Ecological Studies, Vol. 219

Analysis and Synthesis

Edited by

M.M. Caldwell, Washington, USA
G. Heldmaier, Marburg, Germany
R.B. Jackson, Durham, USA
O.L. Lange, Würzburg, Germany
H.A. Mooney, Stanford, USA
E.-D. Schulze, Jena, Germany
U. Sommer, Kiel, Germany

For further volumes see list at the end of the book and:
http://www.springer.com/series/86

Christian Wiencke • Kai Bischof
Editors

Seaweed Biology

Novel Insights into Ecophysiology, Ecology and Utilization

Springer

Editors
Christian Wiencke
Department Seaweed Biology
Section Functional Ecology
Alfred Wegener Institute
Am Handelshafen 12
27570 Bremerhaven
Germany

Kai Bischof
Marine Botany
BREMARE, Bremen Marine Ecology,
Centre for Research and Education
University of Bremen
Leobener Str. NW2
28359 Bremen
Germany

ISSN 0070-8356
ISBN 978-3-642-28450-2 ISBN 978-3-642-28451-9 (eBook)
DOI 10.1007/978-3-642-28451-9
Springer Heidelberg New York Dordrecht London

Library of Congress Control Number: 2012939125

Printed on acid-free paper

Springer is part of Springer Science+Business Media (www.springer.com)

Preface

Two thirds of the world are covered by oceans, whose upper layer is inhabited by photoautotrophic organisms, known as algae. Within coastal ecosystems, marine seaweeds have been identified as a group of organisms of vital importance for ecosystem function. On rocky coasts, they form vast underwater forests of considerable size with a structure similar to terrestrial forests and provide diverse habitats and breeding areas for an uncountable number of organisms including fishes and crustaceans. They are an important food source not only for numerous herbivores, such as sea urchins, gastropods, and chitons, but also for detritivores such as filter feeders and zooplankton, which are feeding on degraded seaweed biomass and on energy-rich spores released in vast quantities from seaweeds. On beaches in some localities large masses of seaweeds are stranded and support meiofauna species.

Although marine seaweeds and seagrasses, altogether known as macrophytes, cover only a minute area of the world's oceans, their production amounts to 5–10% of the total oceanic production. Carbon assimilation of kelps, large brown algae of the order Laminariales, is with 1.8 kg carbon m^{-2} $year^{-1}$ similarly high as that of dense terrestrial forests and even exceeds the primary production of marine phytoplankton up to ten times.

Seaweeds are not only of high ecological, but also of great economic importance. Dried thalli are directly used as human and animal food and also as fertilizer. Extracted seaweed substances are used as stabilizers and stiffeners in food industry, cosmetics, pharmaceutical industry, and biotechnology. In future, aquaculture of seaweeds will certainly strongly intensify, especially in integrated multi-trophic aquaculture systems making use of the waste products or biomass generated by other organisms in the system. Industrial use of seaweeds will also strongly increase as basis for CO_2-neutral production of ethanol and methanol as biofuels.

Seaweeds are exposed to a variety of external factors, which affect their physiological and ecological performance. This suite of factors is, however, not constant, but changes along different timescales. There are short-term daily fluctuations of abiotic factors, especially in low latitudes, whereas in high latitudes seasonal variations are dominating. Long-term changes related to the climate history of the earth extend over thousands and millions of years. But changes do not only have natural reasons. In the younger history of the earth man-made changes became evident on a local and global level. An important local change is for example eutrophication of estuaries, bays, and side oceans. With respect to global changes, the burning of fossil fuels leads on the one hand to global warming through the greenhouse effect and on the other hand to ocean acidification through CO_2-mediated changes on the seawater carbonate system. Stratospheric ozone depletion results in an increase of UV-B radiation at the earth's surface and in the upper layers of the oceans. If the acclimation potential of an individual seaweed species is high, this species might not strongly be affected by external changes. In contrast, in species, which are more strongly (genetically) adapted to a special suite of external factors, the effects will be more prominent. So, on the level of communities variation of each of these factors can change the achieved ecological equilibria between the species, and can finally also prompt economic consequences.

This book is a collection of articles summarizing the advances of seaweed biology achieved within the last decades and also pointing to overlooked treasures. The overall aim of this book is to complement available textbooks for advanced students and young researchers. The book contains 22 chapters, written by experts in the various research areas. The chapters are grouped into five parts.

The first part of the book covers fundamental processes and acclimation strategies of seaweeds toward the abiotic environmental variables. Acclimation to limiting and excessive light conditions is considered in Chap. 1 by *D. Hanelt* and *F.L.-Figueroa*. Changes in pigment composition in response to different underwater spectra and the use of light as environmental signal are also discussed. In Chap. 2 by *I. Gómez* and *P. Huovinen*, the adaptations to incorporate and process dissolved inorganic carbon are summarized. An important part of this chapter deals with morpho-functional aspects of carbon metabolism, in particular with the role of storage carbohydrates, thallus anatomy, and long-distance transport of photoassimilates and patterns of carbon allocation, important features e.g. for supporting seasonal development. *A. Eggert* focuses on phenotypic acclimation to temperature outlined in Chap. 3. Here, disruptive temperature stress and thermal tolerance is another focus, before the prime role of temperature for the determination of geographic distribution is discussed. In this respect, this chapter sets the basis for Chap. 18 by *I. Bartsch* et al. who elaborate on the shift of marine phytogeographic regions under conditions of global warming. In Chap. 4, *F. Gordillo* reviews the relationship between algal nutrition and their environment in order to better understand how seaweeds meet their nutritional needs including the uptake of inorganic carbon. *U. Karsten* (Chap. 5) discusses the effect of salinity stress and desiccation on the physiology of seaweeds and the involved metabolic processes during osmotic acclimation. In Chap. 6, *K. Bischof* and *R. Rautenberger* summarize the processes involved in the generation of reactive oxygen species (ROS) during environmental perturbations, their effects on seaweed performance, and the respective antioxidative strategies against photosynthetically formed ROS. At the end of the chapter pathogen defense through oxidative bursts is discussed.

Part II focuses on the multitude of biotic interactions found in seaweed communities. The first chapter in this part (Chap. 7) by *M.S. Edwards* and *S.D. Connell* addresses competition as a major factor structuring seaweed communities. Along that line *K. Iken* discusses grazer–seaweed interactions as other major drivers for seaweed standing biomass and community composition, for the energy flow through the system, and with respect to higher trophic level predator–prey interactions (Chap. 8). Chapter 9 by *C.D. Amsler* summarizes our present knowledge on seaweed chemical ecology with emphasis on sensory and defensive systems. Hitherto, the microbial community populating seaweeds and, thus, potentially affecting algal growth or secondary epibiosis is still understudied: In Chap. 10 *M. Friedrich* summarizes the advances in metagenomic approaches to address seaweed–bacterial associations. The topic of epibiosis is taken further in Chap. 11 by *P. Potin* who addresses recent insights into the interactions of seaweeds with their epi- and endophytes, as well as with their parasites. The problems related with invasive marine seaweeds are presented in Chap. 12 by *N. Andreakis* and *B. Schaffelke*. The increasing number of marine invasions is mainly due to intensified shipping and global environmental changes. Moreover, many invasive seaweeds are commercially used, but risks are high and strategies are needed to control intentional and accidental introductions.

In Part III, the reader is introduced into structure and function of the main seaweed systems of the world. The first chapter in this part, Chap. 13, by *C. Wiencke* and *C.D. Amsler* focuses on seaweeds and their communities in polar regions. Biodiversity, biogeographical relationships, ecophysiological characteristics of individual species, and ecology of polar seaweed communities are considered here. The cold-temperate seaweed communities of the southern hemisphere are addressed in Chap. 14 by *P. Huovinen* and *I. Gómez*. To our knowledge, this is the first comprehensive overview covering the southwestern and southeastern South American region, the Victoria-Tasmania region, the southern New Zealand region, and the sub-Antarctic Islands region. After an introduction of these regions and their basic abiotic environmental settings, structure and function of the respective communities, as well as biogeographical processes, are discussed. As an example for the warm-temperate region the largely understudied deep-water kelp forests of the Alboran Sea (SW Mediterranean Sea) and the Strait of Gibraltar are presented in Chap. 15 by *A. Flores-Moya* focusing in particular on growth and reproductive strategies in context to the prevailing abiotic factors. The role of seaweeds in tropical marine coastal systems is discussed in Chap. 16 by *A.Y. Mejia* et al. In these systems, seaweeds are not the dominant habitat providers but are important with respect to reduction of nutrients, provision of food, and refuge for predators and prey. However, excessive growth can lead to complete regime shifts, threatening the stability of the entire coastal system. An overview on the ecology of floating seaweeds and their communities is given by *E. Rothäusler* et al. in Chap. 17. Floating seaweeds can bridge large distances especially at high latitudes where the algae can compensate grazer-induced tissue loss by relatively high growth rates at the prevailing temperatures. In this way, seaweed floes may act as important vectors of dispersal for the associated invertebrate fauna.

Part IV contains a compilation on the effects of global and local environmental changes on the performance of seaweeds and their communities. In Chap. 18, *I. Bartsch* et al. identify expected distributional shifts of major biogeographical regions under conditions of global warming using a macro-ecological modeling approach. According to their results, the tropical regions will expand considerably and all other regions, except the Antarctic region, will extend toward the poles. Along with the shift of the biogeographical regions, the community structure at the boundaries will also dramatically change along vast coastlines. Increasing carbon dioxide (CO_2) concentrations in the atmosphere do not only result in global warming, also the pH of the world's oceans is lowered, a process called ocean acidification. This phenomenon is presented in Chap. 19 by *M.Y. Roleda* and *C.L. Hurd*. The consequences of ocean acidification can affect seaweeds from the cellular to the community level. Particularly important are the interactive effects with other factors, e.g., global warming, eutrophication, and ultraviolet radiation (UVR) due to stratospheric ozone depletion. The latter effect on seaweeds is discussed in Chap. 20 by *K. Bischof* and *F.S. Steinhoff* not only with respect to the damaging effects of UVR, but also to the acclimation strategies and the adaptive traits of seaweeds exposed to UVR. Special attention is given here to the microscopic developmental stages of seaweeds. A rather locally acting, but nonetheless very severe environmental impact is eutrophication, which is addressed in Chap. 21 by *M. Teichberg*. This contribution provides an overview over eutrophication-induced formation of macroalgal blooms and their ecological consequences. Moreover, shifts in macroalgal growth in shallow estuaries, coral reefs, and intertidal or subtidal rocky shores are discussed in order to understand how different systems may contrast in response to shifts in top-down versus bottom-up control.

The final part provides a comprehensive overview on recent developments in seaweed aquaculture, industrial applications, and the overall economical importance of seaweeds. This information is incorporated in Chap. 22 by *C.M. Buchholz* et al.

For sure, the present volume of the 'Ecological Studies' presented is far from being completely covering all aspects effective on and affected by seaweeds in its entirety. However, a review on the respective concluding remarks stated by contributing authors at the end of the respective chapters allows for the identification of the most important gaps in knowledge and invaluable insights into future research priorities with respect to seaweed biology.

It is evident that there is still a need for studies on the fundamental adaptational features allowing seaweeds to thrive in their respective environments. This holds especially true for species inhabiting somewhat extreme habitats, like intertidal systems, or polar areas. Newly developed methods in molecular physiology need to be implemented also in seaweed biology on a large scale. In this respect, more studies in gene expression responses following transcriptomic approaches will be promising tools and will revolutionize our understanding of seaweed responses to the environment. *In concreto*, more profound studies on the fundamental strategies of adaptation are urgently needed for example for addressing life strategies of Arctic kelp to understand the triggering of light vs. dark metabolism in extremely seasonal habitats. Hitherto, dark metabolism of kelp, which needs to be sustained under conditions of the polar night and additional sea ice cover, is just fragmentarily understood.

The enforced implementation of molecular tools will also ease our endeavors to predict seaweed responses to environmental change. Along this line molecular analyses should also go beyond gene expression and also aim at the understanding of sensing of environmental cues and identify signal transduction pathways. Genomic, proteomic, and metabolomic studies will become more prominent in seaweed research, facilitated by past, present, and future whole genome sequencing projects. The completion of the *Ectocarpus* genome project in the year 2010 represented a new ignition to seaweed-related research activities, and the currently ongoing *Chondrus* and *Porphyra* genome project will result in another boost in research. Based on their prime ecological significance to coastal ecosystem function, we propose that dominant species of kelp, e.g., *Macrocystis pyrifera, Laminaria hyperborea* should be considered as upcoming candidates in such sequencing projects.

A second priority of future research on seaweeds may be described by the buzzwords "integration" and "interaction": In their environment seaweeds are exposed to a complex set of abiotic and biotic variables, which may change independently or interdependently. Factors may interact synergistically or compensatory. Studies aiming to predict consequences of environmental change in seaweed communities have to consider the interaction of (multiple) stress factors more thoroughly and conceptualise multifactorial experiments. Furthermore, the different life histories of the species under investigation need to be integrated more strongly in environmental stress physiology. In this respect, it is important to identify the life history stage most susceptible to the impinging environmental stressors, as this will represent the bottleneck for reproduction, recruitment, and thus, population structure. Stage-specific acclimation capacities have to be addressed in order to define thresholds of stress, biogeographical boundaries, and the so-called tipping points, which will be important to define with respect to resilience. Furthermore, also biotic factors largely interact, and changes in interspecific interaction may be both the outcome and the cause of environmental stress susceptibility. The multiple interactions between seaweeds and their grazers, foulers, parasites etc. offer a hitherto largely understudied field of interspecific sensing and signaling. Again, these studies should be linked to transcriptomic and metabolomic approaches.

In an era of climate change, these approaches should be integrated by modellers in order to increase the predictability of consequences of environmental disturbance. Climate modellers, physical oceanographers, and seaweed ecophysiologists need to team up in order to outline scenarios of future seaweed community and coastal ecosystem functions. As climate change will strongly affect biogeographical boundaries, future studies on the dynamics, range expansion, connectivity, and ecotype formation of populations will be most important. In a concrete example, there are current indications for ecotype formation in *Laminaria* at Arctic and cold-temperate sites, respectively, which might be reverted in the future due to the predicted process of atlantification of Arctic regions. Improvements in the molecular tools applied in population genetics should then also be used to increase our knowledge in invasive species ecology as well as to understand the significance of kelp rafting to species dispersal.

The improvements in analytical techniques applied in studies on seaweed biology are paralled by the increasing efforts for seaweed uses for economic purposes. Apart from the potentially innumerous pharmacological effects seaweed-derived compounds may confer, the use of seaweeds as bioabsorbers of aquaculture effluents has a great application potential. The increasing applied sector in seaweed biology directly relates to further study questions related to conservation and socio-economy.

Climate change in its different manifestations and facets on global and regional scales, the rapidly increasing anthropogenic pressures on coastal areas, as well as the vastly growing demand for alga-derived products, represent challenges to both human society in general and seaweed biologists in particular. Hitherto, it is commonly accepted that the ecological and economical values of seaweeds can hardly be overestimated. Against this background there is an increasing need for integrated studies on seaweed biology, stretching from molecular physiology to community ecology and even further to societal aspects. Along this line, we hope that this book will further increase the awareness of the enormous ecological significance of seaweeds in coastal environments.

Bremerhaven and Bremen Christian Wiencke and Kai Bischof

Contents

Part I
Fundamental Processes and Acclimation to Abiotic Environmental Variables

Chapter 1
Physiological and Photomorphogenic Effects of Light on Marine Macrophytes

Dieter Hanelt and Félix López Figueroa

1.1 Introduction

The coastal zone of the marine habitat is exposed to considerable diurnal changes in solar radiation due to the position of the sun, clouding and, especially, the tides. Thus, these temporary changes in the costal environment can appear much stronger compared to terrestrial habitats. Whereas algae in air or the upper littoral zone are supplied with sufficient or even with an excess of radiation energy (Wilhelm and Selmar 2011), in deeper regions of the euphotic zone light is clearly limiting for seaweed growth. It is a reason why seaweeds generally need to grow on the continental shelf (epipelagic), down to a maximal depth where the light level is not lower as 0.05% or 0.001% of the surface irradiance in the tropics (Lüning 1990). Moreover, conditions in the coastal areas are quite different compared to the open ocean as in coastal zones generally an inflow of freshwater occurs which can be loaded with sediments. Penetration of solar radiation into the water body is largely determined by scattering and absorption of biological and inorganic material, with higher concentrations in coastal areas. Coastal waters show large temporal changes and regional differences in the concentration of dissolved and particulate matter influencing temporal penetration of solar radiation. This is the basis on which Jerlov (1976) classified marine waters into nine types of coastal and five types of oceanic waters in dependence on the respective transmittance characteristics.

D. Hanelt (✉)
Biocenter Klein Flottbek, University Hamburg, Ohnhorst-Str. 18, 22609 Hamburg, Germany
e-mail: dieter.hanelt@uni-hamburg.de

F.L. Figueroa
Departamento de Ecología, Facultad de Ciencias, Universidad de Málaga, Campus Univ de Teatinos s/n, 29071 Málaga, Spain
e-mail: Felix_Lopez@uma.es

C. Wiencke and K. Bischof (eds.), *Seaweed Biology*, Ecological Studies 219,
DOI 10.1007/978-3-642-28451-9_1, © Springer-Verlag Berlin Heidelberg 2012

1.2 Photosynthesis Under Limiting Light Conditions

Sublittoral species growing in deep water must be adapted to chronic low light. For planktonic algae, generally a lower depth limit of 1% of the surface irradiance is regarded as defining the euphotic zone (Steemann Nielsen 1975), and this limit can be even 0.1% for picoplankton in oceanic waters. Seaweeds of the order *Laminariales* reach their lower growth limit at about 0.6–1.2% of surface light, whereas for some deep growing rhodophytes a minimum of 0.001–0.05% was determined (Lüning 1981, 1990). The photosynthetic rate of different *Laminaria* species exceeds the respiratory rate and thus the compensation point (E_c) at about 5–8 μmol m^{-2} s^{-1}, whereas in deep water red algae an irradiation of about 2 μmol m^{-2} s^{-1} is already sufficient (Lüning 1981). Deepest crustose macroalgae seem to survive at an absolute light minimum of about 0.01 μmol m^{-2} s^{-1} (Littler et al. 1986). These red crustose corallines show a large light absorptance and employ light-harvesting pigments with a high energy-cost in their production per unit light absorption rate in a given underwater spectrum (Raven and Geider 2003). Raven et al. (2000) outlined that it is difficult to explain growth of algae below 0.5 μmol m^{-2} s^{-1} as there are energy-consuming reactions which use an increasing fraction of energy input when photon flux density decreases. Among these processes are redox back reactions of reaction center II, the leakage of H^+ through thylakoid membranes and the turnover of photosynthetic proteins. The first of the two processes limit the rate of linear electron transport and ADP phosphorylation, while the latter consumes ATP. Thus, it is not yet clear how crustose red algae can grow down to 274-m water depth where the average incident photon flux density for 12 h day^{-1} does not exceed 0.02 μmol m^{-2} s^{-1} (Raven and Geider 2003).

Low light adapted species are typically characterized by high photosynthetic efficiencies and very low light compensation (E_c) and saturation points (E_k), ranging between 1 and 15 and between 14 and 52 μmol photons m^{-2} s^{-1}, respectively (Dunton and Jodwalis 1988; Wiencke et al. 1993; Weykam et al. 1996; Brouwer 1996; Eggert and Wiencke 2000).

For a certain time period net photosynthetic rates need to be high enough that seaweeds can promote growth, reproduction, as well as enough energy storage to cope with reduced light availability also during the winter season. Using data on daily changes of in situ irradiance and P–E derived parameters such as photosynthetic capacity (P_{max}), dark respiration, E_c and E_k, it is possible to calculate the daily periods over which carbon (C) assimilation exceeds C losses due to respiratory activity at different water depths. This metabolic C balance is regarded to determine the lower distribution limit of algae. Due to low E_k values, species from King George Island (Antarctica) are exposed during spring–summer for periods $\geq$12 h day^{-1} at about 20 m depths to irradiances above saturation (period of saturation H_{sat}). Under these conditions, species such as *Palmaria decipiens* or *Desmarestia anceps* achieve positive C balances about 3 mg C g^{-1} FW d^{-1}. At 30-m depth daily light availability decreases to values between 7 and 9 h in summer, but C balance is still positive in 4 of 5 studied species (Gómez et al. 1997).

Only in the case of *Desmarestia anceps* from 30 m, a negative C balance was determined, indicating that at this depth the alga is at its lower distribution limit. In contrast, *P. decipiens, Gigartina skottsbergii* and *Trematocarpus antarcticus* and especially *Himantothallus grandifolius* are metabolically able to grow even in deeper waters which mean they have very modest requirements.

Under low light conditions higher pigment content within the thalli was observed as under strong light conditions (Ramus et al. 1976, 1977). The chance of photon absorption increases with increasing photosynthetic antenna size. Algae collected from or transplanted to different water depths show that the content of accessory pigments increases with lower light conditions in deep waters (Lüning 1990), e.g., in green algae the chlorophyll *a:b* ratio decreases demonstrating especially the increase of the antenna size of the photosynthetic apparatus (Yokohama and Misonou 1980). Whereas under low light conditions in deep water a larger antenna size increases the capacity of light absorption, a smaller antenna helps to avoid photoinhibition and photodamage due to over excitation under high light conditions close to the water surface or during emergence at low tide. Under low light conditions the plant invests more energy in the synthesis of light-collecting pigments and in strong light into the synthesis of photosynthetic enzymes, electron chain components as well as photo-protective structures and energy-dissipating mechanisms.

A study of Marquardt et al. (2010) showed that the saturation point E_k of all red algal species tested decreased with increasing depth concomitantly with the decreasing light availability. This may be due to the adjustment of the photosynthetic apparatus itself via changes of the reaction center ratio, changes of the relative size of the light-harvesting complex (LHC) or changes in the relative content of light protective pigments. Changes in thallus morphology are another possibility to achieve acclimation, e.g., change of thickness, branching, length, density of photosynthetic units (Küster et al. 2004). Johansson and Snoeijs (2002) demonstrated by measurements of photosynthesis versus irradiance curves (PE curves) that light-saturated net photosynthetic rates (P_{max}), respiratory rates in darkness (R_d) and the initial slope (α) were strongly related to algal morphology with generally higher values for thinner species. The compensation irradiance (E_c) and saturating irradiance (E_k) were strongly related to water depth with lower values at higher depth. One advantage of thin sheet-like and filamentous species is the capability of fast growth, which is coupled to high photosynthetic rates per unit biomass (Littler et al. 1983; Falkowski and Raven 1997), resembling rather the conditions in shallow waters. In several macroalgae the photosynthetic parameters P_{max} and α are highly dependent on thallus morphology with higher and faster O_2 production rates for thinner and filamentous species, and lower rates for coarser and thicker species when normalized to biomass (measured as dry weight) and opposite when normalized to algal surface area (Johansson and Snoeijs 2002). Similar relationships were described for five green-algal species by Arnold and Murray (1980) as well as by Littler (1980) for 45 species of marine macroalgae from field incubations. Thus, deeper growing algal species are expected to have lower E_c and

E_k as species growing higher up in the littoral. Also, within species E_c can decrease as an acclimation to ambient light with water depth (Gómez et al. 1997).

The ideal strategy for life in deep water is represented by crustose coralline algae. They are well protected against grazing and can survive in spite of slow growth. Their thallus structure represents a horizontal light receiver with none self-shading by a single cell layer which enhances light absorption (Lüning 1990). In dependence to the clarity of water and the annual sum of photosynthetic active radiation impinging at the water surface, the lower depth limit of crustose coralline algae shifts with lower latitude from several meters in cold temperate waters (e.g., 15 m on Helgoland) to several hundred meters in tropical waters (e.g., 268 m Bahamas, Lüning 1990). The annual sum of impinging irradiance must support at least the annual need of energy for maintenance metabolism, measured by the maintenance respiration rate, and guarantee a minimum of energy surplus for establishment of growth and reproduction. Compared to cold-temperate regions, Caribbean algae are able to survive in such extreme depths due to the higher solar irradiance, a 12 hour period of day light and the clear water conditions (Jerlov type I) enables these Caribbean algae to survive in such extreme depths. For algae with a more complex, even erected thallus and the presence of nonphotosynthetic tissue as typically found in kelps, the need for light energy increases and the algae have to grow in more shallow waters as the amount of respiration and self-shading areas increases.

An ideal marker for macroalgal depth distribution seems to be the respective stable carbon isotope composition. Rapid carbon assimilation under high photon fluence rates leads to 13 C enrichments, probably due to extracellular and/or intracellular isotopic disequilibria resulting in a trend toward more positive carbon isotope values with increasing photon fluence rates (Wiencke and Fischer 1990). The pattern of isotope composition of algae grown at different depth was found in sediment trap samples from the 2,000-m deep King George Basin off the Antarctic Peninsula. It also revealed a strong contribution of seaweeds to the total organic carbon pool of the deeper basin waters in spring and summer (Fischer and Wiencke 1992).

In conclusion, distribution of marine macrophytes to the lower light limit which accommodates biomass production depends mainly on the minimal energy input (Kirst and Wiencke 1995). This is dependent on the annual fluence or minimum light level occurring in the respective depth for the maintenance of existing plant material and a surplus for growth and reproduction. Due to seasonal changes, algae in low light habitats have to live for long periods each year at photon fluence rates which do not cover their energy needs. Then, the photosynthetic activity in high light periods of the year (e.g., summer season) needs to be high enough to produce sufficient resources to endure periods with light conditions generally below the compensation point (E_c). Light saturation of growth in seaweeds is fortunately lower than those for photosynthesis (Lüning 1990) so that algae can produce enough reserve materials under favorable light conditions. Changes in the water transparency may shift the lower light limit so that the algal distribution pattern could be affected by anthropogenic changes of the light transmittance of the water body.

1.3 Photosynthesis Under Excessive Light Conditions

In contrast to the conditions at the lower algal distribution limit, photosynthetic organisms close to the water surface are exposed to high irradiances which may even exceed their light energy requirement for photosynthetically provided anabolism. Generally, these photosynthetic organisms do not suffer from energy shortage, but instead they have to invest in proteins and cellular components which protect them from potential damage by excess supply of energy (Wilhelm and Selmar 2011). As consequence a reduction of photosynthetic activity, called photoinhibition, may occur (Powles 1984; Krause 1988; Krause and Weis 1991). As defined by Franklin et al. (2003) photoinhibition is a generic outcome of the failure of photoprotection to mitigate photoinactivation. This may result under high light conditions in which damage of reaction center proteins exceeds photorepair in the complex molecular structure of photosystem II (PSII). Photoacclimation is the adjustment in structure and function of the photosynthetic apparatus that avoid or rapidly repair damage and thus mitigate (chronic) photoinhibition. Photoprotection (also called dynamic photoinhibition) refers to processes that decrease excitation transfer to the reaction centers; these processes are mainly operative in the antenna complexes. Loss of PS II reaction center function is called photoinactivation (formerly called chronic photoinhibition), often accompanied by loss of reaction center protein D_1, which can be repaired under dim light conditions. Under excessive light conditions a significant degradation of the reaction center protein (D_1) of PS II was found (Ohad et al. 1984; Mattoo et al. 1984). Nonfunctional photosynthetic centers are able to protect active centers against continuous damaging effects of excess absorbed light energy by harmless energy dissipation, mainly by heat dissipation (Guenther and Melis 1990; Öquist and Chow 1992; Critchley and Russell 1994). Photodamage occurs if the rate of the D_1-protein damage exceeds the rate of its repair process, leading to a breakdown of the D_1-protein pool (Aro et al. 1993). Photoprotection or dynamic photoinhibition is important for the regulation of quantum yield of photosynthesis, namely by the xanthophyll cycle in the chloroplasts of green and brown algae (Demmig-Adams and Adams 1992). Quantum yield of photosystem II (especially the in vivo fluorescence ratio F_v/F_m which represents photosynthetic efficiency after dark acclimation) is diminished reversibly by increasing thermal energy dissipation under strong light conditions. Carotenoids control this harmless dissipation of excessively absorbed light energy by, e.g., de-epoxidation of violaxanthin to zeaxanthin (Ruban et al. 2007) a process catalyzed by a thylakoid-bound deepoxidase that is activated by a low lumenal pH due to a high electron transport rate (Pfündel and Dilley 1993). Zeaxanthin may dissipate excessively absorbed light energy harmlessly as heat. The process is reversed by the epoxidation of zeaxanthin to violaxanthin in dim light (Xanthophyll-cycle). The low pH induces also protonation of the photosystem II protein, PSBS, which induces a conformational change to specific chlorophyll- and carotenoid-binding LHC proteins (Peers et al. 2009). LHC may aggregate leading to, e.g., PSBS–LHC–zeaxanthin complex in some green algae. As a consequence, the exciton transfer from exited chlorophylls to other chlorophylls is disabled and the LHCs are converted to heat dissipaters (Niyogi et al. 2005). Dissipation of excess excitation energy occurs by a charge transfer mechanism involving a carotenoid radical cation (Ahn et al. 2008) and/or by chlorophyll-to-carotenoid energy transfer (Ruban et al. 2007). The photosynthetic systems of algae share many central functions with land plants. But recently, an ancient light-harvesting protein (LHCSR) was described in *Chlamydomonas* which is involved in fast regulation of algal photosynthesis (Peers et al. 2009). LHCSRs are absent in vascular land plants, but present in a variety of photosynthetic organisms, such as diatoms that show an extremely high nonphotochemical energy-quenching capacity included in the photoprotection mechanism (Eberhard et al. 2008; Peers et al. 2009). LHCSR transcripts accumulate under environmental conditions known to induce photo-oxidative stress, including deprivation of carbon dioxide, sulfur, or iron, as well as high light (Peers et al. 2009). In algae LHCSR orthologues are missing only in Rhodophytes (and cyanobacteria), which dissipate excess light energy from phycobilisomes by a mechanism distinct from the typical, above-described, energy-dissipating mechanism (energetic fluorescence-quenching mechanism, q_E) (Wilson et al. 2006).

The short-term acclimation of plants to high irradiances and its relation to photosystem II photochemistry and fluorescence emission were reviewed in detail by Dau (1994a, b). Moreover, a general overview of photoinhibition, its molecular aspects and its mechanisms in the field are given by several articles in the book edited by Baker and Bowyer (1994), and the effects of a changing irradiance environment, especially on marine macrophyte physiology, were also reviewed (Franklin and Forster 1997; Häder and Figueroa 1997; Bischof et al. 2000a; Wiencke et al. 2007).

In the marine habitat macrophytes are exposed to considerable diurnal changes of the impinging photon fluence rates due to the position of the sun, clouding and, especially, the tides (Hanelt and Nultsch 2002). Therefore, at midday benthic marine algae, which grow normally underwater at dim light conditions, can be exposed to extremely high irradiances on sunny days during low tide in the intertidal. As a consequence, light energy is excessively absorbed by the photosynthetic apparatus and, hence, the extent of its photodamage increases. One of the damaging processes is the production of highly reactive oxygen species, which attack target molecules such as the D_1-protein, chlorophylls and unsaturated fatty acids (Asada and Takahashi 1987; see Chap. 6 by Bischof and Rautenberger). The damage to the photosystem is counteracted by a repair process that involves partial disassembly of inactive PSII, proteolytic degradation of the photodamaged reaction center protein (D_1) and cotranslational insertion of newly synthesized D_1 into PS II, also called D_1 repair cycle (Aro et al. 1993; Barber and Andersson 1992). Permanent photodamage occurs when scavenging of oxygen radicals by superoxide dismutase, hydrogen peroxidase or catalase is insufficient. Under light stress conditions, the concentration of active oxygen is increased either by higher production rates or by insufficient capacity of the oxygen-scavenging systems. The reactive oxygen species seem to induce inhibition of repair processes by suppression of the de novo synthesis of proteins and, in particular, of the D1 protein that are required for the repair of PSII (Nishiyama et al. 2006; Murata et al. 2007). In addition, H_2O_2 can inhibit CO_2 fixation by 50% because of the oxidation of the thiolmodulated enzymes of the Calvin cycle even at low concentrations (e.g., 10 mM) (Foyer and Shigeoka 2011). A charge recombination between the PSII primary electron acceptor in its semiquinone state and the oxidized chlorophyll donor lead to accumulation of excited triple chlorophylls (Eberhard et al. 2008). These triplet chlorophyll molecules can react with oxygen and generates harmful singlet oxygen, causing damage to PSII and other cell constituents. At the photosystem I acceptor side the redox potential under strong supersaturating light is high enough to reduce molecular oxygen to form O_2-superoxide radicals. Quenching of these radicals by carotenoids and ascorbate can protect the photosynthetic apparatus against such damage.

Nultsch et al. (1987) showed that, depending on the duration and the fluence rate of the excessive irradiation, a decrease in both the photosynthetic efficiency and capacity in the brown alga *Dictyota dichotoma* was caused by photoinhibition. Recovery in a subsequent dim light period was very fast so that in this species the photosynthetic efficiency recovered from inhibition of 80% already within 1 h. Moreover, the action spectrum of photoinhibition revealed that the photoinhibitory radiation was absorbed by all photosynthetic pigments, and that photosystem II is the main site of the photoinhibitory process. According to the definition by Franklin et al. (2003) this species shows a high potential for photoprotection.

Similiar results were presented with the red alga *Polyneura hilliae* (Nultsch et al. 1990). However, in contrast to *Dictyota*, the red algal species recovered very slowly after strong photoinhibition. In this red alga only a partial recovery was observed in complete darkness, and the recovery phase is clearly biphasic. A first fast phase occurs already in darkness, whereas the subsequent slow phase requires dim light. The potential for photoprotection in the red algal species was low compared to *Dictyota*. The above described kinetics indicates that apparently two different molecular mechanisms are involved in photoinhibition. The discrimination of dynamic and chronic photoinhibition recommended by Osmond (1994) is ingenious, according to photoprotection and photoinactivation. Studies of the kinetics revealed that especially during a fast light acclimation the reaction can be described combining two different processes of a slow fraction (P_{slow}) with a slow rate constant (k_{slow}) and a fast fraction (P_{fast}) with a fast rate constant (k_{fast}) at a given time (t).

1. The phase of inhibition is described by:
   $y_{\mathrm{Inh}} = P_{\mathrm{fast}} \bullet e^{(-k_{\mathrm{fast}} \bullet t)} + P_{\mathrm{slow}} \bullet e^{(-k_{\mathrm{slow}} \bullet t)}$
   at which $F_{\mathrm{v}}/F_{\mathrm{m}} = P_{\mathrm{fast}} + P_{\mathrm{slow}}$ at $t = 0$

2. The phase of recovery is described by:
   $y_{\mathrm{Rec}} = F_{\mathrm{v}}/F_{\mathrm{m}} - \left(P_{\mathrm{fast}} \bullet e^{(-k_{\mathrm{fast}} \bullet t)} + P_{\mathrm{slow}} \bullet e^{(-k_{\mathrm{slow}} \bullet t)}\right)$

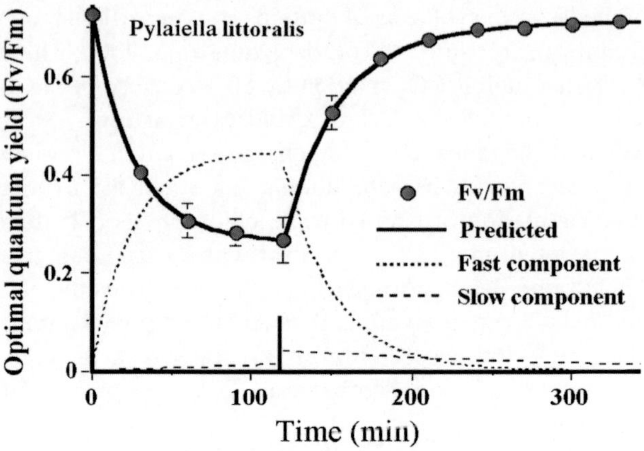

Fig. 1.1 Course of a typical optimal quantum yield curve of the brown alga *Pylaiella littoralis* during an inhibition phase of 120 min with an irradiance of 500 μmol m^{-2} s^{-1} and a subsequent period of recovery under 10 μmol m^{-2} s^{-1} white light. Points show yield measurements (F_v/F_m) dependent on the time course. The *dashed line* shows the kinetics of the slow process, the *dotted line* those of the fast process, and the *solid line* the kinetics of the predicted process combined by the slow and the fast kinetics. Mean values ± SD, $n = 3$, $r^2 = 0.999$ (after Hanelt 1998)

This explains why the course of recovery shows generally two phases (Hanelt et al. 1997a, b, c). Using these equations the data of the optimal quantum yield can be well fitted with a coefficient of determination of $r^2 \geq 0.99$ (Fig. 1.1).

In conclusion, the fast process could represent photoprotection or the activity of the xanthophyll cycle and the slow process the course of photoinactivation or the occurrence of D_1 damage and/or inactivation of reaction centers. Algae which are adapted to dim light conditions show a higher fraction of the slow process, whereas the reaction of algae adapted to strong light conditions mainly consist of the fast reaction e.g., Fig. 1.1 (Hanelt 1998).

First field experiments on photoinhibition in macroalgae were done by Huppertz et al. (1990). They showed that the midday depression of photosynthesis was clearly caused by photoprotection. Specimens of the brown alga *Fucus serratus* living in intertidal rock pools and, hence, are covered by water during low tide, showed a considerable decrease of photosynthetic activity under excessive natural light conditions. The photoinhibitory state is conserved by moderate desiccation in air in this species because the thalli show the same state of photoinhibition after reabsorbing water as measured before they became uncovered during falling tide. Thereafter it was demonstrated that the photosynthetic oxygen production measured at nonsaturating fluence rates and the optimal quantum yield (F_v/F_m) show an approximately inverse course in comparison with the fluence rate of daylight during the day (Hanelt 1992; Hanelt et al. 1993).

According to the recent review of Murchie and Niyogi (2011) the mechanisms included in photoprotection regulate absorption and dissipation of light energy. The immobility of plants does not allow leaving temporarily high light-exposed areas, but they have the ability to regulate the amount of absorbed sunlight by changes in leaf area, leaf angle, chloroplast movement and, on a molecular level, through acclimatory adjustments in LHC antenna size. Excessively absorbed light can be dissipated via several routes, including thermal dissipation. A number of other reactions within the chloroplast can act as photochemical sinks for excess electrons, and there are efficient antioxidant systems for the removal of reactive oxygen species which are produced under high energy load of the system.

1.4 Effect of Ultraviolet Radiation on Zonation of Macroalgae

In most studies on marine macrophytes, there is common sense that the sensitivity of photosynthesis to ultraviolet radiation (UVR) is a function of vertical zonation of the species (Larkum and Wood 1993; Dring et al. 1996; Bischof et al. 1998a; see Chap. 20 by Bischof and Steinhoff). Moreover, Maegawa et al. (1993) regard solar UVR as a major factor controlling the upper zonation limit of red macroalgae on the shore. The potential of UV to inhibit photosynthesis of algae was first demonstrated by Jones and Kok (1966). Results of Larkum and Wood (1993) indicated that increasing UV levels of the solar radiation can cause similar effects comparable to high PAR in all types of aquatic plants. In the field, high irradiances of PAR are generally accompanied by higher UV-radiation. The mechanisms are, however, likely to be different. UV radiation cannot be regarded as an "excessive energy input" in a proper sense. Its maximal irradiance is much smaller than that of PAR, and the UV wavebands do not contribute significant energy supply for photosynthetic chemistry. UV exhibits adverse effects on photosynthesis in a more direct way, such as its waveband with high energy content is absorbed by aromatic and sulfhydryl-containing biomolecules, causing a direct molecular damage (Vass 1997). The UV-B inhibition spectrum corresponds much more with the spectral absorption by DNA and proteins than with photosynthetic pigments or the action spectrum of photoinhibition (Jones and Kok 1966; Setlow 1974; Nultsch et al. 1987; Hanelt et al. 1992). Numerous studies have shown that recovery from photoinhibition is delayed after exposure to additional UV-B irradiation (see the review of Häder and Figueroa 1997). In contrast, Flores-Moya et al. (1999) demonstrated that in the marine macroalga *Dictyota dichotoma* a delay of recovery of photoinhibition is observed if the natural UV-B wavelength range is removed from the solar spectrum, in specimens collected from a high UV environment. This was later confirmed under simulated sunlight conditions with different aquatic plants in New Zealand (Hanelt et al. 2006) or in field studies with natural sun radiation (Hanelt and Roleda 2009). Positive effects of UV-B on growth and abundance in phytoplankton are also reported by Thomson et al. (2008); some taxa were most abundant in treatments of intermediate fluxes of UV-B radiation. This suggests that moderate UV-B irradiances may enhance protection from PAR and/or UV-A. In *Synechocystis*, UV-B radiation accompanied by low intensity visible light enhanced synergistically protein-repair capacity, which provides

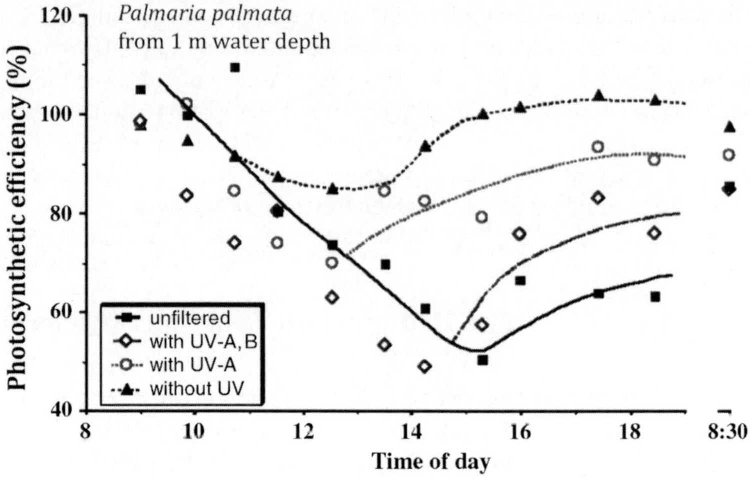

Fig. 1.2 Photosynthetic efficiency of the red alga *Palmaria palmata* during the course of the day. The day was cloudy and the fluence rate decreased continuously after 14:10 h. The photosynthetic efficiencies of the nonphotoinhibited controls, collected the day before from a depth of 1 m, were measured by the fluorescence ratio F_v/F_m and standardized to 100%. The *solid fitted line* indicates the inhibition caused by unfiltered sun radiation. The *broken* and the *dotted lines* represent the inhibition caused by radiation depleted of different UV wavelength ranges. Next morning (8:30), photosynthesis was measured again (after Hanelt et al. 1997a)

protection against photodamage (Sicora et al. 2003). Generally, field experiments at latitudes with a low UV environment (e.g., polar region) do never show UV-B repair effects in macroalgae (Hanelt et al. 1997a, b, c; Bischof et al. 2000b). This indicates a special adaptation of organisms to UV-B when they live in an environment with high solar radiation under high UV pressure. However, algae growing in deeper, UV-protected locations on the shore are more sensitive to the natural UVR as for instance the red alga *Palmaria palmata* (Hanelt et al. 1997a). This species shows a reaction which is clearly dependent on the different UVR ranges. By cutting off successively the shorter wavelengths ranges, the degree of photoinhibition decreases and recovery commences earlier during the course of the day (Fig. 1.2).

Various factors contribute to the individual sensitivity of algae toward solar radiation, e.g., growth depth (Sagert et al. 1997; Dring et al. 1996; Bischof et al. 1998b), season (Gómez et al. 1995a), position within and below the canopy (Stengel and Dring 1998), life history stage as well as the different parts and sizes of the thalli (Gómez et al. 1995b; Dring et al. 1996; Hanelt et al. 1997b; Karsten and Wiencke 1999). Thus, acclimation to the ambient radiation plays an important role in photosynthesis of marine macroalgae. A study on three abundant brown algal species from Spitsbergen (*Saccharina latissima, Alaria esculenta,* and *Saccorhiza dermatodea*) shows that individual photosynthetic performance reflects changing light climate in accordance with depth (Bischof et al. 1998b). Photosynthetic acclimation was found for both ambient PAR and UVR. Exposure to artificial UVR reduces photosynthetic rates in deep water plants significantly while photosynthesis in the same species collected from shallow waters is unaffected. To cope with higher irradiance levels in shallow waters a capability for fast recovery from light stress is one prerequisite (Hanelt 1998).

The acclimation potential of maximal quantum yield of photosynthesis to changing radiation conditions was studied in detail in the Arctic/cold temperate brown alga *Alaria esculenta* (Bischof et al. 1999). In this species, acclimation to changing radiation conditions occurs within very few days. This is of great ecological importance as algae, subjected to 6 months of darkness during the Arctic winter and sea ice with snow cover shielding the algae from solar radiation in spring, become suddenly exposed to high radiation, and thus also to UV, as soon as sea ice breaks up. *A. esculenta* shows two different responses involved in the acclimation of maximal quantum yield of photosynthesis. At first, after a few days of exposure to artificial UVR, the recovery from induced photoinhibitory processes proceeds significantly faster; later, the degree of photoinhibition decreases. This implies again that different molecular mechanisms are involved in photoacclimation. In conclusion, photosynthesis of macroalgae from the intertidal zone is rather resistant to natural UV-B radiation. Algae from the upper sublittoral seem to be able to acclimate rapidly to fast changes of solar irradiance, consequently reducing the adverse effects of UVR exposure or even need UV-B as induction for repair processes (Hanelt and Roleda 2009). Deep water algae react highly sensitive to UVR (Bischof et al. 2000b), but, due to the absorption of the water body, UV-B is no natural component of the ambient light at higher depths. However, it is important to note that UVR may also exert adverse effects on the algae, e.g., growth rate reduction and/or reproductive success which is not reflected by a reduced photosynthetic activity.

Two aspects have received little attention so far. In many Polar species growth rates are highest in Spring (Wiencke 1990a, b; see also Chap. 13 by Wiencke and Amsler), which are also affected by UVR, and not only photosynthesis of young thallus parts (Dring et al. 1996; Wood 1987). Moreover, some species from the Arctic partly reproduce in spring. During this time, algal spores were found to be the most light-sensitive life history stage of the studied brown algae and are strongly affected by increased UV-B radiation, both in respect to their photosynthetic performance and their susceptibility to DNA damage (Wiencke et al. 2000). As has been widely publicized, increased UV-B due to ozone depletion occurs mainly in the Polar Spring due to atmospherical and geographical reasons. Therefore, Polar species will be most exposed to the anthropogenic increase of UV-B radiation.

1.5 Light Absorption and Light Spectrum

The color of the algae is mainly based on the accessory photosynthetic pigments of the LHC. Engulfment of a cyanobacterium or eukaryotic microalga by a process called endosymbiosis is a cause of physiological, structural and genetic adaptation of the different types of chloroplasts in evolution. The brown color typical for the

Phaeophyceae is caused by light absorption of the carotenoid fucoxanthin, and the red color of the Rhodophyceae by the phycobiliproteins. If the LHC consists mainly of chlorophylls, the color is green, as is typical for the Chlorophyceae. The different absorption characteristics of the respective LHC correspond to different underwater light spectra. Different light qualities occur within the water body by absorption and scattering of light (Jerlov 1976) especially with increasing water depth and/or turbidity. Generally, the blue-green waveband penetrates deepest into the water body as the shorter and longer wavelength are more absorbed by the water molecules or scattered by particles. Photosynthetic apparatus of algae has adapted to these different wavebands. The occurrence of the various types of pigments in the LHC and their arrangement in both photosystems are responsible for different photosynthetic efficiencies of different spectral wavebands which affect photosynthetic activity. The action spectrum of photosynthesis of red algae shows a so-called blue and red drop, first described by Haxo and Blinks (1950). Green light is best absorbed by the phycobilines so it shows the highest photosynthetic rates in red algae. In contrast blue and red/far red light does not induce high electron transport rates because chlorophyll molecules act as the main antenna pigment in photosystem I so that the reaction center of PS I is primarily activated (Butler 1978). Blue and far red light induces principally charge separation in the reaction center of PS I and cyclic electron transport around PS I. As charge separation of PS II in red algae is not induced to an equal amount, linear electron transport rate is small and oxygen production rate is low (Haxo and Blinks 1950; Hanelt et al. 1992).

Engelmann (1883, 1884) pointed to the fact that most green algal species occur in the eulittoral and upper sublittoral, whereas brown algae grow often in deeper zones and many red algae can be characterized as deep water species. In deep water, where blue-green light prevails, the red pigments of the Rhodophyceae allow an efficient absorption (Biebl 1962). However, this is only partly valid because the absorption characteristic depends also on several other factors, especially the thallus morphology. It applies more to coastal waters than to oceanic waters (Larkum et al. 1967). If the algal thallus is thick enough it appears nearly black and absorbs light over the whole spectral range, as typical for brown kelps (Lüning 1990). In addition accessory pigments do enhance light absorption in the blue-green range and examples of green algae growing in deep waters are not rare (Dring 1981, 1982; Ramus 1981). Anyhow, the deepest algae found are crustose red algae with a quite low light demand (Littler et al. 1986), which is not only due to their low growth rate and special morphology (one absorption layer) but also due to their capability to use the impinging photons very efficiently in the blue-green wavelength range (Hanelt et al. 2003). In vivo absorptance of thin and thick algal thalli of the different pigment groups was investigated by Lüning and Dring (1985). Their study demonstrated that a good correlation exists between spectral thallus absorption and action spectrum of photosynthesis, as well as a greater thallus thickness supports photosynthetic activity also when wavebands are slightly absorbed. Leukart and Lüning (1994) demonstrated in several red algal species that growth rate and photosynthesis depends on the light quality during culture and on the pigment content under these conditions. The light requirements were lowest in green light for all red algae investigated. The action spectra of growth followed the photosynthetic action spectra, with maximum efficiencies in the green wavebands, corresponding to the wavelength distribution occurring in deep coastal waters. This points to the importance of light quality for survival at low photon fluence rates and corroborates the findings of Harder and Bederke (1957), Beer and Levy (1983) and Glover et al. (1986, 1987).

1.6 Light as an Environmental Signal

The color of light can induce photomorphogenetic effects, enzyme activity induction or controls the life cycle of the algae. Light is not only the primary energy source but it also provides them with information to modulate developmental processes such as phototaxis of swarmers, phototrophic reactions, chloroplast movement, shade avoidance, circadian rhythms, etc. (Lüning 1990; Kleine et al. 2007). Plants can detect almost all facets of light, including direction, duration and wavelength using three major classes of photoreceptors: the red/far-red light-absorbing phytochromes which are only proven to be a sensor in green algae (Dring 1988; Rüdiger and López-Figueroa 1992), the blue/UV-A light-absorbing cryptochromes and phototropins, and UV-B-sensing UV-B receptors (Chen et al. 2004). A phytochrome-like protein was described in all pigment groups of the macroalgae; however, red/far-red forms were isolated only from green algae (López-Figueroa et al. 1989, 1990). Cryptochromes seem to be widespread in the group of Phaeophyta and, possibly, among chromophyte algae in general (Dring 1988). The photoreceptors perceive light signals and initiate intracellular signaling pathways involving proteolytic degradation of signaling components and large reorganization of the transcriptional program to modulate plant growth and development (Chen et al. 2004). Nitrate reductase activity in green algae and biliprotein accumulation in some red algae may be stimulated by blue or green light, and an interaction with phytochrome like photoreceptors was indicated (López-Figueroa and Rüdiger 1991; López-Figueroa and Niell 1991). Blue light, similar to low light, induces an increase in the number of pigment systems per electron transfer chain in green algae, whereas red light blocks chlorophyll b synthesis and leads to a decreased light-harvesting system together with an increase in the number of reaction centers per electron transfer chain (Senger et al. 2002). The latter equates an adaptation to strong light conditions. The opposite behavior of algae and higher plants to red or blue light corresponds to the different spectral conditions in their habitat. In deep water regions algae grow under blue light conditions whereas higher plants are exposed to a higher fraction of red light. However, Senger et al. (2002) come to the conclusion that the phylogenetic relationship is the major factor for this difference in the light adaptation between algae and higher plants rather than a long-term adaptation to the environment.

The phototactic response of swarmers of the brown algae *Scytosiphon lomentaria* and *Petalonia fascia* causes a photoaccumulation at a peak of 450 nm, and no effect was found at a wavelength above 550 nm, as it is typical in a blue light response (Flores-Moya et al. 2002). In *Ectocarpus siliculosus* the waveband between 430 and 450 nm was found to be the most effective for photoaccumulation of the male gametes (Müller et al. 1987), also shown by the action spectrum of their phototactic response (Kawai et al. 1990). The action spectrum of the chloroplast movement of the brown alga *Dictyota dichotoma* resembles the absorption spectrum of the blue light photoreceptor cryptochrom and serves as a control of the light absorptance in brown seaweeds (Hanelt and Nultsch 1989). In addition a blue light receptor is also responsible for the egg release in *D. dichotoma* (Lüning 1990). Phototrophic reactions have been observed, e.g., in the zygotes of *Fucales*, haptera of *Laminariales* or rhizoids of kelps, and even more examples on light control are given in the seaweed book of Lüning (1990).

Circadian and circannual rhythms in algae are also controlled by light that time metabolic, physiological and/or behavioral events to occur at optimal phases of the daily or annual cycle. Eukaryotic algae serve for long time as model system to study circadian rhythms. (Suzuki and Johnson 2001). For example, UV shows the strongest effect of destruction at sunset and the early night when UV levels are naturally low and, hence, the control due to the daily clock enhances the fitness during the day at higher UV levels (Nikaido and Johnson 2000). Photoperiodic time measurements, i.e., the detection of the lengths of day or night, is the ability to sense the season of the year and to respond appropriately so as to adapt to seasonal changes in the environment (Lüning and tom Dieck 1989). The red Antarctic alga *Palmaria decipiens* shows a seasonal pattern of photosynthetic activity and pigment synthesis (Lüder et al. 2001; see also Chap. 13 Wiencke and Amsler). Maximal electron transport rates of photosynthesis and pigment content increased in *P. decipiens* during mid-autumn and winter. Highest photosynthetic capacity and size and number of phycobilisomes were observed in spring according to sea ice break up so that photosynthetic performance was most effective during clear water conditions. In summer the photosynthetic apparatus degrades and the life strategy of the algae was defined as season anticipator (Lüder et al. 2001). This is contrary to species (e.g., *Iridaea cordata* or *Adenocystis utricularis*) where photosynthesis and growth follow the seasonally increasing daylight period and are defined as season responders with opportunistic life strategy. After an artificially prolonged dark period of 6 months *P. decipiens* lost its ability to photosynthesize with degradation of its light harvesting antennae, the phycobilisomes, and probably through degradation of the reaction centers of photosystem II. Re-illumination within 1 day induced fast accumulation of chlorophyll a and enables active photosynthesis, which indicates rapid repair mechanisms (Lüder et al. 2002). This shows how light controls the activity of the photosynthetic apparatus. The formation of erect thalli in red and brown algae, from a prostrate system (a crust or branching filaments), is commonly controlled by the photoperiod, and either gametogenesis or sporogenesis may also occur in response to changes in daylength (Dring 1988). In spring the erect siphonous thallus of *Scytosiphon* and the folios thalli of the green *Monostroma* appear in the cold temperate eulittoral zone, whereas they survive the summer to winter month as crustaceous or respective *Codiolum* phase

(Lüning 1990). Short-day responses of *Laminaria hyperborea* and *Constantinea subulifera* result in the initiation of a new blade in the autumn or winter while the rate of growth of the young blades of a kelp species (*Pleurophycus gardneri*) is stimulated by long-day conditions (Dring 1988). In contrast, photoperiodism in unicellular algae is unknown and seasonal behavior shall be governed by ecological factors, such as temperature, irradiance and nutrient levels (Lüning 1990).

1.7 Conclusive Remarks

Distribution of marine macrophytes in deep water with low light depends mainly on the minimal energy input. The annual fluence or minimum light level occurring in the respective depth for maintenance of existing plant material must have a surplus for growth and reproduction. During winter seasons algae in low light habitats have to live for long periods each year at photon fluence rates which do not cover their energy needs so that the energy must be chemically stored in the bright light seasons. As light saturation of growth is fortunately lower than those for photosynthesis, deep water algae can still produce enough reserve materials under favorable light conditions. In contrast, at the water surface, irradiance can be too high and the algae have to avoid photodamage by additional investment in proteins and cellular components for protection. These algae exhibit strong and fast regulation of photosynthesis and perform photoprotective mechanisms, i.e., harmless heat dissipation.

However, light does not only serve as a source for energy in photoautotrophic organisms, but is also an environmental signal for regulation and development, inducing changes in photoperiodism (dependence on daylength), photomorphogenesis (dependence on spectral range) and phototropism. In these cases the signal character of light is evident from the low light requirement to induce a reaction. Special photoreceptors are the sensors in the signal chain, whereas the energy to perform the respective reaction is mostly provided by photosynthesis and thus channeled from the energy absorbed by photosynthetic pigments. The variety of the sensors and their responses in algae seem to be more varied than those among flowering plants.

References

Ahn TK, Avenson TJ, Ballottari M, Cheng YC, Niyogi KK, Bassi R, Fleming GR (2008) Architecture of a charge-transfer state regulating light harvesting in a plant antenna protein. Science 320:794–797

Arnold KE, Murray SN (1980) Relationships between irradiance and photosynthesis for marine benthic green algae (Chlorophyta) of differing morphologies. J Exp Mar Biol Ecol 43:183–192

Aro EM, Virgin I, Andersson B (1993) Photoinhibition of photosystem II. Inactivation, protein damage and turnover. Biochim Biophys Acta 1143:113–134

Asada K, Takahashi M (1987) Production and scavenging of active oxygen in photosynthesis. In: Kyle DJ, Osmond CB, Arntzen CJ (eds) Photoinhibition, topics in photosynthesis, vol 9. Elsevier, Amsterdam, pp 89–109

Baker NR, Bowyer JR (eds) (1994) Photoinhibition of photosynthesis: from the molecular mechanisms to the field. BIOS Scientific, Oxford, p 471

Barber J, Andersson B (1992) Too much of a good thing: light can be bad for photosynthesis. Trends Biochem Sci 17:61–66

Beer S, Levy I (1983) Effects of photon fluence rate and light spectrum composition on growth, photosynthesis and pigment relations in *Gracilaria* sp. J Phycol 19:516–522

Biebl R (1962) Seaweeds. In: Lewin RA (ed) Physiology and biochemistry of algae. Academic, New York, pp 799–815

Bischof K, Hanelt D, Wiencke C (1998a) UV-radiation can affect depth-zonation of Antarctic macroalgae. Mar Biol 131:597–605

Bischof K, Hanelt D, Tüg H, Karsten U, Brouwer PEM, Wiencke C (1998b) Acclimation of brown algal photosynthesis to ultraviolet radiation in Arctic coastal waters (Spitsbergen, Norway). Polar Biol 20:388–395

Bischof K, Hanelt D, Wiencke C (1999) Acclimation of maximal quantum yield of photosynthesis in the brown alga Alaria esculenta under high light and UV radiation. Plant Biol 1:435–444

Bischof K, Hanelt D, Wiencke C (2000a) UV-effects on photosynthesis and related enzyme reactions of marine macroalgae. Planta 211:555–562

Bischof K, Hanelt D, Wiencke C (2000b) UV-radiation and Arctic marine macroalgae. In: Hessen D (ed) UV-radiation and Arctic ecosystems, vol 153, Ecological Studies series. Springer, New York, pp 227–244

Brouwer PEM (1996) In situ photosynthesis and estimated annual production of the red alga *Myriogramme mangini* in relation to underwater irradiance at Signy island (Antarctica). Antarct Sci 8:245–252

Butler WL (1978) Energy distribution in the photochemical apparatus of photosynthesis. Annu Rev Plant Physiol 29:345–378

Chen M, Chory J, Fankhauser C (2004) Light signal transduction in higher plants. Annu Rev Genet 38:87–117

Critchley C, Russell AW (1994) Photoinhibition of photosynthesis in vivo: the role of protein turnover in photosystem II. Physiol Plant 92:188–196

Dau H (1994a) Short-term adaptation of plants to changing light intensities and its relation to photosystem II photochemistry and fluorescence emission. J Photochem Photobiol B Biol 26:3–27

Dau H (1994b) Molecular mechanisms and quantitative models of variable photosystem II fluorescence. Photochem Photobiol 60:1–23

Demmig-Adams B, Adams WW III (1992) Photoprotection and other responses of plants to high light stress. Annu Rev Plant Physiol Plant Mol Biol 43:599–626

Dring MJ (1981) Photosynthesis and development of marine macrophytes in natural sun light spectra. In: Smith H (ed) Plants and the day light spectrum. Academic, London, pp 297–314

Dring MJ (1982) The biology of marine plants. Edward Arnold, London, p 199

Dring MJ (1988) Photocontrol of development in algae. Annu Rev Plant Physiol Plant Mol Biol 39:157–174

Dring MJ, Makarov V, Schoschina E, Lorenz M, Lüning K (1996) Influence of ultraviolet-radiation on chlorophyll fluorescence and growth in different life-history stages of three species of *Laminaria* (Phaeophyta). Mar Biol 126:183–191

Dunton KH, Jodwalis CM (1988) Photosynthetic performance of *Laminaria solidungula* measured in situ in the Alaskan High Arctic. Mar Biol 98:277–285

Eberhard S, Finazzi G, Wollmann FA (2008) The dynamics of photosynthesis. Annu Rev Genet 42:463–515

Eggert A, Wiencke C (2000) Adaptation and acclimation of growth and photosynthesis of five Antarctic red algae to low temperatures. Polar Biol 23:609–618

Engelmann TW (1883) Farbe und assimilation. Botan Zentr 41:1–29

Engelmann TW (1884) Untersuchungen über die quantitativen beziehungen zwischen absorption des lichtes und assimilation in pflanzenzellen. Botan Zentr 42:82–95

Falkowski PG, Raven JA (1997) Aquatic photosynthesis. Blackwell Scientific, Oxford

Fischer G, Wiencke C (1992) Stable carbon isotope composition, depth distribution and fate of macroalgae from the Antarctic Peninsula region. Polar Biol 12:341–348

Flores-Moya A, Hanelt D, López-Figueroa F, Altamirano M, Viñegla B, Salles S (1999) Solar UV-B radiation shows beneficial effects on recovery of inhibited photosynthesis in the brown alga *Dictyota dichotoma*. J Photochem Photobiol B Biol 49:129–135

Flores-Moya A, Posudin YI, Fernández JA, Figueroa FL, Kawai H (2002) Photomovement of the swarmers of the brown algae *Scytosiphon lomentaria* and *Petalonia fascia*: effect of photon irradiance, spectral composition and UV dose. J Photochem Photobiol B Biol 66:134–140

Foyer CH, Shigeoka S (2011) Understanding oxidative stress and antioxidant functions to enhance photosynthesis. Plant Physiol 155:93–100

Franklin LA, Forster RM (1997) The changing irradiance environment: consequences for marine macrophyte physiology, productivity and ecology. Eur J Phycol 32:207–232

Franklin LA, Osmond CB, Larkum AWD (2003) Photoinhibition, UV-B and algal photosynthesis. In: Larkum AWD, Douglas SE, Raven JA (eds) Photosynthesis in algae. Kluwer, Dordrecht, pp 351–384

Glover HE, Keller MD, Guillard RRL (1986) Light quality and oceanic ultraphytoplankters. Nature 319:142–143

Glover HE, Keller MD, Spinrad RW (1987) Effects of light quality and intensity on photosynthesis and growth of marine eukaryotic and prokaryotic phytoplankton clones. J Exp Mar Biol Ecol 105:137–159

Gómez I, Thomas DN, Wiencke C (1995a) Longitudinal profiles of growth, photosynthesis and light independent carbon fixation in the Antarctic brown alga *Ascoseira mirabilis*. Bot Mar 38:157–164

Gómez I, Wiencke C, Weykam G (1995b) Seasonal photosynthetic characteristics of *Ascoseira mirabilis* (Ascoseirales, Phaeophyceae) from King George Island, Antarctica. Mar Biol 123:167–172

Gómez I, Weykam G, Klöser H, Wiencke C (1997) Photosynthetic light requirements, metabolic carbon balance and zonation of sublittoral macroalgae from King George Island (Antarctica). Mar Ecol Prog Ser 148:281–293

Guenther JE, Melis A (1990) The physiological significance of photosystem II heterogeneity in chloroplasts. Photosynth Res 23:105–109

Häder DP, Figueroa FL (1997) Photoecophysiology of marine macroalgae. Photochem Photobiol 66:1–14

Hanelt D (1992) Photoinhibition of photosynthesis in marine macrophytes of the South Chinese Sea. Mar Ecol Prog Ser 82:199–206

Hanelt D (1998) Capability of dynamic photoinhibition in marine macroalgae is related to their depth distribution. Mar Biol 131:361–369

Hanelt D, Nultsch W (1989) Action spectrum of phaeoplast displacement from the dark to the low intensity arrangement in the brown alga *Dictyota dichotoma*. J Photochem Photobiol B Biol 4:111–121

Hanelt D, Nultsch W (2002) Photoinhibition in seaweeds. In: Heldmaier G, Werner D (eds) Environmental signal processing and adaptation. Springer, Berlin, pp 141–168

Hanelt D, Roleda M (2009) UV-B radiation may ameliorate photoinhibition in specific shallow-water tropical marine macrophytes. Aquatic Bot 91:6–12

Hanelt D, Huppertz K, Nultsch W (1992) Photoinhibition of photosynthesis and its recovery in red algae. Bot Acta 105:278–284

Hanelt D, Huppertz K, Nultsch W (1993) Daily course of photosynthesis and photoinhibition in marine macroalgae investigated in the laboratory and in the field. Mar Ecol Prog Ser 97:31–71

Hanelt D, Melchersmann B, Wiencke C, Nultsch W (1997a) Effects of high light stress on photosynthesis of polar macroalgae in relation to depth distribution. Mar Ecol Prog Ser 149:255–266

Hanelt D, Wiencke C, Nultsch W (1997b) Influence of UV-radiation on the photosynthesis of Arctic macroalgae in the field. J Photochem Photobiol B Biol 38:40–47

Hanelt D, Wiencke C, Karsten U, Nultsch W (1997c) Photoinhibition and recovery after high light stress in different developmental and life-history stages of *Laminaria saccharina* (Phaeophyta). J Phycol 33:387–395

Hanelt D, Wiencke C, Bischof K (2003) Photosynthesis in marine macroalgae. In: Larkum AWD, Douglas SE, Raven JA (eds) Photosynthesis in algae, vol 14. Kluwer, Dordrecht, pp 413–435

Hanelt D, Hawes I, Rae R (2006) Reduction of UV-B radiation causes an enhancement of photoinhibition in high light stressed aquatic plants from New Zealand lakes. J Photochem Photobiol B Biol 84:89–102

Harder R, Bederke B (1957) Über Wachstumsversuche mit Rot- und Grünalgen (*Porphyridium cruentum, Trailliella intricata, Chlorella pyrenoidosa*) in verschiedenfarbigem, energiegleichem Licht. Arch Mikrobiol 28:153–172

Haxo FT, Blinks LR (1950) Photosynthetic action spectra of marine algae. J Gen Physiol 33:389–442

Huppertz K, Hanelt D, Nultsch W (1990) Photoinhibition of photosynthesis in the marine brown alga Fucus serratus as studied in field experiments. Mar Ecol Prog Ser 66:175–182

Jerlov NG (1976) Marine optics. Elsevier, Amsterdam

Johansson G, Snoeijs P (2002) Macroalgal photosynthetic responses to light in relation to thallus morphology and depth zonation. Mar Ecol Prog Ser 244:63–72

Jones LW, Kok B (1966) Photoinhibition of chloroplast reactions I. Kinetics and action spectra. Plant Physiol 41:1037–1043

Karsten U, Wiencke C (1999) Factors controlling the formation of UV-absorbing mycosporine-like amino acids in the marine red alga *Palmaria palmata* from Spitsbergen (Norway). J Plant Phys 155:407–415

Kawai H, Müller DG, Fölster E, Häder DP (1990) Phototactic response in the gametes of the brown alga, *Ectocarpus siliculosus*. Planta 182:292–297

Kirst GO, Wiencke C (1995) Ecophysiology of polar algae. J Phycol 31:181–199

Kleine T, Kindgren P, Benedict C, Hendrickson L, Strand A (2007) Genome-wide gene expression analysis reveals a critical role for cryptochrome1 in the response of *Arabidopsis* to high irradiance. Plant Physiol 144:1391–1406

Krause GH (1988) Photoinhibition of photosynthesis an evaluation of damaging and protective mechanisms. Physiol Plant 74:566–574

Krause GH, Weis E (1991) Chlorophyll fluorescence and photosynthesis: the basics. Annu Rev Plant Physiol Plant Mol Biol 42:313–349

Küster A, Schaible R, Schubert H (2004) Light acclimation of the charophyte Lamprothamnium papulosum. Aquat Bot 79:111–124

Larkum AWD, Wood WF (1993) The effect of UV-B radiation on photosynthesis and respiration of phytoplankton, benthic macroalgae and seagrasses. Photosynth Res 36:17–23

Larkum AWD, Drew EA, Crossett RN (1967) The vertical distribution of attached marine algae in Malta. J Ecol 55:361–371

Leukart P, Lüning K (1994) Minimum spectral light requirements and maximum light levels for long-term germling growth of several red algae from different water depths and a green alga. Eur J Phycol 29:103–112

Littler MM (1980) Morphological form and photosynthetic performances of marine macroalgae: tests of a functional/form hypothesis. Bot Mar 22:161–165

Littler MM, Littler DS, Taylor PR (1983) Evolutionary strategies in a tropical barrier reef system: functional-form groups of marine macroalgae. J Phycol 19:229–237

Littler MM, Littler DS, Blair SM, Norris JM (1986) Deep water plant communities from an uncharted seamount off San Salvador island, Bahamas: distribution, abundance and primary productivity. Deep Sea Res 33:881–892

López-Figueroa F, Niell FX (1991) Photocontrol of chlorophyll and biliprotein synthesis in seaweeds: possible photoreceptors and ecological considerations. Scientia Marina 55:519–527

López-Figueroa F, Rüdiger W (1991) Stimulation of nitrate net uptake and reduction by red and blue light and the reversion by far-red light in the green alga *Ulva rigida*. J Phycol 27:389–394

López-Figueroa F, Lindemann P, Braslavsky SE, Schaffner K, Schneider-Poetsch HA, Rüdiger W (1989) Detection of a phytochrome -like protein in macroalgae. Bot Acta 102:178–180

López-Figueroa F, Lindemann P, Braslavsky SE, Schaffner K, Schneider-Poetsch HA, Rüdiger W (1990) Detection of some conserved domains in phytochrome-like protein from algae. J Plant Physiol 136:484–487

Lüder UH, Knoetzel J, Wiencke C (2001) Acclimation of photosynthesis and pigments to seasonally changing light conditions in the endemic Antarctic red alga *Palmaria decipiens*. Polar Biol 24:598–603

Lüder UH, Wiencke C, Knoetzel J (2002) Acclimation of photosynthesis and pigments during and after six months of darkness in *Palmaria decipiens* (Rhodophyta): A study to simulate Antarctic winter sea ice cover. J Phycol 38:904–913

Lüning K (1981) Light. In: Mj W, Lobban CS (eds) The biology of seaweeds. Blackwell, Oxford, pp 326–355

Lüning K (1990) Seaweeds: their environment, biogeography and ecopysiology. Wiley, New York

Lüning K, Dring MJ (1985) Action spectra and spectral quantum yield of photosynthesis in marine macroalgae with thin and thick thalli. Mar Biol 87:119–129

Lüning K, tom Dieck I (1989) Environmental triggers in algal seasonality. Bot Mar 32:389–397

Maegawa M, Kunieda M, Kida W (1993) The influence of ultraviolet radiation on the photosynthetic activity of several red algae from different depths. Jpn J Phycol 41:207–214

Marquardt R, Schubert H, Varela DA, Huovinen P, Henríquez L, Buschmann AH (2010) Light acclimation strategies of three commercially important red algal species. Aquaculture 299:140–148

Mattoo AK, Hoffman-Falk H, Marder JB, Edelman M (1984) Regulation of protein metabolism: coupling of photosynthetic electron transport in vivo degradation of the rapidly metabolised 32-kilodalton protein of the chloroplast membranes. Proc Natl Acad Sci USA 81:1380–1384

Müller DG, Maier I, Müller H (1987) Flagellum autofluorescence and photoaccumulation in heterokont algae. Photochem Photobiol 46:1003–1008

Murata N, Takahashi S, Nishiyama Y, Allakhverdiev SI (2007) Photoinhibition of photosystem II under environmental stress. Biochim Biophys Acta 1767:414–421

Murchie EH, Niyogi KK (2011) Manipulation of photoprotection to improve plant photosynthesis. Plant Physiol 155:68–92

Nikaido SS, Johnson CH (2000) Daily and circadian variation in survival from ultraviolet radiation in Chlamydomonas reinhardtii. Photochem Photobiol 71:758–765

Nishiyama Y, Allakhverdiev SI, Murata N (2006) A new paradigm for the action of reactive oxygen species in the photoinhibition of photosystem II. Biochim Biophys Acta 1757:742–749

Niyogi KK, Li XP, Rosenberg V, Jung HS (2005) Is PsbS the site of non-photochemical quenching in photosynthesis? J Exp Bot 56:375–382

Nultsch W, Pfau J, Materna-Weide M (1987) Fluence and wavelength dependence of photoinhibition in the brown alga Dictyota dichotoma. Mar Ecol Prog Ser 41:93–97

Nultsch W, Pfau J, Huppertz K (1990) Photoinhibition of photosynthetic oxygen production and its recovery in the subtidal red alga *Polyneura hilliae*. Bot Acta 103:62–67

Ohad I, Kyle DJ, Arntzen CJ (1984) Membrane protein damage and repair: Removal and replacement of inactivated 32-kilodalton polypeptides in chloroplast membranes. J Cell Biol 99:481–485

Öquist G, Chow WS (1992) On the relationship between the quantum yield of photosystem II electron transport, as determined by chlorophyll fluorescence and the quantum yield of CO_2-dependent O_2 evolution. Photosynth Res 33:51–62

Osmond CB (1994) What is photoinhibition? Some insights from comparisons of shade and sun plants. In: Baker NR, Bowyer JR (eds) Photoinhibition of photosynthesis, from the molecular mechanisms to the field. BIOS Scientific, Oxford, pp 1–24

Peers G, Truong TB, Ostendorf E, Busch A, Elrad D, Grossman AR, Hippler M, Krishna K, Niyogi K (2009) An ancient light-harvesting protein is critical for the regulation of algal photosynthesis. Nature 462:518–522

Pfündel EE, Dilley RA (1993) The pH dependence of violaxanthin deepoxidation in isolated pea chloroplasts. Plant Physiol 101:65–71

Powles SB (1984) Photoinhibition of photosynthesis induced by visible light. Annu Rev Plant Physiol 35:15–44

Ramus J (1981) The capture and transduction of light energy. In: Lobban CS, Wynne MJ (eds) The biology of seaweeds. University of California Press, Berkely, pp 458–492

Ramus J, Beale SI, Mauzerall D, Howard KL (1976) Changes in photosynthetic pigment concentration in seaweeds as a function of water depth. Mar Biol 37:223–229

Ramus J, Lemons F, Zimmerman C (1977) Adaptation of light-harvesting pigments to downwelling light and the consequent photosynthetic performance of the eulittoral rockweeds *Ascophyllum nodosum* and *Fucus vesiculosus*. Mar Biol 42:293–303

Raven JA, Geider RJ (2003) Adaptation, acclimation and regulation in algal photosynthesis. In: Larkum AWD, Douglas SE, Raven JA (eds) Photosynthesis in algae. Kluwer, Dordrecht, pp 385–412

Raven JA, Kübler JE, Beardall J (2000) Put out the light, and then put out the light. J Mar Biol Assoc UK 80:1–25

Ruban AV, Berera R, Ilioaia C, van Stokkum IH, Kennis JT, Pascal AA, van Amerongen H, Robert B, Horton P, van Grondelle R (2007) Identification of a mechanism of photoptotective energy dissipation in higher plants. Nature 450:575–578

Rüdiger W, López-Figueroa F (1992) Photoreceptors in algae. Photochem Photobiol 55:949–954

Sagert S, Forster RM, Feuerpfeil P, Schubert H (1997) Daily course of photosynthesis and photoinhibition in Chondrus crispus (Rhodophyta) from different shore levels. Eur J Phycol 32:363–371

Senger H, Humbeck K, Schiller H (2002) Light adaptation of the photosynthetic apparatus of green algae. In: Werner D, Heldmaier G (eds) Environmental signal processing and adaptation. Springer, Berlin, pp 71–86

Setlow RB (1974) The wavelengths in solar radiation effective in producing skin cancer: a theoretical analysis. Proc Nat Acad Sci USA 71:3363–3366

Sicora C, Máté Z, Vass I (2003) The interaction of visible and UV-B light during photodamage and repair of photosystem II. Photosynth Res 75:127–137

Steemann Nielsen E (1975) Marine photosynthesis. Elsevier, Amsterdam

Stengel D, Dring M (1998) Seasonal variation in the pigment content and photosynthesis of different thallus regions of *Ascophyllum nodosum* (Fucales, Phaeophyta) in relation to position in the canopy. Phycologia 37:259–268

Suzuki L, Johnson CH (2001) Algae know the time of day: circadian and photoperiodic programs. J Phycol 37:933–942

Thomson PG, Davidson AT, Cadman N (2008) Temporal changes in effects of ambient UV radiation on natural communities of Antarctic marine protists. Aquat Microb Ecol 52:131–147

Vass I (1997) Adverse effects of UV-B light on the structure and function of the photosynthetic apparatus. In: Pessarakli M (ed) Handbook of photosynthesis. Marcel Dekker, New York, pp 931–949

Weykam G, Gómez I, Wiencke C, Iken K, Klöser H (1996) Photosynthetic characteristics and C:N ratios of macroalgae from King George Island (Antarctica). J Exp Mar Biol Ecol 204:1–22

Wiencke C (1990a) Seasonality of brown macroalgae from Antarctica -a long-term culture study under fluctuating Antarctic daylength. Polar Biol 10:589–600

Wiencke C (1990b) Seasonality of red and green macroalgae from Antarctica -a long-term culture study under fluctuating Antarctic daylength. Polar Biol 10:601–607

Wiencke C, Fischer G (1990) Growth and stable carbon isotope composition of cold-water macroalgae in relation to light and temperature. Mar Ecol Prog Ser 65:283–292

Wiencke C, Rahmel J, Karsten U, Weykam G, Kirst GO (1993) Photosynthesis of marine macroalgae from Antarctica: light and temperature requirements. Bot Acta 106:77–87

Wiencke C, Gómez I, Pakker H, Flores-Moya A, Altamirano M, Hanelt D, Bischof K, Lopez-Figueroa F (2000) Impact of UV radiation on viability, photosynthetic characteristics and DNA of brown algal zoospores: Implications for depth zonation. Mar Ecol Prog Ser 197:217–229

Wiencke C, Clayton MN, Gómez I, Iken K, Lüder UH, Amsler CD, Karsten U, Hanelt D, Bischof K, Dunton K (2007) Life strategy, ecophysiology and ecology of seaweeds in polar waters. Rev Environ Sci Biotechnol 6:95–126

Wilhelm C, Selmar D (2011) Energy dissipation is an essential mechanism to sustain the viability of plants: The physiological limits of improved photosynthesis. J Plant Physiol 168:79–87

Wilson A, Ajlanib G, Verbavatzb JM, Vassc I, Kerfeldd CA, Kirilovsky D (2006) A soluble carotenoid protein involved in phycobilisome-related energy dissipation in cyanobacteria. Plant Cell 18:992–1007

Wood WF (1987) Effect of solar ultraviolet radiation on the kelp *Ecklonia radiata*. Mar Biol 96:143–150

Yokohama Y, Misonou T (1980) Chlorophyll a:b ratios in marine benthic algae. Jpn J Phycol 28:219–223

Chapter 2
Morpho-functionality of Carbon Metabolism in Seaweeds

Iván Gómez and Pirjo Huovinen

2.1 Introduction

The process by which inorganic carbon is converted into organic forms driven by the solar energy is unique to photoautotrophic organisms (some types of prokaryotes, cyanobacteria, eukaryotic algae, and plants). Although in many cases the underlying mechanisms and enzymatic machinery associated with carbon metabolism are essentially similar to those found in higher plants, seaweeds show some differences, especially related with carbon acquisition and concentration, biochemical strategies to avoid the oxygen/CO_2 interference of RUBISCO (photorespiration), and C4 metabolism. Other striking characteristic lies in the remarkable versatility of the different biochemical pathways and products that allow seaweeds to operate under resource limitation (e.g., CO_2, light, nutrients) and under changing environments. On the other hand, seaweeds exhibit a great variety of storage carbohydrates (e.g., mannitol, floridoside). These compounds can be remobilized during growth and reproduction, and are also normally involved in a series of reactions associated with osmoregulation and antioxidant activity (see Lobban and Harrison 2000 for an overview).

Seaweeds are multicellular, macroscopic organisms and thus a major aspect of carbon metabolism that makes them different from other groups of algae and cyanobacteria is its dependency on gross morphology. The morpho-functional processes of carbon assimilation occurring in foliose and finely branched seaweeds can be considerably different from those of, e.g., coarsely branched and leathery species (Ramus 1978; Rosenberg et al. 1995). In many large brown algae, seasonal decoupling of carbon assimilation and anabolic processes, long-distance transport of photoassimilates, and biomass formation restricted to meristematic cells are

I. Gómez (✉) • P. Huovinen
Instituto de Ciencias Marinas y Limnológicas, Universidad Austral de Chile, Campus Isla Teja s/n, Casilla 567, Valdivia, Chile
e-mail: igomezo@uach.cl; pirjo_huovinen@uach.cl

C. Wiencke and K. Bischof (eds.), *Seaweed Biology*, Ecological Studies 219,
DOI 10.1007/978-3-642-28451-9_2, © Springer-Verlag Berlin Heidelberg 2012

conceived as fundamental adaptations for optimal allocation of energy in changing environments (Gómez et al. 2007).

In terms of its ecological importance, the photosynthetic transformation of inorganic carbon into organic molecules carried out by benthic seaweeds accounts for an important fraction of the coastal primary production and biomass. Furthermore, the fate of seaweed-derived organic material is central for the higher trophic levels and geochemical processes of the coastal ecosystems (Mann 1973).

2.2 Inorganic Carbon Acquisition

Seaweeds must use CO_2 dissolved in seawater as inorganic carbon source. In the surface water the different species of inorganic carbon (CO_2, HCO_3^-, and CO_3^{2-}) are in equilibrium in the so-called carbonate buffer system (see also Chap. 19 by Roleda and Hurd). For example, at partial pressure of 365 μatm, pH 8.1, 25°C, and salinity of 35 psu, the CO_2 concentration is close to 10.4 μmol kg^{-1}, while HCO_3^- and CO_3^{2-} have values of 1,818 and 272 μmol kg^{-1}, respectively. Thus, only a small fraction (~0.5%) is in the form of CO_2 (Zeebe and Wolf-Gladrow 2001). Diffusion of CO_2 in water (0.16×10^{-4} cm^{-2} s^{-1}) is four orders of magnitude lower than in air (0.16 cm^{-2} s^{-1}), which has important consequences for photosynthesis (Badger and Spalding 2000). In the case of aquatic organisms, the entry of CO_2 into the cell is normally limited by the diffusion boundary layer, whose thickness, and hence its resistance, depends on the form and volume of the alga as well as the speed of the water flow around it. For example, for an aqueous phase system with a diffusion coefficient of 1.5×10^{-5} m^{-2} s^{-1} and a boundary layer of 15 μm, a maximum flux of CO_2 close to 2.6 m^{-2} s^{-1} can be estimated (Falkowski and Raven 1997). Due to these constraints the sole diffusive entry of CO_2 does not support photosynthetic demands and thus algae can suffer carbon limitation. This situation has been documented for some subtidal red algae that apparently rely on CO_2 diffusion as the only mechanism of inorganic carbon uptake (Raven and Beardall 1981; Maberly 1990). The majority of seaweeds, however, have developed the capacity to concentrate CO_2 in order to guarantee an adequate supply to RUBISCO. One of the most efficient carbon concentrating mechanism (CCM) is the active transport not only of CO_2 but also HCO_3^-, which accounts for up to 95% of the total dissolved inorganic carbon in seawater (reviewed by Raven 2010; see also Chap. 4 by Gordillo). In many cases, the HCO_3^- is an intermediate pool, which is converted to CO_2 in the vicinity of RUBISCO (Badger and Price 1994). Similar to strict CO_2 users, there is depth dependence in the ability of seaweeds to use HCO_3^- with a tendency of higher affinity for HCO_3^- in seaweeds occupying the upper littoral zones compared to their counterparts from deeper locations (Sand-Jensen and Gordon 1984; Mercado et al. 1998; Murru and Sandgren 2004).

The principal way by which algae utilize HCO_3^- is through the enzyme carbonic anhydrase (CA). This enzyme catalyzes the interconversion between HCO_3^- and CO_2 in distinct sites outside or inside the cell (Badger and Price 1994; Badger 2003) and its activity has experimentally been demonstrated in different seaweeds (Haglund et al. 1992; Mercado and Niell 1999). CA activities can vary considerably depending on various environmental factors (e.g., temperature, UV radiation) and at different temporal scales (Flores-Moya et al. 1998; Gómez et al. 1998a, b; Choo et al. 2005). Various surveys carried out in North Atlantic (Giordano and Maberly 1989), Mediterranean (Mercado et al. 1998, 2009), southern Chile (Huovinen et al. 2007), and the Arctic (Gordillo et al. 2006) have confirmed that the CA-based inorganic carbon acquisition is broadly extended in seaweeds, suggesting that this metabolic ability is advantageous in coping with changes in the availability of ambient CO_2. Other mechanisms include the nondiffusive incorporation of HCO_3^- via a specific transporter or proton pump, e.g., ATPase, an OH^-/HCO_3^- antiport system, which have been reported, e.g., in the brown alga *Laminaria digitata* (Klenell et al. 2002), and the green alga *Cladophora* (Choo et al. 2005), or an anion exchanger at the plasmalemma as has been postulated for *Ulva* sp. (Drechsler et al. 1994). Many aspects dealing with the nature of the transporter or the unbalance in the electrochemical potential across membranes remain unknown, but apparently its operation does not preclude the action of any intracellular CA (Raven and Lucas 1985).

Unlike terrestrial C3 plants that base their inorganic acquisition on diffusive CO_2 entry, the majority of seaweeds exhibit functional CCMs (Raven 2010). For example, the model of carbon acquisition/assimilation in the Chlorophyta *Ulva* sp. is based on the capacity of this alga to convert HCO_3^- into CO_2 via external CA and also to transport actively HCO_3^- through the plasmalemma (Beer 1996). If one considers that at normal atmospheric CO_2 level (350 ppm, chloroplast flux of CO_2 of about 2.8 mM s^{-1}), the uncatalyzed rate of interconversion of CO_2 to HCO_3^- is 10,000 times slower than the biological flux via CO_2 fixation by RUBISCO, then the action of CA is necessary in seaweeds (Badger and Price 1994). In the case of some intertidal algae such as *Fucus*, CCMs have also been proposed to serve as an inhibitor of the oxygenase activity of RUBISCO (photorespiration) during emersion periods (Kawamitsu and Boyer 1999) (see below). Overall, the ecological significance of these mechanisms in seaweeds, as well as their prevalence in relation to phylogeny and biogeography, has been proposed (Surif and Raven 1990; Raven 1991; Mercado et al. 2009). An increasingly relevant issue is the unpredictable effect of present and future global change-driven increases in CO_2 concentration, which probably have impact on carbon acquisition patterns of seaweeds (Raven et al. 2002; Hurd 2000; Mercado et al. 2009, see Chap. 19 by Roleda and Hurd).

2.3 Photosynthetic Carbon Fixation

2.3.1 Calvin–Benson Cycle and RUBISCO

The process of fixation of CO_2 into ribulose 1,5-bisphosphate (RuBP) to form triose phosphate is denominated photosynthetic carbon reduction cycle or Calvin–Benson cycle and occurs in the chloroplasts of seaweeds. The whole cycle consists of three major reactions: carboxylation, reduction, and regeneration. The carboxylation part can be summarized as:

3 Ribulose-1,5-bisphosphate (RuBP) + $3CO_2$ + H_2O + 2 × 3 Phosphoglycerate (PGA)

The chloroplast enzyme ribulose1,5 bis-phosphate carboxylase/oxygenase (RUBISCO) is central in the carboxylation phase. The enzyme has a complex structure and is formed by eight large catalytic subunits of approximately 53 kDa encoded in the chloroplast genome and eight small 15 kDa mass peptides encoded in the nucleus. Different types of RUBISCO (I, II, III, and IV), which vary in different groups of photosynthetic organisms according to their kinetic characteristics, have been recognized (Tabita et al. 2008). Only variants of the type I have been found in seaweeds (Raven 1997). Although the enzyme is dispersed in the chloroplast stroma, a portion of the RUBISCO is located in proteinaceous bodies denominated pyrenoids, which are present in many seaweed groups (McKay et al. 1991). The role of pyrenoids in carbon metabolism is still not well understood; however, the evidence that other enzymes (e.g., RUBISCO activases) are also located in these bodies suggests its functional involvement in carbon fixation in the chloroplast.

The Calvin–Benson cycle is autocatalytic, i.e., its reactions generate the biochemical intermediates, which enhance the rate of carbon fixation in the case of increase in its concentration. It has been demonstrated that thallus regions of seaweeds with higher relative abundances in RUBISCO exhibit the highest rates of carbon fixation (Cabello-Pasini and Alberte 2001b). On the other hand, activity of RUBISCO is highly dependent on the environmental factors (Raven 1997; Bischof et al. 2002). During the photosynthetic carboxylation, CO_2 is incorporated into the carboxyl group of the RuBP to form an unstable intermediate (enediol) and finally 3-phosphoglycerate (PGA), which is the first and most important compound of the cycle that is labeled in presence of ^{14}C (Calvin 1956; Beer and Israel 1986). In the reduction phase, the main reaction is the conversion of PGA in glyceraldehyde 3-phosphate (GAP) through the use of ATP and NADPH formed in the photochemical reactions by GAP-dehydrogenase (GAP-DH). Due to this dependence, a feedback between both processes has been examined in some seaweeds, mainly in relation with environmental factors that affect thylakoid membranes (Bischof et al. 2002). The third stage of the carbon fixation process is the regeneration that allows maintaining the operation of the cycle through the constant supply of RuBP from molecules of triose phosphate. In the ATP-consuming process, which includes isomerization, condensation, hydrolization, and phosphorylation reactions, three molecules of RuBP are formed from five molecules of triose phosphate (Nelson and Cox 2004).

2.3.2 *Rates of Carbon Fixation*

Rates of RUBISCO-catalyzed carbon fixation measured using radioactive carbon isotopes ($H^{14}CO_3$) vary in seaweeds depending on different environmental and endogenous factors. Values compiled in Table 2.1 indicate that, irrespective of taxa

Table 2.1 Summary of photosynthetic carbon fixation and light-independent carbon fixation (LICF) rates as well as percentage ratios of selected seaweeds from three major taxonomic divisions and biogeographic regions

Species	Location	Thallus part	Light ^{14}C fixation	LICF	Ratio (%)	References
			(µmol ^{14}C h^{-1}) g^{-1} FW			
Chlorophyta						
Cladophora columbina	Pacific Grove, California	Whole thallus	6.6	0.06	0.90	Cabello-Pasini and Alberte (1997)
Codium fragile	Helgoland Island, North Sea	Whole thallus	20.5	0.27	1.31	Kremer (1981)
Codium decorticatum[a]	Atlantic coast of Florida	Whole thallus	20.6	0.1	0.48	Reiskind et al. (1988)
Ulva lactuca	Pacific Grove, California	Whole thallus	27	0.18	0.66	Cabello-Pasini and Alberte (1997)
Ulva lactuca[b]	Helgoland Island, North Sea	Whole thallus	14.6	0.08	0.55	Kremer (1981)
Udotea flabellum	Florida, Gulf of Mexico	Blades	10.1	0.65	6.22	Reiskind et al. (1988)
Rhodophyta						
Delesseria sanguinea[c]	Helgoland Island, North Sea	Whole thallus	7.95	0.06	0.75	Kremer (1981)
Chondracantus canaliculatus	Pacific Grove, California	Whole thallus	15.6	0.07	0.45	Cabello-Pasini and Alberte (1997)
Chondrus crispus	Helgoland Island, North Sea	Whole thallus	4.2	0.18	0.75	Kremer (1981)
Gracilaria lemaneiformis	Pacific Grove, California	Apical branch	12	0.06	0.50	Cabello-Pasini and Alberte (1997)
Iridaea cordata	Pacific Grove, California	Whole thallus	14.4	0.07	0.48	Cabello-Pasini and Alberte (2001a)
Iridaea cordata[d]	King George I., Antarctica (culture)	Whole thallus	2.5–15	0.2–2.1	1.9–17.2	Weykam (1996)
Palmaria decipiens	King George I., Antarctica (culture)	Young blade	11.6	0.20	1.72	Thomas and Wiencke (1991)
Palmaria decipiens[d]	King George I., Antarctica (culture)	Whole thallus	4–13	0.12–1.59	0.7–31.1	Weykam (1996)
Plocamium cartilagineum	Pacific Grove, California	Apical branch	12	0.04	0.33	Cabello-Pasini and Alberte (1997)
Phaeophyceae						
Ascophyllum nodosum	Tay Estuary, East Coast of Scotland	Blades	15	0.37	2.46	Johnston and Raven (1986)
Ascoseira mirabilis	King George I., Antarctica (culture)	Growing zone	11.4	0.7	6.14	Thomas and Wiencke (1991)
Ascoseira mirabilis	King George I., Antarctica (culture)	Apical zone	15.8	1.5	9.50	Gómez et al. (1995a, b)
Desmarestia anceps	King George I., Antarctica (culture)	Whole thallus	29.7	1	3.36	Thomas and Wiencke (1991)
Desmarestia antarctica	King George I., Antarctica (culture)	Young thallus	43.9	1.2	2.73	Thomas and Wiencke (1991)
Egregia menziesii	Pacific Grove, California	Lateral blade	16.2	0.11	0.67	Cabello-Pasini and Alberte (1997)
Eisenia arborea[e]	Helgoland I., North Sea	Blades	18.5	0.71	3.83	Kremer (1981)
Fucus serratus	Helgoland I., North Sea	Growing zone	17.4	0.70	4	Kremer (1981)
Himantothallus grandifolius	King George I., Antarctica (culture)	Young thallus	11	0.40	3.63	Thomas and Wiencke (1991)
Laminaria setchellii	Pacific Grove, California	Growing zone	15	1.92	12.80	Cabello-Pasini and Alberte (2001b)

(continued)

Table 2.1 (continued)

Species	Location	Thallus part	Light ^{14}C fixation (μmol ^{14}C h^{-1}) g^{-1}	LICF FW	Ratio (%)	References
Laminaria digitata	Helgoland I., North Sea	Blade	5	0.38	7.60	Kremer (1981)
Laminaria digitata	Helgoland I., North Sea	Growing zone	5.8	1.61	27.75	Kremer (1981)
Laminaria hyperborea	Helgoland I., North Sea	Growing zone	2.1	1.01	48.09	Kremer (1981)
Laminaria solidungula	Beauford Sea, Alaskan High Arctic	Growing zone	7.6	0.34	4.47	Dunton and Schell (1986)
Lessonia nigrescens[f]	Valdivia, South East Pacific coast	Whole thallus	5.4–47.5	2.3–9.6	20.2–42.3	Gómez et al. (2005)
Lessonia nigrescens	Valdivia, South East Pacific coast	Growing zone	21	3.8	18.09	Gómez et al. (2007)
Macrocystis pyrifera	Corona del Mar, California	Inmature blades	51.1	0.2	0.39	Arnold and Manley (1985)
Pelvetia compressa	Pacific Grove, California	Apical blade	10.3	1.10	10.67	Cabello-Pasini and Alberte (1997)
Saccharina latissima	Helgoland I., North Sea	Growing zone	5.5	0.92	16.72	Kremer (1981)
Stephanocystis osmundacea	Pacific Grove, California	Lower blade	7.2	1.14	15.8	Cabello-Pasini and Alberte (1997)

Values were compiled from studies using ^{14}C labeling. Ratio (%) represents the proportion of LICF relative to photosynthetic carbon fixation

[a]A Chl *a* content of 0.5 mg g^{-1}FW was assumed

[b]A FW:DW ratio of 0.3 was assumed

[c]A FW:DW ratio of 0.2 was assumed

[d]Range from seasonal data

[e]A FW:DW ratio of 0.18 was assumed

[f]Range between different thallus parts

or geographical region, carbon fixation reaches maxima of close to 40–50 μmol $^{14}C\ g^{-1}$ FW h^{-1}. However, age and thallus part can be relevant components of variability. For example, in complex morphs (e.g., some red algae and large brown algae) ^{14}C fixation can considerably increase in mature thallus regions (which attain a well-developed photosynthetic apparatus) compared to meristematic (growing) zones (Küppers and Kremer 1978; Gómez et al. 2007). At a molecular level, the number of active sites (~4–8 mM), the concentrations of CO_2, levels of O_2 (which competes with CO_2), and RuBP are key factors determining the in vivo kinetics of RUBISCO (Woodrow and Berry 1988). In contrast to the terrestrial C4 plants enriching the concentration of CO_2 via decarboxylation of C4-acids (van Caemmerer and Furbank 2003), seaweeds increase the availability of inorganic CO_2 to RUBISCO (and in parallel inhibits the oxygenase activity of the enzyme) through the action of CCMs (Raven 2010).

2.3.3 Photorespiration

The oxygenase property of RUBISCO, mainly of organisms with diffusive entry of CO_2, is a relevant topic in photosynthetic physiology. In fact, RUBISCO catalyzes the competitive oxidation of RuBP by fixation of O_2 to RuBP to form glycolate and PGA, a pathway-denominated C2 oxidative photosynthetic carbon cycle, which coexists with the Calvin–Benson cycle. In strict sense, the term "photorespiration" implies the consumption of O_2 and release of CO_2 in the light and thus the process depends on the CO_2/O_2 balance, the so-called CO_2 compensation. At low partial pressure of CO_2 and high O_2, photosynthetic carbon fixation is competitively inhibited by the oxygenase activity of RUBISCO with formation of CO_2 from the metabolism of glycolate (Raven et al. 2005). Thus, photorespiration is integrated in the whole photosynthetic carbon metabolism (Tolbert 1997).

Photorespiration in seaweeds has been less studied than in other photosynthetic organisms, probably because seaweeds exhibit CCMs. However, an effect of O_2 on carbon fixation has been demonstrated for some representative seaweeds, which exhibit ratios of oxygenase to carboxylase activities between 0.1 and 0.25 (Raven 1997; Giordano et al. 2005). In seaweeds physiologically resembling C3 plants, e.g., understory red algae that acquire carbon via diffusive CO_2 entry, the effects of photorespiration on photosynthetic carbon fixation are higher than in other groups (Raven 2010). In addition, the detection of various enzymes involved in the glycolate metabolism (e.g., P-glycolate phosphatase, glycolate oxidase, and glycolate dehydrogenase) (Gross 1990; Suzuki et al. 1991) as well as some of their products in different seaweeds (Reiskind et al. 1988) suggests that photorespiratory carbon oxidation is widespread in these organisms and in many ways similar to terrestrial plants (Raven 1997). The fate of glycolate in the cell, which includes its oxidation to glyoxylate in peroxisomes and further conversion to amino acids and CO_2, has been studied only in some seaweeds (Iwamoto and Ikawa 1997). Overall, although the implications of the photorespiratory pathway for seaweed ecology and its reason of maintenance along the evolution of algae are not well understood, its expression has been demonstrated and apparently, under certain environmental conditions, it can have consequences for the whole carbon metabolism of algae (Raven 1997).

2.4 Light-Independent Carbon Fixation

Carboxylation is not an exclusive feature of RUBISCO; seaweeds are equipped with a suite of diverse nonphotosynthetic enzymes that, like C4 and CAM in plants, are able to carboxylate and decarboxylate various C3 and C4 compounds. Light-independent carbon fixation (LICF) is also called "dark carbon fixation" or "β-carboxylation," since inorganic carbon is fixed into the β-site of acceptors such as phosphoenolpyruvate (PEP) or pyruvate. Two enzymes, phosphoenolpyruvate carboxylase (PEPC) and phosphoenolpyruvate carboxykinase (PEP-CK), are especially important in seaweeds. The role of PEP-CK, which uses CO_2 as inorganic carbon source, in LICF has been demonstrated for various species of seaweeds, in particular large brown algae (Küppers and Kremer 1978; Johnston and Raven 1986; Cabello-Pasini et al. 2000). In contrast to PEPC, during the PEP-CK catalysis the energy of the phosphorylated group of PEP is saved by phosphorylation of nucleoside diphosphates. The first studies using radiocarbon (^{14}C) in different groups of seaweeds revealed that amino acids such as aspartate, glutamate, citrate, and alanine were primarily ^{14}C labeled (Akagawa et al. 1972; Kremer 1981; Kerby and Evans 1983). The formation of oxalacetic acid (OAA) as a key intermediate of the Krebs cycle, suggested a link with anabolic processes (Kremer 1981). In fact, an apparent function of LICF is the replenishing of carbon via "anaplerotic" reactions, especially when pyruvate is degraded to acetyl-CoA during glycolysis (Kremer 1981). However, LICF reactions do not increase the net fixed carbon but are essential for cell metabolism, i.e., the pathway provides indispensable C4 acids that are not synthesized in the Calvin–Benson cycle.

Like photosynthetic C-fixation rates, LICF rates show considerable variation among different seaweeds; however, there is a tendency of higher values in brown algae compared to Chlorophytes and Rhodophytes (Table 2.1). In Chlorophytes and Rhodophytes, LICF rarely exceeds 1 μmol $^{14}C\ g^{-1}$ FW h^{-1}, which in terms of their contribution to the photosynthetic carbon is normally <1% (Cabello-Pasini and Alberte 1997). In the case of brown algae, values can be considerably higher (up to 9.6 μmol $^{14}C\ g^{-1}$ FW h^{-1}), accounting for up to 48% of the photosynthetic fixation (Kremer 1981). Especially high LICF rates have been reported in growing thallus areas of Laminariales (e.g., *Laminaria* and *Lessonia*) and during the spring/summer season for temperate and cold-temperate species. In the case of temperate red and brown algae, values of LICF can also be important (Cabello-Pasini and Alberte 1997). Carboxylation measured as the activity of PEP-CK is also linked to growth requirements, especially in species with marked seasonality in growth and photosynthetic carbon fixation, e.g., polar seaweeds (Weykam 1996; Weykam et al.

1997; Wiencke et al. 2009). Interestingly, algae from cold regions have exploited very efficiently the potential for LICF as a strategy to minimize the carbon losses due to high respiration and to optimize the supply of carbon skeletons during rapid growth during the short open water season (Drew and Hastings 1992; Gómez and Wiencke 1998) (see below). For example, in the kelp-like Antarctic brown alga *Ascoseira mirabilis*, LICF represents approximately 9.5% of light C-fixation (Gómez et al. 1995a, b), which is comparable to ratios found in species of *Laminaria* (Küppers and Kremer 1978) (Table 2.1). Despite the potential for LICF that seaweeds exhibit, it is not clear whether this pathway may compensate for C losses due to respiration as pointed by Kremer (1981). Thomas and Wiencke (1991) did not conclusively demonstrate its relationship with dark respiration in several Antarctic marine algae. In general, LICF was between 4.9 and 31% of dark respiration in five brown algae and one red alga. In species such as *Himantothallus grandifolius* and *Desmarestia anceps*, low LICF values were coupled with high respiration rates (Thomas and Wiencke 1991). This situation confirms the findings reported in *Ascophyllum nodosum* where a net C loss due to respiration was estimated in the dark (Johnston and Raven 1986). Recent studies revisiting the role of LICF in carbon metabolism of seaweeds have demonstrated that these reactions can be functional to morpho-physiological strategies to cope with, e.g., enhanced solar UV radiation. In blades of *Lessonia nigrescens*, LICF decreased 70% whereas light carbon fixation decreased by 90% under elevated doses of UV-B radiation. This suggests that LICF could be regarded a compensating mechanism necessary to keep physiological performance of algae during severe photodamage (Gómez et al. 2007). The findings that LICF is also well expressed in temperate and polar Rhodophytes such as *Cryptopleura lobulifera*, *Palmaria decipiens*, and *Iridaea cordata* (Thomas and Wiencke 1991; Weykam 1996; Weykam et al. 1997; Cabello-Pasini and Alberte 1997) open questions related with its involvement in morpho-functional processes that allow these organisms to cope with stressful conditions. Involvement of LICF as a mechanism to reduce photorespiration has only been reported in the Chlorophyte *Udotea* (Reiskind et al. 1988). For most of seaweed groups, especially green and red algae, data on LICF are lacking and thus further studies are required in order to outline accurate conclusions on the significance of this pathway for the ecology of seaweeds.

2.5 Morpho-functional Aspects of Carbon Metabolism

Carbon metabolism in seaweeds is integrated in multicellular organization that in many groups exhibits several plant-like traits. Although seaweeds do not display the structural complexity of vascular plants, the integration of form and function is an important factor even in the simplest groups, e.g., uncorticated filaments and sheet-like species. Thus, gross morphology of seaweeds has been related with ecophysiological adaptations (especially photosynthetic performance and carbon production) in response to abiotic and biotic determinants (Littler and Littler 1980;

Steneck and Watling 1982). It is now well established that thallus morphology defines much of the carbon physiology of large brown and red algae as well as some siphonal Chlorophytes.

2.5.1 The Role of Storage Carbohydrates

Storage carbohydrates of seaweeds are normally formed in the chloroplasts. Starch, a common compound of Chlorophyta and plants, is an insoluble storage polysaccharide constituted by units of ADP-glucose, which is synthesized from the triose phosphate generated in the Calvin–Benson cycle. Alternatively, triose phosphate can be exported via an antiport system to the cytosol to form sucrose. A significant part of the pool of sucrose is recycled to RuBP and thus the formation of starch or sucrose in the cell is a highly regulated process, closely synchronized with the carbon requirements of the Calvin–Benson cycle (Nelson and Cox 2004). The D-glucose monomers are linked to form branched polymers of starch composed of two types of chains: α 1,4-D-glucans (amylose) and/or additional α-1,6-D-glucans (amylopectin). In Rhodophyta, carbohydrates synthesized from carbon fixation are stored as floridean starch, which is characterized by α 1,4-D and α 1,6-glucans. The brown algae have storage laminaran (β-D-glucopyranose), a combination of soluble and insoluble chains of the type β-1,3 and β-1,6-D-glucans. Seaweeds contain also important amounts of low-molecular-weight compounds such as sucrose (green algae), mannitol (which can form part of laminaran chains of brown algae), and floridoside (red algae). These compounds are not only reserve products but also have a series of intracellular functions (e.g., osmolytes; see also Chap. 5 by Karsten) or are also precursors of cell wall polysaccharides (reviewed by Craigie 1974; Wëiwer et al. 2008).

Seasonal variation in major organic compounds, especially carbohydrates, is well known since 60 years and primarily based on studies of large brown algae, especially Laminariales and Fucales (Black 1950; Haug and Jensen 1954; Jensen and Haug 1956) and some Rhodophyta (Dawes et al. 1974). Although these changes were related to gradients of environmental variations (salinity, temperature, light, etc.), the importance of these compounds in life strategy, morpho-functional processes, and stress tolerance mechanisms was addressed later. Only after the classic works by Chapman and Craigie (1977, 1978), the relationship between nutrient availability, growth, photosynthesis, and organic composition in Laminariales could be comprehensively understood. In these seaweeds, degradation of storage carbohydrates, which are built up in summer (when net photosynthetic C assimilation occurs), supplies the energy requirements for growth during high nutrient availability in winter-early spring (Hatcher et al. 1977). In species such as *Saccharina latissima*, mannitol and laminaran vary from total absence in winter (4.5 and 0% DW, respectively) to high values close to 26% DW in summer (Black 1950). This strategy is extreme in the Arctic species *Laminaria solidungula*, which grow only in darkness in winter powered by the carbohydrates (laminaran/mannitol) synthesized during the previous season. About

25% of the original carbon content of the thallus is depleted during the dark winter period while it completes nearly 90% of its annual linear growth (Dunton and Schell 1986). In Antarctic brown algae, due to the seasonally constant levels of nutrients, the dynamics of synthesis and utilization of storage carbohydrates mostly depend on the availability of light. In Antarctic Desmarestiales and Ascoseirales, depletion of laminaran during spring and summer results in increases of mannitol, suggesting that these compounds support requirements during lamina elongation (Drew and Hastings 1992; Gómez and Wiencke 1998; Wiencke et al. 2009).

In cold-temperate and Arctic Laminariales, the significant direct relation between seasonal changes of amino acids to mannitol and the inverse relationship between amino acids and laminaran content confirms that N availability regulates remobilization of stored carbon (Lüning et al. 1973; Küppers and Kremer 1978; Cagné et al. 1982). In the northern hemisphere, high ambient N supply in winter exceeds substantially the N requirements for protein and amino acid synthesis and free N is accumulated in surplus (Chapman and Craigie 1977). In the case of Antarctic algae, seasonal changes in mannitol and laminaran seem to be mainly triggered by daylength, as seasonal carbon budget of *Himantothallus grandifolius*, *Ascoseira mirabilis*,, and *Desmarestia menziesii* is not affected by nutrients in summer like in *Laminaria* (Drew and Hastings 1992; Gómez and Wiencke 1998; Gómez et al. 1995b; Gómez and Wiencke 1998). As for kelps, Antarctic algae suffer a photosynthetic carbon deficit during the growth period, i.e., carbon losses due to anabolism and dark respiration exceed photosynthetic carbon fixation, which may be compensated by reutilization of storage carbohydrates (Gómez and Wiencke 1998). In any case, daylength-dependent variations of storage carbohydrates have been documented in cultured *Laminaria hyperborea*, a species exposed generally to a severe *N* limitation in summer (Schaffelke 1995). Furthermore, exposure to constant short day alters the seasonal growth cycle of *Laminaria digitata* by preventing the decrease of growth rates in summer (Gómez and Lüning 2001), suggesting that not only nutrients but also photoperiodic responses are crucial in carbon metabolism in this group of algae.

Figure 2.1 describes a model of synthesis, remobilization, and utilization of storage carbohydrates during the growth phase for Laminariales. Laminaran, which is built up mainly in distal regions of various species of kelps (Küppers and Kremer 1978; Lüning 1979; Cabello-Pasini and Alberte 2001a), is degraded with release of mannitol, which is transported (along with various amino acids) via translocation to the meristematic region (see below). Mannitol enters in the LICF pathway as a precursor of 3-phosphoglycerate (PGA), which is transformed to phosphoenol pyruvate (PEP) in the mitochondria. One molecule of mannitol generates two molecules of PEP: one can be converted to oxaloacetate (OAA) by the enzyme PEP-CK following the biosynthetic pathway in the Krebs cycle while the other is transformed to acetyl-CoA by the successive action of pyruvate kinase (PK) and pyruvate dehydrogenase (PD) with loss of one molecule of CO_2, which is saved by PEP-CK (Kremer 1981).

The glycolysis of mannitol, as well as the ^{14}C labeling of carboxylic diacids (e.g., malate, aspartate) in the meristematic zones of the blade, has unequivocally confirmed that growth in large brown algae can be supported by nonphotosynthetic,

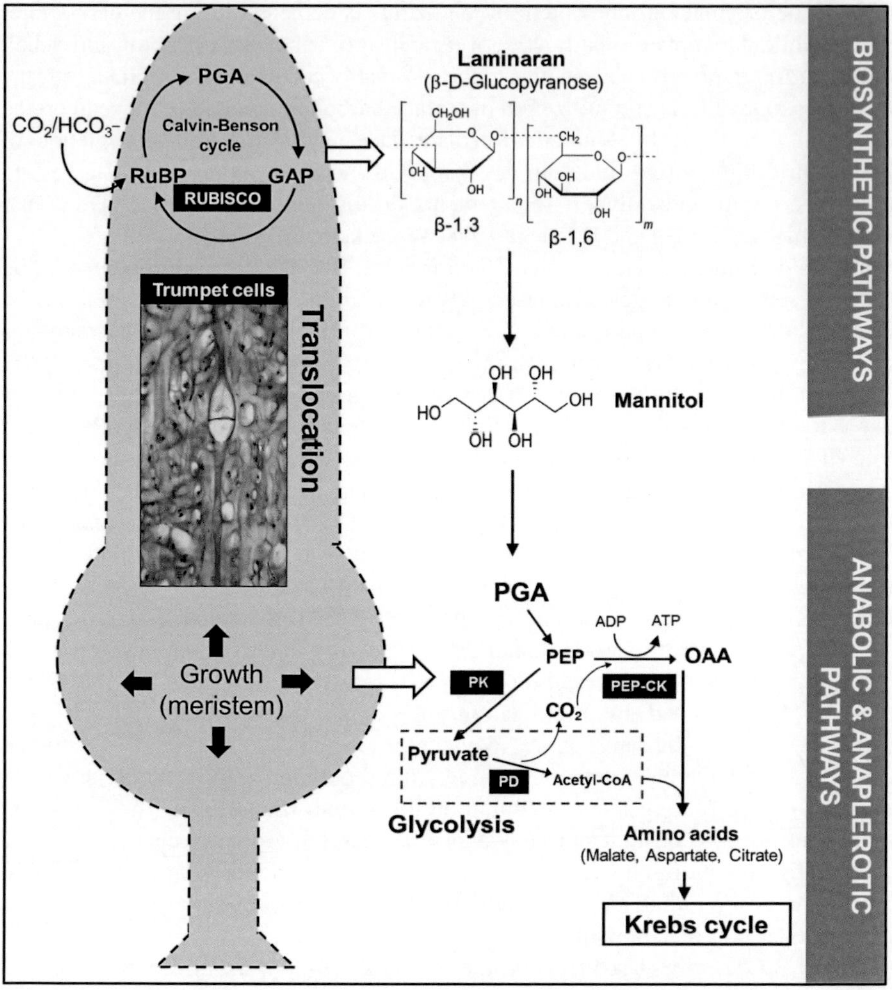

Fig. 2.1 Schema indicating the morpho-functional processes during active growth in Laminariales. In distal thallus regions, photosynthetic carbon fixation via RUBISCO and biosynthetic processes in the Calvin–Benson cycle are connected through translocation of low-weight molecular sugars (e.g., mannitol) with anaplerotic processes mediated by LICF in the intercalary meristem. *PGA*, 3-phosphoglycerate; *GAP*, glyceraldehyde-3-phosphate; *PK*, Pyruvate kinase; *PD*, Pyruvate dehydrogenase; *PEP-CK*, Phosphoenolpyruvate carboxykinase; *PEP*, Phospoenolpyruvate; *RuBP*, ribulose 1,5-bisphosphate

anaplerotic reactions (Kremer 1981; Cabello-Pasini and Alberte 2001b). In Rhodophyta, anaplerotic consumption of storage carbohydrates has been less studied. In *Gelidium coulteri* grown under *N* starvation, rates of LICF were relatively low (1–7%), floridoside was rapidly consumed in glycolysis, and an important fraction of the ^{14}C labeling was allocated to amino acids of the tricarboxylic acids cycle (Macler 1986). This pattern has been found in other Rhodophyta and apparently indicates that, while present, the LICF pathway has much less importance in the whole carbon assimilation in temperate algae (Cabello-Pasini and Alberte 2001b). However, results from the two contrasting polar red algae, the endemic *Palmaria decipiens* and the widely distributed *Iridaea cordata*, indicated that accumulation of floridean starch allows the first species to overwinter in darkness. In the case of *Iridaea*, use of floridean starch to power growth was less marked. In both cases, LICF accounting for up to 9% of the total carbon assimilation was active during a part of the dark period (Weykam et al. 1997), suggesting that nonphotosynthetic carbon metabolism in Rhodophyta may have similar ecophysiological importance in extreme environments as has been reported for large brown algae (Wiencke et al. 2009).

2.5.2 Thallus Anatomy and Long-Distance Transport of Photoassimilates

Unlike vascular plants, where long-distance transport represents the exchange of resources between genuine, highly differentiated tissues, the transport of substances in seaweeds, called translocation, is normally a strategy to redistribute via mass flux diverse organic compounds towards zones of high metabolic activity (Lüning et al. 1973; Schmitz 1981). In Laminariales and Fucales, carbon metabolism is spatially separated in carbon "source" and "sink" regions (Küppers and Kremer 1978; Arnold and Manley 1985; Cabello-Pasini and Alberte 2001a). As is shown in Fig. 2.1 for Laminariales, photoassimilates are stored in the mature, commonly distal regions of the algae, and then transported as mannitol and amino acids to the meristematic region. This morpho-functional arrangement is a consequence of the allometric growth and the action of an intercalary meristem that normally results in tissues with different metabolic activity.

Different types of sieve elements, such as the "trumpet" cells, have been identified in members of Laminariales and Phyllariaceae (Buggeln 1983; Schmitz 1981; Gómez et al. 2007). These structures are formed by specialized, normally vacuolated cells that are longitudinally arranged in the medulla or below the cortex in parenchymatous and peudo-parenquimatous thalli. Cell lengths measured in different Laminariales vary considerably and can reach several millimeters. Most of the sieve tubes end in the so-called sieve plates, which present pores of up to 100 nm diameter permitting the connection between adjacent cells (Schmitz 1990).

Apart from Laminariales, other groups of seaweeds have been shown to have sieve elements or at least a translocation function has been hypothesized. This is the case of the Antarctic genera *Ascoseira* and *Himantothallus*, whose advanced structural organization resembles that of *Laminaria* species from the northern Hemisphere. For example, members of the Ascoseirales are characterized by a strap-like lamina with an intercalary basally located meristem forming new tissue during each growth phase. Thus, the blade in this species is formed by tissues differing in age and developmental stage (Gómez et al. 1995a; Gómez et al. 1996). Histological studies have revealed the presence of medullar structures denominated "conducting channels". Apparently, putative translocation could occur only in young plants, as early "conducting channels" are metabolically active, possess plasmodesmata, and contain relatively few physodes (Clayton and Ashburner 1990). Long-distance transport of substances has also been documented in members of the brown algal orders Scytosiphonales (Guimaraes et al. 1986), Desmarestiales (Moe and Silva 1981; Wiencke and Clayton 1990), and Fucales (Moss 1983). In Rhodophyta, evidence for translocation of photoassimilates using ^{14}C labeling has been obtained in *Polysiphonia* sp. (Wetherbee 1979), *Delesseria sanguinea* (Hartman and Eschrich 1969), and *Gracilaria cornea* (Gonen et al. 1996). Although the structures and probably the mechanisms of translocation in red algae are different compared to brown algae, a relationship between carbon fixation and translocation has been clearly demonstrated in *Gracilaria* (Gonen et al. 1996).

2.5.3 Patterns of Carbon Allocation

Large and complex seaweeds show a differential allocation of carbon fixation products along the thallus. Various brown algal genera such as *Sargassum* (Gorham and Lewey 1984), *Macrocystis* (Wheeler and North 1981; Gerard 1982), *Lessonia* (Percival et al. 1983; Westermeier and Gómez 1996), *Durvillaea* (Cheshire and Hallam 1985; Lawrence 1986; Gómez and Westermeier 1995), and *Desmarestia* (Carlberg et al. 1978) show longitudinal variation in organic composition. Primarily, changes in carbon allocation can be directly caused by differential capacity for carbon uptake among parts of thallus. Using $^{13}C/^{12}C$ ratios ($\delta\ ^{13}C$), it was possible to identify active HCO_3^- uptake sites along the thallus of Antarctic seaweeds correlated to growth activity (Wiencke and Fischer 1990, 1992). For example, $\delta^{13}C$ values between -12 and -16.8% (indicating ^{13}C enrichment) were measured in new blade regions of *Ascoseira mirabilis* during high irradiances and summer daylength (Gómez 1997). Apparently, enhanced carboxylation rates during high light compensate for the energy costs of active HCO_3^- incorporation by decreasing the C supply via diffusive CO_2 entry, and thus the heavier C isotope is preferentially assimilated (Kübler and Raven 1994; Raven et al. 1995). On the other hand, changes in light use and carbon fixation efficiency along with increasing thallus size and age affect the carbon uptake and allocation. In cultures of *Desmarestia menziesii*, $\delta\ ^{13}C$ values $>-29\%$ were found in small algae, but with increasing size, $\delta\ ^{13}C$ signatures increased accordingly (-32%) (Gómez 1997).

Hydrodynamic processes regulate also the allocation of photoassimilated carbon in the thallus. In many large brown algae, carbon (normally in the form of structural carbohydrates) is preferentially allocated in the basal structures, which are biomechanically designed to attach algae to the substrate and to withstand drag forces from water movement (Hurd 2000). In the fucoid *Durvillaea antarctica*, characterized by large and floating laminar blades, 85% of the total energy contents from organic compounds is allocated to fronds, while the rest is allocated to the holdfast (Lawrence 1986). In the case of the kelp *Postelsia palmaeformis*, 63% of the total organic carbon is allocated to holdfast and stipe, which is in line with a gross morphology designed to resist the direct impact of waves (Lawrence and McClintock 1988). In the southern kelp *Lessonia nigrescens,* carbon allocation changed with increasing size and age: adult plants deposited higher proportion of energy in the holdfast and stipes than young plants with important consequences for population density and local demography (Westermeier and Gómez 1996).

Differential allocation of organic carbon along the thallus can also be functional to withstand other environmental stressors (Wakefield and Murray 2009). The optimal defense theory (ODT) has been proposed to understand the mechanisms that control the interaction between algal allocation of organic compounds and the action of, e.g., herbivores (Cronin and Hay 1996). For example, secondary metabolites are normally allocated in the structures with high fitness value, probably where the investment in energy is higher (Pansch et al. 2008). In brown algae, phlorotannins may function as deterrents for many grazers, but additionally, due to their primary role as cell wall precursors, these compounds act also as cell-wall hardening, conferring mechanical resistance and toughness (Lucas et al. 2000). In the intertidal kelp *Lessonia nigrescens*, holdfast and stipes contain higher concentrations of phlorotannins, and hence are better defended than transient fronds (Gómez et al. 2005; Gómez and Huovinen 2010, see also Chap. 8 by Iken).

Carbon fixation, biomass, and overall the sum of the morpho-functional processes of seaweeds define much of the primary productivity and energy fluxes in the coastal ecosystems (Mann 1973). Due to their size and patterns of substrate occupation, seaweeds represent habitat for other organisms and also modify the physical and chemical environment (Jackson 1998; Delille et al. 2000). In coastal areas of cold-temperate and polar regions, seaweeds can account for >50% of the total fixed carbon (Gattuso et al. 2006). The outcome of seaweed carbon metabolism is transferred to the food web not only via direct consumption by herbivores but also as secondary product via detritus in near shore (Duggins et al. 1989) and abyssal areas (Wiencke and Fisher 1992).

2.5.4 Concluding Remarks

Overall, carbon metabolism of seaweeds is highly versatile and has allowed these organisms to thrive in all types of habitats and environmental conditions. However, despite the considerable advances in our knowledge on mechanisms and pathways, the ecological consequences of many photosynthetic adaptations are not well understood. For example, carbon acquisition patterns in relation with morpho-physiological processes, biomass allocation, reproduction, and development of adult and early stages of seaweeds have been hitherto overlooked (Raven 2003). The significance of endogenous processes regulating carbon metabolism, which has been studied in few brown algae, is also a relevant topic that should be expanded to other seaweed groups in order to gain insights into possible evolutionary issues (Schmid et al. 1996). Probably, one of the most important issues in seaweed physiology will be the understanding of the complex interaction between the expression and modulation of carbon metabolism, and processes underlying large biogeographical patterns of seaweeds. For many aspects related to regulation of carbon metabolism in seaweeds, proteomic and genomic studies are urgently needed and fundamental in order to understand the role of seaweeds in present and future scenarios of global change.

The consequences of increasing CO_2 levels and related phenomena such as the ocean warming and ozone depletion on seaweed ecophysiology have begun to be explored and different surveys describe various biological and geochemical scenarios modified by anthropogenic activities (Israel and Einav 2010). Due to the dependence of RUBISCO on CO_2, it has been postulated that photosynthetic organisms will respond positively to present and future increases in atmospheric CO_2 (Amthor 1995). However, some experimental evidence suggests that large increases in photosynthetic carbon fixation are not expected mainly because most of the studied seaweeds exhibit CCMs (Gao et al. 1993; Beer and Koch 1996; Israel and Hophy 2002). Apparently, the focus could be on the increased competiveness of algae without or with poorly developed CCM, which will depend on a series of other environmental and geographic factors (Raven et al. 2002). Overall, although the studies compiled here give important insights into the potential responses of organisms, the central question whether the mechanisms exploited by seaweeds today allow them to adapt to future scenarios remains open.

References

Akagawa H, Ikawa T, Nisizawa K (1972) Initial pathway of dark $^{14}CO_2$-fixation in brown algae. Bot Mar 15:119–125

Amthor JS (1995) Terrestrial higher-plant response to increasing atmospheric [CO_2] in relation to the global carbon cycle. Global Change Biol 1:243–274

Arnold KE, Manley SL (1985) Carbon allocation in *Macrocystis pyrifera* (Phaeophyta): intrinsic variability in photosynthesis and respiration. J Phycol 21:154–167

Badger M (2003) The roles anhydrases in photosynthetic CO_2 concentrating mechanisms. Photosynth Res 77:83–94

Badger MR, Price GD (1994) The role of carbonic anhydrase in photosynthesis. Annu Rev Plant Physiol Plant Mol Biol 45:369–392

Badger MR, Spalding MH (2000) CO_2 acquisition, concentration and fixation in cyanobacteria and algae. In: Leegood RC, Sharkey TD, von Caemmerer S (eds) Photosynthesis: physiology and metabolism. Kluwer, Dordrecht, pp 369–397

Beer S (1996) Photosynthetic utilization of inorganic carbon in *Ulva*. Sci Mar 60:125–128

Beer S, Israel A (1986) Photosynthesis of *Ulva* sp. III. O_2 effects, carboxylase activities, and the CO_2 incorporation pattern. Plant Physiol 81:937–938

Beer S, Koch E (1996) Photosynthesis of marine macroalgae and seagrasses in globally changing CO_2 environments. Mar Ecol Prog Ser 141:199–204

Bischof K, Kräbs G, Wiencke C, Hanelt D (2002) Solar ultraviolet radiation affects the activity of ribulose- 1,5-bisphosphate carboxylase-oxygenase and the composition of photosynthetic and xanthophyll cycle pigments in the intertidal green alga *Ulva lactuca* L. Planta 215:502–509

Black WAP (1950) The seasonal variation in weight and chemical composition of the common British Laminariaceae. J Mar Biol Assoc UK 29:45–72

Buggeln RG (1983) Photoassimilate translocation in brown algae. Prog Phycol Res 2:283–332

Cabello-Pasini A, Alberte RS (1997) Seasonal patterns of photosynthesis and light-independent carbon fixation in marine macrophytes. J Phycol 33:321–329

Cabello-Pasini A, Alberte RS (2001a) Expression of carboxylating enzymes in *Laminaria setchellii* (Phaeophyceae). Phycologia 40:351–358

Cabello-Pasini A, Alberte RS (2001b) Enzymatic regulation of photosynthetic and light-independent carbon fixation in *Laminaria setchelli* (Phaeophyta), *Ulva lactuca* (Chlorophyta) and *Iridaea cordata* (Rhodophyta). Rev Chil Hist Nat 74:229–236

Cabello-Pasini A, Smith GJ, Alberte RS (2000) Phosphoenol pyruvate carboxykinase from the marine diatom *Skeletonema costatum* and the phaeophyte *Laminaria setchelli* I. Isolation and biochemical characterization. Bot Mar 43:559–568

Cagné JA, Mann KH, Chapman ARO (1982) Seasonal patterns of growth and storage in *Laminaria longicruris* in relation to differing patterns of availability of nitrogen in the water. Mar Biol 69:91–101

Calvin M (1956) The photosynthetic carbon cycle. J Chem Soc 1956:1895–1915

Carlberg GE, Percival E, Rhaman A (1978) Carbohydrates of the seaweeds *Desmarestia ligulata* and *D. firma*. Phytochemisty 17:1289–1292

Chapman AR0, Craigie JS (1977) Seasonal growth in *Laminaria longicruris*: relations with dissolved inorganic nutrients and internal reserves of nitrogen. Mar Biol 40:161–164

Chapman AR0, Craigie JS (1978) Seasonal growth in *Laminaria longicruris*: relations with reserve carbohydrate storage and production. Mar Biol 46:209–213

Cheshire AC, Hallam ND (1985) The environmental role of alginates in *Durvillaea potatorum* (Fucales, Phaeophyta). Phycologia 24:147–153

Choo KS, Nilsson J, Pedersén M, Snoeijs P (2005) Photosynthesis, carbon uptake and antioxidant defense in two coexisting filamentous green algae under different stress conditions. Mar Ecol Prog Ser 292:127–138

Clayton MN, Ashburner CM (1990) The anatomy and ultrastructure of "conducting channels" in *Ascoseira mirabilis* (Ascoseirales, Phaeophyceae). Bot Mar 33:63–70

Craigie JS (1974) Storage products. In: Stewart WDP (ed) Algal physiology and biochemistry. Blackwell Scientific, Oxford, pp 206–235

Cronin G, Hay ME (1996) Within-plant variation in seaweed palatability and chemical defenses: optimal defense theory versus the growth-differentiation balance hypothesis. Oecologia 105:361–368

Dawes CJ, Lawrence JM, Cheney DP, Mathieson AC (1974) Ecological studies of floridian *Eucheuma* (Rhodophyta, Gigartinales). III. Seasonal variation of carrageenan, total carbohydrate, protein, and lipids. Bull Mar Sci 24:286–299

Delille B, Delille D, Fiala PC, Frankignoulle M (2000) Seasonal changes of pCO_2 over a subantarctic *Macrocystis* kelp belt. Polar Biol 23:706–716

Drechsler Z, Sharkia R, Cabantchik ZL, Beer S (1994) The relationship of arginine groups to photosynthetic HCO_3^- uptake in *Ulva* sp. mediated by a putative anion exchanger. Planta 194:250–255

Drew EA, Hastings RM (1992) A year-round ecophysiological study of *Himantothallus grandifolius* (Desmarestiales, Phaeophyta) at Signy Island, Antarctica. Phycologia 31:262–277

Duggins DO, Simenstad CA, Estes JA (1989) Magnification of secondary production by kelp detritus in coastal marine ecosystems. Science 245:170–173

Dunton KH, Schell DM (1986) Seasonal carbon budget and growth of *Laminaria solidungula* in the Alaskan High Arctic. Mar Ecol Prog Ser 31:57–66

Falkowski PG, Raven J (1997) Aquatic photosynthesis. Blackwell, Oxford

Flores-Moya A, Gómez I, Viñegla B, Altamirano M, Pérez-Rodríguez E, Maestre C, Caballero RM, Figueroa FL (1998) Effects of solar radiation on the endemic Mediterranean red alga *Rissoella verruculosa*: Photosynthetic performance, pigment content and the activities of enzymes related to nutrient uptake. New Phytol 139:673–683

Gao KY, Aruga Y, Asada K, Kiyohara M (1993) Calcification in the articulated coralline algae *Corallina pilulifera*, with special reference to the effect of elevated CO_2 concentration. Mar Biol 117:129–132

Gattuso J-P, Gentili B, Duarte CM, Kleypas JA, Middelburg JJ, Antoine D (2006) Light availability in the coastal ocean: impact on the distribution of benthic photosynthetic organisms and contribution to primary production. Biogeosciences Discuss 3:895–959

Gerard VA (1982) Growth and utilization of internal nitrogen reserves by the giant kelp *Macrocystis pyrifera* in a low-nitrogen environment. Mar Biol 66:27–35

Giordano M, Maberly SC (1989) Distribution of carbonic anhydrase in British marine macroalgae. Oecologia 81:534–539

Giordano M, Beardall J, Raven JA (2005) CO_2 concentrating mechanisms in algae: mechanisms, environmental modulation, and evolution. Annu Rev Plant Biol 56:99–131

Gómez I (1997) Life strategy and ecophysiology of Antarctic macroalgae. Rep Polar Res 238:1–99

Gómez I, Huovinen P (2010) Induction of phlorotannins during UV exposure mitigates inhibition of photosynthesis and DNA damage in the kelp *Lessonia nigrescens*. Photochem Photobiol 86:1056–1063

Gómez I, Lüning K (2001) Constant short day treatment of outdoor-cultivated *Laminaria digitata* prevents summer drop in growth rates. Eur J Phycol 36:391–395

Gómez I, Westermeier R (1995) Energy contents and organic constituents in Antarctic and south Chilean marine brown algae. Polar Biol 15:597–602

Gómez I, Wiencke C (1998) Seasonal changes in C, N, and major organic compounds and their significance to morpho-functional processes in the endemic Antarctic brown alga *Ascoseira mirabilis*. Polar Biol 19:115–124

Gómez I, Thomas DN, Wiencke C (1995a) Longitudinal profiles of growth, photosynthesis and light independent carbon fixation in the Antarctic brown alga *Ascoseira mirabilis*. Bot Mar 38:157–164

Gómez I, Wiencke C, Weykam G (1995b) Seasonal photosynthetic characteristics of *Ascoseira mirabilis* (Ascoseirales, Phaeophyceae) from King George Island, Antarctica. Mar Biol 123:167–172

Gómez I, Wiencke C, Thomas DN (1996) Variations in photosynthetic characteristic of the Antarctic marine brown alga *Ascoseira mirabilis* Skottsberg in relation to age and size. Eur J Phycol 31:167–172

Gómez I, Weykam G, Wiencke C (1998a) Photosynthetic metabolism and major organic compounds in the marine brown alga *Desmarestia menziesii* from King George Island (Antarctica). Aquatic Bot 60:105–118

Gómez I, Pérez-Rodríguez E, Viñegla B, Figueroa FL, Karsten U (1998b) Effects of solar radiation on photosynthesis, UV absorbing compounds and enzyme activities of the green alga *Dasycladus vermicularis* from southern Spain. J Photochem Photobiol B Biol 47:46–57

Gómez I, Ulloa N, Orostegui M (2005) Morpho-functional patterns of photosynthesis and UV sensitivity in the kelp *Lessonia nigrescens* (Laminariales, Phaeophyta). Mar Biol 148:231–240

Gómez I, Orostegui M, Huovinen P (2007) Morpho-functional patterns of photosynthesis in the South Pacific kelp *Lessonia nigrescens*: effects of UV radiation on ^{14}C fixation and primary photochemical reactions. J Phycol 43:55–64

Gonen Y, Kimmel E, Tel-Or E, Friedlander M (1996) Intercellular assimilate translocation in *Gracilaria cornea* (Gracilariaceae, Rhodophyta). Hydrobiologia 326(327):421–428

Gordillo FJL, Aguilera J, Jiménez C (2006) The response of nutrient assimilation and biochemical composition of Arctic seaweeds to a nutrient input in summer. J Exp Bot 57:2661–2671

Gorham J, Lewey SA (1984) Seasonal changes in the chemical composition of *Sargassum muticum*. Mar Biol 108:103–107

Gross W (1990) Occurrence of glycolate oxidase and hydroypyruvate reductase in *Egreggia menziesii* (Phaeophyta). J Phycol 26:381–383

Guimaraes SMPB, Braga MRA, Cordeiro-Marinho M, Pedrini AG (1986) Morphology and taxonomy of *Jolyna laminarioides*, a new member of the Scytosiphonales (Phaeophyceae) from Brazil. Phycologia 25:99–108

Haglund K, Björk M, Ramazanov Z, García-Reina G, Pedersén M (1992) Role of carbonic anhydrase in photosynthesis and inorganic carbon assimilation in the red alga *Gracilaria tenuistipitata*. Planta 187:275–281

Hartman T, Eschrich W (1969) Stofftransport in Rotalgen. Planta 85:303–312

Hatcher BG, Chapman AR0, Mann KH (1977) An annual carbon budget for the kelp *Laminaria longicruris*. Mar Biol 44:85–96

Haug A, Jensen A (1954) Seasonal variations in chemical composition of *Alaria esculenta*, *Laminaria saccharina*. *Laminaria hyperborea* and *Laminaria digitata* from northern Norway, Norw Inst Seaweed Res Rep No, 4

Huovinen P, Gómez I, Orostegui M (2007) Patterns and UV sensitivity of carbonic anhydrase and nitrate reductase activities in south Pacific macroalgae. Mar Biol 151:1813–1821

Hurd CL (2000) Water motion, marine macroalgae physiology and production. J Phycol 36:453–472

Israel A, Einav R (eds) (2010) Seaweeds and their role in globally changing environments. Cellular origin, life in extreme habitats and astrobiology, vol 15, 1st edn. Springer, New York, p 480

Israel A, Hophy M (2002) Growth, photosynthetic properties and Rubisco activities and amounts of marine macroalgae grown under current and elevated seawater CO_2 concentrations. Global Change Biol 8:831–840

Iwamoto K, Ikawa T (1997) Glycolate metabolism and subcellular distribution of glycolate oxidase in *Patoglossum pacificum* (Phaeophyceae, Chromophyta). Phycol Res 45:77–83

Jackson GA (1998) Currents in the high drag environment in a coastal kelp stand off California. Cont Shelf Res 17:1913–1918

Jensen A, Haug A (1956) Geographical and seasonal variation in the chemical composition of *Laminaria hyperborea* and *Laminaria digitata* from the Norwegian coast. Norw Inst Seaweed Res Rep No 14

Johnston AM, Raven JA (1986) Dark carbon fixation studies on the intertidal macroalga *Ascophyllum nodosum* (Phaeophyta). J Phycol 22:78–83

Kawamitsu Y, Boyer JS (1999) Photosynthesis and carbon storage between tides in a brown alga, *Fucus vesiculosus*. Mar Biol 133:361–369

Kerby WN, Evans LV (1983) Phosphoenolpyruvate carboxykinase activity in *Ascophyllum nodosum* (Phaeophyceae). J Phycol 19:1–3

Klenell M, Snoeijs P, Pedersén M (2002) The involvement of a plasma membrane H^+-ATPase in the blue-light enhancement of photosynthesis in *Laminaria digitata* (Phaeophyta). J Phycol 38:1143–1149

Kremer BP (1981) Metabolic implications of non-photosynthetic carbon fixation in brown algae. Phycologia 20:242–250

Kübler E, Raven JA (1994) Consequences of light limitation for carbon acquisition in three rhodophytes. Mar Ecol Prog Ser 110:203–209

Küppers U, Kremer BP (1978) Longitudinal profiles of carbon dioxide fixation capacities in marine macroalgae. Plant Physiol 62:49–53

Lawrence JM (1986) Proximate composition and standing crop of *Durvillaea antarctica* (Phaeophyta) in the Bay of Morbihan, Kerguelen (South Indian Ocean). Mar Ecol Prog Ser 33:1–5

Lawrence JM, McClintock JB (1988) Allocation of organic material and energy to the holdfast, stipe, and fronds in *Postelsia palmaeformis* (Phaeophyta: Laminariales) on the California coast. Mar Biol 99:151–155

Littler MM, Littler DS (1980) The evolution of thallus form and survival strategies in benthic marine macroalgae: field and laboratory tests of a functional form model. Am Nat 116:25–44

Lobban CS, Harrison PJ (2000) Seaweed ecology and physiology. Cambridge University Press, Cambridge

Lucas PW, Turner IM, Dominy NJ, Yamashita N (2000) Mechanical defenses to herbivory. Ann Bot 86:913–920

Lüning K (1979) Growth strategies of three *Laminaria* species (Phaeophyceae) inhabiting different depth zones in the sublittoral region of Helgoland (North Sea). Mar Ecol Prog Ser 1:195–207

Lüning K, Schmitz K, Willenbrink K (1973) CO_2 fixation and translocation in benthic marine algae III. Rates and ecological significance of translocation in *Laminaria hyperborea* and *L. saccharina*. Mar Biol 12:275–281

Maberly SC (1990) Exogenous source of inorganic carbon for photosynthesis by marine macroalgae. J Phycol 26:439–449

Macler BA (1986) Regulation of carbon flow by nitrogen and light in the red alga, *Gelidium coulteri*. Plant Physiol 82:136–141

Mann KH (1973) Seaweeds: their productivity and strategy for growth. Science 182:975–981

McKay RML, Sarah I, Gibbs P, Vaughn KC (1991) RuBisCo activase is present in the pyrenoid of green algae. Protoplasma 162:38–45

Mercado JM, Niell FX (1999) Carbonic anhydrase activity and use of HCO_3^- in *Bostrychia scorpioides* (Ceramiales, Rhodophyceae). Eur J Phycol 34:13–19

Mercado JM, Gordillo FJL, Figueroa FL, Niell FX (1998) External carbonic anhydrase and affinity for inorganic carbon in intertidal macroalgae. J Exp Mar Biol Ecol 221:209–220

Mercado JM, de los Santos CB, Pérez-Lloréns JL, Vergara JJ (2009) Carbon isotopic fractionation in macroalgae from Cádiz Bay (Southern Spain): Comparison with other bio-geographic regions. Estuar Coast Shelf Sci 85:449–458

Moe RL, Silva PC (1981) Morphology and taxonomy of *Himantothallus* (including *Phaeoglossum* and *Phyllogigas*) an Antarctic member of the Desmarestiales (Phaeophyceae). J Phycol 17:15–29

Moss BL (1983) Sieve elements in the Fucales. New Phytol 93:431–433

Murru M, Sandgren CD (2004) Habitat matters for inorganic carbon acquisition in 38 species of red macroalgae (Rhodophyta) from Puget Sound, Washington, USA. J Phycol 40:837–845

Nelson DL, Cox MN (2004) Lehninger, principles of biochemistry, 4th edn. W.H. Freeman, New York

Pansch C, Gómez I, Rothäusler E, Veliz K, Thiel M (2008) Defense of vegetative vs. reproductive blades of the Pacific kelps *Lessonia nigrescens* and *Macrocystis integrifolia*: A test on species-specific defense strategies. Mar Biol 155:51–62

Percival EE, Venegas MJ, Weigel H (1983) Carbohydrates of the brown seaweed *Lessonia nigrescens*. Phytochemistry 22:1429–1432

Ramus J (1978) Seaweed anatomy and photosynthetic performance: the ecological significance of light guides, heterogeneous absorption and multiple scatter. J Phycol 14:352–362

Raven JA (1991) Implications of inorganic C utilization: ecology, evolution and geochemistry. Can J Plant Physiol 69:908–924

Raven JA (1997) Putting the C in phycology. Eur J Phycol 32:319–333

Raven JA (2003) Inorganic carbon concentrating mechanisms in relation to the biology of algae. Photosynth Res 77:155–171

Raven JA (2010) Inorganic carbon acquisition by eukaryotic algae: four current questions. Photosynth Res 106:123–134

Raven JA, Beardall J (1981) Carbon dioxide as the exogenous inorganic carbon source for *Batrachospermum* and *Lemanea*. Br Phycol J 16:165–172

Raven JA, Lucas J (1985) The energetics of carbon acquisition. In: Lucas WJ, Berry JA (eds) Inorganic carbon uptake by aquatic photosynthetic organisms. Am Soc Plant Physiol, pp 305–324

Raven JA, Walker DI, Johnston AM, Handley LL, Kübler JE (1995) Implications of ^{13}C natural abundance measurements for photosynthetic performance by marine macrophytes in their natural environment. Mar Ecol Prog Ser 123:193–205

Raven JA, Johnston AM, Kübler JE, Korb RE, McInroy SG, Handley LL, Scrimgeour CM, Walker DI, Beardall J, Clayton MN, Vanderklift M, Chudek JA, Fredriksen S, Dunton KH (2002) Seaweeds in cold seas: evolution and carbon acquisition. Ann Bot 90:525–536

Raven JA, Ball LA, Beardall J, Giordano M, Maberly SC (2005) Algae lacking carbon concentrating mechanisms. Can J Bot 83:879–890

Reiskind JB, Seamon PT, Bowes G (1988) Alternative methods of photosynthetic carbon assimilation in marine macroalgae. Plant Physiol 87:686–692

Rosenberg G, Littler DS, Littler MN, Oliveira EC (1995) Primary production and photosynthetic quotients of seaweeds from Sao Paulo State, Brazil. Bot Mar 38:369–378

Sand-Jensen K, Gordon DM (1984) Differential ability of marine and freshwater macrophytes to utilize HCO_3^- and CO_2. Mar Biol 80:247–253

Schaffelke B (1995) Storage carbohydrates and abscisic acid contents in *Laminaria hyperborea* are constrained by experimental daylengths. Eur J Phycol 30:313–317

Schmid R, Mills JA, Dring MJ (1996) Influence of carbon supply on stimulation of light saturated photosynthesis by blue light in *Laminaria saccharina:* implications for the mechanism of carbon acquisition in higher brown algae. Plant Cell Environ 19:383–391

Schmitz K (1981) Translocation. In: Lobban CS, Wynne MJ (eds) The biology of seaweeds. University of California Press, Berkeley, pp 534–558

Schmitz K (1990) Algae. In: Behnke H-D, Sjolund RD (eds) Sieve elements. Comparative structure, induction and development. Springer, New York, pp 1–18

Steneck RS, Watling L (1982) Feeding capabilities and limitation of herbivorous mollusks—a functional—group approach. Mar Biol 68:299–319

Surif MB, Raven JA (1990) Photosynthetic gas exchange under emersed conditions in eulittoral and normally submersed members of the Fucales and the Laminariales: interpretation in relation to C isotope ratio and N and water use efficiency. Oecologia 82:68–80

Suzuki K, Iwamoto K, Yokoyama S, Ikawa T (1991) Glycolate-oxidizing enzymes in algae. J Pycol 27:492–498

Tabita FR, Satagopan S, Hanson TE, Kreel NE, Scott SS (2008) Distinct form I, II, III, and IV Rubisco proteins from the three kingdoms of life provide clues about Rubisco evolution and structure/function relationships. J Exp Bot 59:1515–1524

Thomas DN, Wiencke C (1991) Photosynthesis, dark respiration and light independent carbon fixation of endemic Antarctic macroalgae. Polar Biol 11:329–337

Tolbert NE (1997) The C2 oxidative photosynthetic carbon cycle. Annu Rev Plant Physiol Plant Mol Biol 48:1–25

van Caemmerer S, Furbank RT (2003) The C4 pathway: an efficient CO_2 pump. Photosynth Res 77:191–207

Wakefield RL, Murray SN (2009) Factors influencing food choice by the seaweed-eating marine snail *Norrisia norrisi* (Trochidae). Mar Biol 130:631–642

Wëiwer M, Sherwood T, Linhardt RJ (2008) Synthesis of floridoside. J Carbohydr Chem 27:420–427

Westermeier R, Gómez I (1996) Biomass, energy contents and major organic compounds in the brown alga *Lessonia nigrescens* (Laminariaes, Phaeophyceae) from Mehuín, South Chile. Bot Mar 39:553–559

Wetherbee R (1979) "Transfer connections": specialized pathways for nutrient translocation in red alga? Science 204:858–859

Weykam G (1996) Photosynthetic characteristics and life strategies of Antarctic macroalgae. Rep Polar Res 192:1–132

Weykam G, Thomas DN, Wiencke C (1997) Growth and photosynthesis of the Antarctic red algae *Palmaria decipiens* (Palmariales) and *Iridaea cordata* (Gigartinales) during and following extended periods of darkness. Phycologia 36:395–405

Wheeler PA, North WJ (1981) Nitrogen supply, tissue composition and frond growth rates for *Macrocystis pyrifera* off the coast of Southern California. Mar Biol 64:59–69

Wiencke C, Clayton MN (1990) The anatomy, life history and development of the Antarctic brown alga *Phaeurus antarcticus* (Desmarestiales, Phaeophyceae). Phycologia 29:303–315

Wiencke C, Fischer G (1990) Growth and stable carbon isotope composition of cold-water macroalgae in relation to light and temperature. Mar Ecol Prog Ser 65:283–292

Wiencke C, Fischer G (1992) Stable carbon isotope composition, depth distribution and fate of macroalgae from the Antarctic Peninsula region. Polar Biol 12:341–348

Wiencke C, Gómez I, Dunton K (2009) Phenology and seasonal physiological performance of polar seaweeds. Bot Mar 52:585–592

Woodrow IE, Berry JA (1988) Enzymatic regulation of photosynthetic CO_2 fixation in C3 plants. Annu Rev Plant Physiol Plant Mol Biol 39:533–594

Zeebe RE, Wolf-Gladrow DA (2001) CO_2 in seawater: equilibrium, kinetics, isotopes. Els Oceanogr Ser 65:346

Chapter 3
Seaweed Responses to Temperature

Anja Eggert

3.1 Introduction

"Why don't seaweeds spread beyond their present boundaries along an uninterrupted rocky coastline?", asked Breeman in 1988. Two principal aspects play a central role in shaping biogeographical distribution patterns: temperature-dependent effects on performance (e.g., growth, photosynthesis) and temperature tolerance (i.e., survival). The temperature responses of species are often correlated with the local thermal environments, i.e., species are locally adapted, but may vary seasonally or among populations or life stages due to phenotypic plasticity. Accordingly, it is necessary to differentiate three types of temperature responses: (1) genetic adaptation to local conditions, (2) phenotypic acclimation in response to variation of environmental conditions, and (3) short-term physiological regulation. The responses take place over different timescales: seconds to minutes (regulation), hours to days (acclimation), and up to thousands of millions of years (adaptation). This chapter reviews the temperature responses of seaweeds and their biogeographical implications.

3.2 Local Temperature Adaptation of Growth and Photosynthesis

The effect of temperature on performance traits, such as growth and photosynthesis, is typically visualized using temperature–response curves. Both growth and photosynthetic rates of seaweeds increase with temperature, plateau at a maximal level,

A. Eggert (✉)
Leibniz Institute for Baltic Sea Research Warnemünde, Section Physical Oceanography and Instrumentation, Theoretical Oceanography and Numerical Modelling, Seestrasse 15, 18119 Rostock, Germany
e-mail: anja.eggert@io-warnemuende.de

C. Wiencke and K. Bischof (eds.), *Seaweed Biology*, Ecological Studies 219,
DOI 10.1007/978-3-642-28451-9_3, © Springer-Verlag Berlin Heidelberg 2012

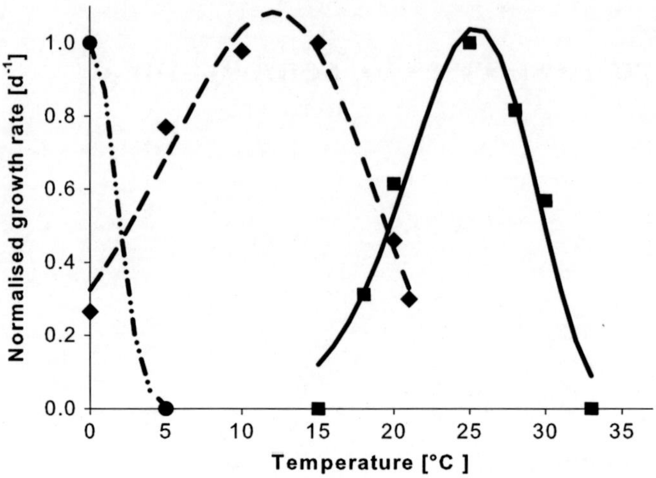

Fig. 3.1 Typical normalized growth temperature–response curves of the stenothermal polar species *Gymnogongrus skottsbergii* (Antarctica, *circles*, Eggert and Wiencke 2000), the eurythermal temperate species *Laminaria digitata* (Helgoland, *diamonds*, tom Dieck 1992), and the tropical species *Wurdemannia miniata* (St. Croix, Caribbean, *squares*, Pakker and Breeman 1996). The temperature–response curves were estimated by fitting the experimental data to the function developed by Blanchard et al. (1996). The temperature optima of the three response curves are 0, 12, and 25°C and the 80%-"performance breadths" are −1–1°C, 7–16°C, and 22–28°C for the Antarctic, temperate, and tropical species, respectively

and then rapidly decline near the upper critical temperature. These temperature–response curves can be characterized quantitatively by estimating various descriptive statistics (e.g., the minimum, maximum, optimum temperature, the performance breadth). Eggert et al. (2003a) introduced the temperature ranges at which 80% and 20% of the maximal growth rate occurs as arbitrary values for "good" and for "sufficient" growth, respectively. This approach intrinsically bypasses the interpretation of the optimum temperature, which is usually a meaningless parameter in the discussion of local temperature adaptation. The "performance breadth" (or alternatively the 80% or 20% temperature ranges) can be interpreted for the concept of eurythermy and stenothermy. Eurythermal seaweeds have a broad "performance breadth" and typically grow in environments with large fluctuations in seawater temperature. In contrast, stenothermal species have a narrow performance breadth and are found in regions with small temperature fluctuations.

Both the "performance breadth" and the position of the growth temperature–response curves of seaweeds typically correlate with the local temperature regime, i.e., reflect local temperature adaptation (Fig. 3.1). The endemic Antarctic species have the longest cold water history of at least 14 Ma (Briggs 1995; Crame 1993; see also Chap. 13 by Wiencke and Amsler) which in turn determines their worldwide lowest temperature requirements (Wiencke et al. 1994; see also Chap. 18 by Bartsch et al.). The red macroalgae *Gigartina skottsbergii* and *Ballia callitricha* grow in a very narrow temperature range between 0 and 5°C (Bischoff-Bäsmann and Wiencke 1996; Eggert and Wiencke 2000). Other Antarctic species with a distribution extending to Tierra del Fuego and subantarctic islands have more eurythermal temperature characteristics. For example, *Gymnogongrus antarcticus* and *Phyllophora ahnfeltioides* grow between 0 and 10°C (Bischoff-Bäsmann and Wiencke 1996; Eggert and Wiencke 2000). Compared to the Antarctic, seaweeds from the Arctic have a shorter cold water history of about 3 Ma (Briggs 1995) which explains the less cold-adapted and more eurythermal characteristics of Arctic species. For example, sporophytes of the kelp *Laminaria solidungula* grow between 0 and 15°C with an optimum at 5–10°C (tom Dieck 1992). Temperate species exhibit the widest performance breadth as they experience largest seasonal temperature changes. Cold-temperate Northeast Pacific and North Atlantic species grow between 0 and 18(20)°C with optima between 5 and 15°C (e.g., Bolton and Lüning 1982; Lüning and Freshwater 1988; tom Dieck 1992; Wiencke et al. 1994), while warm-temperate Atlantic species grow at up to 23–24°C and have slightly elevated optima (e.g., tom Dieck and Oliveira 1993). Additionally, life history stages (i.e., macrothalli/microthalli) of temperate seaweeds with a heteromorphic life history often have different temperature–response curves leading to an overall eurythermal temperature adaptation of the species (tom Dieck 1993). Tropical seaweeds from the Indo-West Pacific have the highest temperature optima of growth at 25–30°C (Pakker and Breeman 1996; Bischoff-Bäsmann et al. 1997).

Similar to growth, photosynthetic temperature responses also reflect adaptation to the local temperature regimes. Seaweeds from colder environments reach higher photosynthetic rates at low temperatures, while seaweeds native to warmer environments exhibit superior photosynthetic rates at higher temperatures. Optimum temperatures for photosynthesis are lowest (10–20°C) in Antarctic macroalgae (Wiencke et al. 1993; Eggert and Wiencke 2000), intermediate (20–25°C) in cold-temperate to Arctic species, and highest (25–35°C) in warm-temperate to tropical species (Terrados and Ros 1992). Thus, temperature optima of photosynthesis are situated well above the temperature optima of growth (Davison 1987; Kübler et al. 1991; Eggert and Wiencke 2000; see also Chap. 13 by Wiencke and Amsler). This shows that temperature effects on a specific physiological process (i.e., photosynthesis in this case) do not necessarily correspond to the temperature–growth pattern as growth integrates the effect of temperature on the total metabolism.

As a consequence of the long cold water history of the Antarctic Ocean, very specific adaptations have evolved in endemic Antarctic species. Eastman (1993) detected "DNA decay" in Antarctic fish, i.e., loss of genetic information not required for life at ambient temperatures (Hoffmann and Willi 2008). Ice-binding proteins have been found in Antarctic sea ice diatoms as a very specific adaptation to the very low, freezing temperatures in this extreme habitat. The extracellular proteins are associated with the diatom community and they serve to prevent freezing injury (Janech et al. 2006). However, it has not been investigated whether these or other types of low-temperature adaptation have evolved in polar seaweeds as well.

Temperature limitations of growth and photosynthesis are a direct effect of temperature sensitivity of the main cellular components, i.e., proteins and membranes. Evolutionary adaptation to local temperatures on the cellular level can involve one or more of the three strategies: quantitative (changing the concentrations of enzymes and/or reactants); qualitative (using a protein variant/ isozyme with different thermal characteristics); or modulation (modifying the protein environment to minimize the impact of temperature change) (Hochachka and Somero 2002). The most important adjustments include changes in enzyme concentration, changes in primary structure affecting the free energy of activation, and modification to both membrane properties and intracellular milieu. These responses are genotypic, but similar changes may be induced by acclimation on a daily or seasonal timescale. The energetic costs of the different strategies of temperature adaptation vary (Clarke 2003). If protein variants/isozymes vary by only a few amino acids, it costs an organism effectively no more ATP to make a molecule of one variant than it does to make another. However, energetic costs become relevant when a particular protein is required in larger amounts, or is turned over faster. Trade-offs with other metabolic processes and changes in energy budgets need to take place. These aspects have been analyzed in polar ectothermic marine animals (Pörtner et al. 2005), but remain to be investigated in polar seaweeds.

Specific knowledge of thermal physiology of proteins is limited to a very small subset of proteins. Clear patterns of adaptive variation have been discovered in structural and functional properties of proteins from ectothermic marine animals adapted to different temperatures (Somero 2004). Studies of dehydrogenase enzymes have demonstrated that a single amino acid substitution is sufficient to cause temperature-adaptive changes in function and stability and define species geographic boundaries (e.g., Fields and Houseman 2004; Dong and Somero 2009). Studies on temperature adaptation are missing for seaweeds and are scarce for unicellular algae. Descolas-Gros and De Billy (1987) showed that the RuBisCO in marine Antarctic diatoms effectively binds CO_2 only at low temperatures. They describe qualitative changes, i.e., an isozyme with modified kinetic properties which allows the maintenance and regulation of RuBisCO activity in Antarctic diatoms at low temperatures (minimum K_m value at 4.5°C for Antarctic species compared to 20°C for temperate species). In contrast, Antarctic *Chloromonas* species did not show an increase in activity at low temperatures (0–5°C) compared to temperate species, which was counterbalanced by increasing enzyme concentration, i.e., quantitative adjustments (Devos et al. 1998).

In addition to proteins, several physical properties of membranes are temperature sensitive, including permeability characteristics, fluidity, and membrane phase state (for reviews, see Murata and Los 1997; Los and Murata 2004). All are crucial to many membrane functions, including the activities of enzymes. A number of studies focused on the effects of low temperature and clearly demonstrated that membrane fluidity decreases with a decrease in temperature (Szalontai et al. 2000; Inaba et al. 2003). The effects of high temperature on the physical state of membranes have also been studied, albeit less intensively (Carratù et al. 1996;

Vigh et al. 1998). High temperatures cause an increase in the fluidity of membranes, which ultimately leads to a disintegration of the lipid bilayer.

The impairment of photosynthesis at low temperature appears to be primarily related to impaired synthesis and functioning of photosynthetic pigment–protein complexes (Robertson et al. 1993; Nie et al. 1995) and reduced activities of key enzymes in the Calvin cycle (Holaday et al. 1992; Huner et al. 1993; Kingston-Smith et al. 1997). There are also several target sites for the impairment of photosynthesis at high temperatures, such as the CO_2 fixation system, photophosphorylation, the electron transport chain, and the oxygen evolving complex (Sharkey 2005; Allakhverdiev et al. 2008). Especially various parameters of fast chlorophyll fluorescence transients of photosystem II, such as the maximum and effective quantum yield (Fv/Fm, Φ PSII), the minimum fluorescence (Fo), and photochemical and non-photochemical quenching of chlorophyll fluorescence, have been widely used to monitor temperature-induced changes of photosynthetic activity in seaweeds (e.g., Antarctic *Palmaria decipiens*, Arctic *Fucus distichus*: Becker et al. 2009, warm-temperate *Valonia utricularis*: Eggert et al. 2003b, tropical to warm-temperate *Laurencia* spp.: Padilla-Gamiño and Carpenter 2007).

3.3 Phenotypic Temperature Acclimation of Growth and Photosynthesis

Most seaweeds have the ability to acclimate growth and photosynthesis in response to changes in ambient temperature, both to daily temperature changes and on a seasonal timescale. Beneficial phenotypic acclimation (*sensu* Leroi et al. 1994) is the improvement of a thermal trait (e.g., growth, photosynthesis) at the respective ambient temperature that allows to maximize performance over a broad temperature range. The potential for temperature acclimation varies between species and is expected to be higher in eurythermal than in stenothermal species. Accordingly, seaweeds native to habitats with large annual temperature variations typically display a stronger ability to acclimate than species from habitats with more stable seasonal regimes. Seasonal acclimation of photosynthesis has been described for seaweeds from the temperate regions (Davison 1987; Kübler and Davison 1995; Pfetzing et al. 2000; Eggert et al. 2006; Padilla-Gamiño and Carpenter 2007). Likewise, acclimation has been described in a number of intertidal species (Smith and Berry 1986; Kim et al. 2009; Henkel et al. 2009). In contrast, a limited acclimation potential has been described for Antarctic and tropical species (Eggert and Wiencke 2000; Eggert et al. 2006).

The fact that temperature changes can induce cellular acclimation responses indicates that temperature is sensed and that the temperature signal is immediately transduced into the cell. Membrane fluidity, protein conformation, cytoskeleton depolymerization, and metabolic reactions have all been identified to be temperature sensors (Horvath et al. 1998; Los and Murata 2004). Enzyme adjustments to temperature occur constantly as temperature changes on different timescales,

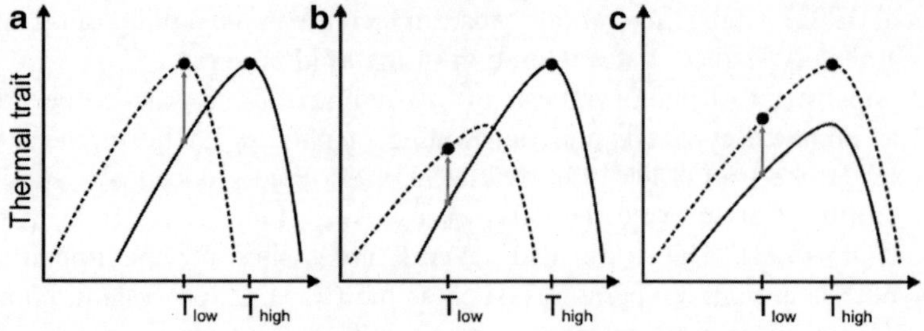

Fig. 3.2 Two individual temperature–response curves of organisms grown at a high temperature T_{high} (*solid line*) and at a low temperature T_{low} (*dashed line*). The *circles* mark the physiological rate at the respective growth temperature. The *gray arrows* indicate (partial) acclimation at T_{low}. (**a**) "Complete low-temperature acclimation", (**b**) "Beneficial acclimation hypothesis", and (**c**) "Colder is better hypothesis"

e.g., diurnally or seasonally. The quantitative and/or qualitative metabolic adjustments of proteins that allow temperature compensation are similar to the genotypic evolutionary adjustments (Clarke 2003). Also, membranes can undergo changes in both lipid composition and degree of unsaturation of fatty acids in response to temperature changes. In this way, they can maintain a relatively constant fluidity and can alter the state transition temperature (Los and Murata 2004). These homeoviscous changes are very important in establishing the appropriate membrane physical state for some enzymes at different temperatures.

Phenotypic acclimation can be tested experimentally by comparing temperature–response curves (short-term effects) of organisms grown at different temperatures (long-term effects). The following three parameters of the short-term temperature–response curves are the most important: maximum rate, temperature optimum, and performance breadth. Both a shift in the response curve (including temperature optimum) and increase in maximum rate may be beneficial and maximize performance over a broad temperature range. Various acclimation hypotheses are discussed in evolutionary biology (Deere and Chown 2006). "Complete temperature compensation" maintains the physiological rates as the temperature changes (Fig. 3.2). However, partial compensation is apparently more common and is predicted from the "Beneficial acclimation hypothesis" caused by some co-variation between optimum temperature and maximum rate and some overlap of the performance curves (Fig. 3.2b). Temperate populations of *Valonia utricularis* (Chlorophyta, Eggert et al. 2006) and *Asparagopsis taxiformis* (Rhodophyta, Padilla-Gamiño and Carpenter 2007) displayed seasonal (partial) temperature acclimation of photosynthesis, whereas tropical individuals of both species did not. Alternatively, the hypothesis "Colder is better" predicts that an organism acclimated to a low-temperature environment will always have a performance advantage over another organism held at a warmer environment (Fig. 3.2c). In this case, the optimal temperature does not change, but the performance curve overall increases. This type of acclimation of photosynthesis has been shown for the Antarctic seaweeds *Ballia callitricha* and *Gymnogongrus antarcticus* (Eggert and

Wiencke 2000) and was also found for cold-temperate specimens of the kelp *Saccharina latissima* (Davison et al. 1991) and the red macroalga *Chondrus crispus* (Kübler and Davison 1995).

Photosynthetic adjustments in response to a change in growth temperature take place in order to counteract the imbalance between photosynthetic energy production and cellular energy consumption (Wilson et al. 2003a). Low-temperature acclimated *Chlorella vulgaris* cells exhibit high chlorophyll *a*:*b* ratio, low light-harvesting complex polypeptide abundance, and an increased zeaxanthin content (Wilson and Huner 2000). Machalek et al. (1996) describe in *Saccharina latissima* a decrease in concentration of major light-harvesting pigments when grown at suboptimal temperatures, and thermal acclimation in *Chondrus crispus* is characterized by variations in antenna size rather than by reaction center densities (Kübler and Davison 1995). Notably, not necessarily all components of the photosynthetic apparatus are sensitive to a temperature change. Wilson et al. (2003a) describe that shifting *C. vulgaris* cells from 5 to 27°C did not induce changes in the accumulation of RuBisCO, a key enzyme of photosynthesis utilizing ATP and NADPH to reduce CO_2 in the Calvin cycle. However, this result contradicts observations by Savitch et al. (1996) and the amount of RuBisCO was also significantly higher in the kelp *Saccharina latissima* when grown at 5°C compared to 17°C-grown algae (Machalek et al. 1996). Acclimation of photosynthesis to high temperatures primarily involves changes in lipid composition of thylakoid membranes and the adjustment of photosystem II thermostability, which could be enhanced either directly through conformational changes of the photosystem or indirectly via a carotenoid-dependent modulation of membrane fluidity (Havaux and Tardy 1996).

It appears that there is more than one cellular signaling mechanism involved in temperature acclimation. There is consensus that the redox state of the plastoquinone pool acts as a sensor of imbalances in the electron transport and is involved in regulating the gene expression of a large number of genes required for photosynthesis (Pfannschmidt et al. 1999). In addition, the *trans*-thylakoid pH gradient, biosynthetic precursors of chlorophyll acting as potential signaling molecules, and reactive oxygen species (see also Chap. 6 by Bischof and Rautenberger) acting as second messenger type molecules have also been shown to be involved (Wilson and Huner 2000; Wilson et al. 2003b). The response of thermal acclimation of photosynthesis appears to be highly comparable to that described for photoacclimation (Huner et al. 1998). However, temperature and light have been shown to have interactive effects on photosynthesis and its regulation (Savitch et al. 1996; Gray et al. 1997).

Today molecular tools such as cDNA and oligonucleotide DNA microarrays ("gene chips") are used for examining environmentally induced changes in the transcriptome. Insights into the role of phenotypic plasticity in transcriptional processes are beginning to be achieved in a variety of species, including Antarctic fishes and invertebrates (Hofmann et al. 2000; Clark and Peck 2009). One major obstacle to studies in seaweeds has been the lack of genomic information, which is now increasing. The genome of the brown macroalga *Ectocarpus siliculosus* has been sequenced by the French sequencing center Genoscope (see http://www.cns.fr)

and the Joint Genome Institute (http://www.jgi.doe.gov/) is presently sequencing the complete nuclear genome of *Porphyra umbilicalis*, which is a large eukaryotic genome (haploid genome 5–270 mb; reviewed in Gantt et al. 2010). However, notably large expressed sequence tag (EST) projects on the red alga *Porphyra yezoensis* (Nikaido et al. 2000; Asamizu et al. 2003), but also smaller EST projects on, e.g., the red alga *Chondrus crispus* (Collén et al. 2006), the brown alga *Laminaria digitata* (Roeder et al. 2005) and the green alga *Ulva linza* (Stanley et al. 2005), allow the utilization of powerful tools of functional genomics.

3.4 Disruptive Temperature Stress and Thermal Tolerance

Heat stress and cold stress (including freezing) cause damage to seaweeds and are referred to "disruptive stress" *sensu* Davison and Pearson (1996) due to adverse conditions beyond phenotypic temperature acclimation to suboptimal temperatures. The timescale of disruptive temperature stress is particularly relevant. Organisms may cope temporarily (timescale of hours) with strong temperature stress and subsequently recover from damage at optimal conditions (Eggert et al. 2003b). But on a longer timescale (timescale of weeks) and/or increasingly stressful conditions, the organisms experience progressively more impaired cellular processes until the minimum and maximum temperatures for survival are reached. The degree of physiological dysfunction becomes very severe at these cardinal temperatures and cell death ultimately follows. The individual temperature tolerance of seaweed species defines the minimum and maximum temperatures for survival. Upper and lower temperature limits for survival with biogeographical implications are typically assessed using incubation periods of 2–8 weeks and a recovery phase of 2 weeks at optimal temperatures (e.g., Wiencke et al. 1994; Eggert et al. 2003a).

Short-term thermal stress is severest for benthic algae in the high intertidal, where factors causing desiccation and/or critical temperatures determine the upper limit for growth (Davison and Pearson 1996). Seaweeds that grow in shallow tide pools or are even exposed to air during tidal emersion may regularly experience abrupt temperature changes of 10–20°C (Helmuth and Hofmann 2001). Also, freezing is an important stress for polar and cold-temperate intertidal algal communities (Pearson et al. 2000). Tolerance ranges (i.e., the range between upper and lower lethal temperatures) have been found to be broader in intertidal seaweeds occupying the upper shore than in species from the subtidal (Einav et al. 1995; Stengel and Dring 1997; Martone et al. 2010). Short-term thermal stress on the organism level seems ecologically less relevant for seaweeds growing in the subtidal where algae are virtually always submerged.

Seaweeds growing in the center of their geographic distribution show seasonal variation in biomass that is typically not directly controlled by seasonal temperature stress, but most importantly by the variability of the light regime and the availability of nutrients (Kain 1989; Wiencke et al. 2009). However, thermal stress may limit seaweed growth in all populations growing near distribution boundaries that are set by lethal temperature limits. Notably these boundary populations are threatened by global warming, an issue that is addressed in Chap. 18 by Bartsch et al. Heat stress may limit seaweed growth in tropical regions in summer when high temperatures exceed the upper thermal tolerance limit of seaweeds, which is <35°C in many strictly tropical seaweeds. Pakker et al. (1995) describe experimentally determined upper tolerance limits of tropical Caribbean seaweeds at 30°C or 33°C, i.e., very close to the local summer temperatures of 30°C. Populations of the fucoid seaweed *Sargassum lapazeanum* from the Gulf of California also showed maximal mortality rates coinciding with highest values of seawater temperatures in summer (29–30°C, Rivera and Scrosati 2006). On shallow reef flats in the southern Red Sea, temperatures may even exceed 34°C for prolonged periods in summer and seaweeds strongly decrease in biomass or disappear during the hot season (Ateweberhan et al. 2005). The tropical green macroalga *Cladophora submarina* is restricted in the western Atlantic to the Caribbean and Bermuda. Its northward extension is prevented by lethal, low winter temperatures <15°C (Cambridge et al. 1987). This species may experience severe cold stress very close to the lethal limits for 1–2 weeks during cold winters at the Florida Keys, when temperatures in the shallow bays may fall below 15°C caused by intrusion of polar air masses.

On the cellular level, heat stress is known to affect membrane-associated processes as high temperatures cause fluidization of membranes and finally disintegration of the lipid bilayer (Los and Murata 2004). Also, protein stability and function are impaired and cause decreased enzyme activities or even enzyme inactivation. Moreover, membrane and protein damage trigger the production of reactive oxygen species, which in turn inhibit the de novo proteins synthesis (Larkindale et al. 2005). Specifically with respect to photosynthesis, there are three major heat-sensitive sites: photosystem II with the oxygen evolving complex, ATP generating ATP synthase, and enzymes of the Calvin–Benson cycle (Allakhverdiev et al. 2008). The primary event of cold stress is the formation of lipid gel phases in cell membranes. When a model membrane enters a phase-separated state in which gel and liquid-crystalline phases coexist, the membrane becomes permeable to small electrolytes. This permeability results in the disruption of ion gradients, across the membrane, that are essential for the maintenance of cellular activities (Nishida and Murata 1996). Changes in the ultrastructure of chloroplasts were detected in the green macroalga *Valonia utricularis* when exposed to cold stress (Eggert 2002). Chloroplasts were less dense packed in the warm-temperate Mediterranean isolate and became particularly disorganized and more swollen in the tropical Indian Ocean isolate when algae were grown at 15 and 18°C, respectively (Fig. 3.3). Disorganization of chloroplasts reduces light-harvesting capacity and ultimately photosynthetic activity (Ciamporova and Trginova 1996). As temperature decreases to freezing temperatures, ice forms in the intercellular spaces (Tomashow 1998). The accumulation of ice would result in the physical disruption of the cells. However, freezing injury results primarily from cellular dehydration rather than from direct mechanical damage by ice crystals. Freeze-induced dehydration causes denaturation of proteins, precipitation of various molecules, and multiple forms of membrane lesions (Steponkus and Webb 1992).

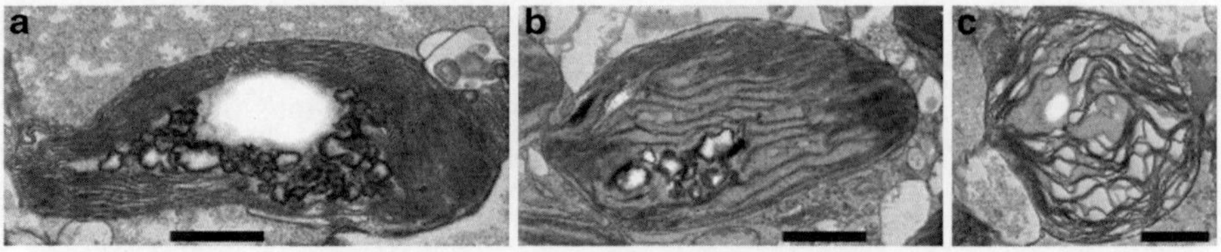

Fig. 3.3 Electron micrographs of cross-sections of *Valonia utricularis* cells, showing chloroplast ultrastructure of the cells. Scale bars measure 1 μm. (**a**) Warm-temperate, Mediterranean isolate grown at optimal temperature (25°C), (**b**) warm-temperate, Mediterranean isolate grown at slight cold stress (15°C), (**c**) tropical, Indian Ocean isolate grown at severe cold stress (18°C). Cells were fixed in situ with 6.0% (w/v) glutaraldehyde in 0.1 M Na-cacodylate buffer (pH 7.2) overnight at 0°C, washed three times with 0.1 M Na-cacodylate buffer (pH 7.2), and postfixed (15 min at room temperature) with 1.5% (w/v) $KMnO_4$. Cells were washed with distilled water and stained in 1.0% (w/v) uranyl acetate overnight. Cells were then dehydrated in an upgrade series of ethanol and embedded in Epon 812 (for more details, see Eggert 2002)

All these types of damage cause reduced photosynthesis and carbon assimilation and ultimately cell dysfunction and cell death.

However, many seaweeds, particularly from temperate regions, can evolve greater resistance to temperature stress via increased tolerance or activation of recovery mechanisms. Both types can serve to extend their temperature range for survival during acute temperature stress. For instance, Davison (1987) reports an acquired high-temperature tolerance of photosynthesis in sporophytes of *Saccharina latissima* grown at 10–20°C compared to specimens grown at 0–5°C due to changes in the thermal stability of the photosystem II electron transport system. Similarly, *Chondrus crispus* grown at summer seawater temperatures (20°C) maintains constant rates of light-saturated photosynthesis at 30°C for 9 h, while photosynthesis of 5°C-grown algae declined rapidly within 10 min following exposure to 30°C (Kübler and Davison 1993). Cold-acclimated (15°C) *Valonia utricularis* exhibited a faster recovery kinetics from chilling-induced photoinhibition than specimens grown at 25°C (Eggert et al. 2003b), which was related to a faster recovery from chronic photoinhibition.

Increased thermal tolerance involves a large suite of processes that modify the cellular metabolism. Again, cellular and molecular investigations in seaweed species are scarce, but the following examples cover a wide range of processes: (1) Heat shock proteins are accumulated as they function as molecular chaperones that protect cellular proteins from protein mis-folding and degradation by environmental stress, including cold and heat stress (Sorensen et al. 2003). Henkel et al. (2009) suggest a high thermal tolerance of the invasive kelp species *Undaria pinnatifida* in the northeast Pacific (California) due to a high expression of the hsp70 gene which encodes heat shock protein 70. (2) The unsaturation of fatty acids increases membrane fluidity and such an increase is necessary if cells are to tolerate cold stress and to survive low temperatures (Murata and Los 1997). Increasing proportions of polyunsaturated fatty acids and, consequently, a higher degree of unsaturation in *Caulerpa racemosa* from the northern Adriatic coincided with a sharp decrease in water temperature in winter (Blažina et al. 2009). By this increased low-temperature tolerance, this species can successfully survive the winter without visible necrotic parts or chilling injuries. (3) As thermal stress disrupts the cellular homeostasis of cells and enhances the production of reactive oxygen species, antioxidants are crucial for increased stress tolerance (Mittler 2002; see also Chap. 6 by Bischof and Rautenberger). Collén et al. (2007) investigated the effects of heat stress on the transcriptome of *Chondrus crispus* using cDNA microarrays. They showed that high-temperature stress caused large changes in gene expression and resource allocation, including antioxidant proteins and detoxifying enzymes. (4) Compatible solutes are accumulated, which are low-molecular-mass, highly soluble organic compounds and are particularly known to serve as osmolytes in salt-stressed cells, including seaweeds (Yancey 2005). The metabolites in seaweeds can include betaines, polyols, and sugars, such as mannitol and trehalose, and amino acids, such as proline (Kirst 1990; see also Chap. 5 by Karsten). Compatible solutes are known to exert multiple protective functions in the metabolism and can also act as antioxidants and can stabilize proteins upon heat stress. Even though similar physiological functions are reasonable to assume in seaweeds as well, they are still to be experimentally proven (Eggert and Karsten 2010).

3.5 Temperature Control of Biogeographical Distribution Boundaries

Geographical distributions of seaweeds are typically delimited at certain seawater isotherms. Based on the historical concept of Setchell (1920) and modified by Lüning (1990), local winter and summer isotherms (i.e., long-term mean monthly temperatures of February and August, respectively) of sea surface temperatures are used for delimitation of biogeographical regions. Seven biogeographical regions are typically recognized in marine phycogeography (for details, see Chap. 18 by Bartsch et al.).

According to van den Hoek (1982a, b) and Breeman (1988), three fundamentally different types of geographic seaweed boundaries can be distinguished: (1) lethal boundaries, i.e., high or low temperatures prevent survival of the hardiest life history stage, (2) reproduction boundaries, i.e., high or low temperatures prevent completion of the life history, (3) growth boundaries, i.e., temperature requirements for sufficient increase of population size. Lethal boundaries are determined by the capacity of the species to survive during the unfavorable season. More specifically, southern lethal temperature boundaries of cold-adapted species (of the Northern hemisphere) are set at the August isotherm of the seawater and the northern lethal limit of warm-adapted species (of the Northern Hemisphere) at the February isotherm. Temperature limits for growth or reproduction operate during the favorable season. Accordingly, southern growth/reproduction limits in the Northern Hemisphere are set at the February isotherm and northern boundaries are determined by the August isotherm. Moreover, seaweed distribution is affected by temperature–daylength interactions (Dring 1984; Molenaar 1996). Optimum temperatures for growth and reproduction or lethal temperatures for the non-hardiest life stage are typically irrelevant in explaining geographic distribution. Suboptimal or sublethal temperatures often prevail at species distribution boundaries. For instance, in case of the tropical to warm-temperate species *Cladophora albida*, which has optimum growth temperatures as high as 25–30°C, temperatures are even suboptimal for year-long growth over a large part of the distribution range at the NW Atlantic coast (Cambridge et al. 1984). Also, many seaweeds from the Arctic grow suboptimally during most of the year and they have a wider distribution in cold-temperate regions (Gómez et al. 2009).

Temperature ranges for survival, growth, and reproduction have been determined experimentally for a large number of seaweed species during the last three decades. Supporting the concepts described above, large-scale biogeographic distribution patterns of many seaweed species were explained by combining the species' thermal traits with local seawater temperatures. Species with similar thermal responses and the same types of distribution boundaries have been assigned to phytogeographic distribution groups (c.f. van den Hoek 1982a, 1982b; Breeman 1988; Wiencke et al. 1994 for designation of groups).

Ecotypic differentiation in thermal responses would be expected in broadly distributed seaweed species, living in a strong thermal gradient. Genetically distinct temperature ecotypes are locally adapted to the particular temperature regime (ecotype definition *sensu* Turesson 1922) and may influence the location of geographic boundaries in a number of species. On the other hand, populations may not be distinct from each other and temperature requirements change gradually over the distribution area, i.e., variation is of an ecoclinal nature. The recognition of ecotypes is intrinsically linked to the definition of species (e.g., Hey 2006), a subject that requires molecular approaches as many seaweed morphospecies are composed of biologically and genetically distinct lineages (Andreakis et al. 2007; Wattier and Maggs 2001; Boedeker et al. 2008). Studies on local adaptation and adaptive evolution in new environments of seaweeds that include molecular data still remain few (Johansson et al. 2003; Bergström et al. 2005; Pereyra et al. 2009; Verbruggen et al. 2009). Most studies on ecotypic differentiation in seaweeds were done with biogeographical isolates, grown under the same laboratory conditions in "common-environment experiments." The temperature responses of thermal traits (i.e., upper and lower survival temperatures, growth range, temperature requirements for reproduction, and photosynthesis) are typically determined and compared with the annual temperature regime of the respective habitats. Especially, the annual growth yield (potential monthly growth rate) is a useful parameter to evaluate local temperature adaptation as it allows to combine the seasonal temperature range with the temperature dependence of growth (e.g., applied in Breeman and Pakker 1994; Bischoff and Wiencke 1995; Eggert et al. 2003a). It is concluded from these "common-environment studies" that observed differences between isolates have a genetic basis and the results provide evidence for intra-specific differentiation. Thermal ecotypes or ecoclinal variation has been described for seaweed species with a disjunct distribution or with wide distributional ranges (Bolton 1983; Bischoff and Wiencke 1995; Pakker and Breeman 1996; Eggert et al. 2003a). In contrast, Pakker et al. (1996) did not find ecotypic differentiation in tropical to subtropical seaweed species with disjunct eastern and western Atlantic populations, despite the large geographic distance. It was suggested that trans-Atlantic dispersal events may be responsible for the lack of ecotype formation. Molecular studies have confirmed the occurrence of long-distance dispersal across the central Atlantic Ocean for a number of seaweed species (Kooistra et al. 1992; Pakker et al. 1996; Pakker and Breeman 1996).

The potential for recent/ongoing geographic differentiation is especially rich on the north-western Atlantic coast, where spatial variability in the environment is strong, with a steep latitudinal gradient in temperature. Differentiation along the north-western Atlantic coast has been found in several marine macrophytes (e.g., *Saccharina latissima*: Gerard and Du Bois 1988). However, the semi-enclosed marginal Mediterranean in the eastern Atlantic also facilitates thermal isolation and selection pressure, processes which appeared especially important during Pleistocene ice ages. Similar marginal seas do not exist in the recent north-western Pacific, even though the Japan Sea might temporarily have also functioned as a warm water pocket during glaciations (van den Hoek and Chihara 2000). *Valonia utricularis*, *Cladophora albida*, and *C. vagabunda* present three examples in which Pacific northern boundaries are situated much further south than Atlantic northern boundaries and which showed ecotypic differences (Breeman et al. 2002; Eggert et al. 2003a).

Biogeographical patterns have long been studied in a rather descriptive way, disregarding that thermal traits are susceptible to evolutionary forces. The rising of molecular data and phylogenetic reconstruction in phycology during the last two decades has led to a number of new research questions related to evolutionary dynamics in biogeography on a species-by-species basis (Avise 2000). Molecular studies put forward evolutionary scenarios to explain the recent distribution of species by considering historical events such as glacial refugia (e.g., *Palmaria palmata*: Provan et al. (2005), *Fucus spiralis*/*F. vesiculosus* species complex: Coyer et al. 2011), vicariance, e.g., due to the rise of the Panamanian Isthmus (*Halimeda* species: Kooistra et al. 2002), or recover historical patterns of seaweed dispersal (*Bostrychia radicans*/*B. moritziana*: Zuccarello et al. 2006). However, many historical biogeographic studies consist of subjective interpretations of genetic and physiological differentiation among geographic isolates. A model-based inference approach is relatively recent and so far not applied for algae (Ree and Sanmartín 2009). Furthermore, sophisticated ecological niche models have been applied in recent years to predict geographical areas with suitable habitat for the species (e.g., Guisan and Thuiller 2005; Raxworthy et al. 2007). So far, species distribution modeling has focussed on terrestrial organisms. Notable exceptions are the study by Graham et al. (2007) which used a niche model to predict kelp habitats in deep tropical waters and the work by Verbruggen et al. (2009) on evolutionary niche dynamics in the green seaweed *Halimeda*. Finally, Tyberghein et al. (2011) presented the data package Bio-ORACLE (Ocean Rasters for Analysis of CLimate and Environment) that permits the development of accurate distribution modeling of marine seaweeds.

3.6 Conclusion

Local temperature adaptation and phenotypic acclimation of performance traits (i.e., growth and photosynthesis) and temperature tolerance ranges (i.e., survival) have been experimentally determined for a wide variety of seaweed species from virtually all biogeographical regions during the last three decades. Large-scale biogeographic distribution patterns of many seaweed species could be explained by combining these thermal traits with local seawater temperatures. In contrast to the vast knowledge on temperature responses of seaweeds on the organism level, studies on temperature effects on cellular components (i.e., proteins and membranes) and environmentally induced changes in transcriptional processes still remain scarce. One major obstacle has been the lack of genomic information of seaweed species, which is now increasing. Furthermore, sophisticated ecological niche models have been applied in the recent years to predict geographical areas with suitable environmental conditions for marine seaweed species.

References

Allakhverdiev SI, Kreslavski VD, Klimov VV, Los DA, Carpentier R, Mohanty P (2008) Heat stress. An overview of molecular responses in photosynthesis. Photosynth Res 98:541–550

Andreakis N, Procaccini G, Maggs C, Kooistra WHCF (2007) Phylogeography of the invasive seaweed *Asparagopsis* (Bonnemaisoniales, Rhodophyta) reveals cryptic diversity. Mol Ecol 16:2285–2299

Asamizu E, Nakajima M, Kitade Y, Saga N, Nakamura Y, Tabata S (2003) Comparison of RNA expression profiles between the two generations of *Porphyra yezoensis* (Rhodophyta), based on expressed sequence tag frequency analysis. J Phycol 39:923–930

Ateweberhan M, Bruggemann JH, Breeman AM (2005) Seasonal patterns of biomass, growth and reproduction in *Dictyota cervicornis* and *Stoechospermum polyplioides* (Dictyotales, Phaeophyta) on a shallow reef flat in the southern Red Sea (Eritrea). Bot Mar 48:8–17

Avise JC (2000) Phylogeography: the history and formation of species. Harvard University Press, Cambridge

Becker S, Walter B, Bischof K (2009) Freezing tolerance and photosynthetic performance of polar seaweeds at low temperatures. Bot Mar 52:609–616

Bergström L, Tatarenkov A, Johannesson K, Jonsson RB, Kautsky L (2005) Genetic and morphological identification of *Fucus radicans* sp. nov. (Fucales, Phaeophyceae) in the brackish Baltic Sea. J Phycol 41:1025–1038

Bischoff B, Wiencke C (1995) Temperature adaptation in strains of the amphi-equatorial green alga *Urospora penicilliformis* (Acrosiphoniales): biogeographical implications. Mar Biol 122:681–688

Bischoff-Bäsmann B, Wiencke C (1996) Temperature requirements for growth and survival of Antarctic Rhodophyta. J Phycol 32:525–535

Bischoff-Bäsmann B, Bartsch I, Xia B, Wiencke C (1997) Temperature responses of macroalgae from the tropical island Hainan (P.R. China). Phycol Res 45:91–104

Blanchard G, Guarini J-M, Richard P, Gros P, Mornet F (1996) Quantifying the short-term temperature effect on light-saturated photosynthesis of intertidal microphytobenthos. Mar Ecol Prog Ser 134:309–313

Blažina M, Ivesa L, Najdek M (2009) *Caulerpa racemosa*: adaptive varieties studied by fatty acid composition (Northern Adriatic Sea, Vrsar, Croatia). Eur J Phycol 44:183–189

Boedeker C, Farr TJ, Nelson WA (2008) Comparative culture experiments with filamentous members of the Bangiales (Rhodophyta) from New Zealand: insight into ecological adaptation and biogeography. Phycol Res 56:183–192

Bolton JJ (1983) Ecoclinal variation in *Ectocarpus siliculosus* (Phaeophyceae) with respect to temperature growth optima and survival limits. Mar Biol 73:131–138

Bolton JJ, Lüning K (1982) Optimal growth and maximal survival temperatures of Atlantic *Laminaria* species (Phaeophyta) in culture. Mar Biol 66:89–94

Breeman AM (1988) Relative importance of temperature and other factors in determining geographic boundaries of seaweeds: experimental and phonological evidence. Helgol Meeresunters 42:199–241

Breeman AM, Pakker H (1994) Temperature ecotypes in seaweeds: adaptive significance and biogeographic implications. Bot Mar 37:171–180

Breeman AM, Oh Y, Hwang M, van den Hoek C (2002) Evolution of temperature responses in the *Cladophora vagabunda* complex and the *C. albida/sericea* complex (Chlorophyta). Eur J Phycol 37:45–58

Briggs JC (1995) Global biogeography. Elsevier, Amsterdam

Cambridge ML, Breeman AM, van Oosterwijk R, van den Hoek C (1984) Temperature responses of some North Atlantic *Cladophora* species (Chlorophyceae) in relation to their geographic distribution. Helgol Meeresunters 38:349–363

Cambridge ML, Breeman AM, Kraak S, van den Hoek C (1987) Temperature responses of tropical to warm-temperate *Cladophora* species in relation to their distribution in the North Atlantic Ocean. Helgol Meeresunters 41:329–354

Carratù L, Franceschelli S, Pardini CL, Kobayashi GS, Horvath I, Vigh L, Maresca B (1996) Membrane lipid perturbation modifies the set point of the temperature of heat shock response in yeast. Proc Natl Acad Sci USA 93:3870–3875

Ciamporova M, Trginova I (1996) Ultrastrucutre of chloroplasts in leaves and of plastids in root tips of two maize lines differing in chilling tolerance. Biologia Bratislava 51:441–447

Clark MS, Peck LS (2009) HSP70 heat shock proteins and environmental stress in Antarctic marine organisms: a mini-review. Mar Genom 2:11–18

Clarke A (2003) Costs and consequences of evolutionary temperature adaptation. Trends Ecol Evol 18:573–581

Collén J, Roeder V, Rousvoal S, Collin O, Kloareg B, Boyen C (2006) An expressed sequence tag analysis of thallus and regeneration protoplasts of *Chondrus crispus* (Gigartinales, Rhodophyceae). J Phycol 42:104–112

Collén J, Guisle-Marsollier I, Léger JJ, Boyen C (2007) Response of the transcirptome of the intertidal red seaweed *Chondrus crispus* to controller and natural stresses. New Phytol 176:45–55

Coyer JA, Hoarau G, Costa JF, Hogerdijk B, Serrão EA, Billard E, Valero M, Pearson A, Olsen JL (2011) Evolution and diversification within the intertidal brown macroalgae *Fucus spiralis/F. vesiculosus* species complex in the North Atlantic. Mol Phylogenet Evol 58:283–296

Crame JA (1993) Latitudinal range fluctuations in the marine realm through geological time. Trends Ecol Evol 10:1096–1111

Davison IR (1987) Adaptation of photosynthesis in *Laminaria saccharina* (Phaeophyta) to changes in growth temperature. J Phycol 23:273–283

Davison IR, Pearson GA (1996) Stress tolerance in intertidal seaweeds. J Phycol 32:197–211

Davison IR, Greene RM, Podolak EJ (1991) Temperature acclimation of respiration and photosynthesis in the brown alga *Laminaria saccharina*. Mar Biol 110:449–454

Deere JA, Chown SL (2006) Testing the beneficial acclimation hypothesis and its alternatives for locomotor performance. Am Nat 168:630–644

Descolas-Gros C, De Billy G (1987) Temperature adaptation of RuBP carboxylase: kinetic properties in marine Antarctic diatoms. J Exp Mar Biol Ecol 108:147–158

Devos N, Ingouff M, Loppes R, Matagne R (1998) Rubisco adaptation to low temperatures: a comparative study in psychrophilic and mesophilic unicellular algae. J Phycol 34:655–660

Dong Y, Somero GN (2009) Temperature adaptation of cytosolic malate dehydrogenases of limpets (genus Lottia): differences in stability and function due to minor changes in sequence correlate with biogeographic and vertical distributions. J Exp Biol 212:169–177

Dring MJ (1984) Photoperiodism and phycology. In: Round F, Chapman DJ (eds) Progress in phycological research, vol 3. Biopress, Bristol, pp 159–192

Eastman JT (1993) Antarctic fish biology: evolution in a unique environment. Academic, San Diego

Eggert A (2002) Thermal ecotypes in the green macrophyte *Valonia utricularis*. PhD thesis, University of Groningen, The Netherlands, pp 213

Eggert A, Karsten U (2010) Low molecular weight carbohydrates in red algae—an ecophysiological and biochemical perspective. Red algae in the genomic age, cellular origin, life in extreme habitats and astrobiology, 13(5):443–456

Eggert A, Wiencke C (2000) Adaptation and acclimation of growth and photosynthesis of five Antarctic red algae to low temperatures. Polar Biol 23:609–618

Eggert A, Burger EM, Breeman AM (2003a) Ecotypic differentiation in thermal traits in the tropical to warm-temperate green macrophyte *Valonia utricularis*. Bot Mar 46:69–81

Eggert A, van Hasselt PR, Breeman AM (2003b) Chilling-induced photoinhibition in nine isolates of *Valonia utricularis* (Chlorophyta) from different climate regions. J Plant Physiol 160:881–891

Eggert A, Visser RJW, van Hasselt PR, Breeman AM (2006) Differences in acclimation potential of photosynthesis in seven isolates of the tropical to warm temperate macrophyte *Valonia utricularis* (Chlorophyta). Phycologia 45:546–556

Einav R, Breckle S, Beer S (1995) Ecophysiological adaptation strategies of some intertidal marine macroalgae of the Israeli Mediterranean coast. Mar Ecol Prog Ser 125:219–228

Fields PA, Houseman DE (2004) Decreases in activation energy and substrate affinity in cold-adapted A4-lactate dehydrogenases: evidence from the Antarctic notothenioid fish *Chaenocephalus aceratus*. Mol Biol Evol 21:2246–2255

Gantt E, Berg M, Bhattacharya D, Blouin NA, Brodie JA, Chan CX, Collén J, Cunningham FX, Gross J, Grossman AR, Karpowicz S, Kitade Y, Klein A, Levine IA, Lin S, Lu S, Lynch M, Minocha SC, Müller K, Neefus CD, De Oliveira MC, Rymarquis L, Smith A, Stiller JW, Wu W-K, Yarish C, Zhuang YY, Brawley SH (2010) *Porphyra*: complex life histories in a harsh environment. *P. umbilicalis*, an intertidal red alga for genomic analysis. In: Seckbach J, Chapman D (eds) Red algae in genomic age. (volume 13 of cellular origins, life in extreme habitats and astrobiology). Springer, Berlin

Gerard V, Du Bois K (1988) Temperature ecotypes near the southern boundary of the kelp *Laminaria saccharina*. Mar Biol 97:575–580

Gómez I, Wulff A, Roleda MY, Huovinen P, Karsten U, Quartino ML, Dunton K, Wiencke C (2009) Light and temperature demands of marine benthic microalgae and seaweeds in polar regions. Bot Mar 52:593–608

Graham MH, Kinlan BP, Druehl LD, Garske LE, Banks S (2007) Deep-water kelp refugia as potential hotspots of tropical marine diversity and productivity. Proc Nat Acad Sci USA 104:16576–16580

Gray GR, Chauvin LP, Sarhan F, Huner NPA (1997) Cold acclimation and freezing tolerance: A complex interaction of light and temperature. Plant Physiol 114:467–474

Guisan A, Thuiller W (2005) Predicting species distribution: offering more than simple habitat models. Ecol Lett 8:993–1009

Havaux M, Tardy F (1996) Temperature-dependent adjustment of the thermal stability of photosystem II *in vivo*: possible involvement of xanthophylls cycle pigments. Planta 198:324–333

Helmuth BST, Hofmann GE (2001) Microhabitats, thermal heterogeneity, and patterns of physiological stress in the rocky intertidal zone. Biol Bull 201:374–384

Henkel SK, Kawai H, Hofmann GE (2009) Interspecific and interhabitat variation in hsp70 gene expression in native and invasive kelp populations. Mar Ecol Prog Ser 386:1–13

Hey J (2006) On the failure of modern species concepts. Trends Ecol Evol 21:447–450

Hochachka PW, Somero GN (2002) Biochemical adaptation: mechanism and process in physiological evolution. Oxford University Press, New York

Hoek C van den, Chihara M (2000) Delimitation of the genus *Cladophora*, reflexions on phylogeny of *Cladophora*, geographic distribution. Koyama H, Kitayama T (ed.) National Science Museum Monographs No. 19, National Science Museum, 14–36

Hoffmann AA, Willi Y (2008) Detecting genetic responses to environmental change. Nat Rev Genet 9:421–432

Hofmann GE, Buckley BA, Airaksinen S, Keen JE, Somero GN (2000) Heat-shock protein expression is absent in the Antarctic fish *Trematomus bernacchii* (Family Nototheniidae). J Exp Biol 203:2331–2339

Holaday A, Martindale W, Alred R, Brooks A, Leegood R (1992) Changes in activities of enzymes of carbon metabolism in leaves during exposure of plants to low temperature. Plant Physiol 98:1105–1114

Horvath I, Glatz A, Varvasovszki V, Torok Z, Pali T, Balogh G, Kovacs E, Nadasdi L, Benko S, Joo F, Vigh L (1998) Membrane physical state controls the signaling mechanism of the heat shock response in *Synechocystis* PCC 6803: identification of hsp17 as a 'fluidity gene. Proc Natl Acad Sci USA 95:3513–3518

Huner N, Öquist G, Hurry V, Krol M, Falk S, Griffith M (1993) Photosynthesis, photoinhibition and low temperature in cold tolerant plants. Photosynth Res 37:19–39

Huner N, Öquist G, Sarhan F (1998) Energy balance and acclimation to light and cold. Trends Plant Sci 3:224–230

Inaba M, Suzuki I, Szalontai B, Kanesaki Y, Los DA, Hayashi H, Murata N (2003) Gene-engineered rigidification of membrane lipids enhances the cold inducibility of gene expression in *Synechocystis*. J Biol Chem 278:12191–12198

Janech MG, Krell A, Mock T, Kang J-S, Raymond JA (2006) Ice-binding proteins from sea ice diatoms (Bacillariophyceae). J Phycol 42:410–416

Johansson G, Sosa PA, Snoeijs P (2003) Genetic variability and level of differentiation in North Sea and Baltic Sea populations of the green alga *Cladophora rupestris*. Mar Biol 142:1019–1027

Kain JM (1989) The seasons in the subtidal. Br Phycol J 24:203–215

Kim JK, Kraemer GP, Yarish C (2009) Comparison of growth and nitrate uptake by New England *Porphyra* species from different tidal elevations in relation to desiccation. Phycol Res 57:152–157

Kingston-Smith A, Harbinson J, Williams J, Foyer C (1997) Effect of chilling on carbon assimilation, enzyme activation, and photosynthetic electron transport in the absence of photoinhibition in maize leaves. Plant Physiol 114:1039–1046

Kirst GO (1990) Salinity tolerance of eukaryotic algae. Annu Rev Plant Physiol Plant Mol Biol 41:21–53

Kooistra WHCF, Stam WT, Olsen JL, van den Hoek C (1992) Biogeography of *Cladophoropsis membranacea* based on comparisons of nuclear rDNA ITS sequences. J Phycol 28:660–668

Kooistra WHCF, Coppejans EGG, Payri C (2002) Molecular systematics, historical ecology, and phylogeography of *Halimeda* (Bryopsidales). Mol Phylogenet Evol 24:121–138

Kübler JE, Davison IR (1993) High-temperature tolerance of photosynthesis in the red alga *Chondrus crispus*. Mar Biol 117:327–335

Kübler JE, Davison IR (1995) Thermal acclimation of light-use characteristics of *Chrondrus crispus* (Rhodophyta). Eur J Phycol 30:189–195

Kübler J, Davison IR, Yarish C (1991) Photosynthetic adaptation to temperature in the red algae *Lomentaria baileyana* and *Lomentaria orcadensis*. Br Phycol J 26:9–19

Larkindale J, Mishkind M, Vierling E (2005) Plant responses to high temperature. In: Jenks MA, Hasegawa PM (eds) Plant abiotic stress. Blackwell Scientific Publications, Oxford

Leroi AM, Bennett AF, Lenski RE (1994) Temperature acclimation and competitive fitness: an experimental test of the beneficial acclimation assumption. Proc Nat Acad Sci USA 91:1917–1921

Los DA, Murata N (2004) Membrane fluidity and its roles in the perception of environmental signals. Biochim Biophys Acta Biomembr 1666:142–157

Lüning K (1990) Seaweeds: their environment, biogeography, and ecophysiology. Wiley, New York

Lüning K, Freshwater W (1988) Temperature tolerance of northeast Pacific marine algae. J Phycol 24:310–315

Machalek KM, Davison IR, Falkowski PG (1996) Thermal acclimation and photoacclimation of photosynthesis in the brown alga *Laminaria saccarina*. Plant Cell Environ 19:1005–1016

Martone PT, Alyono M, Stites S (2010) Bleaching of an intertidal coralline alga: untangling the effects of light, temperature, and desiccation. Mar Ecol Prog Ser 416:57–67

Mittler R (2002) Oxidative stress, antioxidants and stress tolerance. Trends Plant Sci 7:405–410

Molenaar F (1996) Seasonal growth and reproduction of North Atlantic red seaweeds. PhD dissertation, Groningen, The Netherlands, p 111

Murata N, Los D (1997) Membrane fluidity and temperature perception. Plant Physiol 115:875–879

Nie G, Robertson E, Fryer M, Leech R, Baker N (1995) Response of the photosynthetic apparatus in maize leaves grown at low temperature on transfer to normal growth temperature. Plant Cell Environ 18:1–12

Nikaido I, Asamizu E, Nakajima M, Nakamura Y, Saga N, Tabata S (2000) Generation of 10,154 expressed sequence tags from a leafy gametophyte of a marine red alga, *Porphyra yezoensis*. DNA Res 7:223–227

Nishida I, Murata N (1996) Chilling sensitivity in plants and cyanobacteria: the crucial contribution of membrane lipids. Annu Rev Plant Physiol Plant Mol Biol 47:541–568

Padilla-Gamiño JL, Carpenter RC (2007) Seasonal acclimatization of *Asparagopsis taxiformis* (Rhodophyta) from different biogeographic regions. Limnol Oceanogr 52:833–842

Pakker H, Breeman AM (1996) Temperature responses of tropical to warm-temperate Atlantic seaweeds. II. Evidence for ecotypic differentiation in amphi-Atlantic-Tropical-Mediterranean species. Eur J Phycol 31:133–141

Pakker H, Breeman AM, Prud'homme van Reine WF, van den Hoek C (1995) A comparative study of temperature responses of Caribbean seaweeds from different biogeographic groups. J Phycol 31:499–507

Pakker H, Breeman AM, Prud'homme van Reine WF, van Oppen MJH, van den Hoek C (1996) Temperature responses of tropical to warm-temperate Atlantic seaweeds. I. Absence of ecotypic differentiation in amphi-Atlantic tropical-Canary Islands species. Eur J Phycol 31:123–132

Pearson G, Kautsky L, Serrão E (2000) Recent evolution in Baltic *Fucus vesiculosus*: reduced tolerance to emersion stresses compared to intertidal (North Sea) populations. Mar Ecol Prog Ser 202:67–79

Pereyra RT, Bergström L, Kautsky L, Johannesson K (2009) Rapid speciation in a newly opened postglacial marine environment, the Baltic Sea. BMC Evol Biol 9:70

Pfannschmidt T, Nilsson A, Allen JF (1999) Photosynthetic control of chloroplast gene expression. Nature 397:625–628

Pfetzing J, Stengel DB, Cuffe MM, Savage AV, Guiry MD (2000) Effects of temperature and prolonged emersion on photosynthesis, carbohydrate content and growth of the brown intertidal alga *Pelvetia canaliculata*. Bot Mar 43:399–407

Pörtner HO, Storch D, Heilmayer O (2005) Constraints and trade-offs in climate-dependent adaptation: energy budgets and growth in a latitudinal cline. Sci Mar 69:271–285

Provan J, Wattier RA, Maggs CA (2005) Phylogeographic analysis of the red seaweed *Palmaria palmata* reveals a Pleistocene marine glacial refugium in the English Channel. Mol Ecol 14:793–803

Raxworthy CJ, Ingram CM, Rabibisoa N, Pearson RG (2007) Applications of ecological niche modeling for species delimitation: a review and empirical evaluation using day Geckos (*Phelsuma*) from Madagascar. Syst Biol 56:907–923

Ree RH, Sanmartín I (2009) Prospects and challenges for parametric models in historical biogeographical inference. J Biogeogr 36:1211–1220

Rivera M, Scrosati R (2006) Population dynamics of *Sargassum lapazeanum* (Fucales, Phaeophyta) from the Gulf of California, Mexico. Phycologia 45:178–189

Robertson E, Baker N, Leech R (1993) Chloroplast thylakoid protein changes induced by low growth temperature in maize revealed by immunocytology. Plant Cell Environ 16:809–818

Roeder V, Collén J, Rousvoal S, Corre E, Leblanc C, Boyen C (2005) Identification of stress gene transcripts in *Laminaria digitata* (Phaophyceae) protoplast cultures by expressed sequence tag analysis. J Phycol 41:1227–1235

Savitch LV, Maxwell DP, Huner NPA (1996) Photosystem II excitation pressure and photosynthetic carbon metabolism in *Chlorella vulgaris*. Plant Physiol 111:127–136

Setchell WA (1920) The temperature interval in the geographical distribution of marine algae. Science 53:187–190

Sharkey TD (2005) Effects of moderate heat stress on photosynthesis: importance of thylakoid reactions, Rubisco deactivation, reactive oxygen species, and thermotolerance provided by isoprene. Plant Cell Environ 28:269–277

Smith CM, Berry JA (1986) Recovery of photosynthesis after exposure of intertidal algae to osmotic and temperature stresses: comparative studies of species with differing distributional limits. Oecologia 70:6–12

Somero GN (2004) Temperature adaptation of proteins: searching for basic "strategies". Comp Biochem Physiol 139:321–333

Sorensen JG, Kristensen TN, Loeschke V (2003) The evolutionary and ecological role of heat shock proteins. Ecol Lett 6:1025–1037

Stanley MS, Perry RM, Callow JA (2005) Analysis of expressed sequence tags from the green alga *Ulva linza* (Chlorophyta). J Phycol 41:1219–1226

Stengel DB, Dring MJ (1997) Morphology and in situ growth rates of plants of *Ascophyllum nodosum* (Phaeophyta) from different shore levels and responses of plants to vertical transplantation. Eur J Phycol 32:193–202

Steponkus PL, Webb MS (1992) Freeze-induced dehydration and membrane destabilization in plants. In: Somero G, Osmond B (eds) Water and life: comparative analysis of water relationships at the organismic, cellular and molecular level. Springer, Berlin, pp 338–362

Szalontai B, Nishiyama Y, Gombos Z, Murata N (2000) Membrane dynamics as seen by Fourier transform infrared spectroscopy in a cyanobacterium, *Synechocystis* PCC 6803. The effects of lipid unsaturation and the protein-to-lipid ratio. Biochim Biophys Acta 1509:409–419

Terrados J, Ros J (1992) The influence of temperature on seasonal variation of *Caulerpa prolifera* (Forsskal) Lamouroux photosynthesis and respiration. J Exp Mar Biol Ecol 162:199–212

tom Dieck I (1992) North Pacific and North Atlantic digitate *Laminaria* species (Phaeophyta): hybridization experiments and temperature responses. Phycologia 31:147–163

tom Dieck I (1993) Temperature tolerance and survival in darkness of kelp gametophytes (Laminariales, Phaeophyta): ecological and biogeographical implications. Mar Ecol Prog Ser 100:253–264

tom Dieck I, Oliveira EC (1993) The section Digitate of the genus *Laminaria* (Phaeophyta) in the northern and southern Atlantic: crossing experiments and temperature responses. Mar Biol 115:151–160

Tomashow MF (1998) Role of cold-responsive genes in plant freezing tolerance. Plant Physiol 118:1–7

Turesson G (1922) The genotypical response of the plant species to the habitat. Hereditas 3:211–350

Tyberghein L, Verbruggen H, Pauly K, Troupin C, Mineur F, De Clerck O (2011) Bio-ORACLE: a global environmental dataset for marine species distribution modelling. Global Ecol Biogeogr. doi:10.1111/j.1466-8238.2011.00656.x

van den Hoek C (1982a) Phytogeographic distribution groups of benthic marine algae in the North Atlantic Ocean: a review of experimental evidence from life history studies. Helgol Meeresunters 35:153–214 van den Hoek C (1982b) The distribution of benthic marine algae in relation to the temperature regulation of their life histories. Biol J Linn Soc 18:81–144

Verbruggen H, Tyberghein L, Pauly K, Vlaeminck C, van Nieuwenhuyze K, Kooistra WHCF, Leliaert F, De Clerck O (2009) Macroecology meets macroevolution: evolutionary niche dynamics in the seaweed *Halimeda*. Global Ecol Biogeogr 18:393–405

Vigh L, Maresca B, Harwood JL (1998) Does the membrane's physical state control the expression of heat shock and other genes? Trends Biochem Sci 23:369–374

Wattier R, Maggs CA (2001) Intraspecific variation in seaweeds: The application of new tools and approaches. Adv Bot Res 35:171–212

Wiencke C, Rahmel J, Karsten U, Weykam G, Kirst G (1993) Photosynthesis of marine macroalgae from Antarctica: Light and temperature requirements. Bot Acta 106:78–87

Wiencke C, Bartsch I, Bischoff B, Peters AF, Breeman AM (1994) Temperature requirements and biogeography of Antarctic, Arctic and amphiequatorial seaweeds. Bot Mar 37:247–259

Wiencke C, Gómez I, Dunton K (2009) Phenology and seasonal physiological performance of polar seaweeds. Bot Mar 52:585–592

Wilson KE, Huner NPA (2000) The role of growth rate, redox-state of the plastoquinone pool and the trans-thylakoid ΔpH in photoacclimation of *Chlorella vulgaris* to growth irradiance and temperature. Planta 212:93–102

Wilson KE, Król M, Huner NPA (2003a) Temperature-induced greening of *Chlorella vulgaris*: The role of the cellular energy balance and zeaxanthin-dependent nonphotochemical queniching. Planta 217:616–627

Wilson KE, Sieger SM, Huner NPA (2003b) The temperature-dependent accumulation of Mg-protophorphyrin IX and reactive oxygen species in *Chlorella vulgaris*. Physiol Planta 119:126–136

Yancey PH (2005) Organic osmolytes as compatible, metabolic and counteracting cytoprotectants in high osmolarity and other stresses. J Exp Biol 208:2819–2830

Zuccarello GC, Buchanan J, West JA (2006) Increased sampling for inferring phylogeographic patterns in *Bostrychia radicans*/*B. moritziana* (Rhodomelaceae, Rhodophyta) in the eastern USA. J Phycol 42:1349–52

Chapter 4
Environment and Algal Nutrition

Francisco J.L. Gordillo

4.1 Introduction

The mechanisms by which macroalgae take up nutrients from the external medium and use them to form biomolecules have been long and widely studied (e.g., Raven 1984; Lobban and Harrison 1994). In this chapter, the focus is on the ecophysiology of these mechanisms for the main nutrients C, N, and P. Particular details on uptake and utilization mechanisms are included when considered ecophysiologically relevant. Physical aspects such as water movement and seasonality are included, nutrient availability is discussed in terms of environmental conditioning of N and P supply for storage and growth, and different species-specific metabolic strategies are presented. Some attention has been paid to the interference of N assimilation with C assimilation pathways. Carbon assimilation is treated more in-depth in Chap. 2 by Gómez and Huovinen. For the interactive effects of ocean acidification on seaweeds, see Chap. 19 by Roleda and Hurd.

Nutrient uptake and assimilation mechanisms are the link between the resources externally available and the demands for growth. However, the environment where macroalgae develop is characterized by strong fluctuations in a number of relevant factors including nutrients, so that macroalgae show some plasticity in terms of resource management and internal composition of the biomass, thus rapidly acclimating to the nutritional conditions. Algae can switch their photosynthetic energy investment from C to N and P acquisition during the course of the day, or from summer to winter. Most remarkably, different mechanisms of resource utilization for different species coexist at the same time in the same community, and this strategy of partitioning brings a more effective resource use, i.e., better transfer to the whole food web.

F.J.L. Gordillo (✉)
Departamento de Ecología, Universidad de Málaga, Málaga, Spain
e-mail: gordillo@uma.es

C. Wiencke and K. Bischof (eds.), *Seaweed Biology*, Ecological Studies 219,
DOI 10.1007/978-3-642-28451-9_4,

The different processes involved in nutrient acquisition follow the terms as used by Berges (1997). Accordingly, uptake is defined as removal of a given molecule or ion from the environment and its physical transport into the algal cell. In macroalgae and higher plants, a further distinction between transport into intercellular spaces and transport into cells themselves may be useful (Redinbaugh and Campbell 1991), but this difference is not considered here. The term assimilation is reserved for the processes by which internalized inorganic molecules are used to form small organic molecules, such as amino acids, and incorporation is defined as the process by which nutrient-containing organic molecules are combined to form macromolecules, such as proteins and nucleic acids, thus participating in growth. These definitions are functional and pragmatic; for example, disappearance of nitrate from the medium can be easily measured, and a separation of small organic and inorganic molecules is relatively straightforward. Note, however, that the enzymatic components of each process are not specified.

4.2 Meeting Nutritional Needs

Nutrient uptake mechanisms must be optimally evolved in order to meet the nutritional needs of a given species in its environment. Algae use light energy to fix carbon (C) and combine the carbon with elements such as nitrogen (N) and phosphorus (P) at relatively constant stoichiometric ratios. C/N/P ratios of unicellular algae are typically around 106:16:1, the so-called Redfield ratio (Redfield et al. 1963), while the C/N/P ratios of benthic plants are larger and more variable, with a median C/N/P ratio of 550:30:1, called the Atkinson ratio (Atkinson and Smith 1983). Falkowski (2000) proposed that the origin of the constant Redfield ratio is a result of a limitation intrinsic to protein synthesis. Atkinson and Smith (1983) proposed that the higher C/N ratio of benthic plants is related to a requirement of structural strength. Baird and Middleton (2004) and Raven and Kübler (2002) discussed the physical limits of nutrient uptake and metabolic rate, respectively, in relation to light absorption. In addition, variability can be further explained by different metabolic strategies and storage capacities (Fujita 1985; Pedersen and Borum 1996).

Regarding uptake kinetics, the extracellular concentrations of many of the chemical species that are required by intracellular assimilation processes are often much lower than the half-saturation value of the transport and/or assimilating enzyme for that chemical species, potentially greatly limiting the rate at which nutrients could be assimilated and used in growth (e.g., Raven et al. 2008). While this situation could, in the general case, be remedied by increasing the quantity of the assimilatory enzyme, the mismatch between the external concentration and the half-saturation value of the assimilatory enzyme can be extreme.

4.2.1 Extracting Nutrients from their Environment

Macroalgae inhabit mostly aquatic environments, and although intertidal species are also exposed to emersion and direct contact with air, nutrients for macroalgal growth are mostly taken from the surrounding water bodies. Benthic communities may be limited by the transport of nutrients through a diffusive boundary layer (Baird and Atkinson 1997; see also Chap. 19 by Roleda and Hurd), by uptake kinetics on the plant surface, or by a combination of the two (Bilger and Atkinson 1995). When limited by transport through a diffusive boundary layer, nutrient uptake rate remains a function of surface roughness and water velocity (Baird and Middleton 2004).

The relatively high viscosity and density of seawater result in drag and acceleration forces on macroalgae that are considerably greater than those experienced by terrestrial plants and restrict the upper size limit to which macroalgae can grow (Gaylord et al. 1994). Rates of molecular diffusion of essential nutrients such as carbon dioxide (CO_2) are 10,000 times slower in seawater than in air (Denny 1990, 1993). Despite this, high production rates of macroalgae indicate the evolution of mechanisms able to overcome this apparently restrictive fluid environment. The main hydrodynamic habitats of macroalgae have been discussed in detail by Hurd (2000).

Intertidal macroalgae are exposed to air and water. On wave-exposed coasts, intertidal and shallow subtidal macroalgae will experience the full force of breaking waves. This multidirectional flow results in the rapid back and forth movement of the thalli, imparting large drag and/or acceleration forces. Intertidal macroalgae also grow in wave-sheltered sites such as bays or estuaries that are protected from ocean waves, and water motion arises predominantly from smaller wind-generated surface waves and the rise and fall of the tide. Some intertidal macroalgae are submerged in rock pools that are isolated from the ocean at low tide. Subtidal macroalgal communities can also be classified as wave-sheltered and -exposed, but subtidal macroalgae do not experience the full force of breaking waves, and the velocities they encounter will often be lower than those of intertidal sites. Relatively calm conditions can occur at both intertidal and subtidal sites that are wave-exposed.

Macroalgae can store nitrogen and phosphorus on a short- (days, e.g., *Chaetomorpha* sp.; McGlathery et al. 1996) or long- (weeks–months, e.g., *Macrocystis*; Brown et al. 1997) term basis. The productivity of a macroalga with a large nitrate storage is, therefore, unlikely, to be limited by short-term reductions (seconds–hours; see discussion below on timescales of velocity fluctuations) in nitrate mass transfer due to slow mainstream velocities. Similarly, photosynthesis can be maintained in the absence of DIC for up to 6 h in *Ulva* (Koch 1993) and 24 h in *Ascophyllum nodosum* (Ryberg et al. 1990) through the utilization of stored DIC. Therefore, under slow mainstream velocities, the degree to which macroalgal production will be reduced will depend on their physiological status, the timescale over which the diffusion boundary layer forms relative to its nutrient requirement, and the ability of the algae to adapt physiologically to the reduced nutrient flux (Hurd 2000).

However, many macroalgae have nutrient sources other than the mainstream seawater which may alleviate nutrient limitation even in slow flows. Benthic algal mats that form in areas of reduced mainstream velocities take up nutrients from the sediment. In fact, macroalgae and benthic microalgae are key factors controlling the fluxes of inorganic and organic N fluxes between the sediment and the water column (Tyler et al. 2003). The high concentration gradient between the sediment and the algal mat may lead to a high nutrient flux (Larned and Atkinson 1997 and references therein). Kamer et al. (2004) found that the relative importance of the water column vs. the sediment as a source of nutrients for *Ulva intestinalis* varied with the magnitude of the different source. Additionally, ammonium and urea are excreted by invertebrates living within the boundary layer on the macroalgal surface (Probyn and Chapman 1983; Taylor and Rees 1998). Gerard and Mann (1979) attributed the high growth rates of a wave-sheltered *Laminaria hyperborea* population to the provision of ammonium from epiphytic bryozoans, an idea supported by the high levels of nitrogen found directly beneath bryozoan colonies on *Macrocystis integrifolia* (Hurd et al. 1994). Similarly, macroalgae may obtain carbon dioxide released from respiration by bryozoans (Muñoz et al. 1991; Mercado et al. 1998).

Another physical determinant on macroalgal nutrition is the algal surface area to biovolume ratio (SA:V), especially at low concentrations (Rees 2007). Hein et al. (1995) suggested that N and P uptake rate and maximum capacity (V_{max}) in algae is, at least in part, linked to their SA:V, so that algae with low SA:V show lower uptake efficiency and V_{max} values than algae with high SA:V at both low and high nutrient concentrations. Rees (2003), however, found no relationship between the affinity constant (K_m) and SA:V for nitrate, ammonium, and phosphate uptake in macroalgae. Light reaching macroalgae is (mainly) a two-dimensional phenomenon while nutrient uptake and use is three-dimensional. Baird and Middleton (2004) studied the relationship between light and N requirements and suggested that for benthic plants the rates of light absorption to nutrient uptake varies depending on both plant morphology and hydrodynamic conditions.

4.2.2 Neutral vs. Ionic Nutrients

Most nutrients are in ionic form, impermeable to the lipidic plasma-membrane. Main exceptions are inorganic carbon in the form of CO_2 and ammonia, NH_3, which can penetrate the cell with fluxes of 7–18 mm s^{-1} and 6–7 mm s^{-1}, respectively (Sültemeyer and Rinast 1996; Ritchie 1987; Ritchie and Gibson 1987) due to their neutral nature. However, the proportion of inorganic carbon in the form of CO_2 and that of ammonia in the form of NH_3 are generally much lower than their ionized forms ($HCO_3^-/CO_3^=$ and NH_4^+ respectively) in the range of pH values where most macroalgae grow. For ionic forms, facilitated transport is necessary; in many cases, this involves energy-demanding mechanisms. However, this does not mean that facilitated transport mechanisms involve a lesser cost-effective cell; on the contrary, these mechanisms allow for the optimization of resources being invested. As an example, Raven et al. (2008) made a theoretical calculation on the transport and assimilation costs of nonfacilitated vs. facilitated mechanisms for N in the form of NH_4^+. In the case of the passive entry of NH_4^+, it requires the use of 560 times the N and NADPH to synthesize the relevant proteins, and 1.61 times the ATP to make the proteins and in running the transport and assimilation processes than in the case of facilitated uptake and assimilation of NH_4^+. These outcomes clearly show an optimized input of resources (N and energy) in achieving the outcome of a given rate of NH_4^+ assimilation in the case of a very low external concentration of NH_4^+. However, for some seaweeds, the relationship between uptake rate and concentration is a combination of a hyperbolic component due to active uptake and a linear diffusive component (Taylor and Rees 1998). These authors showed that seaweeds from New Zealand have lower ammonium uptake rates than their northern hemisphere counterparts, and suggested that uptake was due largely to passive diffusion of NH_3.

4.2.3 N vs. P Limitation

Under conditions of coastal eutrophication (see also Chap. 21 by Teichberg), fast-growing, ephemeral macroalgae bloom at the expense of slow-growing, perennial macroalgae. As mentioned above, this change in community composition has been explained by a differential ability to exploit and utilize inorganic nitrogen among macroalgae with different growth strategies. It is generally assumed that inorganic N availability is the main control for macroalgal growth in temperate coastal areas (Nixon and Pilson 1983; Oviatt et al. 1995; Howarth et al. 2000); however, some coastal areas have been identified as phosphorus- rather than nitrogen-limited. In tropical latitudes, P limitation might be established as carbonate sediments derived from coral reefs may sequester phosphate (Lapointe et al. 1992; McGlathery et al. 1994; Lapointe and Bedford 2010; see also Chap. 16 by Mejia et al.), although other studies have shown exceptions to this situation (Larned 1998; Fong et al. 2001; Elser et al. 2007). Teichberg et al. (2010) made a global survey of the growth response of *Ulva* spp. to experimental N or P enrichment across temperate and tropical sites and found that N and P limitation of growth was linked directly to nutrient availability and not to geographic or latitudinal differences as had been previously suggested (Howarth 2008). Teichberg et al. (2010) indicated that it is likely that ambient N:P ratios may be useful to predict nutrient limitation in bloom-forming species, despite the variability in the growth responses, uptake affinities, and tissue N and P storage capacities among macroalgal taxa (Fujita 1985; Pedersen and Borum 1996; Fong et al. 2003). Barile (2004) found that macroalgae with high uptake affinities for DIN (*Ulva lactuca*, *Chaetomorpha linum*, *U. intestinalis*, *Caulerpa* spp., and *Gracilaria tikvahiae*) were N-limited in subtropical coastal waters of east-central Florida, where water N:P was on average 8:1. In southeastern Brazil, macroalgal growth was P-limited where ambient N:P were greater than 16:1 (Lourenço et al. 2005). In contrast, in Waquoit Bay estuaries, where the water N:P ratio was approximately 3:1 during the growing season, *G. tikvahiae* (Teichberg et al. 2008) and *U. lactuca* (Teichberg et al. 2010) were N-limited. Lapointe and Bedford (2010) suggested that the blooms of the invasive *Caulerpa brachypus f. parvifolia* occurring at coral reefs off southeast Florida (USA) were P-limited due to high biomass N:P ratio despite relatively high ambient P.

Several studies have examined P-uptake kinetics in macroalgae (e.g., Wallentinus 1984; Hurd and Dring 1990; Gordillo et al. 2002; Runcie et al. 2004), some of them comparing growth-related P requirements to uptake kinetics and storage capacity (e.g., Björnsäter and Wheeler 1990; Lavery and McComb 1991). Pedersen et al. (2010) found that tissue P-concentrations, the balance between P-uptake and requirements, and the P-storage capacity varied systematically among the three groups of algae (i.e., algae with fast, intermediate, and slow growth).

4.2.4 Considering C as Just Another Nutrient

Traditionally, C has been considered as a special case when treated together with nutrients, since the term nutrient is reserved mainly for N and P (e.g., Rees 2003), and to a lesser extent S, Fe, Ca, and others, the latter set commonly being referred to as micronutrients. This might be due to the majority of nutrient studies being carried out in terrestrial plants, where C is taken from the air and the rest from soil by roots in most cases. C is usually considered to be at saturating concentrations. Although this is a pragmatic consideration for many studies on terrestrial plants nutrition, the uptake and assimilation dynamics of C in aquatic plants share many characteristics with other nutrients; i.e., it is essential for the cell, its utilization needs photosynthetic energy, there are specific transporters in the cell surface, it can be stored intracellularly and used later on when other restricting conditions ease, and its availability can limit growth.

The vast majority of algal primary production involves the C_3 pathway (the photosynthetic carbon reduction cycle or Calvin–Benson cycle) for DIC acquisition, fixing DIC directly via ribulose bisphosphate carboxylase oxygenase (Rubisco). However, Rubisco has a relatively low affinity for CO_2 and, for most species, it is consequently less than half saturated under current ambient CO_2 levels. The biochemical properties of Rubisco imply that for autotrophs dependent on diffusive CO_2 entry, the physiology of CO_2 assimilation is characterized by inherent inefficiencies, such as significant inhibition of CO_2 fixation by oxygen, high CO_2 compensation points, and low affinities for external CO_2.

An alternative source of inorganic carbon available is HCO_3^-. This is the most abundant form and can be converted to CO_2 to be used by Rubisco by means of both spontaneous or catalyzed dehydration. Spontaneous HCO_3^- dehydration kinetics in seawater are described in Johnson (1982). The rate of spontaneous CO_2 formation from HCO_3^- dehydration is clearly below the photosynthetic demand for most algae. Most algae meet their Ci demands by the operation of so-called carbon concentrating mechanisms (CCMs). The functional characteristics of these mechanisms have been reviewed by Giordano et al. (2005) (see also Chap. 19 by Roleda and Hurd and Chap. 2 by Gomez and Huovinen).

C and N Temporal Partition: As mentioned above, facilitated transport and assimilation mechanisms are cost-effective, optimizing molecules and energy invested in the process. This does not mean that the amount of energy invested in these mechanisms is a marginal proportion of that produced during photosynthesis (in addition to C assimilation in the Calvin–Benson cycle). On the contrary, they account for quite a significant amount of photosynthetic energy producing values for the photosynthetic quotient (PQ, number of C molecules fixed per O_2 evolved) lower than 1. In *Ulva rigida*, Gordillo et al. (2003) calculated that the amount of photosynthetically derived energy used in processes other than C fixation (operation of CCMs and other uptake and biosynthesis mechanisms) can account for nearly as much as 50%. When cultured at high pCO_2, CCMs were repressed and the proportion dropped to 28% despite the fact that in this species, the nitrate reductase activity (NR) was enhanced by CO_2 under nonlimiting NO_3^- supply. This suggests that in this species CCM could well account for >22% of the gross photosynthetic energy. Since CCM activity and C fixation need to be coupled, it is reasonable to question whether energy-demanding nutrient uptake and assimilation (other than C, mainly N) would benefit from any uncoupling mechanism.

Gordillo et al. (2002) suggested that in several brown algae, during the diel cycle, the reduction of nutrient uptake at the end of the dark period and the beginning of the light period may indicate a mechanism in which cellular energy is differentially distributed between nitrate and phosphate uptake mechanisms on one side and carbon uptake and fixation on the other, depending on the time of the day (Fig. 4.1). This strategy has been suggested by Falkowski (1975), Keller and Paerl (1980), Lean et al. (1982), and Turpin (1983) for some microalgal species. According to Gordillo et al. (2002), both the temporal partitioning of C and nutrients, and the ability to maintain nutrient uptake in darkness are consistent with a mechanism that reacts to the changing availability of these nutrients in an intertidal habitat. Gevaert et al. (2007) found in *Ulva* that rates of biosynthetic processes such as ammonium uptake are highest in the morning. Consistent with this hypothesis, the maximum rate of ammonium assimilation in *U. pertusa* peaked in the morning and coincided with low levels of the photosynthetic product sucrose, which peaked in the afternoon. There was a diurnal cycle in the rate of ammonium uptake and assimilation in light and dark, but the amplitude was much greater for assimilation than uptake. Moreover, these authors suggest that net ammonium assimilation only occurs during the day in *U. pertusa* and that two major roles for diurnal cycles are minimization of interspecific competition for resources and reduced metabolic costs. Further mechanistic models for C:N partition can be found in Flynn et al. (2001) and Wirtz and Pahlow (2010).

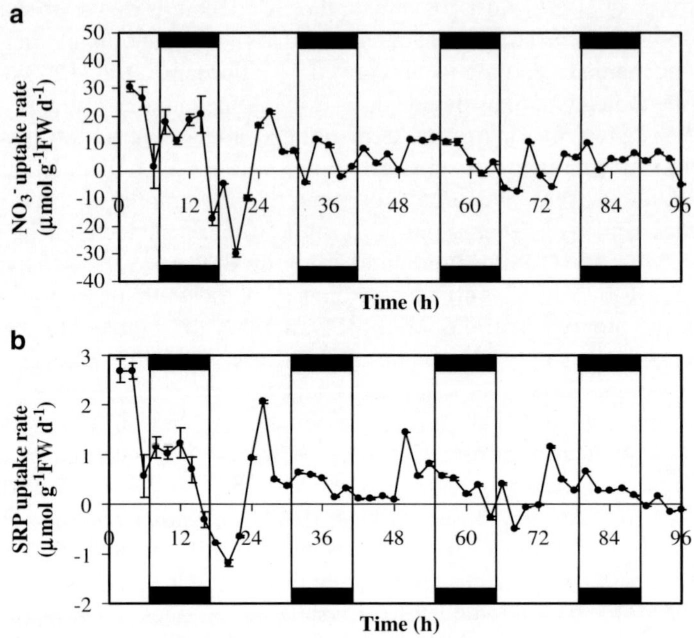

Fig. 4.1 Diel patterns of nitrate (**a**) and phosphate (**b**) uptake by *Fucus serratus*. *Black rectangles* represent the dark period [photoperiod 14:10 (L:D)]. Taken from Gordillo et al. (2002)

4.3 Uncoupling Uptake and Incorporation

Uptake and incorporation might be temporally and spatially uncoupled. In this context even storage in the vacuoles could be considered as spatial reallocation; however, it is more pragmatic to use the term "spatial uncoupling" when resources are reallocated in different parts of the thallus. In turn, storage in vacuoles is considered as a temporal uncoupling since nutrients can be immediately made available for assimilation and incorporation processes occurring in the cell whenever they are needed. According to Fujita (1985), differences in nutrient storage capacity between species may be important in structuring algal assemblages in environments with episodic N supply.

One of the most remarkable examples of temporal uncoupling is that found in polar kelps and kelp-like species (see also Chap. 13 by Wiencke and Amsler). The two polar coastal regions share some major characteristics, including low temperature and light regimes with complete darkness for months in winter and continuous sunlight for months in summer, as well as seasonal melting events of large ice masses. There are also key factors, however, differentiating both systems. In the Arctic, nutrients drop below detection limits during spring, and remain low until autumn mainly due to phytoplanktonic blooms occurring right after ice breakup (Aguilera et al. 2002). The Antarctic, however, is considered virtually unlimited by nutrients all year-round, and growth of algal species inhabiting the Antarctic is considered to be light-driven. This main difference in nutrient availability drives the internal regulation and strategies of light harvesting and nutrient assimilation. Although there are no kelps in the Antarctic, kelp-like *Himantothallus grandifolius* (Desmarestiales) is endemic to the Antarctic and dominates large coastal areas, but its growth is largely restricted to the ice-free months in spring (Drew and Hastings 1992; see also Chap. 13 by Wiencke and Amsler). On the other hand, the kelp *Laminaria solidungula*, endemic to the Arctic, stops its growth during N-depleted summer and accumulates carbon skeletons that are used for growth later in winter when N becomes available (Henley and Dunton 1995). According to Korb and Gerard (2000), *L. solidungula* can be considered a "storage specialist" for nitrogen assimilation. Its high V_{max} for nitrate uptake, coupled with low growth rates, allows Arctic plants to take advantage of seasonally elevated concentrations and accumulate large internal pools of nitrate and organic N-reserves. In the Antarctic, *H. grandifolius* does not accumulate nitrate, so that N-uptake characteristics are better adapted to optimize energy consumption rather than N-assimilation. Preferential use of NH_4^+ over NO_3^- displayed by *H. grandifolius* (but not by *L. solidungula*) would add evidence of an energy-saving strategy (Korb and Gerard 2000; see also Chap. 2 by Gómez and Huovinen, Chap. 13 by Wiencke and Amsler). It is then expected that these two different strategies will ultimately determine their ability to face a given environmental change.

Traditionally, it was thought that this seasonal growth pattern observed in Arctic kelps was a direct consequence of N availability (Chapman and Lindley 1980) as well as light availability (Henley and Dunton 1997). However, it has become apparent that, at least in some species, this pattern is under the control of an endogenous free-running circannual rhythm entrained by a critical minimum daylength in autumn (Lüning 1991; Schaffelke and Lüning 1994). This suggests that the addition of nitrate-N to summer N-limited kelps would have only a marginal effect on growth and biochemical composition, presumably due to the prevailing internal clock (Henley and Dunton 1997). Gordillo et al. (2006) confirmed that the effects of nutrient enrichment on biomass composition play only a marginal role in a number of species from Kongsfjord (Svalbard). When thalli of 21 species were collected in summer and incubated for 2 days under nitrate and phosphate enrichment, the C:N ratio was only affected by 7% on average. Lack of N accumulation and enhanced photosynthetic ability is in agreement with the internal clock prevailing over external nutrient conditions, which in summer would promote active photosynthesis over nutrient use.

4.4 Different Ways of N Utilization

Sources of N for macroalgal growth are mostly in the form of nitrate and ammonium, and the metabolism of inorganic N by algae is known to be regulated by both N forms (Solomonson and Barber 1990). Furthermore, different species show different preferences and dynamics for each N form. The use of nitrate has a clear advantage over ammonium since it can be stored in a vacuole with a concentration factor (internal:external) of >1,000 (e.g., Gordillo et al. 2002), while high concentrations of ammonium are toxic, acting as an uncoupler (Krogmann et al. 1959). On the other hand, ammonium has an energetic advantage over nitrate (but this is a minor issue when light is saturating). As a cation, less energy is required for transport, and it can be immediately used for amino acid synthesis via glutamine synthetase/ glutamine–oxoglutarate aminotransferase (GS/GOGAT), while nitrate must be reduced to nitrite by the highly regulated enzyme nitrate reductase, and then nitrite must be further reduced to ammonium by the nitrite reductase, investing eight electrons per NO_3^- molecule being reduced to NH_4^+ overall (e.g., Raven 1984).

For most algae, ammonium is the preferred inorganic N form. According to Rees (2003), ammonium uptake is more efficient than nitrate uptake in macroalgae. In addition, the presence of ammonium can suppress the uptake and assimilation of less reduced forms of N, mainly nitrate, in phytoplankton (Dortch 1990) and macroalgae (DeBoer 1981; Harrison et al. 1986; Naldi and Wheeler 2002). Elevated ammonium concentrations can also result in a decline in the activity of the primary nitrate-assimilating enzyme, nitrate reductase (NR) in microalgae (Berges et al. 1995; Vergara et al. 1998), and in a brown macroalga, *Giffordia mitchellae* (Weidner and Kiefer 1981). Ammonium was taken up at a higher rate than nitrate when presented alone (Lotze and Schramm 2000; Nishihara et al. 2005) or simultaneously (Naldi and Wheeler 2002; Cohen and Fong 2004).

4.4.1 Seaweed Diversity Linked to Nutrient Source

Ammonium is primarily associated with local scale regeneration, whereas nitrate availability is predominantly determined by oceanographic processes in the adjacent nearshore and/or runoff from land. The physiological mechanisms by which each form is used differ substantially, and seaweeds with different morphologies, physiological capabilities, and life-history strategies might differ in their abilities to utilize nitrate vs. ammonium (e.g., Naldi and Wheeler 1999). However, Bracken and Stachowicz (2006) found that, in diverse seaweeds assemblages, the uptake of either nitrate or ammonium alone was equal to the average of the component monocultures, but when nitrate and ammonium were available simultaneously, total N uptake by the assemblage was 22% higher than the average from monocultures because different species were complementary in the use of the nitrogen forms. These results provide a mechanistic link between N use and the diversity of primary producers in marine ecosystems. Previously, Bracken and Nielsen (2004) found that in nitrate-exhausted intertidal pools, slow-growing species tolerant to low nitrogen availability were joined by fast-growing species with higher nitrogen requirements due to increased ammonium loadings coming from macroinvertebrates. A fourfold increase in the ammonium loading rate was associated with a doubling in the number of macroalgal species.

Other factors such as herbivory greatly influence diversity. However, niche partitioning by the differential N use can reduce competition and lead to greater N fluxes into the intertidal food web. The effect of resource-based niche partitioning on diversity has been also suggested in terrestrial ecosystems (McKane et al. 2002). The nutritional status of aquatic primary producers has been found to determine the efficiency of energy transfer through food chains (the so-called food chain efficiency, FCE), with the carryover effects of algal quality across trophic levels (Dickman et al. 2008).

4.4.2 Slow-Growing vs. Fast-Growing Species

Aquatic plant communities are formed by a number of species representing various growth strategies and life forms, and the contribution of the different plant types to total autotrophic biomass and production is regulated by several factors that influence growth and loss processes. The distribution of macroalgae along nutrient gradients is typically characterized by slow-growing species in nutrient-poor regions or seasons, while fast-growing, ephemeral species dominate under nutrient-rich conditions (e.g., Kautsky et al. 1986; Taylor et al. 1995; Borum and Sand-Jensen 1996; see also Chap. 21 by Teichberg). The association between high nutrient availability and increased dominance of ephemeral macroalgae suggests that fast-growing species require high nutrient inputs to sustain growth, while large, slow-growing species are better adapted to nutrient-poor conditions. In support of this idea, Pedersen and Borum (1996) demonstrated that ephemeral macroalgae suffered from N limitation during longer periods of low N availability in summer than did slow-growing species. The observed variation in N limitation among algae of different growth strategy could partly be explained by species-specific differences in N requirements and storage capacity (Pedersen and Borum 1997). As an example, Lartigue and Sherman (2006) found that, under fluctuating NO_3^- discharges to an estuary, total N in the nitrophillic fast-growing *Ulva lingulata* ranged from 2.6% to 6.4% dry weight, while the slower growing *Gelidium pusillum* showed no significant variation in total N content (3.8–4.1% dw). Additionally, Pedersen and Borum (1996) also indicated that, compared to ephemeral species, slow-growing macroalgae were better able to meet their N requirements by exploiting low external N concentrations. Other studies have associated a higher position in the intertidal with higher internal nitrate pools (Phillips and Hurd 2003), higher NR activities (Murthy et al. 1986), and elevated nutrient uptake capacity, which may compensate for periods of emersion (Hurd and Dring 1990; Phillips and Hurd 2004). However, this association is not always valid, and contrasting results have also been found (Young et al. 2009).

4.5 Use of Isotope Discrimination

The stable nitrogen isotope ^{15}N discrimination ($\partial^{15}N$) has been frequently used to determine nitrogen sources in ecosystem studies (Peterson and Fry 1987; McKinney et al. 2001; Cole et al. 2004; Lapointe 2004). The underlying assumptions are that the $\partial^{15}N$ of sources are known and that primary producers take up ^{15}N in proportion to availability in a predictable manner (Cifuentes et al. 1988). The natural abundance of ^{15}N in many sources has been measured, including fertilizer (0‰), sewage (10‰–20‰), soil (5‰), and rainfall (−3‰ to −5‰; Heaton 1986). Groundwater, depending on the amount of microbial processing that selects for ^{14}N and leaves the remaining N in ^{15}N, can exceed 30‰ (Page 1995). Cohen and Fong (2005) showed that *Ulva intestinalis* did not differentially select ^{14}N over ^{15}N in a range of N concentrations and isotopes proportions. Rather, the species took up ^{14}N and ^{15}N in direct proportion to the amount in the source, suggesting that this alga may be useful for determining N sources in estuaries. However, uptake preference for ^{15}N-ammonium rather than ^{15}N-nitrate can result in differences in tissue $\partial^{15}N$ that may confound its use as an indicator. For example, if ammonium from sewage with a low $\partial^{15}N$ and nitrate from groundwater with a high $\partial^{15}N$ co-occur in an estuary and the algae take up more ammonium, then it will appear that the sewage is a more important source relative to groundwater even though it may not be. Teichberg et al. (2010) showed that $\partial^{15}N$ of macroalgae is generally heavier where DIN concentrations are higher due to increased N loads associated with wastewater rather than other N sources in many coastal waters. Fast-growing algae, such as *Ulva* spp. and *Gracilaria* spp., are better at reflecting the $\partial^{15}N$ values of the N source in a shorter period of time (Aguiar et al. 2003; Teichberg et al. 2008), but this might not be the case when examining slower growing algae such as *Fucus vesiculosus* (Deutsch and Voss 2006).

4.6 Aquaculture

Seaweeds have an essential role for maintenance and expansion of sustainable marine aquaculture (Chopin et al. 2001; Neori 2008; see also Chap. 22 by Buchholz et al.). The integration of seaweeds to marine animal cultures has long been recognized as the most promising approach to reduce the excess nutrients released by aquaculture activity, due to the high nutrient uptake efficiency of macroalgae, their fast growth rates, and economical and practical aspects of their cultivation (Vandermeulen and Gordin 1990; Shpigel et al. 1993; Chopin et al. 2001; Hernández et al. 2006; Lüning and Pang 2003; Neori et al. 2004). Research and development of integrated mariculture, recently known as integrated multi-trophic aquaculture (IMTA), has advanced over the last 20 years (Neori et al. 2004), although it still needs to be implemented at a larger scale by the modern global industry (Hernández et al. 2002). Some of the state-of-art seaweed-based integrated aquaculture has now been focused on the cultivation of high-value seaweed species which have not been previously used in integrated systems, and on improving tank-culture success for seaweed yields and for reduction of undesirable effects of intensive fish mariculture (e.g., Mata et al. 2006; Hernández et al. 2006; Flower and Neori 2008).

The majority of the research on integrated mariculture is based on the integration of biofiltering seaweed and fish aquaculture. There are also studies focussed on the integration of seaweed and shrimp culture (e.g., Jones et al. 2001; Sato et al. 2006a, b; Lombardi et al. 2006) and human sewage (Gordillo et al. 2002). Thus, treatment with macroalgae has now become an alternative to earlier adoptions of microalgae, but care must be taken when extrapolating pilot plant results to a larger scale. The factors for considering the advantages of using a given species over others should include that they are abundant in the system during the whole year and are easier to maintain in long-term cultures. For example, for human sewage treatment Gordillo et al. (2002) showed that the intertidal species *Fucus vesiculosus* and *F. serratus* have the ability to adapt to high nutrient concentrations and low salinity and are, therefore, suitable candidates for wastewater treatment, while the subtidal *Laminaria digitata* would be a less suitable species since it did not stand long periods under reduced salinity despite its N accumulating ability.

4.6.1 Nutrient Uptake at High Concentrations

Frequently, when a given resource is available at concentrations well above the $K_{0.5}$ of the high affinity uptake mechanism, a second uptake mechanism can be observed showing uptake rates above the V_{max} of the former (e.g., Fujita 1985; Gordillo et al. 2002, Fig. 4.2). Nonsaturating and biphasic uptake kinetics have been observed in coastal phytoplankton assemblages, which normally bloom under conditions of high nutrient fluxes and turbulent mixing and are considered to be mechanisms of adaptation to such environments (Collos et al. 1997; Lomas and Glibert 1999). Furthermore, nonsaturating nutrient uptake kinetics have been reported for several macroalgae species by Wallentinus (1984) and Fujita (1985). Biphasic kinetics was observed in the red alga *Gracilaria gaditana* when inorganic carbon uptake was investigated at high CO_2 levels (Andría et al. 1999). Gordillo et al. (2002) evidenced that the onset of a second mechanism in *Fucus serratus* dealing with high external phosphate is apparently different from the high affinity one operating at low concentrations (Fig. 4.2). These adaptations would make the N:P ratio for uptake drop from values around 30 at relatively subsaturating nutrient levels to values more in accordance with internal inorganic N:P ratios observed when internal pools are saturated. Since common assay concentrations are rarely above 30 μM for both N and P, Lomas and Glibert (1999) have suggested, from their results with phytoplankton, that more nonsaturating and biphasic uptake kinetics may be found in algae of coastal areas if researchers use a wider range of concentrations in their assays.

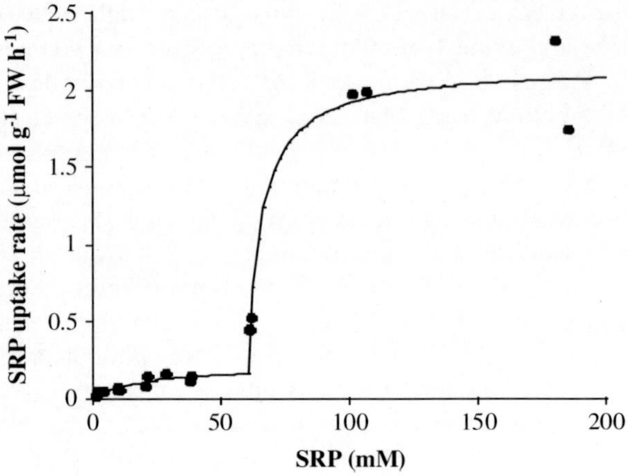

Fig. 4.2 Phosphate uptake rate as a function of external phosphate concentration (soluble reactive phosphate, SRP), fitted to two different Michaelis–Menten equations. Taken from Gordillo et al. (2002)

4.7 Conclusions

The coastal environments are characterized by strong and rapid changes in abiotic factors, among which nutrients load play a key role. Evolution has provided macroalgae with different mechanisms that allow for optimization of resources available. Regarding nutrient acquisition, this implies not only the enzymes for uptake and assimilation. A wide variety of combinations are displayed by the different species. Some are storage specialists; others are more restricted to external availability. Periods of intense nutrient uptake do not necessarily mean faster growth, and during the course of the day, algae can switch their photosynthetic energy investment from C to N and P acquisition. Overall, the major output is that these different mechanisms for the different species coexist at the same time in the same community, and this strategy partitioning allows a more effective resource use, and an increased availability of nutrients to the rest of the food web.

Future research is needed to quantify the benefits of the coexistence of mixed strategies, and how are these combinations going to be affected by the modification of other abiotic factors such as temperature and salinity, the latter being particularly relevant for threatened ice-driven ecosystems such as the Arctic and the Antarctic. Many ongoing studies are focusing on the modification of geographical distribution boundaries for different ecologically relevant species, commonly being related to changes in water temperature and currents pattern. However, the result might also depend on the ability of a given species to deal with a new nutrient regime. The incorporation of this approach to geographically relevant studies will add knowledge on the mechanisms determining the distribution of species.

References

Aguiar AB, Morgan JA, Teichberg M, Fox S, Valiela I (2003) Transplantation and isotopic evidence of the relative effects of ambient an internal nutrient supply on the growth of *Ulva lactuca*. Biol Bull 205:250–251

Aguilera J, Bischof K, Karsten U, Hanelt D, Wiencke C (2002) Seasonal variation in ecophysiological patterns in macroalgae from an Arctic fjord II. Pigment accumulation and biochemical defense systems against high light stress. Mar Biol 140:1097–1106

Andría JR, Pérez-Lloréns JL, Vergara JJ (1999) Mechanisms of inorganic carbon acquisition in *Gracilaria gaditana* nom. prov. (Rhodophyta). Planta 208:564–573

Atkinson MJ, Smith SV (1983) C:N:P ratios of benthic marine plants. Limnol Oceanogr 28:568–574

Baird ME, Atkinson MJ (1997) Measurement and prediction of mass transfer to experimental coral reef communities. Limnol Oceanogr 42:1685–1693

Baird M, Middleton J (2004) On relating physical limits to the carbon: nitrogen ratio of unicellular algae and benthic plants. J Mar Syst 49:169–175

Barile PJ (2004) Evidence of anthropogenic nitrogen enrichment of the littoral waters of east central Florida. J Coast Res 20:1237–1245

Berges J (1997) Algal nitrate reductases. Eur J Phycol 32:3–8

Berges JA, Cochlan WP, Harrison PJ (1995) Laboratory and field responses of algal nitrate reductase to diel periodicity in irradiance, nitrate exhaustion, and the presence of ammonium. Mar Ecol Prog Ser 124:259–269

Bilger RW, Atkinson MJ (1995) Effects of nutrient loading on mass-transfer rates to a coral community. Limnol Oceanogr 40:279–289

Björnsäter BR, Wheeler PA (1990) Effect of nitrogen and phosphorus supply on growth and tissue composition of *Ulva fenestra* and *Enteromorpha intestinalis* (Ulvales, Chlorophyta). J Phycol 26:603–611

Borum J, Sand-Jensen K (1996) Is total primary production in shallow coastal marine waters stimulated by nitrogen loading? Oikos 76:406–410

Bracken MES, Nielsen KJ (2004) Diversity of intertidal macroalgae increases with nutrient loading by invertebrates. Ecology 85:2828–2836

Bracken M, Stachowicz J (2006) Seaweed diversity enhances nitrogen uptake via complementary use of nitrate and ammonium. Ecology 87:2397–2403

Brown MT, Nyman MA, Keogh JA, Chin NKM (1997) Seasonal growth of the giant kelp *Macrocystis pyrifera* in New Zealand. Mar Biol 129:417–424

Chapman ARO, Lindley JE (1980) Seasonal growth of *Laminaria solidungula* in the Canadian high arctic in relation to irradiance and dissolved nutrients concentrations. Mar Biol 57:1–5

Chopin T, Buschmann AH, Halling C, Troell M, Kautsky N, Neori A, Kraemer GP, Zertuche-González JA, Yarish C, Neefus C (2001) Integrating seaweeds into aquaculture systems: a key towards sustainability. J Phycol 37:975–986

Cifuentes LA, Sharp JH, Fogel ML (1998) Stable carbon and nitrogen isotope biogeochemistry in Delaware estuary. Limnol Oceanog 33:1102–1115

Cohen RA, Fong P (2004) Nitrogen uptake and assimilation in *Enteromorpha intestinalis*: using ^{15} N to determine preference during simultaneous pulses of nitrate and ammonium. J Exp Mar Biol Ecol 309:67–77

Cohen R, Fong P (2005) Experimental evidence supports the use of δ^{15}N content of the opportunistic green macroalga *Enteromorpha intestinalis* (Chlorophyta) to determine nitrogen sources to estuaries. J Phycol 41:287–293

Cole ML, Valiela I, Kroeger KD, Tomasky GL, Cebrian J, Wigand C, McKinney RA, Grady SP, Carvalho de Silva MH (2004) Assessment of a ∂^{15} N isotopic method to indicate anthropogenic eutrophication in aquatic ecosystems. J Environ Qual 33:124–132

Collos Y, Vaquer A, Bibent B, Slawyk G, García N, Souchu P (1997) Variability in nitrate uptake kinetics of phyto- plankton communities in a Mediterranean coastal lagoon. Estuar Coast Shelf Sci 44:369–375

DeBoer JA (1981) Nutrients. In: Lobban CS, Wynne MJ (eds) The biology of seaweeds. Blackwell Scientific Publications, Oxford

Denny MW (1990) Terrestrial versus aquatic biology: the medium and its message. Am Zool 30:111–121

Denny MW (1993) Air and water. Princeton University Press, NJ, USA

Deutsch B, Voss M (2006) Anthropogenic nitrogen input traced by means of ∂^{15} N values in macroalgae: results from in-situ incubation experiments. Sci Total Environ 366:799–808

Dickman EM, Newell JM, Gonzalez MJ, Vanni MJ (2008) Light, nutrients, and food-chain length constrain planktonic energy transfer efficiency across multiple trophic levels. Proc Nat Acad Sci 105:18408–18412

Dortch Q (1990) The interaction between ammonium and nitrate uptake in phytoplankton. Mar Ecol Prog Ser 61:183–201

Drew EA, Hastings RM (1992) A year-round ecophysiological study of *Himantothallus grandifolius* (Desmarestiales, Phaeophyta) at Signy Island, Antarctica. Phycologia 31:262–277

Elser JJ, Bracken MES, Cleland EE, Gruner DS, Harpole WS, Hillebrand H, Ngai JT, Seabloom EW, Shurin JB, Smith JE (2007) Global analysis of nitrogen and phosphorus limitation of primary producers in freshwater, marine, and terrestrial ecosystems. Ecol Let 10:1135–1142

Falkowski PA (1975) Nitrate uptake in marine phytoplankton: comparison of half-saturation constants from seven species. Limnol Oceanogr 20:412–416

Falkowski PG (2000) Rationalizing elemental ratios in unicellular algae. J Phycol 36:3–6

Flower EM, Neori A (2008) Effect of water aeration and nutrient load level on biomass yield. N uptake and protein content of the seaweed Ulva lactuca cultured in seawater tanks. J Appl Phycol. doi:10.1007/s10811-007-9300-6

Flynn KJ, Marshall H, Geider RJ (2001) A comparison of two N-irradiance interaction models of phytoplankton growth. Limnol Oceanogr 46:1794–1802

Fong P, Kamer K, Boyer KE, Boyle KA (2001) Nutrient content of macroalgae with differing morphologies may indicate sources of nutrients for tropical marine systems. Mar Ecol Prog Ser 220:137–152

Fong P, Boyer KE, Kamer K, Boyle KA (2003) Influence of initial tissue nutrient status of tropical marine algae on response to nitrogen and phosphorus additions. Mar Ecol Prog Ser 262:111–123

Fujita RM (1985) The role of nitrogen status in regulating transient ammonium uptake and nitrogen storage by macroalgae. J Exp Mar Biol Ecol 92:283–301

Gaylord B, Blanchette CA, Denny MW (1994) Mechanical consequences of size in wave-swept algae. Ecol Monogr 64:287–313

Gerard VA, Mann KH (1979) Growth and production of *Laminaria longicruris* (Phaeophyta) populations exposed to different intensities of water movement. J Phycol 15:33–41

Gevaert F, Barr NG, Rees TAV (2007) Diurnal cycle and kinetics of ammonium assimilation in the green alga *Ulva pertusa*. Mar Biol 151:1517–1524

Giordano M, Beardall J, Raven JA (2005) CO_2 concentrating mechanisms in algae: mechanisms, environmental modulation, and evolution. Annu Rev Plant Biol 56:99–131

Gordillo F, Dring M, Savidge G (2002) Nitrate and phosphate uptake characteristics of three species of brown algae cultured at low salinity. Mar Ecol Prog Ser 234:111–118

Gordillo F, Figueroa F, Niell F (2003) Photon-and carbon-use efficiency in *Ulva rigida* at different CO_2 and N levels. Planta 218:315–322

Gordillo FJL, Aguilera J, Jiménez C (2006) The response of nutrient assimilation and biochemical composition of Arctic seaweeds to a nutrient input in summer. J Exp Bot 57:2661–2671

Harrison PJ, Druehl LD, Lloyd KE, Thompson PA (1986) Nitrogen uptake kinetics in three year-classes of *Laminaria groenlandica* (Laminariales, Phaeophyta). Mar Biol 93:29–35

Heaton THE (1986) Isotopic studies of nitrogen pollution in the hydrosphere and atmosphere: a review. Chem Geol 59:87–102

Hein M, Pedersen MF, Sand-Jensen K (1995) Size dependent nitrogen uptake in micro and macroalgae. Mar Ecol Prog Ser 118:247–253

Henley WJ, Dunton KH (1995) A seasonal comparison of carbon, nitrogen and pigment content in *Laminaria solidungula* and *Laminaria saccharina* (Phaeophyta) in the Alaskan Arctic. J Phycol 31:325–331

Henley WJ, Dunton KH (1997) Effects of nitrogen supply and continuous darkness on growth and photosynthesis of the Arctic kelp *Laminaria solidungula*. Limnol Oceanogr 42:209–216

Hernández I, Martinez-Aragón JF, Pérez-Llorens JL, Vázquez R, Vergara JJ (2002) Biofiltering efficiency in removal of dissolved nutrients by three species of estuarine macroalgae cultivated with sea bass (*Dicentrarchus labrax*) waste waters 1 Phosphate. J Appl Phycol 14:365–374

Hernández I, Pérez-Pastor A, Vergara JJ, Martínez-Aragón JF, Fernández-Engo MA, Pérez-Llorens JL (2006) Studies on the biofiltration capacity of *Gracilariopsis longissima*: from micro- scale to macroscale. Aquaculture 252:43–53

Howarth RW (2008) Coastal nitrogen pollution: a review of sources and trends globally and regionally. Harmful Algae 8:14–20

Howarth RW, Anderson D, Cloern J, Elfring C, Hopkinson C, Lapointe B, Malone T, Marcus N, McGlathery K, Sharpley A, Walker D (2000) Nutrient pollution of coastal rivers, bays, and seas. Iss Ecol 7:1–15

Hurd C (2000) Water motion, marine macroalgal physiology, and production. J Phycol 36:453–472

Hurd CL, Dring MJ (1990) Phosphate uptake by intertidal algae in relation to zonation and season. Mar Biol 107:281–289

Hurd CL, Durante KM, Chia FS, Harrison PJ (1994) Effect of bryozoan colonization on inorganic nitrogen acquisition by the kelps *Agarum fimbriatum* and *Macrocystis integrifolia*. Mar Biol 121:167–173

Johnson KS (1982) Carbon dioxide hydration and dehydration kinetics in seawater. Limnol Oceanogr 27:849–855

Jones AB, Dennison WC, Preston NP (2001) Integrated treatment of shrimp effluent by sedimentation, oyster filtration and macroalgal absorption: a laboratory scale study. Aquaculture 193:155–178

Kamer K, Fong P, Kennison RL, Schiff K (2004) The relative importance of sediment and water column supplies of nutrients to the growth and tissue nutrient content of the green macroalga *Enteromorpha intestinalis* along an estuarine resource gradient. Aquatic Ecol 38:45–56

Kautsky N, Kautsky H, Kautsky U, Waern M (1986) Decreased depth penetration of *Fucus vesiculosus* (L.) since the 1940's indicates eutrophication of the Baltic Sea. Mar Ecol Prog Ser 28:1–8

Keller P, Paerl H (1980) Physiological adaptation and response to environmental stress during an N_2 fixing *Anabaena* bloom. Appl Environ Microbiol 40:587–595

Koch EW (1993) The effect of water flow on photosynthetic processes of the alga *Ulva lactuca* L. Hydrobiology 260(261):457–462

Korb RE, Gerard VA (2000) Nitrogen assimilation characteristics of polar seaweeds from differing nutrient environment. Mar Ecol Prog Ser 198:83–92

Krogmann DW, Jagendorf AT, Avron M (1959) Uncouplers of spinach chloroplast photosynthetic phosphorylation. Plant Physiol 34:272–277

Lapointe BE (2004) Anthropogenic nutrient enrichment of seagrass and coral reef communities in the lower Florida Keys: discrimination of local versus regional nitrogen sources. J Exp Mar Biol Ecol 308:23–58

Lapointe B, Bedford B (2010) Ecology and nutrition of invasive *Caulerpa brachypus* f. *parvifolia* blooms on coral reefs off southeast Florida, USA. Harmful Algae 9:1–12

Lapointe BE, Littler MM, Littler DS (1992) Nutrient availability to marine macroalgae in siliciclastic versus carbonate-rich coastal waters. Estuar 15:75–82

Larned ST (1998) Nitrogen- versus phosphorus-limited growth and sources of nutrients for coral reef macroalgae. Mar Biol 132:409–421

Larned ST, Atkinson MJ (1997) Effects of water velocity on NH_4 and PO_4 uptake and nutrient-limited growth in the macroalga *Dictyosphaeria cavernosa*. Mar Ecol Prog Ser 157:295–302

Lartigue J, Sherman TD (2006) A Field Study of Nitrogen Storage and Nitrate Reductase Activity in the Estuarine Macroalgae *Enteromorpha lingulata* (Chlorophyceae) and *Gelidium pusillum* (Rhodophyceae). Estuar Coast 29:699–708

Lavery PS, McComb AJ (1991) The nutritional ecophysiology of *Chaetomorpha linum* and *Ulva rigida* in Peel inlet, Western Australia. Bot Mar 34:251–260

Lean DRS, Murphy TP, Pick FR (1982) Photosynthetic response of lake plankton to combined nitrogen enrichment. J Phycol 18:509–521

Lobban CS, Harrison PJ (1994) Seaweed ecology and physiology. Cambridge University Press, New York

Lomas MW, Glibert PM (1999) Temperature regulation of nitrate uptake: a novel hypothesis about nitrate uptake and reduction in cool-water diatoms. Limnol Oceanogr 44:556–572

Lombardi JV, Marques HLA, Pereira R, Lima T, Barreto OJS, Paula EJ (2006) Cage polyculture of the Pacific white shrimp Litopenaeus vannamei and the Philippines seaweed *Kappaphycus alvarezii*. Aquaculture 258:412–425

Lotze HK, Schramm W (2000) Ecophysiological traits explain species dominance patterns in macroalgal blooms. J Phycol 36:287–295

Lourenço S, Barbarino E, Nascimento A, Paranhos R (2005) Seasonal variations in tissue nitrogen and phosphorus of eight macroalgae from a tropical hypersaline coastal environment. Crypt Algol 26:355–371

Lüning K (1991) Circannual growth rhythm in a brown alga, *Pterygophora californica*. Bot Act 104:157–162

Lüning K, Pang S (2003) Mass cultivation of seaweeds: current aspects and approaches. J Appl Phycol 15:115–119

Mata L, Silva J, Schuenhoff A, Santos R (2006) The effects of light and temperature on the photosynthesis of *Asparagopsis armata* tetrasporophyte (*Falkenbergia rufolanosa*), cultivated in tanks. Aquaculture 252:12–19. doi:10.1016/j.aquaculture.2005.11.045

McGlathery KJ, Marino R, Howarth RW (1994) Variable rates of phosphate-uptake by shallow marine carbonate sediments-mechanisms andecological significance. Biogeochem 25:127–146

McGlathery KJ, Pedersen MF, Borum J (1996) Changes in intracellular nitrogen pools and feedback controls on nitrogen uptake in *Chaetomorpha linum* (Chlorophyta). J Phycol 32:393–401

McKane RB, Johnson LC, Shaver GR, Nadelhoffer KJ, Rastetter EB, Fry B, Giblin AE, Kielland K, Kwiatkowski BL, Laundre JA, Murray G (2002) Resource-based niches provide a basis for plant species diversity and dominance in arctic tundra. Nature 415:68–71

McKinney RA, Nelson WG, Charpentier MA, Wigand C (2001) Ribbed mussel nitrogen isotope signatures reflect nitrogen sources in coastal salt marshes. Ecol Appl 11:203–214

Mercado JM, Carmona R, Niell FX (1998) Bryozoans increase available CO_2 for photosynthesis in *Gelidium sesquipedale* (Rhodophyceae). J Phycol 34:925–927

Muñoz J, Cancino JM, Molina MX (1991) Effect of encrusting bryozoans on the physiology of their algal substratum. J Mar Biol Ass UK 71:877–882

Murthy MS, Rao AS, Reddy ER (1986) Dynamics of nitrate reductase activity in two intertidal algae under desiccation. Bot Mar 29:471–474

Naldi M, Wheeler PA (1999) Changes in nitrogen pool in *Ulva fenestrata* (Chlorophyta) and *Gracilaria pacifica* (Rhodophyta) under nitrate and ammonium enrichment. J Phycol 35:70–77

Naldi M, Wheeler PA (2002) N-15 measurements of ammonium and nitrate uptake by *Ulva fenestrata* (chlorophyta) and *Gracilaria pacifica* (rhodophyta): Comparison of net nutrient disappearance, release of ammonium and nitrate, and N-15 accumulation in algal tissue. J Phycol 38:135–144

Neori A (2008) Essential role of seaweed cultivation in integrated multi-trophic aquaculture farms for global expansion of mariculture: an analysis. J Appl Phycol 20:567–570. doi:10.1007/s10811-007-9206-3

Neori A, Chopin T, Troell M, Buschmann AH, Kraemer GP, Halling C et al (2004) Integrated aquaculture: rationale, evolution and state of the art emphasizing seaweed biofiltration in modern maricultura. Aquaculture 231:361–391
Nishihara GN, Terada R, Noro T (2005) Effect of temperature and irradiance on the uptake of ammonium and nitrate by *Laurencia brongniartii* (Rhodophyta, Ceramiales). J Appl Phycol 17:371–377
Nixon SW, Pilson MEQ (1983) Nitrogen in estuarine and coastal marine ecosystems. In: Carpenter EJ, Capone DG (eds) Nitrogen in the marine environment. Academic, New York
Oviatt C, Doering P, Nowicki B, Reed L, Cole J, Frithsen J (1995) An ecosystem level experiment on nutrient limitation in temperate coastal marine environments. Mar Ecol Prog Ser 116:171–179
Page HM (1995) Variation in the natural abundance of ^{15} N in the halophyte, *Salicornia virginica*, associated with groundwater subsidies of nitrogen in a southern California salt marsh. Oecology 104:181–188
Pedersen MF, Borum J (1996) Nutrient control of algal growth in estuarine waters. Nutrient limitation and the importance of nitrogen requirements and nitrogen storage among phytoplankton and species of macroalgae. Mar Ecol Prog Ser 142:261–272
Pedersen M, Borum J (1997) Nutrient control of estuarine macroalgae: growth strategy and the balance between nitrogen requirements and uptake. Mar Ecol Prog Ser 161:155–163
Pedersen M, Borum J, Fotel F (2010) Phosphorus dynamics and limitation of fast-and slow-growing temperate seaweeds in Oslofjord, Norway. Mar Ecol Prog Ser 399:103–115
Peterson BJ, Fry B (1987) Stable isotopes in ecosystem studies. Annu Rev Ecol Syst 18:293–320
Phillips JC, Hurd CL (2003) Nitrogen ecophysiology of intertidal seaweeds from New Zealand: N uptake, storage and utilisation in relation to shore position and season. Mar Ecol Prog Ser 264:31–48
Phillips JC, Hurd CL (2004) Kinetics of nitrate, ammonium, and urea uptake by four intertidal seaweeds from New Zealand. J Phycol 40:534–545
Probyn TA, Chapman ARO (1983) Summer growth of *Chordaria flagelliformis* (O. F. Muell.) C. Ag.: physiological strategies in a nutrient stressed environment. J Exp Mar Biol Ecol 73:243–271
Raven JA (1984) Energetics and transport in aquatic plants. AR Liss, New York
Raven JA, Kübler JE (2002) New light on the scaling of metabolic rate with the size of algae. J Phycol 38:11–16
Raven J, Giordano M, Beardall J (2008) Insights into the evolution of CCMs from comparisons with other resource acquisition and assimilation processes. Physiol Plant 133:4–14
Redfield AC, Ketchum BH, Richards FA (1963) The influence of organisms on the composition of sea-water. In: Hill N (ed) In the sea, 2nd edn. Wiley, New York
Redinbaugh MG, Campbell WH (1991) Higher plant responses to environmental nitrate. Physiol Plant 82:640–650
Rees TAV (2003) Safety factors and nutrient uptake by seaweeds. Mar Ecol Prog Ser 263:29–42
Rees TAV (2007) Metabolic and ecological constraints imposed by similar rates of ammonium and nitrate uptake per unit surface area at low substrate concentrations in marine phytoplankton and macroalgae. J Phycol 43:197–207
Ritchie RJ (1987) The permeability of ammonia, methylamine and ethylamine in the charophyte *Chara corallina* (C. australis). J Exp Bot 38:67–76
Ritchie RJ, Gibson J (1987) Permeability of ammonia, methylamine and ethylamine in the cyanobacterium Synechococcus R-2 (*Anacystis nidulans*). J Membr Biol 95:131–142
Runcie JW, Ritchie RJ, Larkum AWD (2004) Uptake kinetics and assimilation of phosphorus by *Catenella nipae* and *Ulva lactuca* can be used to indicate ambient phosphate availability. J Appl Phycol 16:181–194
Ryberg H, Axelsson L, Carlberg S, Larsson C, Uuisitalo J (1990) CO_2 storage and CO2 concentration in brown seaweeds I. Occurrence and ultrastructure. Curr Res Phot 4:517–520

Sato K, Ueno Y, Egashira R (2006a) Uptake of nitrate-nitrogen in intensive shrimp culture ponds by sterile *Ulva* sp. J Chem Eng Jpn 39:1128–1131. doi:10.1252/jcej.39.1128

Sato K, Eksangsri T, Egashira R (2006b) Ammonia-nitrogen uptake by seaweed for water quality control in intensive mariculture ponds. J Chem Eng Jpn 39:247–255. doi:10.1252/jcej.39.247

Schaffelke B, Lüning K (1994) A circannual rhythm controls seasonal growth in the kelps *Laminaria hyperborea* and *L. digitata* from Helgoland (North Sea). Eur J Phycol 29:49–56

Shpigel M, Neori A, Popper DM, Gordin H (1993) A proposed model for "environmentally clean" land based culture of fish, bivalves and seaweeds. Aquaculture 117:115–128. doi:10.1016/0044-8486 (93)90128-L

Solomonson LP, Barber MJ (1990) Assimilatory nitrate reductase: functional properties and regulation. Annu Rev Plant Physiol Plant Mol Biol 41:225–253

Sültemeyer D, Rinast KA (1996) The CO_2 permeability of the plasma membrane of *Chlamydomonas reinhardtii*: mass spectrometric O^{18} exchange measurements from (CO_2)-C^{13}-O^{18} in suspensions of carbonic anhydrase-loaded plasma-membrane vesicles. Planta 200:358–368

Taylor RB, Rees TAV (1998) Excretory products of mobile epifauna as a nitrogen source for seaweeds. Limnol Oceanogr 43:600–606

Taylor DI, Nixon SW, Granger SL, Bucley BA, McMahon JP, Lin HJ (1995) Responses of coastal lagoon plant communi- ties to different forms of nutrient enrichment, a mesocosm experiment. Aquat Bot 52:19–34

Teichberg M, Fox SE, Aguila C, Olsen YS, Valiela I (2008) Macroalgal response to experimental nutrient enrichment in shallow coastal waters: growth, internal nutrient pools, and isotopic signatures. Mar Ecol Prog Ser 368:117–126

Teichberg M, Fox S, Olsen Y, Valiela I, Martinett P, Iribarne O, Muto E, Petti M, Corbisier T, Soto Jiménez M (2010) Eutrophication and macroalgal blooms in temperate and tropical coastal waters: nutrient enrichment experiments with *Ulva spp*. Glob Change Biol 16:2624–2637

Turpin DH (1983) Ammonium induced photosynthetic suppression in ammonium limited *Dunaliella tertiolecta* (Chlorophyta). J Phycol 19:70–76

Tyler AC, McGlathery KJ, Anderson IC (2003) Benthic algae control sediment-water column fluxes of organic and inorganic nitrogen compounds in a temperate lagoon. Limnol Oceanogr 48:2125–2137

Vandermeulen H, Gordin H (1990) Ammonium uptake using *Ulva* (Chlorophyta) in intensive fishpond systems. Mass-culture and treatment of effluent. J Appl Phycol 2:363–374. doi:10.1007/ BF02180927

Vergara JJ, Berges JA, Falkowski PG (1998) Diel periodicity of nitrate reductase activity and protein levels in the marine diatom *Thalassiosira weissflogii* (Bacillariophyceae). J Phycol 34:952–961

Wallentinus I (1984) Comparisons of nutrient uptake rates for Baltic macroalgae with different thallus morphologies. Mar Biol 80:215–225

Weidner M, Kiefer H (1981) Nitrate reduction in the marine brown algae *Giffordia mitchellae* (Harv.) Ham. Z Pflazenphysiol Bd 104:341–351

Wirtz K, Pahlow M (2010) Dynamic chlorophyll and nitrogen: carbon regulation in algae optimizes instantaneous growth rate. Mar Ecol Prog Ser 402:81–96

Young E, Berges J, Dring M (2009) Physiological responses of intertidal marine brown algae to nitrogen deprivation and resupply of nitrate and ammonium. Physiol Plant 135:400–441

Chapter 5
Seaweed Acclimation to Salinity and Desiccation Stress

Ulf Karsten

Dedicated to my mentor Professor Dr. Gunter O. Kirst on the occasion of his 70th birthday.

5.1 Variability of Salinity in Seaweed Habitats

Salinity is a technical term in marine sciences and represents the sum of all dissolved salts of a defined body of water. Therefore, marine biologists for a long time used the expression *promille* (‰) or *parts per thousands* (ppt) to describe salinity concentration. In contrast, since 1978 oceanographers defined salinity in the Practical Salinity Scale as dimensionless Practical Salinity Units (PSU) which is the conductivity ratio of a seawater sample to a standard KCl solution. In October 2010, the Intergovernmental Oceanographic Commission (IOC), International Association for the Physical Sciences of the Oceans (IAPSO), and the Scientific Committee on Oceanic Research (SCOR) jointly adopted a new standard for the calculation of the thermodynamic properties of seawater. This new standard, called TEOS-10, uses Absolute Salinity S_A (mass fraction of salt in seawater) to describe the salt content of seawater. Ocean salinities now have units of g dissolved salts kg^{-1} water (http://www.teos-10.org; Wright et al. 2010). Therefore, all salinities throughout the text are expressed as Absolute Salinity S_A.

The chemical composition of the dissolved salts is relatively constant throughout the open oceans due to intensive mixing, and it varies only between 33 and 37 S_A, gradually decreasing from the subtropics toward the tropics and polar seas. In contrast, salinity strongly varies in nearshore waters and estuaries where river freshwater mixes with marine water bodies. Here horizontal and vertical gradients between 0 and 33 S_A can be measured. The degree of salt dilution in estuaries

U. Karsten (✉)
Institute of Biological Sciences, Applied Ecology, University of Rostock,
Albert-Einstein-Strasse 3, 18059 Rostock, Germany
e-mail: ulf.karsten@uni-rostock.de

C. Wiencke and K. Bischof (eds.), *Seaweed Biology*, Ecological Studies 219,
DOI 10.1007/978-3-642-28451-9_5, © Springer-Verlag Berlin Heidelberg 2012

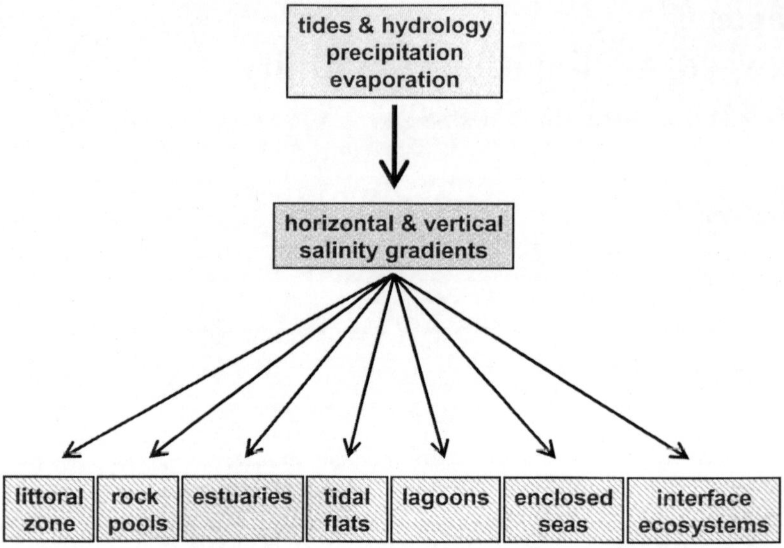

Fig. 5.1 Illustration of the abiotic factors influencing salt concentration resulting in horizontal and vertical salinity gradients in various aquatic systems

largely depends on the rate and volume of freshwater runoff, and can be dramatic. The water of the Amazon, for example, can be recognized at the surface of the open Atlantic Ocean for approximately 300 km (Ryther et al. 1967). Since the geochemistry of the river's catchment area is typically highly variable as well, it strongly affects besides the salt concentration also the salt composition in estuaries. In addition, tidal flows, hydrological conditions, wind, precipitation, and evaporation strongly influence salt concentration of the respective water bodies (Fig. 5.1). Consequently, salinity is typically a local rather than a global parameter and can be highly variable in coastal regions.

Since most seaweeds are sessile organisms that preferentially grow attached to hard substrata such as rocks, gravel, or as epiphytes on salt marsh plants, on mangroves, and on mussel colonies, they are mainly confronted with salinity fluctuations and desiccation when exposed at low tides. At lowest water levels, hyposaline conditions may be present due to the mixing of seawater with rain, snow, or melt water, while hypersaline stress may occur due to evaporation during high insolation in summer or freezing-out of freshwater in winter. In addition, in estuaries and fjords which often exhibit extensive seaweed communities (Schramm and Nienhuis 1996), rivers or freshwater runoff mix with seawater resulting in diurnally and seasonally fluctuating salinity gradients. In Arctic waters, seaweed species can be strongly affected by melt water influx and calving glaciers (Hanelt et al. 2001; Karsten et al. 2003a).

Most seaweeds such as kelps inhabit the sublittoral which is characterized by quite stable environmental factors, and hence these more deepwater plants only rarely experience salinity and desiccation stress. One of these occasions, however, is spring low tides where sublittoral seaweed species may be exposed for some time to salinity changes and desiccation (Lüning 1990). Besides the intertidal zone and estuaries, there are other habitats occupied by seaweeds which are affected by salinity changes or gradients (Fig. 5.1). Rock pools at low tide often exhibit strong salinity fluctuations due to precipitation or evaporation. Seaweeds living on tidal flats (e.g., on mussel beds) or in interface ecosystems (salt marshes, mangroves) are regularly affected by the tidal flows and hence to a combination of salinity and desiccation stress. In mangroves epiphytic red algae exposed to the prop roots can be observed completely covered with salt crystals. Semi-enclosed seas with only small connections to the open oceans mostly exhibit strong horizontal salinity gradients or hypersaline conditions depending on the hydrology of the catchment area and precipitation. The best studied semi-enclosed system is the Baltic Sea with its strong, but stable horizontal salinity gradient from almost freshwater conditions in the Northeast to brackish/marine conditions in the western part. Since the largest part of the Baltic Sea shows a salinity between 4 and 7 S_A only, biodiversity of the aquatic flora (incl. seaweeds) and fauna is generally strongly reduced, because this salinity range (ß-mesohalinum 5–10 S_A) is too low for marine species and too high for freshwater organisms (Remane 1940). In warm-temperate to subtropical coastal regions such as South Australia, lagoons are typical shallow water systems that are separated from the open ocean by some form of barrier (e.g., sand spit) resulting in often hypersaline conditions because of the high prevailing evaporation. A typical example is the Coorong lagoon in South Australia with salinities $>100\ S_A$, where high abundances of charophyte algae can occur (Bisson and Kirst 1983). Marine and brackish waters are consequently almost infinitely variable in the amplitude and frequency of their saline changes, which of course has various consequences for the physiological performance of seaweeds.

5.2 Effects of Salinity Stress and Desiccation on the Physiology of Seaweeds

Since decades, the effects of salinity and desiccation on the ecophysiological performance of seaweeds have been described to outline the species-specific width of tolerance, i.e., upper and lower limits for survival, as well as mechanisms for acclimation (Kirst 1990 and references therein). Nevertheless, both salinity stress and desiccation reflect two different forms of water deprivation. While under salt stress conditions, seaweed cells are still in full contact to liquid water of diminished water potential, desiccation leads to strong cellular dehydration. Typical physiological parameters which have been studied under such stress conditions include rate of survival, growth, photosynthesis, respiration, and reproduction.

5.2.1 Rate of Survival

The rate of survival can be easily estimated by following seaweed growth or photosynthesis activity under defined stress conditions, or using classical vital staining techniques for individual cells such as Evans Blue (Russell 1985; Thomas and Kirst 1991). Only dead cells with damaged membranes will take up this stain and hence appear blue under the microscope. These days a range of commercially available fluorescence dyes such as the so-called Live–Dead kits are available which in combination with an epifluorescence microscope allow the evaluation and visualization of even smallest damages on membrane integrity.

The available data on cell viability in relation to salinity indicate that members of eulittoral green algal genera such as *Ulva* (incl. *Enteromorpha*), *Acrosiphonia*, *Ulothrix*, *Cladophora*, and *Prasiola* are generally much less affected than sublittoral red and brown seaweeds (Russell 1985, 1987, Kain and Norton 1990; Thomas and Kirst 1991; Karsten et al. 1991a). While even under triple seawater concentration >75% of green algal cells remain intact, mortality in sublittoral brown algae such as *Phaeurus antarcticus* increased to almost 100% (Karsten et al. 1991a). This rather euryhaline feature of eulittoral green seaweeds may be considered as important environmental factor for controlling vertical species zonation on the shore (Russell 1987).

5.2.2 Growth

From an ecological perspective, growth represents the most relevant physiological process to describe the performance of seaweeds in their respective habitat because it optimally integrates all positive and negative environmental effects on the organism and hence reflects the acclimation potential (Gustavs et al. 2009). A careful analysis of growth patterns allows the evaluation of salinity tolerance limits, growth optima, and acclimation abilities of individual seaweed species, and, thus facilitates the interpretation of natural distributions. Knowledge of physiological limits is necessary to estimate and understand distribution limits within an unknown community, i.e., in a competitive situation. Additionally, the shape of the growth rate curve in dependence of any abiotic factor characterizes a seaweed species as steno- or euryoecious, or in case of salinity as steno- or euryhaline, which definitively affects its competitive strength.

Until now, numerous ecophysiological studies have been undertaken on the salinity effects on growth in seaweeds (Jacob et al. 1991; Karsten et al. 1991a, 1993b, 1994, 1996b; Karsten and West 1993; Kirst 1990 and references therein, Thomas and Kirst 1991). All these papers clearly indicate that eulittoral seaweeds from polar to tropical regions grow between low salinities (5–10 S_A) and double seawater concentration (66–68 S_A), in most cases with conspicuous optima under normal seawater conditions. The supralittoral *Prasiola crispa* ssp. *antarctica* even grows between 0.3 and 105 S_A (Jacob et al. 1991). In contrast to the generally broad salinity tolerance of upper-shore seaweeds, those from the sublittoral usually exhibit the narrowest tolerance limits (Russell 1987). Salinities commonly encountered in areas of abundance are most favorable for growth (Bird et al. 1979; Bird and McLachlan 1986). This is well reflected in the growth pattern of the green alga *Ulva pertusa* from an eelgrass bed in a semi-protected bay at the southwest coast of Korea (Choi et al. 2010). *Ulva pertusa* exhibited optimum growth at 20 S_A, a situation encountered in the field during the rainy season when this species often forms blooms in eelgrass beds. Growth under hyposaline conditions, i.e., in aquatic systems such as estuaries or the eastern part of the Baltic Sea, may be governed by the availability of certain inorganic ions which increase the lower tolerance limits of seaweeds. In this context, the presence of Ca^{2+} plays an essential role in cell signaling, as structural component of seaweed cell walls and membranes, and as cation to balance organic anions in the plant vacuole (Kauss 1987; Tazawa et al. 1987; Verret et al. 2010). Whole seaweed thalli have been reported to exhibit differential salinity tolerances. Particularly young apical growing parts of species of the genera *Cladophora*, *Ceramium*, *Phycodrys*, and *Plumaria* are more sensitive to hyposaline conditions than older, basal parts (Russell 1987). Kirst (1990) speculated that this observation is a secondary effect of Ca^{2+} availability, as particularly fast-growing cells depend on this cation, for example, for cell wall formation.

If not only growth but also survival is considered, most seaweeds show a remarkable physiological potential. While the red alga *Porphyra umbilicalis* exhibits optimum growth between 7 and 52 S_A, it survived without cell division even in sixfold fully marine salinities for 2 weeks (Wiencke and Läuchli 1980). Similar observations have been reported in the studies of elongation growth in the siphonous green alga *Valonia macrophysa* (Gutknecht et al. 1978). Near the limits of salinity tolerance, growth of most studied seaweeds is typically strongly reduced or even completely inhibited in order to funnel all available metabolic energy into the process of osmotic adjustment, which guarantees survival under fluctuating salinities. Besides these energetic considerations, high inorganic ion concentrations under salt stress conditions exert negative, i.e., inhibiting effects on seaweed growth. This is reflected in conspicuous changes in size and morphology of seaweeds under long-term salt stress (Russell 1987).

If additional environmental factors that greatly affect the growth rate of seaweeds as well are included in the investigation of the growth–salinity relationship (e.g., radiation (UV), temperature, etc.), the emerging picture becomes very complex. Therefore, only a few macroalgal species have been investigated in this respect, such as the green algae *Cladophora glomerata* and *C. rupestris* (Thomas et al. 1988), the red alga *Polysiphonia lanosa* (Reed 1983), or the kelp species *Laminaria groenlandica* and S*accharina latissima* (Druehl 1967). Low salinity may be compromised by temperature as shown for North Pacific *L. groenlandica* which cannot tolerate the combination of low salinity and high temperature conditions encountered in areas subjected to snow-melt runoff, whereas *S. latissima* can. Both species, however, do well in areas subjected to winter rain runoff where cold conditions prevail (Druehl 1967). All available data indicate that growth of seaweeds primarily depends on radiation and temperature conditions, and the more these environmental factors approach species-specific optimum requirements, the broader is the salinity range tolerated. In addition, if seaweeds are exposed to desiccation during low tides, growth is completely inhibited. The physiological strategy is to cope with and survive this stress condition by maximum reduction of all metabolic activities.

5.2.3 Photosynthesis and Respiration

Besides growth, photosynthesis and respiration are two central physiological processes in seaweeds which are strongly affected by salinity changes (Kirst 1990). Under extreme hypo- or hypersaline conditions, photosynthesis and respiration are typically completely inhibited and in many cases confirm the growth–salinity relationships described above. Whether salinity stress is extreme depends on the habitat and vertical zonation of the respective seaweed species, because sublittoral taxa are generally much more sensitive and hence stenohaline than their eulittoral, mainly euryhaline, counterparts (Russell 1987). A relatively salt-insensitive photosynthesis and respiration seem to be a prerequisite for the successful occupation of the eulittoral habitat and may ensure long-term survival and reproduction under large amplitudes of salinity in combination with other environmental factors (Gessner and Schramm 1971). A similar relation regarding photosynthesis as a function of desiccation was reported in various seaweeds from different tidal heights (Wiltens et al. 1978). In addition, photosynthesis and respiration exhibit different responses under a range of salinities as documented for a set of eulittoral green algae from Antarctica (Karsten et al. 1991a; see also Chap. 13 by Wiencke and Amsler). While most investigated species showed optimum photosynthesis at 34 S_A and decreasing rates between 7 and 17 S_A as well as between 51 and 68 S_A, respiration was much less affected. Similarly, under desiccation photosynthesis of seaweeds is much more affected than respiration (Wiltens et al. 1978).

The photosynthetic and respiratory responses following exposure to moderate and high changes in salinity are inconsistent among seaweeds. Frequently, a transient stimulation of respiration and a stimulatory or inhibitory effect on photosynthesis have been observed (Kirst 1990). The time required for a more or less complete recovery is species specific and lasted for several hours to days for seaweeds (Kirst 1990).

Using chlorophyll fluorescence kinetics, the underlying processes leading to salt-induced inhibition of photosynthesis were studied in various eulittoral green and red alga (*Prasiola*, *Ulva*, *Porphyra*) (Wiltens et al. 1978; Fork and Öquist 1981; Satoh et al. 1983; Smith et al. 1986). These investigations were aimed primarily at measuring the effects of desiccation on photosynthesis. The two stresses (increasing salinity and desiccation) are comparable since they result in a reduction of the cellular water potential. During desiccation, however, cellular ionic concentrations increase and the ion ratios remain constant. In contrast, during salinity stress algal cells may not only increase ionic concentrations but also undergo changes in ion ratios owing to selective uptake. This has to be taken into account when comparing the results obtained with species under salt or desiccation stress.

The primary photosynthetic mechanism is affected at the electron transport stage between PS I and PS II. The sensitive site in *Porphyra* and *Ulva* species is most likely between plastoquinone and P 700 (Wiltens et al. 1978). In *Porphyra perforata* there seem to be at least three sites in the photosynthetic apparatus that are inhibited by high salinity, namely the photoactivation of electron flow on the reducing side of PS I, the electron flow on the water side of PS II and the transfer of light energy between the pigment complexes (Satoh et al. 1983). These authors concluded that a free electron flow at all three sites is essential to avoid photodamage through chronic photoinhibition, which will occur if only one site is blocked because of, e.g., the accumulation of highly reactive oxygen species(ROS; see also Chap. 6 by Bischof and Rautenberger).

More recent data on cyanobacteria indicate a salt-induced inactivation of both oxygen evolution in PSII and electron transport in PSI (Allakhverdiev and Murata 2008). The site of inactivation is the electron-donating side of PSII, i.e., the oxygen-evolving machinery, due to the influx of uncontrolled Na^+ and Cl^- ions with resultant dissociation of extrinsic proteins from photosystems (Allakhverdiev and Murata 2008).

5.3 Processes of Osmotic Acclimation

Hypersaline conditions in the external medium affect seaweeds in two ways: first, the water potential is strongly reduced leading to dehydration of cells, and secondly, high concentrations of some inorganic ions exert toxic effects on cellular metabolism. Similarly, desiccation stress will also result in strong dehydration of cells. Since both hypersalinity and desiccation affect the internal osmotic potential, the acclimation responses of seaweeds are comparable.

Water is taken up in all living cells by osmosis, which is driven by the water potential gradient, i.e., only if the intracellular water potential is lower than in the external medium there is a water influx. Consequently, marine seaweeds have to create an internal osmotic potential higher than that of seawater to gain and retain constant water content of the cells, which is necessary to maintain turgor as the driving force for growth. Seaweeds typically respond to external salinity changes with osmotic acclimation processes which involve the control of cytoplasmic and vacuolar concentrations of osmotically active compounds (Kirst 1990). Therefore, osmotic acclimation is a fundamental mechanism of salinity tolerance of these plants that conserves the stability of the intracellular milieu (homeostasis), and it is essential for maintaining an efficient functional state in the cells (Kirst 1990). Most seaweeds use inorganic ions and small organic osmolytes to create the internal osmotic potential.

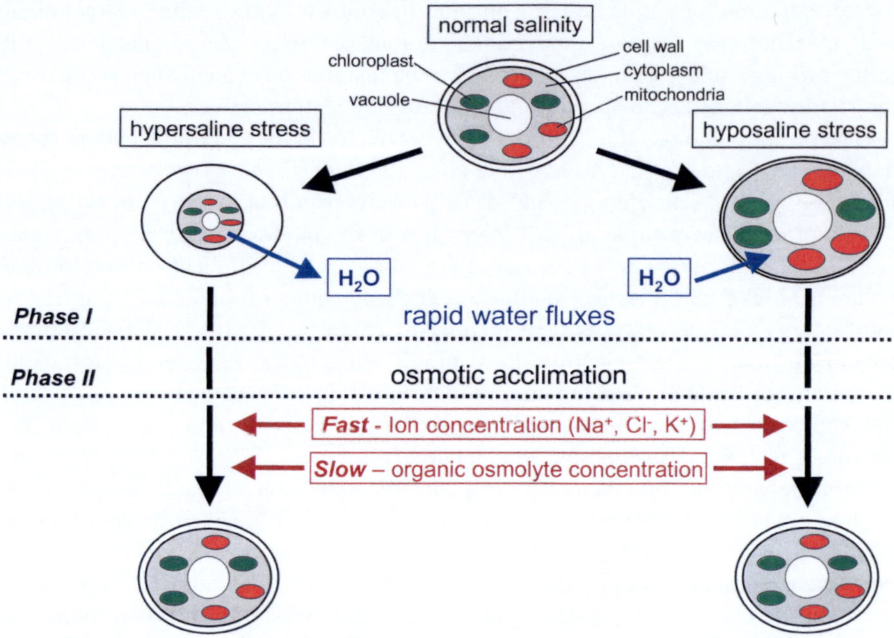

Fig. 5.2 Schematic summary of the processes of osmotic acclimation in seaweeds after hypersaline and hyposaline stress leading to a new steady state

The response of seaweeds to salinity changes is a well-regulated biphasic process. Phase I is characterized by rapid changes in turgor pressure caused by massive water fluxes in or out of the organism following the osmotic gradient. Phase II represents the process of osmotic acclimation, i.e., adjustment of the intracellular concentrations of the osmotically active inorganic ions and organic osmolytes until a new steady state is achieved (Fig. 5.2). Both phases are part of a feedback loop that comprises the osmotic acclimation.

The water fluxes during phase I are rapid processes with half times in seaweeds lasting from minutes to hours (Zimmermann and Steudle 1978), resulting in water influx under hyposaline conditions or water efflux during hypersaline stress, both of which mitigate at least transiently the osmotic stress (Fig. 5.2). Most importantly, these processes follow passive "osmometer" behavior, and depend on physico-chemical properties of the cell-wall membrane complex such as the hydraulic conductivity (water permeability, water channels). So far it has been assumed that all phase I processes are not under metabolic control (Kirst 1990), but since the discovery of the so-called aquaporins as membrane-located water channels it seems that cells can control their water content (Törnroth-Horsefield et al. 2006). These channels are widely distributed in all kingdoms of life. They form tetramers in the plasmalemma, and facilitate the transport of water through the lipophilic membrane. Probably, many water channels at the cytoplasmic membrane open for short time during salt stress, and not only water but also small water-soluble solutes are taken up or lost during phase I.

In contrast to the phase I, phase II is a rather slow process in seaweeds lasting from several hours up to 2–3 days (Kirst 1990, Karsten et al. 1996b). As a result, the internal osmotic potential becomes adjusted by changing the concentrations of ions and organic osmolytes to restore the original turgor pressure. These processes are under direct metabolic control.

5.3.1 Inorganic Ions

To adjust inorganic ion concentrations to homeostatic conditions, ions are actively extruded under hypersaline conditions but must be imported under hyposaline stress. The main ions involved in osmotic acclimation are K^+, Na^+, Cl^- and to a lesser extent sulfate, nitrate, or phosphate. The ionic composition of algal cells and in particular of their vacuoles varies widely depending on cellular number and volume of vacuoles (Kirst 1990). In most seaweeds, the Cl^- concentration usually follows the fluctuation in salinity. With respect to the Na^+ content, particularly siphonous green algae such as members of the genera *Caulerpa*, *Halimeda* or *Bryopsis* with their huge vacuolar system accumulate this cation. In contrast, other seaweeds with small vacuoles and high cytoplasmic fraction prefer to accumulate K^+ under salinity stress (Kirst 1990). In contrast to the toxic properties of Na^+ (and Cl^-), K^+ is fully compatible to metabolic activities, although an explanation for this fact is still missing (Maathuis and Amtmann 1999). Both cations exhibit similar physico-chemical properties, as the smaller Na^+ together with its rather large hydration shell mimics the size of K^+. Consequently, uptake systems for K^+ have difficulties discriminating between both ions, and high external Na^+ amounts may result in K^+ deficiency. Inside the cell, Na^+ can compete for K^+-binding sites of proteins, which contribute to their stabilization, resulting in the inhibition of K^+-dependent metabolic processes (Hagemann 2011). Therefore, all organisms tend to ensure a defined usually high K^+/Na^+ ratio in the cytoplasm (Maathuis and Amtmann 1999).

During osmotic adjustment, the changes in and control of ion composition are achieved by regulating the activity of specific transport systems, which are particularly well studied in cyanobacteria (Hagemann 2011) and probably act also in seaweeds. Ion concentrations in algae are mainly regulated by ion-selective carriers driven by the membrane potential (Gimmler 2000). In addition, facilitated diffusion via ion-selective channels may play a role during rapid changes and recovery of ionic composition (Kirst 1990).

The generally low cytoplasmic Na^+ concentrations observed in most seaweeds clearly indicate that active sodium ion export mechanisms exist in these plants. The complete genome sequence of the cyanobacterial strain *Synechocystis* 6803 showed six different genes, annotated as Na^+/H^+ antiporters (Kaneko et al. 1996), of which at least three were functioning (Inaba et al. 2001). The occurrence of rather large gene families for Na^+/H^+ antiporters clearly point to a group of proteins that fulfill many important functions in osmotic acclimation of cyanobacteria and probably of seaweeds as well. Various studies showed that proton-pumping activity can be directly switched on in salt-treated cyanobacteria due to increased respiratory activity or because of stimulated activities of specific ATPases (Wiangnon et al. 2007, Hagemann 2011). Other Na^+-export systems such as primary active sodium pumps or transporters might exist in seaweeds, but experimental proof is still lacking.

The much higher concentration of intracellular K^+ than in the external medium clearly points to an active uptake of this important cation from the seaweed exterior. In the genome sequences of cyanobacteria, various K^+ transporters and putative K^+ channels have been assigned as potential candidates for performing the uptake of this cation (Hagemann 2011), and more recently it was reported that all cyanobacteria possess the structural genes for a functional ATP-dependent K^+ transport system consisting of a K^+ permease, an ATPase that provides the energy, and a structural stabilisator (Ballal et al. 2007). Most interestingly, this complex K^+ transport system requires but does not transport Na^+ (Matsuda et al. 2004). Such a direct activation by the presence of Na^+ would nicely explain how the K^+ transport activity is rapidly enhanced under salt stress.

In contrast to Na^+ and K^+ transport mechanisms much less is known about Cl^- extrusion/uptake systems. Although Cl^- was involved in the osmotic acclimation of various eulittoral green algae from Antarctica, the concentration remained relatively low even under hypersaline treatments (Karsten et al. 1991b). This is in accordance with the related temperate green algae such as *Ulva prolifera* (as *Enteromorpha prolifera*) (Young et al. 1987). Except *Acrosiphonia arcta*, all other green algal species studied from Antarctica have typically small cells and a large cytoplasmic:vacuolar ratio (Karsten et al. 1991b). They resemble cytoplasm-rich microalgae, which also tend to maintain low Cl^- values in their cytoplasm (Dickson and Kirst 1986). Cl^- like Na^+ has adverse effects on many enzymes (Gimmler et al. 1984). In addition, ribosomes of plants are not functional in the presence of high Cl^- contents (Ritchie 1988).

In the green alga *Acetabularia* spp. a negative membrane potential was measured and interpreted as primarily caused by an electrogenic Cl^- pump (Wendler et al. 1983). Nevertheless, in eukaryotic algae and cyanobacteria Cl^- transport is still badly understood. In contrast, in some bacteria and other eukaryotes such as invertebrates and humans, various genes encoding Cl^- channels or Cl^-/H^+ exchangers have been described (Jentsch 2008). The underlying proteins assemble to dimers, with each monomer containing an ion translocation pathway (Jentsch 2008). From the available data on these organisms it is reasonable to assume similar Cl^- transport systems for seaweeds.

Although K^+, Na^+ and Cl^- represent the major inorganic ions involved in osmotic acclimation, some seaweeds such as *Laminaria digitata* use NO_3^- to satisfy osmotic and nitrogen requirements (Davison and Reed 1985a). However, there are strong seasonal changes in the cytoplasmic composition of major inorganic and organic osmolytes in this kelp. *Laminaria digitata* accumulates high nitrate concentrations in spring. During summer, this anion is completely metabolized, and the gap in the osmotic potential is filled through the biosynthesis and accumulation of the polyol mannitol. This seasonal increase in mannitol concentration compensates for the intracellular decrease in nitrate rather than for changes in external salinity (Davison and Reed 1985a).

In some green seaweeds such as *Codium decorticatum* (Bisson and Gutknecht 1975), *Acrosiphonia arcta* and *Ulva rigida* (Karsten et al. 1991b) SO_4^{2-} plays an important role as an osmolyte. Some brown seaweeds of the genus *Desmarestia* even contain this anion in the form of free sulfuric acid inside the vacuole resulting in a pH of 1–2 (Eppley and Bovell 1958; Anderson and Velimirov 1982).

The energy requirements for ion transport in seaweed species such as *Ulva lactuca* are equivalent to 10–30% of the energy provided by respiration (Ritchie 1988). It seems that all inorganic ion transporters in seaweeds are generally quite fast systems with much lower requirements for metabolic energy compared with the cost for biosynthesis or degradation of organic osmolytes (Kirst 1990). The formation of an intracellular osmotic potential of 1 osmol kg^{-1} requires 1.2 mol ATP when created by KCl uptake only, but 66 mol ATP when based on sorbitol or mannitol biosynthesis.

5.3.2 Organic Osmolytes

The main physiological strategy of all seaweeds studied so far under saline conditions is to keep particularly the Na^+ and Cl^- concentrations in the cytoplasm as low as possible. Since protein and organelle function (e.g., ribosomes, mitochondria), enzyme activity, membrane integrity and structural macromolecules in seaweeds are adversely affected by these ion concentrations (Kirst 1990), it is now generally accepted that the biosynthesis and accumulation of organic osmolytes in the cytoplasm, although energetically costly, permits the generation of low water potentials without incurring metabolic damage (Karsten et al. 1996a; Yancey 2005; Eggert and Karsten 2010).

In many, but not all, cases the organic osmolytes are identical with the main photosynthetic product, and thus there are preferences in different seaweed groups. Polyols such as mannitol are characteristic for most Phaeophyceae, while the Chlorophyta contain typically sucrose, some taxa in addition to proline, glycine-betaine, or dimethylsulphoniumpropionate (DMSP) (Kirst 1990). Although mannitol is present as the sole osmotically significant low-molecular weight organic solute in most brown seaweeds, some species are capable of synthesizing a second polyol (Davison and Reed 1985b). The top-shore Phaeophyceae *Pelvetia canaliculata* contains the heptitol volemitol as an additional intracellular organic solute (Kremer 1976), while the lower shore brown alga *Himanthalia elongata* from the Northern hemisphere as well as various Southern hemisphere Fucales (e.g., *Bifurcariopsis capensis*, *Hormosira banksii*, *Xiphophora chondrophylla*) accumulate the hexitol altritol as isomeric solute in addition to mannitol (Chudek et al. 1984; Wright et al. 1987).

In the Rhodophyta a multitude of low-molecular-weight carbohydrates have been identified in recent years (Eggert and Karsten 2010). The main photosynthetic product in members of most orders of red seaweeds is clearly the heteroside floridoside (α-D-galactopyranosyl-(1-2)-glycerol), while most members of the Ceramiales (Florideophyceae) generally synthesize and accumulate instead of floridoside, the chemically related digeneaside (α-D-mannopyranosyl-(1-2)-glycerate) (Kremer 1978).

Already Lindberg (1955) provided evidence that, in addition to floridoside, members of the Bangiales (Bangiophyceae) contain an isomeric form of floridoside, isofloridoside (α-D-galactopyranosyl-(1-1)-glycerol). Wickberg (1958) later reported isofloridoside in *Porphyra umbilicalis* as a uniform mixture of D- and L-forms. More recently, the chemical structures and configurations of all three heterosides from *Porphyra perforata* were investigated, and the occurrence of floridoside along with both D- and L-isofloridoside was verified (Meng et al. 1987), while the strong involvement of all these compounds in the process of osmotic acclimation was experimentally proven in *P. columbina* (Karsten et al. 1993a). More interesting is the observation of different heteroside patterns in *Porphyra* species from different biogeographic regions in Europe, Africa, North America, Asia, and Australia (Karsten 1999). The composition of the three compounds varied among the species studied. In *P. columbina* from Australia, L-isofloridoside was always quantitatively dominant, while floridoside was the major component in *P. dioica* from the North Sea. D-Isofloridoside was usually present in small concentrations, except in *P. perforata* from the Pacific coast of the USA where it occurred in equal concentrations along with floridoside and L-isofloridoside (Karsten 1999). These results point to species-specific different enzymatic activities of the underlying anabolic pathways, which are, however, not completely understood.

The consistent presence of digeneaside together with a new compound was noted in some members of the genus *Hypoglossum* (Delesseriaceae, Ceramiales), and a chemical survey in members of this taxon resulted in the identification of digalactosylglycerol (2,3-dihydroxypropyl (α-D-galactopyranosyl)-(1→6)-ß-D-galactopyranoside), which was strongly involved in osmotic acclimation (Karsten et al. 2005). Surprisingly, digalactosylglycerol has never been reported for any other seaweed species before, even though it most probably represents the main photosynthetic product in *Hypoglossum barbatum* and *H. heterocystideum* (Karsten et al. 2005).

The disaccharide trehalose was detected in several members of the Ceramiales. While some taxa of this order such as *Aglaothamnion* spp. exhibited only trehalose, others such as *Delesseria sanguinea* showed trehalose together with digeneaside. Although more recent data indicate that only a few Ceramialean taxa are capable of synthesizing trehalose (Karsten et al. 2007), the summarized results in the review of Craigie (1974) point to other Florideophycean species also forming this disaccharide. Therefore, this compound may be more widely distributed among the Rhodophyta than currently thought.

Most interesting is the fact that in some Ceramiales such as in the mangrove-associated genera *Bostrychia* and *Caloglossa* as well as in some early diverging red algal lineages the polyols mannitol, sorbitol, and dulcitol can be found (Karsten and

West 1993; Karsten et al. 1994, 1999, 2003b; Eggert and Karsten 2010), which are otherwise uncommon for red algae.

With the exception of digeneaside that plays no more than a minor role in osmotic acclimation of Rhodophyta (Karsten et al. 2005; Eggert and Karsten 2010), all other low-molecular-weight carbohydrates act as organic osmolytes. Since these organic compounds can be accumulated and tolerated at high intracellular concentrations, and permit the generation of low water potentials without incurring metabolic damage (Yancey 2005), the term "compatible solute" was introduced by Brown and Simpson (1972). In general, the intracellular concentrations of these organic osmolytes are actively adjusted by photosynthesis driven de novo biosynthesis or by remobilization of storage products and are directly proportional to external salinity (Kirst 1990).

Although differing in their chemical structure, compatible solutes in seaweeds have some features in common: they are typically highly soluble, in most cases have no net charges at physiological pH, and are non-inhibitory at high concentrations (Kirst 1990; Karsten et al. 1996a). The interactions of these organic compounds with intracellular macromolecules are not completely understood and several mechanisms have been suggested. Bisson and Kirst (1995) discussed different models to explain the protection of enzyme systems: (1) binding of the solute to the protein, (2) colligative action of the solute, (3) buffering of potentially damaging changes in solution properties, (4) inhibition of conformational changes resulting in formation of inter- or intramolecular disulfide bridges, and (5) preferential exclusion of the solute from the protein surface. These models can be basically summarized into two types: (1) those that hypothesize direct solute–protein interactions and (2) those that postulate that protein stability is mediated by solute-induced changes in water structure (Roberts 2005; Yancey 2005). However, there is little experimental evidence in seaweeds for any of these models.

Desiccation tolerant seaweeds such as supralittoral green algal *Prasiola* or brown algal *Pelvetia* species typically exhibit morphological adaptations such as thick cell walls that protect against rapid water loss and the capability to synthesize and accumulate polyols as "water-keeping" substances (Jacob et al. 1991, 1992).

5.3.3 Biosynthesis of Organic Osmolytes

Kremer and Kirst (1981) showed that exogenously applied inorganic ^{14}C is rapidly taken up by various red algae and assimilated into floridoside, which is thus acting as a major photosynthetic product. The biosynthesis of floridoside is initiated by a condensation reaction of l-glycerol-3-P and UDP-galactose resulting in floridoside-P. This reaction is mediated by a respective heteroside-P synthase (Kremer and Kirst 1981). Floridoside-P is subsequently de-phosphorylated by a specific phosphatase. In this anabolic pathway l-glycerol-3-P serves as precursor, and the condensation reaction takes place at the C-2 position of glycerol. In contrast, the biosynthesis of D- and L-isofloridoside in the Bangiales is unknown.

The biosynthesis of digeneaside in red algae is initiated by a condensation reaction of L-glycerate-P and UDP-mannose resulting in digeneaside-P. This reaction is mediated by digeneaside-P synthase, and followed by a specific digeneaside phosphatase which de-phosphorylates digeneaside-P (Kirst 1990).

All members of the mangrove red algal genus *Caloglossa* synthesize the polyol mannitol under salinity stress (Karsten and West 1993). The biosynthesis of mannitol was experimentally verified for the first time in *Caloglossa leprieurii* (Karsten et al. 1997). This species exhibits four enzymes that control the size of the mannitol pool. Mannitol-1-P dehydrogenase (Mt1PDH; EC 1.1.1.17) reduces fructose-6-P to mannitol-1-P. Mannitol-1-phosphatase (Mt1Pase; EC 3.1.3.22) subsequently dephosphorylates mannitol-1-P and releases mannitol in the anabolic pathway. The catabolic pathway includes the conversion of mannitol to fructose by mannitol dehydrogenase (MtDH; EC 1.1.1.67) and further to fructose-6-P by hexokinase (HK; EC 2.7.1.1.). Both pathways involved in mannitol metabolism are known as the so-called mannitol cycle (Karsten et al. 1997).

A current study on the key enzyme mannitol-1-P dehydrogenase in the brown alga *Ectocarpus siliculosus* took advantage of the recently published genome (Cock et al. 2010; Rousvoal et al. 2011). Applying a biochemical and for the first time a genomic approach, the latter authors documented a salt-induced gene expression and upregulation for this enzyme, i.e., hypersaline conditions stimulated the formation of mannitol. Similarly, Iwamoto et al. (2003) documented the biochemical and kinetic properties of purified mannitol-1-P dehydrogenase from *Caloglossa continua*, and also reported strong enzyme regulation by salinity.

In contrast to the mannitol metabolism, the biosynthesis of sorbitol and dulcitol as well as of trehalose in red seaweeds is unstudied. Nevertheless, a trehalose-6-phosphate synthase gene was recently screened out from a large DNA fragment library constructed from *Porphyra yezoensis* (Dai et al. 2004), indicating the genotypic presence of a trehalose biosynthesis key enzyme. This example strongly supports the usefulness of various recent genomic projects on different seaweed taxa, which will give a deeper look into the molecular mechanisms of biosynthesis and regulation of organic osmolytes.

5.3.4 Antioxidants

Since seaweeds perform oxygenic photosynthesis using water as an electron donor they steadily release molecular oxygen, which can be accumulated and easily chemically converted to potentially damaging reactive oxygen species (ROS). The sources and production sites of ROS are mainly related to photosynthetic activities, such as pseudocyclic photophosphorylation and the Mehler reaction, which stimulate the accumulation of hydrogen peroxide (Asada 1994; see Chap. 6 by Bischof and Rautenberger). Besides these internal processes, formation of ROS might also be induced under hypersaline conditions as reported for *Ulva fasciata* (Sung et al. 2009). These authors undertook a gene expression study and documented a salt-induced upregulation of various antioxidant enzymes such as superoxide dismutase, ascorbate peroxidase, glutathione reductase, and catalase, which efficiently detoxified ROS.

5.3.5 Ultrastructural Changes

In many seaweeds, the separation of the plasmalemma from the cell wall rather than cell volume changes under hypersaline conditions is the most damaging aspect of plasmolysis (Reed 1990). However, in some species such as the red alga *Caloglossa leprieurii*, the elastic cell walls accommodated marked shrinkage of the protoplasm when treated with hypersaline media resulting in a concomitant swelling and hence prevention of plasmolysis (Fig. 5.2, Mostaert and King 1993). These authors suggested that the enormous flexibility of the cell wall thickness may be a protective mechanism in response to fluctuating salinities.

Besides the adverse ionic influences on particular sites of the photosynthetic machinery, the cellular ultrastructure may also be affected by salt stress resulting in reduced photosynthesis and respiration. In *Porphyra umbilicalis* changes in the thylakoid structure of the chloroplasts after salt treatment have been described (Wiencke 1982). Hypersaline conditions typically result in cell shrinkage due to water loss with thylakoids and membranes appressed, and under hyposaline conditions cellular organelles are typically swollen. This disturbance of fine structure may cause a disruption of energy transfer in photosynthetic electron flow (Satoh et al. 1983). In contrast to chloroplasts, mitochondrial cristae ultrastructure is less affected by salinity changes (Kirst 1990), and this well explains the better functioning of respiration compared to photosynthesis under the respective stress.

Another ultrastructural observation is the formation of vacuoles after hypo- or hypersaline stress in seaweeds such as *Porphyra umbilicalis*, which under normal conditions contain only very few of these organelles (Wiencke and Läuchli 1980). The vacuoles formed under hypersaline conditions may serve as compartments to sequester metabolically toxic ions, mainly Na^+ and Cl^- (Wiencke et al. 1983). Concomitantly, the fine structure of the tonoplast changes is probably related to the intensity of ion transport across it (Wiencke and Läuchli 1983).

5.3.6 Self Protection

In addition to active processes, which compensate for osmotic stress, *Laminaria* blades form multiple-layered, mat-like canopies at neap tides, protecting the individual thalli against desiccation and salinity changes (Lüning 1990). Macroalgal canopies of *Ulva* sp. in the upper littoral zone in southern Spain form under emersed conditions sheet-like, multiple-layered structures in which the top layer usually bleaches due to strong insolation, desiccation, and other abiotic stresses, thereby providing photoprotection and moisture for subcanopy thalli (Bischof et al. 2002).

5.4 Salinity Ecotypes

Habitats with strong horizontal salinity gradients such as estuaries or the Baltic Sea promote genetic adaptation and eventually speciation. Indeed, the geographic semi-isolation from the Atlantic and the steep horizontal salinity gradient of the Baltic Sea have caused local genetic adaptation in several seaweeds and contributed to the formation of so-called ecotypes. Ecotypes (*sensu* Turesson, 1922) are genetically distinct populations of a species that are locally adapted to particular environmental conditions. The ability of populations to adapt to a novel habitat is a key component of the speciation process. Ecotypes are recognized as physiologically distinct variants of a species. The concept of ecotypes is widely accepted in relation to ecophysiological diversity in seaweeds when considering salinity stress responses. Several Baltic seaweeds were physiologically and morphologically compared with their marine counterparts and a number of salinity ecotypes have been described such as *Ceramium strictum* (Rueness and Kornfeldt 1992), *Delesseria sanguinea* and *Membranoptera alata* (Rietema 1993), and *Fucus vesiculosus* (Nygard and Dring 2008). But data from other regions also clearly point to strong ecotypic differentiation between populations of the same seaweed species such as estuarine and marine isolates of *Pylaiella littoralis* (Bolton 1979), *Polysiphonia lanosa* (Reed 1984), *Bostrychia radicans*, and *B. moritziana* (Karsten et al. 1993b, 1994). Although the kelp genus *Laminaria* is stenohaline (Bartsch et al. 2008), ecotypic differentiation in terms of growth under different salinities has been reported in North Atlantic populations of *Saccharina latissima* originating from Long Island Sound, New York, and Cape Neddick, Maine (Gerard et al. 1987), and might be considered as a mechanism to adapt to locally environmentally unfavorable conditions. However, in most seaweeds the underlying molecular mechanisms of ecotypic differentiation are unstudied.

One exception is *Bostrychia tenuissima*, which is restricted to Southern Australia and New Zealand. Previous studies have revealed two distinct patterns in the presence of osmotically active polyols. Southern Australian populations only have sorbitol whereas northern Australian populations have in addition to sorbitol also dulcitol. These polyol patterns led to speculation on ecotypic differentiation in both population types (Karsten et al. 1995). Using molecular approaches, a 100% congruence was found between polyol patterns and three plastid haplotypes observed among all isolates studied, which experimentally proved for the first time a genetic basis for different ecotypes of the same seaweed species (Zuccarello et al. 1999).

5.5 Outlook

Since the review of Kirst (1990) on salinity tolerance in algae not much scientific progress has been made in seaweeds. In strong contrast, in cyanobacteria many molecular mechanisms involved in osmotic acclimation have been successfully addressed (Hagemann 2011). The main reason for this huge gap in knowledge between both organism groups is related to available genome information. Without genomic data or other modern approaches such as metabolomics or proteomics, it is still difficult to evaluate molecular mechanisms, such as ion transport across membranes, biosynthesis of organic osmolytes, gene expression, and regulation. Therefore, the establishment of more genomes of model seaweeds is urgently needed to get a fundamental understanding of salinity and desiccation stress responses. A first step has been recently taken with the brown alga *Ectocarpus siliculosus* (Cock et al. 2010), so that deeper insight into the molecular biology of osmotic acclimation can be expected in the near future.

Acknowledgments The author likes to thank many of his colleagues for excellent collaboration on salinity stress in seaweeds over the last two decades, particularly Robert King, Gunter Kirst, Annika Mostaert, John West, and Christian Wiencke, as well as the Deutsche Forschungsgemeinschaft for funding various projects.

References

Allakhverdiev SI, Murata N (2008) Salt stress inhibits photosystems II and I in cyanobacteria. Photosynth Res 98:529–539

Anderson RJ, Velimirov B (1982) An experimental investigation of the palatability of kelp bed algae to the sea urchin *Parechinus angulosus* Leske. P.S.Z.N.I. Mar Ecol 3:357–373

Asada K (1994) Production and action of active oxygen species in photosynthetic tissues. In: Foyer CH, Mullineaux PM (eds) Causes of photooxidative stress and amelioration of defence systems in plants. CRC, Boca Raton, pp 77–104

Ballal A, Basu B, Apte SK (2007) The Kdp-ATPase system and its regulation. J Biosci 32:559–568

Bartsch I, Wiencke C, Bischof K, Buchholz CM, Buck BH, Eggert A, Feuerpfeil P, Hanelt D, Jacobsen S, Karez R, Karsten U, Molis M, Roleda MY, Schubert H, Schumann R, Valentin K, Weinberger F, Wiese J (2008) The genus *Laminaria* sensu lato: recent insights and developments. Eur J Phycol 43:1–86

Bird CJ, McLachlan J (1986) The effect of salinity on distribution of species of *Gracilaria* (Rhodophyta, Gigartinales): an experimental assessment. Bot Mar 29:231–238

Bird NL, Chen LCM, McLachlan J (1979) Effects of temperature, light and salinity on growth in culture of *Chondrus crispus*, *Furcellaria lumbricalis*, *Gracilaria tikvahiae* (Gigartinales, Rhodophyta), and *Fucus serratus* (Fucales, Phaeophyta). Bot Mar 22:1–27

Bischof K, Peralta G, Kräbs G, van den Poll WH, Perez-Llorens JL, Breeman AM (2002) Effects of solar UV-B radiation on canopy structure of *Ulva* communities from southern Spain. J Exp Bot 379:2411–2421

Bisson MA, Gutknecht J (1975) Osmotic regulation in the marine alga, *Codium decorticatum*. I. Regulation of turgor pressure by control of ionic composition. J Membr Biol 37:85–98

Bisson MA, Kirst GO (1983) Osmotic adaptations of charophyte algae in the Coorong, South Australia and other Australian lakes. Hydrobiology 105:45–51

Bisson MA, Kirst GO (1995) Osmotic acclimation and turgor pressure regulation in algae. Naturwissenschaften 82:461–471

Bolton JJ (1979) Estuarine adaptation in populations of *Pilayella littoralis* (L.) Kjellm. (Phaeophyta, Ectocarpales). Estuar Coast Mar Sci 9:273–280

Brown AD, Simpson JR (1972) Water relations of sugar-tolerant yeasts: the role of intracellular polyols. J Gen Microb 72:589–591

Choi TS, Kang EJ, Kim JH, Kim KY (2010) Effect of salinity on growth and nutrient uptake of *Ulva pertusa* (Chlorophyta) from an eelgrass bed. Algae 25:17–26

Chudek JA, Foster R, Davison IR, Reed RH (1984) Altritol in the brown alga *Himanthalia elongata*. Phytochemistry 23:1081–1082

Cock JM, Sterck L, Rouze P, Scornet D, Allen AA, Amoutzias G, Anthourad V, Artiguenave F, Aury JM, Badger JH et al (2010) The *Ectocarpus* genome and the independent evolution of multicellularity in the brown algae. Nature 465:617–621

Craigie JS (1974) Storage products. In: Stewart WDP (ed) Algal physiology and biochemistry. Blackwell, Oxford, pp 206–235

Dai YD, Zhao G, Jin SX, Ji LY, De LD, Man LW, Wang B (2004) Construction and characterization of a bacterial artificial chromosome library of marine macroalga *Porphyra yezoensis* (Rhodophyta). Plant Mol Biol Rep 22:375–386

Davison IR, Reed RH (1985a) Osmotic adjustment in *Laminaria digitata* (Phaeophyta) with particular reference to seasonal changes in internal solute concentrations. J Phycol 21:41–50

Davison IR, Reed RH (1985b) The physiological significance of mannitol accumulation in brown algae: the role of mannitol as a compatible cytoplasmic solute. Phycologia 24:449–457

Dickson DMJ, Kirst GO (1986) The role of dimethylsulphoniopropionate, glycine betaine and homarine in the osmoacclimation of *Platymonas subcordiformis*. Planta 155:409–415

Druehl LD (1967) Distribution of two species of *Laminaria* as related to environmental factors. J Phycol 3:103–108

Eggert A, Karsten U (2010) Low molecular weight carbohydrates in red algae – an ecophysiological and biochemical perspective. In: Seckbach J, Chapman D, Weber A (eds) Cellular origins, life in extreme habitats and astrobiology red algae in the genomics age. Springer, Berlin, pp 445–456

Eppley RW, Bovell CR (1958) Sulphuric acid in *Desmarestia*. Biol Bull 115:101–106

Fork DC, Öquist G (1981) The effects of desiccation on excitation energy transfer at physiological temperatures between the two photosystems of the red alga *Porphyra perforata*. Z Pflanzenphysiol 104:385–393

Gerard VA, Dubois K, Greene R (1987) Growth responses of two *Laminaria saccharina* populations to environmental variation. Hydrobiology 151(152):229–232

Gessner F, Schramm W (1971) Salinity: plants. In: Kinne O (ed) Marine ecology, vol 1(2), Environmental factors. Wiley Interscience, London, pp 705–1083

Gimmler H (2000) Primary sodium plasma membrane ATPases in salttolerant algae: facts and fictions. J Exp Bot 348:1171–1178

Gimmler H, Kaaden R, Kirchner U, Weyand A (1984) The chloride sensitivity of *Dunaliella parva* enzymes. Z Pflanzenphysiol 114:131–50

Gustavs L, Schumann R, Eggert A, Karsten U (2009) In vivo growth fluorometry: accuracy and limits of microalgal growth rate measurements in ecophysiological investigations. Aquat Microb Ecol 55:95–104

Gutknecht J, Hastings DF, Bisson MA (1978) Ion transport and turgor pressure regulation in giant algal cells. In: Giebisch G, Tosteson DC, Ussing HH (eds) Membrane transport in biology, vol 3, Transport across biological membranes. Springer, Berlin, pp 25–74

Hagemann M (2011) Molecular biology of cyanobacterial salt acclimation. FEMS Microbiol Rev 35:87–123

Hanelt D, Tüg H, Bischof K, Groß C, Lippert H, Sawall T, Wiencke C (2001) Light regime in an Arctic fjord: a study related to stratospheric ozone depletion as a basis for determination of UV effects on algal growth. Mar Biol 138:649–658

Inaba M, Sakamoto A, Murata N (2001) Functional expression in *Escherichia coli* of low-affinity and high-affinity $Na^+(Li^+)/H^+$ antiporters of *Synechocystis*. J Bacteriol 183:1376–1384

Iwamoto K, Kawanobe H, Ikawa T, Shiraiwa Y (2003) Characterization of salt-regulated mannitol-1-phosphate dehydrogenase in the red alga *Caloglossa continua*. Plant Physiol 133:893–900

Jacob A, Kirst GO, Wiencke C, Lehmann H (1991) Physiological responses of the Antarctic green alga *Prasiola crispa* ssp. *antarctica* to salinity stress. J Plant Physiol 139:57–62

Jacob A, Wiencke C, Lehmann H, Kirst GO (1992) Physiology and ultrastructure in the green alga *Prasiola crispa* from Antarctica. Bot Mar 35:297–303

Jentsch TJ (2008) CLC chloride channels and transporters: from genes to protein structure, pathology and physiology. Crit Rev Biochem Mol 43:3–36

Kain JM, Norton TA (1990) Marine ecology. In: Cole KM, Sheath RG (eds) Biology of the red algae. Cambridge University Press, Cambridge, pp 377–422

Kaneko T, Sato S, Kotani H et al (1996) Sequence analysis of the genome of the unicellular cyanobacterium *Synechocystis* sp. strain PCC 6803. II. Sequence determination of the entire genome and assignment of potential protein-coding regions. DNA Res 3:109–136

Karsten U (1999) Seasonal variation in heteroside concentrations of field-collected *Porphyra* species (Rhodophyta) from different biogeographic regions. New Phytol 143:561–571

Karsten U, West JA (1993) Ecophysiological studies on six species of the mangrove red algal genus *Caloglossa*. Aust J Plant Physiol 20:729–739

Karsten U, Wiencke C, Kirst GO (1991a) The effect of salinity changes upon the physiology of eulittoral green macroalgae from Antarctica and Southern Chile. I. Cell viability, growth, photosynthesis and dark respiration. J Plant Physiol 138:667–673

Karsten U, Wiencke C, Kirst GO (1991b) The effect of salinity changes upon the physiology of eulittoral green macroalgae from Antarctica and Southern Chile. II. Intracellular inorganic ions and organic compounds. J Exp Bot 245:1533–1539

Karsten U, Barrow KD, King RJ (1993a) Floridoside, L-isofloridoside, and D- isofloridoside in the red alga *Porphyra columbina*. Plant Physiol 103:485–491

Karsten U, West JA, Ganesan EK (1993b) Comparative physiological ecology of *Bostrychia moritziana* (Ceramiales, Rhodophyta) from freshwater and marine habitats. Phycologia 32:401–409

Karsten U, West JA, Zuccarello G, Kirst GO (1994) Physiological ecotypes in the marine red alga *Bostrychia radicans* (Ceramiales, Rhodophyta) from the east coast of the USA. J Phycol 30:174–182

Karsten U, Bock C, West JA (1995) Low molecular weight carbohydrate patterns in geographically different isolates of the eulittoral red alga *Bostrychia tenuissima* from Australia. Bot Acta 108:321–326

Karsten U, Barrow KD, Nixdorf O, King RJ (1996a) The compability of unusual organic osmolytes from mangrove red algae with enzyme activity. Aust J Plant Physiol 23:577–582

Karsten U, Koch S, West JA, Kirst GO (1996b) Physiological responses of the eulittoral macroalga *Stictosiphonia hookeri* (Rhodomelaceae, Rhodophyta) from Argentina and Chile: salinity, light and temperature acclimation. Eur J Phycol 31:361–368

Karsten U, Barrow KD, Nixdorf O, West JA, King RJ (1997) Characterization of mannitol metabolism in the mangrove red alga *Caloglossa leprieurii* (Montagne) J. Agardh Planta 201:173–178

Karsten U, West JA, Zuccarello GC, Nixdorf O, Barrow KD, King RJ (1999) Low molecular weight carbohydrate patterns in the Bangiophyceae (Rhodophyta). J Phycol 35:967–976

Karsten U, Dummermuth A, Hoyer K, Wiencke C (2003a) Interactive effects of ultraviolet radiation and salinity on the ecophysiology of two Arctic red algae from shallow waters. Polar Biol 26:249–258

Karsten U, West JA, Zuccarello GC, Engbrodt R, Yokoyama A, Hara Y, Brodie J (2003b) Low molecular weight carbohydrates of the Bangiophycidae (Rhodophyta). J Phycol 39:584–589

Karsten U, Michalik D, Michalik M, West JA (2005) A new unusual low molecular weight carbohydrate in the red algal genus *Hypoglossum* (Delesseriaceae, Ceramiales) and its possible function as osmolyte. Planta 222:319–326

Karsten U, Görs S, Eggert A, West JA (2007) Trehalose, digeneaside and floridoside in the Florideophyceae (Rhodophyta) – a re-evaluation of its chemotaxonomic value. Phycologia 46:143–150

Kauss H (1987) Some aspects of calcium-dependent regulation in plant metabolism. Annu Rev Plant Physiol 38:47–72

Kirst GO (1990) Salinity tolerance of eukaryotic marine algae. Annu Rev Plant Physiol Plant Mol Biol 41:21–53

Kremer BP (1976) Distribution and biochemistry of alditols in the genus *Pelvetia* (Phaeophyceae, Fucales). Br Phycol J 11:239–243

Kremer BP (1978) Patterns of photoassimilatory products in Pacific Rhodophyceae. Can J Bot 56:1655–1659

Kremer BP, Kirst GO (1981) Biosynthesis of 2-O-D-glycerol-D-galactopyranoside (floridoside) in marine Rhodophyceae. Plant Sci Lett 23:349–357

Lindberg B (1955) Low-molecular carbohydrates in algae. XI. Investigation of *Porphyra umbilicalis*. Acta Chem Scand 9:1097–1099

Lüning K (1990) Seaweeds: their environment, biogeography, and ecophysiology. Wiley, New York

Maathuis FJM, Amtmann A (1999) K^+ nutrition and Na^+ toxicity: the basis of cellular K^+/Na^+ ratios. Ann Bot 84:123–133

Matsuda N, Kobayashi H, Katoh H, Ogawa T, Futatsugi L, Nakamura T, Bakker EP, Uozumi N (2004) Na^+-dependent K^+ uptake Ktr system from the cyanobacterium *Synechocystis* sp. PCC 6803 and its role in the early phases of cell adaptation to hyperosmotic shock. J Biol Chem 279:54952–54962

Meng J, Rosell KG, Srivastava LM (1987) Chemical characterization of floridosides from *Porphyra perforata*. Carbohyd Res 161:171–180

Mostaert AS, King RJ (1993) The cell wall of the halotolerant red alga *Caloglossa leprieurii* (Montagne) J. agardh (Ceramiales, Rhodophyta) from freshwater and marine habitats: effect of changing salinity. Crypt Bot 4:40–46

Nygard CA, Dring MJ (2008) Influence of salinity, temperature and dissolved inorganic carbon and nutrient concentration on the photosynthesis and growth on *Fucus vesiculosus* from Baltic and Irish Seas. Eur J Phycol 43:253–262

Reed RH (1983) The osmotic response of *Polysiphonia lanosa* from marine and estuarine sites: evidence for incomplete recovery of turgor. J Exp Mar Biol Ecol 68:69–93

Reed RH (1984) The effects of extreme hyposaline stress upon *Polysiphonia lanosa* (L.) Tandy from marine and estuarine sites. J Exp Mar Biol Ecol 76:131–144

Reed RH (1990) Solute accumulation and osmotic adjustment. In: Cole KM, Sheath RG (eds) Biology of the red algae. Cambridge University Press, Cambdrige, pp 147–170

Remane A (1940) Einführung in die zoologische Ökologie der Nord- und Ostsee. In: Grimpe G, Wagler E (eds) Die Tierwelt der Nord- und Ostsee, 1a. – Akademische Verlagsgesellschaft Leipzig, pp 1–238

Rietema H (1993) Ecotypic differences between baltic and north-sea populations of *Delesseria sanguinea* and *Membranoptera alata*. Bot Mar 36:15–21

Ritchie RJ (1988) The ionic relations of *Ulva lactuca*. J Plant Physiol 33:183–192

Roberts MF (2005) Organic compatible solutes of halotolerant and halophilic microorganisms. Saline Systems 1:5. doi:10.1186/1746-1448-1-5

Rousvoal S, Groisillier A, Dittami SM, Michel G, Boyen C, Tonon T (2011) Mannitol-1-P dehydrogenase activity in *Ectocarpus siliculosus*, a key role for mannitol synthesis in brown algae. Planta 233:261–273

Rueness J, Kornfeldt RA (1992) Ecotypic differentiation in salinity responses of *Ceramium strictum* (Rhodophyta) from Scandinavian waters. Sarsia 77:207–212

Russell G (1985) Recent evolutionary changes in the algae of the Baltic Sea. Br Phycol J 20:87–104

Russell G (1987) Salinity and seaweed vegetation. In: Crawford RMM (ed) The physiological ecology of amphibious and intertidal plants. Blackwell, Oxford, pp 35–52

Ryther JH, Menzel DW, Corwin N (1967) Influence of the Amazon river outflow on the ecology of the western tropical Atlantic. I. Hydrography and nutrient chemistry. J Mar Res 25:69–83

Satoh K, Smith CM, Fork DC (1983) Effects of salinity on primary processes of photosynthesis in the red alga *Porphyra perforata*. Plant Physiol 73:643–647

Schramm W, Nienhuis PH (1996) Marine benthic vegetation. Recent changes and the effects of eutrophication. Ecol Stud Anal Synth 123:1–470

Smith CM, Satoh K, Fork DC (1986) The effects of osmotic tissue dehydration and air drying in morphology and energy transfer in two species of *Porphyra*. Plant Physiol 80:843–847

Sung MS, Hsu YT, Hsu YT, Wu TM, Lee TM (2009) Hypersalinity and hydrogen peroxide upregulation of gene expression of antioxidant enzymes in *Ulva fasciata* against oxidative stress. Mar Biotechnol 11:199–209

Tazawa M, Shimmen T, Mimura T (1987) Membrane control in the Characeae. Annu Rev Plant Physiol 38:95–117

Thomas DN, Kirst GO (1991) Salt tolerance of *Ectocarpus siliculosus* (Dillw.) Lyngb.: comparison of gametophytes, sporophytes and isolates of different geographic origin. Bot Acta 104:26–36

Thomas DN, Collins JC, Russell G (1988) Interactive effects of temperature and salinity upon net photosynthesis of *Cladophora glomerata* and *C. rupestris*. Bot Mar 31:73–77

Törnroth-Horsefield S, Wang Y, Hedfalk K, Johanson U, Karlsson M, Tajkhorshid E, Neutze R, Kjellbom P (2006) Structural mechanism of plant aquaporin gating. Nature 439:688–694

Turesson G (1922) The genotypical response of the plant species to the habitat. Hereditas 3:211–350

Verret F, Wheeler G, Taylor AR, Farnham G, Brownlee C (2010) Calcium channels in photosynthetic eukaryotes: implications for evolution of calcium-based signalling. New Phytol 187:23–43

Wendler S, Zimmermann U, Bentrup FW (1983) Relationship between cell turgor pressure, electrical membrane potential, and chloride efflux in *Acetabularia mediterranea*. J Membr Biol 72:75–84

Wiangnon K, Raksajit W, Incharoensakdi A (2007) Presence of a Na^+-stimulated P-type ATPase in the plasma membrane of the alkaliphilic halotolerant cyanobacterium *Aphanothece halophytica*. FEMS Microbiol Lett 270:139–145

Wickberg B (1958) Synthesis of 1-glyceritol D-galactopyranosides. Acta Chem Scand 12:1187–1201

Wiencke C (1982) Effect of osmotic stress on thylakoid fine structure in *Porphyra umbilicalis*. Protoplasma 111:215–220

Wiencke C, Läuchli A (1980) Growth, cell volume, and fine structure of *Porphyra umbilicalis* in relation to osmotic tolerance. Planta 150:303–311

Wiencke C, Läuchli A (1983) Tonoplast fine structure and osmotic regulation in *Porphyra umbilicalis*. Planta 159:342–346

Wiencke C, Stelzer R, Läuchli A (1983) Ion compartmentation in *Porphyra umbilicalis* determined by electron-probe X-ray microanalysis. Planta 159:336–341

Wiltens J, Schreiber U, Vidaver W (1978) Chlorophyll fluorescence induction: an indicator of photosynthetic activity in marine algae undergoing desiccation. Can J Bot 56:2787–2794

Wright PJ, Clayton MN, Chudek JA, Foster R, Reed RH (1987) Low molecular weight carbohydrates in marine brown algae from the southern hemisphere: the occurrence of altritol in *Bifurcariopsis capensis*, *Hormosira banksii*, *Notheia anomala* and *Xiphophora chondrophylla*. Phycologia 26:429–434

Wright DG, Pawlowicz R, McDougall TJ, Feistel R, Marion GM (2010) Absolute salinity, "density salinity" and the reference-composition salinity scale: present and future use in the seawater standard TEOS-10. Ocean Sci Discuss 7:1559–1625

Yancey PH (2005) Organic osmolytes as compatible, metabolic and counteracting cytoprotectants in high osmolarity and other stresses. J Exp Biol 208:2819–2830

Young AJ, Collins JC, Russell G (1987) Solute regulation in the euryhaline marine alga *Enteromorpha prolifera* (O.F. Mull) J. Ag. J Exp Bot 38:1298–1308

Zimmermann U, Steudle E (1978) Physical aspects of water relations of plant cells. Adv Bot Res 6:45–117

Zuccarello GC, West JA, Karsten U, King RJ (1999) Molecular relationships within *Bostrychia tenuissima* (Rhodomelaceae, Rhodophyta). Phycol Res 47:81–85

Chapter 6
Seaweed Responses to Environmental Stress: Reactive Oxygen and Antioxidative Strategies

Kai Bischof and Ralf Rautenberger

6.1 Introduction

In an oxygenated environment virtually all metabolic processes involving the transfer of electrons bear the potential hazard of oxygen radical formation. In biological systems the photosynthetic and respiratory electron transport chains (ETCs) located in the thylakoid membranes of chloroplasts and in the inner mitochondrial membrane, respectively, are prominent sites for the evolvement of oxidative stress because of the omnipresence of molecular oxygen (O_2). Reactive oxygen species (ROS) are generally considered as intermediates of O_2 with an unpaired electron; this applies to the superoxide radical ($\bullet O_2^-$), singlet oxygen ($^1O_2$), and the hydroxyl radical ($\bullet OH$). Additionally, due to its high reactivity resulting in major oxidative functions, hydrogen peroxide (H_2O_2) is commonly also included in the group of ROS.

Oxidative stress is considered to be a major component in the cellular stress response of any organism exposed to any kind of environmental stress. As in particular the photosynthetic ETC, and here especially the pseudo-cyclic electron flow (also referred to as "Mehler reaction" or "photoreduction of oxygen") passing electrons from PSI to molecular oxygen, is one of the most important sources for the formation of oxygen radicals, the ROS evolving pathways in plants and algae have been studied quite intensively (Asada and Takahashi 1987; Polle 1996; Ledford and Niyogi 2005). The Mehler reaction is typically active as a photoprotective

---

K. Bischof (✉)
Department of Marine Botany, University of Bremen, Leobener Str. NW2, 28359 Bremen, Germany
e-mail: kbischof@uni-bremen.de

R. Rautenberger
Department of Marine Botany, Bremen Marine Ecology, University of Otago, 464 Great King Street, Dunedin 9016, New Zealand
e-mail: ralf.rautenberger@otago.ac.nz

C. Wiencke and K. Bischof (eds.), *Seaweed Biology*, Ecological Studies 219,
DOI 10.1007/978-3-642-28451-9_6,

mechanism under conditions of restricted electron drainage, resultant from e.g., saturated photosynthetic secondary reactions (i.e., Calvin–Benson cycle) and accumulation of reduced NADP (NADPH + H^+). These conditions are typically induced under excessive irradiance conditions. Nowadays, it is common knowledge in plant sciences that a sudden exposure to high irradiance of photosynthetically active radiation (PAR: 400–700 nm) may impair or even destruct the photosynthetic apparatus, particularly in low light adapted plants (Andersson et al. 1992; Osmond 1994). Excessive radiation conditions may thus result in an inhibition of photosynthesis, called photoinhibition (Aro et al. 1993; Osmond 1994; Hanelt 1996; Franklin et al. 2003; see Chap. 1 by Hanelt and Figueroa). Furthermore, studies on intertidal seaweeds have shown that besides excessive radiation most other environmental factors leading to physiological stress, like e.g., freezing, desiccation, hypo- and hypersalinity, heavy metals, as well as wounding may accelerate the cellular production of ROS (Benet et al. 1994; Collén and Davison 1999a, b; Burritt et al. 2002; Lu et al. 2006; Pereira et al. 2009; Wu et al. 2009a, b; see also Chap. 5 by Karsten). In general, if accumulation of ROS exceeds the scavenging capacity of enzymatic and nonenzymatic antioxidant systems, remaining non-detoxified ROS may inhibit photosynthesis and become auto-destructive to cells due to the oxidation of lipids, proteins, and nucleic acids. Consequently, the ability to control oxidative stress, i.e., to limit the generation of ROS effectively, is an important feature of organisms inhabiting variable environments. It is, thus, of vital importance to e.g., intertidal macroalgae to suppress or remove stress-induced ROS in order to thrive in their challenging habitat [see Davison and Pearson (1996) for review]. Particularly in the intertidal, macroalgae are subjected to a highly variable environment, which requires permanent adjustments of metabolic rates and protective strategies to keep oxidative stress at minimum, nondestructive levels. Most intertidal macroalgae are exposed to large fluctuation in irradiance levels, i.e., when low tide coincides with high solar radiation conditions around noon (Davison and Pearson 1996). Furthermore, drastic changes in (air and water) temperatures, salinity, desiccation, as well as mechanical forcings represent significant stress factors to intertidal macroalgae with implications to the respective reactive oxygen metabolism (Collén and Davison 1999a; Burritt et al. 2002; Sung et al. 2009).

Besides its major role in stress responses and as mediator of cellular damage, ROS also have been identified forming an important component in signal transduction pathways (Mackerness et al. 2001) and even interspecific interactions. The function in pathogen defence or fouling control via oxidative bursts has been proven by the light-independent release of H_2O_2 triggered by membrane-bound NADPH oxidases (Potin 2008). The present contribution, however, focuses on the stress-mediated induction of ROS, forced by the respective abiotic environment and highlights adaptive and acclimatory traits in macroalgae to cope with oxidative stress in particular with respect to superoxide dismutase (SOD) activity. For extended information on the significance of reactive oxygen to seaweed physiology and biotic interactions, the reader is referred to the recent and very comprehensive reviews by Dring (2005) and Potin (2008).

6.2 Stress-Induced ROS Production and Its Effects on Seaweed Performance

Being photoautotrophic organisms conducting oxygenic photosynthesis, macroalgae display the same major generation sites for oxygen radicals as identified for higher plants. Any malfunction of photosynthesis triggered by an unfavorable environment will finally be reflected by an increase in the level of internally generated ROS. As photosynthetic primary reactions are directly related to the generation of oxygen radicals, hampered dissipation of excessive PAR has been identified as an important inducer of ROS formation in plants (Asada and Takahashi 1987; Asada 1999; Bischof et al. 2002, 2003; see also Chap. 1 by Hanelt and Figueroa). Excessively absorbed light energy may result in an over-excitation of chlorophyll *a* molecules present in the photosynthetic reaction centers, which may then pass into the so-called triplet state. The critical characteristic of this process is that triplet chlorophyll (^{3}Chl) may subsequently promote the formation of $^1O_2$, which is highly reactive with any kind of cellular component (see Ledford and Niyogi 2005). Higher plants as well as algae try to avoid formation of ^{3}Chl by dissipation of excess radiation energy by the xanthophyll cycle, in which energy is dissipated harmlessly as heat by the interconversion of specialized xanthophylls located in the antennae (see Jahns et al. 2009). In addition, ^{3}Chl may be deactivated by carotenoid associated to the photosynthetic reaction centers (β-carotene and lutein).

A second and quantitatively by far the most important source of ROS formation in photosynthesis is the formation of $O_2^{\bullet-}$ in the process of Mehler reaction or photoreduction of oxygen: At a high reduction state of ferredoxin (e.g., under high light conditions, or in case electrons are not drained off at sufficient pace under varying stress conditions), electrons coming from PSI might be transferred to O_2 rather than to ferredoxin, which generates $O_2^{\bullet-}$ (Asada 1999). This ROS may subsequently reduce metal ions like Fe^{3+} and Cu^{2+} univalently. The $O_2^{\bullet-}$, however, is quenched via the enzyme SOD, yielding O_2 and H_2O_2. The latter, highly reactive compound may also either react with the formerly reduced metal ions ("Fenton reaction": Fe^{2+} and Cu^+) or with another $O_2^{\bullet-}$ ("Haber–Weiss reaction"). In both cases, the extremely reactive •OH is formed (see Dring 2005). This aggressive compound initiates free radical cascades and may denature proteins, nucleic acids, and peroxidise lipids, potentially resulting in biochemical dysfunctions and structural damages such as membrane leakage. As plants do not have specific protective agents against •OH, the only means of protection is to scavenge both the $O_2^{\bullet-}$ and H_2O_2 as fast as possible and to suppress metal ion reduction (Asada 1999). Moreover, H_2O_2 may also harm multiple biological components, and therefore, it has to be detoxified as well. This is achieved by the enzyme ascorbate peroxidase (APX). In order to minimize destructive effects by H_2O_2 via free diffusion across biological membranes, APX is located closely to SOD in the neighborhood of PSI as well. Its cofactor, ascorbate, is regenerated by the vitamin's two oxidized forms: monodehydroascorbate (MDHA) and dehydroascorbate (DHA). The enzymes glutathione reductase (GR), monodehydroascorbate reductase (MDHAR), and dehydroascorbate reductase (DHAR) as well as glutathione are involved in the regenerating process often called ascorbate–glutathione (AsA–GSH) cycle. In contrast to higher plants, seaweeds are additionally able to detoxify H_2O_2 by excretion out of the cell into their environment (Ross and Van Alstyne 2007). It has been shown for the chlorophyte *Ulva rigida* that H_2O_2 excreted from the thallus can reach concentrations up to 4.0 μM in the surrounding seawater (Collén et al 1995; Collén and Pedersén 1996). Catalase (CAT) is another enzyme that cleaves H_2O_2 to oxygen and water, but it seems that this enzyme exclusively detoxifies H_2O_2 produced by glycolate oxidase in peroxisomes during photorespiration (Gross 1993). Collén and Davison (2001) concluded from their studies on the seaweed *Fucus vesiculosus* that CAT activity is regulated independently of the other components of the reactive oxygen metabolism like SOD and GR and, thus, of light. Because photorespiration is low at low temperatures, an increased CAT activity is unlikely to be detected at suboptimal seawater temperatures. In *Ulva rigidia*, no significant CAT activities have been detected when exposed to several high light environments and only APX increased its activity (Collén et al. 1995). Peroxidase enzymes (PX) need the presence of a reducing substrate e.g., ascorbate and glutathione which are regenerated in the AsA–GSH cycle. Another PX is glutathione peroxidase (GPX), which has been intensively studied in mammalian tissues and higher plants, but much less is still known for this enzyme in seaweeds (Eshdat et al. 1997; Contreras et al. 2009).

Interestingly, the above-mentioned enzymes for oxidative stress management (SOD and CAT) have hardly been reported for kelps (i.e., the representatives of the brown algal order Laminariales). A single DNA sequence likely encoding for SOD was isolated from the gametophyte of *Laminaria digitata* (Crepineau et al. 2000; see Bartsch et al. 2008), but the further analysis of 1985 gene transcripts did not reveal the expression of any of these enzymes in *L. digitata* (Roeder et al. 2005). Instead, it is apparent that the stress-induced expression of bromoperoxidases is involved in oxidative stress management. These enzymes are typically catalyzing the oxidation of halide ions to hypohalous acid, in the required presence of H_2O_2. It was shown that haloperoxidases in *Laminaria* are mainly located near the cuticle in the external cortex region of the thallus (Almeida et al. 2001) and around the mucilaginous channels specifically in *L. hyperborea*. In *Saccharina latissima* and *L. digitata* bromoperoxidases were only found being active in the blade (Jordan et al. 1991; Mehrtens and Laturnus 1997). Oxidative stress in kelps was shown to increase halomethane production (Palmer et al. 2005) and the inhibition of photosynthetic electron transport. Consequently, elevated H_2O_2 production via the Mehler reaction reduced the halogenation process (Goodwin et al. 1997). Haloperoxidase is thus considered as another key enzyme for oxidative stress management in kelps, and in this respect further studies on other ecologically important species of seaweeds are urgently needed (see Bartsch et al. 2008).

Next to ROS-scavenging enzymes there is a suite of a large number of different nonenzymatic intermediate metabolites present which confer antioxidative properties and thus may additionally or alternatively be employed to ROS scavenging in macroalgae, like e.g., ascorbate, glutathione, carotenoids, etc. Today's definition of antioxidants also includes polyphenols, vitamin A and retinoids, vitamin E (α-tocopherol), and various sulfur compounds. Hitherto, various phenolic compounds, carotenoids, as well as mycosporine-like amino acids (MAAs) (Karsten et al. 1998; Aguilera et al. 2002b; Schoenwaelder 2002) have been identified as potential antioxidants from Arctic macroalgae. However, currently little information is available from field studies on the antioxidative potential of Antarctic macroalgae. Interestingly, some MAAs which have received significant research attention due to their UV-screening capacities have also been found to reveal some antioxidative potential (de la Coba et al. 2009), which presumably makes them part of a general stress response in red macroalgae (see also Chap. 20 by Bischof and Steinhoff). Again, there is evidence that the cellular concentration and composition of antioxidative compounds in seaweeds are highly species-specific and dependent on the external environmental forcings (Collén and Davison 1999a, b; Lohrmann et al. 2004).

Although low surrounding temperatures can lead to adverse effects on the metabolic functioning, they also may modulate the impact of high irradiances quite substantially. It has been shown that the combination of high PAR at low water temperature may confer adverse conditions to Antarctic macroalgae (Becker et al. 2010). The reason behind is likely to be a temperature-dependent reduction in photosynthetic secondary reactions (i.e., the Calvin–Benson cycle) resulting in a reduced reoxidation of NADPH, and consequently in an elevated reduction state of ferredoxin (Asada 1999). However, while photosynthetic secondary reactions are strongly temperature controlled, primary reactions are not. Thus, high irradiances of PAR will pump electrons into the photosynthetic ETC and, at the high reduction state of ferredoxin, a considerably higher share of electrons will be fuelled into the Mehler reaction, yielding an increased concentration of $O_2^{\bullet -}$. These environmental conditions are frequently observed in shallow waters of polar and cold-temperate regions (Collén and Davison 1999b, 2001; Becker et al. 2009, 2010). The impact of temperature on the reactive oxygen metabolism was studied by Collén and Davison (2001) in the cold-temperate intertidal brown alga *Fucus vesiculosus*. Here, particularly the increase in the ROS-scavenging activity of SOD has been shown as a characteristic feature, which was related to experimental temperature fluctuations. This, however, might also be highly significant in terms of responses to seasonal changes in the abiotic environment, which are most pronounced in cold-temperate regions.

For completeness, it should be noted that apart from the chloroplasts, other important ROS generation sites in plants are the inner mitochondrial membrane. There, superoxide radicals might be generated, e.g., by the NADPH-dependent electron transport involving the cytochrome P-450. In peroxisomes and glyoxysomes, H_2O_2 formed via enzymatic reactions (by glycolate oxidase and acyl-CoA oxidase) is rapidly degraded by CAT to oxygen and water (see Dring 2005). However, with respect to the quantitative significance toward ROS generation under abiotic environmental stress these processes seem to be of minor importance in plants and seaweeds. The generation of oxidative stress has been largely attributed to the result of a stress-induced impairment/restriction of electron flow, as it may be evidently the case under e.g., excess radiation, temperature stress, and nutrient limitation. The induction of ROS formation by mechanical stress, however, has been a matter of debate and has now been more attributed as an oxidative burst response, similar to the responses upon wounding of the thallus (Collén and Pedersén 1994; Ross et al. 2005; see Dring (2005) and Potin (2008) for review).

There is a multitude of adverse effects that ROS may confer to biological components. One prominent target of ROS is biological membranes, i.e., the lipids, which are differentially sensitive depending on their respective saturation state. ROS-induced peroxidation of membrane lipids results in the formation of aldehydes, e.g., malondialdehyde as readily shown for a variety of marine macroalgae under excessive radiation conditions, both in the PAR and ultraviolet (UV) range (Bischof et al. 2002, 2003, 2006; Dummermuth et al. 2003). Due to their complex structures, proteins are also highly susceptible to the interaction with ROS. A multitude of ROS impacts on proteins have been described, interfering on all levels, the primary, secondary, and tertiary structure of proteins, e.g., by specific interaction with the respective amino acids involved based on their differential substitutes (see Dring 2005). For example, in enzymes ROS may impact on active centers, and in particular ROS-induced impairments of iron–sulfur centers have been described to result in inactivation of enzyme function (see Dring 2005).

An important, visual result of oxidative stress is "photobleaching": the loss of photosynthetic pigments like chlorophyll due to their photo-oxidation. This phenomenon has been frequently observed in seaweeds exposed to high PAR (Bischof et al. 2002, 2006). In example, loss in pigmentation in response to steep microscale gradients of solar exposure has been studied within green macroalgal mats (Vergara et al. 1998; Bischof et al. 2002). Supported by high nutrient loads, the green macroalga *Ulva rotundata* may form thick mats consisting of multiple thallus layers. Within these assemblages a steep gradient of irradiance persists, which results in different physiological effects. Whereas the pigments in the top layers (canopy) are directly exposed to solar radiation and become visibly photobleached, subcanopy algae suffer from light limitation. Under the high and prolonged radiation conditions, the canopy algae may become completely bleached within 2 days of exposure. The drastic loss in chlorophyll is accompanied by an increase in malondialdehyde levels, which is an indicator for light stress-induced lipid peroxidation (Bischof et al. 2002, 2003). High production of malondialdehyde was also observed in the top layers of mats of another green seaweed, *Chaetomorpha linum*, however, with not as pronounced changes in chlorophyll concentration as in the latter species (Bischof et al. 2006).

6.2.1 Seaweed Exposure to Heavy Metal Stress and Reactive Oxygen Metabolism

Very recent studies have highlighted the particular significance of heavy metal exposure for the induction of oxidative stress in seaweeds. In natural seawater, concentrations of heavy metals are usually very low. Because copper (Cu^{2+}) concentration ranges between 0.05 and 3 g L^{-1} (Contreras et al. 2005, 2009), it is a micronutrient, which is used as cofactors for enzymes (e.g., oxidases, SODs) and ETC components (e.g., plastocyanin). However, due to industrial activities such as copper mining, Cu^{2+} concentrations may rise significantly when highly contaminated wastewaters are drained off into coastal areas. Ecologically, such an impact reduces species richness as well as changes the structures of coastal communities. Consequently, only opportunistic macroalgae, characterized as highly copper tolerant, are able to grow under these extreme conditions. For example, copper-impacted coastal areas with 25 g L^{-1} and higher in northern Chile are inhabited by only two macroalgal species (*Ulva compressa*, *Scytosiphon lomentaria*), although *Lessonia nigrescens* dominates related, but pristine, ecosystems (Contreras et al. 2009).

Macroalgae, independently of their heavy metal-tolerance, are able to accumulate copper in their tissues as a function of its concentration in seawater (Ratkevicius et al. 2003; Andrade et al. 2006). Tissues of *U. compressa* form copper-impacted sites; an approx. 40-fold higher copper content was measured than in conspecifics from pristine habitats (Ratkevicius et al. 2003). These free copper ions (Cu^{+}, Cu^{2+}), which are not bound in protein complexes in the cell, are able to increase ROS production substantially by their involvement in the Haber–Weiss and the Fenton reaction (Pinto et al. 2003). The highly reactive hydroxyl radical (•OH), which is a product of both reactions, changes and inactivates biological macromolecules by oxidation and peroxidation. Therefore, macroalgae exposed to copper excess showed increase in lipid peroxidation, indicating cell membrane damage. If oxidative stress is not sufficiently detoxified by the cellular ROS-scavenging system, it may consequently lead to cell death.

However, the antioxidant defence system is also activated by copper excess. A general response of macroalgae seems to be the stimulation of those enzymes that are involved in scavenging of lipid and fatty acid peroxidation. In copper-tolerant macroalgae (*U. compressa*, *Ulva fasciata*, *S. lomentaria*), activities of APX and peroxiredoxin (PRX) detoxifying H_2O_2 and fatty acid hydroperoxides increased significantly as a result of copper excess. This response is attributable to enhanced gene expression providing new enzymes rather than modulation of already existing enzymes, which represents an efficient acclimation mechanism to cope with copper stress (Contreras-Porcia et al. 2011). In *U. fasciata*, ROS acts as a signal for upregulation of genes encoding for antioxidant defence enzymes (FeSOD, APX, GR) because their transcript levels increased by exogenously applied H_2O_2 concentrations above 0.2 mM (Wu et al. 2009a, b). Moreover, induction of transcription of antioxidant enzyme genes by copper led to an increase in activities of MnSOD, FeSOD, APX, CAT, and GR to cope with copper stress successfully. However, at high copper concentrations (above 20 μM $CuSO_4$), copper might be responsible for a selective inhibition of APX and CAT, which resulted in an insufficient ROS detoxification with enhanced H_2O_2 accumulation and lipid peroxidation (Wu and Lee 2008). However, one must take into account that these mechanisms of copper acclimation are species-specific rather than a general response.

Different efficiencies of buffering oxidative stress-induced by copper excess (20–100 g L^{-1}) might play a crucial role for the absence of the copper-sensitive *L. nigrescens* and the presence of the copper-tolerant *S. lomentaria* in copper-enriched coastal areas in Chile (Contreras et al. 2009). H_2O_2 formed right after copper stress was completely scavenged in both species but much slower in *L. nigrescens* than in *S. lomentaria*. Additionally, lipoperoxide production detected in both species over a 96 h period might be a result from the activation of a lipoxygenase (LOX) that only uses arachidonic acid as substrate. Both species show different responses to copper excess. A stable lipoperoxide production in *S. lomentaria* which was reached early after exposure to copper excess could be ascribed to the constant LOX activity as well as to an efficient ROS scavenging by increased activities of APX and GPX. These enzymatic responses seem to be crucial for macroalgal survival since potentially toxic peroxides are converted into nontoxic compounds. In contrast, continuously increasing accumulation of lipoperoxides in *L. nigrescens* over the entire period might be a consequence of the biphasic boosts in LOX activity. Additionally, inactivation of GPX and APX possibly due to direct inhibition by copper might be a reason for a further accumulation of fatty acid hydroperoxides and H_2O_2 at high copper concentrations. Impaired detoxification may trigger the production of the highly reactive, short-chain aldehydes such as 4-hydroxy-2(E)-nonenal (HNE) which are cytotoxic, genotoxic, and mutagenic due to its electrophilic properties (Guéraud et al. 2010). Dysfunctions of proteins and destruction/changes in the fine structures of mitochondria and chloroplasts in *L. nigrescens* are a consequence of the accumulation of lipoperoxides. Thus, an efficient detoxifying mechanism seems to be crucial for survival of macroalgae in copper-impacted habitats.

6.2.2 A Peculiar Antioxidant System in Kelp

An unusual antioxidant system that is based on the enzymatic oxidation of accumulated iodide (I^-) has been recently described to be present in the sporophytes of the kelp *L. digitata* (Küpper et al. 2008). Two types of vanadium-dependent haloperoxidases (vHPOs) can be found in this life stage to mediate the oxidation of halides (X^-) under consumption of H_2O_2 to hypohalous acids (HXO), which may subsequently react to form iodocarbons such as polyhalomethans (CH_2I_2):

Bromoperoxidases

$$Br^- + H_2O_2 \rightarrow HBrO + OH^- \text{ and}$$
$$I^- + H_2O_2 \rightarrow HIO + OH^-$$

Iodoperoxidase

$$I^- + H_2O_2 \rightarrow HIO + OH^-$$

Although multimeric aggregates of several bromoperoxidases (vBrPOs) oxidize both iodide (I^-) and bromide (Br^-) ions, these enzymes are not involved in the iodide antioxidant system because of their inefficient catalysis toward iodide (see below) as well as their various locations in the cell (Colin et al. 2003). Numerous studies did demonstrate that iodide-oxidizing enzymes, iodoperoxidases (vIPOs), play the central role in the iodine metabolism rather than vBrPOs. These homodimeric enzymes being encoded by a multigenic family are responsible for both iodide uptake and ROS scavenging. They have a strict specificity to iodide because of their special halide binding site topology and the fine-tuning of the vanadate cofactor electronegativity (Colin et al. 2005; Pacios and Gálvez 2010). Iodoperoxidases have a lower affinity to iodide (K_m,I^- = 2.5 mM) and, more important, a higher catalytic turnover rate of iodide (k_{cat},I^- = 462 s^{-1}) than vBrPOs (K_m,I^- = 18.1 mM, k_{cat},I^- = 38 s^{-1}). Moreover, a seven times higher specific activity toward iodide of the purified iodoperoxidases compared to vBrPOs (cf. 1,200 vs. 180 U mg^{-1}) represents a high efficiency of iodide incorporation and the crucial role in iodine metabolism (Colin et al. 2003).

When *L. digitata* is unstressed and submerged during high tide, iodide is taken up from seawater (0.3 μM) probably by facilitated diffusion down its concentration gradient to the apoplast (Küpper et al. 1998). There, the iodoperoxidases mediates the oxidation of iodide. However, this process requires a steady flow of H_2O_2 into the apoplast, which was estimated to as small as approx. 5 μM (Küpper et al. 1998). So far, the source of H_2O_2 is not absolutely clear yet, but it is supposed to be produced by cell wall oxidases or membrane-bound enzymes with an extracellular domain. Intracellular sources of H_2O_2 might also be responsible for the functioning of the iodoperoxidase. However, it lacks detailed studies, which investigate the role of vIPOs under, for example, high light conditions when H_2O_2 is produced by APX or the photorespiratory pathway in response to excessively absorbed light energy.

Depending on the age of the sporophyte and the regions of the thallus, iodide may accumulate between 13,000 and even 150,000-fold relative to natural seawater in the apoplast of *L. digitata* (Küpper et al. 1998). In contrast to these vast pools, iodide contents in related *Ectocarpus* and *Fucus* species are only a fraction of that one found in Laminariales (Saenko et al. 1978; Cock et al. 2010). This substantial difference can be ascribed to the lack of iodoperoxidases in species of these genera. Only bromoperoxidases being characterized by their substrate preference to bromide over iodide fix comparable small amounts of iodide. This is supported by the recent analysis of the genome sequence of *Ectocarpus siliculosus*, which revealed that only one haloperoxidase, a vBrPO, is encoded (Cock et al. 2010).

In the apoplast, iodide is non-covalently associated with biomolecules such as phenolic compounds, polysaccharides, and proteins (Verhaeghe et al. 2008; Küpper et al. 2008). This unique type of accumulation ensures that iodide can be readily and rapidly mobilized for chemical defence and under oxidative stress conditions during high tide (Verhaeghe et al. 2008). It was demonstrated by extracellular addition of elicitors (e.g., oligoguluronates) mimicking pathogen attack and oxidants (e.g., H_2O_2) to *L. digitata* that iodide is released into the seawater within seconds after addition. This response represents a very fast mechanism to acclimate to an external abiotic and biotic stressor. However, because these high iodide concentrations are not directly toxic to both pro- and eukaryotes (Küpper et al. 2008), iodocarbons with high toxicity might also be used to prevent from microbial assault. In fact, iodocarbons such as CH_2I_2 (diiodomethane) and CH_3I (iodomethane) were emitted after oligoguluronate treatment (Palmer et al. 2005). Moreover, oxidative burst in the apoplast is triggered by the recognition of oligoguluronates (Küpper et al. 2001) and bacterial lipopolysaccharides (Küpper et al. 2006). Superoxide anions probably produced by an oxidase with a flavoprotein subunit (i.e., NADPH oxidase) in this process become subsequently converted to H_2O_2 that can diffuse into the seawater. There, H_2O_2 may affect settling of microorganisms on the thallus surface. In the apoplast where organic antioxidants such as ascorbate and glutathione are absent, accumulated iodide is released under oxidative stress conditions to detoxify ROS. Enzymatically, vIPO catalyzes the removal of H_2O_2 by reoxidation of iodide. Furthermore, due to fast, nonenzymatic reactions of iodide with O_3, $^1O_2$, and superoxide anions, these ROS can be scavenged efficiently, which makes iodide the best antioxidant available in the apoplast of *L. digitata* (Küpper et al. 2008). At the end of the oxidative stress period, the iodide remaining in the surrounding seawater is taken up again to the apoplasm where it is re-associated with biomolecules (Küpper et al. 2008).

During low tide events when the thallus is uncovered by water (desiccation stress) and simultaneously exposed to high photon fluxes (high light stress) and O_3 concentrations (ozone stress), iodide may have a significant impact on atmospheric chemistry. Molecular iodine (I_2) is released directly from the surface into the atmosphere above the kelp beds by the abiotic reaction of O_3 with accumulated iodide. Then, I_2 is photolytically (impact of high light) broken down to the iodine radical ($I^\bullet$) on the thallus surface. The $I^\bullet$ reacts with O_3 in the troposphere to IO (iodine oxides) providing a precursor for aerosol particles (>3 nm) that may lead to cloud condensation nuclei formation. Particle formation seems to be a result from O_3-scavenging reactivity of iodide on the thallus surface (Küpper et al. 2008) rather than by the iodine atom-releasing breakdown of iodocarbons as presumed earlier (O'Dowd et al. 2002). So, taken all these impacts together, the iodine metabolism in kelps has not only a great impact on ROS scavenging in the apoplast and the extracellular defence but also on atmospheric chemistry.

6.3 Antioxidative Strategies in Seaweeds on Seasonal and Spatial Scales

For all seaweeds living in highly variable environments, it is essential to apply protective and detoxifying mechanisms to scavenge and to minimize the adverse effects of oxygen radicals, which are resultants from abiotic environmental stress (Collén and Davison 1999a, b; Rijstenbil et al. 2000; Bischof et al. 2003).

As already indicated above, macroalgae apply a multitude of enzymatic and nonenzymatic tools to reduce the impact of ROS on their biological/cellular structures. Studies on Arctic and cold-temperate macroalgae demonstrated the plasticity of the

ROS-scavenging mechanisms in order to protect against cytotoxic reactive oxygen. These mechanisms include the inducible increase in activities of the radical scavenging enzymes SOD, APX, GR as well as the cellular concentration of antioxidants, e.g., glutathione, ascorbate, carotenoids (Aguilera et al. 2002b; Dummermuth et al. 2003).

SOD has often been referred to as "the first line of defence" against ROS. However, in most of the studies conducted on SOD activity in macroalgae so far, it was not discriminated between the different isoforms of SOD, the cytosolic Cu/Zn-SOD or the Fe-SOD located in chloroplasts, neglecting the assumption that it is indeed the combination and interaction of different isoforms of SOD characterizing it as the central protection mechanisms against $O_2^{\bullet -}$-mediated damage (Pinto et al. 2003; see Dring 2005). Variation in total SOD activity has been found among different seaweed species and also with respect to the respective growth site, i.e., abiotic stress exposure along latitudinal or vertical gradients (Aguilera et al. 2002b). In general, *inter*specific differences in SOD activity are mostly related to differences in the respective abiotic environment of the species under investigation, depending on latitudinal differences, but also the shore level, which is typically populated by the respective species. In their survey on general antioxidative properties of polar macroalgae, Aguilera et al. (2002b) reported on an interspecific depth-dependent antioxidative pattern: species from the upper shore are generally characterized by higher SOD activities than those from deeper growth sites as a likely consequence of high light exposure.

Latitudinal differences in SOD activity were found in the closely related green macroalgae *Ulva bulbosa* (now also referred to as *U. hookeriana*) and *Ulva clathrata* from Antarctic and sub-Antarctic regions, respectively (Rautenberger and Bischof 2006). Total SOD activities in *U. bulbosa* were significantly higher (86–89 Units mg^{-1} protein) compared to those in *U. clathrata* (58–63 Units mg^{-1} protein). Apparently, these differences based on the respective geographical dispersal of these species are resultant from differences in water temperature and radiation regimes, modulated by latitude. The Antarctic ecotype of *U. bulbosa* is permanently exposed to lower water temperatures (around 0 °C) than *U. clathrata* isolated from waters in which temperatures never decrease below 5 °C. Exposure to lower water temperatures may permanently cause a higher background level of ROS in *U. bulbosa* which requires a more efficient detoxification system, reflected by the higher SOD activity. Furthermore, *U. bulbosa*'s lower susceptibility to enhanced irradiances of UV-B radiation at 0 °C might be a consequence of a more efficient photoprotection, possibly achieved by higher SOD activities as well.

On a smaller spatial scale, i.e., with respect to growth at different shore levels, the related intertidal rhodophytes *Mastocarpus stellatus* and *Chondrus crispus* (Gigartinales) have been tested for their respective total scavenging activity of SOD, which was found to be clearly related to general stress tolerance (freezing, high light, temperature) and tidal height (Collén and Davison 1999c; Bischof et al. 2000; Lohrmann et al. 2004). Also, different intertidal congener species of the brown alga *Fucus* display differential overall activities of SOD, positively correlated with the respective position of the respective species on the shore (Collén and Davison 1999a, b).

6.3.1 SOD Activity in the Brown Seaweed Desmarestia anceps *Along a Depth Gradient: A Case Study in Antarctica*

In line with *inter*specific variation of SOD activity along gradients of different spatial scales, also *intra*specific variations have been found, reflecting the specific abiotic conditions in individuals of species growing over pronounced latitudinal and/or vertical ranges.

The variation of SOD activity with depth was studied in an ecologically important brown seaweed from Antarctica (Rautenberger and Bischof, previously unpublished data): The endemic Antarctic phaeophyte *Desmarestia anceps* colonizes water depths between 5 and 30 m in coastal ecosystems (see also Chap. 13 by Wiencke and Amsler). Occurring across such a wide range within the sublittoral zone, this species is subjected to a pronounced vertical gradient of habitat stability. Thus, individuals from highest positions on the shore are regularly exposed to frequent and strong changes in abiotic conditions, such as solar radiation, whereas specimens from deeper waters inhabit a much more stable environment. It was, thus, expected that *D. anceps* has a high potential to respond to environmental changes at the individual growth sites by metabolic adjustments to the respective in situ conditions.

Overall, the species-specific capacity to detoxify ROS with respect to environmental stability, also including the exposure to detrimental UV-B-radiation, is a major factor determining the vertical distribution patterns of seaweeds on the shore (Bischof et al. 1998; Aguilera et al. 2002b; see Chap. 20 by Bischof and Steinhoff). In *D. anceps*, site-specific variation of total SOD activity as a function of water depth was investigated in adult sporophytes. Individuals were collected at 5.5, 9.0, and 13.5 m water depth, near the Antarctic scientific base Jubany (62.23°S, 58.63°W) at King George Island (South Shetland Islands, Antarctic Peninsula) in January 2005. Specimens were cultivated in air-bubbled basins in the laboratory at 15 μmol m^{-2} s^{-1} PAR (16:8 h light/dark) and 2 $\pm$ 1°C for 10 days. Changes in SOD activity in response to exposure to enhanced ultraviolet radiation were studied by supplementing the PAR regime with 15.3 W m^{-2} UV-A and 2.2 W m^{-2} UV-B. For exposure, Petri dishes containing the specimens were completely covered by cutoff filters in order to expose experimental individuals to a defined spectral composition: PAR alone (400–700 nm), PAR + UV-A (320–700 nm), and PAR + UV-A + UV-B (280–700 nm). Consequently, the integral UV-B dose over the entire experimental period of 4 h was 31.8 kJ m^{-2} and the three radiation treatments were used to study *D. anceps*' potential to acclimate to enhanced UV-B radiation by SOD adjustment. SOD activities were determined according to a modified xanthine/xanthine oxidase assay of McCord and Fridovich (1969) and Aguilera et al. (2002b) at 25 °C. Initial total SOD activities (PAR alone) were related to the original water depths where the individuals of *D. anceps* have grown before collection (Fig. 6.1a). Individuals obtained from 5.5 m exhibited very high activities of 2,760 $\pm$ 2,283 Units mg protein^{-1}. Although SOD activities of 1,389 $\pm$ 559 Units mg protein^{-1} determined in individuals from 9.0 m showed a trend to lower values, there was no statistical difference in SOD activity in specimens from both depths. However,

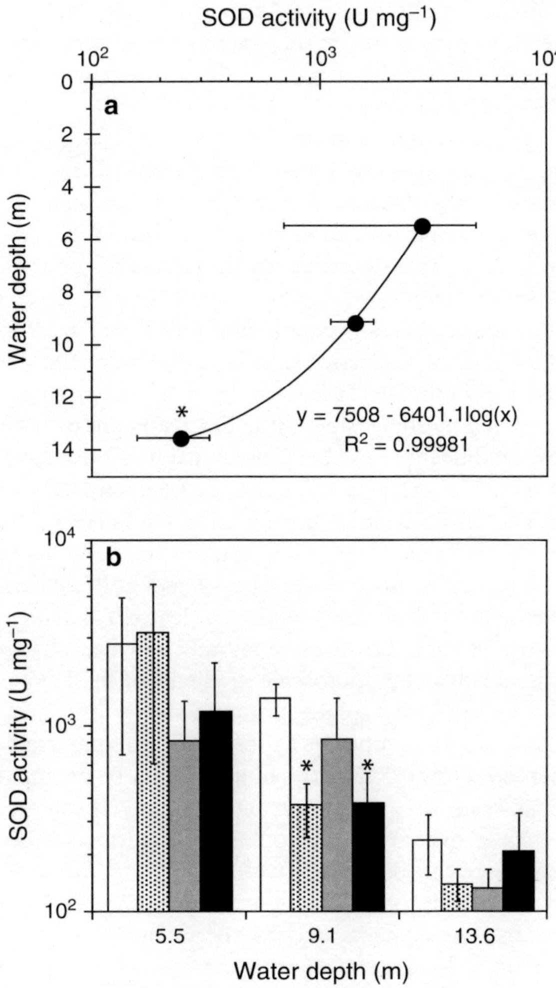

Fig. 6.1 (**a**) Total activities of superoxide dismutase (SOD) in individuals of *Desmarestia anceps* collected from 5.5, 9.0, and 13.5 m water depth before UV exposure. Activities were fitted according to a logarithmical function. (**b**) Changes of SOD activities after 4 h of exposure to PAR alone (control, *dotted columns*), PAR + UV-A (*grey columns*), and PAR + UV-A + UV-B (black columns) in comparison to initial SOD activities before UV exposure (*white columns*). Error bars characterize standard deviations. The *asterisks* represent statistical differences between SOD activities of individuals from different depths (**a**) or light treatments (**b**). Note that the SOD activities are plotted logarithmically

individuals from the deepest position tested (13.5 m) revealed significantly lower SOD activities of 241 $\pm$ 184 Units mg protein^{-1}. SOD activities of *D. anceps* from each water depth did not change in the course of the 4 h exposure to either UV-A or UV-B radiation compared to samples under PAR alone (Fig. 6.1b).

Thus, the results presented suggest that *D. anceps*' capacity to detoxify the superoxide radical formed under conditions of environmental stress (here: artificial UV radiation) is strongly related to environmental variability, i.e., according to original growth depth: high SOD activities may protect shallow water individuals of *D. anceps* efficiently against ROS induced by environmental stress. The intraspecific adjustment of SOD activity thus seems to represent a crucial mechanism of acclimation along environmental gradients. However, although UV-B radiation is regarded to be an important factor determining macroalgal vertical distribution

(Bischof et al. 1998; see Chap. 20 by Bischof and Steinhoff), this factor did not seem to play a major role in ROS formation under the experimental conditions tested, since SOD activities did not increase, even not in individuals from the deepest collection site. Also, results obtained do not point to a highly flexible response of SOD adjustment, at least not within the 4 h of experimental exposure. Thus, data obtained rather reflect long-term acclimatory adjustments, potentially allowing the efficient detoxification of light-induced superoxide in individuals from the upper and mid sublittoral (5.5 and 9.0 m), where irradiance conditions may increase and may become harmful to photosynthesis on sunny days, as soon as noon coincides with low tides. In contrast, low SOD activities shown for the deep-water individuals (13.5 m water depth) reflect the larger degree of habitat stability as e.g., the extent of attenuation of solar radiation by the water column above and the in situ radiation climate. However, high SOD activities of *D. anceps* are presumably achieved by increasing the cellular content of the scavenging enzyme rather than by modification to the protein itself. Consequently, higher energetic costs for protein biosynthesis might then be compensated by higher photosynthetic activities as demonstrated in *D. anceps* from the upper sublittoral.

As outlined above, the variation in SOD activity described in individuals of *D. anceps* resulted from site-specific differences in habitat stability and light attenuation along the depth gradient, in the present case approximately 10 m. Other studies, however, explored whether changes in SOD activities are also significant along microscale gradients of light exposure, e.g., within algal canopies where irradiance might be diminished within just a few centimeters. In fact, in dense algal assemblages exposed to high solar irradiance, distinct responses of SOD activities toward variation in light quantity and quality have been observed in green algal mats of *U. rotundata* and *C. linum* (Bischof et al. 2002, 2003, 2006). Results obtained over microscale gradients confirmed the findings on *D. anceps* that overall SOD activity represents well the in situ radiation climate and thus the vertical position of respective specimens, but does not exhibit changes in activity and thus flexible adjustments under experimental conditions. These adjustments seem to require a longer exposure to the respective environmental conditions in order to become significant. Thus, the studies mentioned above did not exhibit changes in SOD activity along diurnal or tidal cycles.

6.3.2 Seasonal Changes in SOD Activity in Seaweeds from an Arctic Fjord

The ecological implications of adjustments of the antioxidant metabolism become apparent when considering pronounced *seasonal* changes in abiotic conditions at the natural growth sites of seaweeds. In particular, polar regions represent a good example for such studies because the abiotic environment of macroalgae changes more drastically throughout the seasons than it does in temperate or tropical regions. The modulation of antioxidant enzyme activities with respect to changing radiation and temperature regimes in the course of the seasons has been observed for a number of macroalgal species from an Arctic fjord system: Kongsfjorden, Svalbard (Aguilera et al. 2002a). It was striking that macroalgae growing subtidally, such as the chlorophyte *Monostroma* sp. and the rhodophyte *Palmaria palmata*, showed patterns of antioxidant enzyme activity modulation, which were corresponding to seasonal changes but also to their respective growth site and, furthermore, life strategy. In *Monostroma* sp. and *P. palmata*, SOD activities closely reflected the particular PAR and UV radiation conditions in Kongsfjorden: SOD activities were low as long as the coastline was still covered by sea ice until the middle of June. Thus, these subtidal organisms were exposed to prolonged conditions of extreme low light. The sea ice breakup may then occur rapidly within very few days and is mainly triggered by changes in the hydrographic conditions of the fjord (Svendsen et al. 2002). Accordingly, seaweeds which were exposed to darkness or very low light conditions during the polar night and under the prolonged sea ice cover may then rapidly be exposed to high irradiances of PAR and UV radiation because of the subsequent very clear water conditions. Therefore, high radiation is able to penetrate deeply into the water column and thus even affects subtidal macroalgae significantly. *Monostroma* sp. responded to the breakup of sea ice with more than a doubling of total SOD activity within very few days and with a continued increase as long as the seawater remained clear. Simultaneously, the detected loss of GR activity might either be a result of increased CAT activities or a consequence of direct radiation damage (Aguilera et al. 2002a). *Palmaria palmata* even quadruplicates SOD activity in summer, but this response was much slower than in *Monostroma sp*. Moreover, this increase was still ongoing throughout the summer season and may thus reflect an acclimation pattern to the season. In contrast, SOD activity in *Monostroma sp.* peaked just before the onset of snow-melt inshore and consequently the runoff of turbid freshwater. These events resulted in an increase of suspended material in the water column, leading to a strong decrease in water transparency and thus decline in in situ light exposure. Consequently, SOD activities might well be reduced again under subsequent turbid water conditions. Overall, this very flexible seasonal response in SOD activity of *Monostroma sp.* is in line with the purely opportunistic life strategy of this ephemeric green alga. In contrast to both species described above, the shallow water red alga *Devaleraea ramentacea* did not exhibit marked variation of SOD activity in the course of the seasons. It rather maintained high SOD activities throughout the year, even under low light conditions under sea ice cover. This is typically characterizing *D. ramentacea* as a shallow water species and is in line with the generally reduced habitat stability, being exposed to more pronounced variation in physicochemical variables in shallow waters.

In summary, high interspecific variation in SOD activity has been demonstrated in a multitude of studies, reflecting species-specific differences with respect to the respective growth site along latitudinal and depth gradients, with respect to site-specific seasonal variation, as well as differences with respect to developmental stages, age, and tissue type under investigation. Oxidative stress is also to be considered as an important component of cellular signal transduction pathways (Mackerness et al. 2001), although there is only scant information obtained from seaweeds in this respect. As a consequence of oxidative stress, most organisms, and seaweeds as well, typically increase the expression of specific cell repair-related proteins. Protoplasts of *L. digitata* sporophytes have been shown to massively upregulate heat-shock proteins HSP-70 and HSP-90, which play a crucial role in the recovery of cells from stress, in the prevention of protein aggregation and in the refolding of denaturated proteins (Roeder et al. 2005).

6.4 Oxidative Bursts: Reactive Oxygen and Interspecific Interaction

The significance of reactive oxygen radicals is by no means restricted to be resultant from abiotic stress resulting in a malfunction of photosynthesis. A fast, light-independent induction of massive ROS formation, which is referred to as an "oxidative burst", has been recognized as a tool involved in interspecific interaction like e.g., pathogen defence in marine macroalgae (see Potin 2008 and Weinberger and Potin 2010 for review). In these complex systems, oxygen radicals ($O_2^{\bullet-}$, H_2O_2, or $\bullet OH$) are transiently formed by the infected or assaulted alga in response to stimuli originating from either the pathogen and parasite or the host's cell wall. These stimuli are often represented by chemical elicitors, as extensively studied in a few red and brown macroalgal species within the last decade.

Chondrus crispus exhibits oxidative bursts upon challenging with its frequent parasite, the green alga *Acrochaete operculata* (Weinberger et al. 2005a; see Dring 2005 and Weinberger 2007 for review). The mechanisms behind ROS and, more specifically, H_2O_2 production by the host upon infection by a parasite is well understood in the *Chondrus*/*Acrochaete* system (Weinberger et al. 2005a): L-asparagine, which is released from the parasite serves as substrate for L-amino oxidase in the apoplasm of the host. Thus, *C. crispus* is able to perceive microbe-associated molecular patterns (MAMPs) which are also referred to as general or exogenous elicitors. The excreted H_2O_2 will result in the inhibition of the parasite's settlement, but may also be quenched again by the presence of host-specific peroxidases.

Red and brown marine macroalgae are able to perceive pathogen-induced molecular patterns (PIMPs) or endogenous elicitors. These elicitors are degradation products of their own cell wall due to the action of lytic enzymes secreted by attacking organisms (Weinberger 2007). In a study in which 17 red macroalgal species of Gracilariaceae were exposed to agar oligosaccharides it was shown that two different mechanisms of defence-related responses to this chemical elicitor exist in this family: (1) oligoagar-activated oxidative burst and (2) oxidation of agar oligosaccharides. In nature, agar oligosaccharides arise when pathogenic organisms cleave the cell wall matrices of a red macroalga enzymatically. In laboratory studies however, this elicitor is used to add to algal media to study their defence responses.

In contrast, oligocarrageenans, which are chemically similar to oligoagars, did not lead to any defence response in *Gracilaria* sp.*"dura"*[1] (Weinberger et al. 1999).

In *Gracilaria* (*Gracilaria "sensu stricto"*) and the closely related *Hydropuntia* species (Gurgel and Fredericq 2004), agar oligosaccharides induce a fast and transient oxidative burst in which relatively high amounts of H_2O_2 (>100 nmol g^{-1} min^{-1}) are released into the medium after the first challenge. It is assumed that agar oligosaccharides are presumably perceived by cell membrane-bound, specific receptors which may trigger a subsequent signal transduction pathway. In fact, the lack of oxidative burst due to the use of the protein kinase inhibitor staurosporine demonstrates that phosphorylation of signal molecules and/or proteins seems to be crucial to relay the "signal of an attack" toward a H_2O_2 producing agent. A NADPH oxidase located in the plasma membrane appears to be activated by the signal transduction pathway. Inhibitor studies with DPI (diphenylene iodonium), which is a specific inhibitor of NADH- and NADPH-dependent enzymes (Weinberger et al. 2005b), showed that H_2O_2 was not released in species of these both genera. These findings suggest that ROS production for pathogen defence by an activation of a NADPH oxidase requires phosphorylation events in a signal transduction pathway which is triggered by oligoagar-specific receptors.

An alternative mechanism leading to strong H_2O_2 release in response to agar oligosaccharides has been shown to be present in species related to *Gracilaria chilensis* (*G. chilensis* clade) and those of the genus *Gracilariopsis* (Weinberger et al. 2010). In these species, fast H_2O_2 production (<100 nmol g^{-1} min^{-1}) results from direct oxidation of agar oligosaccharides by an agar oligosaccharide oxidase. Inhibitor studies have proven that neither NADPH oxidases nor kinase-mediated phosphorylation events are involved in H_2O_2 release. Moreover, oxidation products of agar oligosaccharides are able to activate expression of genes related to agar oligosaccharide oxidase within the first 24 h after the challenge (Weinberger et al. 2010). In contrast to the members of the *Gracilaria "sensu lato"* major clade (*Gracilaria "sensu stricto"* and *Hydropuntia*) which show a refractory state of oxidative burst after first challenge, H_2O_2 is repeatedly released without any delay after a previous challenge in species of the "*G. chilensis* clade" and the genus *Gracilariopsis*.

A comparative study of agar oligosaccharide responses within the family of the Gracilariaceae revealed that the activation of a NADPH oxidase may be a relatively "newly" acquired feature because it is only present in the *Gracilaria "sensu lato"* major clade (Weinberger et al. 2010). It was speculated by the authors of the study that the activation of a NADPH oxidase can be coincident either with the acquisition of agar oligosaccharide-specific receptors or the signaling pathway that interlinks the receptors with the oxidase (Weinberger et al. 2010). In contrast, direct oxidation of agar oligosaccharides by agar oligosaccharide oxidase seems to be widely distributed in Gracilariaceae because of its activity in the whole

[1] *Gracilaria conferta* from Israel being used in the study of Weinberger et al. (1999) was recently identified as *G.* sp. "*dura*" and can be classified as a *Hydropuntia* species. However, species belonging to the *G. "sensu lato"* major cluster show the same response as *G.* sp. "*dura*" from Israel (Weinberger et al. 2010).

family. Thus, two mechanisms of H_2O_2 release seem to be coexist in the *Gracilaria* "*sensu lato*" major clade. This pattern reveals that defence response by direct oxidation of agar oligosaccharides is more ancient than oxidative burst triggered by oligoagar (Weinberger et al. 2010).

ROS formation due to perception of PIMPs also occurs in brown macroalgae. In Laminariales, oxidative burst is triggered by recognition of specific degradation products of alginates, being the major polysaccharides in their cell walls. The addition of oligoguluronates, which are homopolymeric blocks of poly-α-1,4-L-guluronic acid, to *L. digitata* induces a transient and strong release of H_2O_2 into the medium within a few minutes. Similar to the *Gracilaria*"*sensu lato*" major clade, the oxidative burst is a result from signal transduction chains triggered by the recognition of the elicitor. The activation of phospholipases and kinases as well as ion fluxes across the plasma membrane is apparently involved in the signaling pathways (Küpper et al. 2001). Moreover, perception of bacterial lipopolysaccharides (i.e., MAMPs) may also induce early events of oxidative burst and fatty acid oxidation (Küpper et al. 2006).

An analysis of the pattern of gene expression in response to oligoguluronate elicitation showed that 50 genes were highly regulated within the first 24 h in *L. digitata* (Cosse et al. 2009). The number of differentially expressed genes which were upregulated was maximal (38 genes) after 6 h and decreased continuously after 12 h (21 genes) and 24 h (12 genes) after challenge. However, six of these differentially expressed genes remained upregulated over the first 24 h. By the use of various inhibitors affecting signal transduction and ROS formation, the expression pattern of ten genes was affected differently. Three trends of changes in transcript levels observed led to the conclusion that three different signal transduction pathways might be involved in oligoguluronate-induced oxidative burst and set the stage for subsequent gene regulation (Cosse et al. 2009). A coexistence of multiple pathways leading to H_2O_2 release due to PIMP perception was shown in the *Gracilaria*"*sensu lato*" major clade as well (Weinberger et al. 2010). One signal transduction pathway appears to be dependent on early phosphorylation events by kinases and has a great impact on expression of *vIPO1* (vanadium-dependent iodoperoxidase 1), *trx* (thioredoxin like 5), and *hsp70* (heat-shock protein 70) genes. A second signaling pathway seems to be fully controlled by the oxidative burst: H_2O_2 production regulates transcript levels of *vBPO3* (vanadium-dependent bromoperoxidase 3) and *g6pd* (glucose-6-phosphate dehydrogenase). The third signal transduction chain apparently depends on both ROS production (oxidative burst) and the activation of putative MAP kinases or other downstream kinases. This pathway led to an upregulation of the *vIPO3* gene whereas the *vBPO1* gene was concurrently downregulated. Based on this finding, Cosse et al. (2009) hypothesized that ROS produced by oxidative burst might also be an internal signal to induced expression of genes with defence-related function. The rapid induction of the *vIPO3* gene encoding a novel vanadium-dependent iodoperoxidase upon oligoguluronate elicitation suggests that the iodine metabolism might be specifically activated on the molecular level during early defence response (Cosse et al. 2009). This is in line with the fast and massive iodide release into the medium after elicitation (Küpper et al. 2008). The simultaneous repression of the *vBPO1 gene* coding for a vanadium-dependent bromoperoxidase indicates a tightly coordinated regulation of these two distinct haloperoxidase gene families (Cosse et al. 2009). This antagonistic response on transcription level might lead to the conclusion that oligoguluronate-induced oxidative burst switches the iodine metabolism on to achieve a more efficient and effective pathogen defence (Cosse et al. 2009).

Stress- and defence-related signals might be responsible for communication between marine macroalgae, e.g., to transmit warning messages to their neighboring conspecifics, as recently demonstrated for *L. digitata* (Thomas et al. 2011). It was shown that elicited sporophytes have a great impact on defence responses to oligoguluronate in un-elicited neighboring conspecifics. It seems that elicited sporophytes release signaling agents (of unknown nature so far) into the surrounding water which are integrated by their neighboring conspecifics. As a consequence of the perception of the putative waterborne signal, *L. digitata*'s capacity to detoxify ROS has potentiated: the pattern of oxidative burst and emission of volatile organic compounds (VOCs) changed as well as genes regulated in response to oligoguluronate become faster induced. This effect resembles the *priming* in terrestrial plants where species that had been in contact with signals borne off infested neighbors show earlier and stronger effects in defence response (Thomas et al. 2011). The less intense oxidative burst in primed sporophytes represented by a significantly earlier decrease in H_2O_2 release after oligoguluronate elicitation may be attributed to a faster triggering of ROS detoxification processes. Lower emission of VOCs might also reflect an enhanced ROS detoxification mechanism. Moreover, the higher and faster elicitation-dependent upregulation of genes involved in ROS management (*prx*, peroxiredoxin, and *msr*, methionine sulfoxide reductase) also represents a more efficient antioxidative response early after elicitation. These responses prepare the primed sporophyte to a better response to subsequent challenges without directly triggering defence responses (Thomas et al. 2011).

In this way, the generation and scavenging of ROS does also play a central role in biotic interactions, which certainly needs to be explored in much more detail.

6.5 Summary

Overall, the formation and effective detoxification of ROS represents a crucial feature in seaweed responses to the abiotic and biotic environment, and it still requires more intense research activities. Hitherto, the mechanisms of ROS-mediated signal transduction pathways in seaweeds are just basically studied. It should also be emphasized that based on the multitude of chemically mediated biotic interactions of seaweeds with their environment, the significance of ROS for grazing and pathogen defences is a challenging but extremely exciting field of future research. With respect to the importance of the oxygen radical metabolism to stress physiology of seaweeds, a general link between ROS formation and habitat stability of the respective growth site is well established. The formation of ROS as resultant from photosynthetic malfunctions, and thus rising from stress generated by extreme radiation salinity and temperature (e.g., freezing) regimes, is apparent, but still the interaction of environmental factors with respect to potential synergistic or antagonistic effects as well as features of cross adaptation needs to be addressed in the future. Moreover, in particular those abiotic factors, which are not directly related to photosynthetic performance, such as mechanical stress, need to be studied with respect to their significance to ROS generation. Considering the generation of ROS as some kind of general stress response of seaweeds under environmental constrains, the need for easy-to-apply as well as for sensitive biochemical assays in order to characterize oxidative stress burdens under in situ conditions becomes apparent. In this respect, interdisciplinary projects between seaweed ecologists and biochemists arc particularly promising and should be considered as an essential approach for upcoming studies.

References

Aguilera J, Bischof K, Karsten U, Hanelt D, Wiencke C (2002a) Seasonal variation in ecophysiological patterns in macroalgae from an Arctic fjord II. Pigment accumulation and biochemical defence systems against high light stress. Mar Biol 140:1087–1095

Aguilera J, Dummermuth A, Karsten U, Schriek R, Wiencke C (2002b) Enzymatic defences against photooxidative stress induced by ultraviolet radiation in Arctic marine macroalgae. Polar Biol 25:432–441

Almeida M, Filipe S, Humanes M, Maia MF, Melo R, Severino N, da Silva JAL, da Silva J, Wever R (2001) Vanadium haloperoxidases from brown algae of the Laminariaceae family. Phytochemistry 57:633–642

Andersson B, Salter AH, Virgin I, Vass I, Styring S (1992) Photodamage to photosystem II-primary and secondary events. J Photochem Photobiol B Biol 15:15–31

Andrade S, Contreras L, Moffett J, Correa J (2006) Kinetics of copper accumulation in *Lessonia nigrescens* (Phaeophyceae) under conditions of environmental oxidative stress. Aquat Toxicol 78:398–401

Aro EM, Virgin I, Andersson B (1993) Photoinhibition of photosystem II. Inactivation, protein damage and turnover. Biochim Biophys Acta 1143:113–134

Asada K (1999) The water-water cycle in chloroplasts: Scavenging of active oxygens and dissipation of excess photons. Annu Rev Plant Physiol Plant Mol Biol 50:601–639

Asada K, Takahashi M (1987) Production and scavenging of active oxygen in photosynthesis. In: Kyle DJ, Osmond CB, Arntzen CJ (eds) Topics in photosynthesis 9. Elsevier Science Publishers, Amsterdam, pp 89–109

Bartsch I, Wiencke C, Bischof K, Buchholz CM, Buck BH, Eggert A, Feuerpfeil P, Hanelt D, Jacobsen S, Karez R, Karsten U, Molis M, Roleda MY, Schumann R, Schubert H, Valentin K, Weinberger F, Wiese J (2008) The genus *Laminaria sensu lato*: recent insights and developments. Eur J Phycol 43:1–86

Becker S, Walter B, Bischof K (2009) Freezing tolerance and photosynthetic performance of polar seaweeds at low temperatures. Bot Mar 52:609–619

Becker S, Graeve M, Bischof K (2010) Photosynthesis and lipid composition of the Antarctic endemic rhodophyte *Palmaria decipiens*: effects of changing light and temperature levels. Polar Biol 33:945–955

Benet H, Argall E, Asensi A, Kloareg B (1994) Protoplast regeneration from gametophytes and sporophytes of some species in the order Laminariales (Phaeophyceae). Protoplasma 199:39–48

Bischof K, Hanelt D, Wiencke C (1998) UV-radiation can affect depth-zonation of Antarctic macroalgae. Mar Biol 131:597–605

Bischof K, Kraebs G, Hanelt D, Wiencke C (2000) Photosynthetic characteristics and mycosporine-like amino acids under UV radiation: a competitive advantage of *Mastocarpus stellatus* over *Chondrus crispus* at the Helgoland shoreline? Helgol Mar Res 54:47–52

Bischof K, Peralta G, Kräbs G, van de Poll WH, Wiencke C, Perez-Llorens JL, Breeman AM (2002) Effects of solar UV-B radiation on canopy formation of natural *Ulva* communities from Southern Spain. J Exp Bot 53:2411–2421

Bischof K, Janknegt PJ, Buma AGJ, Rijstenbil JW, Peralta G, Breeman AM (2003) Oxidative stress and enzymatic scavenging of superoxide radicals induced by solar UV-B radiation in *Ulva* canopies from southern Spain. Sci Mar 67:353–359

Bischof K, Rautenberger R, Brey L, Pérez-Lloréns JL (2006) Physiological acclimation to gradients of solar irradiance within mats of the filamentous green macroalga *Chaetomorpha linum* from southern Spain. Mar Ecol Prog Ser 306:165–175

Burritt DJ, Larkindale J, Hurd CL (2002) Antioxidant metabolism in the intertidal red seaweed *Stictosiphonia arbuscula* following desiccation. Planta 215:829–838

Cock J, Sterck L, Rouze P, Scornet D, Allen A, Amoutzias G, Anthouard V, Artiguenave F, Aury JM, Badger J, Beszteri B, Billiau K, Bonnet E, Bothwell J, Bowler C, Boyen C, Brownlee C, Carrano C, Charrier B, Cho G, Coelho S, Collen J, Corre E, Da Silva C, Delage L, Delaroque N, Dittami S, Doulbeau S, Elias M, Farnham G, Gachon C, Gschloessl B, Heesch S, Jabbari K, Jubin C, Kawai H, Kimura K, Kloareg B, Küpper F, Lang D, Le Bail A, Leblanc C, Lerouge P, Lohr M, Lopez P, Martens C, Maumus F, Michel G, Miranda-Saavedra D, Morales J, Moreau H, Motomura T, Nagasato C, Napoli C, Nelson D, Nyvall-Collen P, Peters A, Pommier C, Potin P, Poulain J, Quesneville H, Read B, Rensing S, Ritter A, Rousvoal S, Samanta M, Samson G, Schroeder D, Segurens B, Strittmatter M, Tonon T, Tregear J, Valentin K, von Dassow P, Yamagishi T, Van de Peer Y, Wincker P (2010) The *Ectocarpus* genome and the independent evolution of multicellularity in brown algae. Nature 465:617–621

Colin C, Leblanc C, Wagner E, Delage L, Leize-Wagner E, Van Dorsselaer A, Kloareg B, Potin P (2003) The brown algal kelp *Laminaria digitata* features distinct bromoperoxidase and iodoperoxidase activities. J Biol Chem 278:23545–23552

Colin C, Leblanc C, Michel G, Wagner E, Leize-Wagner E, Van Dorsselaer A, Potin P (2005) Vanadium-dependent iodoperoxidases in *Laminaria digitata*, a novel biochemical function diverging from brown algal bromoperoxidases. J Biol Inorg Chem 10:156–166

Collén J, Davison IR (1999a) Reactive oxygen metabolism in intertidal *Fucus* spp (Phaeophyceae). J Phycol 35:62–69

Collén J, Davison IR (1999b) Production and damage of reactive oxygen in intertidal *Fucus* spp (Phaeophyceae). J Phycol 35:54–61

Collén J, Davison IR (1999c) Stress tolerance and reactive oxygen metabolism in the intertidal red seaweeds *Mastocarpus stellatus* and *Chondrus crispus*. Plant Cell Environ 22:1143–1151

Collén J, Davison IR (2001) Seasonality and thermal acclimation of reactive oxygen metabolism in *Fucus vesiculosus* (Phaeophyceae). J Phycol 37:474–481

Collén J, Pedersén M (1994) A stress-induced oxidative burst in *Eucheuma platycladum* (Rhodophyta). Physiol Planta 92:417–422

Collén J, Pedersén M (1996) Production, scavenging and toxicity of hydrogen peroxide in the green seaweed *Ulva rigida*. Eur J Phycol 31:265–271

Collén J, Jiménez del Río M, García-Reina G, Pedersén M (1995) Photosynthetic production of hydrogen peroxide by *Ulva rigida* C. Ag. (Chlorophyta). Planta 196:225–230

Contreras L, Moenne A, Correa J (2005) Antioxidant responses in *Scytosiphon lomentaria* (Phaeophyceae) inhabiting copper-enriched coastal environments. J Phycol 41:1184–1195

Contreras L, Mella D, Moenne A, Correa JA (2009) Differential responses to copper-induced oxidative stress in the marine macroalgae *Lessonia nigrescens* and *Scytosiphon lomentaria* (Phaeophyceae). Aquat Toxicol 94:94–102

Contreras-Porcia L, Dennett G, González A, Vergara E, Medina C, Correa J, Moenne A (2011) Identification of copper-induced genes in the marine alga *Ulva compressa* (Chlorophyta). Mar Biotechnol 13:544–556

Cosse A, Potin P, Leblanc C (2009) Patterns of gene expression induced by oligoguluronates reveal conserved and environment-specific molecular defense responses in the brown alga *Laminaria digitata*. New Phytol 182(1):239–250

Crepineau F, Roscoe T, Kaas R, Kloareg B, Boyen C (2000) Characterisation of complementary DNAs from the expressed sequence tag analysis of life cycle stages of *Laminaria digitata* (Phaeophyceae). Plant Mol Biol 43:503–513

Davison IR, Pearson GA (1996) Stress tolerance in intertidal seaweeds. J Phycol 32:197–211

de la Coba F, Aguilera J, Figueroa FL, de Gálvez MV, Herrera E (2009) Antioxidant activity of mycosporine-like amino acids isolated from three red macroalgae and one marine lichen. J Appl Phycol 21:161–169

Dring MJ (2005) Stress resistance and disease resistance in seaweeds: the role of reactive oxygen metabolism. advances in botanical research, vol 43. Academic, New York, pp 175–207

Dummermuth AL, Karsten U, Fisch FM, König GM, Wiencke C (2003) Responses of marine macroalgae to hydrogen-peroxide stress. J Exp Mar Biol Ecol 289:103–121

Eshdat Y, Holland D, Faltin Z, Ben-Hayyim G (1997) Plant glutathione peroxidases. Physiol Plant 100:234–240

Franklin LA, Osmond CB, Larkum AWD (2003) Photoinhibition, UV-B and Algal Photosynthesis. In: Larkum AWD, Douglas SE, Raven JA (eds) Photosynthesis in algae. Kluwer Academic, Dordrecht, pp 351–384

Goodwin KD, North WJ, Lidstrom ME (1997) Production of bromoform and dibromomethane by giant kelp: Factors affecting release and comparison to anthropogenic bromine sources. Limnol Oceanogr 42:1725–1734

Gross W (1993) Peroxisomes in algae: their distribution, biochemical function and phylogenic importance. In: Round FE, Chapman DJ (eds) Progress in phycological research, vol 9. Biopress Bristol, UK, pp 47–78

Guéraud F, Atalay M, Bresgen N, Cipak A, Eckl P, Huc L, Jouanin I, Siems W, Uchida K (2010) Chemistry and biochemistry of lipid peroxidation products. Free Radic Res 44:1098–1124

Gurgel CFD, Fredericq S (2004) Systematics of the Gracilariaceae (Gracilariales, Rhodophyta): a critical assessment based on *rbcL* sequence analyses. J Phycol 40:138–159

Hanelt D (1996) Photoinhibition and photosynthesis in marine macroalgae. Sci Mar 60:243–248

Jahns P, Latowski D, Strzalka K (2009) Mechanism and regulation of the violaxanthin cycle: The role of antenna proteins and membrane lipids. Biochim Biophys Acta 1787:3–14

Jordan P, Kloareg B, Vilter H (1991) Detection of vanadate-dependent bromoperoxidases in protoplasts from the brown algae *Laminaria digitata* and *L. saccharina*. J Plant Physiol 137:520–524

Karsten U, Sawall T, Hanelt D, Bischof K, Figueroa FL, Flores-Moya A, Wiencke C (1998) An inventory of UV-absorbing mycosporine-like amino acids in macroalgae from polar to warm-temperate regions. Bot Mar 41:443–453

Küpper F, Schweigert N, Ar Gall E, Legendre J-M, Vilter H, Kloareg B (1998) Iodine uptake in Laminariales involves extracellular, haloperoxidase-mediated oxidation of iodide. Planta 207:163–171

Küpper F, Kloareg B, Guern J, Potin P (2001) Oligoguluronates elicit an oxidative burst in the brown algal kelp *Laminaria digitata*. Plant Physiol 125:278–291

Küpper F, Gaquerel E, Boneberg E, Morath S, Salaün J, Potin P (2006) Early events in the perception of lipopolysaccharides in the brown alga *Laminaria digitata* include an oxidative burst and activation of fatty acid oxidation cascades. J Exp Bot 57:1991–1999

Küpper FC, Carpenter LJ, McFiggans GB, Palmer CJ, Waite TJ, Boneberg EM, Woitsch S, Weiller M, Abela R, Grolimund D, Potin P, Butler A, Luther GW, Kroneck PMH, Meyer-Klaucke W, Feiters MC (2008) Iodide accumulation provides kelp with an inorganic antioxidant impacting atmospheric chemistry. Proc Natl Acad Sci USA 105(19):6954–6958

Ledford HK, Niyogi KK (2005) Singlet oxygen and photo-oxidative stress management in plants and algae. Plant Cell Environ 28:1037–1145

Lohrmann NL, Logan BA, Johnson AS (2004) Seasonal acclimatisation of antioxidants and photosynthesis in *Chondrus crispus* and *Mastocarpus stellatus*, two co-occurring red algae with differing stress tolerances. Biol Bull 207:225–232

Lu IF, Sung MS, Lee TM (2006) Salinity stress and hydrogen peroxide regulation of antioxidant defense system in *Ulva fasciata*. Mar Biol 150:1–15

Mackerness SAH, John CF, Jordan B, Thomas B (2001) Early signalling components in ultraviolet-B responses: distinct role for different reactive oxygen species and nitric oxide. FEBS Lett 489:237–242

McCord JM, Fridovich I (1969) Superoxide dismutase—an enzymic function for erythrocuperein (hemocuprein). J Biol Chem 244:6049–6055

Mehrtens G, Laturnus F (1997) Halogenating activity in an Arctic population of brown macroalga *Laminaria saccharina* (L) Lamour. Polar Res 16:19–25

O'Dowd C, Jimenez J, Bahreini R, Flagan R, Seinfeld J, Hameri K, Pirjola L, Kulmala M, Jennings S, Hoffmann T (2002) Marine aerosol formation from biogenic iodine emissions. Nature 417:632–636

Osmond CB (1994) What is photoinhibition? Some insights from comparisons of shade and sun plants. In: Baker NR, Bowyer NR (eds) Photoinhibition of photosynthesis, from the molecular mechanisms to the field. BIOS Scientific Publications, Oxford, pp 1–24

Pacios L, Gálvez O (2010) Active site, catalytic cycle, and iodination reactions of vanadium iodoperoxidase: a computational study. J Chem Theor Comput 6:1738–1752

Palmer CJ, Anders TL, Carpenter LJ, Küpper FC, McFiggans GB (2005) Iodine and halocarbon response of *Laminaria digitata* to oxidative stress and links to atmospheric new particle production. Environ Chem 2:282–290

Pereira P, de Pablo H, Rosa-Santos F, Pacheco M, Vale C (2009) Metal accumulation and oxidative stress in *Ulva* sp. substantiated by response integration into a general stress index. Aquat Toxicol 91:336–345

Pinto E, Sigaud-Kutner T, Leitão MAS, Okamoto OK, Morse D, Colepicolo P (2003) Heavy metal-induced oxidative stress in algae. J Phycol 39:1008–1018

Polle A (1996) Mehler reaction: friend or foe in photosynthesis. Bot Acta 109:84–89

Potin P (2008) Oxidative burst and related responses in biotic interactions of algae. In: Amsler CD (ed) Algal chemical ecology. Springer, Berlin, pp 245–272

Ratkevicius N, Correa J, Moenne A (2003) Copper accumulation, synthesis of ascorbate and activation of ascorbate peroxidase in *Enteromorpha compressa* (L.) Grev. (Chlorophyta) from heavy metal-enriched environments in northern Chile. Plant Cell Environ 26:1599–1608

Rautenberger R, Bischof K (2006) Impact of temperature on UV-susceptibility of two *Ulva* (Chlorophyta) species from Antarctic and Subantarctic regions. Polar Biol 28:988–996

Rijstenbil JW, Coelho SM, Eijsackers M (2000) A method for the assessment of light-induced oxidative stress in embryos of fucoid algae via confocal laserscan microscopy. Mar Biol 137:763–774

Roeder V, Collén J, Rousvoal S, Corre E, Leblanc C, Boyen C (2005) Identification of stress gene transkripts in *Laminaria digitata* (Phaeophyceae) protoplast cultures by expressed sequence tag analysis. J Phycol 41:1227–1235

Ross C, Van Alstyne KL (2007) Intraspecific variation in stress-induced hydrogen peroxide scavenging by the ulvoid macroalga *Ulva lactuca*. J Phycol 43:466–474

Ross C, Küpper FC, Vreeland V, Waite JH, Jacobs RS (2005) Evidence of a latent oxidative burst in relation to wound repair in the giant unicellular chlorophyte *Dasycladus vermicularis*. J Phycol 41:531–541

Saenko G, Kravtsova Y, Ivanenko V, Sheludko S (1978) Concentration of iodine and bromine by plants in the seas of Japan and Okhotsk. Mar Biol 47:243–250

Schoenwaelder MEA (2002) The occurrence and cellular significance of physodes in brown algae. Phycologia 41:125–139

Sung MS, Hsu YT, Hsu YT, Wu TM, Lee TM (2009) Hypersalinity and hydrogen peroxide upregulation of gene expression of antioxidant enzymes in *Ulva fasciata* against oxidative stress. Mar Biotechnol 11:199–209

Svendsen H, Beszczynska-Møller A, Hagen JO, Lefauconnier B, Tverberg V, Gerland S, Ørbaek JB, Bischof K, Papucci C, Zajaczkowski M, Azzolini R, Bruland O, Wiencke C, Winther JG, Dallmann W (2002) The physical environment of Kongsfjorden–Krossfjorden, an Arctic fjord system in Svalbard. Polar Res 21:133–166

Thomas F, Cosse A, Goulitquer S, Raimund S, Morin P, Valero M, Leblanc C, Potin P (2011) Waterborne signaling primes the expression of elicitor-induced genes and buffers the oxidative responses in the brown alga *Laminaria digitata*. PLoS ONE 6(6):e21475. doi:10.1371/journal.pone.0021475

Vergara JJ, Sebastian M, Perez-Llorens JL, Hernandez I (1998) Photoacclimation of *Ulva rigida* and *U. rotundata* (Chlorophyta) arranged in canopies. Mar Ecol Prog Ser 165:283–292

Verhaeghe E, Fraysse A, Guerquin-Kern J-L, Wu T-D, Devés G, Mioskowski C, Leblanc C, Ortega R, Ambroise Y, Potin P (2008) Microchemical imaging of iodine distribution in the brown alga *Laminaria digitata* suggests a new mechanism for its accumulation. J Biol Inorg Chem 13:257–269

Weinberger F (2007) Pathogen-induced defense and innate immunity in macroalgae. Biol Bull 213(3):290–302

Weinberger F, Potin P (2010) Red Algal Defenses in the Genomics Age. In: Seckbach J, Chapman DJ (eds) Red Algae in the Genomic Age. Cellular origin, life in extreme habitats and astrobiology, vol 13. Springer, New York, pp 457–477

Weinberger F, Friedlander M, Hoppe H-G (1999) Oligoagars elicit a physiological response in *Gracilaria conferta*(Rhodophyta). J Phycol 35:747–755

Weinberger F, Leonardi P, Miravalles A, Correa JA, Lion U, Kloareg B, Potin P (2005a) Dissection of two distinct defense-related responses to agar oligosaccharides in *Gracilaria chilensis* (Rhodophyta) and *Gracilaria conferta* (Rhodophyta). J Phycol 41:863–873

Weinberger F, Pohnert G, Berndt ML, Bouarab K, Kloareg B, Potin P (2005b) Apoplastic oxidation of L-asparagine is involved in the control of the green algal endophyte *Acrochaete operculata* Correa & Nielsen by the red seaweed *Chondrus crispus* Stackhouse. J Exp Bot 56:1317–1326

Weinberger F, Guillemin M-L, Destombe C, Valero M, Faugeron S, Correa JA, Pohnert G, Pehlke C, Kloareg B, Potin P (2010) Defense evolution in the Gracilariaceae(Rhodophyta): substrate-regulated oxidation of agar oligosaccharides is more ancient than the oligoagar-activated oxidative burst. J Phycol 46:958–968

Wu T-M, Lee T-M (2008) Regulation of activity and gene expression of antioxidant enzymes in *Ulva fasciata* Delile (Ulvales, Chlorophyta) in response to excess copper. Phycologia 47:346–360

Wu TM, Hsu YT, Lee TM (2009a) Effects of cadmium on the regulation of antioxidant enzyme activity, gene expression, and antioxidant defenses in the marine macroalga *Ulva fasciata*. Bot Stud 50:25–34

Wu T-M, Hsu Y-T, Sung M-S, Lee T-M (2009b) Expression of genes involved in redox homeostasis and antioxidant defence in a marine macroalga *Ulva fasciata* by excess copper. Aquat Toxicol 94:275–285

Part II
Biotic Interactions

Chapter 7
Competition, a Major Factor Structuring Seaweed Communities

Matthew S. Edwards and Sean D. Connell

7.1 Competition Theory

Competition for shared resources has long been viewed as an important structuring agent in natural communities (Gause 1934; Strong 1980). Generally defined as the simultaneous use of limited resources by two or more organisms, competition can negatively impact population growth rates and influence community-level processes such as predation and/or how species respond to disturbances (Roughgarden 1979). This can occur between individuals of the same species or among individuals of different species. Early mathematical models such as those posed by Lotka (1925) and Volterra (1926) and later demonstrated by Gause (1934) suggest that when competition between two species is asymmetrical, one species will eventually drive the other to local extinction in their "struggle for existence." However, these models do not include the effects of other factors (e.g., predation, herbivory, environmental heterogeneity, or extreme environmental conditions) which often mediate the strength and outcome of competitive interactions (reviewed in Olson and Lubchenco 1990). In fact, even though two species rely on a shared resource, they may not compete if that resource is not limited or if the effects of other factors such as grazing or disturbances are substantially greater. However, when the shared resource is limited and competition for it is strong relative to other factors, the outcome of competition may vary as a function of both inter- and intraspecific interactions (Creese and Underwood 1982), especially as they relate to each species' own carrying capacity. Here, one species may drive

M.S. Edwards (✉)
Department of Biology, San Diego State University, 5500 Campanile Dr., San Diego, CA 92182, USA
e-mail: edwards@sciences.sdsu.edu

S.D. Connell
Southern Seas Ecology Laboratories, School of Earth and Environmental Sciences, University of Adelaide, DP 418, Adelaide 5005, Australia

C. Wiencke and K. Bischof (eds.), *Seaweed Biology*, Ecological Studies 219,
DOI 10.1007/978-3-642-28451-9_7, © Springer-Verlag Berlin Heidelberg 2012

the other to local extinction if the strength of their competitive abilities is asymmetrical, but they exhibit similar carrying capacities. Conversely, if the carrying capacities of the two species are substantially different, the species with the larger carrying capacity can drive the other species to local extinction even if their competitive abilities are equal.

While numerous examples of competition among individuals and/or species exist, debate over the general importance of competition as a structuring agent in natural communities remains. On the one side, Weins (1977) argues that competition among many species is actually rare in nature and that its importance as an agent of natural selection is therefore minimal. On the other hand, Diamond (1978) supports the view of Darwin (1859), who suggested that competition between closely related species is a deterministic factor in natural selection. In fact, Darwin wrote "As the species of the same genus usually have, though by no means invariably, much similarity in habits and constitution, and always in structure, the struggle will generally be more severe between them, if they come into competition with each other, than between species of distant genera." In discussing competition among macroalgae specifically, Miller (1967) notes that competition can occur via direct interactions in which one organism and/or species prevents the other from access to a limited resource (interference competition) or via indirect interactions where the species exploit a shared resource but do not directly interact with each other (exploitative competition). However, given that different algal traits such as rapid and/or indeterminate growth, large thallus size, and resistance to unfavorable environmental conditions can be associated with different competitive interactions, it may be difficult to identify if certain interactions are due to interference, exploitation, or a combination of the two (Olson and Lubchenco 1990). This problem can be exacerbated considering that different ecologists will often use different methodologies in their studies of competition and thus the results of their studies may vary simply due to the experimental approached used (e.g., Underwood and Fairweather 1986), and that experiments aimed at identifying the responses to competition may actually reflect only indirect evidence that it has actually occurred (e.g., Connell 1980). Consequently, demonstrating that competition is important to structuring natural communities requires carefully planned combinations of field observations to identify the patterns coupled with directed manipulative experiments to determine the processes behind them. These methods often involve removing one or more (e.g., the competitive dominant) species(s) and monitoring responses in the other species under consideration. In fact, a large number of studies have used these methods in studies of competition within macroalgal communities and have largely demonstrated the importance of both interspecific and intraspecific interactions as structuring agents. In this chapter, we review some of the main developments associated with these concepts and provide examples from case studies that demonstrate their importance. We focus our discussion on competition among benthic macroalgae, although it should be noted that among bloom-forming microalgae, both exploitative (i.e., for shared limited resources) and interference (i.e., via allelopathy) competition can be important in structuring planktonic communities, especially when the blooms are fully developed (e.g., Solé et al. 2005).

7.2 Competition Among Marine Macroalgae

7.2.1 Overview

The distribution of plants is regulated by both physical and biological factors, though the relative importance of these factors is notoriously variable through time and space, and is particularly dependent on the scale at which they are examined (Dayton and Tegner 1984; Edwards 2004). While plant populations may persist within a given ecosystem due to predation on herbivores as postulated by Hairston et al. (1960), in the absence of herbivory, plants would increase in number and ultimately compete among themselves for light, space, and nutrients. In fact, competition has been predicted to be the most important factor regulating the distribution and abundance of plant species in the absence of environmental stressors (Grime 1974, 1977, 1979). It is considered crucial in setting the latitudinal range limits (e.g., Edwards and Hernández-Carmona 2005; see also Chap. 3 by Eggert and Chap. 18 by Bartsch et al.) and tidal distributions (e.g., Hawkings and Hartnoll 1985) of some macroalgae, and regulating patterns of their succession following disturbances in others (e.g., Sousa 1979). However, the mere presence of a competitor may not be sufficient to cause strong competitive effects. Variation in morphology may have variable effects on the availability of and/or access to resources. For example, the exclusion of erect species from the understory within Australian *Ecklonia radiata* kelp forests depends on the morphology of the algae. Shorter and more flexible forms exclude erect taxa by shading and whiplashing the substratum (Connell 2003a; Irving and Connell 2006), whereas longer and more rigid forms do not exclude erect taxa from the understory which is strongly affected by shade (Kennelly 1987a). These differences have profound effects on the understory communities across Australia as 5,000 km of coastline across the Leeuwin Current (west and south coast) are largely structured by the short-flexible forms whereas the east coast (East Australian Current) understory is structured by the long-rigid forms (Connell and Irving 2009). These mechanisms match closely with variation in understory communities across temperate Australia (Irving et al. 2004). Furthermore, interactions among macroalgal holdfasts may result in strong interspecific competition, as seen in some intertidal red algae in Chile, but little-to-no intraspecific competition, as the holdfasts of the same species tend to coalesce (Santelices et al. 2003). What this means for their populations, however, is uncertain. One thing that is clear is that marine macroalgae are susceptible to competition as space, light, and nutrients are often in short supply in coastal ecosystems (reviewed in Carpenter 1990).

7.2.2 Competition for Light

Light is the primary resource requirement for algal growth, reproduction, and survival. Both the quantity (irradiance) and spectral quality (wavelength) are important in determining the distribution of many algal species, especially along depth gradients in the ocean (Kitching 1941; Saffo 1987). For example, Novaczek (1984) observed that the lower depth limit of *Eckonia radiata* off the coast of New Zealand was set by minimum light requirements of ca 40 mol photons m^{-2} d^{-1}. Conversely, Graham (1996) and Fejtek et al. (2011) found that the shallow limits of *Macrocystis pyrifera* and *Pelagophycus porra* (respectively) were set, at least in part, by high irradiance (PAR) effects on their microscopic gametophytes. Further, high UV-B irradiance appears to be especially important in reducing spore survival and consequently inhibiting algal recruitment in shallow water (e.g., Wiencke et al. 2000; Roleda et al. 2005; see also Chap. 1 by Hanelt and Figueroa and Chap. 20 by Bischof and Steinhoff). With regard to spectral quality, Lüning (1981) examined the effects of blue light on sexual reproduction in gametophytes of *Saccharina latissima* near Helgoland (North Sea) and observed that egg release by the gametophytes was reduced by 50% when they were held under only 1.4 μmol photons m^{-2} s^{-1} of blue light ($\lambda = 449$ nm) for 45 min. However, blue light may also have positive effects, such as enhancing nitrate (e.g., Aparicio et al. 1976) and carbon (e.g., Schmid and Dring 1996) uptake and storage as seen in some green and brown algae, respectively. While irradiance and spectral quality may be important in controlling recruitment, growth, survival, and reproduction in many algae, photoperiod may be as important in controlling the timing of their growth and reproduction. For example, Lüning (1986, 1990, 1994) and Lüning and Kadel (1993) observed that new frond formation in several brown algae is regulated by changes in photoperiod, resulting in circannual rhythms and synchronized seasonal growth. Further, Edwards (1998) observed that recruitment in the brown alga *Desmarestia ligulata* in central California, USA, was closely tied to increase in day length during the early spring, but this was limited to areas where the dominant kelp canopies had been removed either experimentally or by winter storms. In summary, light quantity and quality appear important to macroalgal physiology and thus factors that reduce access to light may be integral in establishing spatial and temporal patterns in their distribution and abundance, especially in the ocean where irradiance rapidly diminishes with increasing depth.

In addition to natural attenuation of light in the ocean (e.g., Kirk 1992), light is absorbed and/or scattered by the macroalgae themselves, placing them in both direct and indirect competition with each other. As a result, many species of macroalgae have developed functional morphology forms that allow them to more effectively capture light for photosynthesis in their respective habitats (Vadas and Steneck 1988). These morphologies range from single cells to multicellular parenchymatous thalli with complex tissue differentiation. Within this later group, many species have evolved morphologies that elevate their photosynthetic blades above their competitors. For example, Kitching (1941) observed that on the shores of Carsaig Island, Scotland, the dominant macroalga *Laminaria digitata* possesses erect nonflexible stipes that hold its blades above the substrate in order to more effectively capture light when occurring in the subtidal where light can be limiting, but flexible stipes that allow its blades to lay prostrate in the water when occurring in the intertidal where light was otherwise abundant but desiccation may be an issue. Similarly, many subtidal kelps produce canopies that are either buoyed at the surface by gas-filled pneumatocysts or are supported above the substrate by thick woody stipes (e.g., Abbott and Hollenberg 1976). These canopies place their photosynthetic blades above those of their competitors and enabling them to more effectively capture light before it reaches their competitors (Horn 1971; Hay 1986). This structure results in a rich mosaic of benthic light regimes below the canopies (Gerard 1984) to which macroalgae can "photoadapt" by altering the amount and/or arrangement of their photosynthetic pigments (Ramus 1981; Rosenberg and Ramus 1982). Combined with other factors such as substrate topography, hydrodynamic forces, and grazing, this can result in benthic algal communities being highly variable at a number of spatial scales (Dayton et al. 1984, 1992, 1999; Foster and VanBlaricom 2001; Edwards 2004).

Much in the way competitive dominance by canopy-forming plants plays a structuring role in terrestrial forest communities (e.g., Brokaw and Scheiner 1989; Connell 1989; Spies and Franklin 1989; Hubbell et al. 1999), shading from kelp canopies can be the determining factor in regulating benthic macroalgal populations (Reed and Foster 1984; Kennelly 1989; Edwards 1998; Connell 2003b). For example, it is well understood that in the terrestrial forests, the removal of only a few canopy dominants can prevent the competitive exclusion of many understory species and thereby promote greater diversity (Whitmore 1989; Brokaw and Scheiner 1989). These understory species can be grouped into three general categories based on how they respond to canopy shading. This "ecological response group" approach has been widely used by terrestrial plant ecologists to classify individual species according to how they respond to changes in their light environment and has been very useful in studies on canopy shading (e.g., Collins et al. 1985; Whitmore 1989; Kursar and Coley 1999). The logic behind the response-group approach is straightforward; canopy removal should elicit positive responses in recruitment and/or growth for species that require high irradiances, negative responses for species that require low irradiances, and little-to-no responses for species adapted to a variety of irradiances (Brokaw and Scheiner 1989; Spies and Franklin 1989; Whitmore 1989; Kursar and Coley 1999). Those species, then, that remain rare under dense canopies but rapidly recruit into canopy gaps or following widespread canopy removal have typically been referred to as "light-adapted," "gap-requiring," or "shade-intolerant" and are often considered to represent fugitive species, while those species that occur under forest canopies but do not markedly respond to canopy loss have typically been referred to as "light-flexible" or "shade-tolerant" species and are often considered to represent climax species (Whitmore 1989; Hay 1994; Clark et al. 2004). Whereas light-flexible species are generally numerically more abundant than light-adapted species both under canopies and in canopy gaps, their ability to withstand low light environments may result in a decreased ability to rapidly respond to sudden increases in light (Canham 1989). As a consequence, light-adapted species, through greater recruitment and growth, typically dominate areas following canopy loss.

In temperate marine communities, where shading by thick canopies of kelps and rockweeds (Orders Laminariales and Fucales, respectively) can regulate understory algal abundance in shallow (<30 m) water (Pearse and Hines 1979; Foster 1982;

Reed and Foster 1984; Kennelly 1987a; Harrold et al. 1988; Dayton et al. 1992), the removal of the dominant canopies typically results in increased bottom light and a corresponding increase in the abundance of opportunistic species (Dayton et al. 1984, 1992; Reed and Foster 1984; Kennelly 1987b; Cecchi and Cinelli 1992; Graham 1996; Edwards 1998). In fact, North et al. (1986) conclude that light is the primary factor regulating species abundances within coastal forests, a claim that has been experimentally tested in numerous studies via experimental canopy removal. For example, Ambrose and Nelson (1982) observed that removal of the invasive *Sargassum muticum* at Santa Catalina Island, USA, resulted in reduced recruitment of the giant kelp *Macrocystis pyrifera* by reducing benthic irradiance. Likewise, Reed and Foster (1984) found that removal of the *Macrocystis pyrifera* surface canopy and the subsurface *Pterygophora californica* canopy in a central California, USA, kelp forest resulted in increased recruitment of understory algae, as well as the kelps themselves. Similarly, Clark et al. (2004) replicated Reed and Foster's canopy clearings at three areas in the same central California kelp forest and followed changes to the understory algae for a period of 2 years. Their clearing design (Fig. 7.1) allowed for the simultaneous testing of the individual and combined effects of shading from both surface and subsurface canopies on understory algal assemblages, and from an opportunistic alga, *Desmarestia ligulata*, that recruited into the clearings in very high abundances. Their results indicated that while understory algae did respond to the canopy clearings, the low abundances of individual species and the small magnitude of each species' response compared to their natural temporal and spatial variability made detecting canopy effects difficult. However, when understory species were grouped together in ecological response groups, they were able to detect otherwise cryptic increases in some (i.e., light-adapted) species as much as 1 year earlier than when each species was examined individually. The exception to this was the opportunistic brown alga *Desmarestia ligulata* which showed dramatic rapid increases within the canopy clearings. Edwards (1998) examined this further and found that while *Desmarestia ligulata* remained in low abundances under existing canopies, it recruited in high abundance in the spring and ultimately reached high bottom cover in areas where the canopies were removed (Fig. 7.2). Similar patterns have been observed for *Desmarestia ligulata* in Point Loma, CA, following canopy removal by winter storms (Dayton et al. 1984) and in three central California kelp forests characterized by different hydrodynamic conditions and canopy covers (Foster 1982). Dayton et al.'s work further described that disturbance to the dominant kelp canopies resulted in variation in the benthic light regimes and a corresponding mosaic of understory algal patches. Some of these patches were able to persist for extended periods of time and competitively exclude or delay recovery of the otherwise dominant kelps. The effects of this canopy shading, however, are not limited to interspecific interactions but also impact individuals of their own species via intraspecific interactions. For instance, shading from the dominant *Macrocystis pyrifera* canopies also inhibits recruitment and growth of their own juvenile sporophytes (Anderson and North 1969; Reed and Foster 1984; Dean et al. 1989). Similar negative effects of intraspecific competition were reported by Neushul and

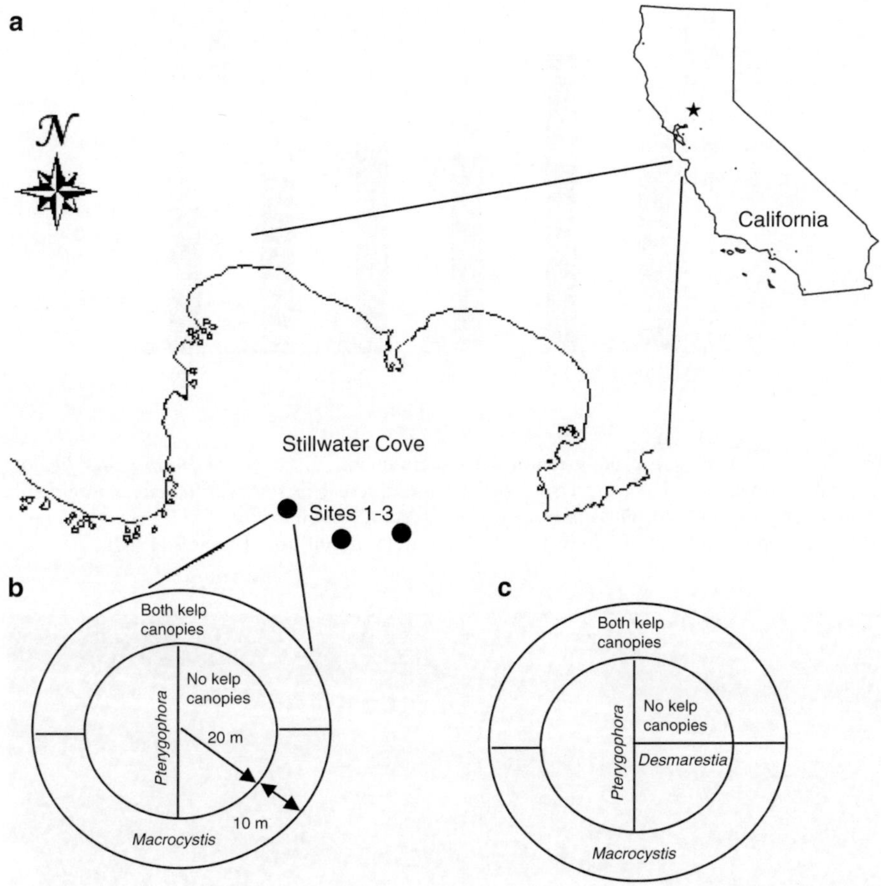

Fig. 7.1 Diagram of kelp canopy clearing experiment used by Clark et al. (2004) showing (**a**) location of the three replicate sites in Stillwater Cove, CA, USA. (**b**) Blocked canopy clearing design showing the 20 m radius inner circles and 10 m radius outer tori from which kelp canopies were removed either individually or in combination. Treatments include "No canopies" = both *Macrocystis* and *Pterygophora* removed, "*Pterygophora* canopy" = *Macrocystis* removed and *Pterygophora* left unmanipulated, "*Macrocystis* canopy" = *Pterygophora* removed and *Macrocystis* left unmanipilated, and "Both canopies" = both kelp species left unmanipulated (also considered as control treatment). (**c**) Canopy clearings showing the placement of the *Desmarestia* removal treatment following dense *Desmarestia* recruitment. Diagram reproduced from Clark et al. (2004)

Harger (1985) and Reed (1987). Consequently, as the dominant kelps begin to recover following a disturbance, they often recruit in much higher densities than can be sustained when they are adults and thus undergo strong intraspecific competition (i.e., density dependence) that results in self-thinning of the population (Fig. 7.3). However, large variability exists in the interpretation of how important density dependence is to algal populations, in part due to differences in the experimental methodologies used to examine it (Flores-Moya et al. 1996; Scrosati 2005).

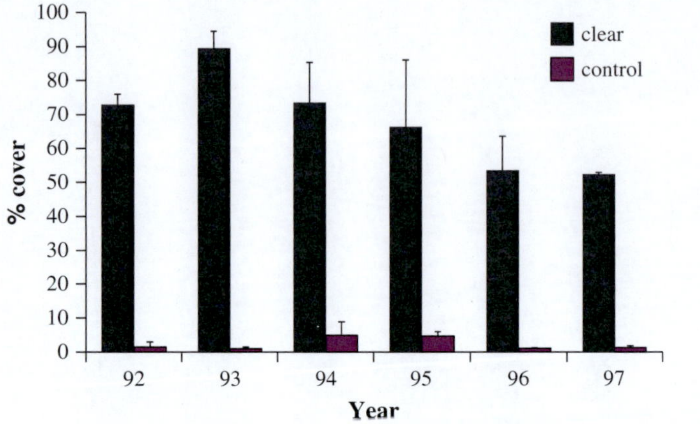

Fig. 7.2 Effects of kelp canopy removal on recruitment (% cover) of *Desmarestia ligulata* sporophytes (means + 1standard error). *Black bars* represent recruitment into areas where kelp canopies had been cleared, and *white bars* represent recruitment into areas where the kelp canopies were unmanipulated (controls) from 1992 to 1997. Reproduced from Edwards (1998)

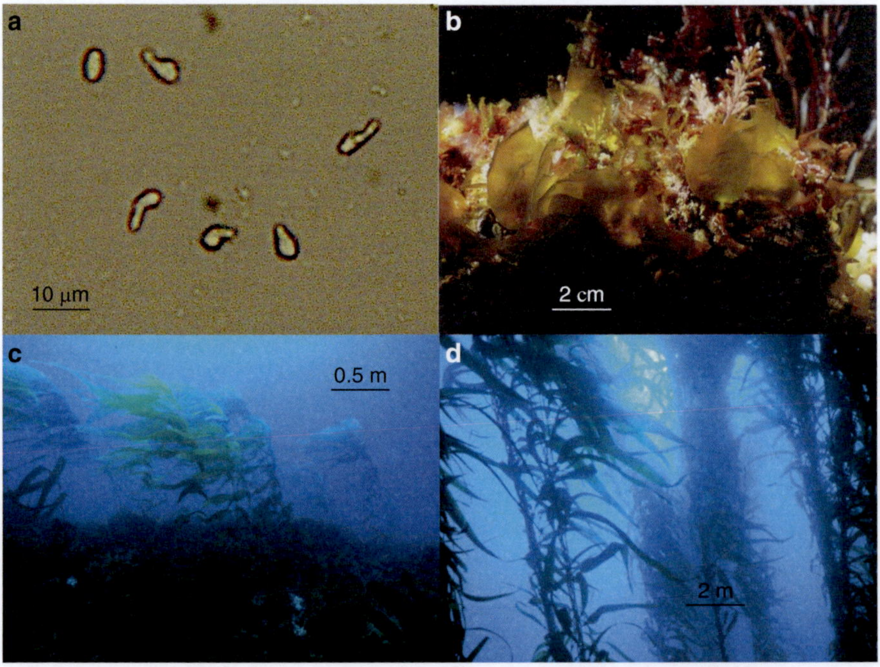

Fig. 7.3 Photographs of giant kelp (*Macrocystis pyrifera*) showing relative densities of different ages as the population self-thins. **a** = young sporophytes (photo taken on a microscope slide), (**b**) recruits, (**c**) juveniles, (**d**) adults. Note the differences in scales showing that younger smaller stages are much more densely packed

Regardless, it is clear that self-thinning is important to many seaweed populations, though this may be more important for unitary species such as *Macrocystis pyrifera* than for clonal species such as *Mastocarpus papillatus* where self-thinning during the growing season is not expected even under high density (Flores-Moya et al. 1996; Scrosati 2006).

While studies along the west coast of North America have shown dramatic responses by the turf-forming algae to kelp canopy clearings, results from other areas of the world are more variable. For example, clearings to the surface and subsurface kelp canopies within *Nereocystis leutkeana* forests in coastal Alaska resulted in complete overgrowth by subsurface kelps within 1 year, while similar clearings within the *Eualaria fistulosa* forest throughout the Aleutian Archipelago resulted in little to no macroalgal recruitment over 1-year period (B. Konar et al. unpublished data). While the reason for this is unclear, we believe it is primarily due to variation in identities and life histories of the species involved (i.e., whether they are annual or perennial), oceanographic conditions (i.e., temperature, nutrient, and wave conditions), grazing pressure, and idiosyncrasies associated with the experiments themselves (i.e., when the clearings were done, how large and thorough they were). However, even with this variability, certain similarities exist among geographic areas. For example, throughout much of Australia, the conditions created by subtidal kelp canopies include reductions in irradiance, sedimentation, and the abundance of turf-forming algae (Kennelly 1987a). Consequently, the kelp canopies maintain encrusting coralline algae from overgrowth by turf-forming algae through shade (Kennelly 1987a; Connell 2003b) and possibly through reduced accumulation of sediments (Melville and Connell 2001). Thus, the clearance of kelp canopies generally leads to the overgrowth of encrusting algae by turf-forming algae, which then trap sediments (Kennelly and Underwood 1993; Connell 2005). Conversely, recovery of the kelp canopies invariably leads to the reverse effect (e.g., Kennelly 1987b), demonstrating that the co-occurrence of canopy and encrusting coralline algae is not just a spatial and temporal coincidence, but rather one driven by the state of the dominant canopy-forming species. Given the strength of association across regions and different taxa in Australia, there appears to be considerable predictive capacity in understanding how the understory macroalgae benefit from the physical light environment created by the canopies (e.g., Bertness and Leonard 1997). These patterns are consistent with the relative abundance of encrusting and turf-forming algae, where the former are relatively abundant in locations of low light and sedimentation (Steneck 1986, 1997), while the latter monopolize locations of high light and high sedimentation (Airoldi 1998).

7.2.3 Competition for Space

On rocky surfaces, access to adequate space is crucial for the attachment and survival of organisms (e.g., Connell 1961). Except for a few examples of free-living algal (e.g., some species in genera *Ulva* (formerly *Enteromorpha*) and

Sargassum, and several species of coralline algae called rhodoliths), most marine macroalgae are attached to rocky reef, other algae, invertebrates, or sediments. This results in their interacting for attachment sites in two dimensions rather than three and thus greatly increases the chances that they will compete for access to the substrate. This will also occur between algae and sessile invertebrates such as is commonly observed on degraded coral reefs where algae are considered competitively dominant to coral larvae and new recruits, but competitively subordinate to coral adults (reviewed in McCook et al. 2001; see also Chap. 16 by Mejia et al. and Chap. 18 by Bartsch et al.). As a consequence, algal proliferation often occurs in areas where coral adults have died, which in turn prevents coral recruitment and slows community recovery (e.g., Kuffner et al. 2006). However, while algal–invertebrate competition is common, we do not address it further here but rather focus our attention to competition between and within algal species.

Although algae may compete strongly for attachment space on the substrate, when examined closely, the actual area of attachment to the substrate is often very small compared to the overall size of the algal thallus. For example, on coral reefs, the actual attachment site of the dominant algae can be as small as 10–100 μm with lots of bare space present between the holdfasts (Hackney et al. 1989). Further, Dayton (1975) found that less than 20% of rocky reef in a subtidal temperate kelp forest was occupied by the kelp holdfasts even though the kelp canopies covered the majority of the reef. Consequently, competition among macroalgae for space, per se, may not be a limiting factor in recruitment in many locations (reviewed in Carpenter 1990). However, given that the thalli of many algae tend to lie prostrate on the substrate or physically abrade the substrate as they move in the waves (Velimirov and Griffiths 1979; Witman 1987; Kennelly 1989), they can prevent settlement and/or growth of other algae on the substrate via interference (abrasion) and exploitative (shading) competition (Duggins and Dethier 1985). Removal of the dominant species, then, generally results in an increase in other macroalgae as observed for intertidal *Hedophyllum sessile* communities in the San Juan Islands, off the coast of Washington State, USA (Duggins and Dethier 1985). Further, some of the strongest competitive interactions in the southern hemisphere tend to be created by large canopy-forming algae (e.g., intertidal *Durvillaea antarctica*; subtidal *Ecklonia radiata*). These individuals exclude many understory species via their fronds which shade or whiplash the substratum (Kennelly 1987b). However, these effects may vary geographically at a number of spatial scales (Connell and Irving 2009) and among different benthic species (Kennelly 1989). Likewise, in the Aleutian Archipelago, physical abrasion from the brown alga *Desmarestia viridis* is an important factor keeping other algae from establishing patches on the tops of pinnacles (Konar and Estes 2003). Physical abrasion by kelp fronds can also prevent recruitment of sessile invertebrates (Connell 2003a) which otherwise monopolize space under lower light and sedimentation such as observed on South Australian coasts (Irving and Connell 2006) or of motile invertebrates that would otherwise graze on the algae (Konar and Estes 2003). Hence, body size often serves as the best measure of the potential for algae and other types of sessile species to exert strong community-wide effects.

Exceptions to the notion that macroalgae do not compete strongly for space have been observed in turf-forming species that can occupy nearly 100% of the substrate in some places and thereby prevent settlement of other algae (e.g., Hruby and Norton 1979), and in areas dominated by encrusting algae which can compete intensely for space (Steneck and Paine 1986). On many temperate coasts, encrusting coralline algae can also occupy up to 80% of hard substrate, dominating space beneath macroalgal canopies (Irving et al. 2004). With the creation of available space (e.g., through storm damage), however, turfs rapidly colonize the space and overgrow the coralline crusts (Russell and Connell 2005; Worm et al. 1999). The morphology of these filamentous turfs allows them to then trap sediment (Airoldi 1998; Gorgula and Connell 2004) and subsequently inhibit the recruitment of canopy-forming species in localities from a diversity of locations around the world (Devinny, 1978; Gorman and Connell 2009). In addition, recruitment of the larger macroalgae can occur on the thalli of the turf algae, though the added hydrodynamic drag on the turf holdfast can result in increased removal rates from waves and thus low survival (Edwards unpublished data). In an example of competition among three species of intertidal crustose and turf algae, Padilla (2003) observed coralline turf algae were better competitors than the upright forms for space, as they were more susceptible to grazing and desiccation. In contrast, two upright forms were inferior competitors for space but were more resistant to grazing and desiccation. Thus, the coralline algae coexisted on the rocky shore, though in different microhabitats. Furthermore, in many temperate systems, if the dominant kelp canopies are excluded from a reef, turf algae may increase in abundance over time and ultimately become the dominant space occupiers, especially in areas where the canopies have been removed (e.g., Edwards 1998; Connell 2003b).

Competition for space itself can be strongly mediated by both physical (disturbance) and biological (competition) factors. For example, Kastendiek (1982) observed three species of algae at Santa Catalina Island, CA, USA, competing strongly with each other but that the nature of this competition differed greatly between species pairs, each of which was influenced by physical disturbances differently. Specifically, the kelp *Eisenia arborea* forms canopies that excludes (or reduces its abundance) the brown alga *Halidrys dioica*, thereby allowing the turf alga *Pterocladia capillacea* to dominate the substrate under the canopy. However, if the *Eisenia arborea* canopy is removed, *Halidrys dioica* recruits in high abundances and dominates the substrate, thus excluding *Pterocladia capillacea* from the area. This switch between competition for light and competition for space appears to be driven by disturbance to the dominant *Eisenia arborea* canopy. Similarly, competition for space may be strongly dependent on species identity and whether the interaction is interspecific or intraspecific. For example, Santelices et al. (2003) observed that in the intertidal region of Chile, when the holdfasts of the intertidal red algae *Mazzaella laminarioides* and *Nothogenia fastigiata* come into contact, one outcompetes the other resulting in its death. However, when the holdfasts of either species come into contact with a member of their own species, the holdfasts coalesce and both individuals survive, suggesting that the populations might be chimeric. Finally, competition for space may be integral in regulating patterns of zonation in the intertidal. Much in the way that Connell (1961) concluded that the upper and lower limits of species distributions in the intertidal may be set by competition and/or predation, Hawkings and Hartnoll (1985) concluded that bands of both dense and sparse macroalgal populations across intertidal shores of the United Kingdom were set by competitive interactions. Specifically, when fronds of the brown algae *Fucus* and *Laminaria* were removed, all other algal species occurring immediately below them were able to colonize further up in the intertidal zone. Similarly, competition may also set the lower limit of those species, as they were also able to colonize the lower parts of the shore following the removal of their competitors.

7.2.4 Competition for Nutrients

Algae require adequate carbon, nitrogen, and phosphorus for growth and survival (see also Chap. 4 by Gordillo), and competition among seaweeds for them has been observed in numerous studies. While the majority of studies on the effects of nutrient limitation have been done for freshwater phytoplankton communities, and have generally reported strong competition for nutrients, their conclusions vary depending on the specific species involved and complex synergies associated with multiple resources (e.g., nitrogen, phosphorus, and silica) that fluctuate independently with one another (e.g., Titman 1976). Unlike freshwater ecosystems, nitrogen (namely nitrate NO_3^-) is the primary limiting nutrient in the ocean, though phosphorus may occasionally be in limited supply. The primary source of nitrogen input into most coastal ecosystems is upwelling of cold nutrient-rich water from below the thermocline, though terrestrial inputs from runoff may also be important in some locations (Gorman et al. 2009). Algae take up the nitrogen directly from the water column via mass transport and thus uptake rates are strongly influenced by thallus morphology ("functional form," surface/volume ratios), tissue and seawater nitrogen content (concentration gradients), and hydrodynamic flow which in turn regulate boundary layer formation around the algal thallus (Hurd et al. 1997). Consequently, during periods of low nutrients and low hydrodynamic flow, algae can compete intensely for nutrients. For example, Dayton and Tegner (1984) report that along southern California, USA, the dominant current flow is caused by longshore currents (see also Winant and Bratovich 1981). While internal waves may be important in cross-shore delivery of nutrients, the longshore currents are the primary source of nutrient delivery into the kelp forest. However, drag imposed on the moving water by the kelp thalli can reduce current speeds within the kelp forest by as much as 33% resulting in reduced nutrient delivery to the inner parts of the forest. Further, the nutrients are taken up by those kelps near the periphery of the forest resulting in further nutrient depletion within the forest. In fact, Jackson and Winant (1983) concluded that nutrient uptake within the Point Loma, CA, USA, kelp forests can be 23% per hour, resulting in the water being nutrient depleted by the time it reaches the interior of the kelp forest, though this may be partially ameliorated by cross-shore delivery of nutrients via internal waves. Consequently, while the individuals near the edge of the forests may be healthy, individuals in the interior can become nutrient stressed and appear less healthy during periods of low nutrient availability, though this varies considerably with forest size (Bernstein and Jung 1979; Dayton et al. 1984). This nutrient limitation may also be important to reproduction in many species. For example, Reed (1990) concluded that both interspecific and intraspecific competition between the microscopic gametophytes of *Macrocystis pyrifera* and *Pterygophora californica* was intense when they settled at high densities but not when they settled at low densities. He attributed this to competition for nutrients even though the culture media was changed regularly. In contrast, Vadas (1972) found no evidence of competition among the gametophytes of *Nereocystis luetkeana*, though this may have been because his overall densities were substantially lower than those used by Reed. Further, Carney and Edwards (2010) observed that delaying development as 1–2 celled gametophytes for 1–3 months can ameliorate the negative effects of intraspecific competition among kelp gametophytes, though she did not test how this impacts eventual sporophyte recruitment.

7.3 Competition in Complex Environments

7.3.1 Synergistic Effects of Different Factors

While it is true that limitation of one or more resources can be important to the recruitment, survival, and reproduction in many macroalgae, synergistic effects among numerous environmental factors may result in substantial variability in how species respond to these limitations. For example, Fujita (1985a, b) reports that when nutrients are plentiful, the green algae *Ulva* spp. (formerly *Enteromorpha*) are able to outcompete the red algae *Gracilaria tikvahiae* for space in the intertidal due to its greater uptake and growth rates. However, *Gracilaria tikvahiae* possesses greater nutrient storage capabilities and therefore outcompetes *Ulva* during periods when nutrients are limited. Likewise, on the New England, USA coast, where the brown alga *Fucus vesiculosus* is the dominant competitor for space over *Ulva* spp. in the absence of grazers, *Ulva* spp. is the competitive dominant species when grazers are present. This occurs because the grazers remove the fast growing young stages of *Ulva* spp., thus allowing the slower growing grazer-resistant *Fucus* to grow. Along the Pacific coast of Baja California, MEX, the giant kelp *Macrocystis pyrifera* forms a surface canopy above the stipitate kelp *Eisenia arborea* and thus outcompetes *Eisenia arborea* for light. However, *Eisenia arborea* is more resistant to low nutrient conditions such as those observed during El Niño Southern Oscillation (ENSO) events (Hernández-Carmona et al. 2001) and thus is able to persist during periods when *Macrocystis pyrifera* dies. Following the ENSO, *Eisenia arborea* is able to form dense subsurface canopies that exclude *Macrocystis pyrifera* recruitment and thus prevent recovery of its populations for several years (Edwards and Hernández-Carmona 2005). Likewise, Dayton et al. (1984) observed similar patterns between *Macrocystis pyrifera* the subsurface kelps *Pterygophora californica* and *Laminaria farlowii* following the 1982–1984 ENSO in a southern California kelp forest. Here, the subsurface kelps were more resistant to the large ocean waves associated with this disturbance and thus exhibited greater survival. Much as with *Eisenia arborea* in Baja California, these kelps were able to delay *Macrocystis pyrifera* recruitment, resulting in longer term changes to the understory algal patch dynamics within the kelp forest. Similar changes to competitive hierarchies have been observed in other areas of the world in response to environmental factors. For example, although *Ecklonia radiata* is generally competitively dominant to turf algae along the South Australia coast, nutrient discharge to the normally oligotrophic waters causes turfs to dominate the space and inhibit recruitment of *Ecklonia radiata* (Gorman and Connell 2009), resulting in a competitive outcome that can persist for several decades (Connell et al. 2008). Finally, Duggins and Dethier (1985) note that while the intertidal kelp *Hedophyllum sessile* outcompetes other algae for space in the intertidal areas of the San Juan Islands, its competitive abilities are regulated by grazing from the chiton *Kathrina tunicata* and by wave exposure, both of which act to remove *Hedophyllum sessile* and create bare space for other algae to recruit. Taken together, it is clear that the competitive abilities of at least some macroalgae can be a trade-off with features that are adapted to the environment. For example, while the giant kelp *Macrocystis pyrifera* develops large canopies above its neighbors and thus outcompetes them for light, its large size and high growth rate make it more susceptible to disturbance from large waves and low nutrients than many of its neighbors (Dayton and Tegner 1984). In their review on competition in seaweeds, Olson and Lubchenco (1990) write, "Because the traits of seaweeds have evolved in a context of multiple selection pressures, the consequences of traits will vary with environmental conditions." Consequently, a comprehensive investigation of competition in heterogeneous environments requires studying how competition operates under different combinations of environmental conditions.

7.3.2 The Importance of Competition in Maintaining Marine Forests

Marine forests form diverse biological habitats and largely persist by facilitating their own recruitment through the competitive exclusion of opportunistic turf-forming algae (Dayton et al. 1984; Reed and Foster 1984; Connell 2005; Irving and Connell 2006). When kelp canopies are lost, however, turfs rapidly colonize space and inhibit the recruitment of juvenile kelp and reformation of kelp forests (Kennelly 1987a; Gorman and Connell 2009). Under conditions of elevated nutrients, these naturally ephemeral turfs persist in fragmented forests (Airoldi et al. 2008; Gorman et al. 2009) to cause intergenerational decline and collapse of the kelp community (Connell et al. 2008). The competitive dominance of kelps over turfs is a key interaction that enables kelp forests to regenerate and persist, but this dominance may be reversed under conditions that favor the persistence of turfs (Gorman and Connell 2009). As a consequence, it is clear that variation in algal assemblages within and among kelp forests can be attributed partly to differences in the primary canopy-disturbing agents, ocean waves, and temperature (Foster 1975; Cowen et al. 1982; Foster 1982; Dayton et al. 1984, 1992; Seymour et al. 1989; Breda and Foster 1985; Harrold et al. 1988; Graham 1997). This, however, can be highly variable especially on a seasonal scale. For instance, along parts of the west coast of North America, large ocean waves associated with winter storms remove large quantities of kelp canopies, typically resulting in maximum kelp canopies in summer and minimum canopies in winter (Kimura and Foster 1984; Reed and Foster 1984; Dayton et al. 1992). Consequently, competitive interactions between the canopy-forming and understory species can vary temporally, further increasing the overall variation in kelp forest community composition (Pearse and Hines 1979; Gerard 1984; Kennelly 1987a, b; Harrold et al. 1988; Dayton et al. 1999). For example, Edwards and Hernández-Carmona (2005) found that along the coast of Baja California Sur, MEX, the dominant *Macrocystis pyrifera* canopies typically reduce the abundance of the understory *Eisenia arborea* canopies. However, all the *Macrocystis pyrifera* canopies were lost during the 1997–1998 ENSO, resulting in a dramatic increase in *Eisenia arborea* canopies that were able to prevent *Macrocystis pyrifera* recovery at its southern range limit for up to 20 years.

7.3.3 Competition in the Face of Climate Change

The role of global environmental change in driving change to macroalgal communities in marine ecosystems has received heightened attention (e.g., Wernberg et al. 2010, 2011). While range contractions and extensions have been forecast as a function of changing climate (e.g., temperature), the effects on competitive interactions among species are less clear. Unlike kelps, many turf-forming species are ephemeral and require increased resource availability to enable their physiology and life history to be competitively superior to perennial species (Airoldi et al. 2008). Coastal waters that facilitate the expansion of turf-forming algae (e.g., enhanced CO_2, temperature, and nutrients; Gorgula and Connell 2004; Connell and Russell 2010) tend to maintain their dominance of space at the expense of canopy-forming algae (e.g., Gorman and Connell 2009). Much of the global research effort into forecasting the effects of climate change focuses on the direct effects, thereby overlooking indirect effects (e.g., competitive effects). As the name implies, "indirect" effects are not as simple to identify as "direct" effects and can often yield "unexpected results" (Wootton 1994). They are unanticipated because the impact of one species on another (i.e., competitors) requires knowledge of a third species or mediating component is poorly understood. Research into ocean acidification (see Chap. 19 by Roleda and Hurd) is a good case in point. Marine waters absorb approximately 30% of the anthropogenic-derived CO_2 from the earth's atmosphere and the resulting ocean acidification (Feely et al. 2004; Orr et al. 2005) and direct negative effects on calcifying algae. The difficulty is that for noncalcifying algae, elevated inorganic carbon has positive effects on some species and not others, and that these nonuniform effects among alternate species (review by Gao and McKinley 1994) have relatively unexplored consequences to competitive hierarchies, particularly if growth is limited by sources of inorganic carbon.

If competitive interactions are reduced by the increasingly novel conditions brought by human activities, then the persistence of entire biological communities is increasingly likely to be disrupted. Such phase shifts are not uncommon, but anticipating them has been problematic because many involve indirect effects for which the impact of one species (e.g., turfs) on another (e.g., kelp) requires knowledge of a third element that is inadequately understood (e.g., synergies among pollutants). The strength of competition (e.g., dominance of kelp over turfs) is important to understand because it identifies the conditions in which turfs expand. The mere presence of foundation species may not always maintain their competitive strength over turfs. As canopies are thinned, reduced in size, or fragmented, their positive effects are reduced and the associated environmental conditions become more similar to those experienced outside the canopy (Jackson and Winant 1983; Bruno and Bertness 2001). Under these conditions, turfs expand to dominate space and inhibit the recruitment of kelp (Gorman and Connell 2009; Connell and Russell 2010), leading to phase shifts over multiple generations (Connell et al. 2008). Therefore, the maintenance of intact populations of foundation species seems critical to the strength of interaction (i.e., continued dominance over kelp inhibitors) and ensuing maintenance of community structure and function (Bruno and Bertness 2001). The competitive balance between species is often shifted by the abiotic environment which alters the relative abilities of algae to recruit to and retain space. Hence climate change seems set to bring change to competitive processes as we currently understand them.

7.4 Concluding Remarks

In conclusion, competition within and among marine macroalgae is a deterministic influence in establishing patterns of biogeography, regulating growth and reproduction, and in maintaining populations at or below their carrying capacities. It is important to setting species range limits, governing how populations will respond to disturbances, and in structuring coastal ecosystems. However, the direction, strength, and importance of these competitive interactions vary considerably with species identity, the location where these interactions take place, and with changes to the physical and biological environment. Many species have evolved morphologies that enable them to better access the resources, and thus outcompete their neighbors, but these morphologies also may make them more susceptible to environmental stresses, thereby temporarily altering competitive hierarchies and ameliorating the effects of limited resources. Even so, numerous studies have demonstrated that competition can be the most important factor in regulating marine macroalgal population and community dynamics. However, while our understanding of these interactions has grown considerably since Lotka (1925), Volterra (1926), and Gause (1934) wrote their seminal works, information on how these interactions will work in the future remains lacking. Perhaps the most important next step is to assess how these competitive interactions will change under environmental variability, such as predicted with climate change. Understanding this may better inform us of how macroalgal communities will likely be structured in the future.

References

Abbott IA, Hollenberg GJ (1976) Marine algae of California. Stanford University Press, Stanford, CA

Airoldi L (1998) Roles of disturbance, sediment stress, and substratum retention on spatial dominance in algal turf. Ecology 79:2759–2770

Airoldi L, Balata D, Beck MW (2008) The gray zone: relationships between habitat loss and marine diversity and their applications in conservation. J Exp Mar Biol Ecol 366:8–15

Ambrose RF, Nelson BV (1982) Inhibition of giant kelp by an introduced brown alga. Bot Mar 25:265–267

Anderson EK, North WJ (1969) Light requirements of juvenile and microscopic stages of giant kelp, *Macrocystis*. Proc Int Seaweed Symp 6:3–15

Aparicio PJ, Roldán JM, Calero F (1976) Blue light photoreactivation of nitrate reductase from green algae and higher plants. Bioch Biophys Res Comm 70:1071–1977

Bernstein BB, Jung N (1979) Selective pressures and coevolution in a kelp canopy in southern California. Ecol Monogr 49:335–355

Bertness MD, Leonard GH (1997) The role of positive interactions in communities: lessons from intertidal habitats. Ecology 78:1976–1989

Breda VA, Foster MS (1985) Composition, abundance, and phenology of foliose red algae associated with two central California kelp forests. J Exp Mar Biol Ecol 94:15–130

Brokaw NVL, Scheiner SM (1989) Species composition in gaps and structure of tropical forests. Ecology 70:538–541

Bruno J, Bertness M (2001) Habitat modification and facilitation in benthic marine communities. In: Bertness M, Gaines S, Hay M (eds) Marine community ecology. Sinauer Associates, Sutherland, pp 201–220

Canham CD (1989) Different responses to gaps among shade-tolerant tree species. Ecology 70:569–576

Carney LT, Edwards MS (2010) Role of nutrient fluctuations and delayed development in gametophyte reproduction by *Macrocystis pyrifera* (Phaeophyceae) in southern California. J Phycol 46:987–996

Carpenter RC (1990) Competition among macroalgae: a physiological perspective. J Phycol 26:6–12

Cecchi LB, Cinelli F (1992) Canopy removal experiments in *Cystoseira*-dominated rockpools from the Western coast of the Mediterranean (Ligurian Sea). J Exp Mar Biol Ecol 15:569–583

Clark RP, Edwards MS, Foster MS (2004) Effects of shade from multiple kelp canopies on an understory algal assemblage. Mar Ecol Prog Ser 267:107–119

Collins BS, Dunne KP, Pickett TA (1985) Responses of forest herbs to canopy gaps. In: Pickett ST, White PS (eds) The ecology of natural disturbance and patch dynamics. Academic, Orlando, pp 218–234

Connell JH (1961) The influence opf interspecific competition and other factors on the distribution of the barnacle *Chthamalus stallatus*. Ecology 42:710–723

Connell JH (1980) Diversity and the coevolution of competitors, or the ghost of competition past. Oikos 35:131–138

Connell JH (1989) Some processes affecting the species composition in forest gaps. Ecology 70:560

Connell SD (2003a) Negative effects overpower the positive of kelp to exclude invertebrates from the understorey community. Oecologia 137:97–103

Connell SD (2003b) The monopolization of understorey habitat by subtidal encrusting coralline algae: a test of the combined effects of canopy-mediated light and sedimentation. Mar Biol 142:1065–1071

Connell SD (2005) Assembly and maintenance of subtidal habitat heterogeneity: synergistic effects of light penetration and sedimentation. Mar Ecol Prog Ser 289:53–61

Connell SD, Irving AD (2009) The subtidal ecology of rocky coasts: local-regional-biogeographic patterns and their experimental analysis. In: Witman JD, Kaustuv R (eds) Marine macroecology. University of Chicago Press, Chicago, pp 392–417

Connell SD, Russell BD (2010) The direct effects of increasing CO2 and temperature on non-calcifying organisms: increasing the potential for phase shifts in kelp forests. Proc R Soc B-Biol Sci 277:1409–1415

Connell SD, Russell BD, Turner DJ, Shepherd SA, Kildea T, Miller DJ, Airoldi L, Cheshire A (2008) Recovering a lost baseline: missing kelp forests from a metropolitan coast. Mar Ecol Prog Ser 360:63–72

Cowen RK, Agegian CR, Foster MS (1982) The maintenance of community structure in a central California giant kelp forest. J Exp Mar Biol Ecol 64:189–201

Creese RG, Underwood AJ (1982) Analysis of inter- and intra-specific competition amongst limpets with different methods of feeding. Oecologia 53:337–346

Darwin C (1859) On the origin of species by means of natural selection, or the preservation of favoured races in the struggle for life. John Murray, London

Dayton PK (1975) Experimental evaluation of ecological dominance in a rocky intertidal algal community. Ecol Monogr 45:137–159

Dayton PK, Tegner MJ (1984) The importance of scale in community ecology: a kelp forest example with terrestrial analogs. In: Price PW, Slobodchikoff CN, Gaud WS (eds) A new ecology: novel approaches to interactive systems. Wiley, New York

Dayton PK, Currie V, Gerrodette TB, Keller D, Rosenthal R, VanTresca D (1984) Patch dynamics and stability of some California Kelp Communities. Ecol Monogr 54:253–289

Dayton PK, Tegner MJ, Parnell PE, Edwards PB (1992) Temporal and spatial patterns of disturbance and recovery in a kelp forest community. Ecol Monogr 62:421–445

Dayton PK, Tegner MJ, Edwards PB, Riser KL (1999) Temporal and spatial scales of kelp demography: the role of oceanographic climate. Ecol Monogr 69:219–250

Dean TA, Thies K, Lagos SL (1989) Survival of juvenile giant kelp: the effects of demographic factors, competitors, and grazers. Ecology 70:483–495

Devinny YS (1978) Ordination of seaweed communities: environmental gradients at Punta Banda, Mexico. Bot Mar 21:357–363

Diamond JM (1978) Niche shifts and the rediscovery of intraspecific competition. Am Sci 66:322–331

Duggins DO, Dethier MN (1985) Experimental studies of herbivory and algal competition in a low intertidal habitat. Oecologia 67:183–191

Edwards MS (1998) Effects of long-term kelp canopy exclusion on the abundance of the annual alga *Desmarestia ligulata* (Light F). J Exp Mar Biol Ecol 228:309–326

Edwards MS (2004) Estimating scale dependency in disturbance impacts: El Niños and giant kelp forests in the Northeast Pacific. Oecologia 138:436–447

Edwards MS, Hernández-Carmona G (2005) Delayed recovery of giant kelp near its southern range limit in the North Pacific following El Niño. Mar Biol 147:273–279

Feely RA, Sabine CL, Lee K, Berelson W, Kleypas J, Fabry V, Millero FJ (2004) Impact of anthropogenic CO_2 on the $CaCO_3$ system in the oceans. Science 305:362–366

Fejtek SM, Edwards MS, Kim KY (2011) Elk kelp, *Pelagophycus porra,* distribution limited due to susceptibility of microscopic stages to high light. J Exp Mar Biol Ecol 396:194–20

Flores-Moya A, Ferández JA, Niell FX (1996) Growth pattern, reproduction and self-thinning in seaweeds: a re-evaluation in reply to Scrosati. J Phycol 32:767–769

Foster MS (1975) Regulation of algal community development in the *Macrocystis pyrifera* forest. Mar Biol 32:331–342

Foster MS (1982) The regulation of macroalgal associations in kelp forests. In: Srivastava L (ed) Synthetic and degradative processes in marine macrophytes. Walter de Gruyter Co., Berlin

Foster MS, VanBlaricom GR (2001) Spatialvariation in kelp forest communities along the Big Sur coast of central California USA. Cryptogamie Algol 22:173–186

Fujita RM (1985a) The role of nitrogeno suplí variability in regulating nitrogen uptake by macroalgae and in structuring a macroalgal community. Ph.D. thesis, Boston University, Massachusetts, p 143

Fujita RM (1985b) The role of nitrogen status in regulating transient ammonium uptake and nitrogen storage by macroalgae. J Exp Mar Biol Ecol 92:283–301

Gao K, McKinley KR (1994) Use of macroalgae for marine biomass production and CO_2 remediation. A review. J Appl Phycol 6:45–60

Gause GF (1934) The struggle for existence. Hafner Publ. Co, New York, p 163

Gerard VA (1984) The light environment in a giant kelp forest: influence of *Macrocystis pyrifera* on spatial and temporal variability. Mar Biol 84:189–195

Gorgula SK, Connell SD (2004) Expansive covers of turf-forming algae on human-dominated coast: the relative effects of increasing nutrient and sediment loads. Mar Biol 145:613–619

Gorman D, Connell SD (2009) Recovering subtidal forests on human-dominated landscapes. J Appl Ecol 46:1258–1265

Gorman D, Russell BD, Connell SD (2009) Land-to-sea connectivity: linking human-derived terrestrial subsidies to subtidal habitat change on open rocky coasts. Ecol Appl 19:1114–1126

Graham MH (1996) Effect of high irradiance on recruitment of the giant kelp Macrocystis. J Phycol 32:903–906

Graham MH (1997) Factors determining the upper limit of giant kelp, *Macrocystis pyrifera* Agardh, along the Monterey Peninsula, central California, USA. J Exp Mar Biol Ecol 218:127–149

Grime JP (1974) Vegetation classification by reference to strategies. Nature 250:26–31

Grime JP (1977) Evidence for the existence of three primary strategies in plants and its relevance to ecological and evolutionary theory. Am Nat 111:1169–1194

Grime JP (1979) Plant strategies and vegetation processes. Wiley, New York, p 222

Hackney JM, Carpenter RC, Adey WH (1989) Characteristic adaptations to grazing among algal turfs on a Caribbean coral reef. Phycologia 28:109–119

Hairston NG, Smith FE, Slobodkin LB (1960) Community structure, population control, and competition. Am Nat 94:421–425

Harrold CJ, Watanabe J, Lisin S (1988) Spatial variation in the structure of kelp forest communities along a wave exposure gradient. P.S.Z.N.I. Mar Ecol 9:131–156

Hawkings SJ, Hartnoll RG (1985) Grazing on intertidal algae by marine invertebrates. Ocean Mar Biol Annu Rev 21:195–282

Hay ME (1986) The functional morphology of turf-forming seaweeds: persistence in stressful marine habitats. Ecology 62:739–750

Hay ME (1994) Species as 'noise' in community ecology: do seaweeds block our view of the kelp forest? Trends Ecol Evol 9:414–416

Hernández-Carmona G, Robledo D, Serviere-Zaragoza E (2001) Effect of nutrient availability on *Macrocystis pyrifera* recruitment survival near its southern limit of Baja California. Bot Mar 44:221–229

Horn HS (1971) The adaptive geometry of trees. Princeton University Press, Princeton, NJ, 144

Hruby T, Norton TA (1979) Algal colonization on rocky shores in the Firth of Clyde. J Ecol 67:65–77

Hubbell SP, Foster RB, O'Brien ST, Harms KE, Condit R, Welchsler B, Wright SJ, Loo se Lao S (1999) Light-gap disturbances, recruitment limitation, and tree diversity in a neotropical forest. Science 283:554–557

Hurd CL, Stevens CL, Laval BE, Lawrence GA, Harrison PJ (1997) Visualization of seawater flow around morphologically distinct forms of giant kelp *Macrocystis integrifolia* from wave-sheltered and exposed sites. Limnol Oceanogr 42:156–163

Irving AD, Connell SD (2006) Physical disturbance by kelp abrades erect algae from the understorey. Mar Ecol Prog Ser 324:127–137

Irving AD, Connell SD, Gillanders BM (2004) Local complexity in patterns of canopy-benthos associations produce regional patterns across temperate Australasia. Mar Biol 144:361–368

Jackson GA, Winant CD (1983) Effects of a kelp forest on coastal currents. Cont Shelf Res 2:75–80

Kastendiek J (1982) Competitor-mediated coexistence: interactions among three species of benthic macroalgae. J Exp Mar Biol Ecol 62:201–210

Kennelly SJ (1987a) Physical disturbances in an Australian kelp community II. Effects on understorey species due to differences in kelp cover. Mar Ecol Prog Ser 40:155–165

Kennelly SJ (1987b) Physical disturbances in an Australian kelp community I. Temporal effects. Mar Ecol Prog Ser 40:145–153

Kennelly SJ (1989) Effects of canopies on understory species due to shade and scour. Mar Ecol Progr Ser 50:215–224

Kennelly SJ, Underwood AJ (1993) Geographic consistencies of effects of experimental physical disturbance on understory species in sublittoral kelp forests in central New South Wales. J Exp Mar Biol Ecol 168:35–58

Kimura S, Foster MS (1984) The effects of harvesting *Macrocystis pyrifera* on the algal assemblage in a giant kelp forest. Hydrobiology 116(117):425–428

Kirk JTO (1992) The nature and measurement of the light environment in the ocean. In: Falkowski A (ed) Primary productivity and biogeochemical cycles in the sea. Plenum, New York, pp 9–29

Kitching JA (1941) Studies in sublittoral ecology: III. *Laminaria* Ffrest on the west coast of Scotland; a study of zonation in relation to wave action and illumination. Biol Bull 80:324–337

Konar BK, Estes JA (2003) The stability of boundary regions between kelp beds and deforested areas. Ecology 84:174–185

Kuffner IB, Walters LJ, Becerro MA, Paul VJ, Ritson-Williams R, Beach KS (2006) Inhibition of coral recruitment by macroalgae and cyanobacteria. Mar Ecol Prog Ser 323:107–117

Kursar TA, Coley PD (1999) Contrasting modes of light acclimation in two species of the rain forest understory. Oecologia 121:489–498

Lotka AJ (1925) Elements of physical biology. Williams & Wilkins, Baltimore

Lüning K (1981) Egg release in gametophytes of *Laminaria saccharina*: induction by darkness and inhibition by blue light and U.V. Br Phycol J 16:379–393

Lüning K (1986) New frond formation in *Laminaria hyperborean* (Phaeophyta): a photoperiodic response. Br Phycol J 21:269–273

Lüning K (1990) Circannual growth rhythm in a brown alga, *Pterygophora californica*. Bot Acta 104:157–162

Lüning K (1994) When do algae grow? The third founders lecture. Eur J Phycol 29:61–67

Lüning K, Kadel P (1993) Daylength range for circannual rhythmicity in *Pterygophora californica* (Alariaceae, Phaophyta) and synchronization of seasonal growth by daylength cycles in several other brown algae. Phycologia 32:379–387

McCook LJ, Jompa J, Diaz-Pulido G (2001) Competition between corals and algae on coral reefs: a review of evidence and mechanisms. Coral Reefs 19:400–417

Melville AJ, Connell SD (2001) Experimental effects of kelp canopies on subtidal coralline algae. Aust Ecol 26:102–108

Miller RS (1967) Pattern and process in competition. Adv Ecol Res 4:1–74

Neushul M, Harger BWW (1985) Studies of biomass yield from a near-shore macroalgal test farm. J Solar Energ Eng 107:93–96

North WJ, Jackson GA, Manley SL (1986) *Macrocystis* and its environment, knowns and unknowns. Aquat Bot 26:9–26

Novaczek (1984) Response of Eklonia radiata (Laminariales) to light at 15 °C with reference to the field light budget at Goat Island Bay, New Zealand. Mar Biol: 263–272

Olson AM, Lubchenco J (1990) Competition in seaweeds: linking plant traits to competitive outcomes. J Phycol 26:1–6

Orr JC, Fabry VJ, Aumont O, Bopp L, Doney SC, Feely RA, Gnanadesikan A, Nicolas Gruber N, Ishida A, Joos F, Key RM, Lindsay K, Maier-Reimer E, Matear R, Patrick Monfray P, Mouchet A, Najjar RG, Plattner GK, Rodgers KB, Sabine CL, Sarmiento JL, Schlitzer R, Slater RD, Totterdell IJN, Weirig MF, Yamanaka Y, Yool A (2005) Anthropogenic ocean acidification over the twenty-first century and its impacts on calcifying organisms. Nature 437:681–686

Padilla DK (2003) The importance of form: differences in competitive ability, resistance to consumers and environmental stress in an assemblage of coralline algae. J Exp Mar Biol Ecol 79:105–127

Pearse JS, Hines AH (1979) Expansion of central California kelp forests following the mass mortality of sea urchins. Mar Biol 51:83–91

Ramus J (1981) The capture and transduction of light energy. In: Lobban C, Wynne M (eds) The biology of seaweeds. Blackwell, Oxford Ramus

Reed DC (1987) Factors affecting the production of sporophylls in the giant kelp *Macrocystis pyrifera* (L.) C.Ag. J Exp Mar Biol Ecol 113:61–69

Reed DC (1990) The effects of variable settlement and early competition on patterns of kelp recruitment. Ecology 71:776–787

Reed DC, Foster MS (1984) The effects of canopy shading on algal recruitment and growth in a giant kelp forest. Ecology 65:937–948

Roleda MY, Wiencke C, van de Poll WH, Gruber A (2005) Sensitivity of Laminariales zoospores from Helgoland (North Sea) to ultraviolet and photosynthetically active radiation: implications for depth distribution and seasonal reproduction. Plant Cell Environ 28:466–479

Rosenberg G, Ramus J (1982) Ecological growth strategies in the seaweeds *Gracilaria folifera* (Rhodophyceae) and *Ulva* sp. (Chlorophyceae): photosynthesis and antenna composition. Mar Ecol Prog Ser 8:233–241

Roughgarden J (1979) Theory of population genetics and evolutionary ecology: an introduction. Macmillan, London

Russell BD, Connell SD (2005) A novel interaction between nutrients and grazers alters relative dominance of marine habitats. Mar Ecol Prog Ser 289:5–11

Saffo MB (1987) New light on seaweeds. Bioscience 37:654–664

Santelices B, Aedo D, Hormazábal M, Flores V (2003) Field testing of inter- and intraspecific coalescence among mid-intertidal red algae. Mar Ecol Prog Ser 250:91–103

Schmid R, Dring MJ (1996) Blue light and carbon acquisition in brown algae: an overview and recent developments. Sci Mar 60:115–124

Scrosati R (2005) Review of studies on biomass-density relationship (including self-thinning lines) in seaweeds: Main contributions and persisting misconceptions. Phycol Res 53:224–233

Scrosati R (2006) Crowding in clonal seaweeds: does self-thinning occur in *Mastocarpus papillatus* shortl;y before stand biomass peaks? Aquat Bot 84:233–238

Seymour RJ, Tegner MJ, Dayton PK, Parnell PE (1989) Storm wave induced mortality of giant kelp, *Macrocystis pyrifera*, in southern California. Estuar Coast Shelf Sci 28:277–292

Solé J, Garía-Ladona E, Ruardij P, Estrada M (2005) Modelling allelopathy among marine algae. Ecol Model 183:373–384

Sousa WP (1979) Experimental investigations of disturbance and ecological succession in a rocky intertidal algal community. Ecol Monogr 49:227–254

Spies TA, Franklin JF (1989) Gap characteristics and vegetation response in coniferous forests of the Pacific Northwest. Ecology 70:543–545

Steneck RS (1986) The ecology of coralline algal crusts: convergent patterns and adaptive strategies. Annu Rev Ecol Syst 17:273–303

Steneck RS (1997) Crustose corallines, other algal functional groups, herbivores and sediments: complex interactions along reef productivity gradients. Proc 8th Int Coral Reef Symp 1:695–700

Steneck RS, Paine RT (1986) Ecological and taxonomic studies of shallow-water encrusting Corallinaceae (Rhodophyta) of the boreal northeastern Pacific. Phycologia 25:221–240

Strong DR (1980) Null hypotheses in ecology. Synthese 43:271–285

Titman D (1976) Ecological competition between algae: experimental confirmation of resource-based competition theory. Science 192:463–465

Underwood AJ, Fairweather PG (1986) Intertidal communities: do they have different ecologies or ecologists? Proc Ecol Soc Aust 14:7–16

Vadas R (1972) Ecological implications of culture studies on *Nereocystis luetkeana*. J Phycol 8:196–203

Vadas RL, Steneck RS (1988) Zonation of deep water algae ion the Gulf of Maine. J Phycol 24:338–346

Velimirov B, Griffiths CL (1979) Wave-induced kelp movement and its importance for community structure. Bot Mar 22:169–172

Volterra V (1926) Variazioni e fluttuazioni del numero d'individui in specie animali conviventi. Mem R Accad Naz dei Lincei Ser. VI, 2

Weins JA (1977) On competition and variable environments. Am Sci 65:590–597

Wernberg T, Thomsen MA, Tuya F, Kendrick GA, Staehr PA, Toohley BD (2010) Decreasing resilience of kelp beds along a latitudinal temperature gradient: potential implications for a warmer future. Ecol Lett 13:685–694

Wernberg T, Russell BC, Moore PJ, Ling SD, Smale DA, Campbell A, Coleman MA, Steinberg PD, Kendrick GA, Connell SD (2011) Impacts of climate change in a global hotspot for temperate marine biodiversity and ocean warming. J Exp Mar Biol Ecol 400:7–16

Whitmore TC (1989) Canopy gaps and the two major groups of forest trees. Ecology 70:569–576

Wiencke C, Gómez I, Pakker H, Flores-Moya A, Altamirano M, Hanelt D, Bischof K, Figueroa FL (2000) Impact of UV radiation on viability, photosynthetic characteristics and DNA of brown algal zoospores: implications for depth zonation. Mar Ecol Prog Ser 197:217–229

Winant CD, Bratovich AW (1981) Temperature and currents on the southern California shelf: a description of the variability. Phys Oceanogr 11:71–86

Witman JD (1987) Subtidal coexistence: storms, grazing, mutualism, and the zonation of kelps and mussels. Ecol Monogr 57:167–187

Wootton JT (1994) The nature and consequences of indirect effects in ecological communities. Annu Rev Ecol Syst 25:443–466

Worm B, Lotze HK, Boström C, Engkvist R, Labanauskas V, Sommer U (1999) Marine diversity shift linked to interactionsamong grazers, nutrients and propagule banks. Mar Ecol Prog Ser 185:309–314

Chapter 8
Grazers on Benthic Seaweeds

Katrin Iken

8.1 Common Classification Systems of Grazers

Seaweed grazers can be classified in several ways, most commonly by their feeding mode, their size, or their diet specificity, with continua rather than distinct groups within each of these classifications (Table 8.1). These classifications are not only of practical use for researchers but also have many ecological implications.

First, grazers can be distinguished by functional feeding mode, such as grazing, browsing, rasping, biting, tearing, etc. Grazer mouthparts can be specifically adapted to feeding on seaweeds, and even on particular algal functional groups with different morphological thallus characteristics (thickness, branching, calcification; Littler and Littler 1980). For example, gastropod grazing has been distinguished based on specific radula types (Steneck and Watling 1982), herbivorous fishes can develop specialized mouthparts such as the "beak" in parrot fish, and the Aristotle's lantern of sea urchins is particularly suitable for grazing on tough kelps. The functional group approach is intriguing to ecologists because it may allow detection of generalizable and predictable patterns in ecosystem functioning rather than focusing on small differences among individual species. This approach, however, may also overemphasize the functional redundancy in a system by obscuring small but ecologically important differences in food selection or assimilation within a particular functional group of grazers. For example, herbivorous coral reef fishes classified within the same functional group displayed strong species-specific food selectivity among algal species, suggesting a low degree of functional redundancy within this feeding group and that species-specific herbivore effects can be missed when looking at functional groups alone (Mantyka and Bellwood 2007).

K. Iken (✉)
School of Fisheries and Ocean Sciences, University of Alaska Fairbanks, Fairbanks, AK 99775, USA
e-mail: kbiken@alaska.edu

C. Wiencke and K. Bischof (eds.), *Seaweed Biology*, Ecological Studies 219,
DOI 10.1007/978-3-642-28451-9_8, © Springer-Verlag Berlin Heidelberg 2012

Table 8.1 Common classifications of seaweed grazers

Grazer taxon	Feeding mode	Size group	Diet specificity	Typical habitat	Selected examples
Amphipods	Biting Tearing	Mesograzers	Specialists–generalists	Tropical rocky shores/coral reefs Temperate rocky shores Polar rocky shores	*Pseudoamphithoides* feeds selectively on and builds domiciles from the chemically defended brown alga *Dictyota* (Hay et al. 1990a) *Amphitoe* grazing is driven by food choice and habitat quality of host algae (Duffy and Hay 1991), and can control brown algal biomass and competitive dominance in temperate Atlantic reefs (Duffy and Hay 2000) *Aora* is a generalist feeder on taxonomically and morphologically diverse macroalgae in New Zealand (Taylor and Brown 2006) Amphipods occur in high abundance on Antarctic macroalgae (Huang et al. 2007) and *Gondogeneia* selectively feeds on the red alga *Palmaria* (Amsler et al. 2009; Aumack et al. 2010)
Isopods	Biting Tearing	Mesograzers	Specialists	Intertidal/shallow subtidal temperate fucoid communities	*Idotea* influences biomass and distribution of *Fucus* in the Baltic Sea (Engkvist et al. 2000) and of *Ascophyllum* in the North Atlantic (Viejo and Åberg 2003) *Idotea* controls epiphyte growth on *Fucus* induced by high nutrient loads (Worm et al. 2000; Orav-Kotta and Kotta 2004)
Polychaetes	Biting Tearing	Mesograzers	Generalists	Temperate rocky shores	Herbivorous polychaetes, mostly *Platynereis*, preferentially feed on filamentous and branched algae in the Mediterranean (Antoniadou and Chintiroglou 2006)
Prosobranch gastropods	Rasping Browsing	Mostly mesograzers	Generalists	Temperate rocky intertidal Temperate kelp beds Polar rocky shores	Selective grazing of the periwinkle *Littorina* in rocky tidepools controls abundance of the dominant space competitor *Ulva* (Lubchenco 1978)

					Lacuna grazing at high densities can affect kelp populations and survival (Chenelot and Konar 2007) The abalone *Haliotis* preferentially consumes kelp (mostly drift) in New Zealand (Cornwall et al. 2009) *Laevilacunaria* is a generalist grazer on Antarctic seaweeds, but also depends on epiphytic microalgae (Iken 1999)
Opisthobranch gastropods (Ascoglossa, Anaspidea)	Biting Rasping	Mostly mesograzers	Specialists	Tropical coral reefs	The ascoglossan *Costasiella* selectively feeds on the chemically defended green alga *Avrainvillea* (Hay et al. 1990b) The anaspidean *Aplysia* performs best when feeding on members of the red alga *Portieria* with moderate chemical defense levels (Ginsburg and Paul 2001)
Crabs	Biting Browsing	Macrograzers	Generalists	Tropical rocky shores	*Grapsus* preferentially feeds on filamentous algae; it supplements its diet with animal items when seasonally forced to feed on low nutritional encrusting algae (Kennish 1996)
Sea urchins	Biting Tearing	Macrograzers	Specialists–generalists	Temperate kelp beds Tropical coral reefs	*Strongylocentrotus* grazing causes phase shifts from kelp beds to urchin barrens in the absence of top predators in the Aleutian Islands (Estes and Duggins 1995) *Diadema* grazing removes macroalgae from coral reefs, thus facilitating space for coral growth and recruitment (Carpenter and Edmunds 2006)
Fishes (e.g., Scaridae Pomacentridae Acanthuridae	Biting Tearing Browsing	Macrograzers	Specialists–generalists Specialists–generalists Specialists–generalists	Tropical coral reefs Temperate kelp beds Polar rocky shores	Damselfishes defend territories on coral reefs to "garden" algae on which they feed, thus influencing reef algal and overall communities (Jones et al. 2006)

(continued)

Table 8.1 (continued)

Grazer taxon	Feeding mode	Size group	Diet specificity	Typical habitat	Selected examples
Siganidae Blenniidae) (e.g., Blenniidae Girellidae Kyphosidae Stichaeidae Scorpididae) (Nototheniidae)					Reef fishes of the same family or functional group can exhibit strong selectivity for particular reef algae, and thus their grazing effect on the reef community may differ (Mantyka and Bellwood 2007) Selective grazing on *Ulva* by the abundant blenny *Scartichthys* in the Chilean rocky intertidal impacts overall community structure (Ojeda and Muñoz 1999) *Notothenia* feeds selectively on some of the available seaweeds in Antarctica, depending to ~40% on algal diet overall (Iken et al. 1997)
Reptiles (Iguana)	Biting Tearing	Macrograzers	Generalists	Tropical systems	The marine iguana *Amblyrhynchus* in the Galapagos feeds on intertidal and subtidal seaweeds, with algal abundance- and temperature-dependent feeding rates to optimize food intake (Shepherd and Hawkes 2005)
(Green turtle)	Biting Tearing	Macrograzers	Generalists	Tropical systems	Green turtles mostly feed on seagrasses but also consume significant portions of seaweeds, possibly selectively; on an individual basis they may even specialize on an algal over seagrass diet (reviewed in Bjorndal 1997)
Sirenians	Biting Tearing Browsing	Macrograzers	Generalists	Tropical systems	Dugongs feed mostly on seagrasses but can consume considerable amounts of seaweeds either opportunistically or more targeted when seagrass diet is sparse (Whiting 2002)
Birds	Tearing Browsing	Macrograzers	Generalists	Temperate and tropical systems	Brent geese grazing can significantly deplete green algal biomass but has little long-term effects on these fast-growing algae (Rowcliffe et al. 2001)

Second, grazers can be separated by size, with a main separation between meso- and macrograzers (Table 8.1); virtually nothing is known about the impact of the smallest group, micrograzers (<0.1 cm), on seaweeds. Mesograzers are small (mostly 0.1–2.5 cm), and can include juveniles of some larger species, with typically small spatial feeding ranges (Hay et al. 1987; Arrontes 1999). Mesograzers can occur in very high densities on their hosts and their grazing effects are often specific to the algal host rather than the entire seaweed community. Conversely, macrograzers are larger (>2.5 cm) with intermediate to large feeding ranges. The voracious feeding activity of some macrograzers can denude large seaweed stands and consequently influence overall community structure.

A third way of classifying grazers is by their diet specificity, distinguishing specialists feeding on one or few seaweed species from generalists feeding on a large variety of seaweed species and types (Table 8.1). Most marine herbivores tend to be generalists, and many also incidentally ingest some animal material. Feeding selectivity can be driven by a number of factors, among others predation pressure on the herbivore. For example, small, sedentary amphipod mesograzers can specialize to live and feed on chemically defended seaweeds to gain protection against predators (Hay et al. 1989a, b). Food selection can also be driven by the nutritional content of the seaweed. Some crustaceans have higher reproductive fitness when feeding on mixed algal species or on higher quality species (Hemmi and Jormalainen 2002). Specialization of herbivores on a single or few select algal species can also be due to the structural habitat an alga provides the herbivore. For example, the isopod *Idotea baltica* prefers feeding on phlorotannin-rich seaweeds although this causes reduced growth rates, likely because the stable habitat structure of these algae affords protection from predators (Jormalainen et al. 2001a).

8.2 Digestive Adaptations in Grazers

One of the “problems” of seaweed food is its low nutritional quality and the restricted digestibility and assimilation (Montgomery and Gerking 1980). Seaweeds have a high structural carbohydrate content with limited enzymatic digestibility by herbivores, low nitrogen content compared to animal diet, tough tissues, and sometimes metabolites that reduce digestion (e.g., phlorotannins; Targett and Arnold 2001; Amsler and Fairhead 2006). Hence, while seaweeds are often relatively easy to acquire as food because of their abundance and stationary distribution, processing large amounts of algal food to satisfy metabolic needs of the grazer poses many challenges. Digestion and assimilation of seaweeds are best, albeit incompletely, known for herbivorous fishes (for reviews, see Horn 1989; Clements et al. 2009). Adaptations in fishes often include a gizzard or other grinding mechanism and/or acidic stomach environment to mechanically break up and/or chemically lyse the food. They also typically possess a long alimentary tract to optimize assimilatory surface area and gut retention time, and hindgut fermentation of complex carbohydrates through microorganisms. Some of these gut features may not occur at all in some species and can be compensated for by strong expression of the other characteristics. For example, high enzymatic activity compensates for a short, stomachless, alkaline digestive tract in the silverside *Atherinops affinis* (Horn et al. 2006). Much less is known about the physiological processes associated with digestion and assimilation in invertebrate herbivores feeding on seaweeds (Targett and Arnold 2001). Many invertebrate guts contain an array of proteinase, amylase, and lipase enzymes. Similar to herbivorous fishes, microflora is supposed to assist with carbohydrate polymer breakdown, while algal lysis with gut acids may be a less likely mechanism in invertebrate herbivores. The field of seaweed digestion by invertebrate herbivores still requires much research so that ecological and evolutionary frameworks of grazer–macroalgal interactions can be more fully developed.

8.3 Grazer Impacts on Seaweeds and Seaweed Communities

Macrograzer feeding can remove large amounts of seaweed biomass, with strong ecosystem-wide effects such as trophic cascades best known from temperate kelp forests and tropical coral reefs (Pinnegar et al. 2000). Sea urchin grazing on kelps can be one cause for phase shifts from kelp beds with an abundance of kelps, other fleshy algae, and associated invertebrates and fishes to an urchin barren state, where few seaweeds except for encrusting corallines prevail (Steneck et al. 2002). For example, in the Alaskan Aleutian Islands, these phase shifts are closely linked to the presence/absence of sea otters, the keystone species that preys on urchin grazers (Estes and Duggins 1995). Under reduced predation pressure in the absence of sea otters, urchins grow unchecked and overgraze the kelp, causing urchin barrens. The barren phase is stable over long time periods, until a major disturbance, e.g., reintroduction of a predator or disease, reduces sea urchin numbers and kelps can grow back into stable kelp beds. Besides the indirect benefits to kelp forests of sea otter predation on urchins, it also provides positive feedbacks for the otters, which will find more nutritious prey in the complex kelp forests. Trophic cascades including grazing by urchins and/or herbivorous fishes also are well documented in tropical coral reef community shifts between coral-dominated and seaweed-dominated phases (Mumby et al. 2006). Grazing removes seaweeds from coral reefs, thus facilitating space for coral growth and recruitment. When grazing is reduced, such as after the mass die-off of the urchin *Diadema antillarum* in the Caribbean in 1983 (Carpenter 1990) or by overfishing, competitively dominant seaweeds smother corals and dominate the reef community (see Chap. 16 by Mejia et al.). A greater diversity of reef grazers is typically more effective in removing algal biomass, thus increasing overall coral reef resilience (e.g., Duffy et al. 2003).

The feeding activity of small mesograzers is typically nonlethal to large seaweeds, but their grazing activity can still affect host fitness (e.g., reduce growth and reproductive output), and influence overall community structure (Arrontes 1999). The overall community effects of algal biomass removal by mesograzers

Table 8.2 Mesograzer effects on macroalgae and macroalgal communities

Grazing effect	Selected example
No effect on macroalgal growth	At natural abundances, amphipod grazers do not reduce growth in *Sargassum* in Australia (Poore et al. 2009)
Reduction of macroalgal growth	Limpets limit macroalgal growth on exposed rocky intertidal shores throughout Europe, with overall stronger grazing effects on macroalgal abundance as well as spatial variability in assemblages in the north than the south (Coleman et al. 2006)
Reduction of macroalgal reproduction	Selective grazing of periwinkles on meristematic tissue reduces growth and reproduction in the rockweed *Fucus* (Van Alstyne 1990)
Removal of propagules	Mesoherbivores are effective in controlling bloom-forming *Ulva* (as *Enteromorpha*) in the Baltic Sea by grazing on early life-stage propagules (Lotze and Worm 2000)
Algal distribution patterns	Limpet grazing controls seaweed growth at mid-intertidal levels in rocky shores in Spain (Arrontes et al. 2004), but periwinkles do not have the same effect in Sweden (Lindegarth et al. 2001)
Selective grazing	Selective grazing on annual seaweeds in the Baltic Sea releases long-lived fucoids from interspecific competition and maintains overall complex communities (Worm et al. 1999)
Damage to macroalgal tissue causes secondary effects	Grazing of the small gastropod *Lacuna vincta* weakens tissue in the kelp *Nereocystis luetkeana* resulting in increased breaking and mortality through hydrodynamics (Duggins et al. 2001)
Importance of herbivore diversity	Higher diversity of grazers removes more algal epiphytes from seagrasses than similar herbivore abundances of lesser diversity, with additional indirect effects such as increased secondary production in the system (Duffy et al. 2003)
	Diversity and productivity of rocky intertidal algal communities is not linearly related to gastropod grazer diversity, but different grazer species have idiosyncratic effects (O'Connor and Crowe 2005)
Importance of herbivore identity	Selective grazing of different species of herbivorous reef fishes causes different impacts on the algal community, which can also vary depending on the successional stage of the algal community (Burkpile and Hay 2010)
Importance of herbivore life stage and gender	In situ feeding of male and female *Idotea* isopods on *Fucus* differs with males preferring apical portions and females preferring the less nutritious but better protected basal portions of the algae, likely as a risk avoidance strategy (Jormalainen et al. 2001b)

are dependent on many complex interactions. These include feeding rate and mode of the dominant grazer, growth rates of the algae, competitive interactions with and among other community members, as well as interactive effects with the physical environment (Hawkins and Hartnoll 1983; Coleman et al. 2006; Jenkins et al. 2008; see Table 8.2 for examples for the following effects). Even if mesograzer consumption of adult seaweeds is negligible, strong community-structuring effects may occur in the case of selective removal of algal propagules. Selective grazing reduces seaweed interspecific competition, influencing overall community composition and algal distribution patterns. Mesograzing at high densities can damage valuable parts of the algal thallus such as reproductive or meristematic regions, facilitate access for other grazers or pathogens, or weaken the thallus and render the seaweed prone to breakage. Seaweed biomass reduction may be greater with increased herbivore diversity, in addition to high abundance, because the complementary feeding modes and preferences of different grazers are more effective in removing algal biomass (Table 8.2). Conversely, while there is little directed research yet, meta-analysis indicates that higher algal diversity may lead to decreased grazer effects (Hillebrand and Cardinale 2004). It was argued that, if more algal species are present, there is a higher likelihood that the community contains unpalatable species, or that a more diverse system may have increased indirect and facilitative effects that may benefit seaweeds overall (Hillebrand and Cardinale 2004). Similar to macrograzers, mesograzers themselves are often under strong control of predators, which can efficiently decimate grazer densities and thus indirectly enhance seaweed growth. For example, micro-carnivorous fishes efficiently prey upon mesograzers in a Californian kelp bed and enhance kelp frond density (Davenport and Anderson 2007).

Among other environmental variables, mesograzers may be particularly susceptible to hydrodynamic forces that can dislodge them from their host alga, thus reducing the effects that grazing has on seaweed communities. Nutrient levels, e.g., due to eutrophication in coastal waters ("bottom-up" effects), may partially counteract grazer-induced "top-down" effects on seaweed diversity and production. The relative importance of grazer versus nutrient effects on seaweed communities is highly context dependent; nutrient effects seem to be strongest in naturally high-nutrient systems but less important in oligotrophic systems, and they may also vary based on algal functional group (e.g., foliose, filamentous, crustose; Burkpile and Hay 2006).

Occasionally, herbivore feeding can also benefit seaweeds and communities (Hay et al. 2004). A common effect of mesograzers is the removal of filamentous epiphytes from a host, possibly affording better light and nutrient acquisition to the host plant by reducing algal epiphyte biomass (Duffy 1990). Also, coralline algae have higher accretion rates when grazed because the positive effects of grazing, such as the removal of competitively superior fleshy algae, outweigh the grazing damage to the coralline (Littler et al. 1995). Grazing, in general, cannot only open new space for algal recruits, but in the case of grazing gastropods, the pedal mucus trail left on the rocky surface can effectively trap spores and other seaweed propagules and support their germination (Santelices and Bobadilla 1996). Considering the typically high abundance of gastropods in rocky intertidal systems, this may enhance new seaweed recruitment (and new food resources for the grazer!) into an area where seaweeds are under strong grazing pressure. Grazing may also enhance algal productivity and algal recruitment even if algal biomass is reduced. For example, it has been hypothesized that either the reduction of self-shading or the increase in nutrients from herbivore excretions may have positive effects on seaweed productivity (Carpenter 1986). Lastly, the preferential feeding of amphipods on reproductive algal tissue can lead to an increased release and recruitment of spores (Buschmann and Vergara 1993).

8.4 Seaweed Defenses Against Grazing

Algal defenses against herbivores present the other side of the coin in grazer–seaweed interactions, where defenses presumably have developed in response to selective pressures of grazers. The large array of algal defense mechanisms ranges from tolerance and fast replacement of lost tissue, to temporal and spatial escapes, to structural and chemical defenses. Several different defense strategies may be employed simultaneously in some seaweeds, and many of these defenses may also have functions other than herbivore deterrence (e.g., antifouling, antimicrobial defense, allelopathic functions). These defenses are commonly distinguished as non-coexistence and coexistence strategies, with the latter encompassing structural and chemical defenses (e.g., Lubchenco and Gaines 1981; Duffy and Hay 1990; Table 8.3).

In non-coexistence strategies the encounter of seaweeds with herbivores is minimized. These "strategies" do not imply a specific, directed behavioral response in seaweeds but likely are a result of chance or are evolutionary developments from situations where spatial and temporal occurrence of seaweeds coincided with minimal herbivore activity. Overall, grazing pressure is variable over time and space, and seaweeds can escape grazing by occurring in places, or by completing important life history phases such as initial growth or reproduction during times, where and when herbivores are not present or active. These strategies are most effective at driving co-evolution in systems where temporal and spatial distribution of herbivores is predictable. Escape in size occurs when seaweeds outgrow the sporeling stage as a bottleneck of high grazer susceptibility and reach a size where they are unaffected by the specific feeding mode of the abundant herbivore. Lastly, macroalgae also can escape herbivores by association with other organisms that provide them chemical or structural protection from grazers (for examples of the above see Table 8.3).

Coexistence strategies lower the herbivore's attraction to a seaweed because of specific structural or chemical traits of the alga. The most common example of structural defenses, also referred to as mechanical or morphological defenses, involves the inclusion of calcium carbonate into algal cells (Paul and Hay 1986). In their functional-form model, Steneck and Watling (1982) suggest that calcareous seaweeds, specifically crustose coralline forms, have the highest grazer resistance of all algal forms, and grazing on calcareous forms requires specialized mouthparts in the herbivore. The main herbivores able to graze on coralline algae are chitons, limpets, sea urchins, and parrot fishes. Coralline algae often are abundant despite high herbivore densities, demonstrating that herbivore damage is minimal.

Table 8.3 Algal defenses against grazing

Defense mechanism	Selected example
Non-coexistence strategies	
Escape in time	The highly palatable green alga *Ulva* forms early spring blooms from an overwintering propagule bank in the Baltic Sea, thus avoiding intense herbivore pressure, which efficiently suppresses *Ulva* later in the season (Lotze et al. 2000)
Escape in space	Turf algae on the Great Barrier Reef thrive in regions of reduced herbivory, but are readily consumed when transplanted to regions of high herbivore abundance (Bonaldo and Bellwood 2010)
Escape in size	The chiton *Katharina* is very effective in removing new recruits of the low intertidal kelp *Hedophyllum* in the Northeast Pacific but does not graze on the adults (Markel and DeWreede 1998)
Associational defense	The palatable red alga *Gracilaria* is protected from sea urchin grazing when it occurs associated with the unpalatable brown alga, *Sargassum,* in the temperate Atlantic (Pfister and Hay 1988)
	In tropical coral reefs, seaweeds persist even under high herbivore abundances when they are covered with epiphytic cyanobacteria (Fong et al. 2006)
Coexistence strategies	
Structural defense	*Fucus* produces adventitious branches after grazing damage, which act as structural defenses against further grazing, although high phlorotannin content in these branches may also present a chemical defense (Van Alstyne 1989)
	Padina on coral reefs exhibit morphological plasticity with an erect, foliose morphology in the absence of grazers and a more resistant turf morphology under high grazing pressure (Lewis et al. 1987)
Chemical defense	
Tropical systems	A large variety of green, red, and brown seaweeds deter feeding in many invertebrate and vertebrate grazers using a rich arsenal of terpene and acetogenin metabolites (see Table 2.1 in Pereira and da Gama 2008). For example, the Brazilian brown alga *Stypopodium zonale* (Dictyotales) produces a diterpene mixture that chemically differs depending on location, and some compounds of the mix deter urchin and crab feeding more than others (Soares et al. 2003)
Temperate systems	While generally less rich in natural products active against grazers than tropical algae, temperate seaweeds are often chemically defended, with phlorotannins in brown algae being the most-studied group (Jormalainen and Honkanen 2008). However, phlorotannins are not always the deterrent compound; in *Fucus vesiculosus*, a galactolipid and uncharacterized non-phenolic compounds are active in deterring an urchin grazer (Deal et al. 2003)
Polar systems	Extracts of 20 macroalgae along the Western Antarctic Peninsula deter sympatric grazers, although responsible compounds are often unidentified (Amsler et al. 2005). Phlorotannins are abundant in most of the dominant brown algae and some deter feeding in sea star, amphipod, and/or fish grazers (Iken et al. 2007, 2011)

Many coralline algae also contain defensive chemicals in addition to calcification. For example, chemical defenses instead of calcification were the driving mechanism for reduced grazing of urchins, amphipods, and parrot fishes on the tropical coralline green algae, *Rhipocephalus* and *Udotea* (Hay et al. 1994). Calcification in seaweeds can also be paired with additional non-coexistence (avoidance) strategies. One example is the tropical green coralline alga *Halimeda*, which produces new, non-calcified segments only at night when herbivorous reef fishes are inactive (Hay et al. 1988). The price for herbivore resistance of calcareous seaweeds is very low growth rates, which prevent coralline algae from easily replacing tissue that is lost through grazing. A different example of structural defense is referred to as bet hedging and can occur in heteromorphic seaweeds. Here, a fleshy, upright form has high photosynthetic and growth rates but is susceptible to grazers, while a crustose form is more grazer resistant at the expense of reduced growth rates (e.g., Slocum 1980). With this strategy, the crustose form prevails in situations of high grazer abundance while the upright form occurs when herbivores are fewer or absent. Another type of structural (mechanical) defense is where water motion creates a whiplash effect of some kelps, which will prevent urchins from entering kelp beds to graze (Konar 2000).

Coexistence strategies using chemical defenses are probably the best-studied area of seaweed defenses against herbivores, and a recent book provides an excellent review of the topic (Amsler 2008). These defenses also are covered from a mechanistic perspective in Chap. 9 by Amsler, while they are viewed in this chapter from a strategic and community perspective. Marine seaweeds from all latitudinal ranges and ecosystem types produce a wide variety of secondary metabolites that function as herbivore deterrents (Maschek and Baker 2008; see Table 8.3 for examples from different latitudinal systems). These chemical defenses mostly reduce the palatability or nutritional quality of the seaweed, or can also reduce the fitness and survivorship of the herbivore through toxic effects. The functions and mechanisms of the multitude of defensive chemical metabolites have evoked a number of predictive models. Many of these models are based on the assumption that the production, storage, or degradation of defensive metabolites is costly, and that implementation of defenses is thus under regulation to optimize energy use in seaweeds (Pavia and Toth 2008). Models of seaweed chemical defenses that are mostly based on resource availability are reviewed elsewhere in detail (Cronin 2001; Amsler and Fairhead 2006; Pavia and Toth 2008). One model, the Optimal Defense Theory (ODT; Rhoades 1979), is specifically based on herbivory as the selective driver of defensive strategies. The ODT predicts that allocation of defensive compounds to specific parts within the alga is in direct proportion to the value of the thallus part for the seaweed's fitness as well as to the risk of attack, and in inverse relation to the cost of defense production. Within the framework of the ODT, induction of these defenses should occur during times when the risk of herbivore attack is greatest and unpredictably variable. The preferential allocation of defensive chemicals and induction of defenses after initial damage in seaweeds has been abundantly tested and reported in the literature and much, although not unanimous, support has been found (see Toth and Pavia 2007 for review). In part the debate about differential defense allocation to the most valuable thallus parts may be hampered by the definition of "valuable." One particularly well-studied system of the ODT is phlorotannin allocation in brown algae, even though the defensive role of phlorotannins has been controversial and seems to be context-dependent on herbivore and environment (Amsler and Fairhead 2006). Using sensitivity analysis, which determines the relative contribution of different life processes to overall population growth, Pavia et al. (2002) related life processes to different thallus parts in the fucoid *Ascophyllum nodosum*, and thus determined fitness values of different thallus parts. Phlorotannin concentrations and resistance to gastropod grazers were positively correlated to these fitness values in accordance to the ODT.

8.5 Effects of Seaweed Chemical Defenses on Herbivores and Communities

Algal chemical defenses can indirectly influence herbivore populations by affecting their overall food consumption and lead to reduced growth or reproduction. Metabolites can also directly affect herbivore survival, although practical separation of indirect and direct effects is often difficult. Phlorotannins, common metabolites in brown algae, have the potential to bind to digestive enzymes as well as with plant nutrients and thus limit assimilation efficiencies in the grazer (Targett and Arnold 2001). For example, the assimilation efficiency in fishes was greatly reduced when feeding on a phlorotannin-rich diet (Boettcher and Targett 1993), and egg production in periwinkles was reduced by phlorotannins (Toth et al. 2005). Similarly, the terpenoid metabolites produced by the brown alga *Dictyota* significantly reduced growth, survival, and reproduction in some grazing amphipods (Cruz-Rivera and Hay 2003). These fitness effects can depend on metabolite concentration and on the specific herbivore species, as there is considerable variation in the effects of the same algal metabolite on even closely related herbivores. Conversely, some mesograzer populations can ultimately benefit from seaweed defenses by gaining protection from predators when living associated with a chemically defended host, even if this comes with reduced fitness as a trade-off (Hay et al. 1988, 1989a, b).

Relatively little understanding exists on the structuring role of seaweed chemical defenses on overall communities beyond the specific interactions between a particular seaweed and herbivore (Hay 2009). Chemically well-defended seaweeds can be very abundant and even dominate communities, independent of latitude or algal group; this attests to the competitive advantage chemical defenses may provide. On tropical reefs in Brazil with high grazing pressure, members of the tropical green algae *Bryopsis* and *Caulerpa,* the red alga *Laurencia,* and the brown alga *Dictyota* chemically deter herbivore grazing and also are highly abundant (Marques et al. 2006; Villaça et al. 2010). *Fucus vesiculosus* is the dominant habitat-providing seaweed along rocky shores in the Baltic Sea, and phlorotannins act as effective defenses in the adult seaweed against most grazers (Jormalainen and Honkanen 2008). Antarctic shallow waters are dominated by large brown algae of the order Desmarestiales, which are unpalatable to most sympatric grazers. In some cases, this may be due to relatively high phlorotannin contents (Amsler et al. 2005; Iken et al. 2011). Chemical defenses in seaweeds may also play a role in the invasive potential of a species, which typically have significant effects on the structure and biodiversity of the native community. For example, the invasive, chemically defended *Caulerpa taxifolia* is largely unpalatable and presents poor habitat to a variety of native grazers in temperate Australia, making it unlikely that grazing could control the abundance or spreading of the invasive species (Gollan and Wright 2006). Chemical defense in seaweeds also can have community-wide effects by altering trophic dynamics, and can even change the outcome of trophic cascades. In Australia, the chemically defended red alga *Delisea pulchra* survived intense urchin grazing in a kelp bed, which led to an intermediate stable community state containing grazer-resistant foliose algae between the typical alternate states of kelp beds and urchin barrens (Wright et al. 2005). Waterborne cues released by herbivores grazing on macroalgae may function as a form of chemical communication with neighboring con-specific algae to increase their defense levels (Toth and Pavia 2007). However, these cues may act in a broader sense as "infochemicals" when signaling to other community members. Cues released from herbivore-algal grazing may attract potential predators of the herbivores, which would benefit the seaweed. Both a fish and a crab predator were significantly attracted to cues released by *Ascophyllum nodosum* when grazed by periwinkles, thus inducing top-down control of the herbivore by the two predators (Coleman et al. 2007). This shows that such waterborne cues may act as a seaweed distress signal, a mechanism well known from terrestrial plant–insect interactions (Dicke 2009). In a different example, cues exuded by the chemically defended *Delisea pulchra* in Australian kelp beds induced settlement of urchin larvae, although the chemically defended alga is unpalatable to the adult urchins (Williamson et al. 2004). Adult urchins will instead feed on other seaweeds within the community, thus providing a competitive advantage to *D. pulchra*. Simultaneously, urchins gain associational protection against fish predators from the chemically defended alga in their habitat.

8.6 Climate Change Effects on Seaweed–Herbivore Interactions

An emerging, yet timely, consideration is the effects that climatic changes will have on the interactions between seaweeds and grazers. Metabolic theory predicts rising temperatures to generally increase metabolic rates in both seaweeds and herbivores (see also Chap. 3 by Eggert and Chap. 18 by Bartsch et al.). Manipulative tank experiments, however, indicated that temperature rise has a proportionally larger effect on grazers, which outweigh any increased algal growth. This resulted in an overall reduction of algal biomass (O'Connor 2009). Conversely, no change in herbivore impact at higher temperatures was seen in an in situ experiment in a rocky intertidal in California (Morelissen and Harley 2007). It was hypothesized that multiple abiotic drivers interacted with varying effects on biotic interactions, partially offsetting and ameliorating each other, thus not leading to a clearly detectable temperature effect. It is yet uncertain how simultaneous changes in light and nutrient regimes with temperature increases would influence seaweed–grazer relationships. But it seems that any climate-driven temperature effects on species-specific seaweed–grazer interactions may not propagate linearly through the food web, and depend, among others, on the body size, diversity, diet specificity, and connectivity of members of the food web (Petchey et al. 2010; see also Chap. 19 by Roleda and Hurd).

The seaweed–grazer effects of other variables subject to global change, such as sedimentation or ocean pH, are poorly understood. Many coastal regions experience an increase in sedimentation, mostly due to coastal erosion and, in cold-temperate and polar regions, the increased melting of glaciers (Bogen and Bønsnes 2003). Regions under strong sediment influence are characterized by lowered seaweed and invertebrate density and diversity (Spurkland and Iken 2011). How this translates into specific effects on seaweed–herbivore interactions is not yet known. Similarly, it is known that increased ocean acidification reduces the abundance and fitness of crustose coralline algae on coral reefs and may enhance fleshy algal growth, but the specific impacts of pH on grazers and on seaweed–grazer relationships are unknown (Kuffner et al. 2008).

8.7 Conclusions

Despite the many advances achieved in the broad field of grazer–seaweed interactions, many if not most topics require more comprehensive investigation to reach a more complete understanding. Generalizations made on patterns or interactions between grazers and seaweeds are often based on few individual studies and need to be more thoroughly tested with different species and at multiple locations. This will allow us to better understand the variability of individual patterns within a (possibly) general concept. Obviously, globally occurring changes in climate and ocean chemistry are likely to have effects on the seaweed–grazer systems as we have started to understand them so far. Aside from metabolic changes in both grazers and seaweeds, marine systems are expected to undergo "functional homogenization," where numerous specialist species fulfilling various functions are replaced with few generalist species assuming most ecosystem functions (Clavel et al. 2011). This homogenization will inevitably have important effects on the often species-specific nuances in seaweed–grazer interactions, evoking important questions of the role of the diversity of these myriad interactions in ecosystem resiliency and functioning.

Acknowledgements I am grateful to Brenda Konar for productive discussions and thoughtful comments on the manuscript. I also thank Charles Amsler, Gunilla Toth, and Tania Spurkland for helpful comments that greatly improved the manuscript. Preparation of the manuscript was supported by NaGISA, the nearshore field project of the Census of Marine Life program, award #2008-12-1 by the Alfred P. Sloan Foundation.

References

Amsler CD (2008) Algal Chemical Ecology. Springer, Berlin, p xviii, 313

Amsler CD, Fairhead VA (2006) Defensive and sensory chemical ecology of brown algae. Adv Bot Res 43:1–91

Amsler CD, Iken K, McClintock JB, Amsler MO, Peters KJ, Hubbard JM, Furrow FB, Baker BJ (2005) Comprehensive evaluation of the palatability and chemical defense of subtidal macroalgae from the Antarctic Peninsula. Mar Ecol Prog Ser 294:141–159

Amsler CD, Amsler MO, McClintock JB, Baker BJ (2009) Filamentous algal endophytes in macrophytic Antarctic algae: prevalence in hosts and palatability to mesoherbivores. Phycologia 48:324–334

Antoniadou C, Chintiroglou C (2006) Trophic relationships of polychaetes associated with different algal growth forms. Helgol Mar Res 60:39–49

Arrontes J (1999) On the evolution of interactions between marine mesoherbivores and algae. Bot Mar 42:137–155

Arrontes J, Arenas F, Fernandez C, Rico JM, Oliveros J, Martinez B, Viejo RM, Alvarez D (2004) Effects of grazing by limpets on mid-shore species assemblages in northern Spain. Mar Ecol Prog Ser 277:117–133

Aumack CF, Amsler CD, McClintock JB, Baker BJ (2010) Chemically mediated resistance to mesoherbivory in finely branched macroalgae along the western Antarctic Peninsula. Eur J Phycol 45:19–26

Bjorndal KA (1997) Foraging ecology and nutrition of sea turtles. In: Lutz PL, Musick JA (eds) The biology of sea turtles. CRC, Boca Raton, pp 199–232

Boettcher AA, Targett NM (1993) Role of polyphenolic molecular size in reduction of assimilation efficiency in *Xiphister mucosus*. Ecology 74:891–903

Bogen J, Bønsnes TE (2003) Erosion and sediment transport in high Arctic rivers, Svalbard. Polar Res 22:175–189

Bonaldo RM, Bellwood DR (2010) Spatial variation in the effects of grazing on epilithic algal turfs on the Great barrier reef, Australia. Coral Reefs: DOI 10.1007/s00338-010-0704-4

Burkpile DE, Hay ME (2006) Herbivore vs. nutrient control of marine primary producers: context-dependent effects. Ecology 87:3128–3139

Burkpile DE, Hay ME (2010) Impact of herbivore identity on algal succession and coral growth on a Caribbean reef. PLoS One 5(1):e8963

Buschmann AH, Vergara PA (1993) Effects of rocky intertidal amphipods on algal recruitment: a field study. J Phycol 29:154–159

Carpenter RC (1986) Partitioning herbivory and its effects on coral reef algal communities. Ecol Monogr 56:345–363

Carpenter RC (1990) Mass mortality of *Diadema antillarum* I. Long-term effects on sea urchin population-dynamics and coral reef algal communities. Mar Biol 104:67–77

Carpenter RC, Edmunds PJ (2006) Local and regional scale recovery of *Diadema* promotes recruitment of scleractinian corals. Ecol Lett 9:271–280

Chenelot H, Konar B (2007) *Lacuna vincta* (Mollusca, Neotaenioglossa) herbivory on juvenile and adult *Nereocystis luetkeana* (Heterokontophyta, Laminariales). Hydrobiology 583:107–118

Clavel J, Julliard R, Devictor V (2011) Worldwide decline of specialist species: toward a global functional homogenization? Front Ecol Environ 9:222–228

Clements KD, Raubenheimer D, Choat JH (2009) Nutritional ecology of marine herbivorous fishes: ten years on. Funct Ecol 23:79–92

Coleman RA, Underwood AJ, Benedetti-Cecchi L, Åberg P, Arenas F, Arrontes J, Castro J, Hartnoll RG, Jenkins SR, Paula J, Della Santina P, Hawkins SJ (2006) A continental scale evaluation of the role of limpet grazing on rocky shores. Oecologia 147:556–564

Coleman RA, Ramchunder SJ, Davies KM, Moody AJ, Foggo A (2007) Herbivore-induced infochemicals influence foraging behavior in two intertidal predators. Oecologia 151:454–463

Cornwall CE, Phillips NE, McNaught DC (2009) Feeding preference of the abalone *Haliotis iris* in relation to macroalgal species, attachment, accessibility and water movement. J Shellfish Res 28:589–597

Cronin G (2001) Resource allocation in seaweeds and marine invertebrates: chemical defense patterns in relation to defense theory. In: McClintock JB, Baker BJ (eds) Marine Chemical Ecology. CRC, Boca Raton, FL, pp 325–354

Cruz-Rivera E, Hay ME (2003) Prey nutritional quality interacts with chemical defenses to affect consumer feeding and fitness. Ecol Monogr 73:483–506

Davenport AC, Anderson TW (2007) Positive indirect effects of reef fishes on kelp performance: the importance of mesograzers. Ecology 88:1548–1561

Deal MS, Hay ME, Wilson D, Fenical W (2003) Galactolipids rather than phlorotannins as herbivore deterrents in the brown seaweed *Fucus vesiculosus*. Oecologia 136:107–114

Dicke M (2009) Behavioural and community ecology of plants that cry for help. Plant Cell Environ 32:654–665

Duffy JE (1990) Amphipods on seaweeds: partners or pests? Oecologia 83:267–276

Duffy JE, Hay ME (1990) Seaweed adaptations to herbivory. BioScience 40:368–375

Duffy JE, Hay ME (1991) Food and shelter as determinants of food choice by an herbivorous marine amphipod. Ecology 72:1286–1298

Duffy JE, Hay ME (2000) Strong impacts of grazing amphipods on the organization of a benthic community. Ecol Monogr 70:237–263

Duffy JE, Richardson JP, Canuel EA (2003) Grazer diversity effects on ecosystem functioning in seagrass beds. Ecol Lett 6:637–645

Duggins D, Eckman JE, Siddon CE, Klinger T (2001) Interactive roles of mesograzers and current flow in survival of kelps. Mar Ecol Prog Ser 223:143–155

Engkvist R, Malm T, Tobiasson S (2000) Density-dependent grazing effects of the isopod *Idotea baltica* Pallas on *Fucus vesiculosus* L in the Baltic Sea. Aquat Ecol 34:253–260

Estes JA, Duggins DO (1995) Sea otters and kelp forests in Alaska: generality and variation in a community ecological paradigm. Ecol Monogr 65:75–100

Fong P, Smith TB, Wartian MJ (2006) Epiphytic cyanobacteria maintain shifts to macroalgal dominance on coral reefs following ENSO disturbance. Ecology 87:1162–1168

Ginsburg DW, Paul VJ (2001) Chemical defenses in the sea hare Aplysia parvula: importance of diet and sequestration of algal metabolites. Mar Ecol Prog Ser 215:261–274

Gollan JR, Wright JT (2006) Limited grazing pressure by native herbivores on the invasive seaweed *Caulerpa taxifolia* in a temperate Australian estuary. Mar Freshwater Res 57:685–694

Hawkins SJ, Hartnoll RG (1983) Grazing of intertidal algae by marine invertebrates. Oceanogr Mar Biol Annu Rev 21:195–282

Hay ME (2009) Marine chemical ecology: chemical signals and cues structure marine populations, communities, and ecosystems. Annu Rev Mar Sci 1:193–212

Hay ME, Duffy JE, Pfister CA, Fenical W (1987) Chemical defense against different marine herbivores: are amphipods insect equivalents? Ecology 68:1567–1580

Hay ME, Renaud PE, Fenical W (1988) Large mobile versus small sedentary herbivores and their resistance to seaweed chemical defenses. Oecologia 75:246–252

Hay ME, Paul VJ, Lewis SM, Gustafson K, Tucker J, Trindell RN (1989a) Can tropical seaweeds reduce herbivory by growing at night? Diel patterns of growth, nitrogen content, herbivory, and chemical versus morphological defenses. Oecologia 75:233–245

Hay ME, Pawlik JR, Duffy JE, Fenical W (1989b) Seaweed-herbivore-predator interactions: host-plant specialization reduces predation on small herbivores. Oecologia 81:418–427

Hay ME, Duffy JE, Fenical W (1990a) Host-plant specialization decreases predation on a marine amphipod: an herbivore in plant's clothing. Ecology 71:733–743

Hay ME, Duffy JE, Paul VJ, Renaud PE, Fenical W (1990b) Specialist herbivores reduce their susceptibility to predation by feeding on the chemically defended seaweed *Avrainvillea longicaulis*. Limnol Oceanogr 35:1734–1743

Hay ME, Kappel QE, Fenical W (1994) Synergisms in plant defenses against herbivores: interactions of chemistry, calcification, and plant quality. Ecology 75:1714–1726

Hay ME, Parker JD, Burkepile DE, Caudill CC, Wilson AE, Hallinan ZP, Chequer AD (2004) Mutualism and aquatic community structure: the enemy of my enemy is my friend. Annu Rev Ecol Evol Syst 35:175–197

Hemmi A, Jormalainen V (2002) Nutrient enhancement increases performance of a marine herbivore via quality of its food alga. Ecology 83:1052–1064

Hillebrand H, Cardinale J (2004) Consumer effects decline with prey diversity. Ecol Lett 7:192–201

Horn MH (1989) Biology of marine herbivorous fishes. Oceanogr Mar Biol Annu Rev 27:167–272

Horn MH, Gawlicka AK, German DP, Logothetis EA, Cavanagh JW, Boyle KS (2006) Structure and function of the stomachless digestive system in three related species of New World silverside fishes (Atherinopsidae) representing herbivory, omnivory, and carnivory. Mar Biol 149:1237–1245

Huang YM, Amsler MO, McClintock JB, Amsler CD, Baker BJ (2007) Patterns of gammaridean amphipod abundance and species composition associated with dominant subtidal macroalgae from the western Antarctic Peninsula. Polar Biol 30:1417–1430

Iken K (1999) Feeding ecology of the Antarctic herbivorous gastropod *Laevilacunaria antarctica* Martens. J Exp Mar Biol Ecol 236:133–148

Iken K, Barrera-Oro ER, Quartino ML, Casaux RJ, Brey T (1997) Grazing in the Antarctic fish *Notothenia coriiceps*: evidence for selective feeding on macroalgae. Antarct Sci 9:386–391

Iken K, Amsler CD, Hubbard JM, McClintock JB, Baker BJ (2007) Allocation patterns of phlorotannins in Antarctic brown algae. Phycologia 46:386–395

Iken K, Amsler CD, Amsler MO, McClintock JB, Baker BJ (2011) Field studies on deterrent roles of phlorotannins in Antarctic brown algae. In: Wiencke C (ed) Biology of polar benthic algae. Walter de Gruyter GmbH & Co. KH, Göttingen, pp 121–138

Jenkins SR, Moore P, Burrows MT, Garbary DJ, Hawkins SJ, Ingólfsson A, Sebens KP, Snelgrove PVR, Wethey DS, Woodin SA (2008) Comparative ecology of North Atlantic shores: do differences in players matter for process? Ecology 89:S3–S23

Jones GP, Santana L, McCook LJ, McCormick MI (2006) Resource use and impact of three herbivorous damselfishes on coral reef communities. Mar Ecol Prog Ser 328:215–224

Jormalainen V, Honkanen T (2008) Macroalgal chemical defenses and their roles in structuring temperate marine communities. In: Amsler CD (ed) Algal chemical ecology. Springer, Berlin, pp 57–89

Jormalainen V, Honkanen T, Heikkilä N (2001a) Feeding preferences and performance of a marine isopod on seaweed hosts: cost of habitat specialization. Mar Ecol Prog Ser 220:219–230

Jormalainen V, Honkanen T, Mäkinen A, Vesakoski O (2001b) Why does herbivore sex matter? Sexual differences in utilization of *Fucus vesiculosus* by the isopod *Idotea baltica*. Oikos 93:77–86

Kennish R (1996) Diet composition influences the fitness of the herbivorous crab *Crapsus albolineatus*. Oecologia 105:22–29

Konar B (2000) Seasonal inhibitory effects of marine plants on sea urchins: structuring communities the algal way. Oecologia 125:208–217

Kuffner IB, Andersson AJ, Jokiel PL, Rodgers KS, MacKenzie FT (2008) Decreased abundance of crustose coralline algae due to ocean acidification. Nat Geosci 1:114–117

Lewis SM, Norris JN, Searles RB (1987) The regulation of morphological plasticity in tropical reef algae by herbivory. Ecology 68:636–641

Lindegarth M, Åberg PA, Cervin G, Nilsson PG (2001) Effects of grazing on the structure of mid-shore, intertidal assemblages on moderately exposed rocky shores on the Swedish west coast. Mar Ecol Prog Ser 212:29–38

Littler MM, Littler DS (1980) The evolution of thallus form and survival strategies in benthic marine macroalgae: field and laboratory tests of a functional form model. Am Nat 116:25–44

Littler MM, Littler DS, Taylor PR (1995) Selective herbivore increases biomass of its prey: a chiton-corraline reef-building association. Ecology 76:1666–1681

Lotze HK, Worm B (2000) Variable and complementary effects of herbivores on different life stages of bloom-forming macroalgae. Mar Ecol Prog Ser 200:167–175

Lotze HK, Worm B, Sommer U (2000) Propagule banks, herbivory and nutrient supply control population development and dominance patterns in macroalgal blooms. Oikos 89:46–58

Lubchenco J, Gaines SD (1981) A unified approach to marine plant-herbivore interactions. 1. Populations and communities. Annu Rev Ecol Syst 12:405–437

Lubchenco J (1978) Plant species diversity in a marine intertidal community: importance of herbivore food preference and algal competitive abilities. Am Nat 112:23–39

Mantyka CS, Bellwood DR (2007) Macroalgal grazing selectivity among herbivorous coral reef fishes. Mar Ecol Prog Ser 352:177–185

Markel RW, DeWreede RE (1998) Mechanisms underlying the effect of the chiton *Katharina tunicata* on the kelp *Hedophyllum sessile*: size escapes and indirect effects. Mar Ecol Prog Ser 166:151–161

Marques LV, Villaça R, Pereira RC (2006) Susceptibility of macroalgae to herbivorous fishes at Rocas Atoll, Brazil. Bot Mar 49:379–385

Maschek JA, Baker BJ (2008) The chemistry of algal secondary metabolites. In: Amsler CD (ed) Algal chemical ecology. Springer, Berlin, pp 1–24

Montgomery WL, Gerking SD (1980) Marine macroalgae as foods for fishes: an evaluation of potential food quality. Environ Biol Fish 5:143–153

Morelissen B, Harley CDG (2007) The effects of temperature on producers, consumers, and plant-herbivore interactions in an intertidal community. J Exp Mar Biol Ecol 348:162–173

Mumby PJ, Dahlgren CP, Harborne AR, Kappel CV, Micheli F, Brumbaugh DR, Holmes KE, Mendes JM, Broad K, Sanchirico JN, Buch K, Box S, Stoffle RW, Gill AB (2006) Fishing, trophic cascades, and the process of grazing on coral reefs. Science 311:98–101

O'Connor NE, Crowe TP (2005) Biodiversity loss and ecosystem functioning: distinguishing between number and identity of species. Ecology 86:1783–1796

O'Connor MI (2009) Warming strengthens an herbivore-plant interaction. Ecology 90:388–398

Ojeda FP, Muñoz AA (1999) Feeding selectivity of the herbivorous fish *Scartichthys viridis:* effects on macroalgal community structure in a temperate rocky intertidal coastal zone. Mar Ecol Prog Ser 184:219–229

Orav-Kotta H, Kotta J (2004) Food and habitat choice of the isopod *Idotea baltica* in the northeastern Baltic Sea. Hydrobiology 514:79–85

Paul VJ, Hay ME (1986) Seaweed susceptibility to herbivory: chemical and morphological correlates. Mar Ecol Prog Ser 33:255–264

Pavia H, Toth GB (2008) Macroalgal models in testing and extending defense theories. In: Amsler CD (ed) Algal chemical ecology. Springer, Berlin, pp 147–172

Pavia H, Toth GB, Åberg P (2002) Optimal defense theory: elasticity analysis as a tool to predict intra-plant variation in defenses. Ecology 83:891–897

Pereira RC, da Gama BAP (2008) Macroalgal chemical defenses and their roles in structuring tropical marine communities. In: Amsler CD (ed) Algal chemical ecology. Springer, Berlin, pp 25–55

Petchey OL, Brose U, Rall BC (2010) Predicting the effects of temperature on food web connectance. Philos Trans R Soc B 365:2081–2091

Pfister CA, Hay ME (1988) Associational plant refuges: convergent patterns in marine and terrestrial communities result from differing mechanisms. Oecologia 77:118–129

Pinnegar JK, Polunin NVC, Francour P, Badalamenti F, Chemello R, Harmelin-Vivien ML, Hereu B, Milazzo M, Zabala M, D'Anna G, Pipitone C (2000) Trophic cascades in benthic marine ecosystems: lessons for fisheries and protected-area management. Environ Conserv 27:179–200

Poore AGB, Campbell AH, Steinberg PD (2009) Natural densities of mesograzers fail to limit growth of macroalgae or their epiphytes in a temperate algal bed. J Ecol 97:164–175

Rhoades DF (1979) Evolution of plant chemical defense against herbivores. In: Rosenthal GA, Janzen DH (eds) Herbivores: their interaction with secondary plant metabolites. Academic, New York, pp 1–55

Rowcliffe JM, Watkinson AR, Sutherland WJ, Vickery JA (2001) The depletion of algal beds by geese: a predictive model and test. Oecologia 127:361–371

Santelices B, Bobadilla M (1996) Gastropod pedal mucus retains seaweed propagules. J Exp Mar Biol Ecol 197:251–261

Shepherd SA, Hawkes MW (2005) Algal food preferences and seasonal foraging strategies of the marine iguana, *Amblyrhynchus cristatus*, on Santa Cruz, Galapagos. Bull Mar Sci 77:51–77

Slocum CJ (1980) Differential susceptibility to grazers in two phases of an intertidal alga: Advantages of heteromorphic generations. J Exp Mar Biol Ecol 46:99–110

Soares AR, Teixeira VL, Pereira RC, Villaça R (2003) Variation on diterpene production by the Brazilian alga *Stypopodium zonale* (Dictyotales, Phaeophyta). Biochem Syst Ecol 31:1347–1350

Spurkland T, Iken K (2011) Kelp bed dynamics in estuarine environments in sub-Arctic Alaska. J Coast Res 27:133–143. doi 10.2112/JCOASTRES-D-10-00194

Steneck RS, Watling L (1982) Feeding capabilities and limitation of herbivore molluscs: a functional group approach. Mar Biol 68:299–319

Steneck RS, Graham MH, Bourque BJ, Corbett D, Erlandson JM, Estes JA, Tegner MJ (2002) Kelp forest ecosystems: biodiversity, stability, resilience and future. Environ Conserv 29:436–459

Targett NM, Arnold TM (2001) Effects of secondary metabolites on digestion in marine herbivores. In: McClintock JB, Baker BB (eds) Marine chemical ecology. CRC, Boca Raton, pp 391–411

Taylor RB, Brown PJ (2006) Herbivory in the gammarid amphipod *Aora typica*: relationships between consumption rates, performance and abundance across ten seaweed species. Mar Biol 149:455–463

Toth GB, Pavia H (2007) Induced herbivore resistance in seaweeds: a meta-analysis. J Ecol 95:425–434

Toth GB, Langhamer O, Pavia H (2005) Inducible and constitutive defenses of valuable seaweed tissues: consequences for herbivore fitness. Ecology 86:612–618

Van Alstyne KL (1989) Adventitious branching as a herbivore-induced defense in the intertidal brown alga *Fucus distichus*. Mar Ecol Prog Ser 56:169–176

Van Alstyne KL (1990) Effects of wounding by the herbivorous snails *Littorina sitkana* and *L. scutulata* (Mollusca) on growth and reproduction of the intertidal alga *Fucus distichus* (Phaeophyta). J Phycol 26:412–416

Viejo RM, Åberg P (2003) Temporal and spatial variation in the density of mobile epifauna and grazing damage on the seaweed *Ascophyllum nodosum*. Mar Biol 142:1229–1241

Villaça R, Fonseca AC, Köppe Jansen V, Knoppers B (2010) Species composition and distribution of macroalgae on Atol das Rocas, Brazil, SW Atlantic. Bot Mar 53:113–122

Whiting SD (2002) Rocky reefs provide foraging habitat for dugongs in the Darwin region of northern Australia. Aust Mammal 24:147–150

Williamson JE, Carson DG, de Nys R, Steinberg PD (2004) Demographic consequences of an ontogenetic shift by a sea urchin in response to host plant chemistry. Ecology 85:1355–1371

Worm B, Lotze HK, Boström C, Engkvist R, Labanauskas V, Sommer U (1999) Marine diversity shift linked to interactions among grazers, nutrients and propagule banks. Mar Ecol Prog Ser 185:309–314

Worm B, Lotze HK, Sommer U (2000) Coastal food web structure, carbon storage, and nitrogen retention regulated by consumer pressure and nutrient loading. Limnol Oceanogr 45:339–349

Wright JT, Dworjanyn SA, Rogers CN, Steinberg PD, Williamson JE, Poore AGB (2005) Density-dependent sea urchin grazing: differential removal of species, changes in community composition and alternative community states. Mar Ecol Prog Ser 298:143–156

Chapter 9
Chemical Ecology of Seaweeds

Charles D. Amsler

9.1 Overview

Chemical ecology is the study of chemically mediated interactions between organisms or between organisms and their environment. It is convenient to separate these interactions into sensory chemical ecology and defensive chemical ecology. However, sensory and defensive chemical interactions cannot always be separated as waterborne signals released by seaweeds under attack by herbivores can be sensed by neighboring algae and induce a defensive response in the unattacked neighbors (Toth and Pavia 2000). Moreover, seaweed defenses against biofouling can function by interfering with chemical communication between bacteria on the algal surface (Kjelleberg et al. 1997).

Sensory chemical ecology can be subdivided into chemical communications between organisms, be they intentional or not, and sensing and responding to the abiotic chemical environment of the alga. There is a large body of literature on seaweeds sensing and physiologically responding to their nutrient environment (see Chap. 4 by Gordillo). However, in practice such studies have been outside the scope of "chemical ecology" and this chapter follows the more narrow definition, which does not include physiological responses to nutrients.

Defensive chemical ecology can likewise be subdivided, often by what is being defended against. Most studies of seaweed chemical ecology as a whole involve studies of defenses against predation, which from a mechanistic perspective are an important focus of this chapter. They are discussed with respect to community impacts in Chap. 8 by Iken. Chemical defenses can also be produced in response to organisms that grow on or over seaweeds, referred to as *biofoulers* or *epibionts*, to pathogenic organisms, and to competitors. Production of chemical defenses against competitors is commonly referred to as *allelopathy*. These categories are not

C.D. Amsler (✉)
Department of Biology, University of Alabama at Birmingham, Birmingham, AL 35294-1170, USA
e-mail: amsler@uab.edu

C. Wiencke and K. Bischof (eds.), *Seaweed Biology*, Ecological Studies 219,
DOI 10.1007/978-3-642-28451-9_9, © Springer-Verlag Berlin Heidelberg 2012

clear-cut as biofoulers may also have characteristics of pathogens or competitors, and individual defensive metabolites may have multiple defensive roles (e.g., Schmitt et al. 1995).

It would be impossible to comprehensively review seaweed chemical ecology in a short chapter such as this. Indeed there is a recent book devoted entirely to algal chemical ecology (Amsler 2008a) and it could not be comprehensive with respect to most subjects. Rather, I have attempted to highlight studies that illustrate the various aspects of seaweed chemical ecology while providing references to more detailed reviews.

9.2 Sensory Chemical Ecology

9.2.1 Chemical Communication

Chemical communication is important in sexual reproduction of brown seaweeds. This has received a great deal of research attention and been the subject of numerous reviews (e.g., Maier and Müller 1986; Maier 1995; Amsler and Iken 2001; Pohnert and Boland 2002; Amsler and Fairhead 2006). Comparatively, virtually nothing is known about such phenomena in green or red seaweeds although similar mechanisms are well documented in some freshwater and microscopic green algae (Sekimoto 2005; Amsler 2008b).

Brown seaweeds utilize pheromones released by female gametes both as male gamete attractants and, in some taxa, for stimulating the release of male gametes. Although this behavioral attraction was first observed in the 1800s, the specific pheromones responsible did not begin to be identified until the 1960s (cf. Amsler and Fairhead 2006). The 12 known pheromones from over 60 brown algal species in 13 orders are all C_8 or C_{11} hydrocarbons or epoxides (Maier 1995; Pohnert and Boland 2002). These diffuse rapidly away from the female gametes and the range at which perception by males is possible has been estimated at up to several millimeters (Maier and Müller 1986).

Gamete attraction in brown algae is probably best understood in *Ectocarpus siliculosus*, where a responsible pheromone was first identified (Müller 1967, 1968). Upon settlement, female gametes secrete the C_{11} pheromones ectocarpene and pre-ectocarpene (Müller 1978; Pohnert and Boland 2002), which cause a complex response in male gametes that has been described as a chemo-thigmo-klinokinesis [reviewed by Maier and Müller (1986) and Müller (1989)]. Upon sensing the pheromone, the male gametes remain in contact with a surface (the "thigmo" component) and slow down. As the pheromone concentration increases, the gametes make sharp, nearly U-turns more and more frequently (hence a "chemokinetic" response; cf. Amsler and Iken 2001) and thereby are effectively trapped in the area of the female gamete until they contact it. In contrast, male gametes of the kelp *Laminaria digitata* are able to swim directly toward female gametes releasing the C_{11} pheromone lamoxirene [reviewed by Maier and Müller (1986) and Müller (1989)], a behavior that can be described as a true chemotaxis (cf. Amsler and Iken 2001). Lamoxirene also stimulates release of male gametes from male gametophytes' antheridia (Maier 1982).

Unintentional forms of chemical communication, at least from the perspective of the signal sender, occur between spores of the green seaweed *Ulva* spp. and biofilm bacteria as well as in bacterial stimulation of spore release of the red alga *Acrochaetium* sp. (see also Chap. 11 by Potin and Chap. 10 by Friedrich). It has been known for some time that biofilm composition can influence *Ulva* spore settlement (e.g., Thomas and Allsopp 1983) and that bacterial metabolites are necessary for normal development of *Ulva* germlings (e.g., Provasoli and Pinter 1980). Joint et al. (2000) reported that within biofilms, the spores preferentially settle on or next to bacterial cells. This settlement pattern is the result of spores "eavesdropping" on *N*-acylhomoserine lactone (AHL) signals (Joint et al. 2002; Tait et al. 2005) which are produced by many bacteria for quorum sensing, an intraspecific form of chemical communication involved in multiple bacterial processes (Chhabra et al. 2005). The mechanism is a form of chemokinesis in which the spores drastically reduce their swimming speed when close to the AHL source, but this only occurs if they are in contact with the surface of the substrate (Wheeler et al. 2006).

AHLs can also be involved in seaweed spore release. The filamentous red alga *Acrochaetium* sp. is a common epiphyte on the larger red seaweed *Gracilaria chilensis* and spore release in *Acrochaetium* sp. appears to be induced by AHLs from bacteria growing epiphytically on *G. chilensis* (Weinberger et al. 2007; see also Chap. 11 by Potin). However, no such effect of AHLs was observed with another (although not closely related) filamentous red alga, *Sahlingia subintegra*, which also grows epiphytically on *G. chilensis* (Weinberger et al. 2007).

As discussed later, some seaweeds release waterborne chemical cues when under attack by some (but not all) herbivores. These cue adjacent conspecifics to induce the production of defensive chemical compounds even without being attacked themselves (Toth and Pavia 2000; Toth 2007). Predators of herbivorous snails preying on the seaweeds can also sense these cues, another example of communication between species. Coleman et al. (2007a) reported that fish predators can be attracted by the scent of snails feeding on other organisms but significantly more so if the snails are grazing on *Ascophyllum nodosum*, an alga known to induce its own defenses when neighboring conspecifics are under attack (Toth and Pavia 2000). Crab predators of the snails were attracted by *A. nodosum* that had been grazed by snails even if the snails had been removed (Coleman et al. 2007a).

9.2.2 Environmental Sensing

Seaweed spores are known to respond to physiochemical cues, in particular to surface hydrophobicity. Spores from several species of brown and green algae settle preferentially on hydrophobic surfaces in the laboratory, although the adaptive significance of this in the context of the biofilm environments they settle within in nature is unknown (Amsler 2008b). Interestingly, although *Ulva* spores given the choice between alternating hydrophobic and hydrophilic surfaces settle preferentially on the hydrophobic patches (Callow et al. 2000), when presented with a hydrophobicity gradient they settle more in hydrophilic regions (Chaudhury et al. 2006). In the brown alga *Hincksia irregularis*, surface hydrophobicity interacts with light and light gradients (Greer and Amsler 2002, 2004) and with potential antifouling compounds (Greer et al. 2003) to influence spore settlement patterns.

Some brown algal spores respond to inorganic nutrients in their swimming and settlement behaviors. Spores from several species of kelps (Order Laminariales) demonstrate chemotactic swimming responses toward or away from nutrients (Amsler and Neushul 1989; Fukuhara et al. 2002). Spores themselves probably have no need for inorganic nutrients, but the filamentous gametophyte stage into which they germinate after settlement certainly does. Because these gametophytes are microscopic and confined to the biofilm environment the spore settles in, the nutrient microenvironment in which they settle likely has a profound influence on the ultimate growth and reproduction of the gametophyte. *Macrocystis pyrifera* spores are attracted to ammonia concentrations which stimulate gametophytic growth but not to high concentrations which inhibit it (Amsler and Neushul 1989). The spores are also attracted to low concentrations of iron which stimulate gametophytic reproduction, but not to higher, inhibitory concentrations (Amsler and Neushul 1989). Independent of chemical gradients, and apparently via a somewhat different sensory mechanism, kelp spores also settle more rapidly in the presence of nutrients which are necessary for gametophytic growth and reproduction (Amsler and Neushul 1990).

9.3 Chemical Defense

9.3.1 Defenses Against Predation

Seaweed chemical defenses against herbivores have received the bulk of attention in seaweed chemical ecology. They have been the subject in whole or large part of numerous recent reviews appearing in journals (e.g., Pohnert 2004; Amsler and Fairhead 2006; Ianora et al. 2006; Amsler et al. 2009b; Paul et al. 2011) and in six chapters of a recent book (Amsler 2008a). They are also discussed in Chap. 8 by Iken, particularly with respect to their population or community level importance. This section focuses on selected sets of studies which illustrate defenses from a more mechanistic perspective.

Seaweeds utilize a variety of chemical compounds for defense (Maschek and Baker 2008) and these may be spread throughout the alga or be localized to particular parts of the thallus (Jormalainen and Honkanen 2008). Chemical defenses can be produced constitutively, only when a seaweed is attacked by a herbivore, or some combination thereof. Defenses synthesized only when under attack or in greater amounts when under attack are referred to as *induced defenses*. Alternatively, some defenses, known as *activated defenses*, are produced constitutively but in an inactive or less toxic form and are rapidly changed into the bioactive form upon herbivore attack. There are a number of theories based on terrestrial plants (Stamp 2003; Agrawal 2007) that have been used for predictions of what kinds of defensive compounds seaweeds should produce as well as when and where they should make them. These ideas were recently reviewed by Amsler and Fairhead (2006) and Pavia and Toth (2008).

The red alga *Delisea pulchra* is a well-studied example of constitutive chemical defenses against predation (and against biofoulers, see also Chap. 8 by Iken, Chap. 11 by Potin). *D. pulchra* elaborates a number of halogenated polyketide metabolites known as *furanones* (de Nys et al. 1993) which deter grazing by many, but not all, herbivores (Wright et al. 2004). Interspecific variations in furanone levels are heritable (Wright et al. 2004) and have a cost to the seaweeds in terms of reduced growth and fecundity under some circumstances (Dworjanyn et al. 2006a). That defensive levels vary within a species, are heritable, and come at a cost to the producing organism are assumptions underlying chemical defense theories (Stamp 2003; Agrawal 2007). However, they are rarely documented, particularly in seaweeds, making *D. pulchra* a particularly important model.

Constitutive (and other) defenses are also well studied in a number of brown seaweeds, particularly in members of the Orders Fucales and Dictyotales (Amsler and Fairhead 2006). In the Fucales, one important model species has been *Fucus vesiculosus*, particularly from northern Europe where the defensive compounds appear to be phlorotannins, a class of large, phenolic compounds that are widely distributed in brown algae (Amsler and Fairhead 2006; Jormalainen and Honkanen 2008). Phlorotannin levels in these seaweeds have a heritable component and also vary geographically, over time, and with a variety of environmental parameters, which has facilitated tests of a diverse range of ideas in defensive chemical ecology (Jormalainen and Honkanen 2008). Seaweeds in the Dictyotales utilize a variety of smaller compounds for defense with the best studied examples being terpenoids, particularly in the genus *Dictyota* (Pereira and da Gama 2008). These seaweeds have also been used in a wide variety of studies of how defensive chemicals mediate seaweed–herbivore interactions. For example, they have been used to demonstrate that chemical defenses interact with the nutritional quality of an alga to influence not only how much of the alga is consumed by a herbivore, but also the ultimate impact on herbivore fitness (Cruz-Rivera and Hay 2003). They have also been particularly useful in studies of how some herbivores are able to tolerate algal chemical defenses (Hay 2009; Sotka et al. 2009).

Activated chemical defenses are best understood in several green seaweeds including two siphonous genera: *Halimeda* and *Caulerpa*. *Halimeda* spp. constitutively produce the diterpenoid halimedatetraacetate which itself deters fish grazing, but upon wounding the seaweeds rapidly convert it to an even more potent grazing deterrent, halimedatrial (Paul and Van Alstyne 1992). *Caulerpa* spp. produce the sesquiterpene caulerpenyne constitutively but upon wounding rapidly convert it to several more reactive aldehydes (Jung and Pohnert 2001; Jung et al. 2002). Caulerpenyne is deterrent against some but not all herbivores (Jung et al. 2002; Baumgartner et al. 2009) as it significantly reduces food quality due to protein cross-linking. Due to their chemical characteristics, the aldehydes derived from caulerpenyne activation presumably have greater bioactivity but this has been difficult to test because of their instability. Other green algal species, particularly in *Ulva* and related genera, produce high levels of dimethylsulfoniopropionate (DMSP) which has multiple roles in algae (Van Alstyne 2008). Upon wounding, several algae convert DMSP into dimethylsulfide and acrylic acid, both of which are feeding deterrents to sea urchins and likely act as activated defenses (reviewed by Van Alstyne 2008).

Fucoid brown algae have served as important models in understanding induced defenses in seaweeds, particularly *Ascophyllum nodosum* and *Fucus* spp. which elaborate phlorotannins as chemical defenses (Amsler and Fairhead 2006). However, interactions involving induction of defenses vary with species and even population. *A. nodosum* has been shown to increase phlorotannin production and become more unpalatable when under attack by snails but not by isopods (Pavia and Toth 2000) while in some (but not all) populations *Fucus vesiculosus* induces defenses (measured as palatability) when attacked by either snails or isopods (Rohde et al. 2004). Mechanical wounding alone does not induce defenses in *A. nodosum* (Pavia and Toth 2000) and the induction appears to be a response to digestive enzymes present in the snails' saliva (Coleman et al. 2007b). Snails grazing on *A. nodosum* result in waterborne cues being released that enable neighboring *A. nodosum* individuals to induce defenses even in the absence of being wounded themselves (Toth and Pavia 2000) and also can attract predators of the snails (see also Chap. 8 by Iken). In *F. vesiculosus*, waterborne cues resulting in induction of defenses in neighboring individuals are produced in response to predation by isopods but not snails (Rohde et al. 2004). However, although predation by isopods also directly induces defense production in two other species of *Fucus*, in these other species there is no induction from waterborne cues (Rohde and Wahl 2008).

9.3.2 Defenses Against Pathogens

Seaweed pathogens include bacteria, fungi, and filamentous algal endophytes. These relationships and the seaweeds' chemical defenses against pathogens have been the subject of several recent reviews (Weinberger 2007; Lane and Kubanek 2008; Potin 2008; Goecke et al. 2010; see also Chap. 11 by Potin).

One seaweed defense that has been studied with both bacterial and filamentous algal endophyte pathogens is the production of reactive oxygen species (e.g., hydrogen peroxide). These are produced in response to breakdown products from the algal cell walls during pathogen entry (Weinberger 2007; Potin 2008). Such defenses are referred to as *oxidative burst responses*. In some cases, particularly with bacteria, the ability of the reactive oxygen species to oxidize a wide variety of organic molecules may be enough to overcome the pathogen (e.g., Weinberger et al. 2001). In many others, particularly with endophytes, the oxidative burst also triggers a series of activated and induced chemical and structural defenses to contain the infection (Potin 2008).

Many seaweeds have been shown to produce organic compounds that can inhibit growth of bacteria or other pathogens (Bhadury and Wright 2004; Goecke et al. 2010). However, just because an alga makes a compound that can inhibit a pathogen (or biofouling organism) in vitro does not show that it functions as such in nature. To establish that, it is necessary to know that the compound is localized in a way that the pathogen comes into contact with it at a concentration high enough to be effective. This has been shown with defenses against fungal pathogens in several seaweeds. The brown alga *Lobophora variegata*, or more likely a bacterial symbiont of it, produces lobophoralide, a polycyclic macrolide which can be extracted from its surface at concentrations which inhibit the growth of some higher fungi, including seaweed pathogens (Kubanek et al. 2003). The red alga *Callophycus serratus* produces a suite of diterpene-benzoate macrolides and diterpene-alcohols which also inhibit fungal growth (Lane et al. 2009). In the first study of its kind, desorption electrospray ionization mass spectrometry was used to image and quantify these molecules in patches on the seaweed's surface and at concentrations sufficient to inhibit the pathogenic fungi (Lane et al. 2009).

9.3.3 Defenses Against Biofouling

Chemical defenses against biofouling are like those against pathogens in that just because a seaweed produces a compound that can inhibit fouling in vitro, to show ecological relevance against biofouling one must establish that the compound is present at the surface of the alga at a biologically effective concentration (Steinberg and de Nys 2002). Perhaps the most thoroughly studied example of this in seaweeds is the case of halogenated furanones produced by the red alga *Delisea pulchra* as defenses against both predators (see also Chap. 8 by Iken, Chap. 11 by Potin) and biofoulers. These are present at the cell surface at known concentrations (Dworjanyn et al. 1999). They mimic bacterial AHL quorum sensing molecules and interfere with numerous bacterial processes, including attachment, at the concentration range in which they occur on the algal surface (Kjelleberg et al. 1997; Maximilien et al. 1998). They also inhibit the attachment of propagules from four species of fouling macroalgae at surface-level concentrations (Dworjanyn et al. 2006b). Surface extracts of the brown alga *Dictyota menstrualis* have been shown to contain the diterpenoids pachydictyol A and dictyol E, which are also known to defend against herbivores, and algal surface extracts including these compounds inhibited the settlement of larvae of a fouling bryozoan when coated onto artificial surfaces (Schmitt et al. 1995). Surface extracts of the red alga *Bonnemaisonia hamifera* contain a poly-halogenated 2-heptanone at concentrations that inhibit bacterial biofouling (Nylund et al. 2008). The green alga *Ulva reticulata* appears to release a water soluble antifouling defense. Water collected from the vicinity of the seaweeds in nature inhibits attachment of larvae of fouling bryozoans and hydroids (Harder et al. 2004). Some, but not all, of this could be the result of compounds actually produced by a bacterial species colonizing the surface of the algae (Harder et al. 2004).

9.3.4 Allelopathy

Allelopathy (chemical defense against competitors) is relatively widely studied in freshwater systems including in macrophytic algae, vascular plants, and phytoplankton (Gross 2003; see also Chap. 11 by Potin). In marine systems there have been studies showing allelopathic effects of water-soluble compounds from seaweeds on the growth of phytoplankton in laboratory conditions, but the significance of such interactions in natural populations is unclear. On the other hand, seaweeds often compete with other sessile, macroscopic organisms for space (see Chap. 7 by Edwards and Connell) and the advantages of being able to inhibit the growth of such competitors are obvious. There have been at least two studies of such interactions under ecologically relevant conditions. The red alga *Plocamium hamatum* causes tissue necrosis in soft corals upon physical contact due to allelopathic effects of the monoterpene chloromertensene (de Nys et al. 1991). In a study of the effects of contact by multiple seaweed species on two species of hard corals, Rasher and Hay (2010) reported that in nearly half the interactions, seaweeds caused bleaching (loss of symbiotic microalgae) of the corals. This was due to the physiological stress imposed on the microalgal symbionts (measured as photosynthetic efficiency). In nearly every interaction where the intact seaweed caused this effect, so did lipophilic chemical extracts of the seaweed (Rasher and Hay 2010).

9.4 Outlook

If this chapter were about the chemical ecology of terrestrial organisms rather than of seaweeds, the relative lengths of the above sections would be vastly different. Terrestrial chemical ecology is currently dominated by studies of sensory chemical ecology, including intentional or unintentional communication among or between prey species, their predators, and their predators' predators. Such studies are in their infancy in seaweed chemical ecology and represent important areas for future studies. Likewise, we are just starting to learn about seaweed defenses against pathogens and biofoulers in ecologically relevant contexts. Doing the necessary experiments in an ecologically relevant manner is difficult, but many more such studies are needed to advance the field. Important work is ongoing in all areas touched on in this short review. In each there is much more to be learned and many important questions yet to be asked.

Acknowledgments I am grateful to M. Amsler, K. Iken, J. McClintock, G. Pohnert, and an anonymous reviewer for helpful comments on earlier drafts of this chapter. Manuscript preparation was supported in part by National Science Foundation award ANT-0838773 from the Antarctic Organisms and Ecosystems program.

References

Agrawal AA (2007) Macroevolution of plant defense strategies. Trends Ecol Evol 22:103–109

Amsler CD (2008a) Algal chemical ecology. Springer, Berlin, p xviii, 313

Amsler CD (2008b) Algal sensory chemical ecology. In: Amsler CD (ed) Algal chemical ecology. Springer, Berlin, pp 297–309

Amsler CD, Fairhead VA (2006) Defensive and sensory chemical ecology of brown algae. Adv Bot Res 43:1–91

Amsler CD, Iken KB (2001) Chemokinesis and chemotaxis in marine bacteria and algae. In: McClintock JB, Baker BJ (eds) Marine chemical ecology. CRC, Boca Raton, FL, pp 413–430

Amsler CD, Neushul M (1989) Chemotactic effects of nutrients on spores of the kelps *Macrocystis pyrifera* and *Pterygophora californica*. Mar Biol 102:557–564

Amsler CD, Neushul M (1990) Nutrient stimulation of spore settlement in the kelps *Pterygophora californica* and *Macrocystis pyrifera*. Mar Biol 107:297–304

Amsler CD, Iken K, McClintock JB, Baker BJ (2009b) Defenses of polar macroalgae against herbivores and biofoulers. Bot Mar 52:535–545

Baumgartner FA, Motti CA, de Nys R, Paul NA (2009) Feeding preferences and host associations of specialist marine herbivores align with quantitative variation in seaweed secondary metabolites. Mar Ecol Prog Ser 396:1–12

Bhadury P, Wright PC (2004) Exploitation of marine algae: biogenic compounds for potential antifouling applications. Planta 219:561–578

Callow ME, Callow JA, Ista LK, Coleman SE, Nolasco AC, Lopez GP (2000) Use of self-assembled monolayers of different wettabilities to study surface selection and primary adhesion processes of green algal (*Enteromorpha*) zoospores. Appl Environ Microbiol 66:3249–3254

Chaudhury MK, Daniel S, Callow ME, Callow JA, Finlay JA (2006) Settlement behavior of swimming algal spores on gradient surfaces. Biointerphases 1:18–21

Chhabra SR, Philipp B, Eberl L, Givskov M, Williams P, Camara M (2005) Extracellular communication in bacteria. Top Curr Chem 240:279–315

Coleman R, Ramchunder S, Davis K, Moody A, Foggo A (2007a) Herbivore-induced infochemicals influence foraging behaviour in two intertidal predators. Oecologia 151:454–463

Coleman RA, Ramchunder SJ, Moody AJ, Foggo A (2007b) An enzyme in snail saliva induces herbivore-resistance in a marine alga. Funct Ecol 21:101–106

Cruz-Rivera E, Hay ME (2003) Prey nutritional quality interacts with chemical defenses to affect consumer feeding and fitness. Ecol Monogr 73:483–506

de Nys R, Coll JC, Price IR (1991) Chemically mediated interactions between the red alga *Plocamium hamatum* (Rhodophyta) and the octocoral *Sinularia cruciata* (Alcyonacea). Mar Biol 108:315–320

de Nys R, Wright AD, Konig GM, Sticher O (1993) New halogenated furanones from the marine alga *Delisea pulchra* (cf. *fimbriata*). Tetrahedron 49:11213–11220

Dworjanyn SA, de Nys R, Steinberg PD (1999) Localisation and surface quantification of secondary metabolites in the red alga *Delisea pulchra*. Mar Biol 133:727–736

Dworjanyn SA, de Nys R, Steinberg PD (2006a) Chemically mediated antifouling in the red alga *Delisea pulchra*. Mar Ecol Prog Ser 318:153–163

Dworjanyn SA, Wright JT, Paul NA, de Nys R, Steinberg PD (2006b) Cost of chemical defence in the red alga *Delisea pulchra*. Oikos 113:13–22

Fukuhara Y, Mizuta H, Yasui H (2002) Swimming activities of zoospores in *Laminaria japonica* (Phaeophyceae). Fish Sci 68:1173–1181

Goecke F, Labes A, Wiese J, Imhoff JF (2010) Chemical interactions between marine macroalgae and bacteria. Mar Ecol Prog Ser 409:267–300

Greer SP, Amsler CD (2002) Light boundaries and the coupled effects of surface hydrophobicity and light on spore settlement in the brown alga *Hincksia irregularis* (Phaeophyceae). J Phycol 38:116–124

Greer SP, Amsler CD (2004) Clonal variation in phototaxis and settlement behaviors of *Hincksia irregularis* (Phaeophyceae) spores. J Phycol 40:44–53

Greer SP, Iken KB, McClintock JB, Amsler CD (2003) Individual and coupled effects of echinoderm extracts and surface hydrophobicity on spore settlement and germination in the brown alga *Hincksia irregularis*. Biofouling 19:315–326

Gross EM (2003) Allelopathy of aquatic autotrophs. Crit Rev Plant Sci 22:313–339

Harder T, Dobretsov S, Qian P-Y (2004) Waterborne polar macromolecules act as algal antifoulants in the seaweed *Ulva reticulata*. Mar Ecol Prog Ser 274:133–141

Hay ME (2009) Marine chemical ecology: chemical signals and cues structure marine populations, communities, and ecosystems. Annu Rev Mar Sci 1:193–212

Ianora A, Boersma M, Casotti R, Fontana A, Harder J, Hoffmann F, Pavia H, Potin P, Poulet SA, Toth G (2006) New trends in marine chemical ecology. Est Coast 29:531–551

Joint I, Callow ME, Callow JA, Clarke KR (2000) The attachment of *Enteromorpha* zoospores to a bacterial biofilm assemblage. Biofouling 16:151–158

Joint I, Tait K, Callow ME, Callow JA, Milton D, Williams P, Camara M (2002) Cell-to-cell communication across the prokaryote-eukaryote boundary. Science 298:1207

Jormalainen V, Honkanen T (2008) Macroalgal chemical defenses and their roles in structuring temperate marine communities. In: Amsler CD (ed) Algal chemical ecology. Springer, Berlin, pp 57–89

Jung V, Pohnert G (2001) Rapid wound-activated transformation of the green algal defensive metabolite caulerpenyne. Tetrahedron 57:7169–7172

Jung V, Thibaut T, Meinesz A, Pohnert G (2002) Comparison of the wound-activated transformation of caulerpenyne by invasive and noninvasive *Caulerpa* species of the Mediterranean. J Chem Ecol 28:2091–2105

Kjelleberg S, Steinberg P, Givskov M, Gram L, Manefield M, de Nys R (1997) Do marine natural products interfere with prokaryotic AHL regulatory systems? Aquat Microb Ecol 13:85–93

Kubanek J, Jensen PR, Keifer PA, Sullards MC, Collins DO, Fenical W (2003) Seaweed resistance to microbial attack: A targeted chemical defense against marine fungi. Proc Nat Acad Sci USA 100:6916–6921

Lane AL, Kubanek J (2008) Secondary metabolite defenses against pathogens and biofoulers. In: Amsler CD (ed) Algal chemical ecology. Springer, Berlin, pp 229–243

Lane AL, Nyadong L, Galhena AS, Shearer TL, Stout EP, Parry RM, Kwasnik M, Wang MD, Hay ME, Fernandez FM, Kubanek J (2009) Desorption electrospray ionization mass spectrometry reveals surface-mediated antifungal chemical defense of a tropical seaweed. Proc Nat Acad Sci USA 106:7314–7319

Maier I (1982) New aspects of pheromone-triggered spermatozoid release in *Laminaria digitata* (Phaeophyta). Protoplasma 113:137–142

Maier I (1995) Brown algal pheromones. Prog Phycol Res 11:51–102

Maier I, Müller DG (1986) Sexual pheromones in algae. Biol Bull 170:145–175

Maschek JA, Baker BJ (2008) The chemistry of algal secondary metabolism. In: Amsler CD (ed) Algal chemical ecology. Springer, Berlin, pp 1–24

Maximilien R, de Nys R, Holmstr"m C, Gram L, Givskov M, Crass K, Kjelleberg S, Steinberg P (1998) Chemical mediation of bacterial surface colonisation by secondary metabolites from the red alga *Delisea pulchra*. Aquat Microb Ecol 15:233–246

Müller DG (1967) Ein leichtflüchtiges Gyno-Gamon der Braunalge *Ectocarpus siliculosus*. Naturwissenschaften 54:496–497

Müller DG (1968) Versuche zur Charakterisierung eines Sexuallockstoffes bei der Braunalge *Ectocarpus siliculosus*. I. Methoden, Isolierung, und gaschromatographischer Nachweis. Planta 81:160–168

Müller DG (1978) Locomotive responses of male gametes to the species-specific sex attractant of *Ectocarpus siliculosus* (Phaeophyta). Arch Protistenkd 120:371–377

Müller DG (1989) The role of pheromones in sexual reproduction of brown algae. In: Coleman AW, Goff LJ, Stein-Taylor JR (eds) Algae as experimental systems. Alan R Liss, New York, pp 201–213

Nylund GM, Cervin G, Persson F, Hermansson M, Steinberg PD, Pavia H (2008) Seaweed defence against bacteria: a poly-brominated 2-heptanone from the red alga *Bonnemaisonia hamifera* inhibits bacterial colonisation. Mar Ecol Prog Ser 369:39–50

Paul VJ, Van Alstyne KL (1992) Activation of chemical defenses in the tropical green algae *Halimeda* spp. J Exp Mar Biol Ecol 160:191–203

Paul VJ, Ritson-Williams R, Sharp K (2011) Marine chemical ecology in benthic environments. Nat Prod Rep 28:345–387

Pavia H, Toth GB (2000) Inducible chemical resistance to herbivory in the brown seaweed *Ascophyllum nodosum*. Ecology 81:3212–3225

Pavia H, Toth GB (2008) Macroalgal models in testing and extending defense theories. In: Amsler CD (ed) Algal chemical ecology. Springer, Berlin, pp 147–172

Pereira RC, da Gama BAP (2008) Macroalgal chemical defenses and their roles in structuring tropical marine communities. In: Amsler CD (ed) Algal chemical ecology. Springer, Berlin, pp 25–55

Pohnert G (2004) Chemical defense strategies of marine organisms. Top Curr Chem 239:179–219

Pohnert G, Boland W (2002) The oxylipin chemistry of attraction and defense in brown algae and diatoms. Nat Prod Rep 19:108–122

Potin P (2008) Oxidative burst and related responses in biotic interactions of algae. In: Amsler CD (ed) Algal chemical ecology. Springer, Berlin, pp 245–271

Provasoli L, Pinter IJ (1980) Bacteria induced polymorphism in an axenic laboratory strain of *Ulva lactuca*. J Phycol 16:196–201

Rasher DB, Hay ME (2010) Chemically rich seaweeds poison corals when not controlled by herbivores. Proc Nat Acad Sci USA 107:9683–9688

Rohde S, Wahl M (2008) Antifeeding defense in baltic macroalgae: induction by direct grazing versus waterborne cues. J Phycol 44:85–90

Rohde S, Molis M, Wahl M (2004) Regulation of anti-herbivore defence by *Fucus vesiculosus* in response to various cues. J Ecol 92:1011–1018

Schmitt TM, Hay ME, Lindquist N (1995) Constraints on chemically mediated coevolution: multiple functions for seaweed secondary metabolites. Ecology 76:107–123

Sekimoto H (2005) Plant sex pheromones. Vitamins Hormones 72:457–478

Sotka EE, Forbey J, Horn M, Poore AGB, Raubenheimer D, Whalen KE (2009) The emerging role of pharmacology in understanding consumer-prey interactions in marine and freshwater systems. Integr Comp Biol 49:291–313

Stamp N (2003) Out of the quagmire of plant defense hypotheses. Q Rev Biol 78:23–55

Steinberg PD, de Nys R (2002) Chemical mediation of colonization of seaweed surfaces. J Phycol 38:621–629

Tait K, Joint I, Daykin M, Milton DL, Williams P, Camara M (2005) Disruption of quorum sensing in seawater abolishes attraction of zoospores of the green alga *Ulva* to bacterial biofilms. Environ Microbiol 7:229–240

Thomas RWSP, Allsopp D (1983) The effects of certain periphytic marine bacteria upon the settlement and growth of *Enteromorpha*, a fouling alga. In: Oxley TA, Barry S (eds) Biodeterioration 5. Wiley, New York, pp 348–357

Toth G (2007) Screening for induced herbivore resistance in Swedish intertidal seaweeds. Mar Biol 151:1597–1604

Toth GB, Pavia H (2000) Water-borne cues induce chemical defense in a marine alga (*Ascophyllum nodosum*). Proc Nat Acad Sci USA 97:14418–14420

Van Alstyne KL (2008) Ecological and physiological roles of dimethylsulfoniopropionate and its products in marine macroalgae. In: Amsler CD (ed) Algal chemical ecology. Springer, Berlin, pp 173–194

Weinberger F (2007) Pathogen-induced defense and innate immunity in macroalgae. Biol Bull 213:290–302

Weinberger F, Richard C, Kloareg B, Kashman Y, Hoppe HG, Friedlander M (2001) Structure-activity relationships of oligoagar elicitors toward *Gracilaria conferta* (Rhodophyta). J Phycol 37:418–426

Weinberger F, Beltran J, Correa JA, Lion U, Pohnert G, Kumar N, Steinberg P, Kloareg B, Potin P (2007) Spore release in *Acrochaetium* sp (Rhodophyta) is bacterially controlled. J Phycol 43:235–241

Wheeler GL, Tait K, Taylor A, Brownlee C, Joint I (2006) Acyl-homoserine lactones modulate the settlement rate of zoospores of the marine alga *Ulva intestinalis* via a novel chemokinetic mechanism. Plant Cell Environ 29:608–618

Wright JT, de Nys R, Poore AGB, Steinberg PD (2004) Chemical defense in a marine alga: Heritability and the potential for selection by herbivores. Ecology 85:2946–2959

Chapter 10
Bacterial Communities on Macroalgae

Michael W. Friedrich

10.1 Introduction

Macroalgae are important primary producers in the marine environment, alter the ecosystem by influencing sediment structure and water flow, and provide a habitat for many marine organisms. The surface of macroalgae is an important habitat for microorganisms, and healthy macroalgae are consistently colonized by microbial communities (Bolinches et al. 1988; Jensen et al. 1996). Interactions of epiphytic microorganisms with its algal host range from beneficial (Goecke et al. 2010; Wahl 2008) to disease and decomposition (Largo et al. 1995, 1997, 1999; Vairappan et al. 2001; Wang et al. 2008). Algae-associated bacteria have been shown to enhance algal growth and provide nutrients (Croft et al. 2005, 2006; Keshtacherliebson et al. 1995; Goecke et al. 2010), influence release and settlement of algal spores (Joint et al. 2002, 2007; Weinberger et al. 2007; see also Chap. 9 by Amsler and Chap. 11 by Potin), and either enhance or deter the settlement of invertebrate larvae (Dobretsov and Qian 2002; Huggett et al. 2008; Patel et al. 2003; Steinberg and de Nys 2002). Specific bacteria play an important role in the development of normal morphology in several green algae (Marshall et al. 2006; Matsuo et al. 2003; Nakanishi et al. 1996; Provasoli and Pintner 1980). Moreover, algal-associated bacteria produce toxins, signaling compounds, and secondary metabolites, which represent an interesting reservoir for the discovery of bioactive compounds (Egan et al. 2008). Such intriguing microbial interactions are based on cultured bacteria.

The analysis of microbial communities associated with macroalgae, however, is still in its infancy. Recent comprehensive reviews have examined the body of literature available on microbial communities associated with macroalgae, which is, however, largely built on cultivation-based studies (Goecke et al. 2010;

M.W. Friedrich (✉)
BreMarE - Bremen Marine Ecology, Faculty of Biology/Chemistry University of Bremen, 28334 Bremen, Germany
e-mail: michael.friedrich@uni-bremen.de

C. Wiencke and K. Bischof (eds.), *Seaweed Biology*, Ecological Studies 219,
DOI 10.1007/978-3-642-28451-9_10, © Springer-Verlag Berlin Heidelberg 2012

Olson and Kellogg 2010) with known limitations in adequately describing the composition of natural microbial communities. This chapter focuses on studies analyzing the composition of bacterial communities using molecular biology approaches by sequence information, thereby allowing to identify individual community members based on phylogenetic and functional traits.

10.2 Microbial Communities Associated with Macroalgae

The past two decades have seen tremendous efforts to unravel the diversity of microorganisms in the marine environment (Pernthaler and Amann 2005; Rappe and Giovannoni 2003). Among sessile eukaryotes with microbial colonization, sponges and corals have received the greatest interest, whereas comparatively few studies have analyzed the microbial diversity associated with the surface of macroalgae (Egan et al. 2008), and specifically macroalgae thriving at mesophotic depths are understudied (Olson and Kellogg 2010). Apart from microorganisms of the domain bacteria (see below), protists as well as fungi occur on macroalgae, including a number of pathogenic fungal species (Armstrong et al. 2000; Correa 1997; Raghukumar et al. 1992), but nothing is known about Archaea (Olson and Kellogg 2010).

Traditionally, classification of microorganisms has relied on cultivation, which requires pure cultures for testing physiological, biochemical, and cell biological properties, and suitable media for their growth. Only a small fraction of microorganisms grow on standard media, in some environments this has been estimated to be as low as 0.001–<1%, e.g., in seawater (Amann et al. 1995; Eilers et al. 2000; Fuhrman and Campbell 1998); thus, cultivation of microorganisms may not reflect numerically abundant microbial populations in the habitat they were retrieved from. Until recently, cultivation-based approaches have nevertheless contributed most of our knowledge regarding microorganisms associated with the surface of macroalgae (Goecke et al. 2010). Microscopy and cultivation-based techniques have revealed densities of 10^4–10^7 cells g^{-1} (wet weight) of algal biomass for healthy individuals (Largo et al. 1997; Lewis et al. 1985; Weinberger et al. 1994). Most common isolates retrieved from algal surfaces comprise the genera *Flavobacterium*, *Bacillus*, *Vibrio*, *Pseudomonas*, and *Moraxella* (Bolinches et al. 1988; Chan and Mcmanus 1969; Egan et al. 2008; Lewis et al. 1985). A large number of novel bacteria have been isolated from diverse macroalgal species over the last decade and a number of novel genera especially in the family Flavobacteriaceae have been described (Goecke et al. 2010).

Given the inherent limitations of cultivation-based approaches, cultivation-independent molecular biology approaches are nowadays used to assess the structure of a microbial community (i.e., its diversity and abundance of microbial populations) (Liesack et al. 1997). Cultivation-independent molecular approaches for community study rely on the concept to clone and sequence genes (e.g., the 16S rRNA gene is predominantly used) directly from biomass rather from cultures

(Pace et al. 1986). The molecular tool box is complemented by whole-cell in situ hybridization with rRNA-targeted fluorescently labeled oligonucleotide probes (fluorescent in situ hybridization, FISH) for PCR-independent identification and enumeration (Amann et al. 1995), as well as various PCR-based fingerprinting methods (e.g., DGGE, denaturing gradient gel electrophoresis; T-RFLP, terminal restriction fragment length polymorphism analysis) (Liu et al. 1997; Muyzer et al. 1993). Essentially, sequence-based analysis (i.e., cloning/sequencing) allows to characterize a microbial community at the level of individual phylotypes; however, this approach has been limited in the past by considering insufficient numbers of sequences (and replicate samples) causing under-sampling bias (Green and Bohannan 2006; Hughes et al. 2001). Fingerprinting methods reflect the diversity of the whole PCR amplicon pool, and thus, are powerful tools to compare larger numbers of sample; however, they have limited phylogenetic resolution (Marsh 2005). Therefore, this review focuses mainly on studies that characterized the composition of microbial communities associated with macroalgae using sequence information (Table 10.1).

A number of cultivation studies complemented with molecular cultivation-independent analysis found only little overlap between the identity of isolates and predominant sequences retrieved from the algal host (Bengtsson et al. 2010; Tujula et al. 2010). For example, the strains of the γ-proteobacterial genus *Pseudoalteromonas* have been frequently isolated from Ulvacean algal hosts (Dobretsov and Qian 2002; Egan et al. 2000, 2001; Patel et al. 2003), but were not detected by catalytically amplified reporter deposition (CARD)-FISH on samples of *Ulva australis* (Tujula et al. 2010).

Sequence-based microbial community composition studies (Table 10.1) included a range of macroalgal species representing Rhodophyta ($n = 5$), Phaeophyceae ($n = 4$), Chlorophyta ($n = 6$), and two functionally defined algae from a coral reef, crustose calcifying algae, and turf algae. The majority of studies analyzed up to 300 sequences by conventional Sanger sequencing of clones or DGGE bands, one study analyzed ~900 clones per individual (*Ulva australis*) (Burke et al. 2011b), and one study used pyrosequencing with up to 60,000 sequences from one individual sample (Barott et al. 2011). High-throughput sequencing methods, such as 16S rRNA gene tag pyrosequencing (Sogin et al. 2006), facilitate to retrieve several thousands (to hundreds of thousands) short sequences (100–400 bp).

Among the microbial phyla encountered sequences representing α-Proteobacteria were most numerous across all studies and macroalgal species (Table 10.1). Frequently detected were also bacteria associated with the phyla Bacteroidetes, Planctomycetes, Verrucomicrobia, Cyanobacteria, and δ-Proteobacteria, and on a few algal species Actinobacteria, Chloroflexi, and Firmicutes. Up to 22 different phyla were encountered in a pyrosequencing-based diversity study of four different algae from a coral reef ecosystem (Barott et al. 2011), indicating the high diversity of microbial communities on macroalgae. Planctomycetal sequences were detected on a relatively large number of algae (Table. 10.1), albeit mismatches of the canonical 27F-primer used to most planctomycetal 16S rRNA genes typically result in underrepresentation of Planctomycetes in clone libraries; when using a specific FISH probe,

Table 10.1 16S rRNA gene sequence-based community composition of microorganisms associated with macroalgae

Taxonomic algal group	Algal species (sampling location)	Major bacterial phyla encountered	Representative species	Number of clones/ sequences analyzed	Methods	Experimental details	References
Rhodophyta (Red algae)	*Delisea pulchra* (Bare Island, Botany Bay, Sydney, Australia)	α-, δ- and γ-Proteobacteria, Planctomycetes, Bacteroidetes	n.d.	236/62	CL-RFLP DGGE Direct tissue lysis	Comparison of three host species (sponge, red, and green macroalgae)	Longford et al. (2007)
	Ahnfeltiopsis flabelliformis *Caulacanthus ustulatus*	Cyanobacterium *Acaryochloris* spp.	n.d	–/23	DGGE Bandseq SP Direct tissue lysis	Tracing cyanobacteria on macroalgal surfaces	Ohkubo et al. (2006)
	Porphyra yezoensis (cultivated in the lab)	γ-Proteobacteria, β-Proteobacteria, Bacteroidetes	*Alteromonas* spp.	100/100	CL Direct tissue lysis	Development of a primer for excluding eukaryal templates	Namba et al. (2010)
	Gracilaria vermiculophylla (Kiel fjord, North Sea)	α-Proteobacteria (31–71%) Bacteroidetes (9–41%) Actinobacteria (3–11%) δ-Proteobacteria (~10%) Planctomycetes (~3%)	*Mesorhizobium* spp. (~25%)	60/60[a]	DGGE CL, Cotton swab samples	Comparison of three algae and seawater; summer and winter samples	Lachnit et al. (2011)
Phaeophyta (Brown algae)	*Laminaria saccharina* (Kiel, Baltic Sea; Helgoland, North Sea)	α-Proteobacteria (9 phylotypes), γ-Proteobacteria (6) Bacteroidetes (4)	Erythrobacter spp. Roseobacter spp.	170[b] 112[b]	DGGE CL	Specific microbial populations associated with different parts of the alga, some overlap with bacteria in seawater found	Staufenberger et al. (2008)
	Laminaria hyperborea (Bergen, Norway)	Planctomycetes	RB1 lineage (73–98%)	260	CU SP CL (FISH) scraped biofilm	Over 50% of epiphytic microbial cells (DAPI) were Planctomycetes, seasonal variability	Bengtsson and Ovreas (2010)
	Laminaria hyperborea (Bergen, Norway)	Planctomycetes (47%) α-Proteobacteria (21%) Verrucomicrobia (18%) Cyanobacteria (5%) Bacteroidetes (4%) γ-Proteobacteria (4%)	n.d.	273 bands	DGGE Bdseq, scraped biofilm	Seasonal variation of microbial communities and comparison with seawater	Bengtsson et al. (2010)
	Laminaria hyperborea	α-Proteobacteria	Roseobacter spp.			Follow-up study, proves that cultivation could not retrieve the most relevant bacteria from the natural biofilm	Bengtsson et al. (2011)

	Fucus vesiculosus (Kiel fjord, Germany)	γ-Proteobacteria (22–33%) Bacteroidetes (20–30%) Verrucomicrobia (6–31%) α-Proteobacteria (13–31%) Planctomycetes (3–19%) Cyanobacteria (2–17%)	*Octadecabacter arcticus* (Rhodobacterales) ~20%	60/60[a]	DGGE, CL Cotton swab samples	Comparison of three algae and seawater, summer and winter samples	Lachnit et al. (2011)
	Dictyota bartayresiana (coral reef, Curacao, Netherlands Antilles)	Unknown bacteria (24–38%) Cyanobacteria (18–33%) Proteobacteria (31–34%) Bacteroidetes (10%)	Various unclassified bacteria cyanobacteria	9,500, 16,300	PYR, Direct tissue lysis	Comparison of four algae and reef-building coral spp.	Barott et al. (2011)
Chlorophyta (Green algae)	*Caulerpa taxifolia*	α-Proteobacteria Bacteroidetes δ-Proteobacteria β-Proteobacteria	*Ruegeria* (Rhodobacter group)	50[d]/54	RFLP CL Direct tissue lysis	Geographical different *Caulerpa* specimen were compared from Mediterranean, Tahiti, the Philippines and Australia	Meusnier et al. (2001)
	Ulva compressa *Ulva intestinalis* (Chañaral Bay, Northern Chile)	Bacteroidetes (49%) α-Proteobacteria (15%) Verrucomicrobia (5%) γ-Proteobacteria (9%)	n.d.	79	TRFLP CL Sonication	Three algae and seawater compared, seasonal effects, copper pollution	Hengst et al. (2010)
	Ulva australis (Bare Island, Botany Bay, Sydney, Australia)	α-, δ- and γ-Proteobacteria Planctomycetes Bacteroidetes	n.d.	197/25	CL-RFLP DGGE Direct tissue lysis	Comparison of three host species (sponge, red and green macroalgae)	Longford et al. (2007)
	Ulva australis (Shark Point, Clovelly, NSW, Australia)	α-Proteobacteria (70%) Bacteroidetes (13%)	*Roseobacter* clade (47%)	34/28	DGGE Bdseq cFISH Direct tissue lysis	Diversity and seasonal dynamics, spatial distribution	Tujula et al. (2010)
	Ulva australis (Bare Island, La Perouse and Shark Point, Clovelly, NSW, Australia	α-Proteobacteria (54%) Bacteroidetes (28%) γ-Proteobacteria (8%) Planctomycetes (3%)	Unclassified Rhodobacteraceae (22%)	5293[e]	CL Microbial cells released from algal surface	Taxonomic in-depth analysis, comparison to seawater community	Burke et al. (2011a, b)
	Ulva prolifera (Jiaozhou Bay, Qingdao)	γ-Proteobacteria (23–53%) α-Proteobacteria (28–35%) CFB group (27%)	Alteromonadales/ Glaciecola	192	DGGE CL cells released by shaking	Comparison of thallus-surface bacterial community composition and surrounding seawater of algal bloom region; two time points	Liu et al. (2011)
	Ulva intestinalis (Kiel fjord, Germany)	α-Proteobacteria (41–71%) Bacteroidetes (5–29%) γ-Proteobacteria (7–15%) OD1-ODP11-WS6-TM7 (5–13%)	*Phyllobacterium* (~14%) (Rhizobiales)	60/60[a]	DGGE CL Cotton swab samples	Comparison of three algae and seawater, summer and winter samples	Lachnit et al. (2011)

(continued)

Table 10.1 (continued)

Taxonomic algal group	Algal species (sampling location)	Major bacterial phyla encountered	Representative species	Number of clones/ sequences analyzed	Methods	Experimental details	References
	Halimeda opuntia (coral reef, Curacao, Netherlands Antilles)	Proteobacteria (31–34%) Cyanobacteria (27%) Unknown bacteria (10–12%) Bacteroidetes (15%) Firmicutes (6–8%)	*Cyanobacteria* Group I (8%)	52,100, 44,800[c]	PYR Direct tissue lysis	Comparison of four algae and reef-building coral spp.	Barott et al. (2011)
Functional groups							
	Turf algae (coral reef, Curacao, Netherlands Antilles)	Proteobacteria (49–57%) Unknown bacteria (10%) Cyanobacteria (5–11%) Firmicutes (8–9%) Bacteroidetes (5–13%) Actinobacteria (2–3%)	*Acidovorax* spp. (11%)	34,000 61,000[c]	PYR Direct tissue lysis	Comparison of four algae and reef–building coral spp.	Barott et al. (2011)
	Crustose coralline algae (coral reef, Curacao, Netherlands Antilles)	Proteobacteria (27–32%) Unknown bacteria (25%) Cyanobacteria (12–20%) Firmicutes (10–18%) Bacteroidetes (15%) Chloroflexi (7%)	*Unclassified Bacteria* (11%)	39,600 36,600[c]	PYR Direct tissue lysis	Comparison of four algae and reef-building coral spp.	Barott et al. (2011)

TRFLP, terminal restriction fragment polymorphism analysis; *DGGE*, denaturing gradient gel electrophoresis; *CU*, culturing; *CL*, clone libraries; *CL-RFLP*, representative clones sequenced after restriction fragment length typing of clones; *Bdseq*, DGGE band sequencing; *SP*, specific primers; *cFISH*, catalyzed reporter deposition fluorescence hybridization; PYR, pyrosequencing; n.d., not determined

[a] ~60 Sequences per alga and season analyzed, DNA pooled from three individuals

[b] Clones ($n = 170$) were from the Baltic Sea algal sample, and 112 clones were from the North Sea algal sample; however, a large number of sequences was from chloroplast 16S rRNA genes ($n = 228$)

[c] Each algal species was analyzed from two different coral sites

[d] ~50 clones per host analyzed by RFLP

[e] Clones from six libraries representing six different *Ulva australis* samples; in addition, ten different seawater samples were analyzed

Bengtsson and Ovreas (2010) could show that Planctomycetes dominate in biofilms on *Laminaria hyperborea* (up to 50% of all DAPI counts), and clone libraries generated with a Planctomycete specific primer set revealed a high diversity of this phylum. Thus, Planctomycetes might be underrepresented in clone libraries of microbial communities of macroalgae studied so far. Whereas similar bacterial phyla are frequently encountered (Table 10.1), algae specific populations, i.e., highly similar (>97%) or identical in sequence could not be identified across all algal taxa examined so far. Nevertheless, 16S rRNA gene sequences of the α-proteobacterial Rhodobacterales, Rhizobiales (*Mesorhizobium* spp.), and *Cyanobacterium* spp. were found as most frequent populations on red, brown, and green algal species (Barott et al. 2011; Burke et al. 2011b; Lachnit et al. 2011; Meusnier et al. 2001; Tujula et al. 2010) (Table 10.1).

10.3 Host-Specific Microbial Communities on Macroalgae?

An important open question in alga–microbe interactions is whether these associations are host-specific. A number of studies report that certain microbial populations appear to be unique to the algal host but were absent in clone libraries or DGGE patterns of the surrounding seawater (Bengtsson et al. 2010; Burke et al. 2011b; Lachnit et al. 2011; Liu et al. 2011; Staufenberger et al. 2008), albeit some overlap might exist at high taxonomic levels, i.e., phyla (Longford et al. 2007). The surrounding seawater and sediment are likely sources of microbial recruitment for algal surfaces. However, small microbial populations might be easily overlooked at limited sample size possibly masking the presence of algal colonizers in seawater. For example, a study on the kelp species *Saccharina latissima* (as *Laminaria saccharina*) involved 170 clones from the algal surface but only 31 clones from seawater (Staufenberger et al. 2008); this is especially critical, as the seawater microbial community was more diverse than that of the algal hosts according to DGGE analysis. Nevertheless, the difference in composition between seawater and algae suggests that algae provide at least a niche for a distinct microbiota. In fact, different algal hosts from the same sampling site had highly dissimilar microbial communities, suggesting that each algal host provides a unique niche for microbial colonization (Barott et al. 2011; Lachnit et al. 2011).

A few studies report the existence of a "core" of specific microbial populations on their host algae. For example, on the kelp species *Laminaria hyperborea* a Planctomycete operational taxonomic unit (OTU) was commonly detected (78% of all samples) over all seasons, albeit the OTU was defined relatively coarsely (95% sequence identity; Bengtsson et al. 2010). Among the macroalgae from the Baltic Sea, between 7 and 16% were identified as algal host-specific populations (OTUs defined >99% sequence identity) across seasonal samples but still not all OTUs were consistently present in summer and winter samples (Lachnit et al. 2011). The detection of a core microbial community is further complicated by patterns of different microbial colonization along algal tissues, or might be subject to succession with the age of the individual alga. On young meristems and cauloids of *Saccharina latissima* microbial communities were more similar to each other than aging blade sections of the same individual (Staufenberger et al. 2008), suggesting that tissue age might contribute to microbial community variability. Conversely, no clear differences were found in microbial communities on meristem and lamina of *Laminaria hyperborea* (Bengtsson et al. 2010), indicating that considerable variability exists even between closely related species.

Several lines of evidence suggest that microbe–macroalgae interactions might not be host-specific when 16S rRNA gene phylogeny is the basis for microbial community assessment (Burke et al. 2011b). (1) Microbial communities can vary considerably among individuals of the same algal species. The microbiota on *Ulva australis* individuals from within the same and three different tidal pools, separated by less than 20 m, were found highly variable (up to 40% based on DGGE analysis; Tujula et al. 2010). A large-scale sequencing analysis revealed that less than 20% of sequences (at 97% sequence identity) were shared among six individuals of *Ulva australis* (Burke et al. 2011b). Also, on *Laminaria hyperborea* variability in community composition was detected between sampling locations (Bengtsson et al. 2010). (2) The algal microbiota varies considerably over seasons. Individuals of *Ulva australis* varied as much as 40% in summer (Tujula et al. 2010). Strong seasonal shifts also occurred in the microbiota of *Fucus vesiculosus*, *Gracilaria vermiculophylla*, and *Ulva intestinalis*; however, shifts were consistent at the phylum level and reoccurred in two consecutive years (Lachnit et al. 2011). The considerable variability among individuals of the same species became especially apparent in the study of the green algae *Ulva australis* based on the resolving power of large-scale sequencing (Burke et al. 2011b). It appears that at least in case of *Ulva australis* a core microbiota does not exist and that a large number of microbial populations are capable of colonizing the algal surface (Burke et al. 2011b).

10.4 A Model of Surface Colonization on Macroalgae

The high variability in microbial colonization of algal surfaces observed in a number of studies has resulted in the proposal that a large functional redundancy exists within the colonizing microbiota (Burke et al. 2011b). If true, then phylogenetic aspects, e.g., which microbial populations (or taxa) are present on the alga, would be inferior to the presence (and expression) of specific functions encoded by their genes. A recent metagenomic analysis of the microbiota on *Ulva australis* supports the hypothesis that the algal surface provides a niche for microorganisms carrying specific functional genes rather than belonging to a specific taxonomic group (Burke et al. 2011a). Despite the high taxonomic variability (~only 15–20% similarity between individuals), the functional composition of genes was highly similar (70%) and a core set of functional genes across all members of the microbial community was detected. Core functions detected at higher rate than in the surrounding seawater encompassed genes involved in detection and movement to the host, attachment and biofilm formation, response to the algal host environment, regulatory mechanisms in response to environmental stimuli, defense, and lateral gene transfer. All these functions were regarded as consistent with the ecological role of the *Ulva australis* microbiota (Burke et al. 2011a). Phylogenetic and functional analysis of the metagenome dataset revealed that the core functions are not restricted to a particular taxonomic group, suggesting that different taxa are responsible for the core functions of the microbial community on *Ulva australis*. Based on their results, Burke et al. (2011a, b) proposed a competitive lottery model for community assembly on the surface of *U. australis*. The competitive lottery model, originally proposed for coral reef fish (Sale 1979), integrates functional and random components of community assembly. According to the model, different species but with similar (or identical) functional traits can occupy the same niche in an ecosystem facilitated through the stochastic chance ("lottery") to get to the particular niche space first (Sale 1979). Translated to microbial community structuring, Burke et al. (2011a,b) argue that microbial populations can colonize the surface of an alga provided they have the necessary function in the form of a particular gene (or sets of genes). The assemblage of microbial population might, however, be determined by stochastics ("lottery") rather than controlled by specific host enforced selection mechanisms that would result in a very specific microbiota that is known from other host–microbe systems. Such a model of microbial community structuring implies that genes (or gene clusters) are more important than microbial taxonomic entities ("species"). In microbiology, the species concept is still a matter of debate and especially lateral transfer of genes renders it difficult to define how a microbial species is defined (Rossello-Mora and Amann 2001). Support for functional genes as drivers of community assembly also comes from the observation that the microbiota on *Ulva australis* abundantly harbors genes necessary for lateral gene transfer (Burke et al. 2011a). Analyzing functional genes revealed biological patterns on *Ulva australis* (Burke et al. 2011a) that were not detected by studying 16S rRNA gene-based phylogenetic community composition (Burke et al. 2011b). Similar observations were made in the human microbiome (Turnbaugh et al. 2009). The repercussions of these findings could have far reaching impacts on studies of microbial community ecology eventually shedding light on differences in the ecology of micro- and macroorganisms.

10.5 Conclusions

The molecular microbial ecology of macroalgal hosts is a rather young field with only small datasets available but with large potentials to influence theory building in microbial ecology and host–microbe interactions. Recent, more in-depth molecular sequence analyses have challenged the notion that host-specific microbial taxa colonize macroalgal surfaces. Rather, a model of algal colonization has been proposed in which specific microbial functional genes govern the structuring of microbial communities and not taxonomic affiliations of microbial populations

(Burke et al. 2011a, b). Methodological improvements, here the application of large-scale sequencing approaches and analyses at the level of individual functional and phylogenetic marker genes, have facilitated to gain insight into novel biological patterns. Naturally, the database of microbial communities on algal host species should be broadened to test the proposed functional gene-driven hypothesis of microbial community structuring. Ultimately, identification of important functional traits might also allow unraveling the role of microorganisms in macroalgae–microbe interactions.

Acknowledgements I would like to thank Kai Bischof for the possibility to gain insights into the field of algae–microbe communities and helpful suggestions on algal taxonomy. Manuscript preparation was supported by the University of Bremen.

References

Amann RI, Ludwig W, Schleifer KH (1995) Phylogenetic identification and in situ detection of individual microbial cells without cultivation. Microbiol Rev 59:143–169

Armstrong E, Rogerson A, Leftley JW (2000) The abundance of heterotrophic protists associated with intertidal seaweeds. Estuar Coast Shelf Sci 50:415–424

Barott KL, Rodriguez-Brito B, Janouskovec J, Marhaver KL, Smith JE, Keeling P, Rohwer FL (2011) Microbial diversity associated with four functional groups of benthic reef algae and the reef-building coral *Montastraea annularis*. Environ Microbiol 13:1192–1204

Bengtsson MM, Ovreas L (2010) Planctomycetes dominate biofilms on surfaces of the kelp *Laminaria hyperborea*. BMC Microbiol 10:261

Bengtsson MM, Sjotun K, Ovreas L (2010) Seasonal dynamics of bacterial biofilms on the kelp *Laminaria hyperborea*. Aquat Microb Ecol 60:71–83

Bengtsson MM, Sjotun K, Storesund JE, Ovreas L (2011) Utilization of kelp-derived carbon sources by kelp surface-associated bacteria. Aquat Microb Ecol 62:191–199

Bolinches J, Lemos ML, Barja JL (1988) Population dynamics of heterotrophic bacterial vommunities associated with *Fucus vesiculosus* and *Ulva rigida* in an estuary. Microb Ecol 15:345–357

Burke C, Steinberg P, Rusch D, Kjelleberg S, Thomas T (2011a) Bacterial community assembly based on functional genes rather than species. Proc Natl Acad Sci USA 108:14288–14293

Burke C, Thomas T, Lewis M, Steinberg P, Kjelleberg S (2011b) Composition, uniqueness and variability of the epiphytic bacterial community of the green alga Ulva australis. ISME J 5:590–600

Chan ECS, Mcmanus EA (1969) Distribution characterization and nutrition of marine microorganisms from algae *Polysiphonia Lanosa* and *Ascophyllum Nodosum*. Can J Microbiol 15:409–420

Correa JA (1997) Infectious diseases of marine algae: current knowledge and approaches. In: Round FE, Chapman DJ (eds) Progress in phycological research. Biopress, Bristol, UK, pp 149–180

Croft MT, Lawrence AD, Raux-Deery E, Warren MJ, Smith AG (2005) Algae acquire vitamin B_{12} through a symbiotic relationship with bacteria. Nature 438:90–93

Croft MT, Warren MJ, Smith AG (2006) Algae need their vitamins. Eukaryot Cell 5:1175–1183

Dobretsov SV, Qian PY (2002) Effect of bacteria associated with the green alga *Ulva reticulata* on marine micro- and macrofouling. Biofouling 18:217–228

Egan S, Thomas T, Holmstrom C, Kjelleberg S (2000) Phylogenetic relationship and antifouling activity of bacterial epiphytes from the marine alga *Ulva lactuca*. Environ Microbiol 2:343–347

Egan S, Holmstrom C, Kjelleberg S (2001) *Pseudoalteromonas ulvae* sp nov., a bacterium with antifouling activities isolated from the surface of a marine alga. Int J Syst Evol Microbiol 51:1499–1504

Egan S, Thomas T, Kjelleberg S (2008) Unlocking the diversity and biotechnological potential of marine surface associated microbial communities. Curr Opin Microbiol 11:219–225

Eilers H, Pernthaler J, Gloeckner FO, Amann R (2000) Culturability and in situ abundance of pelagic bacteria from the North Sea. Appl Environ Microbiol 66:3044–3051

Fuhrman JA, Campbell L (1998) Microbial microdiversity. Nature 393:410–411

Goecke F, Labes A, Wiese J, Imhoff JF (2010) Chemical interactions between marine macroalgae and bacteria. Mar Ecol Prog Ser 409:267–299

Green J, Bohannan BJM (2006) Spatial scaling of microbial biodiversity. Trends Ecol Evol 21:501–507

Hengst MB, Andrade S, Gonzalez B, Correa JA (2010) Changes in epiphytic bacterial communities of intertidal seaweeds modulated by host, temporality, and copper enrichment. Microb Ecol 60:282–290

Huggett MJ, Crocetti GR, Kjelleberg S, Steinberg PD (2008) Recruitment of the sea urchin *Heliocidaris erythrogramma* and the distribution and abundance of inducing bacteria in the field. Aquat Microb Ecol 53:161–171

Hughes JB, Hellmann JJ, Ricketts TH, Bohannan BJM (2001) Counting the uncountable: statistical approaches to estimating microbial diversity. Appl Environ Microbiol 67:4399–4406

Jensen PR, Kauffman CA, Fenical W (1996) High recovery of culturable bacteria from the surfaces of marine algae. Mar Biol 126:1–7

Joint I, Tait K, Callow ME, Callow JA, Milton D, Williams P, Camara M (2002) Cell-to-cell communication across the prokaryote-eukaryote boundary. Science 298:1207

Joint I, Tait K, Wheeler G (2007) Cross-kingdom signalling: exploitation of bacterial quorum sensing molecules by the green seaweed Ulva. Philos Trans R Soc Lond B Biol Sci 362:1223–1233

Keshtacherliebson E, Hadar Y, Chen Y (1995) Oligotrophic bacteria enhance algal growth under iron-deficient conditions. Appl Environ Microbiol 61:2439–2441

Lachnit T, Meske D, Wahl M, Harder T, Schmitz R (2011) Epibacterial community patterns on marine macroalgae are host-specific but temporally variable. Environ Microbiol 13:655–665

Largo DB, Fukami K, Nishijima T (1995) Occasional pathogenic bacteria promoting ice-ice disease in the carrageenan producing red algae *Kappaphycus alvarezii* and *Eucheuma denticulatum* (Solieriaceae, Gigartinales, Rhodophyta). J Appl Phycol 7:545–554

Largo DB, Fukami K, Adachi M, Nishijima T (1997) Direct enumeration of bacteria from macroalgae by epifluorescence microscopy as applied to the fleshy red algae *Kappaphycus alvarezii* and *Gracilaria* spp. (Rhodophyta). J Phycol 33:554–557

Largo DB, Fukami K, Nishijima T (1999) Time-dependent attachment mechanism of bacterial pathogen during ice-ice infection in *Kappaphycus alvarezii* (Gigartinales, Rhodophyta). J Appl Phycol 11:129–136

Lewis TE, Garland CD, Mcmeekin TA (1985) The bacterial biota on crustose (nonarticulated) coralline algae from tasmanian waters. Microb Ecol 11:221–230

Liesack W, Janssen PH, Rainey FA, Ward-Rainey NL, Stackebrandt E (1997) Microbial diversity in soil: the need for a combined approach using molecular and cultivation techniques. In: van Elsas JD, Trevors JT, Wellington EMH (eds) Modern soil microbiology. Marcel Dekker, New York, pp 375–439

Liu WT, Marsh TL, Cheng H, Forney LJ (1997) Characterization of microbial diversity by determining terminal restriction fragment length polymorphisms of genes encoding 16S rRNA. Appl Environ Microbiol 63:4516–4522

Liu M, Dong Y, Zhao Y, Zhang GT, Zhang WC, Xiao T (2011) Structures of bacterial communities on the surface of *Ulva prolifera* and in seawaters in an *Ulva* blooming region in Jiaozhou Bay, China. World J Microbiol Biotechnol 27:1703–1712

Longford SR, Tujula NA, Crocetti GR, Holmes AJ, Holmstrom C, Kjelleberg S et al (2007) Comparisons of diversity of bacterial communities associated with three sessile marine eukaryotes. Aquat Microb Ecol 48:217–229

Marsh TL (2005) Culture-independent microbial community analysis with terminal restriction fragment length polymorphism. Environ Microbiol 397:308–329

Marshall K, Joint I, Callow ME, Callow JA (2006) Effect of marine bacterial isolates on the growth and morphology of axenic plantlets of the green alga *Ulva linza*. Microb Ecol 52:302–310

Matsuo Y, Suzuki M, Kasai H, Shizuri Y, Harayama S (2003) Isolation and phylogenetic characterization of bacteria capable of inducing differentiation in the green alga *Monostroma oxyspermum*. Environ Microbiol 5:25–35

Meusnier I, Olsen JL, Stam WT, Destombe C, Valero M (2001) Phylogenetic analyses of *Caulerpa taxifolia* (Chlorophyta) and of its associated bacterial microflora provide clues to the origin of the Mediterranean introduction. Mol Ecol 10:931–946

Muyzer G, Waal EC, Uitterlinden AG (1993) Profiling of complex microbial populations by denaturing gradient gel electrophoresis analysis of polymerase chain reaction-amplified genes coding for 16S rRNA. Appl Environ Microbiol 59:695–700

Nakanishi K, Nishijima M, Nishimura M, Kuwano K, Saga N (1996) Bacteria that induce morphogenesis in *Ulva pertusa* (chlorophyta) grown under axenic conditions. J Phycol 32:479–482

Namba A, Shigenobu Y, Kobayashi M, Kobayashi T, Oohara I (2010) A new primer for 16S rDNA analysis of microbial communities associated with *Porphyra yezoensis*. Fisheries Sci 76:873–878

Ohkubo S, Miyashita H, Murakami A, Takeyama H, Tsuchiya T, Mimuro M (2006) Molecular detection of epiphytic *Acaryochloris* spp. on marine macroalgae. Appl Environ Microbiol 72:7912–7915

Olson JB, Kellogg CA (2010) Microbial ecology of corals, sponges, and algae in mesophotic coral environments. FEMS Microbiol Ecol 73:17–30

Pace NR, Stahl DA, Lane DJ, Olsen GJ (1986) The analysis of natural microbial populations by ribosomal RNA sequences. Adv Microb Ecol 9:1–55

Patel P, Callow ME, Joint I, Callow JA (2003) Specificity in the settlement–modifying response of bacterial biofilms towards zoospores of the marine alga Enteromorpha. Environ Microbiol 5:338–349

Pernthaler J, Amann R (2005) Fate of heterotrophic microbes in pelagic habitats: focus on populations. Microbiol Mol Biol Rev 69:440–461

Provasoli L, Pintner IJ (1980) Bacteria induced polymorphism in an axenic laboratory strain of *Ulva lactuca* (Chlorophyceae). J Phycol 16:196–201

Raghukumar C, Nagarkar S, Raghukumar S (1992) Association of thraustochytrids and fungi with living marine algae. Mycol Res 96:542–546

Rappe MS, Giovannoni SJ (2003) The uncultured microbial majority. Annu Rev Microbiol 57:369–394

Rossello-Mora R, Amann R (2001) The species concept for prokaryotes. FEMS Microbiol Rev 25:39–67

Sale PF (1979) Recruitment, loss and coexistence in a guild of territorial coral-reef fishes. Oecologia 42:159–177

Sogin ML, Morrison HG, Huber JA, Mark Welch D, Huse SM, Neal PR et al (2006) Microbial diversity in the deep sea and the underexplored "rare biosphere". Proc Natl Acad Sci USA 103:12115–12120

Staufenberger T, Thiel V, Wiese J, Imhoff JF (2008) Phylogenetic analysis of bacteria associated with *Laminaria saccharina*. FEMS Microbiol Ecol 64:65–77

Steinberg PD, de Nys R (2002) Chemical mediation of colonization of seaweed surfaces. J Phycol 38:621–629

Tujula NA, Crocetti GR, Burke C, Thomas T, Holmstrom C, Kjelleberg S (2010) Variability and abundance of the epiphytic bacterial community associated with a green marine Ulvacean alga. ISME J 4:301–311

Turnbaugh PJ, Hamady M, Yatsunenko T, Cantarel BL, Duncan A, Ley RE et al (2009) A core gut microbiome in obese and lean twins. Nature 457:480–4U7

Vairappan CS, Suzuki M, Motomura T, Ichimura T (2001) Pathogenic bacteria associated with lesions and thallus bleaching symptoms in the Japanese kelp *Laminaria religiosa* Miyabe (Laminariales, Phaeophyceae). Hydrobiologia 445:183–191

Wahl M (2008) Ecological lever and interface ecology: epibiosis modulates the interactions between host and environment. Biofouling 24:427–438

Wang G, Shuai L, Li Y, Lin W, Zhao XW, Duan DL (2008) Phylogenetic analysis of epiphytic marine bacteria on Hole-Rotten diseased sporophytes of *Laminaria japonica*. J Appl Phycol 20:403–409

Weinberger F, Friedlander M, Gunkel W (1994) A bacterial facultative parasite of Gracilaria-Conferta. Dis Aquat Organ 18:135–141

Weinberger F, Beltran J, Correa JA, Lion U, Pohnert G, Kumar N et al (2007) Spore release in *Acrochaetium* sp (Rhodophyta) is bacterially controlled. J Phycol 43:235–241

Chapter 11
Intimate Associations Between Epiphytes, Endophytes, and Parasites of Seaweeds

Philippe Potin

11.1 Introduction

Most, but not all, seaweeds studied in natural ecosystems are plagued by epiphytes, endophytes or other eukaryotic parasites that cause no or few apparent disease symptoms. Hence, such intimate associations can range from mutualism through commensalism to parasitism in a continuous manner (Correa 1994). In shallow hard-bottom marine habitats, seaweeds compete for substrate, nutrients, and light (see Chap. 7 by Edwards and Connell). A planktonic bank of seaweed spores and propagules is periodically and widely distributed in the coastal zone that is ready to settle in all appropriate microhabitats (Wahl and Mark 1999; Lotze et al. 2000). In this context, epibiotism (using other organisms as substrates) illustrates competition for space (e.g., Menge 1991; Underwood 2000; Wahl and Mark 1999). Epibiotic associations comprise a substrate organism (basibiont) and one or several attached species (epibionts). This chapter focuses on associations between seaweeds, and therefore it corresponds to associations between algal basiphytes and epiphytes. A wide variety of algae colonizes other algae representing a continuum between epiphytism and endophytism. Such interactions begin with nonrandom patterns of settlement and recruitment (Gonzalez and Goff 1989; Pearson and Evans 1989, 1990) and artificial substrata are often preferred over nearby living surfaces for epiphytes (Wahl and Mark 1999) very likely depending on the previous colonization of these surfaces by specific bacterial biofilms (Goecke et al. 2010, see also Chap. 10 by Friedrich). Therefore, host specificity was considered as an exception in such interactions. However, the presence of epiphytes, endophytes, and eukaryotic parasites in marine macrophytes results from complex interactions

P. Potin (✉)
CNRS, Université Pierre et Marie Curie-Paris 6, UMR 7139, Station Biologique BP74, 29682 Roscoff, France
e-mail: philippe.potin@sb-roscoff.fr

C. Wiencke and K. Bischof (eds.), *Seaweed Biology*, Ecological Studies 219,
DOI 10.1007/978-3-642-28451-9_11, © Springer-Verlag Berlin Heidelberg 2012

between hosts and colonizers that were hardly investigated. These interactions have positive effects, such as providing additional ecological niches, food and protective habitat for animals (Pavia et al. 1999; Karez et al. 2000; Viejo and Åberg 2003; Bittick et al. 2010), as well as negative effects, such as competition with the host for light and nutrients (D'Antonio 1985; Cebrian et al. 1999; Honkanen and Jormalainen 2005); increasing drag (D'Antonio 1985; Hemmi et al. 2005); and favoring the attraction of grazers to the host alga (Bernstein and Jung 1979; Wahl and Hay 1995; Wahl et al. 1997). During the last decades, sustained nutrient enrichment from land-based activities resulted in increased biomass of attached epiphytes in many ecosystems (Rönnberg and Bonsdorff 2004; Russell et al. 2005, see also Chap. 21 by Teichberg et al.) and consequences of global warming may also strongly affect these biotic interactions (Eggert et al. 2010). In addition, closed associations between seaweeds are involved in human affairs when they impact the mariculture or the harvesting of wild populations of economically important seaweeds (Friedlander 1992; Critchley et al. 2004; Hurtado et al. 2006; Vairappan 2006; Gachon et al. 2010, see also Chap. 22 by Buchholz et al.) or when opportunistic species foul artificial substrates (Callow and Callow 2011).

This chapter attempts to summarize a part of the literature available on the biology and ecology of such close associations. The particular focus will be on cross-talk mechanisms involved in host specificity, the role of bacterial biofilms in mediating some interactions, as well as the role of host chemical defenses that impact the diversity of associations through allelopathic interactions. In the context of coastal eutrophication and climate change, these subtle relationships between seaweeds and colonizers at the community level and in human affairs, including algal mariculture or the search for environment-friendly antifouling and antimicrobial strategies, will be an important and emerging field of research.

11.2 Ecology and Biology of Close Biotic Interactions of Seaweeds

11.2.1 Epiphytes

Algal epiphytes are ubiquitous and play major ecological functions in coastal benthic communities. Therefore, the available literature on algal epiphytes is abundant but also very diverse with a lack of a comprehensive review. Algal epiphytes have been studied mainly in terms of species diversity and community ecology. However, in comparison with the abundant literature available on seagrasses, considerably less work has concerned the distribution and effects of epiphytic populations on macroalgal hosts (D'Antonio 1985; Rindi and Guiry 2004). The composition of epiphytic assemblages on macrophyte host surfaces displays strong seasonal and spatial variations (Markham 1969; Whittick 1983; Arrontes 1990; Rindi and Guiry 2004; Leonardi et al. 2006). Old or damaged portions of thalli appear more susceptible to epiphyte colonization (Ballantine 1979; Lobban and Baxter 1983; Ducker and Knox 1984; Pearson and Evans 1989, 1990; Arrontes 1990). Patterns of epiphytic assemblages are also strongly dependent on the intensity of grazing by various invertebrates or fishes (Lubchenco and Gaines 1981). The species composition of epiphyte guilds is also conditioned by physical factors, particularly desiccation and wave action (Hayward 1980; Graham and Hanna 1989; Longtin et al. 2009). These factors strongly impact the distribution and stability of suitable substrata and increase selective forces affecting epiphyte growth (Benzing 1987). Despite substantial work on marine epiphytes, their population ecology has received little attention (Harlin 1987; Longtin and Scrosati 2009; Longtin et al. 2009).

Various types of interactions can be established between basiphytes and epiphytes, but only a few studies, based on both wild and laboratory-infected material, report on the contact surface established between epiphytes and their hosts (Rawlence 1972; Rawlence and Taylor 1972; Ducker and Knox 1984; Gonzalez and Goff 1989; Gonzalez et al. 2003; Dawes et al. 2000; Leonardi et al. 2006). The damage caused by an epiphyte to its basiphyte can be highly variable and is mainly influenced by the type of anatomical association and the incidence of the epiphyte (Fletcher 1995). According to Linskens (1963), holo-epiphytes are those attached to the outer layers of the host, whereas amphi-epiphytes are deeply anchored in the tissue of their hosts. Linskens (1963) suggested, however, that the type of anatomical contact is highly variable and determined by the nature of the partners. Epiphytes attach via single cells, filamentous bases, or massive rhizoidal structures. Host–epiphyte interfaces were typed and their relative abundance and temporal variability were monitored to unravel the mechanisms of host recognition and host damage that could explain the loss of crops and the negative effects of epiphytism in *Gracilaria chilensis* farming (Leonardi et al. 2006): Five types of anatomical relationships were detected. *Infection type I* included the epiphytes weakly attached to the surface of the host and not associated with damage of host tissues (i.e., *Hincksia mitchelliae*, *H. granulosa* and *Ectocarpus acutus*). *Infection type II* included those epiphytes strongly attached to the surface of the host but not associated with any host tissue damage (i.e., *Acrochaetium* sp., *Antithamnionella* sp. and *Colpomenia sinuosa*). *Infection type III* included all the epiphytes that penetrated the outer layer of the host wall without damaging its cortical cells (i.e., *Xenococcus* sp. and *Sahlingia subintegra*). *Infection type IV* included epiphytes penetrating deep into the host cell wall, disorganizing the cortical tissue (i.e., *Ulva lactuca* and *Acrosorium corallinarum*). *Infection type V* included epiphytes that penetrated deeply into the cortex, reached the medullary tissue, and caused destruction of the host's cells in the area around the infection (i.e., *Ceramium rubrum* and *Polysiphonia harveyi*). Prevalence varied with time and with infection type, with types II and III reaching up to 80% and 90% of the thalli, respectively. Severity of epiphyte infection was similar to the distribution of infection prevalence, with crustose epiphytes colonizing up to 80% of the host surface.

It was also noted that epiphytes could actually be present on the host in the field, yet in a "dormant" state, and that they later developed to maturity in the richer culture conditions.

Epiphytes are usually defined as organisms that grow on plants, but do not derive nutrients from their hosts (Linskens 1976). However, one of the most extensively studied relationships is the obligate partnership of the red alga *Vertebrata (Polysiphonia) lanosa,* which is a hemiparasite of the fucoid brown alga *Ascophyllum nodosum* (Taylor 1957; Turner and Evans 1977; Garbary et al. 1991; Levin and Mathieson 1991). Perhaps, hemiparasitism is an intermediate step between simple epiphytism and parasitism. The rhizoids of *P. lanosa* penetrate the host and obtain some nutrition from *A. nodosum* (Rawlence and Taylor 1972; Turner and Evans 1977). However, the quantity of carbon obtained by *P. lanosa* is minimal (Harlin and Craigie 1975) and *P. lanosa* is pigmented and capable of photosynthesis (Bidwell 1958; Fralick and Mathieson 1975). Field observations suggest that the distribution and abundance of *V. lanosa* are highly dependent upon the occurrence of injured host tissue and/or number of lateral shoots (Lobban and Baxter 1983; Burke 1986; Pearson and Evans 1989, 1990; Longtin et al. 2009). Recent work by Longtin and Scrosati (2009) in *Ascophyllum nodosum* beds has reinvestigated the various factors that affect the settlement of the red alga *V. lanosa* at the surface of *A. nodosum* including competitive interactions with epiphytic filamentous brown algae (*Elachista fucicola* and *Pylaiella littoralis*) (Longtin et al. 2009)

11.2.2 Pigmented Endophytes

Numerous species of pigmented algal endophytes are associated with red, brown, and green seaweeds (Goff 1982; Correa 1996; Bouarab et al. 2001a; Bown et al. 2003; Eggert et al. 2010). They are generally small, filamentous algae that belong to three main lineages of multicellular algae (Chlorophyta, Phaeophyceae, and Rhodophyta) and inhabit mainly the intercellular spaces of their hosts. They appear to be carbon independent of their hosts (Correa et al. 1988; Correa 1990; Eggert et al. 2010). Therefore, endophytism has arisen independently numerous times throughout the diversification of the multicellular algal lineages and has likely been constrained by selective pressure exerted on epiphytes by grazers (Peters 2003; Amsler et al. 2009; Aumack et al. 2011) or by abiotic conditions and environmental gradients leading to differential levels of desiccation in intertidal seaweeds (Longtin et al. 2009). During the last decade, additional information has become available on the taxonomy, prevalence, and ecology of pigmented endophytes in brown, red, and green algae and some culture and molecular studies have revealed the occurrence of cryptic stages of the life cycles of some important seaweeds such as kelps (Garbary et al. 1999a, b; Hubbard et al. 2003) or green algae (Sussmann and DeWreede 2002) growing endophytically in the tissues of various algal hosts. Difficulties in isolating endophytes into pure culture and the lack of a robust taxonomy constrain our ability to develop a full understanding of their ecology and that of their hosts.

11.2.2.1 Endophytes in Red Algae

The occurrence of epi-endophyte algae in the red algal host *Hymenena falklandica* was recently reported as the first epidemiological study conducted in Argentina (Gauna et al. 2010). The green alga *Epicladia heterotricha* revealed a 100% prevalence of occurrence, particularly in the basal region of the host and low (14%), moderate (28%), and high (58%) severity degrees of infection were registered (Gauna et al. 2010). In contrast, the green epiphyte *Pseudendoclonium submarinum* exhibited a lower frequency of occurrence, close to 3%. The experimental infection of *H. falklandica* by *E. heterotricha* demonstrated that *E. heterotricha* initially behaves as epiphytic but with endophytic filaments growing into the cortex of the host during late infection, while *P. submarinum* was found to be exclusively epiphytic.

A reassessment of the taxonomy of such green algal epi/endophytes has been attempted during the recent years using molecular tools. Their assignment to genus and/or species level has been previously done on the basis of their morphology, but their growth inside an algal host and their simple growth forms, with relatively few reliable diagnostic characters, make their taxonomy problematic (Dixon 1963; White and Boney 1969, 1970; Woelkerling 1971; Bown et al. 2003) and strongly impact the epidemiological studies. Positive identification depends on culturing living isolates to reveal valuable taxonomic characters that are not always apparent in field collected or preserved material (Nielsen and McLachlan 1986; Correa and McLachlan 1991). The relationships between isolates belonging to the green algal genus *Acrochaete* and other chlorophyte taxa were better characterized by the analysis of internal transcribed spacer 2 (ITS2) rDNA sequence variability (Bown et al. 2003). These findings indicate that the taxonomy of *Acrochaete* may need major revision and suggest that species-specific oligonucleotide primers could be used as the basis for a polymerase chain reaction (PCR) assay to assess endophyte diversity within populations of *Chondrus* and related host species (Bown et al. 2003).

11.2.2.2 Endophytes in Brown Algae

Endophyte species of kelps were recently reviewed by Eggert et al. (2010). Some of these endophytes cause only minor changes in their hosts such as tissue darkening, whereas others are known to produce either degradative losses or tumoral lesions accompanied by severe thalli deformations (Apt 1988a; Peters and Schaffelke 1996). Epidemiological studies in populations of kelps demonstrated that infections may affect 100% of the host population (Andrews 1977; Lein et al. 1991; Peters and Schaffelke 1996; Ellertsdottir and Peters 1997). Recent work on the biology and ecology of brown algal endophytes has focused on exploring novel interactions between kelps and brown endophytes in the South-East Pacific Ocean and Antarctic regions, respectively. Along the Pacific coasts of South America, *Macrocystis pyrifera* is infected by *Laminariocolax macrocystis* (Burkhardt and Peters 1998), initially described as *Streblonema macrocystis* (Peters 1991). Recent results showed that galls affecting natural populations of *Lessonia nigrescens* were associated with the infection by a filamentous brown algal endophyte of the genus *Laminariocolax* (Thomas et al. 2009).

Similar patterns of few filamentous algae growing subtidally except within the thalli of the larger chemically defended macroalgae were also observed by Peters (2003) in Antarctica. The endophytes were highly palatable to the amphipod *Gondogeneia antarctica*, suggesting that these filamentous endophytes obtain some protection from mesograzers by growing in association with unpalatable macroalgae (Amsler et al. 2009). These data suggest that macroalgae along the western Antarctic Peninsula rely on grazers to control populations of potentially harmful epiphytes. These results strongly suggest that the chemically defended macroalgal flora lives in mutualism with high densities of mesograzers, providing amphipods with shelter from predation while continually being cleaned of potentially harmful endo/epiphytes (Aumack et al. 2011).

11.2.2.3 Endophytes in Green Algae

Reports on the occurrence of algal endophytes in green algae are scarce. The endophytic filamentous green alga *Acrochaete geniculata* was identified as the causative agent of a disease affecting tank-cultivated *Ulva rigida* (Del Campo et al. 1998). This destructive disease is characterized by green spots, initially located at the basis of the thalli that spread through the host and gradually cause perforations of the frond (Del Campo et al. 1998). Advanced stages of infection result in frond wrinkling and severe tissue loss. Infection is transmitted within 2 weeks when healthy and diseased fronds are cocultivated, whereas previous trials to infect the host in the laboratory with actively reproducing suspensions of the endophyte remained unsuccessful (Del Campo et al. 1998).

11.2.2.4 Life History Stages of Seaweeds as Endophytic Filaments in Other Taxa

Kelp Gametophytes in Red Filamentous Algae

Kelp gametophytes have been extensively studied in the laboratory, but are elusive in nature. They have occasionally been found growing on rocks, and wood, and as epiphytes on crustose coralline algae (Parke 1932; Sakai and Funano 1964; Funano 1969; Kaneko 1973; Kain 1979). Artificial substrata placed in nature have also successfully been utilized for settlement of kelp gametophytes (e.g., Reed et al. 1988). More recently, gametophytes have been found growing endophytically in the cell walls of red algae where they become reproductive and produced juvenile sporophytes (Garbary et al. 1999a, b). Initially, 17 red algal species from western North America were observed with endophytic kelp gametophytes. They were most common in filamentous hosts, and hundreds of gametophytes were observed in many individual host thalli. To date, endophytic gametophytes have only been observed either in situ or in vitro in cultures from the Pacific Ocean (Hubbard et al. 2003). On the one hand, it is possible that the endophytic habit is an accident of spore settlement and that it is of little or no adaptive significance. On the other hand, endophytism may represent a facultative strategy of primary importance to gametophyte growth and reproduction, at least in some species or in some environments. The irregular branching pattern and absence of heterotrichous development in the kelp gametophytes are analogous to that found in many endophytic red algae (e.g., in Acrochaetiales, Garbary et al. 1982) or endophytic green algae [e.g., *Acrochaete*, *Endophyton*, *Phaeophila* (Nielsen and McLachlan 1986)]. These common features suggest a morphological convergence in red, brown, and green algae based on the common adaptation for endophytism. The benefits and drawbacks of the endophytic lifestyle for kelp gametophytes remain unresolved. The small morphology and low density of endophytic gametophytes in culture suggests drawbacks. However, both free-living and endophytic gametophytes can be found in nature (e.g., Kain 1979; Garbary et al. 1999a, b), implying that both lifestyles represent adaptations. Symbiosis could provide protection from grazing and sedimentation, or improve access to light. Fine sediments reduce kelp spore attachment and survival (Devinny and Volse 1978) and the understories of kelp stands experience higher levels of sedimentation than adjacent cleared areas (Eckman et al. 1989; Duggins et al. 1990). Anderson et al. (1997) found that juvenile *Ecklonia* were grazed less when they grew on adult holdfasts. Similarly, red algal hosts may inhibit herbivores from accessing the gametophytes, or the host surface may be more resistant than in epilithic gametophytes (Padilla 1985). It is also possible that endophytes have no better chance of surviving than free-living gametophytes, but the increased diversity of life history strategies may increase the overall likelihood of gametophytes surviving. More recently, molecular and microscopy data revealed that gametophytes of *Alaria*, *Macrocystis,* and *Nereocystis* epi/endophytically grow in the older portions of the thallus of *Lessoniopsis littoralis* (Lane and Saunders 2005). However, developing sporophytes were hardly observed in the field attached to *L. littoralis*, suggesting that endophytic habit is an accident of spore settlement and that it is of little or no adaptive significance.

Green Algal Endophytes as Alternate Life History Phases of *Acrosiphonia* Species

Molecular studies have shown the relationship between the unicellular endophytes ("*Codiolum*" and "*Chlorochytrium*") and their alternative life history stages, the gametophytes of *Acrosiphonia coalita* and *A. arcta* (Sussmann et al. 1999). The population dynamics of the endophyte and their common hosts (*Mazzaella splendens* and "*Petrocelis*") has been explored, showing a seasonal pattern of abundance of the endophyte (Sussmann and DeWreede 2002). A puzzling aspect of the relationship is the presence of many endophytes at the time the host plant is disintegrating. A bet-hedging strategy is proposed for this complex life history; not only are there two different phases adapted to different seasonal conditions, but there are two different hosts (a crustose and a bladed host) for the endophytic sporophyte phase (Sussmann and DeWreede 2002).

11.2.3 Eukaryotic Parasites and Pathogens

In addition with the case of hemiparasitism caused by *V. lanosa* in *A. nodosum* (see above), strict parasitic relationships have been well studied in red algae, but remained nearly unexplored from a physiological point of view in fungal or oomycete colonizers of marine macroalgae.

11.2.3.1 Adelphoparasites and Alloparasites of Red Algae

Some of the algal endophytes in fronds of red algae are nonphotosynthetic and are assumed to obtain carbon from their hosts (Goff 1982). The evolution of these red algal parasites is one of the most fascinating aspects of algal associations: It is common among florideophyte red algae, involving about 8–10% of the species, and has a unique type of development in which parasite nuclei are transferred to host cells and "control" host cell development in a kind of hijacking. It was reinvestigated in the recent years using new molecular tools to better decipher phylogenetic relationships between the host and its parasite (Zuccarello et al. 2004; Kurihara et al. 2010), as well as to generate genomics data on the two partners. A recent work reported the sequencing of the mitochondrial genomes of the free-living gracilarioid *Gracilariopsis andersonii* and its closely related parasite *Gracilariophila oryzoides* to characterize the effect of a parasitic lifestyle on the organellar genomes of red algal parasites (Hancock et al. 2010). This latter paper provides a comprehensive review of the biology of aldelphoparasitism that is summarized below.

Red algal parasites are unusual because the vast majority (80%) of them parasitize species with which they share a recent common ancestor. This strategy has earned them the name "adelphoparasites," from the Greek, adelpho, meaning "kin." Intracellular adelphoparasites are hardly observed in nature, morphologically reduced to a miniature, often colorless pustule and, yet, have independently evolved hundreds of times among the florideophyte red algae. Another part of red algal parasites (20%) are known as alloparasites ("allo" meaning other), which also diverged from free-living species, but then have radiated to exploit more distantly related red algal hosts (Goff et al. 1997; Kurihara et al. 2010). Goff et al. (1997) give two possible scenarios for the origin of parasitic red algae. In one, the parasites are ancestrally epiphytic, later becoming endophytic and eventually parasitic. In the second, the parasites derive directly from spores that lose the ability to survive independently of the parent, following the evolution of a carposporophyte after fertilization in the florideophyte. One possibility is that adelphoparasites arose by the second process while alloparasites arose by the first.

These interactions were extensively studied at a cell biology level by the group of Lynda Goff and Annette Coleman since the 1980s (see Hancock et al. 2010), but little is known about how these algal parasites affect their hosts at an organismic level and even less at the population level. Research has concentrated on assessing changes in the host at the cellular level (Kugrens and West 1973; Goff 1976) or in detecting translocation of substances between partners (e.g., Court 1980; Kremer 1983). Kremer (1983) showed that although there was carbon transfer in the form of the photoassimilate digeneaside and of amino acids from the host *Rhodomela confervoides* to its parasite *Harveyella mirabilis*, carbon fixed by the host was enough to cover the needs of both the parasite and the host. There are, though, indications that parasitized hosts are more susceptible to removal from the population following local tissue destruction and a decrease in growing tips (Goff 1976). Few studies have specifically evaluated the effects of algal parasitic infections on the host, and these demonstrated reduced growth rates and elongation of the host thallus (Nonomura 1979; Apt 1984). It was shown that infections by the red alga *Hypneocolax stellaris* reduced the growth rate of its host *Hypnea musciformis* by 40% in the field and up to 70% in laboratory trials (Apt 1984).

In contrast with the excellent knowledge on the life history strategies of red algal parasites, the molecular and genomic consequences of becoming a parasite remained unknown until the recent work by Hancock et al. (2010). It reports the sequencing of the mitochondrial genomes of the free-living *Gracilariopsis andersonii* and its closely related parasite *Gracilariophila oryzoides*. Whereas the parasite genome is similar to the host in many ways, the genes encoding the essential proteins *atp8* and *sdhc* are pseudogenes in the parasite. The mitochondrial genome of a parasite from a different class of red algae, *Plocamiocolax pulvinata*, has also lost the *atp*8 gene entirely, indicating that this gene is no longer critical in red algal parasite mitochondria. The trend toward the loss of *atp8* is significant considering that rapid rates of evolution are not occurring among the majority of mitochondrial genes. This creates a bit of a paradox because genes considered of vital importance are being lost while, at the same time, there are clearly selective forces maintaining the genome sequence from further losses and accelerated mutational rates. Further analysis of host and parasite nuclear DNA may elucidate some fundamental genomic shifts and molecular changes that have resulted in a parasitic existence (Hancock et al. 2010). In addition, it is interesting to mention that the galactans from *Tikvahiella candida*, an adelphoparasite of *Solieria robusta*, closely resemble those of its host and furnish evidence in support of a close phylogenetic relationship between the two species (Chiovitti et al. 1999). Similarly with the importance of cell wall galactans in the host perception by spores of some green algal endophytes (Bouarab et al. 1999) and oomycete spores (Uppalapati et al. 2001), it is tempting to speculate that part of the host specificity in red alga parasitism has evolved through sophisticated surface signal perception. Future work in this field will undoubtedly beneficiate of comparative transcriptome analysis using RNAseq that will reveal the expression and the regulation of both host and parasite genes in parasitized red algae as recently done in algal oomycete interactions (see below).

11.2.3.2 Fungal and Oomycete Parasites and/or Pathogens

Considering the vast lack of knowledge and the importance to seaweed aquaculture, the significance of oomycetes and fungi infecting marine algae was outlined in two recent reviews (Gachon et al. 2010; Li et al. 2010) and in a book chapter (Strittmatter et al. 2009). A comprehensive report (Li et al. 2010) included 15 species of oomycetes, six species of chytrids, 31 Ascomycota species, and one species of mitosporic fungi. In nature, both the oomycetes and chytrids frequently occurred and induced prevalence of disease, which could destroy the populations of host plants greatly. However, the Ascomycota growing on algae have never occurred as epidemics so far. Further investigations are required in this field, such as performing tests to satisfy Koch's postulates, investigations of host specificity, interactions between host and parasite, and the potential effects of environmental factors on occurrence of a disease.

Despite their small size and ephemeral life stages, filamentous brown algae are plagued by various pathogens, including viruses (Müller et al. 1998), which are not commented in this chapter, which focuses on eukaryotic parasites. In addition, numerous historical records described ectocarpoids with abnormal sporangia or vegetative cells suspected to contain unknown parasites (Rattray 1885; Müller et al. 1998). The oomycete *Eurychasma dicksonii* has been described mainly in wild populations of *Pylaiella littoralis* (Küpper and Müller 1999), but it displays a broad host range and infects various brown algae, including *Ectocarpus* (Müller et al. 1999), in which it was initially described by Wright (1879). *E. dicksonii* belongs to the most early branching clade within the oomycete lineage (Sekimoto et al. 2008; Küpper et al. 2006).

11.3 Cross-Talk Between Host and Colonizers and Defenses Against Colonization

Resistance against epibionts can be generally mediated by either structural or chemical means, with both mechanisms interfering with the settling and penetration of spores or propagules. Algal basiphytes can exhibit a variety of defenses against attachment and colonization (Ducker and Knox 1984), including rapid growth by having ephemeral life histories (den Hartog 1972), changes in pH at the plant surface, and sloughing of outer cell walls *(Ulva,* coralline algae) (Filion-Myklebust and Norton 1981; Johansen 1981; Moss 1982; Steneck 1982; Russell and Veltcamp 1984). *Chondrus crispus* will shed outer layers of cell wall and cuticule glycoproteins (Craigie et al. 1992), whereas *Ascophyllum* produces slime and may lose an entire epidermal layer. *Eucheuma,* a carrageenophyte, has a slimy surface during growth and is free of epiphytes, but when it ceases growth, the plant surface loses that slime and becomes epiphytized (Dawes et al. 2000). Survivorship of epiphyte recruits may also be related to the host, as epiphytes may be less exposed to herbivore pressure when associated with a chemically defended host (Hay 1986; Amsler et al. 2009) and endophytes may survive by avoiding grazing when embedded in host tissues (Amsler et al. 2009). Allelopathy occurs through the exudation of antifouling compounds such as phenolics by the two species of *Sargassum* in the Sargasso Sea (Sieburth and Conover 1965) or iodinated compounds by species of *Laminaria*.

In comparison with epiphytism, alga-parasite or pathogen interactions are generally highly specific and characterized by typical symptoms (for reviews, see Andrews 1976, 1977; Fujita 1990; Goff 1982; Correa 1996; Eggert et al. 2010). Some of the biochemical bases of the responses of seaweeds toward algal endophyte infection have been uncovered recently with either biochemical emphasis (Potin et al. 2002; Weinberger 2007; Cosse et al. 2008; Potin 2008; Weinberger and Potin 2010) or ecological emphasis (Pohnert 2004; Amsler and Fairhead 2006). This contribution does not aim to add one more review on a specific aspect of this research field but rather provide further understanding of the role of cross-talk mechanisms involved in host specificity as well as host chemical defenses that impact the diversity of associations.

11.3.1 Cross-Talk Involved in Host Colonization

Host specificity by several epiphytic or endophytic algal species is a dramatic example of nonrandom settlement and recruitment (Ducker and Knox 1984). In terrestrial plant–fungus mutualistic or pathogenic associations, surface compounds of the host, such as wax and other cell wall compounds, were shown to provide signals to the pathogen in the early stages of colonization. Cell wall polysaccharides from the host are also a source of settlement cues for successful colonization of red seaweeds by swimming spores of pigmented endophytes. Interestingly, the causative agent of green patch disease in the red alga *Mazzaella laminarioides,* the endophytic green alga *Endophyton ramosum,* colonizes carrageenan-producing red algae (carrageenophyte), whereas agar producers (agarophytes) remain uninfected (Sanchez et al. 1996). Surprisingly, another green algal endophytic pathogen, *Acrochaete operculata,* infects carrageenophytes containing lambda-carrageenan more successfully than carrageenophytes containing kappa-carrageenan. When zoospores of *A. operculata* settle and germinate, the vegetative filaments of the parasite completely invade sporophytic fronds of *C. crispus*, which contain lambda-carrageenan. In contrast, gametophytic fronds—containing kappa-carrageenan—are not infected beyond the epidermis and outer cortex (Correa and McLachlan 1991). Recognition by the pathogen of the carrageenan oligomeric degradation products of the host cell walls governs this interaction (Bouarab et al. 1999). Another example of host recognition through cell wall polysaccharides was also partly elucidated in the case of *Olpidiopsis* (*Pythium*) *porphyrae*, the oomycete agent of red rot disease in *Porphyra yezoensis*. Porphyran, the sulfated agar that characterized the cell wall of the red algal Bangiales, was shown to enhance zoospore attachment, encystment, and appressorium formation.

Here again, sulfation is likely an important cue in signal recognition because agarose, a neutral agar fraction, did not enhance appressorium formation. However, *O. porphyrae* ultimately only infects *Porphyra yezoensis* and *Bangia atropurpurea* and protoplast fusion of *P. yezoensis* with the resistant *Porphyra tenuipedalis* resulted in hybrids with reduced susceptibility. Therefore, probably other signals are involved to increase specificity leading to selective infection.

Various oomycetes are able to infect other red algae, but the spectrum of compatible hosts is usually narrow (West et al. 2006). For example, a strain of *Olpidiopsis* sp. from Madagascar that had been isolated from *Bostrychia moritziana* was also able to infect certain isolates of *B. tenella*, *B. radicans,* and *B. radicosa*, but no other species (West et al. 2006). Five other species of *Bostrychia* were ultimately resistant, as were six different species of *Stictosiphonia*, *Lophosiphonia*, *Neosiphonia,* and *Polysiphonia*. Interestingly, *Bostrychia* isolates originating from Madagascar were in most cases susceptible, and strains from other parts of the world mostly resistant, which seemingly indicates the existence of pathovars (strains of the same species with distinctive pathogenicity due to coevolution with different hosts) among algal pathogens.

Surprisingly, basal oomycetes, such as *Eurychasma dicksonii*, which is the most common and widespread eukaryotic pathogen of marine algae (Gachon et al. 2009), appear far less specific than *O. porphyrae*. It occurs in all cold and temperate seas worldwide (Sekimoto et al. 2008). In culture, it infects virtually all species of brown algae tested (more than 40, Müller et al. 1999), including representatives of all major orders of this phylum. qPCR and microscopic observations made on laboratory-controlled cultures revealed that clonal brown algal strains exhibit different levels of resistance against *Eurychasma*, ranging from high susceptibility to complete absence of symptoms (Gachon et al. 2009). In the coevolution race between parasites and hosts, the parasites quickly overcome new resistance strategies by the host. This process should lead to local adaptation, where a parasite population has a higher mean performance on local (sympatric) versus foreign (allopatric) host populations. Hence, this latter study opens the perspective of combining large-scale disease monitoring in the field with laboratory-controlled experiments on the genome model seaweed *Ectocarpus siliculosus* to improve our understanding of host specificity in brown algal diseases (Grenville-Briggs et al. 2011).

11.3.2 "Ménages à trois," Quorum Sensing Mediated Bacterial–Macroalgal Interactions Modulate Colonization

In marine environments, the space competitors include, in addition to macrophytes, sedentary animals and the periphytic community of microalgae, cyanobacteria, heterotrophic microbes, and protists (see Chap. 7 by Edwards and Connell). The bacterial communities that are associated with macroalgae often provide further settlement cues for macrofoulers, which has in particular been studied with coralline crustose algae (Wieczorek and Todd 1998; Steinberg and deNys 2002; see also Chap. 10 by Friedrich). Shifts in the community composition of alga-associated bacterial communities may potentially result in changed availability of such cues and subsequently in changed intensity of fouling (Steinberg and deNys 2002).

Two recent reports show that the main determinant of the composition and structure of epiphytic bacterial communities at the surfaces of various seaweeds is the nature of the algal host (Hengst et al. 2010; Lachnit et al. 2011). Algae clearly shape the structure and composition of microbial communities in their vicinity (Lam and Harder 2007; Lam et al. 2007). On the other hand, they are also affected by these microorganisms. Certain bacteria have been shown to provide their host algae with growth factors, nutrients, or protection from settlement by other micro- or macrofoulers (Armstrong et al. 2001; Dobretsov and Qian 2002; Matsuo et al. 2005; Zheng et al. 2005) and nonpalatable epiphytes potentially protect more palatable hosts from grazing (Karez et al. 2000). Quorum sensing (QS) is a cell-to-cell communication mechanism that allows bacteria to coordinate swarming, biofilm formation, stress resistance, and production of secondary metabolites in response to an excess of the threshold of QS signals (Paul and Ritson-Williams 2008; Dobretsov et al. 2009). Gram-negative bacteria, such as *Pseudomonas* or *Vibrio* strains, produce *N*-acyl homoserine lactones (AHLs) as signaling compounds. *Pseudomonas* spp. are also known to produce diketopiperazines acting as QS signals (Dickschat 2010). The signal molecules γ-butyrolactones and oligopeptides are known to be synthesized by Gram-positive bacteria, e.g., members of the genera *Streptomyces* or *Bacillus* (reviewed by Dobretsov et al. 2009). The interaction between zoospores of eukaryotic green macroalgae (Ulvales) with *Vibrio anguillarum* indicates algal susceptibility to quorum sensing AHL molecules (Joint et al. 2002; Wheeler et al. 2006). Although the specific mechanism regulating these responses to AHLs is not known, it was shown that the AHL molecules affect the calcium influx into the spores of *Ulva* sp., affecting their motility toward the surfaces where they eventually settle (Diggle et al. 2007; Joint et al. 2007). In addition, it has been demonstrated that life cycle completion and spore release in the red epiphytic alga *Acrochaetium* sp. strongly depend on AHLs, which are produced by bacteria associated with the basiphyte *Gracilaria chilensis* (Weinberger et al. 2007). These findings of AHL perception in green and red algae confirm that AHL signaling is more widespread among eukaryotes than previously thought. This indicates a more general importance of the associated microbial communities in interactions with macroalgae.

During the past decade it has been shown that various macroalgae are able to stimulate, inhibit, or inactivate QS signals in bacteria by producing QS inhibitors or analogs thereof (Maximilien et al. 1998; Joint et al. 2007; Kanagasabhapathy et al. 2009). The Australian red macroalga *Delisea pulchra* produces halogenated furanones, structural analogs to *N*-acyl homoserine lactones. These furanones protect the algal surfaces by interfering with AHL-regulated processes and selectively inhibit bacterial colonization and biofilm formation (Maximilien et al. 1998; Rasmussen et al. 2000; Manefield et al. 2002). In addition to the furanones of

D. pulchra, a variety of bacteria and eukaryotes have been shown to produce cyclic dipeptides that can act as AHL mimics and affect QS-regulated behavior in other bacteria (Dobretsov et al. 2009; Dickschat 2010). Recently, Kanagasabhapathy et al. (2009) suggested that certain epibiotic bacteria from the brown macroalga *Colpomenia sinuosa* may play a role in defense mechanisms and suppress the settlement of other competitive bacteria by producing quorum sensing inhibitors (QSI) or QSI-like compounds.

11.3.3 Allelopathic Interactions and Induced Defenses

Allelopathy was initially defined when a plant secretes toxic biochemicals, such as phytotoxic root exudates (Bais et al. 2003) into its environment. These toxic compounds, called allelochemicals, can have adverse effects on the growth of other plants that come in contact with them. Allelopathy provides a strong advantage in the competition for space by interfering with settlement and/or expansive growth of competitors. According to Gross (2003), in aquatic environments, allelopathy could be extended to include negative influences between photosynthetic organisms and other organisms as well as antifouling strategies (see also Chap. 9 by Amsler).

One of the most dramatic and recent examples of allelopathy between macroalgae in temperate systems is seen in the introduced Japanese alga *Bonnemaisonia hamifera,* which has become one of the most common red algae in Sweden during the last decades (Svensson et al. 2009). Its life history involves an alternation between morphologically different gametophytes and tetrasporophytes and it seems that only filamentous tetrasporophytes (formerly described as a distinct species *Trailliella*) are propagating, often growing as an epiphyte on native algae, primarily *Corallina officinalis* and *Furcellaria fastigiata*. *B. hamifera* was coated on surfaces to estimate settling success of native algal spores. Some field experiments also tested whether this substance can be transferred among species by placing exotic adjacent to native algae under simulated wave action. These authors showed that at natural concentrations, 4-bromo-heptanone, the most abundant brominated compounds in the alga, inhibits settlement of spores from native algal species. Furthermore, the substance can be transferred from *B. hamifera* to the native species. The mechanism of transferring 4-bromo-heptanone to other surfaces might enable the algae to "reserve" space for future colonization, as native algae cannot settle on the coated surfaces. This promising study is the first to show that invasions by exotic algae are facilitated by allelopathy and that chemical defenses can be transferred from exotic to native species (Svensson et al. 2009; see also Chap. 12 by Andreakis and Schaffelke). It is in agreement with the Novel Weapons Hypothesis that predicts that exotic species will be successful if they possess biochemicals unique to their new range (Callaway and Ridenour 2004). *B. hamifera* is also found to be released from predators in the archipelago of Tjärnö (Enge et al.

2009) and another polybrominated compound 1,1,3,3-tetrabromo-2-heptanone inhibits also bacterial colonization (Nylund et al. 2008).

Other examples of allelopathic interactions have been also recently reviewed by Jormalainen and Honkanen (2008) in bloom-forming macroalgae, such as *Ulva fenestrata*, *Ulvaria,* and *Pylaiella littoralis* that gain competitive success using toxic exudates against germlings of canopy-forming seaweeds (Nelson et al. 2003; Råberg et al. 2005). Recently also, the invasive red alga *Lophocladia lallemandii* was observed growing as an epiphyte on *C. taxifolia*, and levels of caulerpenyne were found to be higher on *C. taxifolia* that was epiphytized by *L. lallemandii* (Box et al. 2008). Additionally, activities of antioxidant enzymes, catalase and superoxide dismutase, and H_2O_2 production were higher in epiphytized *C. taxifolia* than in thalli that were not epiphytized by *L. lallemandii*, pointing as the evolution of a defensive mechanism against the new invasive algal epiphyte using caulerpynyne, previously proposed as an allelopathic substance in the invasion success of *C. taxifolia* (Pohnert 2004).

In crustose coralline algae also, allelopathy may be an important means to prevent overgrowth by canopy-forming macroalgae (reviewed by Gross 2003). The crustose coralline alga, *Lithophyllum* sp., produces an allelopathic, nonpolar substance that destroys zoospores of the brown alga, *Laminaria religiosa* (Suzuki et al. 1998). Suzuki et al. (1998) suggested that the reduction of epiphytes due to allelopathy may contribute to the predominance of crustose coralline algae in the coastal region of the Northern Japan Sea. In another study, extracts from the crustose coralline alga *Lithophyllum yessoense* inhibited spore settlement of 14, and germination of spores of 13, of the total of 17 tested macroalgal species (Kim et al. 2004). Kim et al. (2004) suggested that *L. yessoense* probably has multiple allelochemicals, including water-soluble exudates, to cope with macroalgal recruits. More recently, the algal spore lytic fatty acid hydroperoxide of heptadeca-5,8,11-trienoic acid (HpDTE: C17:3) was isolated from the crustose coralline seaweed *Lithophyllum yessoense*. HpDTE showed more than 50% lysis at a concentration of 5 $\mu g\ mL^{-1}$ against the spores of three chlorophyte species, nine rhodophytes, four phaeophytes, and the cells of four phytoplankton species. Lysis activity increased with the number of double bonds and carbon atoms in the fatty acid derivatives (Luyen et al. 2008).

The wound response of the red alga *G. chilensis* involves the release of free fatty acids as well as the hydroxylated eicosanoids, 8R-hydroxy eicosatetraenoic acid (8-HETE) and 7S,8R-dihydroxy eicosatetraenoic acid (7,8-di-HETE) (Lion et al. 2006). While the release of free arachidonic acid and subsequent formation of 8-HETE is likely controlled by phospholipase A, 7,8-di-HETE production is independent of this lipase. This dihydroxylated fatty acid might be directly released from galactolipids that contained 8-HETE or 7,8-di-HETE (Lion et al. 2006). In this context, the induced chemical defense response of the red alga *G. chilensis* against epiphytes was also investigated. An extract of an epiphyte challenged alga was shown to trigger a defense response (Weinberger et al. 2011). The hormonally active metabolite(s) could further be purified by reverse phase RP-HPLC. Treatment with the extract or the fraction resulted in pronounced changes of the chemical profile of the alga and in an increased resistance against spores of the epiphytic red alga *Acrochaetium* sp. Semiquantitative RT-PCR and enzyme assays demonstrated that this metabolic response occurs in succession of an upregulation of lipoxygenases and phospholipase A2 activity. Even if this strongly suggests the involvement of regulatory oxylipins, the known plant hormone jasmonic acid and the algal metabolite prostaglandin E2 do not trigger comparable defense responses.

Interestingly also, *C. crispus* gametophytes synthesize ultraviolet absorbing compounds around the sites of penetration of zoospores of the green endophyte *A. operculata*, whereas this response is absent in the sensitive generation (Bouarab et al. 2004). These findings are reminiscent of the deposition of phenolic compounds in higher plant–fungus or oomycete interactions (McLusky et al. 1999). Although the exact structure of these ultraviolet absorbing compounds remains to be elucidated, their aromatic nature suggests that the perception of the pathogen induces a pathway related to the phenylpropanoid metabolism of terrestrial plants. Indeed, *A. operculata* extracts activated two key enzymes of this pathway, shikimate dehydrogenase and phenylalanine ammonia-lyase (Bouarab et al. 2004) that are also activated by oxylipins.

The only very clear case of nonphlorotannin brown algal secondary metabolites acting as antifoulants in nature that we are aware of is in *Dictyota menstrualis* (Order Dictyotales). Schmitt et al. (1995) reported that *D. menstrualis* was noticeably less fouled in nature than other macroalgae and that bryozoan larvae would not settle on *D. menstrualis* in laboratory bioassays even though they would contact its surface and would settle on several other species of brown and red macroalgae.

Extracts of *Lobophora variegata* (Order Dictyotales) from the Bahamas have particularly strong antifungal bioactivity and Lane et al. (2009) identified the cyclic lactone, lobophorolide, as a defensive compound with significant bioactivity at exceptionally low concentrations against both pathogenic and saprophytic sympatric fungi. Although *L. variegata* from 46 of 51 samples collected at ten sites in the Bahamas contained measurable, bioactive concentrations of lobophorolide, it was not present in two collections from the Red Sea even though crude extracts of the Red Sea algae did have antifungal bioactivity. This indicates that although *L. variegata* from different regions may be chemically defended against pathogens, the same specific compounds may not be involved.

Although the vast majority of the investigations into the defensive chemical ecology of macroalgae have focused on roles of secondary metabolites (see Chap. 9 by Amsler), recent work has demonstrated that both red and brown algae employ oxidative burst defenses against pathogens, a defensive mechanism that has been recognized in terrestrial plants for a number of years (reviewed by Lamb and Dixon 1997; Wojtaszek 1997; Mahalingam and Fedoroff 2003; see also Chap. 6 by Bischof and Rautenberger).

This was first discovered in red algae, where cell wall degradation products elicit a rapid release of H_2O_2 that is toxic to epiphytic bacteria (Weinberger et al. 1999, 2001; Weinberger and Friedlander 2000). Küpper et al. (2001) demonstrated an analogous response in *Laminaria digitata* (Order Laminariales), which produces a rapid oxidative burst of H_2O_2 and O_2—in response to oligomeric degradation products of alginate (a major cell wall component of *L. digitata* and other brown algae). Küpper et al. (2001) also reported that H_2O_2 concentrations in the range released by the algae were toxic to pathogenic, alginate-degrading bacteria and, using histological techniques, demonstrated that the response was primarily confined to the meristoderm. These authors also began to dissect the underlying physiological mechanisms and showed that at least some of the mechanisms are likely conserved within the oxidative burst responses of algae, terrestrial plants, and animals. Küpper et al. (2002) surveyed 45 species of brown algae from 11 Orders for constitutive vs. induced production of H_2O_2. They reported that oxidative bursts in response to alginate oligomers were common in sporophytes in the Laminariales and Desmarestiales but not in their filamentous gametophyte stages. One species (*Pylaiella littoralis*) from the Ectocarpales had an oxidative burst response but ten other ectocarpalean species did not. Members of the Fucales had high constitutive production of H_2O_2 but most did not produce additional H_2O_2 in response to alginate oligomers. Küpper et al. (2002) also reported that axenic sporophytes of *Macrocystis pyrifera* (Order Laminariales) were rapidly infected by pathogenic bacteria when the oxidative burst response was blocked with an NAD(P)H oxidase inhibitor but not in controls, and that when nonaxenic *M. pyrifera* or *L. digitata* sporophytes were treated with the inhibitor, they were rapidly attacked by their natural bacterial flora. These authors also demonstrated that the oxidative burst response is important in resistance of both *M. pyrifera* and *L. digitata* to the pathogenic brown algal endophytes, *Laminariocolax tomentosoides* and *Laminariocolax macrocystis*, although the response took 7 days to occur and likely involves elicitation of other structural or chemical defenses (Küpper et al. 2002).

11.4 Impacts of Close Associations

11.4.1 Community Context

The consequences of epibiotism and endobiosis for the host algae are not necessarily negative, but instead depend on the community context as discussed recently in Jormalainen and Honkanen (2008). Examples of positive effects of epibiont removal by grazers are numerous and depend strongly of the type of interactions established among hosts, epibiota, and grazers. In many cases grazers, often gastropods, can drastically counteract the negative influences of epibiotism.

Epibiota on the host alga may modify the host's susceptibility to herbivores such that it would decrease or increase susceptibility to herbivores. Wahl and Hay (1995) took the host species as the reference point, and classified epibiosis-caused decreases in herbivory as "associational resistance" and epibiosis-caused increases in herbivory as "shared doom." The same phenomena have also been called "protective coating" and "co-consumption," respectively (Karez et al. 2000). Both these epibiosis-related modifications of susceptibility to herbivory have been found in macroalgae (critically reviewed by Jormalainen and Honkanen 2008).

The community context where macroalgae are susceptible to multiple natural enemies, including herbivores and epibionts, and where the susceptibility to one enemy may depend on the occurrence of others, implies that the patterns of natural selection for resistance to epibiotism and resistance to herbivory may be correlated. When enemies tend to coexist on the same hosts, generalized defenses, i.e., traits that provide resistance to assorted enemies, are expected to evolve (Rausher 1996). There are some data supporting such multiple functions of macroalgal secondary metabolites: The diterpene alcohols pachydictyol A and dictyol E from *Dictyota menstrualis*, which provide resistance to herbivores (reviewed in Hay and Steinberg 1992), also prevent fouling organisms from colonizing the surface (Schmitt et al. 1995). Similarly, brown algal phlorotannins have been suggested to have multiple functions, e.g., to act both against herbivores and epibiota (reviewed in Amsler and Fairhead 2006). This is also the case of some oxylipins, which are allelopathic substances and were also shown to deter herbivores (Hay et al. 1998).

Epibiosis and grazing are very likely to be intimately interdependent in the influence of both community dynamics and the evolution of resistance traits.

11.4.2 Implications to Seaweed Mariculture

Epiphytes, endophytes, and parasites are well involved in human affairs, as when they increase the cost for mariculture of economically important seaweeds (Friedlander 1992; see also Chap. 22 by Buchholz et al.). Epiphytism has been considered one significant problem in seaweed cultivation (Fletcher 1995; Oliveira et al. 2000; Ask and Azanza 2002; Lüning and Pang 2003). Massive monocultures of algae offer ideal conditions for the growth of epiphytes and propagation of diseases (Wheeler et al. 1981; Fletcher 1995; Lüning and Pang 2003). Epiphytes also produce a series of undesired effects in the production of the host depending on the extent of the contamination (Fletcher 1995). Some important problems with epiphytes have been attributed to (1) competition for nutrients (Buschmann and Gómez 1993), (2) shading (Fletcher 1995), (3) increased weight and drag (Kuschel and Buschmann 1991; Buschmann and Gómez 1993), (4) the possible liberation by the epiphyte of growth inhibiting compounds that affect the host (Santelices and Varela 1993), and (5) a decrease in the reproductive output of the infected host (D'Antonio 1985).

Methods for controlling epiphytes have been described and utilized in cultures of *Gracilaria* (cf. Fletcher 1995). However, many of these methods can only be used in cultivation tanks and are more difficult to apply to open ocean cultivation.

The concern in seaweed mariculture with epiphytism has led to special attention being directed to the selection of resistant strains and to knowledge of the strategies developed by the host to avoid epiphytism (e.g., Santelices and Ugarte 1990; Santelices 1992; Fletcher 1995; Ask and Azanza 2002). For example, the ability to produce compounds, which deter the settlement and growth of epiphytic organisms, has been described in some algae of commercial value (Santelices

1992; Jennings and Steinberg 1997; Harder et al. 2004; Hellio et al. 2004). On the other hand, it has been demonstrated that the capacity to resist an attack depends on the host susceptibility to epiphytes. As mentioned earlier, the presence of different types of carrageenan in the different reproductive phases of *Chondrus crispus* confers a differential adaptation in recognizing an attack by endophytic organisms (Bouarab et al. 1999, 2001b).

Farming of red algal eucheumoids in South-East Asia has been threatened recently by the occurrence of various diseases, mainly "ice-ice", during which parts of the plant turn white from primary or secondary attacks by bacteria and fungi, as well as heavy colonization by endo/epiphytic filamentous red algae (EFAs) of the *Polysiphonia* type, which results in swelling of the host tissue followed by fragmentation and rotting (Critchley et al. 2004). In successive outbreaks of epiphyte infestation, in the Calaguas Islands, the impact of the disease was economically, socially, and environmentally very serious as the farmers became disillusioned and they either moved to other cultivation sites (thus further propagating the epidemy) or returned to the environmentally damaging practices of dynamite or cyanide reef fishing. Production thus declined markedly in this region, from 6,000T per year during 1997–1999 to nil in 2001. At the levels of the Philippines production *Kappaphycus alvarezii* ("Cottonii") farms also are sensitive to smothering by other pest weeds, such as the green algae *Ulva* and *Chaetomorpha* and the brown alga *Hydroclathrus*, as well as to grazing by siganidae herbivory fishes (e.g., rabbit fish) and to marked fluctuations in water temperature due to warm water events. Farmers now tend to grow Cottonii for 45 days (or less) instead of 60 days, resulting in a lower quality of the harvest in terms of biomass and colloid content. The productivity of some *Kappaphycus* farms is decreasing and one estimate indicates that the losses due to diseases now represent about 15% of the potential crop of the Philippines and Indonesia, with a conservative estimated value of US$ 5–10 million to the farmers. That this does not show up yet in the production figures is due to the extension of production areas (i.e., increased hectarage of farming activities, but there is a marked decline in productivity per unit area in certain farms). If the situation is not addressed there is a very real risk that the raw material source in many areas of production could be compromised if a particularly virulent epiphyte/pathogen strain were to arise. In Malaysia also, cultured seaweed became susceptible to epiphytes in the dry seasons between March–June and September–November. Findings revealed *Neosiphonia savatieri* as the dominant infecting epiphyte, representing up to 80–85% of the epiphytes present during peak seasons. Besides *N. savatieri*, *Neosiphonia apiculata*, *Ceramium* sp., *Acanthophora* sp., and *Centroceras* sp. were observed in smaller quantities (Vairappan 2006; Vairappan et al. 2008).

Similarly, there is general agreement that epiphytism is one of the major biological problems in *Gracilaria* farms (Pringle et al. 1989, reviewed by Fletcher 1995) due to the high density of individuals essentially maintained under monoculture conditions. These conditions are known to make the host more susceptible to pests in general and to epiphytes in particular (Friedlander 1992). Competition between hosts and their epiphytes has been demonstrated under natural and artificial conditions of growth (Arrontes 1990; Friedlander and Ben-Amotz 1991;

Svirski et al. 1993), and the extent of the damage is clearly determined by the intensity of the infections (Cancino et al. 1987; Buschmann and Gomez 1993). Farming of *Gracilaria chilensis* during the last 40 years is a clear example of how detrimental epiphytes may become (Kuschel and Buschmann 1991; Pizarro and Santelices 1993).

In addition, negative effects on the economic value of brown and red algae infested by endophytes have been reported (Yoshida and Akiyama 1979; Apt 1988a, b). The endophytic phaeophycean *Laminariocolax* (*Streblonema) aecidioides* causes tissue thickening in the commercially valuable alga *Undaria* sp. (Yoshida and Akiyama 1979), and *Streblonema*-like endophytes are known to produce galls in some algal hosts (Andrews 1977; Apt 1988a).

Estimates of economic losses for Japan and Korea (Gachon et al. 2010) are estimated at about 10% of the annual production (0.15 billion US dollars globally), mostly due to the oomycete pathogens *Olpidiopsis* spp. and *Pythium* spp. (Ding and Ma 2005; Park et al. 2000). However, spatial and interannual variability feature dramatic variations with reductions of 15–20% and 20–30% in yields and value, respectively, in South Korea and losses in Japan rose to 64% in 1993 (Gachon et al. 2010).

Removal of epiphytes can be carried out by physical (mechanical brushing, rapid water movement), chemical (rinsing in chlorine, copper, or changes in pH), and biological means (use of epiphyte herbivores). However, type III to type V epiphytes (see above), as well as endophytes and eukaryotic parasites, cannot be controlled by such treatments and alternative methods based on the current understanding of the genetic and biochemical basis of close associations are required in order to limit their impacts on algal mariculture. Interestingly, a recent study showed that dipping of seedlings in a commercial extract from *Ascophyllum nodosum* (AMPEP), prior to out-planting, was efficient for both improving the daily growth rate and increased productivity of both varieties of *Kappaphycus* and could be used as part of a management protocol to control or reduce the impact of *Neosiphonia* infection in commercial cultivation areas in the sea (Borlongan et al. 2011).

11.4.3 Fouling

The negative effects of intimate associations on macroalgal fitness can be expected to exert a strong pressure of selection for efficient antifouling defenses. Chemical resistance against various epibiotic organisms, e.g., bacteria, microalgae, periphyton, fungi, invertebrate larvae, and epiphytic macroalgae, has been tested for all major macroalgal groups (worth 1.5 billion US dollars globally reviewed in Davis et al. 1989; Steinberg and deNys 2002; Lane and Kubanek 2008).

Although a wealth of examples on antifouling properties of algal extracts exists, the generality of antifouling metabolites has been questioned. The problem with most of the existing studies is that, although they clearly show macroalgae commonly containing metabolites with antifouling properties, it is not known whether the compounds have an antifouling role in the real ecological interaction with their natural enemies (Steinberg et al. 2001; Dworjanyn et al. 2006). The metabolites may have evolved in other contexts, e.g., as antiherbivory compounds, and their antifouling properties may be artifacts generated by breaking up the cell structure and exposing the compounds to fouling organisms. For example, in the red alga *Delesseria sanguinea*, crude extracts decrease settlement in assays where substrates are coated with metabolites even though the intact alga does not inhibit settlement (Nylund and Pavia 2003).

Although the ecologically relevant examples of natural antifoulants are still few, their number is increasing (Lane and Kubanek 2008), indicating that chemical characteristics of macroalgae do affect the epibiotic community. Further understanding of the ecology and evolution of macroalgal antifouling strategies has become increasingly important (Callow and Callow 2011). When a trait provides resistance to biofoulers in nature, it is realistic to assume that the trait has evolved under selection generated by the epibiotic organisms and such adaptation is susceptible of inspiring biomimetic approaches to design new antifouling strategies.

11.5 Summary and Prospects

Altogether, the data described earlier indicate that marine plants are not passive participants in epibiotic interactions and that they can actively alter their susceptibility to various attackers. Macroalgae produce allelopathic metabolites that act against sessile invertebrates, canopy-forming macroalgal competitors, and epibiotic organisms in particular. Although a large body of evidence that shows that algae are endowed with chemical defenses has been available for some time, the idea that many of these defenses are induced following challenge by bioaggressors has emerged only recently. Consequences of epi/endobiotism to the host alga depend also on the community composition and, vice versa, epibiotism may modify the susceptibility of the host to herbivory. However, relatively few studies have tested the responses in ecologically relevant contexts. The scarcity of epidemiological studies in natural populations of algal species is a major gap for the understanding of naturally occurring mechanisms impacting ecosystem functioning. Hopefully, metagenomics and qPCR approaches open the possibility of scoring the presence and relative abundance of several epiphytes and pathogens in parallel, in order to determine the global pressure acting on natural algal populations. A major challenge, which remains for further understanding, is to address whether fouling pressure and pathogen attacks act as a selective force driving population divergence. An approach that could be applied will be to compare the expression of defense-responsive genes in natural populations experiencing different fouling intensities or infectious diseases. Low cost and more accessible Next Generation of Sequencing methods for monitoring gene expression (RNAseq) will help to identify gene networks and the interplay of defense pathways in a growing number of ecologically and/or economically relevant algal models in response to specific pests. Therefore, patterns of differential gene expression will be assessed and will allow identification of major defense-responsive genes whose expression varies in a population-dependent manner. This will link population genetic variation in stress responses with identification of traits under possible selective pressure. The future for these integrated studies includes molecular genetics, microarrays, proteomics, RNAi assays, and knockouts on a limited amount of marine model algae, combined with a quantitative organismal approach.

Acknowledgements This chapter is dedicated to the memories of Pr. Jean-Yves Floc'h (1940–2010) and Dr Jean-Pierre Salaün (1945–2011) wonderful mentors and friends who taught P. Potin so much about the biology and physiology of algae and the biochemistry of algal defenses, respectively. Research on epiphytism by the Potin group and collaborators has been supported by the European Commission INCO-DEV Programme (contribution of the joint research effort INCO-EPIFIGHT ICA4-CT-2001-10021). This study also constitutes a contribution from the Laboratoire International Associé (LIA), Dispersal and Adaptation in Marine Species.

References

Amsler CD, Fairhead VA (2006) Defensive and sensory chemical ecology of brown algae. Adv Bot Res 43:1–91

Amsler CD, Amsler MO, McClintock JB, Baker BJ (2009) Filamentous algal endophytes in macrophytic Antarctic algae: prevalence in hosts and palatability to mesoherbivores. Phycologia 48:324–334

Anderson RJ, Carrick P, Levitt GJ, Share A (1997) Holdfasts of adult kelp Ecklonia maxima provides refuges from grazing for recruitment of juvenile kelps. Mar Ecol Prog Ser 159:265–273

Andrews JH (1976) The pathology of marine algae. Biol Rev 51:211–253

Andrews JH (1977) Observations on pathology of seaweeds in Pacific Northwest. Can J Bot 55:1019–1027

Apt KE (1984) Effects of the symbiotic red alga *Hypneocolax stellaris* on its host *Hypnea musciformis* (Hypneaceae, Gigartinales). J Phycol 20:148–150

Apt KE (1988a) Galls and tumor-like growth on marine macroalgae. Dis Aquat Org 3:211–217

Apt KE (1988b) Etiology and development of hyperplasia induced by Streblonerna sp. (Phaeophyta) of members of the Laminariales (Phaeophyta). J Phycol 24:28–34

Armstrong E, Yan LM, Boyd KG, Wright PC, Burgess JG (2001) The symbiotic role of marine microbes on living surfaces. Hydrobiologia 461:37–40

Arrontes J (1990) Composition, distribution on host and seasonality of epiphytes on three intertidal algae. Botanica Mar 33:205–211

Ask EI, Azanza RV (2002) Advances in cultivation technology of commercial eucheumatoid species: a review with suggestions for future research. Aquaculture 206:257–277

Aumack CF, Amsler CD, McClintock JB, Baker BJ (2011) Changes in amphipod densities among macroalgal habitats in day versus night collections along the Western Antarctic Peninsula. Mar Biol 158:1879–1885

Bais HP, Vepachedu R, Gilroy S, Callaway RM, Vivanco JM (2003) Allelopathy and exotic plant invasion: from molecules and genes to species interactions. Science 301:1377–1380

Ballantine DL (1979) The distribution of algal epiphytes on macrophyte hosts offshore from La Parguera, Puerto Rico. Botanica Mar 22(107–1):11

Benzing DH (1987) Vascular epiphytism: taxonomic participation and adaptive diversity. Ann Mo Bot Gard 74:183–204

Bernstein BB, Jung N (1979) Selective pressures and coevolution in a kelp canopy community in Southern California. Ecol Monogr 49:335–355

Bidwell RGS (1958) Photosynthesis and metabolism of marine algae 11. A survey of rates and products of photosynthesis in $^{14}CO_2$. Can J Bot 36:337–349

Bittick SJ, Bilotti ND, Peterson HA, Stewart HL (2010) Turbinaria ornata as an herbivory refuge for associate algae. Mar Biol 157:317–323

Borlongan IAG, Tibubos KR, Yunque DAT, Hurtado AQ, Critchley AT (2011) Impact of AMPEP on the growth and occurrence of epiphytic Neosiphonia infestation on two varieties of commercially cultivated Kappaphycus alvarezii grown at different depths in the Philippines. J Appl Phycol 23:615–621

Bouarab K, Potin P, Correa J, Kloareg B (1999) Sulfated oligosaccharides mediate the interaction between a marine red alga and its green algal pathogenic endophyte. Plant Cell 11:1635–1650

Bouarab K, Kloareg B, Potin P, Correa JA (2001a) Ecological and biochemical aspects in algal infectious diseases. Cah Biol Mar 42:91–100

Bouarab K, Potin P, Weinberger F, Correa J, Kloareg B (2001b) The Chondrus crispus-Acrochaete operculata host-pathogen association, a novel model in glycobiology and applied phycopathology. J Appl Phycol 13:185–193

Bouarab K, Adas F, Gaquerel E, Kloareg B, Salaün J-P, Potin P (2004) The innate immunity of a marine red alga involves oxylipins from both the eicosanoid and octadecanoid pathways. Plant Physiol 135:1838–1848

Bown P, Plumb J, Sánchez-Baracaldo P, Hayes PK, Brodie J (2003) Sequence heterogeneity of green (Chlorophyta) endophytic algae associated with a population of Chondrus crispus (Gigartinaceae, Rhodophyta). Eur J Phycol 38:153–164

Box A, Sureda A, Terrados J, Pons A, Deudero S (2008) Antioxidant response and caulerpenyne production of the alien Caulerpa taxifolia (Vahl) epiphytized by the invasive algae Lophocladia lallemandii (Montagne). J Exp Mar Biol Ecol 364:24–28

Burke JM (1986) Studies on the relationship between the obligate epiphyte Polysiphonia lanosa (Rhodophyta) and its host Ascophyllum nodosum (Phaeophyta). Honours thesis (Bachelor), St. Xavier University, Anntogonish, Nova Scotia

Burkhardt E, Peters AF (1998) Molecular evidence from nrDNA sequences that *Laminariocolax* (Phaeophyceae, Ectocarpales *sensu lato*) is a worldwide clade of closely related kelp endophytes. J Phycol 34:682–691

Buschmann A, Gomez P (1993) Interaction mechanisms between Gracilaria chilensis (Rhodophyta) and epiphytes. Hydrobiologia 260:345–351

Callaway RM, Ridenour WM (2004) Novel weapons: invasive success and the evolution of increased competitive ability. Front Ecol Environ 2:436–443

Callow JA, Callow ME (2011) Trends in the development of environmentally friendly fouling-resistant marine coatings. Nature Comm 2. doi:10.1038/ncomms1251

Cancino JM, Muñoz J, Muñoz M, Orellana MC (1987) Effects of the bryozoan *Membranipora tuberculata* (Bosc.) on the photosynthesis and growth of Gelidium rex. J Exp Mar Biol Ecol 113:105–112

Cebrian J, Enriquez S, Fortes M, Agawin N, Vermaat JE, Duarte CM (1999) Epiphyte accrual on, Posidonia oceanica (L.) Delile leaves: implications for light absorption. Bot Mar 42:123–12

Chiovitti A, Bacic A, Kraft GT, Craik DJ, Liao M-L (1999) Pyruvated carrageenans from Solieria robusta and its adelphoparasite Tikvahiella candida. Hydrobiologia 398–399:401–409

Correa JA (1990) Pigmented algal endophytes of Chondrus crispus Stackhouse: host specificity, fine structure, and effects on host performance in infections by Acrochaete operculata Correa & Nielsen and A. heteroclada Correa & Nielsen. Ph.D. thesis, Dalhousie University, Halifax, Nova Scotia, Canada

Correa JA (1994) Infections by pigmented algal *endophytes*: misuse of concepts and terminology. Revista Chilena de Historia Natural 67:4–8

Correa JA (1996) Infectious diseases of marine algae: current knowledge and approaches. In: Round FE, Chapman DJ (eds) Progress in phycological research. Biopress Ltd, Bristol, U.K, pp 149–180
Correa JA, McLachlan JL (1991) Endophytic algae of Chondrus crispus (Rhodophyta) III. Host specificity. J Phycol 27:448–459
Correa J, Nielsen R, Grund DW, McLachlan J (1987) Endophytic algae of Irish Moss (Chondrus crispus Stackh.). Proc Int Seaweed Symp 12:223–228
Correa JA, Nielsen R, Grund DW (1988) Endophytic algae of Chondrus crispus (Rhodophyta). II. Acrochaete heteroclada sp. nov., A. operculata sp. nov., and Phaeophila dendroides (Chlorophyta). J Phycol 24:528–539
Correa J, Buschmann AH, Retamales C, Beltran J (1997) Infectious diseases of Mazzaella laminarioides (Rhodophyta): changes in infection prevalence and disease expression associated with season, locality, and within-site location. J Phycol 33:344–352
Cosse A, Leblanc C, Potin P (2008) Dynamic defense of marine macroalgae against pathogens: from early activated to gene-regulated responses. Adv Bot Res 46:221–266
Court GJ (1980) Photosynthesis and translocation studies of Laurencia spectabilis and its symbiont Janczewskia gardneri (Rhodophyceae). J Phycol 16:270–279
Craigie JS, Correa JA, Gordon ME (1992) *Cuticles* from Chondrus crispus (Rhodophyta). J Phycol 28:777–786
Critchley AT, Largo D, Wee W, Bleicher Lhonneur G, Hurtado AQ, Schubert J (2004) A preliminary summary on Kappaphycus farming and the impact of epiphytes. Jpn J Phycol 52:231–232
D'Antonio C (1985) Epiphytes on the rocky intertidal algal, Rhodomela larix (Turner) C. Agardh: negative effects on the host and food for herbivores? J Exp Mar Biol Ecol 86:197–218
Davis AR, Targett NM, McConnell OJ, Young CM (1989) Epibiosis of marine algae and benthic invertebrates: natural products chemistry and other mechanisms inhibiting settlement and overgrowth. Bioorg Mar Chem 3:85–114
Dawes CJ, Teasdale BW, Friedlander M (2000) Cell wall structure of the agarophytes Gracilaria tikvahiae and G. cornea (Rhodophyta) and penetration by the epiphyte Ulva lactuca (Chlorophyta). J Appl Phycol 12:567–575
Del Campo E, García-Reina G, Correa JA (1998) Degradative disease in Ulva rigida (Chlorophyceae) associated with Acrochaete geniculata (Chlorophyceae). J Phycol 34:160–166
den Hartog C (1972) Substratum: multicellular plants. In: Kinne O (ed) Marine ecology, vol I, Environmental factors, Part 3. Wiley, New York, pp 1277–1289
Devinny JS, Volse LA (1978) Effects of sedimentation on the development of Macrocystis pyrifera gametophytes. Mar Biol 48:343–348
Dickschat JS (2010) Quorum sensing and bacterial biofilms. Nat Prod Rep 27:343–369
Diggle SP, Gardner A, West SA, Griffin AS (2007) Evolutionary theory of bacterial quorum sensing: when is a signal not a signal? Philos Trans R Soc B 362:1241–1249
Ding HY, Ma JH (2005) Simultaneous infection by red rot and chytrid diseases in Porphyra yezoensis Ueda. J Appl Phycol 17:51–56
Dixon PS (1963) The Rhodophyta: some aspects of their biology. Ocean Mar Biol Annu Rev 1:177–196
Dobretsov S, Teplitski M, Paul VJ (2009) Mini-review: quorum sensing in the marine environment and its relationship to biofouling. Biofouling 25(5):413–427
Dobretsov SV, Qian P-Y (2002) Effect of bacteria associated with the green alga Ulva reticulata on marine micro- and macrofouling. Biofouling 18:217–228
Ducker SC, Knox RB (1984) Epiphytism at the cellular level, with special reference to algal epiphytes. In: Linskens HF, Heslop-Harrison J (eds) Cellular interactions, vol 17, Encyclopedia of plant physiology, new series. Springer, Berlin, pp 115–133
Duggins DO, Eckman JE, Sewell AT (1990) Ecology of understory kelp environments II. Effects of kelps on recruitment of benthic invertebrates. J Exp Mar Biol Ecol 143:27–45

Dworjanyn SA, de Nys R, Steinberg PD (2006) Chemically mediated antifouling in the red alga Delisea pulchra. Mar Ecol Prog Ser 318:153–163

Eckman JE, Duggins DO, Sewell AT (1989) Ecology of understorey kelp environments I. Effects of kelps on flow and particle transport near the bottom. J Exp Mar Biol Ecol 129:173–187

Eggert A, Peters AF, Küpper FC (2010) The potential impact of climate change on endophyte infections in kelp sporophytes. In: Seckbach J (ed) Seaweeds and their role in globally changing environments. Springer, Berlin, pp 141–154

Ellertsdóttir E, Peters AF (1997) High prevalence of infection by endophytic brown algae in populations of *Laminaria* spp. (Phaeophyceae). Mar Ecol Prog Ser 146:135–143

Enge S, Nylund GM, Harder T et al (2009) A novel chemical releases the invasive red alga Bonnemaisonia hamifera from new enemies. Phycologia 48(4):s. 81

Filion-Myklebust C, Norton T (1981) Epidermis shedding in the brown seaweed *Ascophyllum nodosum* (L.) Le Jolis and its ecological significance. Mar Biol Lett 2:45–51

Fletcher RL (1995) Epiphytism and fouling in Gracilaria cultivation: an overview. J Appl Phycol 7:325–333

Fralick RA, Mathieson AC (1975) Physiological ecology of four Polysiphonia species (Rhodophyta, Ceramiales). Mar Biol 29:29–36

Friedlander M (1992) Gracilaria conferta and its epiphytes: the effect of culture conditions on growth. Bot Mar 35:423–428

Friedlander M, Ben-Amotz A (1991) The effect of outdoor culture conditions on growth and epiphytes of Gracilaria conferta. Aquat Bot 39:315–333

Fujita Y (1990) Diseases of cultivated Porphyra in Japan. In: Akatsuka I (ed) Introduction to applied phycology. SPB Academic, The Hague, pp 177–190

Funano T (1969) Observation on the female gametophytes and the microscopic sporophytes of Laminaria religiosa Miyabe. Rep Hokkaido Fish Exp Stn 10:43–50

Gachon CM, Strittmatter M, Müller DG, Kleinteich J, Küpper FC (2009) Detection of differential host susceptibility to the marine oomycete pathogen Eurychasma dicksonii by real-time PCR: not all algae are equal. Appl Environ Microbiol 75:322–328

Gachon CM, Sime-Ngando T, Strittmatter M, Chambouvet A, Kim GH (2010) Algal diseases: spotlight on a black box. Trends Plant Sci 15:633–640

Garbary D, Burke J, Tian L (1991) The Ascophyllum/Polysiphonia/Mycosphaerella symbiosis II. Aspects of the ecology and distribution of Polysiphonia lanosa in Nova Scotia. Bot Mar 34:391–401

Garbary D, Hansen GI, Scagel RF (1982) The marine algae of British Columbia and northern Washington: division Rhodophyta (red algae), Class Florideophyceae, Orders Acrochaetiales and Nemaliales. Syesis 15(1(Suppl)):1–102

Garbary DJ, Kim KY, Klinger T, Duggins D (1999a) Preliminary observations on the development of kelp gametophytes endophytic in red algae. Hydrobiologia 398(399):247–252

Garbary DJ, Kim KY, Duggins D (1999b) Red algae as hosts for endophytic kelp gametophytes. Mar Biol 135:35–40

Gauna M, Parodi C, Elisa R (2010) Green epi-endophytes in Hymenena falklandica (Rhodophyta) from the Patagonian coasts of Argentina: preliminary observations. Phycol Res 56:1440–1835

Goecke F, Labes A, Wiese J, Imhoff JF (2010) Chemical interactions between marine macroalgae and bacteria Mar. Ecol Progr Ser 409:267–300

Goff LJ (1976) Solitary bodies (S-bodies) in the parasitic red algaHarveyella mirabilis (Choreocolaceae, Cryptonemiales). Protoplasma 89:189–195

Goff LJ (1982) The biology of parasitic red algae. Phycol Res 1:289–370

Goff LJ, Ashen J, Moon D (1997) The evolution of parasites from their hosts: a case study in the parasitic red algae. Evolution 51:1068–1078

Gonzalez MA, Goff LJ (1989) The red algal epiphytes Microcladia coulteri and M. californica (Rhodophyceae, Ceramiaceae) II. Basiphyte specificity. J Phycol 25:558–567

Gonzalez MA, Barrales HL, Candia A, Cid L (2003) Spatial and temporal distribution of dominant epiphytes on Gracilaria from a natural subtidal bed in central-southern Chile. Aquaculture 116:135–148

Grahame J, Hanna FS (1989) Factors affecting the distribution of the epiphytic fauna of *Corallina officinalis* (L.) on an exposed rocky shore. Ophelia 30:113–129

Grenville-Briggs L, Gachon CM, Strittmatter M, Sterck L, Küpper FC, van West P (2011) A molecular insight into algal-oomycete warfare: cDNA analysis of Ectocarpus siliculosus infected with the basal oomycete Eurychasma dicksonii. PLoS One 6(9):e24500

Gross EM (2003) Allelopathy of Aquatic Autotrophs. Crit Rev Plant Sci 22:313–339

Hancock L, Goff LJ, Lane C (2010) Red algae lose key mitochondrial genes in response to becoming parasitic. Genome Biol Evol 2:897–910

Harder T, Dobretsov S, Qian PY (2004) Waterborne polar macromolecules act as algal antifoulants in the seaweed Ulva reticulata. Mar Ecol Prog Ser 274:131–141

Harlin MM, Craigie JS (1975) The distribution of photosynthate in Ascophyllum nodosum as it relates to epiphytic Polysiphonia lanosa. J Phycol 11:109–113

Harlin MM (1987) Allelochemistry in marine macroalgae. Crit Rev Plant Sci 5:237–249

Hay ME (1986) Associational plant defenses and the maintenance of species diversity: turning competitors into accomplices. Am Nat 128:617–648

Hay ME (1996) Marine chemical ecology: what's known and what's next? J Exp Mar Biol Ecol 200:103–134

Hay ME, Steinberg PD (1992) The chemical ecology of plant-herbivore interactions in marine versus terrestrial communities. In: Rosenthal GA, Berenbaum M (eds) Herbivores: their interaction with secondary plant metabolites, vol 2, Ecological and evolutionary processes. Academic, San Diego, pp 371–413

Hay ME, Piel J, Boland W, Schnitzler I (1998) Seaweed sex pheromones and their degradation products frequently suppress amphipod feeding but rarely suppress sea urchin feeding. Chemoecology 8:91–98

Hayward PJ (1980) Invertebrate epiphytes of coastal marine algae. In: Price JH, Irvine DEG, Farnham WF (eds) The shore environment, vol 2, Ecosystems. Academic, New York, pp 761–787

Hellio C, Marechal J-P, Veron B, Bremer G, Clare AS, Le Gal Y (2004) Seasonal variation of antifouling activities of marine algae from the Brittany Coast (France). Mar Biotechnol 6:67–82

Hemmi A, Makinen A, Jormalainen V, Honkanen T (2005) Responses of growth and phlorotannins in Fucus vesiculosus to nutrient enrichment and herbivory. Aquat Ecol 39:201–211

Hengst MB, Andrade S, González B, Correa JA (2010) Changes in epiphytic bacterial communities of intertidal seaweeds modulated by host, temporality, and copper enrichment. Microb Ecol 60:282–90

Honkanen T, Jormalainen V (2005) Genotypic variation in tolerance and resistance to fouling in the brown alga Fucus vesiculosus. Oecologia 144:196–205

Hubbard CB, Garbary DJ, Kim KY, Chiasson DM (2003) Host specificity and growth of kelp gametophytes symbiotic with filamentous red algae (Ceramiales, Rhodophyta). Helgol Mar Res 58:18–25

Hurtado AQ, Critchley AT, Trespoey A, Bleicher Lhonneur G (2006) Occurence of Polysiphonia epiphytes in Kappaphycus farms at Calaguas Is., Camarines Norte Philippines. J Appl Phycol 18:301–306

Jennings JG, Steinberg PD (1997) Phlorotannins versus other factors affecting epiphyte abundance on the kelp Ecklonia radiata. Oecologia 109:461–473

Joint I, Tait K, Callow ME, Callow JA, Milton D, Williams P, Camara M (2002) Cell-to-cell communication across the prokaryote-eukaryote boundary. Science 298:1207

Joint I, Tait K, Wheeler G (2007) Cross-kingdom signalling: exploitation of bacterial quorum sensing molecules by the green seaweed Ulva. Philos Trans R Soc Lond B Biol Sci 36:1223–33

Jormalainen V, Honkanen T (2008) Macroalgal chemical defenses and their roles in structuring temperate marine communities. In: Amsler CD (ed) Algal chemical ecology. Springer, Berlin, pp 75–90

Kain JM (1979) A view of the genus Laminaria. Oceanogr Mar Biol Annu Rev 17:101–161

Kanagasabhapathy M, Yamazaki G, Ishida A, Sasaki H, Nagata S (2009) Presence of quorum-sensing inhibitor-like compounds from bacteria isolated from the brown alga Colpomenia sinuosa. Lett Appl Microbiol 49:573–579

Kaneko T (1973) Morphology of the female gametophyte and young sporophyte of Laminaria japonica var. ochotensis Okamura in nature. Sci Rep Hokkaido Fish Exp Stn 15:1–8

Karez R, Engelbrecht S, Sommer U (2000) "Co-consumption" and "protective coating": two new proposed effects of epiphytes on their macroalgal hosts in mesograzer-epiphyte-host interactions. Mar Ecol Prog Ser 205:85–93

Kim MJ, Choi JC, Kang SE, Cho JY, Jin HJ, Chun BS, Hong YK (2004) Multiple allelopathic activity of the crustose coralline alga Lithophyllum yessoense against settlement and germination of seaweed spores. J Appl Phycol 16:175–179

Kremer BP (1983) Carbon economy and nutrition of the alloparasitic red alga Harveyella mirabilis. Mar Biol 76:231–239

Kugrens P, West JA (1973) The ultrastructure of the alloparasitic red alga Choreocolax polysiphoniae. Phycologia 12:175–186

Küpper FC, Müller DG (1999) Massive occurrence of the heterokont and fungal parasites Anisolpidium, Eurychasma and Chytridium in Pylaiella littoralis (Ectocarpales, Phaeophyceae). Nova Hedwigia 69:381–389

Küpper FC, Kloareg B, Guern J, Potin P (2001) Oligoguluronates elicit an oxidative burst in the brown algal kelp Laminaria digitata. Plant Physiol 125:278–291

Küpper FC, Müller DG, Peters AF, Kloareg B, Potin P (2002) Oligoalginate recognition and oxidative burst play a key role in natural and induced resistance of sporophytes of Laminariales. J Chem Ecol 28:2057–2081

Küpper FC, Maier I, Müller DG, Loiseaux-de Goer S, Guillou L (2006) Phylogenetic affinities of two eukaryotic pathogens of marine macroalgae, Eurychasma dicksonii (Wright) Magnus and Chytridium polysiphoniae Cohn Cryptogam. Algol 27:165–184

Kurihara A, Abe T, Tani M, Sherwood AR (2010) Molecular phylogeny and evolution of red algal parasites: a case study of Benzaitenia, Janczewskia, and Ululania (Ceramiales). J Phycol 46:580–590

Kuschel FA, Buschmann AH (1991) Abundance, effects and management of epiphytism in intertidal cultures of Gracilaria (Rhodophyta) in Southern Chile. Aquaculture 92:7–19

Lachnit T, Meske D, Wahl M, Harder T, Schmitz R (2011) Epibacterial community patterns on marine macroalgae are host-specific but temporally variable. Environ Microbiol 13:655–65

Lam C, Harder T (2007) Marine macroalgae affect abundance and community richness of bacterioplancton in close vicinity. J Phycol 43:874–881

Lam C, Stang A, Harder T (2007) Planktonic bacteria and fungi are selectively eliminated by exposure to marine macroalgae in close proximity. FEMS Microbiol Ecol. doi:10.1111/j.1574-6941.2007.00426.x

Lamb C, Dixon R (1997) The oxidative burst in plant disease resistance. Annu Rev Plant Physiol Plant Mol Biol 48:251–275

Lane AL, Kubanek J (2008) Secondary metabolite defenses against pathogens and biofoulers. In: Amsler CD (ed) Algal chemical ecology. Springer, Berlin, pp 229–243

Lane AL, Nyadong L, Galhena A, Shearer TL, Stout EP, Parry RM, Kwasnik M, Wang MD, Hay ME, Fernandez F, Kubanek J (2009) Surface-mediated antifungal chemical defenses in a tropical seaweed. Proc Natl Acad Sci USA 106:7314–7319

Lane CE, Saunders GW (2005) Molecular investigation reveals epiphytic extrageneric kelp (Laminairales, Phaeophyceae) gametophytes on Lessoniopsis littoralis. Bot Mar 48:426–436

Lein TE, Sjatun K, Wakili S (1991) Mass-occurrence of a brown filamentous endophyte in the lamina of the kelp Laminaria hyperborea (Gunnerus) Foslie along the southwestern coast of Norway. Sarsia 76:187–193

Leonardi PI, Miravalles AB, Faugeron S, Flores V, Beltran J, Correa JA (2006) Diversity, phenomenology and epidemiology of epiphytism in farmed Gracilaria chilensis (Rhodophyta) in northern Chile. Eur J Phycol 41:247–257

Levin P, Mathieson A (1991) Variation in a host–epiphyte relationship along a wave exposure gradient. Mar Ecol Prog Ser 77:271–278

Li W, Zhang T, Tang X, Wang B (2010) Oomycetes and fungi: important parasites on marine algae Acta Oceanol. Sinica 29:74–81

Linskens HF (1963) Beitrag zur Frage der Beziehungen zwischen Epiphyt und Basiphyt bei marinen Algen. Pubbl Staz Zool Napoli 33:274–293

Linskens HF (1976) Specific interactions in higher plants. In: Wood RKS, Graniti A (eds) Specificity in plant diseases. Plenum, New York

Lion U, Wiesemeier T, Weinberger F, Beltrán J, Flores V, Faugeron S, Correa J, Pohnert G (2006) Phospholipases and galactolipases trigger oxylipin-mediated wound-activated defence in the red alga Gracilaria chilensis against epiphytes. Chembiochem 7:457–462

Lobban CS, Baxter DM (1983) Distribution of the red algal epiphyte Polysiphonia lanosa on its brown algal host Ascophyllum nodosum in the Bay of Fundy. Can Bot Mar 26:533–538

Longtin C, Scrosati R (2009) Role of surface wounds and brown algal epiphytes in the colonization of Ascophyllum nodosum (Phaeophyceae) fronds by Vertebrata lanosa (Rhodophyta). J Phycol 45:535–9

Longtin C, Scrosati R, Whalen G, Garbary D (2009) Distribution of algal epiphytes across environmental gradients at different scales: intertidal elevation, host canopies, and host fronds. J Phycol 45:820–7

Lotze HK, Worm B, Sommer U (2000) Propagule banks, herbivory and nutrient supply control population development and dominance patterns in macroalgal blooms. Oikos 89:46–58

Lubchenco J, Gaines SD (1981) A unified approach to marine plant-herbivore interactions I. Populations and communities. Annu Rev Ecol Syst 12:405–437

Lüning K, Pang S (2003) Mass cultivation of seaweeds: current aspects and approaches. J Appl Phycol 15:115–119

Luyen QH, Cho J-Y, Choi J-S, Kang J-Y, Park NG, Hong KI (2008) Isolation of algal spore lytic C17 fatty acid from the crustose coralline seaweed Lithophyllum yessoense. J Appl Phycol 21:423–427

Mahalingam R, Fedoroff NV (2003) Stress response, cell death and signaling: the many faces of ROS. Physiol Plant 119:56–68

Manefield M, Rasmussen T, Kumar N, Hentzer M, Anderson JB, Steinberg PD, Kjelleberg S, Givskov M (2002) Quorum sensing inhibition through accelerated degradation of the LuxR protein by halogenated furanones. Microbiology 148:1119–1127

Markham JW (1969) Vertical distribution of epiphytes on the stipe of Nereocystis luetkeana (Mertens) Postels and Ruprecht. Svesis 2:227–240

Matsuo Y, Imagawa H, Nishizawa M, Shizuri Y (2005) Isolation of an algal morphogenesis inducer from a marine bacterium. Science 307:1598

Maximilien R, de Nys R, Holmstrom C, Gram L, Kjelleberg S, Steinberg PD (1998) Bacterial fouling is regulated by secondary metabolites from the red alga Delisea pulchra. Aquat Microbial Ecol 15:233–246

McLusky SR, Bennett MH, Beale MH, Lewis MJ, Gaskin P, Mansfield JW (1999) Cell wall alteration and localised accumulation of feruloyl-3′-methoxytyramine in onion epidermis at sites of attempted penetration by Botrytis allii associated with actin polarization, peroxidase activity and suppression of flavonoid biosynthesis. Plant J 17:523–534

Menge BA (1991) Relative importance of recruitment and other causes of variation in rocky intertidal community structure. J Exp Mar Biol Ecol 146:69–100

Moss BL (1982) The control of epiphytes by Halidrys siliquosa. Phycologia 21:185–191

Müller DG, Kapp M, Knippers R (1998) Viruses in marine brown algae. Adv Virus Res 50:50–67

Müller DG, Küpper FC, Küpper H (1999) Infection experiments reveal broad host ranges of *Eurychasma dicksonii* (Oomycota) and *Chytridium polysiphoniae* (Chytridiomycota), two eukaryotic parasites in marine brown algae (Phaeophyceae). Phycological Res 47:217–23

Nelson TA, Lee DJ, Smith BC (2003) Are "green tides" harmful algal blooms? Toxic properties of water-soluble extracts from two bloom-forming macroalgae, Ulva fenestrata and Ulvaria obscura (Ulvophyceae). J Phycol 39:874–879

Nielsen R, McLachlan J (1986) Investigations of the marine algae of Nova Scotia XVI. The occurrence of small green algae. Can J Bot 64:808–814

Nonomura AM (1979) Development of Janczewskia morimotoi (Ceramiales, Rhodophyceae). J Phycol 15:154–162

Nylund GM, Pavia H (2003) Inhibitory effects of red algal extracts on larval settlement of the barnacle Balanus improvisus. Mar Biol 143:875–882

Nylund GM, Cervin G, Persson F, Hermansson M, Steinberg PD, Pavia H (2008) Seaweed defence against bacteria: a poly-brominated 2-heptanone from the red alga Bonnemaisonia hamifera inhibits bacterial colonisation. Mar Ecol Progr Ser 369:39–50

Oliveira EC, Alveal K, Anderson R (2000) Mariculture of the agar-producing Gracilarioid red algae. Rev Fish Sci 8:345–378

Padilla DK (1985) Structural resistance of algae to herbivores: a biomechanical approach. Mar Biol 90:103–109

Park CS et al (2000) Comparison of the morphological and physiological features of the red rot disease fungus Pythium sp. isolated from Porphyra yezoensis from Korea and Japan. Fisheries Sci 66:261–269

Parke MW (1932) Port Erin: report of the algologist. Rep Mar Biol Stn Port Erin 45:18–21

Paul V, Ritson-Williams R (2008) Quorum sensing in the marine environment and its relationship to biofouling. Biofouling 25:413–427

Pavia H, Carr H, Åberg P (1999) Habitat and feeding preferences of crustacean mesoherbivores inhabiting the brown seaweed Ascophyllum nodosum (L.) Le Jol. and its epiphytic macroalgae. J Exp Mar Biol Ecol 236:15–32

Pearson CA, Evans LV (1989) Polysiphonia lanosa basiphyte inter-relationships. Br Phycol J 24:310–311

Pearson CA, Evans LV (1990) Settlement and survival of Polysiphonia lanosa (Ceramiales) spores on Ascophyllum nodosum and Fucus vesiculosus (Fucales). J Phycol 26:597–603

Peters AF (1991) Field and culture studies of Streblonema macrocystis sp. nov. (Ectocarpales, Phaeophyceae) from Chile, a sexual endophyte of giant kelp. Phycologla 30:365–377

Peters AF (2003) Molecular identification, taxonomy and distribution of brown algal endophytes with emphasis on species from Antarctica. In: Chapman ARO, Anderson RJ, Vreeland V, Davison IR (eds) In: Proceedings of the 17th international seaweed symposium. pp 293–302. Oxford University Press, New York

Peters AF, Schaffelke B (1996) Streblonema (Ectocarpales. Phaeophyceae) infection in the kelp Laminaria saccharina (Laminariales, Phaeophyceae) in the western Baltic. Hydrobiologia 326(327):111–116

Pizarro AB, Santelices (1993) Environmental variation and large-scale Gracilaria production. In: Chapman ARO, Brown y MT, Lahaye M (eds) Fourteenth International Seaweed Symposium—Kluwer Academic Publishers, Belgium. Hydrobiología 260/261:357–363

Pohnert G (2004) Chemical defense strategies of marine organisms. In: Schulz S (ed) The chemistry of pheromones and other semiochemicals I. Topics in current chemistry, vol 239. Springer, Berlin, pp 179–219

Potin P (2008) Oxidative burst and related responses in biotic interactions of algae. In: Amsler CD (ed) Algal chemical ecology. Springer, Berlin, pp 245–271, 121–146

Potin P, Bouarab K, Salaün J-P, Pohnert G, Kloareg B (2002) Biotic interactions of marine algae. Curr Opin Plant Biol 5:308–317

Pringle JD, James D, Tseng CK (1989) Overview of the workshop on production and utilization of commercial seaweeds—Quingdao, China, 1987. J Appl Phycol 1:83–90

Råberg S, Berger-Jönsson R, Björn A, Graneli E, Kautsky L (2005) Effects of Pilayella littoralis on Fucus vesiculosus recruitment: implications for community composition. Mar Ecol Prog Ser 289:131–139

Rasmussen TB, Manefield M, Andersen JB, Eberl L, Anthoni U, Christophersen C, Steinberg P, Kjelleberg S, Givskov M (2000) How Delisea pulchra furanones affect quorum sensing and swarming motility in Serratia liquefaciens MG1. Microbiology 146:3237–3244

Rattray J (1885) Note on *Ectocarpus*. Trans R Soc Ed 32:589–602

Rausher MD (1996) Genetic analysis of coevolution between plants and their natural enemies. Trends Genet 12:212–217

Rawlence DJ (1972) An ultrastructural study of the relationship between rhizoids of Polysiphonia lanosa (L.) Tandy (Rhodophyceae) and tissue of Ascophyllum nodosum (L.) Le Jolis (Phaeophyceae). Phycologia 11:279–290

Rawlence DJ, Taylor ARA (1972) A light and electron microscopic study of rhizoid development in Polysiphonia lanosa (L.) Tandy. J Phycol 8:15–24

Reed DC, Laur DR, Ebeling AW (1988) Variation in algal dispersal and recruitment: the importance of episodic events. Ecol Monogr 58:321–335

Rindi F, Guiry MD (2004) Composition and spatio temporal variability of the epiphytic macroalgal assemblage of Fucus vesiculosus Linnaeus at Clare Island, Mayo, western Ireland. J Exp Mar Biol Ecol 311:233–252

Rönnberg C, Bonsdorff E (2004) Baltic Sea eutrophication: area-specific ecological consequences. Hydrobiologia 514:227–241

Russell BD, Elsdon TS, Gillanders BM, Connell SD (2005) Nutrients increase epiphyte loads: broad-scale observations and an experimental assessment. Mar Biol 147:551–558

Russell G, Veltcamp CJ (1984) Epiphyte sunrival on skinshedding macrophytes. Mar Ecol Prog Ser 18:149–153

Sakai Y, Funano T (1964) Observation on the female gametophytes and the microscopic sporophytes of Laminaria religiosa from Oshoro Bay, Hokkaido, Japan. Sci Rep Hokkaido Fish Exp Stn 2:16

Sanchez P, Correa JA, Garcia-Reina G (1996) Host-specificity of Endophyton ramosum (Chlorophyta), the causative agent of green patch disease in Mazzaella laminarioides (Rhodophyta). Eur J Phycol 31:173–179

Santelices B, Ugarte R (1990) Ecological differences among Chilean populations of commercial Gracilaria. J Appl Phycol 2:17–26

Santelices B, Varela D (1993) Intra-clonal variation in the red seaweed Gracilaria chilensis. Mar Biol 116:543–552

Schmitt TM, Hay ME, Lindquist N (1995) Constraints on chemically mediated coevolution: multiple functions for seaweed secondary metabolites. Ecology 76:107–123

Sekimoto S, Beakes GW, Gachon CMM, Müller DG, Küpper FC et al (2008) The development, ultrastructural cytology and molecular phylogeny of the basal oomycete Eurychasma dicksonii infecting the filamentous phaeophyte algae Ectocarpus siliculosus and Pylaiella littoralis. Protist 159:299–318

Sieburth JM, Conover JT (1965) Sargassum tannin, an antibiotic which retards fouling. Nature 208:52–53

Steinberg PD, deNys R (2002) Chemical mediation of colonization of seaweed surfaces. J Phycol 38:621–629

Steinberg PD, de Nys R, Kjelleberg S (2001) Chemical mediation of surface colonization. In: McClintock JB, Baker BJ (eds) Marine chemical ecology. CRC, Boca Raton, pp 355–387

Steneck RS (1982) Adaptive trends in the ecology and evolution of crustose coralline algae (Rhodophyta, Corallinaceae) PhD dissertation, John Hopkins University, Baltimore, Maryland

Strittmatter M, Gachon CMM, Küpper FC (2009) Ecology of lower oomycetes. In: S. Kamoun and K. Lamour (eds) "Oomycete genetics and genomics: diversity, plant and animal interactions, and toolbox"

Sussmann A, DeWreede RE (2002) Host specificity of the endophytic sporophyte phase of Acrosiphonia (Codiolales, Chlorophyta) in southern British Columbia. Phycologia 41:169–177

Sussmann A, Mable BK, DeWreede RE, Berbee M (1999) Identification of green algal endophytes as the alternate phase of Acrosiphonia (Codiolales, Chlorophyta) using ITS1 and ITS2 ribosomal DNA sequence data. J Phycol 35:607–614

Suzuki Y, Takabayashi T, Kawaguchi T, Matsunaga K (1998) Isolation of an allelopathic substance from the crustose coralline algae, Lithophyllum spp., and its effect on the brown alga, Laminaria religiosa Miyabe (Phaeophyta). J Exp Mar Biol Ecol 225:69–77

Svensson JR, Nylund GM, Cervin G et al (2009) Japanese macroalgae use chemical warfare to invade Sweden. Phycologia 48(4):s 363

Svirski E, Beer S, Friedlander M (1993) Gracilaria conferta and its epiphytes: (2) interrelationship between the red seaweed and Ulva lactuca. Hydrobiologia 260–261:391–396

Taylor WM (1957) Marine algae of the northeastern coast of North America, Volume 13, Partie 1. University of Michigan Press, New York, p 509

Thomas D, Beltrán J, Flores V, Contreras L, Bollmann E, Correa JA (2009) Laminariocolax sp. (Phaeophyceae) associated with gall developments in Lessonia nigrescens (Phaeophyceae). J Phycol 45:1252–1258

Turner CHC, Evans LV (1977) Physiological studies on the relationship between Ascophyllumn nodosum and Polysiphonia lanosa. New Phytol 79:363–371

Underwood AJ (2000) Experimental ecology of rocky intertidal habitats: what are we learning? J Exp Mar Biol Ecol 250:51–76

Uppalapati SR, Kerwin JL, Fujita Y (2001) Epifluorescence and scanning electron microscopy of host-pathogen interactions between Pythium porphyrae (Peronosporales, Oomycota) and Porphyra yezoensis (Bangiales, Rhodophyta). Bot Mar 44:139–145

Vairappan CS (2006) Seasonal occurrences of epiphytic algae on the commercially cultivated red alga Kappaphycus alvarezii (Solieriaceae, Gigartinales, Rhodophyta). J Appl Phycol 18:611–617

Vairappan CS, Chung CS, Hurtado AQ, Soya FE, Bleicher-Lhonneur G, Critchley A (2008) Distribution and symptoms of epiphyte infection in major carrageenophyte-producing farms. J Appl Phycol 20:477–483

Viejo RM, Åberg P (2003) Temporal and spatial variation in the density of mobile epifauna and grazing damage on the seaweed Ascophyllum nodosum. Mar Biol 142:1229–1241

Wahl M, Hay ME (1995) Associational resistance and shared doom: effects of epibiosis on herbivory. Oecologia 102:329–340

Wahl M, Mark O (1999) The predominantly facultative nature of epibiosis: experimental and observational evidence. Mar Ecol Prog Ser 187:59–66

Wahl M, Hay ME, Enderlein P (1997) Effects of epibiosis on consumer-prey interactions. Hydrobiologia 355:49–59

Weinberger F (2007) Pathogen-induced defense and innate immunity in macroalgae. Biol Bull 213:290–302

Weinberger F, Beltran J, Correa JA, Lion U, Pohnert G, Kumar N, Steinberg P, Kloareg B, Potin P (2007) Spore release in *Acrochaetium* sp. is bacterially controlled. J Phycol 43:235–241

Weinberger F, Friedlander M (2000) Response of Gracilaria conferta (Rhodophyta) to oligoagars results in defense against agar-degrading epiphytes. J Phycol 36:1079–1086

Weinberger F, Potin P (2010) Red algal defenses in the genomics age. In: red algae in the genomic age. Cellular origin, life in extreme habitats and astrobiology, 13(5), Springer, Dordrecht pp 457–477

Weinberger F, Friedlander M, Hoppe HG (1999) Oligoagars elicit a physiological response in Gracilaria conferta (Rhodophyta). J Phycol 35:747–755

Weinberger F, Richard C, Kloareg B, Kashman Y, Hoppe H-G, Friedlander M (2001) Structure activity relationship of oligoagar elicitors toward Gracilaria conferta (Rhodophyta). J Phycol 37:418–426

Weinberger F, Lion U, Delage L, Kloareg B, Potin P, Beltrán J, Flores V, Faugeron S, Correa J, Pohnert G (2011) Up-regulation of lipoxygenase, phospholipase, and oxylipin-production in the induced chemical defense of the red alga Gracilaria chilensis against epiphytes. J Chem Ecol 37:677–686

West JA, Klochkova TA, Kim GH, Loiseaux-de Goer S (2006) Olpidiopsis sp., an oomycete from Madagascar that infects Bostrychia and other red algae: host species susceptibility. Phycol Res 54:72–85

Wheeler GL, Tait K, Taylor A, Brownlee C, Joint I (2006) Acyl-homoserine lactones modulate the settlement rate of zoospores of the marine alga Ulva intestinalis via a novel chemokinetic mechanism. Plant Cell Environ 29:608–618

Wheeler W, NeushulM, Harger B (1981) Development of a coastal marine farm and its associated problems. In: Levring T (ed) Proceedings of the Tenth International Seaweed Symposium. Walter de Gruyter, Berlin, pp 631–636

White EB, Boney AD (1969) Experiments with some endophytic and endozoic Acrochaetium species. J Exp Mar Biol Ecol 3:246–274

White EB, Boney AD (1970) In situ and in vitro studies on some endophytic and endozoic Acrochaetium species. Nova Hedwigia 19:841–881

Whittick A (1983) Spatial and temporal distributions of some dominant epiphytes on the stipes of Laminaria hyperborea (Gunn.) Fosl. (Phaeophyta, Laminariales) in SE Scotland. J Exp Mar Biol Ecol 73:1–10

Wieczorek S, Todd CD (1998) Inhibition and facilitation of settlement of epifaunal marine invertebrate larvae by microbial biofilm cues. Biofouling 12:81–118

Woelkerling WJ (1971) Morphology and taxonomy of the Audouinella complex (Rhodophyta) in southern Australia. Austr J Bot (Suppl. 1):1–91

Wojtaszek P (1997) Oxidative burst: an early plant response to pathogen infection. Biochem J 322:681–692

Wright EP (1879) On a species of *Rhizophydium* parasitic on species of *Ectocarpus*, with notes on the fructification of the *Ectocarpi*. Trans Royal Irish Acad 26:369–379

Yoshida T, Akiyama K (1979) Streblonema (Phaeophyta) infection in the frond of cultivated Undaria (Phaeophyceae). Proc Int Seaweed Symp 9:219–223

Zheng TL, Su JQ, Maskaoui K, Yu ZM, Hu Z, Xu JS, Hong HS (2005) Microbial modulation in the biomass and toxin production of a red-tide causing alga. Mar Pollut Bull 51:1018–25

Zuccarello GC, Debra M, Goff LJ (2004) A phylogenetic study of parasitic genera placed in the family Choreocolacaceae (Rhodophya). J Phycol 40:937–945

Chapter 12
Invasive Marine Seaweeds: Pest or Prize?

Nikos Andreakis and Britta Schaffelke

12.1 Introduction

In recent decades, nonindigenous marine species (NIMS) have become noticeable features of coastal ecosystems worldwide as well as a major environmental problem and a symptom of global change (e.g., Rilov and Crooks 2009; Sellers et al. 2010). The accelerated occurrence of new marine invasions is a function of intensified maritime traffic, coupled with changing environmental conditions in the "recipient" ecosystems. The latter factor may improve the competitiveness of new arrivals in well-established native communities. Transport by ships is the major vector moving marine biota around the globe, mainly via ballast tanks and biofouling on ships' hulls (Flagella et al. 2007). However, the long-range movement of NIMS is also associated with other marine industries such as aquaculture, fisheries, and marine tourism (Hewitt et al. 2009a; Minchin et al. 2009). Global climate change and other regional stressors such as overexploitation of marine resources, marine habitat fragmentation, and coastal pollution are altering the "recipient" ecosystems. This may provide both available space and a suitable ecophysiological window for survival and establishment of NIMS outside the species' native distribution range, or may reveal species' biological traits that are responsible for "invasiveness." Unfortunately, modeled estimates indicate that the rate of biological introductions will continue to increase, in proportion with increasing global trade and economic development (Lin et al. 2007; Meyerson and Mooney 2007). Now and in the future, a major challenge will be to achieve improvements in (1) the early detection of NIMS, (2) the prediction of ecological and socio-economic impacts, and (3) environmental policy and management to prevent new and to control established introductions of NIMS (Lodge et al. 2006; Costello et al. 2007; Rilov and Crooks 2009).

N. Andreakis (✉) • B. Schaffelke
Australian Institute of Marine Science, PMB 3, Townsville 4810, QLD, Australia
e-mail: n.andreakis@aims.gov.au; b.schaffelke@aims.gov.au

C. Wiencke and K. Bischof (eds.), *Seaweed Biology*, Ecological Studies 219,
DOI 10.1007/978-3-642-28451-9_12,

Macroalgae represent approximately 20% of the world's marine introduced species. High-profile seaweed taxa such as *Caulerpa taxifolia* (Vahl) C. Agardh, *Codium fragile* (Suringar) Hariot ssp. *tomentosoides* (Van Goor) Silva, *Sargassum muticum* (Yendo) Fenshold, and *Undaria pinnatifida* (Harvey) Suringar have led to well-documented economic and ecological consequences in their introduction range (Schaffelke et al. 2006; Schaffelke and Hewitt 2007; Williams and Smith 2007). Following establishment, these species are able to directly affect the habitat profile by monopolizing space and acting as ecosystem engineers and can significantly modify the composition of local communities by altering competitive interactions and trophic networks (Wallentinus and Nyberg 2007).

Terms such as "pest" or "invasive" are often used to describe introduced taxa. A "pest" has been defined as an introduced species with a documented negative economic effect (Williamson and Fitter 1996), while an "invasive species" spreads rapidly from its point of introduction and becomes highly abundant (Kolar and Lodge 2001). In this book chapter, we mainly use the term "invasive" because the economic consequences, and hence the pest status, of an introduced species are generally not well understood (reviewed in Schaffelke and Hewitt 2007).

Global climate change plays a pivotal role in the perpetuation and evolution of organisms and populations, for example, by slowly changing ecological niches and triggering evolutionary events at the species and population levels (Howden et al. 2003; Mainka and Howard 2010). However, it is difficult to predict how climate change will influence species introductions in the near future (Occhipinti-Ambrogi 2007). For instance, changes in environmental conditions may turn a species into an invasive species in its introduced or even in its native range by revealing certain biological traits of that species that were previously not expressed and confer invasiveness (Mueller and Hellmann 2008; Willis et al. 2010). Conversely, environmental changes may decrease the suitability of the ecosystem for newly established or even well-adapted invasive marine species (Mainka and Howard 2010). Although population growth and dispersal potential can be modeled, it is nearly impossible to predict whether a newly introduced species will become invasive, or where and when.

12.2 Biological Traits of Invasive Seaweeds

Seaweeds are good candidates for becoming invasive given their potential to survive long-range relocations via maritime transport (Flagella et al. 2007; up to intercontinental scales; see also Chap. 17 by Rothhäusler et al.) and because many species have a set of biological traits that make them highly competitive. Functional traits that are recognized to facilitate successful invasions include:

1. High growth rates and in some cases large individual sizes (most invasive green and brown macroalgae are larger than their native counterparts)
2. Vegetative propagation (e.g., red and green invasive macroalgae such as *Kappaphycus* sp., *Ulva* sp., *Caulerpa* sp.)
3. Spectacular growth strategies (e.g., invasive species such as *Caulerpa* spp. form dense covers of siphonous single-cell individuals several meters long and inhabit

a larger depth range than native species; canopy-forming *Sargassum muticum* reduces native biodiversity in invaded areas)
4. High levels of sexual reproduction and high fecundity (e.g., in the invasive lineage of *Asparagopsis taxiformis*)
5. Parthenogenetic reproduction and broad environmental tolerances (*Codium fragile* ssp. *tomentosoides*)

A quantitative ranking of European introduced and native seaweed species (using comparisons of categories of biological traits such as dispersal capabilities, environmental tolerances, reproductive mode, and size) indicated clearly that introduced species with a number of these biological traits have an increased likelihood of being a successful invader (Nyberg and Wallentinus 2005). In some cases, however, the biological traits identified in important invasive seaweeds are also present in noninvasive co-specific or co-generic relatives (Paula and Eston 1987; Trowbridge 1996; Chapman 1999; Vroom and Smith 2001).

Rapid micro-evolutionary changes are common in invasive species because introduced populations are often subject to founder effects and population bottlenecks, have higher incidence of hybridization (suggested to provide innovative genetic variants), and are exposed to a range of novel selective pressures encountered in the introduction range (Brown and Marshall 1981). Acclimatization, adaptation, and thereafter phenotypic modification may arise in response to new biotic (e.g., competitors, consumers, or parasites) and environmental conditions and the drivers of the functional changes are likely to be genetic (reviewed in Whitney and Gabler 2008). However, the question remains whether well-established gene regulation mechanisms, already present in the genome's background, are simply activated by the local selective pressure or the aforementioned changes are the result of a de novo genomic response.

Published accounts suggest that only a limited number of biologically distinct species within algal orders or just one ESU (evolutionary significant taxonomic unit); among a number of cryptic ESUs within the same morpho-species become suddenly invasive. In plants, the switch to invasiveness has been recently related with differences in ploidy levels suggesting that genetic attributes such as polyploidy and high chromosome counts may be the drivers for this transformation (endangered plants exhibit disproportionally low levels of ploidy and chromosome numbers compared to invasive plant species; Pandit et al. 2011). Similar to hybridization, polyploidy may lead to the production of novel and greater numbers of genetic variants, which increases the probability of a successful invasion. An association between polyploidy and invasiveness has been reported in the red alga *Asparagopsis taxiformis* (Andreakis et al. 2007b, 2009) and should be further explored in other invasive marine algae, given the general propensity for polyploidy in seaweeds (based on nuclear genome size estimates; Kapraun 2005).

It has been debated whether ecosystems exhibiting high species richness are less vulnerable to biological invasions because in theory only a limited, highly specialized, number of empty niches are available for the invader. Furthermore, it has been assumed that polluted "recipient" environments are likely to promote invasive traits in NIMS (Davis et al. 2000; Davis and Pelsor 2001; Dunstan and Johnson 2007; Valentine et al. 2007; Whitney and Gabler 2008). In addition, invasion attempts are more likely to be successful when the environmental conditions in recipient ecosystems are similar to conditions in the ecosystem the invader has evolved. In the latter case, the chances of survival and establishment of invasive seaweeds following introduction are often improved because their natural enemies such as herbivores or parasites are not introduced with them (i.e., the enemy release hypothesis; Cacabelos et al. 2010; Engelen and Henriques 2011). This means that the energy invested into defense in their native environment (e.g., grazing resistance; see Chap. 8 by Iken) can now support growth and biomass production, increasing the competitiveness of the introduced species.

For a successful invasion, biological traits such as broad ecophysiological tolerance and the consequences of differential gene expression are likely to act in synergy under sufficient inoculation pressure and favorable characteristics of the "recipient" ecosystem. The relative importance of each of these traits, however, is poorly understood because they are likely to be species and location-specific and it is impossible to perform controlled experiments at large biogeographic and evolutionary timescales (Schaffelke and Hewitt 2007).

12.2.1 Seaweed Invasions Are Mostly Human-Mediated

The introduction of a seaweed species often comprises a plethora of immigrants of several genetic variants and ecotypes entering the "recipient" environment. Simplistically, settlement and establishment will be feasible if the species' ecophysiological limits for survival match the environmental conditions in the "recipient" environment. While not fully understood for many individual species, the stages of a successful seaweed invasion have been described as: (1) uptake and transport of propagules, (2) survival and release of propagules and establishment of low-density populations, and (3) increased competition with native biota, spread, expansion, and impact of populations of introduced seaweeds (Schaffelke et al. 2006).

The accidental uptake, transport, and release of propagules into the recipient environment are the only stages of the process when human intervention can prevent biological invasions. Marine seaweeds make use of many vectors for dispersal such as specialized floating structures (see Chap. 17 by Rothäusler et al.), gametes or spores for sexual reproduction, fragments of thalli for vegetative propagation, or attachment to long-distance moving organisms (e.g., sea turtles). The dispersal of most invasive seaweeds, however, is associated with human activities, which could be managed (Hewitt et al. 2009a). Biofouling (i.e., the attachment of the seaweed to any part of a vessel or marine equipment) is one of the most common and also oldest mechanisms of human-mediated relocation of seaweeds. The other main vectors are ballast waters, used to stabilize large vessels for long-distance voyages, that transport algal propagules and intentional relocations of aquaculture species (e.g., as fouling on oyster shells or as packing material) consequently responsible for seaweed introductions (Hewitt et al. 2006).

12.3 The Identification of Introduced Seaweeds Is Not Always Easy

Seaweeds have circumnavigated the globe for centuries. Unintentional introductions by humans have occurred since the beginning of naval explorations and this may explain a perceived "cosmopolitan distribution" reported in many "hull-fouling" species (mostly small, ephemeral species that colonize ship's hulls; Godwin 1975; Ruiz and Carlton 2003). A species' geographic distribution can be misinterpreted especially in taxa with few diagnostic morphological characters (e.g., *Ulva lactuca* Linnaeus, *Bryopsis plumosa* (Hudson) C. Agardh) for the following reasons: firstly, most seaweed taxa have been described from a limited number of geographical regions and/or expeditions; taxonomists in the past relied heavily on these already characterized "type species" to identify new taxa. Secondly, although molecular tools are today able to characterize genealogically distinct units, a significant number of taxonomists still rely on the use of morphological characters for a routine identification of species (Sherwood 2007).

The identification of even well-studied introduced seaweeds is not always straightforward and the assessment of an introduction episode is extremely complex. We will discuss here two important reasons. (1) Due to extreme phenotypic plasticity of many seaweed genera, closely related algal species can be morphologically almost identical and therefore difficult to distinguish. Furthermore, cryptic species with separate geographic distributions have been observed within morphologically homogeneous taxa. Cryptic invasion episodes may therefore occur and are likely to lead to (a) misidentified biologically distinct invasive species due to morphological similarities with natives or (b) misidentified genealogically distinct lineages of invasive character found within a single morpho-species complex. Although forensic techniques may provide the necessary tools to achieve a reliable and quick identification of the invader, results are often compromised by the absence of sufficient DNA sequence data for comparisons and species identification (Box 12.1), even in some phylogenetically important groups of seaweeds (e.g., *Ulva* and *Bryopsis*). Furthermore, a species concept to describe the taxonomical units within these groups is often nonexistent. (2) A large number of species are today recognized as "cryptogenic" (i.e., there is no reliable evidence to indicate whether a species is native or introduced, Carlton 1996); because human-mediated relocations of marine biota have occurred since historical times while archeological or paleontological evidence related to the geographical origins of many soft-body marine organisms is largely absent. A number of biogeographical criteria have been proposed to distinguish between native and introduced taxa (Geller et al. 2010) and a species' cryptogenic history can be resolved in some cases where biological invasions have been systematically studied, e.g., *Ulva flexuosa* Wulfen cited as *Enteromorpha lingulata* J Agardh, *Caloglossa leprieurii* (Montagne) G. Martens, *Bostrychia radicans* (Montagne) Montagne in southern Brazil (Neves and Rocha 2008).

The effective management of a biological invasion requires a solid knowledge of (1) which species has been introduced, (2) the source of the introduction, and (3) the

Box 12.1

Two main approaches are used to identify the source of NIMS. The first relies on the assumption that the vector of introduction (e.g., ballast waters of cargo ships along a specific route) is known. Knowledge of the target species' distribution range and the navigation routes can be combined to identify the likely source regions for that species. Vector management can then prevent further introductions (Ruiz and Carlton 2003). Most importantly, biogeoclimatic models can forecast potential zones vulnerable to invasion and predict the range expansion of invasive NIMS by combining information gathered from species distribution, transport vectors, local ecophysiological conditions, and the species-known ecophysiological tolerance ranges (Arrontes 2005).

A second method deploys DNA sequencing markers and geographical information (phylogeographic approach). Specimens are collected, DNA is extracted, and the same (one or more than one) DNA region is sequenced from multiple individuals to infer the genealogical relationships among introduced and putative native populations of the invasive species and reconstruct the spatial history of the invasion (Fig. 12.1). A variety of DNA sequencing markers have been today developed from mitochondrial, plastid, and nuclear genomes (see case studies); however, the successful use of the approach relies on the resolution of the marker and an adequate sampling strategy. Less applied in seaweed invasion biology is the population genetic approach based on allele frequencies of microsatellite markers to assign individuals to a specific population (for reviews of both approaches, see Wares et al. 2005; Muirhead et al. 2008).

extent of the introduced range. This information can be immediately available where algal floras are well described and survey programs for early detection are in place (Anderson 2007). In geographical regions, however, where local seaweed species are poorly explored and historical records such as species lists are limited, a presumably introduced species may either be native or cryptogenic. Visual surveys and identifications based on gross morphology are often sufficient to detect introductions since many NIMS are conspicuous and genetic discontinuities do not always occur within species (O'Doherty and Sherwood 2007; Castelar et al. 2009). However, many algal genera have extreme levels of phenotypic plasticity and species can potentially hybridize with close relatives (e.g., introduced and native populations of *Fucus evanescens* and *F. serratus*; Coyer et al. 2007). For many introduced seaweeds only multidisciplinary approaches, which take into account molecular, morphological, reproductive, and ecophysiological evidence, are able to characterize the introduced taxonomic units. Such approaches provide sound information on the origins and the geographical trajectory of the invasion and are able to predict persistence and expansion potential of introduced populations (Flagella et al. 2010; Pang et al. 2010).

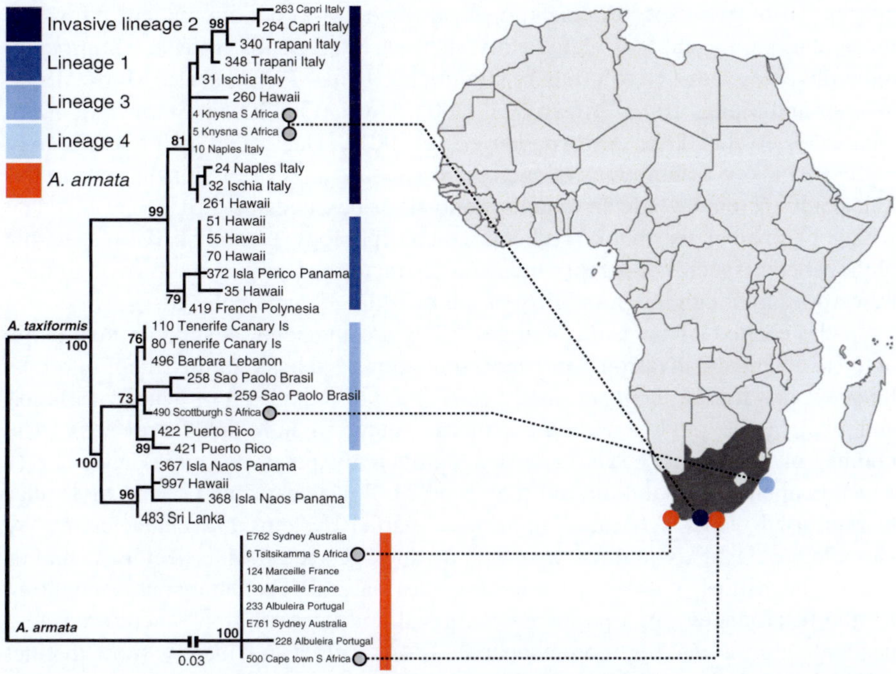

Fig. 12.1 Molecular phylogeographic approach modified from Bolton et al. (2011) to elucidate multiple cryptic introductions of morphologically nearly identical lineages and species (gametophytes and tetrasporophytes) of the red seaweed genus Asparagopsis from specimens collected in South Africa. Both the Indo-Pacific lineage 2 of *A. taxiformis* and *A. armata* constitute notorious invasive species. The highly informative mitochondrial *cox*2–3 intergenic spacer has been deployed in this case. Sequences from previously identified specimens belonging to the five ESUs, collected worldwide and of known origin, are merged with sequences from specimens collected in the study area (*gray circles*) to produce a robust phylogeny and reproduce the two-species genealogical history. From the latter, the identity of each of the target cryptic specimens is revealed. The superimposed geographical information on the phylogram will indicate "which target ESU is where." The level of sequence similarity between sequences of known geographical origin and the target sequences will indicate the source population. Following analyses, unclassified specimens (*gray circles*) are successfully identified (*colored circles*) at the species and lineage levels; lineages and species are finally associated with a geographical location

12.3.1 Relevant Taxonomic Units in Seaweeds Molecular Ecology

Levels and patterns of local marine biodiversity fluctuate, in part due to the exchange of biota among bioregions. In recent years, however, remarkable changes in biodiversity have been revealed at the subspecies level. Molecular techniques in algal systematics are critical to define new relevant units of biological organization for ecological surveys and biodiversity estimates (Box 12.1). These taxonomic units in the vast majority of the cases do not correspond to the traditional species concept (Biological species concept; sensu Mayr 1942) but are recognized within molecular phylogenies as genetically distinct, statistically robust, reciprocally monophyletic clades (Evolutionary Significant Units, ESUs, sensu Moritz 1994; Wattier and Maggs 2001; Provan et al. 2005; Uwai et al. 2006a; Stam et al. 2006; Andreakis et al. 2007a; Verbruggen et al. 2007). The condition of monophyly represents a key assumption for biological classification of organisms; monophyletic clades herein include an organism and all its descendants. ESUs may correlate with geographical distribution and common morphology. Overall, ESUs respect the phylogenetic species concept and are characterized by distinct evolutionary trajectories although the possibility of interbreeding is not excluded.

In the context of invasion biology, ESUs are intended as suitable taxonomic units to elucidate invasions and represent terms such as strains, haplo-groups, lineages, or varieties, conventionally used to express the level of genetic variation and explain the phylogeographic patterns found in introduced seaweeds. The number of uncovered ESUs can vary significantly between studies and largely depends upon the resolution and the speed of "lineage sorting" of the molecular marker used. This is because molecular marker systems are characterized by variable levels of evolutionary speed, distinct genealogical trajectories, and/or uneven intensities of selective pressure; these are all characteristics accounting for the heterogeneity in a phylogenetic signal and its resolution. Therefore, independent information for more than one DNA region and ideally from distinct genomes (nuclear, mitochondrial, and plastid) is necessary to achieve consistent results for delineating ESUs and inferring phylogenies at the sub-species level.

In the last decades, DNA barcoding has been largely used in plant and animal systematics. The method involves sequencing and similarity analysis of a short fragment of DNA from multiple specimens, corresponding to a highly conserved region in the genome such as the subunit one of the mitochondrial gene cytochrome *c* oxidase (COI; Blaxter 2003). However, information coming from a single, extremely short DNA region, compared to the size of a genome, is far from being reliable in modern phylogenetics. This is particularly true in (1) groups of organisms with limited sequence diversity, (2) recently diverged species, (3) hybrids (i.e., offspring of parents belonging to biologically different species, and (4) pseudogenes (i.e., nonfunctional DNA sequences resembling functional genes). The need to employ in a single assay more than one universal biological marker in organism classification can today be satisfied by the increasingly growing number of completely sequenced prokaryotic and eukaryotic genomes. At present, techniques such as DNA microarrays or DNA chips represent the most powerful way to gather as much information as possible from a single individual genome in a single trial in a cost-effective way. In the future, the latter techniques and improvements therein, in combination with advances in data processing and computation power, will permit the inference of phylogenetic relationships among taxonomically robust taxa, ideally at the whole-genome level based on the analysis of their full set of genes (Garzon and Wong 2010).

Fig. 12.2 (**a**) *Sargassum muticum* (photo by Stefan Kraan); (**b**) *Undaria pinnatifida* (Harvey) Suringar (Algaebase, image Ref. 15830, photo by Mike Guiry)

12.3.2 Case Study: Sargassum Muticum

The large brown Japanese seaweed *Sargassum muticum* (Yendo) Fensholt (Fig. 12.2a) is one of the most invasive and widely established algae in Europe and North America (Monteiro et al. 2009; Kraan 2008). Contrary to other brown macroalgae (such as *Undaria pinnatifida* and *Fucus serratus*) no evidence for genetic differentiation has been detected between native Asian and introduced populations in this species. This indicated a number of recent invasion events supported mainly by the transport of Pacific oysters introduced for farming in Canada, UK, and France. Nuclear (ITS2; internal transcribed spacer 2) mitochondrial (*Trn*W-*Trn*I spacer) and plastid (RUBISCO spacer) markers failed to uncover significant levels of DNA polymorphisms in *S. muticum* individuals collected globally, which would be required for the reconstruction of the genetic identity of invasive and native populations in this species. The low levels of genetic variability encountered in the native and introduced range of *S. muticum* together with previous morphological confusions with closely related species prohibited the identification of the source populations and the direction of the introduction episodes along the invasion pathway of this important invasive seaweed (Cheang et al. 2010).

12.3.3 Case Study: **Undaria Pinnatifida**

The brown seaweed *Undaria pinnatifida*, known as the edible *wakame* (Fig. 12.2b), was originally endemic to Japan, Korea, and China where it is industrially cultivated. Firstly observed outside its native distribution range in the Mediterranean Sea (Etang de Thau, French, 1970s) the species has in few years invaded many of the world's oceans from Europe to North America, Australasia, and New Zealand; both by accidental translocation via oyster farming from Japan and intentional introduction for cultivation purposes (Voisin et al. 2005). It was possible to explore and elucidate the invasion history of this species by means of sequence analysis of the mitochondrial *atp*8-*trn*S and *trn*W-*trn*I intergenic spacers, the partial *cox*3 gene and the *tatC-tLeu* gene regions. The latter DNA loci revealed highly differentiated populations, each characterized by low levels of haplotype diversity in the native geographical range of the species (Voisin et al. 2005). On the contrary, where *U. pinnatifida* has been introduced, populations were found to be highly diverse. The so distributed genetic diversity clearly indicated that introduced populations of *U. pinnatifida* are the results of multiple introduction events from more than one native population. Furthermore, the difference in haplotype diversity found in introduced populations sampled from Europe and Australasia indicated differences in the dynamics of the colonization process (Voisin et al. 2005; Uwai et al. 2006a, b).

12.3.4 Case Study: **Asparagopsis Armata**

The red seaweed *Asparagopsis armata* Harvey (Fig. 12.3a) is a genetically homogeneous species originally described from Southern Australia, Tasmania, and New Zealand and first reported in the northern hemisphere from Algeria in 1925 (Horridge 1951). Molecular phylogenies inferred from the D1, D2, and D3 hyper-variable domains of the nuclear large subunit rDNA gene the plastid RUBISCO spacer and the mitochondrial cox2–3 intergenic spacer indicated just a single plastid haplotype distributed worldwide and a limited number of panmictic nuclear and mitochondrial haplotypes in presumably native and introduced populations. *Asparagopsis armata* exhibits great dispersal potential due to a free-floating tetrasporophyte in its life cycle (see below) and propagation via fragmentation and attachment to other floating structures and has always been reported as a recently introduced, potentially invasive, species from the Mediterranean Sea and western European coasts. Given the complete absence of phylogeographic structuring on global and local scales and the inadequate phylogenetic signal of nuclear, mitochondrial, and plastid markers, only general assumptions on the origins of introduced populations and the direction of the spread can be made. Furthermore, significant taxonomic confusion surrounds this species and has been propagated since its first description. This is because no diagnostic morphological characters are known to differentiate between tetrasporophytes (the "*Falkenbergia*" stage, previously described as separate species)

Fig. 12.3 (**a**) *Asparagopsis armata* on *Ulva* (photo by Stefan Kraan; (**b**) unidentified lineage of *Asparagopsis taxiformis* (photo by Mark JA Vermeij)

of *Asparagopsis* species. Although it is easy to identify the macroscopic *A. armata* gametophytes in the field, the identification of cryptic Falkenbergia isolates and their assignment to *A. armata* populations is only feasible with molecular tools or by inference when *Asparagopsis* gametophytes and *Falkenbergia* occur at the same sites.

12.3.5 Case Study: Asparagopsis Taxiformis

The warm temperate to tropical red seaweed *Asparagopsis taxiformis* (Delile) Trevisan de Saint-Léon, 1845, (Fig. 12.3b) consists of four genetically distinct yet morphologically virtually identical mitochondrial lineages, detectable only by molecular tools (the nuclear D1, D2, and D3 hyper-variable domains of the large subunit rDNA gene, the plastid RUBISCO spacer, and the mitochondrial cox2–3 intergenic spacer). One of these, lineage two, is of Indo-Pacific origin and invaded the Mediterranean basin from the Suez Canal (Ní Chualáin et al. 2004; Andreakis et al. 2004, 2007a, b). Phylogeographic patterns inferred from individuals of this species-complex collected globally were suitable to delineate cryptic taxa, infer lineage-specific distribution patterns, and elucidate single or multiple introduction events worldwide. Most importantly, the phylogeographic approach was useful to distinguish between native and invasive cryptic taxa where more than one lineage was found in sympatry. For instance, lineages 1, 2, and 4 co-occur in the Hawaiian archipelago [lineage 2 is considered native and lineage 4 introduced to the islands as recently as in 1991 (Sherwood 2008)]. Lineages 1 and 4 co-occur sympatrically within hundreds of meters in the Pacific side of the Isthmus of Panama and they may represent multiple introduction events. Lineages 2 and 3, the latter of Atlantic distribution, are found in South Africa and are considered introduced and invasive, respectively (Bolton et al. 2011).

Individuals with a *A. taxiformis* morphology collected in unrelated geographic locations have in the past been described as separate species. This observation, together with the multiple ESUs found in *A. taxiformis*, strongly suggests that the taxonomic status of the species sensu stricto needs to be revisited. For instance, one of these lineages may correspond to *A. svedelii* W.R. Taylor (1945), an endemic taxon of the Galápagos Islands, considered synonymous to *A. taxiformis* (Taylor 1945), while cryptic lineage 2 may correspond to *A. sanfordiana* Harvey (Setchell and Gardner 1924), another species considered to be a synonym of *A. taxiformis*.

12.3.6 Case Study: **Codium Fragile *ssp.* Tomentosoides**

A single strain of the green alga *Codium fragile* sensu stricto known as *Codium fragile* ssp. *tomentosoides* (Fig. 12.4a) is recognized as highly invasive. DNA sequence analysis based on plastid microsatellites and the plastid *psb*J-*psb*L, *rpl*16-*Rps*3, and *trn*G(UCC)-5 S DNA regions of individuals collected throughout the strain's known native and introduced distribution range indicated extremely low levels of genetic variation with only four distinct haplotypes encountered in Japan where the strain occurs naturally. Despite the limited levels of genetic variability, the phylogenetic signal was sufficient to uncover at least two separate introductions in the biogeographic history of *C. fragile* ssp. *tomentosoides*: two of these haplotypes were highly structured geographically and were found in introduced populations, one exclusively in the Mediterranean Sea, and the other exclusively in Northwest Atlantic, Northern European, and South Pacific populations (Provan et al. 2005, 2008).

12.3.7 Case Study: **Caulerpa *spp.***

The green algal genus *Caulerpa* comprises morpho-species complexes, species that may become invasive under certain conditions and ESUs known from the literature as "invasive strains" (e.g., described in *C. taxifolia* sensu stricto) or "invasive varieties" [e.g., described in *C. racemosa* (Forsskål) J. Agardh sensu stricto (Stam et al. 2006)]. For instance, a potentially invasive species, *Caulerpa prolifera* (Forsskål) J.V. Lamouroux, is widely distributed in tropical and subtropical seas and has been significantly altering the structure and functioning in its introduced

Fig. 12.4 (**a**) *Codium fragile* subsp. tomentosoides (van Goor) P.C.Silva (Algaebase; image Ref. 17373, photo by Mike Guiry); (**b**) *Caulerpa taxifolia* (photo by John Huisman); (**c**) *C. racemosa* var. cylindracea (Sonder) Verlaque, Huisman and Boudouresque (photo by John Huisman)

habitats (e.g., Gab-Alla 2007). The widespread tropical/subtropical seaweed *Caulerpa taxifolia* (Fig. 12.4b), native to the Atlantic, Indian, and Pacific Oceans, was the first seaweed invasion to draw widespread public attention after its introduction to the Mediterranean Sea in the 1980s, its fast spread, and well-documented negative ecological impacts (reviewed in Schaffelke and Hewitt 2007).

Nuclear molecular phylogenies inferred from sequences of the rDNA-ITS1 + 2 region (ribosomal DNA—internal transcribed spacer 1 and 2) from hundreds of *Caulerpa* individuals collected worldwide were used to delineate species, "varieties," and "strains" of problematic morphological identification. The global phylogenetic reconstruction contributed to the identification of the invasive "aquarium strain" of Australian origin of *C. taxifolia*, which has been introduced at various locations worldwide (Stam et al. 2006). *Caulerpa taxifolia* sensu stricto cannot be distinguished morphologically from this invasive ESU which, following accidental introductions, has successfully established in temperate areas such as the United States, Mediterranean Sea, and southern Australia. Eradication attempts in Australia and the United States (the latter seems to have been successful) and an international ban of this species in the aquarium trade were the initial management responses (Anderson 2007).

The same nuclear marker has been successfully used to uncover the now well-known invasive variety of *Caulerpa racemosa* (*C. racemosa* var. cylindracea (Sonder) Verlaque, Huisman and Boudouresque; Fig. 12.4c) against two other morphologically similar, and apparently noninvasive, sister varieties within the introduced range of the *C. racemosa* complex (*C. racemosa* var. *turbinata–uvifera* (C. Agardh) J. Agardh, *C. racemosa* var. *lamourouxii* (Turner) Weber-van Bosse f. *requienii* (Montagne) Weber-van Bosse; varieties reviewed by Klein and Verlaque 2008). The approach was particularly useful in the Mediterranean basin where other *C. racemosa* varieties have been erroneously misidentified as the invasive variety (Verlaque et al. 2003, Yeh and Chen 2004; Nuber et al. 2007). Although the invasive lineage represents a dominant algal component in the Mediterranean basin, its identity has been definitely confirmed by combining morphological and molecular data in only five of the 13 Mediterranean countries in which *C. racemosa* sensu stricto has been reported so far (France, Italy, Greece, Croatia, and Cyprus; Klein and Verlaque 2008). This ESU has been introduced into the basin from south-western Australia. Due to its well-documented negative ecological impact, this invasion is considered as one of the most serious biological invasions in the history of species introduced into the basin (Klein and Verlaque 2008).

12.3.8 Assessing Seaweed Introductions

In recent years, several conspicuous seaweed invasions have been detected. Our case studies illustrate that introduced seaweeds may be either genetically homogeneous or species consisting of multiple cryptic lineages.

Populations of taxa in the former category are characterized by weak phylogenetic signals and insufficient phylogeographic structuring between the native and introduced geographical range of the invasive species. Such introduced species are only easy to detect based on morphology if detailed floristic records exist in a particular region. It is however impossible to identify the source populations and the subsequent direction of the invasion based on either morphological identifications or genetic approaches.

Morphologically cryptic species often occur within so-called widely distributed taxa characterized by large geographic disjunctions and apparently low levels of connectivity among populations. The morphological delineation of these ESUs is often impossible even for the expert taxonomist; most importantly, it is difficult to distinguish between a recent introduction and the tendency of the species to disperse naturally (the latter may have occurred in timescales of hundreds to thousands of years before present). However, a molecular phylogeographic approach may represent a sufficiently robust strategy to delineate the cryptic taxa and, most importantly, provide information on the origins of the invader and the direction of the invasion process (see Box 12.1; Voisin et al. 2005; Andreakis et al. 2007a, b; Sherwood 2008).

When cryptic taxa remain unrecognized, significant evolutionary and ecological changes of the local marine biodiversity at the subspecies level will remain unknown. Most importantly, cryptic invasions generally escape the critical management mechanisms of early detection, control, and future prevention. Conversely, if cryptic taxa are misidentified, the impacts of invasions can be erroneously overestimated due to the co-occurrence of morphologically similar invasive and noninvasive cryptic ESUs belonging to the same species-complex.

The increased recognition of cryptic ESUs has significantly changed our view of distribution, abundance, ecophysiology, and conservation status of seaweed species. In the light of the high rate with which new ESUs are discovered, firstly, the taxonomic status of several so-called cosmopolitan species has to be reconsidered. Secondly, the newly emerging patterns can be used to formulate hypotheses on diversification rates and to discover biodiversity hotspots. Thirdly, and most importantly, results are essential to develop accurate assumptions on extinction risks and population sizes. Marine science in the post-genomic period with hundreds of genomes sequenced so far requires a convincing yet suitable taxonomic unit. This is of crucial importance to corroborate ecological, biological, and evolutionary interpretations, but also for practical applications such as management of Marine Protected Areas and active bio-invasion control programs.

12.4 Are Introduced Species a Pest or Prize? A Crab's Tale

The Chinese mitten crab is native to the rivers and estuaries of the Yellow Sea in Korea and China, and has been introduced to many regions of the world. The two quotations below illustrate the *conundrum* whether this crab species is a pest or a prize. Counting the number of scientific publications on this species (ISI web of Knowledge, viewed 05 April 2011) suggests the latter, as most publications report on aspects of economic use, e.g., reproductive performance and breeding optimization.

"*The Chinese mitten crab (*Eriocheir sinensis*) is considered one of the world's top 100 invaders by the Invasive Species Specialist Group (ISSG), part of the Species Survival Commission of The World Conservation Union (IUCN). Populations of* E. sinensis *have recently become established on the west coast of the U.S. and now pose a potential threat to the native fishes and invertebrates and to freshwater and brackish estuarine communities*" (Dittel and Epifanio 2009).

"*The Chinese mitten crab (Eriocheir sinensis) is one of the most important cultivated aquatic species in northern China. Significant advances have been made in the crab farming during last years due to its "high market value with declining yields. (...) The annual production of crabs has reached more than 400,000 t valued more than 20 billion RMB in 2005 in China*" (Li et al. 2007).

The food on our tables today comes mostly from a small number of widely cultivated plant and domesticated animal species. Most of them are historical intentional introductions. For example, the earliest archeological records of agriculture in India 2400–1750 BC show crop species of west Asian origin (Hutchinson 1976), while the transport of crop species between Europe, Africa, and Asia peaked with the establishment of direct intercontinental sea trade in the 1400s (National Research Council 1996). There is no evidence that the main crops, such as cereals, oil seeds, fruits, and vegetables, were invasive in their native range. However, over the last 10,000 years modern cultivars (from "*culti*vated *var*iety") have been continuously selected for maximum performance in a range of environments. In addition, modern farming practices such as fertilization, irrigation, crop rotation, the use of pesticides, and genetically assisted breeding have provided extraordinary growth rates and production of high biomass. Interestingly, many invasive seaweed species or ESUs fit the profile of successful agricultural cultivars because they exhibit evolutionary traits for e.g., fast growth, high reproductive capacity, grazing resistance, and wide ecophysiological tolerances to persist in different environments.

12.4.1 Commercially Used Invasive Seaweeds

Seaweeds have been highly valued food resources for thousands of years, especially in Asia. Today, seaweed harvesting and mariculture is an actively growing global industry with a currently estimated total annual value of approximately six billion USD (McHugh 2003). Seaweeds are used mainly as food products for direct human consumption, sources of bulk chemicals such as hydrocolloids and other higher value biochemical compounds, fertilizers, and animal feed additives (see Chap. 22 by Buchholz et al.). The farming of seaweeds has expanded rapidly as demand has outstripped the supply available from natural resources (McHugh 2003).

Many cultivated seaweed species, harvested in their native range, have been intentionally or unintentionally introduced (reviewed in Schaffelke and Hewitt 2007) and are often highly valued for their invasive traits, especially for fast growth rates and biomass accumulation, and often low levels of epiphytism and resistance to herbivores. The commercialization of invasive seaweeds and the exploitation of species-specific invasive traits are often considered to be positive for the economic development of the regional community (Pickering et al. 2007). However, this has to be balanced with the potential negative side effects on the ecological integrity of the "recipient" environments due to accidental release or escape of invasive species, which are difficult to manage. The following sections present brief summaries of the most important commercially used seaweed species, which are also high-profile invaders.

12.4.2 Gracilaria salicornia

Gracilaria species represent well-recognized sources of linear galactan polysaccharide agar, widely used in industry for its excellent gelling or thickening abilities (reviewed in Mehta et al. 2010). Species of the genus *Gracilaria* such as *Gracilaria salicornia* (C. Agardh) Dawson and *G. parvispora* Abbott, a highly valuable alga in the Hawaiian markets, occur widely throughout the tropical Indian and Pacific Oceans (Ryder et al. 2004). Furthermore, *G. debilis* (Forsskål) Børgesen and *G. salicornia* occur naturally at the west coast of India where they represent valuable natural sources for superior quality agars (Oza and Zaidi 2001; Mehta et al. 2010). Several *Gracilaria* species from various sources were intentionally introduced to the Hawaiian Islands for aquaculture development (Russell 1992). *Gracilaria salicornia* has become a successful invader. After escaping from sea farms, *G. salicornia* has invaded surrounding coral reef flats, either as unattached drift or attached populations outcompeting corals, propagated mainly via fragmentation (Smith et al. 2004; Conklin and Smith 2005). *Gracilaria salicornia* displays a variety of biological traits that facilitate high yield cultivation but also make this a potentially invasive species: e.g., high growth rates that can be sustained in variable light regimes, the ability for vegetative dispersal, desiccation tolerance, remarkable tolerance to a wide range of temperatures, salinities, chemical treatments (commercial-grade herbicides and algaecides), and resistance to herbivores (Rodgers and Cox 1999; Smith et al. 2004; Nelson et al. 2009).

12.4.3 Eucheuma *and* Kappaphycus *spp.*

Wild and farmed strains of the carrageenophytes *Eucheuma denticulatum* (Burman) Collins et Harvey, *Kappaphycus alvarezii* (Doty) Doty and *Kappaphycus striatum* (Schmitz) *Doty ex Silva* have been intentionally introduced in more than 20 countries for mariculture purposes (Bulboa and de Paula 2005; Hayashi et al. 2007). These species represent the world's most important kappa-carrageenan producing red algae (Ask and Azanza 2002), a hydrocolloid that is widely used in industry as a gelling and thickening agent. All, however, have been introduced to Kane'ohe Bay in Hawai'i in 1974 for growth experiments and since then accidentally escaped and established highly abundant populations on Hawaii's coral reefs (Conklin and Smith 2005). However, most introductions for cultivation have occurred in the developing world and adverse impacts on the surrounding environment are not well studied (Zemke-White 2004; Bindu and Levine 2011).

Invasive traits that make these species ideal for farming are easy propagation and dispersal by fragmentation and very high growth rate (doubling in size every 15–30 days). The improvement of growth performance in these species is an area of active research (Luhan and Sollesta 2010).

12.4.4 Asparagopsis *Species*

Gametophytes of *Asparagopsis armata* and *A. taxiformis* represent promising sources of pharmaceutical agents of remarkable anti-protozoan, bactericidal, and fungicidal activities (McConnell and Fenical 1977; Woolard et al. 1979; Paul et al. 2006; Salvador et al. 2007; Genovese et al. 2009). In addition, *Asparagopsis* species produce valuable natural secondary metabolites such as compounds with considerable antimicrobial activity against *vibrio harveyi* strains, useful in cosmetics industry and animal aquaculture (Salvador et al. 2007; Manilal et al. 2009); compounds of cytotoxic activity against cancer cells (Zubia et al. 2009) and as preservatives in industrial applications (Kraan and Barrington 2005). Furthermore, "*Falkenbergia*" stages have been successfully cultivated in southern Portugal and can be used as continuous, rapidly growing bio-filter systems for the effluent of commercial fish farms (Schuenhoff et al. 2006; Figueroa et al. 2008; Mata et al. 2010). Introduced in Ireland in 1939, and following population decline due to inadequate temperature for sexual reproduction, *Asparagopsis armata* is now commercially cultivated. The persistence of this species along the Irish West coast seems to be due to continued replenishment from a single aquaculture operation (Kraan and Barrington 2005).

Both *A. taxiformis* and *A. armata* exhibit a number of invasive traits, including vegetative propagation of gametophytes and tetrasporophytes, frequent sexual reproduction of tetrasporophytes, and a suite of halogenated secondary metabolites that increase resistance to herbivory (Boudouresque and Verlaque 2002; Paul et al. 2006; Williams and Smith 2007). In addition, polyploidy has been associated with the extraordinary invasive success of the Indo-Pacific Mediterranean linage 2 of *A. taxiformis* (Andreakis et al. 2009). Furthermore, "*Falkenbergia*" is particularly well suited to human-mediated transport (ballast water) and capable of surviving for long periods as free-floating or loosely attached filaments transported on ships, animals, and other materials in aquaculture (Flagella et al. 2007).

12.4.5 Caulerpa *Species*

Among three genetically distinct ESUs of the *Caulerpa racemosa* species-complex occurring in the Mediterranean Sea, the so-called *C. racemosa* var. *cylindracea* (Sonder) Verlaque, Huisman and Boudouresque, is highly invasive. *Caulerpa* species are greatly valued for human consumption due to their high vitamin and protein content ("sea grapes", *C. racemosa* var *laetevirens*, *C. racemosa*, *C. geminata*, *C. hodgkinsoniae*) or pleasant taste ("green caviar", *C. lentillifera*) and have been identified in several countries as high-value candidates for cultivation and commercial production (Matanjun et al. 2009). *Caulerpa* species also offer promising potential for bioremediation of nutrient-enriched effluent from intensive tropical aquaculture systems because of their high nutrient uptake rates (linked to high growth rates) and the fact that many *Caulerpa* species are native and sufficiently abundant in regions where aquaculture operations are expanding (Paul and de Nys 2008). The downside of the easy cultivation is that several potentially invasive *Caulerpa* species have been commercialized and can be easily purchased from aquarium shops or ordered online. The aquarium industry has enormously facilitated the transportation of these species to distant geographic locations and hence the potential for introductions into natural habitats with serious ecological consequence (Zaleski and Murray 2006; Stam et al. 2006).

The main invasive traits in *Caulerpa* species are high growth rates, propagation via fragmentation (i.e., 1 cm small fragments of *C. taxifolia* are able to give rise to viable single siphonous-cell three-dimensional structures up to 5 m long), broad environmental tolerances, survival out of the water in humid conditions for several days, and in some cases, such as in *Caulerpa racemosa* and *C. taxifolia,* production of secondary metabolites and allelopathic compounds that act as herbivore deterrents and/or inhibit other floral components during space competition (e.g., Meinesz 1999; Vroom and Smith 2001; Gollan and Wright 2006; Wright and Davis 2006; Klein and Verlaque 2008; Raniello et al. 2007).

12.4.6 Ulva *Species*

The cosmopolitan genus *Ulva* Linnaeus, commonly known as the "sea lettuce," is represented by species distributed in all oceans and estuaries of the world, 17 of which have been recorded in the Mediterranean Sea (Guiry and Guiry 2008). Several species in this genus are commercially cultivated for human consumption in Asian countries or as a component in integrated multi-trophic aquaculture systems, e.g., for bioremediation of effluent and as a feed crop for abalone aquaculture in South Africa (Bolton et al. 2009). Furthermore, due to their higher biomass yield and the higher photosynthetic activity compared to the terrestrial crops, *Ulva* spp. are considered as a promising aquatic crop for biofuel production (Bruhn et al. 2011).

Some Ulva species are problematic to identify morphologically due to the extreme phenotypic plasticity in different environments (Hayden and Waaland 2004). Among them, *Ulva ohnoi* Hiraoka et Shimada, a green-tide-forming alga described from Japan, has been recently found in ballast water tanks of container ships in the Mediterranean Sea and may represent a newly introduced species in the basin (Flagella et al. 2010). *Ulva ohnoi* together with *Ulva prolifera* (Müller) J. Agardh were two of the species responsible for the world largest drifting macroalgal bloom in 2008 estimated as approx. one million tons of harvestable biomass (Qingdao, China, before the Olympic sailing competition; Leliaert et al. 2009; Pang et al. 2010).

Ulva species are well known for their broad environmental tolerance, high growth rate, and photosynthetic activity. They represent early successional, opportunistic species, often occurring in degraded or polluted (high nitrogen) environments, mainly due to their simple morphological features and the high reproductive success supported by the ability of gametes and zoospores to photosynthesize (Beach et al. 1995). Furthermore, *Ulva* species can survive up to 10 months in darkness and are therefore easily transported via ballast water tanks (Santelices et al. 2002).

12.4.7 Undaria pinnatifida

The brown seaweed *Undaria pinnatifida*, commonly known as "wakame," is one of the most commonly eaten seaweeds in Japan. Large-scale commercial cultivation throughout Asia is primarily for human consumption; however, recently discovered antiviral compounds offer prospects for further commercial uses (Hemmingson et al. 2006). Invasive biological traits of *U. pinnatifida* are the very high growth rate (1–2 cm per day), high reproductive rate, long-range dispersal by floating reproductive thalli, the capacity for establishing populations in a variety of habitats and surfaces including artificial substrata, and the ability of gametophytes to colonize ship hulls (e.g., Casas et al. 2004; Thornber et al. 2004; Schaffelke et al. 2005; Russell et al. 2008). Due to the high biomass production and canopy-forming large blades, *U. pinnatifida* often outcompetes other algae by shading and space monopolization; however, ecosystem effects seem to vary greatly between introduced locations (reviewed in Schaffelke and Hewitt 2007).

12.4.8 Codium Fragile *Subspecies* Tomentosoides

Codium fragile (Suringar) Hariot sensu stricto is an edible green alga, widely consumed in Korea, China, and Japan. Cultivated in Korea, using seed stock produced from regeneration of isolated utricles and medullary filaments, the species is preferred to seaweeds such as *Porphyra* and *Undaria* because of its higher market value (Hwang et al. 2008). The protein and amino acid composition in this seaweed is similar to cereals like wheat, corn, and oats; its high content of minerals, dietary fiber, and low lipid content make the species a good candidate as a food additive in human and animal nutrition. In addition, bioactive compounds in *Codium fragile* ssp. *tomentosoides*, including anticancer agents and antibiotics are currently under investigation (Ortiz et al. 2009).

However, *Codium fragile* ssp. *tomentosoides* is one of the most invasive seaweeds, with significant ecological effects ranging from direct competition and replacement of native species to major changes at the community level (reviewed in Schaffelke and Hewitt 2007). In Chile, the invasive *Codium* strain competes with the native agarophyte *Gracilaria chilensis* causing potentially negative effects to the coastal economy by threatening the persistence of *G. chilensis* farms (Neill et al. 2006).

Several biological traits, related to life history and ecology of this ESU, are likely to be a reason for its invasive success. Invasive *Codium* benefits from both sexual and parthenogenetic proliferation as well vegetative propagation facilitated via thalli fragmentation and dispersal. Furthermore, colonization and the persistence of populations are facilitated by broad physiological tolerances to temperature and salinity regimes (Bégin and Scheibling 2003).

12.5 Conclusions

The relocation of marine species with invasive potential represents extraordinary challenges for marine ecosystems. In the future, the rate and range of introductions is likely to increase with increasing volume and speed of maritime traffic and with unprecedented changes in "recipient" environments by global climate change and local habitat degradation. Anticipating and controlling potential invasions is easier and more cost-effective than investing in posteriori management such as control and eradication efforts (e.g., Hewitt et al. 2009b).

The invasive seaweed species discussed in this chapter are recognized priority pest species with a high potential for negative ecological and socio-economic impacts (Doelle et al. 2007). However, where these seaweeds are native, many of them have long been exploited as valuable natural resources. The low capital costs of seaweed farming, good economic returns, and high demand for seaweed products in the world's market have led to intentional introductions of potentially invasive, high yield taxa like *Kappaphycus*, *Gracilaria*, and *Undaria* spp. and the expansion of farming areas, especially in developing and low-wage countries (Pickering et al. 2007). The incentives to introduce high-yield taxa are significant and the risks of intentional seaweed introductions have not yet been evaluated with cost-benefit analyses that include ecological and socio-economic consequences. The seaweed industry in the new millennium is predicted to grow with increasing demand for traditional seaweed products and for new applications of seaweeds, e.g., for the production of biofuels and bioactive compounds. Selective breeding and genetic modification to produce better cultivars for these applications is likely to add another dimension of uncertainty. New "super strains" with high invasive potential may be introduced to new locations with unprecedented consequences or may interbreed with con-specific native populations and increase their invasiveness.

The potential for controlled harvest has been suggested as a positive consequence of seaweed introductions, especially in areas where invasive seaweeds have established in high abundance and eradication or ongoing control is not economically feasible. However, stringent harvest regulations must be developed and applied to prevent further translocation and spread of invasive seaweeds into areas previously unaffected (see Schaffelke and Hewitt 2007).

In application of the precautionary principle of environmental management, preference should be given to the development of sustainable commercial seaweed culture in regions where a species (sensu stricto) occurs naturally. Valuable marine cultivars could be certified similar to D.O.P (protected denomination of origin) and D.O.C (controlled origin denomination) standards of agricultural crops, which would confirm that a seaweed has been produced in a specific geographical location. In addition, marine stewardship certifications would attest that a product is produced under certain environmental management and product quality standards. However, for this to happen seaweed products must attain a much higher commercial value than they currently have. In the meantime, a number of international agreements and voluntary codes of practice have been formulated to encourage national governments to restrict and control the intentional import and use of NIMS in aquaculture (reviewed in Hewitt et al. 2006). Also, recommendations for the assessment of proposals for intentional introductions are available to assist industries and regulators, even though quantitative data on ecological and socio-economic impacts of seaweed introductions are scarce. Research describing and assessing past and present NIMS invasions is vital to better predict future invaders, invaded regions, and the associated impacts and to support management and conservation decisions. At present, many questions around seaweed invasions and marine invasions, in general, remain to be answered (see e.g., Johnson 2007).

References

Anderson LWJ (2007) Control of invasive seaweeds. Bot Mar 50:418–437

Andreakis N, Procaccini G, Kooistra WHCF (2004) *Asparagopsis taxiformis* and *Asparagopsis armata* (Bonnemaisoniales, Rhodophyta): genetic and morphological identification of Mediterranean populations. Eur J Phycol 39:273–283

Andreakis N, Procaccini G, Maggs CA, Kooistra WHCF (2007a) Phylogeography of the invasive seaweed *Asparagopsis* (Bonnemaisoniales, Rhodophyta) reveals cryptic diversity. Mol Ecol 16:2285–2299

Andreakis N, Kooistra WHCF, Procaccini G (2007b) Microsatellite markers in an invasive strain of *Asparagopsis taxiformis* (Bonnemaisoniales, Rhodophyta): insights in ploidy levels and sexual reproduction. Gene 406:144–151

Andreakis N, Kooistra WHCF, Procaccini G (2009) High genetic diversity and connectivity in the polyploid invasive seaweed *Asparagopsis taxiformis* (Bonnemaisoniales) in the Mediterranean, explored with microsatellite alleles and multilocus genotypes. Mol Ecol 18:212–226

Arrontes J (2005) A model for range expansion of coastal algal species with different dispersal strategies: the case of Fucus serratus in northern Spain. Mar Ecol Prog Ser 295:57–68

Ask IE, Azanza RV (2002) Advances in cultivation technology of commercial eucheumatoid species: a review with suggestions for future research. Aquaculture 206:257–77

Beach KS, Smith CM, Michael T, Shin HW (1995) Photosynthesis in reproductive unicells of *Ulva fasciata* and *Enteromorpha flexuosa:* implications for ecological success. Mar Ecol Prog Ser 125:229–237

Bégin C, Scheibling RE (2003) Growth and Survival of the Invasive Green Alga *Codium fragile* ssp. *tomentosoides* in Tide Pools on a Rocky Shore in Nova Scotia. Bot Mar 46:404–412

Bindu MS, Levine IA (2011) The commercial red seaweed *Kappaphycus alvarezii*–an overview on farming and environment. J Appl Pcycol 23:789–796

Blaxter M (2003) Counting angels with DNA. Nature 421:122–124

Bolton JJ, Robertson-Andersson DV, Shuuluka D, Kandjengo L (2009) Growing Ulva (Chlorophyta) in integrated systems as a commercial crop for abalone feed in South Africa: a SWOT analysis. J Appl Phycol 21:575–583

Bolton JJ, Andreakis N, Anderson RJ (2011) Molecular evidence for three separate cryptic introductions of the red seaweed *Asparagopsis* (Bonnemaisoniales, Rhodophyta) in South Africa. Afr J Mar Sci. doi:10.2989/1814232X.2011.600339

Boudouresque CF, Verlaque M (2002) Biological pollution in the Mediterranean Sea: invasive versus introduced macrophytes. Mar Poll Bull 44:32–38

Brown AHD, Marshall DR (1981) Evolutionary changes accompanying colonization in plants. Evolution today, proceeding of the second international congress of systematic and evolutionary biology. In: Scudder GGT and Reveal JL (ed.). Carnegie-Mellon University, Pittsburgh, PA, pp 351–363

Bruhn A, Dahl J, Nielsen HB, Nikolaisen L, Rasmussen MB, Markager S, Olesen B, Arias C, Jensen PD (2011) Bioenergy potential of Ulva lactuca: Biomass yield, methane production and combustion. Bioresour Technol 102:2595–2604

Bulboa CR, de Paula EJ (2005) Introduction of non-native species of *Kappaphycus* (Rhodophyta, Gigartinales) in subtropical waters: Comparative analysis of growth rates of *Kappaphycus alvarezii* and *Kappaphycus striatum in vitro* and in the sea in south-eastern Brazil. Phycol Res 53:183–188

Cacabelos E, Olabarria C, Incera M, Troncoso JS (2010) Do grazers prefer invasive seaweeds? J Exp Mar Biol Ecol 393:182–187

Carlton JT (1996) Biological invasions and cryptogenic species. Ecology 77:1653–1655

Casas G, Scrosati R, Piriz ML (2004) The invasive kelp *Undaria pinnatifida* (Phaeophyceae, Laminariales) reduces native seaweed biodiversity in Nuevo Gulf (Patagonia, Argentina). Biol Invasions 6:411–416

Castelar B, Perpetuo Reis R, Moura AL, Kirk R (2009) Invasive potential of *Kappaphycus alvarezii* off the south coast of Rio de Janeiro state, Brazil: a contribution to environmental secure cultivation in the tropics. Bot Mar 52:283–289

Chapman AS (1999) From introduced species to invader: what determines variation in the success of *Codium fragile* ssp. tomentosoides (Chlorophyta) in the North Atlantic Ocean? Helgolander Mee 52:277–289

Cheang CC, Chu KH, Fujita D, Yoshida G, Hiraoka M, Critchley A, Choi HG, Duan D, Serisawa Y, Angr Jr PO (2010) Low genetic variability of sargassum muticum (Phaeophyceae) revealed by a global analysis of native and introduced populations. J Phycol 46:1063–1074

Conklin EJ, Smith JE (2005) Abundance and spread of the invasive red algae, *Kappaphycus* spp., in Kane'ohe Bay, Hawai'i and an experimental assessment of management options. Biol Invasions 7:1029–1039

Costello C, Drake JM, Lodge DM (2007) Evaluating an invasive species policy: Ballast water exchange in the great lakes. Ecol Appl 17:655–662

Coyer JA, Hoarau G, Stam WT, Olsen JL (2007) Hybridization and introgression in a mixed population of the intertidal seaweeds *Fucus evanescens* and *F. serratus*. J Evol Biol 20: 2322–2333

Davis MA, Pelsor M (2001) Experimental support for a resource-based mechanistic model of invisibility. Ecol Lett 4:421–428

Davis MA, Grime JP, Thompson K (2000) Fluctuating resources in plant communities: a general theory of invasibility. J Ecol 88:528–534

Dittel AI, Epifanio CE (2009) Invasion biology of the Chinese mitten crab *Eriochier sinensis*: a brief review. J Exp Mar Biol Ecol 374:79–92

Doelle M, McConnell ML, VanderZwaag DL (2007) Invasive seaweeds: global and regional law and policy responses. Bot Mar 50:438–450

Dunstan PK, Johnson CR (2007) Mechanisms of invasions: can the recipient community influence invasion rates? Bot Mar 50:361–372

Engelen AH, Henriques N (2011) Mesograzers prefer mostly native seaweeds over the invasive brown seaweed *Sargassum muticum*. Hydrobiologia 669:157–165

Figueroa FL, Bueno A, Korbee N, Santos R, Mata L, Schuenhoff A (2008) Accumulation of mycosporine-like amino acids in *Asparagopsis armata* grown in tanks with fishpond effluents of gilthead sea bream *Asparus aurata*. J World Aquacult Soc 39:692–699

Flagella MM, Verlaque M, Soria A, Buia MC (2007) Macroalgal survival in ballast water tanks. Mar Pollut Bull 54:1395–401

Flagella MM, Andreakis N, Hiraoka M, Verlaque M, Buia MC (2010) Identification of cryptic Ulva species (Chlorophyta, Ulvales) transported by Ballast water. J Biol Res-Thessalon 13:47–57

Gab-Alla AAFA (2007) Ecological study on community of exotic invasive seaweed *Caulerpa prolifera* in suez canal and its associated macro invertebrates. J Appl Sci 7:679–686

Garzon HM, Wong TY (2010) DNA chips for species identification and biological phylogenies. Nat Comput. doi:10.1007/s11047-010-9232-y

Geller JB, Darling JA, Carlton JT (2010) Genetic perspectives on marine biological invasions. Annu Rev Mar Sci 2:367–394

Genovese G, Tedone L, Hamann MT, Morabito M (2009) The Mediterranean red alga *Asparagopsis*: a source of compounds against *Leishmania*. Mar Drugs 7:361–366

Godwin H (1975) The history of the British flora, 2nd edn. Cambridge University Press, Cambridge

Gollan JR, Wright JT (2006) Limited grazing pressure by native herbivores on the invasive seaweed *Caulerpa taxifolia* in a temperate Australian estuary. Mar Freshwater Res 57:685–694

Guiry MD, Guiry GM (2008) AlgaeBase. (http://www.algaebase.org) World-wide electronic publication, National University of Ireland, Galway [updated: continuously]. Available from: http://www.algaebase.org/search/species/

Hayashi L, Jose de Paula E, Chow F (2007) Growth rate and carrageenan analysis in four strains of *Kappaphycus alvarezii* (Rhodophyta, Gigartinales) farmed in the subtropical waters of São Paulo State, Brazil. J Appl Phycol 19:393–399

Hayden HS, Waaland JR (2004) A molecular systematic study of *Ulva* (Ulvaceae, Ulvales) from the northern Pacific. Phycologia 43:364–382

Hemmingson JR, Furneaux FR, Thompson K (2006) Structure and antiviral activity of the galactofucan sulfates extracted from *Undaria pinnatifida* (Phaeophyta). J Appl Phycol 18:185–193

Hewitt CL, Campbell ML, Gollasch S (2006) Alien species in aquaculture. Considerations for responsible use, IUCN. Gland, Switzerland, p 32

Hewitt CL, Gollasch S, Minchin D (2009a) The vessel as a vector—biofouling, ballast water and sediments. In: Rilov R, Crooks JA (eds) Biological invasions in marine ecosystems. Springer, Berlin, pp 117–132

Hewitt CL, Everett RA, Parker N, Campbell ML (2009b) Marine bioinvasion management: structural framework. In: Rilov R, Crooks JA (eds) Biological invasions in marine ecosystems. Springer, Berlin, pp 327–334

Horridge GA (1951) Occurrence of *Asparagopsis armata* Harvey on the Scilly Isles. Nature 167:732–733

Howden M, Hughes L, Dunlop M, Zethoven I, Hilbert D, Chilcott C (2003) *Climate Change Impacts On Biodiversity In Australia*, Outcomes of a workshop sponsored by the Biological Diversity Advisory Committee, 1–2 October 2002, Commonwealth of Australia, Canberra

Hutchinson JFRS (1976) India: local and introduced crops. Philos Trans R Soc London 275:129–141

Hwang EK, Baek JM, Park CS (2008) Cultivation of the green alga, *Codium fragile* (Suringar) Hariot, by artificial seed production in Korea. J Appl Phycol 20:469–475

Johnson CR (2007) Seaweed invasions: conclusions and future directions. Bot Mar 50:451–457

Kapraun DF (2005) Nuclear DNA content estimates in multicellular green, red and brown algae: phylogenetic considerations. Ann Bot 95:7–44

Klein J, Verlaque M (2008) The *Caulerpa racemosa* invasion: a critical review. Mar Poll Bull 56:205–225

Kolar CS, Lodge DM (2001) Progress in invasion biology: predicting invaders. Trends Ecol Evol 16:199–204

Kraan S (2008) *Sargassum muticum* (Yendo) Fensholt in Ireland: an invasive species on the move. J Appl Phycol 20:825–832

Kraan S, Barrington KA (2005) Commercial farming of *Asparagopsis armata* (Bonnemaisoniaceae, Rhodophyta) in Ireland, maintenance of an introduced species? J Appl Phycol 17:103–110

Leliaert F, Zhang X, Ye N, Malta EJ, Engelen AH, Mineur F (2009) Identity of the Qingdao algal bloom. Phycol Res 57:147–151

Li X, Dong S, Lei Y, Li Y (2007) The effect of stocking density of Chinese mitten crab *Eriocheir sinensis* on rice and crab seed yields in rice–crab culture systems. Aquaculture 273:487–493

Lin W, Zhou G, Cheng X, Xu R (2007) Fast economical development accelerates biological invasions in China. PLoS One 2(11):e1208. doi:10.1371/journal.pone.0001208

Lodge DM, Williams S, MacIsaac HJ, Hayes KR, Leung B, Reichard S, Mack RN, Moyle PB, Smith M, Andow DA, Carlton JT, McMichael A (2006) Biological invasions: recommendations for U.S. policy and management. Ecol Appl 16:2035–2054

Luhan MRJ, Sollesta H (2010) Growing the reproductive cells (carpospores) of the seaweed, *Kappaphycus striatum*, in the laboratory until outplanting in the field and maturation to tetrasporophyte. J Appl Phycol 22:579–585

Mainka SA, Howard GW (2010) Climate change and invasive species: double jeopardy. Integr Zool 5:102–111

Manilal A, Sujith S, Kiran GS, Selvin J, Shakir C, Gandhimathi R, Panikkar MVN (2009) Biopotentials Of Seaweeds Collected From Southwest Coast Of India. J Mar Sci Technol 17:67–73

Mata L, Schuenhoff A, Santos R (2010) A direct comparison of the performance of the seaweed biofilters, *Asparagopsis armata* and *Ulva rigida*. J Appl Phycol 22:639–644

Matanjun P, Mohamed S, Mustapha NM, Muhammad K (2009) Nutrient content of tropical edible seaweeds, *Eucheuma cottonii*, *Caulerpa lentillifera* and *Sargassum polycystum*. J Appl Phycol 21:75–80

Mayr E (1942) Systematics and the origin of species. Columbia University Press, New York

McConnell O, Fenical W (1977) Halogen chemistry of the red alga *Asparagopsis*. Phytochemistry 16:367–374

McHugh DJ (2003) A guide to the seaweed industry, FAO Fisheries Technical Paper. No. 441, Rome, pp 105

Mehta GK, Meena R, Prasad K, Ganesan M, Siddhanta AK (2010) Preparation of galactans from *Gracilaria debilis* and *Gracilaria salicornica* (Gracilariales, Rhodophyta) of Indian waters. J Appl Phycol 22:623–627

Meinesz A (1999) Killer algae: a true tale of biological invasion. University of Chicago Press, Chicago, 360

Meyerson LA, Mooney HA (2007) Invasive alien species in an era of globalization. Front Ecol Environ 5:199–208

Minchin D, Gollasch S, Cohen AN, Hewitt CL, Olenin S (2009) Characterizing vectors of marine invasions. In: Rilov R, Crooks JA (eds) Biological invasions in marine ecosystems. Springer, Berlin, pp 109–116

Monteiro C, Engelen AH, Serrão EA, Santos R (2009) Habitat differences in the timing of reproduction of the invasive alga *Sargassum muticum* (Phaeophyta, sargassaceae) over tidal and lunar cycles. J Phycol 45:1–7

Moritz C (1994) Defining "evolutionarily significant units" for conservation. Trends Ecol Evol 9:373–375

Mueller JM, Hellmann JJ (2008) An assessment of invasion risk from assisted migration. Conserv Biol 22:562–567

Muirhead JR, Gray DK, Kelly DW, Ellis SM, Heath DD, Macisaac HJ (2008) Identifying the source of species invasions: sampling intensity vs. genetic diversity. Mol Ecol 17:1020–1035

National Research Council (1996) Lost Crops of Africa. Volume 1: Grains. E-book, available via http://www.nap.edu/openbook.php?record_id=2305&page=R2. Cited 11 May 2011

Neill PE, Alcalde O, Faugeron S, Naverrete SA, Correa JA (2006) Invasion of *Codium fragile* subspecies tomentosoides in northern Chile: A new threat for *Gracilaria* farming. Aquaculture 259:202–210

Nelson SG, Glenn EP, Moore D, Ambrose B (2009) Growth and distribution of the Macroalgae *Gracilaria salicornia* and *G. parvispora* (Rhodophyta) established from aquaculture introductions at Moloka'i, Hawai'i. Pac Sci 63:383–396

Neves CS, Rocha RM (2008) Introduced and cryptogenic species and their management in Paranaguá Bay, Brazil. Braz Arch Biol Technol 51:623–633

Ní Chualáin F, Maggs CA, Saunders GW, Guiry MD (2004) The invasive genus *Asparagopsis* (Bonnemaisoniaceae, Rhodophyta): molecular systematics, morphology and ecophysiology of *Falkenbergia* isolates. J Phycol 40:1112–1126

Nuber N, Gornik O, Lauc G, Bauer N, Zuljevic´ A, Papes D, Zoldos V (2007) Genetic evidence for the identity of *Caulerpa racemosa* (Forsska°l) J. Agardh (Caulerpales, Chlorophyta) in the Adriatic Sea. Eur J Phycol 42:113–120

Nyberg CD, Wallentinus I (2005) Can species traits be used to predict marine macroalgal introductions? Biol Invasions 7:265–279

O'Doherty DC, Sherwood AR (2007) Genetic Population Structure of the Hawaiian Alien Invasive Seaweed *Acanthophora spicifera* (Rhodophyta) as Revealed by DNA Sequencing and ISSR Analyses. Pac Sci 61:223–233

Occhipinti-Ambrogi A (2007) Global change and marine communities: Alien species and climate change. Mar Poll Bull 55:342–352

Ortiz J, Uquiche E, Robert P, Romero N, Quitral V, Llantén C (2009) Functional and nutritional value of the Chilean seaweed *Codium fragile*, *Gracilaria chilensis* and *Macrocystis pyrifera*. Eur J Lipid Sci Tech 111:320–327

Oza RM, Zaidi SH (2001) A revised checklist of Indian marine algae. CSMCRI, Bhavnagar, Gujarat, pp 24–29

Pandit MK, Pocock MJO, Kunin WE (2011) Ploidy influences rarity and invasiveness in plants. J Ecol doi:10.1111/j.1365-2745.2011.01838.x

Pang SJ, Liu F, Shan FL, Xu N, Zhang ZH (2010) Tracking the algal origin of the *Ulva* bloom in the Yellow Sea by combination of molecular, morphological and physiological analysis. Mar Environ Res 69:207–215

Paul NA, de Nys R (2008) Promise and pitfalls of locally abundant seaweeds as biofilters for integrated aquaculture. Aquaculture 281:49–55

Paul NA, de Nys R, Steinberg PD (2006) Chemical defence against bacteria in the red alga *Asparagopsis armata*: linking structure with function. MEPS 306:87–101

Paula EJ, Eston VR (1987) Are there other *Sargassum* species potentially as invasive as *S. muticum*? Bot Mar 30:405–410

Pickering TD, Skelton P, Sulu RJ (2007) Intentional introductions of commercially harvested alien seaweeds. Bot Mar 50:338–350

Provan J, Murphy S, Maggs CA (2005) Tracking the invasive history of the green alga *Codium fragile* subsp. *tomentosoides*. Mol Ecol 14:189–194

Provan J, Booth D, Todd NP, Beatty GE, Maggs CA (2008) Tracking biological invasions in space and time: elucidating the invasive history of the green alga *Codium fragile* using old DNA. Diversity Distrib 14:343–354

Raniello R, Mollo E, Lorenti M, Gavagnin M, Buia MC (2007) Phytotoxic activity of caulerpenyne from the Mediterranean invasive variety of Caulerpa racemosa: a potential allelochemical. Biol Invasions 9:361–368

Rilov G, Crooks JA (2009) In: Biological invasions in marine ecosystems: Ecological, management and geographical perspectives. Rilov and Crooks (ed.). Springer, Berlin, pp 619–626

Rodgers SK, Cox EF (1999) Rate of spread of introduced Rhodophytes *Kappaphycus alvarezii*, *Kappaphycus striatum*, and *Gracilaria salicornica* and their current distributions in Kane'ohe Bay, O'ahu, Hawai'i. Pac Sci 53:232–241

Ruiz GM, Carlton CT (2003) Invasive species: vectors and management strategies. Island, Washington DC

Russell DJ (1992) The ecological invasion of Hawaiian reefs by two marine red algae, *Acanthophora spicifera* (Vahl) Boerg. and *Hypnea musciformis* (Wulfen) J. Ag., and their association with two native species, Laurencia nidifica J. Ag. and *Hypnea cervicornis*. J Ag ICES Mar Sci Symp 194:110–125

Russell L, Hepburn C, Hurd C, Stuart M (2008) The expanding range of *Undaria pinnatifida* in southern New Zealand: distribution, dispersal mechanisms and the invasion of wave-exposed environments. Biol Invasions 10:103–115

Ryder E, Nelson S, Glenn E, Nagler P, Napolean S, Fitzsimmons K (2004) Review: production of *Gracilaria parvispora* in two-phase polyculture systems in relation to nutrient requirements and uptake. Bull Fish Res Agen 1:71–76

Salvador N, Garreta AG, Lavelli L, Ribera MA (2007) Antimicrobial activity of Iberian macroalgae. Sci Mar 71:101–113

Santelices B, Aedo D, Hoffman A (2002) Banks of microscopic forms and survival to darkness of propagules and microscopic stages of macroalgae. Rev Chil Hist Nat 75:547–555

Schaffelke B, Hewitt CL (2007) Impacts of introduced seaweeds. Bot Mar 50:397–417

Schaffelke B, Campbell ML, Hewitt CL (2005) Reproductive phenology of the introduced kelp *Undaria pinnatifida* (Phaeophyta, Laminariales) in Tasmania, Australia. Phycologia 44:84–94

Schaffelke B, Smith JE, Hewitt CL (2006) Introduced macroalgae—a growing concern. J Appl Phycol 18:529–541

Schuenhoff A, Mata L, Santos R (2006) The tetrasporophyte of *Asparagopsis armata* as a novel seaweed biofilter. Aquaculture 252:3–11

Sellers E, Simpson A, Curd-Hetrick S (2010) List of Invasive Alien Species (IAS) Online Information Systems. A 'living document' based on a preliminary draft document, prepared for the Experts Meeting Towards the Implementation of a Global Invasive Species Information Network (GISIN), Baltimore, Maryland, USA, 6–8 April 2004. http://www.gisinetwork.org/Documents/draftiasdbs.pdf

Setchell WA, Gardner NL (1924) New marine algae from the Gulf of California. Proc Calif Acad Sci 12:695–949

Sherwood AR (2007) Where are we now regarding Hawaiian stream algal systematics? (a suspiciously cosmopolitan flora). Bishop Mus Bull Cult Environ Stud 3:195–206

Sherwood AR (2008) Phylogeography of *Asparagopsis* (Bonnemaisoniales, Rhodophyta) in the Hawaiian Islands: two mtDNA markers support three separate introductions. Phycologia 47:79–88

Smith JE, Hunter CL, Conklin EJ, Most EJ, Sauvage R, Squair T, Smith C, Celia M (2004) Ecology of the invasive red alga *Gracilaria salicornia* (Rhodophyta) on O'ahu, Hawai'i. Pac Sci 58:325–343

Stam WT, Olsen JL, Zaleski SF, Murray SN, Brown KR, Walters LJ (2006) A forensic and phylogenetic survey of *Caulerpa* species (Caulerpales, Chlorophita) from the Florida coast, local aquarium shops, and e-commerce: establishing a proactive baseline for early detection. J Phycol 42:1113–1124

Taylor WR (1945) Pacific marine algae of the Allan Hancock Expeditions to the Galapagos Islands. Allan Hancock Pacific Expeditions 12: i-iv, 1–528, 3 figs, 100 pls

Thornber CS, Kinlan BP, Graham MH, Stachowicz JJ (2004) Population ecology of the invasive kelp *Undaria pinnatifida* in California: environmental and biological controls on demography. MEPS 268:69–80

Trowbridge CD (1996) Introduced versus native subspecies of *Codium fragile*: How distinctive is the invasive subspecies tomentosoides? Mar Biol 126:193–204

Uwai S, Yotsukura N, Serisawa Y, Muraoka D, Hiraoka M, Kogame K (2006a) Intraspecific genetic diversity of *Undaria pinnatifida* in Japan, based on the mitochondrial *cox3* gene and the ITS1 of nrDNA. Hydrobiologia 553:345–356

Uwai S, Nelson W, Neill K, Wang WD, Aguilar-Rosas LE, Boo SM, Kitayama T, Kawai H (2006b) Genetic diversity in *Undaria pinnatifida* (Laminariales, Phaeophyceae) deduced from mitochondria genes—origins and succession of introduced populations. Phycologia 45:687–695

Valentine JP, Magierowski RH, Johnson CR (2007) Mechanisms of invasion: establishment, spread and persistence of introduced seaweed populations. Bot Mar 50:351–360

Verbruggen H, Leliaert F, Maggs CA, Shimada S, Schils T, Provan J, Booth D, Murphy S, De Clerck O, Littler DS, Littler MM, Coppejans E (2007) Species boundaries and phylogenetic relationships within the green algal genus *Codium* (Bryopsidales) based on plastid DNA sequences. Mol Phyl Evol 44:240–254

Verlaque M, Durand C, Huisman JM, Boudouresque CF, Le Parco Y (2003) On the identity and origins of the Mediterranean invasive *Caulerpa racemosa* (Caulerpales, Chlorophyta). Eur J Phycol 38:325–339

Voisin M, Engel CR, Viard F (2005) Differential shuffling of native genetic diversity across introduced regions in a brown alga: Aquaculture vs. maritime traffic effects. Proc Natl Acad Sci 102:5432–5437

Vroom P, Smith CM (2001) The challenge of siphonous green algae. Am Sci 89:525–531

Wallentinus I, Nyberg CD (2007) Introduced marine organisms as habitat modifiers. Mar Poll Bull 55:323–332

Wares JP, Hughes AR, Grosberg RK (2005) Mechanisms that drive evolutionary change: insights from species introductions and invasions. In: Species invasions, insights into ecology, evolution, and biogeography. Sax DF, Stachowicz JJ, Gaines SD (eds.). Sinauer & Associates, Sunderland, pp 229–257

Wattier R, Maggs CA (2001) Intraspecific variation in seaweeds: the application of new tools and approaches. Adv Bot Res 35:171–212

Whitney KD, Gabler CA (2008) Rapid evolution in introduced species, 'invasive traits' and recipient communities: challenges for predicting invasive potential. Diversity Distrib 14: 569–580

Williams SL, Smith JE (2007) A global review of the distribution, taxonomy, and impacts of introduced seaweeds. Annu Rev Ecol Evol Syst 38:327–359

Williamson M, Fitter A (1996) The varying success of invaders. Ecology 77:1661–1666

Willis CG, Ruhfel BR, Primack RB, Miller-Rushing AJ, Losos JB, Davis CC (2010) Favorable climate change response explains non-native species' success in Thoreau's Wwoods. PLoS ONE 5(1):e8878. doi:10.1371/journal.pone.0008878

Woolard FX, Moore RE, Roller PP (1979) Halogenated acetic and acrylic acids from the red alga *Asparagopsis taxiformis*. Phytochemistry 18:617–620

Wright JT, Davis AR (2006) Demographic feedback between clonal growth and fragmentation in an invasive seaweed. Ecology 87:1744–1754

Yeh WL, Chen GY (2004) Nuclear rDNA and internal transcribed spacer sequences clarify Caulerpa racemosa vars. From other Caulerpa species. Aquat Bot 80:193–207

Zaleski SF, Murray SN (2006) Taxonomic diversity and geographic distributions of aquarium-traded species of *Caulerpa* (Chlorophyta: Caulerpaceae) in southern California, USA. MEPS 314:97–108

Zemke-White WL (2004) Assessment of the current knowledge on the environmental impacts of seaweed farming in the tropics. Marine Science into the New Millennium: New Perspectives and Challenges. In: Proceedings of the Asia-Pacific Marine Science and Technology Conference, 12–16 May Kuala Lumpur 2002, Malaysia, pp 465–476

Zubia M, Fabre MS, Kerjean V, Deslandes E (2009) Antioxidant and cytotoxic activities of some red algae (Rhodophyta) from Brittany coasts (France). Bot Mar 52:268–277

Part III
Structure and Function of the World's Main Seaweed Systems

Chapter 13
Seaweeds and Their Communities in Polar Regions

Christian Wiencke and Charles D. Amsler

13.1 Introduction

The natural environment of polar seaweeds is characterized by strong seasonal light conditions and constant low temperatures (Zacher et al. 2011). At the northern and southern distribution limits of seaweeds in the Arctic (80°N) and Antarctic (77°S), the polar night lasts for about 4 months. At lower latitudes, e.g., the northern border of the Antarctic region, at King George Island (South Shetland Islands; 62°S) daylength varies between 20 h in summer and 5 h in winter. Sea-ice cover extends the period of hibernal darkness. If the ice is covered by snow under-ice irradiance can be reduced to <2% of the surface value. Extremely low irradiances ≤6.5 μmol photons m^{-2} s^{-1} have been measured even in midsummer (June) below a sea-ice cover and a snow layer of 30 cm in June on Arctic Spitsbergen (Hanelt et al. 2001). So, seaweeds can be exposed for long periods to darkness and very low light conditions. However, when the sea-ice breaks up solar radiation can penetrate deeply into the relatively clear water. In October/November 1993, average midday irradiances as high as 70 μmol photons m^{-2} s^{-1} were measured in 30 m water depth at King George Island (Gómez et al. 1997). At Signy Island (South Orkney Islands), the mean 1% depth of surface photosynthetically active radiation (PAR) is at about 29 m (Brouwer 1996a). Ultraviolet (UV) radiation as well can go down to considerable depths. In coastal areas, the 10% level for UV-A/UV-B radiation can be as deep as 7.1/4.3 m water depth, respectively, in summer (Richter et al. 2008).

C. Wiencke (✉)
Department Seaweed Biology, Section Functional Ecology, Alfred Wegener Institute for Polar and Marine Research, Am Handelshafen 12, 27570 Bremerhaven, Germany
e-mail: christian.wiencke@awi.de

C.D. Amsler
Department of Biology, University of Alabama at Birmingham, Birmingham, AL 35294-1170, USA
e-mail: amsler@uab.edu

C. Wiencke and K. Bischof (eds.), *Seaweed Biology*, Ecological Studies 219,
DOI 10.1007/978-3-642-28451-9_13,

These relatively high light conditions prevail until phytoplankton blooms are formed and turbid melt-water carrying fine sediments and detritus flows into the sea (Drew and Hastings 1992). Thus, favorable light conditions for algal growth are present for only 2–3 months per year.

In contrast to the strong seasonality of the light conditions, seawater temperatures in the sublittoral vary only slightly between −1.8°C in winter and 2.2°C in summer in the Antarctic Peninsula region (Drew and Hastings 1992). At the boundary of the temperate region, maximum temperatures can reach 5°C in the Antarctic and 8–10°C in the Arctic (Wiencke et al. 2007). Temperature variation is, however, stronger in eulittoral and supralittoral communities. On King George Island, temperatures can go up to 14°C in tide pools while seaweeds exposed to air in the supralittoral experience temperatures up to nearly 30°C in summer and down to −27°C in winter (Zacher et al. 2011).

Although the current water temperatures in both polar regions are similar, the Arctic Oceans differ considerably in their cold-water history and their genesis. Whereas the water temperatures in the Southern Ocean have been low for 14 million years, glaciation and a winter sea-ice cover did not develop earlier than two million years in the Arctic (Crame 1993; Zachos et al. 2001; Zacher et al. 2011). Other differences between both polar regions refer to the connection with the temperate regions and to the nutrient levels in the seawater. Whereas the Antarctic region is strongly isolated, the Arctic region is continuously connected to temperate coasts. With respect to the nutrients, levels of nitrate and phosphate are high throughout the year in the Southern Ocean (Drew and Hastings 1992; Ducklow et al. 2007), while there is a strong seasonal variation of these levels in the Arctic. Here, nutrient levels are usually high during winter only and drop suddenly after sea-ice breakup (Chapman and Lindley 1980; Aguilera et al. 2002).

Seaweeds growing in such an extreme environment have to be adapted to these conditions. Their seasonal development must be tuned to the strong seasonality of the light conditions. Species from the Arctic must furthermore be adapted to the seasonal changes of the nutrient concentrations. An important prerequisite for seaweeds from polar waters is their capability to utilize the—for most parts of the year—prevailing low light conditions and, on the other hand, to sustain and use as efficient as possible the high light conditions during spring for photosynthesis.

The difference in the cold-water history of both polar regions has forced seaweeds in the Antarctic to reduce their temperature demands considerably compared to species from the Arctic (Gómez et al. 2011). Moreover, in conjunction with the strong geographic isolation of the Antarctic region, this has had great effects on biodiversity: Whereas in the Arctic only few endemic species occur, endemism is high in the Southern Ocean (Wulff et al. 2011). These environmental forcings influenced also the biotic interactions between seaweeds and their associated organisms. As we will discuss below, all these factors are important determinants of the functioning of polar seaweed communities.

13.2 Biodiversity and Biogeographical Relationships of Antarctic and Arctic Seaweeds

As mentioned above, both polar regions differ considerably with respect to their genesis and cold-water history. Antarctica is an ice-covered continent surrounded by the Southern Ocean without any land connection to temperate regions since the Mesozoic (Lawver et al. 1985). The Antarctic Circumpolar Current (ACC), which started to develop at about 32 Ma and became persistent around 15 Ma, amplifies this separation further (Hommersand et al. 2011). Driven by westerly winds, it flows clockwise around the continent. In contrast, the Arctic Ocean is a "mediterranean" sea surrounded by continental land masses and is continuously connected to the temperate coasts of Eurasia and America. The Arctic Polar Front is discontinuous and there is a strong influx of warm North Atlantic water into the Arctic through the Fram Strait between Spitsbergen and Greenland (Zacher et al. 2011). In contrast, the Antarctic region is sharply delimited to the north by the forceful Antarctic Polar Front. A second feature differentiating the two polar regions is their cold-water history. Whereas Antarctica became glaciated 14 million years ago (Crame 1993), a perennial ice cover over the Arctic Ocean did not develop before 0.7–2.0 Ma (Clarke 1990). So, seawater temperatures have been low for a much longer time in the Antarctic compared to the Arctic (Zacher et al. 2011).

These differences have resulted in strongly different biodiversity patterns in both polar regions. In the Antarctic region, 35% of the species are endemic, whereas only few endemic Arctic species have been identified so far (Wiencke and Clayton 2002; Wulff et al. 2011). In the Antarctic, 44% of the Heterokontophyta (Phaeophyceae and Chrysophyceae), 36% of the Rhodophyta, and 18% of the Chlorophyta are endemic and the number of endemic species is continuously increasing. Only recently, Hommersand et al. (2011) described four new genera and five new endemic red algal species. There is also one endemic order, the brown algal order Ascoseirales. Conspicuous and ecologically important endemic species include the brown algae *Himantothallus grandifolius*, *Desmarestia anceps*, *D. menziesii*, *Cystosphaera jacquinotii*, *Ascoseira mirabilis*, and the red algae *Myriogramme manginii*, *Georgiella confluens*, *Phycodrys antarctica*, *Trematocarpus antarcticus*, and others. The red algae *Palmaria decipiens* and *Iridaea cordata* occur from the Ross Sea in the south to a few sub-Antarctic islands in the north (Wiencke and Clayton 2002).

In contrast to the high degree of endemism in Antarctica, only very few seaweed species endemic to the Arctic have been detected (Wilce 1990; Wilce et al. 2009). These include the brown algae *Chukchia pedicellata*, *C. endophytica*, *Punctaria glacialis*, *Platysiphon verticillatus*, the red alga *Petrocelis polygyna*, and the green alga *Acrosiphonia incurva*. Most species from the Arctic have a distribution that extends well into the temperate region, e.g., the red algae *Devaleraea ramentacea*, *Turnerella pennyi*, *Dilsea integra*, and *Pantoneura baerii*. This is also the case for the kelp *Laminaria solidungula*, but this species is regarded as a true Arctic species as its occurrence in the temperate region is restricted to cold, deep waters (Müller et al. 2011).

A general characteristic of the floras of both polar regions is their low species richness. In the Antarctic, about 130 species have been documented (Wulff et al. 2011) and in the Arctic about 150 species have been recorded (Wilce 1994). These numbers will likely increase with increased exploration of these remote areas of our planet. In Antarctica, species richness is highest in the Antarctic Peninsula region and lowest in the southernmost part of the Ross Sea at 77°S (Wiencke and Clayton 2002). In the Arctic, species richness peaks with about 70 species around Svalbard, followed by the Canadian High Arctic (mainly Baffin Bay area) with 55 species. The Russian Arctic principally harbors an impoverished Svalbard flora. Similarly, species richness decreases in eastern Greenland from the south to the north. Whereas the seaweed flora in southernmost Greenland is relatively rich (Pedersen 1976), in the Alaskan Beaufort Sea shelf west of the Canadian Arctic Archipelago only 15 species were recorded (Dunton and Schonberg 2000; Wulff et al. 2011).

Two other characteristics of the Antarctic seaweed flora must be mentioned here: First, a conspicuous feature of the Antarctic seaweed flora, compared to temperate regions, is the scarcity of small macroalgal epiphytes. Such epiphytes are, however, not absent. Rather, they occur as endophytes in larger seaweeds (Peters 2003; see also Chap. 11 by Potin and Chap. 9 by Amsler). Second, the order Desmarestiales, which ecologically replaces the Laminariales (kelps) in Antarctica, is considered to have its evolutionary center in the Southern Ocean and subsequently radiated into the Northern hemisphere (Peters et al. 1997).

In contrast to the earlier assumption that Arctic seaweeds are mostly of Atlantic origin (Wilce 1990; Dunton 1992), recent molecular data obtained on six red algal species and of *Laminaria solidungula* from the Arctic indicate that Arctic and Atlantic subarctic species are evolutionary based in the North Pacific (Adey et al. 2008). Another example of an Arctic invader from the Pacific may be the green alga *Acrosiphonia arcta* (van Oppen et al. 1994). The species is—as is also *Desmarestia viridis/confervoides*—a good example for the evolution and spread of bipolar distributed species. According to molecular studies, the biogeographic disjunctions of both species date back to the maximum of the Würm/Wisconsin glaciation 18,000 years ago (van Oppen et al. 1993). The dispersal phases were presumably microscopic stages which, due to a high temperature tolerance, were able to cross the tropics (Peters and Breeman 1993; Bischoff and Wiencke 1995; see also Chap. 18 by Bartsch et al.).

13.3 Physiological Adaptations to the Environment

13.3.1 Seasonal Development and Physiological Performance

In polar regions, adaptation to the strong seasonality of the light regime is one of the most important prerequisites for the ecological success of seaweeds (Wiencke et al. 2011). There are two different growth strategies classified as season anticipators and season responders *sensu* Kain (1989). The season anticipators begin growing under short-day conditions in late winter/spring, often under the sea-ice. Some species even reproduce in winter. Maximal growth rates occur in spring. Growth and reproduction in these species are presumably based on circannual rhythms and photoperiodism, synchronized or triggered by daylength, as shown for some Antarctic (Wiencke 1990a; Wiencke et al. 1996) and for many cold-temperate species (Lüning 1988, 1991; Lüning and Kadel 1993; Schaffelke and Lüning 1994). The season responders, in contrast, start growth and reproduction later, not before favorable light conditions are present in spring and summer. Species of this group react directly to the primary factors in their environment and show an opportunistic life strategy (Wiencke 1990a).

Typical Antarctic season anticipators are the brown algae *Desmarestia menziesii*, *D. anceps*, *D. antarctica*, *Himantothallus grandifolius*, *Ascoseira mirabilis* (Wiencke 1990a; Drew and Hastings 1992; Gómez et al. 1995, 1996; Gómez and Wiencke 1997) and the red algae *Palmaria decipiens*, *Delesseria salicifolia*, *Gymnogongrus antarcticus*, *G. turquetii*, *Hymenocladiopsis crustigena*, *Trematocarpus antarcticus*, and *Phyllophora ahnfeltioides* (Wiencke 1990b; Weykam et al. 1997; Dummermuth and Wiencke 2003). Typical Antarctic season responders are *Adenocystis utricularis* (Wiencke 1990a), the red algae *Iridaea cordata* (Weykam et al 1997) and *Gigartina skottsbergii* (Wiencke 1990b), and the green algae *Ulva hookeriana* and *Acrosiphonia arcta* (Wiencke 1990b).

In contrast to the Southern Ocean in the Arctic there is not only a strong seasonal variation of the light regime, but also of the levels of the macronutrients nitrate and phosphate, which are high in winter and low in summer (see Chap. 4 by Gordillo). Arctic season anticipators like *Laminaria solidungula* and *Saccharina latissima* (*L. saccharina*) take full advantage of these conditions (Chapman and Lindley 1980; Dunton 1985). New blades start to form in *L. solidungula* in fall under decreasing daylengths. Optimum growth rates occur in late winter/spring under thick ice. In comparison, *S. latissima* grows mostly during a brief period in late spring, when the first light penetrates into the water during sea-ice breakup.

As most endemic Antarctic and Arctic seaweeds are season anticipators, this type of life strategy is therefore regarded as the typical adaptation to the seasonally changing conditions in polar seas. In contrast, season responders are mostly distributed also in the adjacent temperate regions. Another difference between both groups is their depth distribution. Polar season anticipators occur almost exclusively in the sublittoral, whereas many polar season responders can grow also in the eulittoral.

As for biomass formation, photosynthetic performance shows also a strong seasonal pattern (Wiencke et al. 2011). In large brown algal season anticipators, photosynthetic rates are highest in late winter/spring (Drew and Hastings 1992; Gómez et al. 1995). As shown in several kelp and kelp-like species, respiration rates increase indicating growth activity in the basal meristem powered by remobilization of carbohydrates from the distal thallus part. An extreme example is *Laminaria solidungula*, in which the mobilization occurs during the 9 months period of darkness under the ice, when the alga completes over 90% of its annual growth.

During this time, up to 30% of its original total carbon content is depleted before photosynthetic production begins in spring. During summer, high photosynthetic rates are used for carbon storage and not for biomass formation (Dunton and Schell 1986; see also Chap. 2 by Gómez and Huovinen).

In the Antarctic red algal season anticipator *Palmaria decipiens*, photosynthetic rates are—like the growth rates—highest in spring. There is a positive correlation between the phycobilin content, photosynthetic capacity and efficiency, which are highest in fall, winter, and spring. During summer, the alga reduces the photosynthetic apparatus to a minimum (Lüder et al. 2001a). The presence of two phycobilisome forms with different aggregation states has been regarded as special advantage for a rapid acclimation to changing environmental light conditions (Lüder et al. 2001b; see also Chap. 1 by Hanelt and Figueroa).

The effect of darkness on physiological performance has been studied in *Palmaria decipiens* and *Iridaea cordata*. In *P. decipiens*, the light harvesting phycobilisomes and later, the chl *a* containing inner antennae are degraded during long-term exposure to darkness. After 6 months, the alga has lost its ability to photosynthesize. Following reexposure to light, pigments are rapidly synthesized and after a week photosynthesis recovers to normal levels (Lüder et al. 2002; Weykam et al. 1997). In contrast, the season responder *I. cordata* maintains a functional photosynthetic apparatus during dark-exposure and is therefore better suited to grow in places with less predictable light conditions (Weykam et al. 1997; see also Chap. 1 by Hanelt and Figueroa).

Overall, seasonal development and physiological performance exhibit many similarities with temperate seaweeds. Although there is no unique mechanism occurring only in polar species, their efficient adaptations to low light, however, allow Arctic and Antarctic species to thrive with great success in polar waters.

13.3.2 Radiation Climate and Depth Zonation

In polar regions, the radiation climate imposes severe constraints not only with respect to seasonal light availability but also with respect to the irradiance level in different water depths ultimately determining seaweed zonation. As polar algae are mainly sublittoral, low light tolerance is a prerequisite for distribution down to great depths. This becomes obvious when the minimum light requirement for completion of the life history is considered, which is lower in polar seaweeds compared to temperate, morphologically similar species. For Antarctic Desmarestiales, the minimum annual light demand is 31 moles photons m^{-2} (Wiencke 1990a) and for *Laminaria solidungula* 45–49 moles m^{-2} (Chapman and Lindley 1980; Dunton 1990). In contrast, *L. hyperborea* from the North Sea requires 71 moles m^{-2} per year (Lüning and Dring 1979).

Another prerequisite for algal life in polar waters is a capacity to tolerate long periods of darkness. Various polar seaweeds tolerate darkness for up to 18 months (tom Dieck 1993; Wiencke 1990a). Growth in the microscopic stages of Antarctic seaweeds is light-saturated at photon fluence rates as low as 4–12 μmol photons m^{-2} s^{-1} (Wiencke 1990a). In young sporophytes of Antarctic Desmarestiales the values are somewhat higher, at 15–20 μmol photons m^{-2} s^{-1} (Wiencke and Fischer 1990).

As with growth, the light demands for photosynthesis are also very low. Species from both polar regions show a high photosynthetic efficiency (α), low respiratory rates, low saturation points for photosynthesis (E_k), and low compensation points for photosynthesis (E_c; Gómez et al. 2011). E_k values range between 3 and 100 μmol photons m^{-2} s^{-1}, and E_c values between <1 and 15 μmol photons m^{-2} s^{-1}, values usually lower than for species from temperate regions (Lüning 1990).

Generally, E_k values for photosynthesis are higher than the irradiances required for saturation of growth. This represents an important ecological advantage for coping with the strong fluctuations of incident irradiance during the open water period. While growth is saturated at low irradiances, the photon fluence rates above the saturation point for growth can be used for purposes other than growth, e.g., for formation of storage compounds. For example, Antarctic seaweeds growing at depths below 20 m are often exposed to irradiances around 80 μmol photons m^{-2} s^{-1} during late winter/spring (Gómez et al. 1997) allowing considerable carbon fixation, fuelling—aside from growth—other metabolic processes.

The light requirements for photosynthesis are an important factor for the determination of zonation patterns. If one relates the daily light course of the irradiance to the E_k value, it is possible to estimate the average daily period of light saturation, called H_{sat}. The obtained metabolic daily carbon balance is regarded as a physiological indicator for the ability to live in deep waters. *Laminaria solidungula* in the Alaskan Beaufort Sea at 70°N was for example exposed in 1986 to total H_{sat} periods of 148 h (Dunton 1990), corresponding to an average daily H_{sat} of 3 h. For five red and brown algae from King George Island (Antarctica), H_{sat} determined during the clear water period in spring decreases with depth from values close to 14 h at 10 m to values between 7 and 12 h at 30 m depths (Gómez et al. 1997). For the red algal species *Palmaria decipiens*, *Trematocarpus antarcticus*, and *Gigartina skottsbergii* the carbon balance was between 1.7 and 2.5 mg C g FW $^{-1}$ d^{-1} at 10 m depth and between 0.6 and 0.8 mg C g FW $^{-1}$ d^{-1} at 30 m depths, setting the lower depth limit at >30 m. For the kelp-like brown alga *Himantothallus grandifolius*, the daily carbon balance varied between 0.6 and 1.0 mg C g FW $^{-1}$ d^{-1} over the studied range of 10–30 m, indicating that this species can potentially occur even deeper, which is actually the case. In contrast, in the brown alga *Desmarestia anceps* the negative (!) carbon balance of −1.9 mg C gFW^{-1} d^{-1} limits the alga to depths of about 30 m at this location (Gómez et al. 1997). Based on photosynthetic measurements, the lower depth distribution limit of the red alga *Myriogramme manginii* at Signy Island (South Orkney Islands) has been predicted to be at approx. 23 m water depth (Brouwer 1996a).

Polar seaweeds are not only strongly shade-adapted but can also cope with high light conditions in summer because of their ability for dynamic photoinhibition, a photoprotective mechanism, by which excessive energy absorbed is rendered harmless by thermal dissipation (see Chap. 1 by Hanelt and Figueroa).

Deep water species are strongly photoinhibited when exposed to high light conditions and recover during subsequent exposure to favorable light only slightly and slowly (Hanelt 1998; Karsten et al. 2001), whereas in eulittoral and upper sublittoral species the decrease in photosynthetic activity is less pronounced and usually a strong and quick recovery is recorded.

Similarly, UV radiation is also regarded as key factor affecting the depth zonation of seaweeds (Karsten et al. 2011). As explained in Chap. 20 by Bischof and Steinhoff, UV-B radiation has damaging effects on various cellular structures and processes, among those on the DNA and photosynthesis. However, damage can be repaired and there are also protective mechanisms to prevent damage. The result can be impaired growth or impaired reproductive capacity. The life-history stages of seaweeds most susceptible to UV radiation are spores. Field experiments have clearly shown that the upper depth distribution limit of Arctic kelps is determined by the UV susceptibility of their spores (Wiencke et al. 2006). Consequently, the succession of polar seaweed communities also depends on the UV radiation regime as explained below. Certainly, metabolic carbon balance and UV radiation are not the only factor controlling the lower and upper depth distribution limits. Substrate, ice abrasion, and competition play besides other factors important roles.

13.3.3 Temperature Requirements and Geographic Distribution

Polar seaweeds are well adapted to the low seawater temperatures, and Antarctic seaweeds more strongly than Arctic species due to the longer cold-water history of the Southern Ocean (see Chap. 18 by Bartsch et al.). Sporophytes of endemic Antarctic Desmarestiales, for example, grow up to 5°C and exhibit upper survival temperatures (USTs) of 11–13°C. Their gametophytes grow up to 10 or 15°C with USTs between 15 and 18°C (Wiencke et al. 1994). The Antarctic red algae *Georgiella confluens*, *Gigartina skottsbergii*, and *Plocamium cartilagineum* grow at 0°C, but not at 5°C and have USTs as low as 7–11°C (Bischoff-Bäsmann and Wiencke 1996). Antarctic cold-temperate species, especially from the eulittoral, are characterized by higher temperature ranges (Wiencke et al. 1994; Gómez et al. 2011).

In contrast, sporophytes of the endemic Arctic kelp *Laminaria solidungula* grow up to temperatures of 15°C with optimum growth rates at 5–10°C and an UST of 16°C. The gametophytes of this species exhibit an UST of 20°C (Bolton and Lüning 1983; tom Dieck 1992). No data are available on the temperature demands of other endemic Arctic species. The Arctic cold-temperate red alga *Devaleraea ramentacea* grows at temperatures up to 10°C and exhibits USTs of 18–20°C (Novaczek et al. 1990; Bischoff and Wiencke 1993). Clearly, more data are needed also on the temperature dependence of other processes in the life history.

The strong adaptation of Antarctic seaweeds to low temperatures is also reflected in their photosynthetic performance. Maximum photosynthetic rates of endemic Antarctic species are at 0°C in a similar range compared to temperate species measured at higher temperatures. Moreover, the temperature optima for photosynthesis in the few tested Antarctic species are well below values determined in temperate species (Gómez et al. 2011). Likewise, the optimum temperature for photosynthesis in zoospores of *Alaria esculenta*, *Laminaria digitata*, and *Saccharina latissima* from Arctic Spitsbergen is with 7–13°C relatively low (Roleda 2009). In contrast, the optima for respiration are clearly located at higher temperatures. Photosynthesis:respiration (P:R) ratios in Antarctic species are highest at the lowest tested temperature (0°C) and decrease with increasing temperatures due to different Q_{10} values for photosynthesis (1.4–3.5) and respiration (2.5–5.1; Gómez et al. 2011). The high P:R ratios at low temperatures explain the high growth rates of polar species at low temperatures, which in the end determine the geographic distribution.

The northern distribution of endemic Antarctic species is often limited by the temperature demands for growth. Endemic Antarctic Desmarestiales for example occur only south of the Antarctic Polar Front in areas with maximum temperatures ≤5°C allowing sufficient growth of their sporophytes (Wiencke et al. 1994; Müller et al. 2011; see also Chap. 18 by Bartsch et al.). The southern distribution of Arctic-North Atlantic species is often limited both by the USTs and the upper limit of gametogenesis (van den Hoek 1982a, b). In the West Atlantic distribution limits are determined by lethal, high summer temperatures, whereas in the East Atlantic they are determined by high winter temperatures inhibiting reproduction. Examples for species from this group are *Laminaria digitata*, *Chorda filum*, and *Halosiphon tomentosus*.

During the ice ages, both the Arctic and the Antarctic were not hospitable for seaweeds. In the southern hemisphere, sub-Antarctic islands and the southern tip of South America have probably served as refugia. Migration of species from Antarctica to South America and vice versa probably took place along the Scotia Arc (Wiencke et al. 1994). In the northern hemisphere, the distribution area of Atlantic seaweeds was extremely reduced during the ice ages. In the north, their distribution was limited by the glaciers and in the south by the 10–15°C winter isotherm, their southern reproduction boundary. This most probably explains the present depauperate flora in the North West Atlantic. A comparable situation did not exist in the Pacific, probably a major reason for the richness of the cold North Pacific (Wiencke et al. 1994). The temperature decrease during the ice ages allowed polar species to extend their distribution limits toward the equator. Some species even crossed the equator during the maximum of the last glaciations (see Chap. 18 by Bartsch et al.; Wiencke et al. 1994).

13.3.4 Effect of Salinity, Temperature, and Desiccation on Supra- and Eulittoral Seaweeds

Information on osmotic acclimation of supra- and eulittoral seaweeds during salinity changes is available only for green algae from the Antarctic. Eulittoral species survive salinities between 7 and 102 psu and grow between 7 and 68 psu.

The supralittoral species *Prasiola crispa* has even a broader growth range between 0.3 and 105 psu (Karsten et al. 1991a, b; Jacob et al. 1991). The general mechanisms of osmotic acclimation are the same as also in species from other regions (Kirst 1990; Wiencke et al. 2007; see Chap. 5 by Karsten). Major inorganic osmolytes are potassium, sodium, and chloride. These osmolytes are used during short-term osmotic regulation. Long-term osmotic stress is counterbalanced by various compatible organic solutes, among them β-dimethylsulphoniumproprionate (DMSP), the imino acid proline, and sucrose (Karsten et al. 1991b; Jacob et al. 1991).

Decreasing temperatures also strongly stimulate the biosynthesis and accumulation of DMSP in Antarctic green algae stabilizing the structure of the enzymes lactate dehydrogenase and malate dehydrogenase of *Acrosiphonia arcta* (Karsten et al. 1996). Ice-binding proteins (IBP) that modify the shape of growing intracellular ice crystals during freezing were recently detected in Antarctic *Prasiola* and sea-ice diatoms (Raymond and Fritsen 2001). Presumably, IBPs prevent damage to membranes by the inhibition of the recrystallization of ice (Raymond and Knight 2003).

The low temperatures in the eulittoral and supralittoral represent a challenge to algal physiology because they are often combined with high irradiances in summer. At low temperatures enzyme activities and turnover velocity of the D1 reaction center protein in photosystem II are reduced (Andersson et al. 1992; Aro et al. 1993), which may result in increased electron pressure in photosynthesis and ultimately in the generation of reactive oxygen species (Dring 2006). The consequences of increased oxidative stress are chronic photoinhibition/photoinactivation, bleaching of photosynthetic pigments, peroxidation of membrane lipids, and enhanced degradation of D1 protein (see Chap. 6 by Bischof and Rautenberger). Eulittoral species, such as the green algae *Urospora penicilliformis* and *Ulva hookeriana*, the red alga *Porphyra endiviifolia*, and the brown alga *Adenocystis utricularis* may overcome radiation stress at low temperatures by their ability for dynamic photoinhibition/photoprotection, which proceeds much faster than in sublittoral algae (Hanelt et al. 1994, 1997). In the upper sublittoral red alga *Palmaria decipiens*, for example, there was a persistent impairment of photosynthetic activity at 0°C combined with irradiances of 400 μmol photons $m^{-2} s^{-1}$, but not at 200 μmol photons $m^{-2} s^{-1}$. In contrast, photosynthesis was not impaired under both light intensities at 8°C (Becker et al. 2011).

The effects of freezing in combination with high light intensities on photosynthesis were studied in the Arctic eulitttoral brown alga *Fucus distichus*. There was a marked decrease of optimum quantum yield with decreasing temperatures down to −10/−15°C and a rapid recovery as soon as the temperatures increased again (Becker et al. 2011). While photosynthetic activity was zero at 15°C in this species, in the Antarctic green alga *Prasiola* photosynthetic activity was observed down to −15°C (Becker 1982). The physiological basis of this extraordinary capacity is not known.

Desiccation is a strong stress parameter in supralittoral species (see also Chap. 5 by Karsten). Thalli of *Prasiola crispa* lose 75% of the cellular water during the first 6 h of exposure to air (Jacob et al. 1992). Irreversible damage occurred after a water loss of more than 90%. Growth rates after reimmersion in seawater depend on the thallus water content and the length of the desiccation period. Few ultrastructural changes were found after desiccation, but the very thick cell walls of the species and the absence of vacuoles are regarded as prerequisites for surviving periods of desiccation (Jacob et al. 1992).

13.4 Ecology of Polar Seaweed Communities

13.4.1 Biomass, Depth Distribution, and Productivity

As a result of the harsh environmental conditions, most seaweeds of the polar regions occur almost entirely in the sublittoral. In Antarctica, seaweeds dominate shallow benthic communities on hard substrates along the northern portion of the western Antarctic Peninsula and adjacent islands, often covering over 80% of the bottom with standing biomass levels in the range of 5–10 wet kg m^{-2}, which is comparable to temperate kelp forests (Amsler et al. 1995; Brouwer et al. 1995; Quartino et al. 2001; Quartino and Boraso de Zaixso 2008). Low intertidal pools and crevices in the upper sublittoral are frequently colonized by the red algae *Iridaea cordata* and *Palmaria decipiens*. Below this zone several species of large, perennial brown algae are particularly abundant. *Desmarestia menziesii* (shallower) and/or *D. anceps* (deeper) typically dominate in waters from 2 to 3 m down to approximately 10–20 m with *D. anceps* sometimes the dominant species to 30 m or more. *Himantothallus grandifolius* typically dominates from whatever depth at which the *Desmarestia* spp. thin out down to 40 m or greater (DeLaca and Lipps 1976; Zielinski 1990; Klöser et al. 1994; Amsler et al. 1995; Brouwer et al. 1995; Quartino et al. 2001) with productivity estimates in the northern part of its range estimated as 16–56 g C m^{-2} $year^{-1}$ (Drew and Hastings 1992). *Cystosphaera jacquinotii* often co-dominates with *H. grandifolius* or *D. anceps* below 20 m in relatively wave-exposed areas (Zielinski 1990; Chung et al. 1994; Amsler personal observations). These brown algal stands are accompanied by various undergrowth species, besides the mentioned *P. decipiens* and *I. cordata*, the red algae *Gigartina skottsbergii, Myriogramme manginii, M. smithii, Trematocarpus antarcticus, Gymnogongrus turquetii, Georgiella confluens*, and others (Wiencke and Clayton 2002; Wulff et al. 2011). In total, over 100 species of seaweeds are found in shallow water along the Antarctic Peninsula (Moe and DeLaca 1976; Klöser et al. 1996; Wiencke and Clayton 2002).

This general pattern of seaweed dominance is reported from numerous locations along the western Antarctic Peninsula region from its northeastern limit at Signy Island (60°S latitude) south to Anvers Island (64°S; e.g., Neushul 1965; Delépine et al. 1966; DeLaca and Lipps 1976; Lamb and Zimmerman 1977; Rakusa-Suszczewski and Zielinski 1993; Amsler et al. 1995; Brouwer et al. 1995; Klöser et al. 1996; Quartino et al. 2001). However, the limited reports of seaweed abundance and distribution on the southern half of the western Antarctic Peninsula describe a drop off in both species richness and biomass as one moves south from Anvers Island (DeLaca and Lipps 1976; Moe and DeLaca 1976; Barnes and Brockington 2003; Bowden 2005). Moe and DeLaca (1976) hypothesized that the decreases in biomass and diversity in more southerly sites are largely due to increases in annual ice cover. This relative decrease in biomass and coverage is not unique to the southern Antarctic Peninsula since the same qualitative patterns are described at other locations of similar latitude throughout most of the coast of Antarctica (e.g., Zaneveld 1966; Dhargalkar et al. 1988; Kirkwood and Burton 1988; Johnston et al. 2007). However, even at these locations seaweeds can dominate their communities. For example in East Antarctica, Irving et al. (2005) reported that *Himantothallus grandifolius* covered approximately 80% of the bottom at 12 m depth at two sites with the red seaweed, *Palmaria decipiens*, covering nearly all the remaining substrate and in another area Johnston et al. (2007) reported 80–90% cover of *Desmarestia* sp. and *P. decipiens* at 6 m depth and approximately 80% cover of *H. grandifolius* at 12 m.

In the southern-most Antarctic seaweed habitats in the Ross Sea, communities dominated by the frondose red seaweeds *Iridaea cordata* and *Phyllophora antarctica* and the crustose coralline alga *Phymatolithon foecundum* have been studied at Terra Nova Bay (74°, 40′ S; reviewed by Cattaneo-Vietti et al. 2000; Cormaci et al. 2000) and in McMurdo Sound (77° 40–50′ S Miller and Pearse 1991). *I. cordata* is more abundant in the 2–10 m depth range with densities up to 4,000 individuals m^{-2} and wet biomass levels up to 3.5 kg m^{-2} (Cattaneo-Vietti et al. 2000). *P. antarctica* dominates from 10 to 20 m depth with densities up to 10,000 individuals m^{-2}, wet biomass levels up to 1.6 kg m^{-2}, and percent cover up to 90% of the benthos (Miller and Pearse 1991; Cattaneo-Vietti et al. 2000). *P. foecundum* includes several coralline alga species previously reported from the Ross Sea (Alongi et al. 2002) and occurs to depths of 70 m (Cormaci et al. 2000). In McMurdo Sound, it is present at most sites which have been observed by divers whereas the foliose red algae occur at only a minority of locations (Miller and Pearse 1991; Amsler, personal observations). Both *P. foecundum* (Miller and Pearse 1991) and *I. cordata* (R. Robbins, personal communication) occur at Cape Armitage in McMurdo Sound, which is the southern-most point in the world where open ocean occurs over waters shallow enough to support seaweeds.

In the Arctic, several studies have documented seaweed communities growing on boulders as well as smaller rock and shell hard substrates in the Alaskan Beaufort Sea. These communities are dominated by the kelp *Laminaria solidungula* along with lesser numbers of the kelps *Saccharina latissima* and *Alaria esculenta*, six species of foliose red seaweeds, and crustose coralline red algae (Dunton et al. 1982; Busdosh et al. 1983). Wet biomass values for the kelps range from means of 0.067 kg m^{-2} where only 10–25% of the bottom is covered by rock to 0.26 kg m^{-2} where rock covers >25% of the bottom (Dunton et al. 1982). Kelp densities are higher in summer than winter, with maximum densities of 6–30 individuals m^{-2} in patches with high percentages of rock (Busdosh et al. 1983). The total mean wet biomass of red seaweeds is approximately 0.12 kg m^{-2} (Dunton et al. 1982).

L. solidungula productivity in this system has been estimated as 6–10 g C m^{-2} $year^{-1}$ (Dunton and Schell 1986) compared to an estimate of up to 20 g C m^{-2} $year^{-1}$ in the Canadian High Arctic (Chapman and Lindley 1980). In this area, however, there are no intertidal algae and seaweeds are restricted to small areas on hard substrata protected from ice scour (Dunton and Schonberg 2000).

At another Arctic location, at Hansneset in the middle part of Kongsfjorden in Svalbard, seaweed communities are better developed and include a greater diversity of species. The intertidal zone supports *Fucus distichus* and smaller brown algae while the shallow subtidal down to 2.5 m is characterized by several annual or pseudo-perennial seaweeds (Hop et al. 2002; Wulff et al. 2011). Important species are the green algae *Acrosiphonia* spp., *Spongomorpha* spp., and the brown algae *Chordaria flagelliformis* and *Chorda filum* (Wulff et al. 2011). The zone between 5 and 15 m is dominated by kelps including *Saccorhiza dermatodea*, *Alaria esculenta*, *Laminaria digitata*, and *Saccharina latissima*, along with a diversity of smaller red, green, and brown seaweeds (Wiencke et al. 2004). Overall, over 70 seaweed species occur here with wet biomass values ranging from 0.02 to 21 kg m^{-2} (Hop et al. 2002). Below this zone of kelps and kelp-like species *Desmarestia aculeata* and *D. viridis* can be very frequent (Beuchel and Gulliksen 2008). At greater depths red algae dominate the community, in particular *Coccotylus truncatus* and *Phycodrys rubens*, the latter growing down to as deep as 60 m (Wiencke, unpublished data). The endemic Arctic *Laminaria solidungula* occurs in deep water in the inner (colder) part of Kongsfjorden or as undergrowth species in the kelp forest. The inner part of the fjord is characterized by high sedimentation rates (Svendsen et al. 2002) and consequently by a much lower biodiversity. The described zonation pattern is in principle also typical for the Russian Arctic coast, eastern Greenland, and the Canadian high Arctic (Wulff et al. 2011).

The few studies on succession in polar seaweed communities were recently summarized by Campana et al. (2011). Succession starts with rapid colonizers, e.g., diatoms and ephemeral seaweeds and goes on with a recruitment of annual and perennial seaweed species in late winter–spring. Colonization is severely affected by physical disturbance and seasonal changes in abiotic conditions including UV stress (see Chap. 20 by Bischof and Steinhoff; Karsten et al. 2011) as well as by biotic factors, such as grazing (Zacher et al. 2007).

13.4.2 Elemental and Nutritional Content

Compared to seaweeds in almost all other parts of the world, Antarctic seaweeds have relatively low Carbon to Nitrogen (C:N) elemental ratios and relatively high tissue nitrogen levels (Dhargalkar et al. 1987; Weykam et al. 1996; Dunton 2001; Peters et al. 2005). This indicates that they are unlikely to be growth limited by nitrogen, which is probably a result of the relatively high nitrogen and other nutrient levels present in Antarctic coastal waters throughout the year (Weykam et al. 1996;

Peters et al. 2005; see Chap. 4 by Gordillo). Not surprisingly then, protein levels in Antarctic macroalgae are also higher than reported for most other places in the world (Peters et al. 2005). Since protein levels are important determinants of the nutritional value and palatability of seaweeds to herbivores (e.g., Horn and Neighbors 1984), this suggests that were it not for the common occurrence of chemical defenses in these algae they would likely be more valuable foods for herbivores than seaweeds from other regions. In the only study we are aware of in the Arctic, *L. solidungula* and *S. latissima* were shown to also have relatively low C:N ratios early in the growing season which increased as stored nitrogen was used for biomass gain during the summer (Henley and Dunton 1995).

13.4.3 Defenses Against Herbivory and Biofouling in Polar Seaweeds

Many seaweeds deter herbivores with organic compounds that make them unpalatable and/or interfere with the animals' digestion, and similar compounds can also be used to deter the settlement of biofouling spores or larvae (see also Chap. 8 by Iken and Chap. 9 by Amsler). In addition or as an alternative to chemical defenses, seaweeds can also decrease their palatability to herbivores by increasing their physical toughness (e.g., Lowell et al. 1991).

Chemical defenses against herbivores are very common in Antarctic seaweeds (Amsler et al. 2005a, 2008, 2011; Aumack et al. 2010). All of the dominant brown seaweeds along the western Antarctic Peninsula described and most of the common red seaweeds in the subtidal community elaborate chemical defenses against herbivores. Although physical toughness may play a role in some species, particularly against amphipods, chemical defenses appear to be far more important overall in deterring herbivory (Amsler et al. 2005a, 2008, 2011). Hence, although these communities are similar to temperate kelp forests in being dominated by large, perennial brown seaweeds, they differ from many such communities in that the seaweeds are able to resist herbivory directly rather than relying on top-down control of herbivores by their predators (e.g., Elner and Vadas 1990; Estes and Duggins 1995).

Both of the dominant red seaweeds in the Ross Sea, *I. cordata* and *P. antarctica*, are also chemically defended against a sea urchin which is the most obvious potential algal consumer in that system (Amsler et al. 1998). However, the sea urchins preferentially cover themselves with detached seaweeds when they are available, which in turn provides a physical barrier for defense against their main predator, large sea anemones (Amsler et al. 1999). This benefits the seaweeds since it keeps a large population of drift algae in the photic zone where they continue to photosynthesize and reproduce (Amsler et al. 1999; Schwarz et al. 2003).

The palatability of 19 seaweeds to herbivorous sea urchins and amphipods has been examined in the Arctic at Svalbard (Wessels et al. 2006). Of these, 17 seaweed species were unpalatable. By comparing assays on fresh seaweed thallus with artificial foods which removed physical defenses, it was concluded that 15 of the 17 seaweeds rely primarily on physical toughness to deter herbivores and that only two elaborate chemical defenses (Wessels et al. 2006). In one of these chemically defended seaweeds, *Desmarestia viridis*, the defensive compound has been shown to be sulfuric acid (Molis et al. 2008). *D. viridis* provides associational defense to other seaweeds growing near because it so strongly repels sea urchins (Molis et al. 2008).

To date, no studies—we are aware of—have examined polar seaweed chemical defenses against biofouling in an ecologically relevant context (see also Chap. 9 by Amsler). However, Antarctic seaweeds do produce compounds which are toxic to epiphytic diatoms in vitro (Amsler et al. 2005b; Iken et al. 2011; Sevak 2010). Grazing amphipods appear to be very important in reducing diatom fouling of these seaweeds in nature (see also Chap. 8 by Iken). However, when held in experimental mesocosms without amphipods, one of four species, *Desmarestia anceps,* still had only very low diatom coverage even though the other three species were heavily fouled (Aumack et al. 2011). *D. anceps* is one of the many species which have anti-diatom bioactivity in vitro, and it is possible that some of these compounds are also playing a role in controlling fouling in nature.

13.4.4 Trophic Interactions

Antarctic seaweeds appear to play important roles in benthic food webs in both shallow and deeper waters along the northern portion of the Antarctic Peninsula and they are likely to be of particular importance to benthic detrital food chains. A majority of the mid-water and, in particular, deep water carbon flux along the western Antarctic Peninsula can be composed of seaweed material, which also forms a significant part of the sediments (Liebezeit and von Bodungen 1987; Reichardt 1987; Fischer and Wiencke 1992). Seaweeds are also important in Antarctic coastal and maritime detrital food chains along the western Antarctic Peninsula (Zielinski 1981; Dawson et al. 1985). Drift seaweeds can be abundant in pockets on the coastal sea floor (Neushul 1965; Brouwer 1996b; Amsler, personal observations). A major portion of the primary production of *Himantothallus grandifolius* is lost as blade erosion and, in contrast to phytoplankton, this carbon input to the detrital communities continues throughout the year (Dieckmann et al. 1985). Although living, drift *Desmarestia anceps* decays fairly slowly via fragmentation, carbon from dead *D. anceps* is recycled relatively quickly (Brouwer 1996b). Stable isotope techniques have shown that macroalgal carbon makes important contributions to invertebrates via the detrital food webs in both shallow hard bottom communities where they grow (Dunton 2001) and in nearby, soft bottom communities (Corbisier et al. 2004) along the western Antarctic Peninsula. In McMurdo Sound, *Phyllophora antarctica* appears to be a similar, important source of carbon to hard and soft bottom communities, likely via the detrital food web (Norkko et al. 2004).

Antarctic seaweeds also provide food and cover to large numbers of invertebrates and fishes (e.g., DeLaca and Lipps 1976; Richardson 1977; Iken 1999; Huang et al. 2007; Zamzow et al. 2011). Amphipods occur on these seaweeds at exceptionally high densities. Richardson (1977) reported over 10,000 amphipods per single *Desmarestia anceps* while Amsler et al. (2008), combining datasets from the same community collected by Amsler et al. (1995) and Huang et al. (2007), estimated densities of amphipods in solid stands of *D. menziesii* at over 300,000 individuals m^{-2} of the benthos with densities in solid stands of *D. anceps* and *Plocamium cartilagineum* of approximately 30,000 individuals m^{-2}. All the seaweed species that have been reported with very high amphipod densities are also chemically defended against amphipod herbivory (Amsler et al. 2005a; Huang et al. 2007; Aumack et al. 2010) and as such are almost certainly not being consumed by the amphipods. Instead, the amphipods are grazing epiphytic diatoms and other macroalgae, as well as filamentous algal epiphytes and emergent filaments from algal endophytes (Aumack et al. 2011). Removing these biofouling algae undoubtedly benefits the host seaweeds. The amphipods also benefit from associating with the chemically defended seaweeds as they are much less likely to be consumed by omnivorous fish (Zamzow et al. 2010). Therefore, the seaweeds, which are the dominant, habitat-forming organisms in the community and the amphipods, which are by far the most numerous animals in the community, exist in a community-wide, mutualistic relationship.

Filamentous algae growing as epiphytes are very rare along the western Antarctic Peninsula (Peters 2003). However, filamentous algae growing as endophytes within the larger, chemically defended seaweeds are unusually common, which has probably been selected for as a refuge from the dense amphipod assemblage (Peters 2003; Amsler et al. 2009; see also Chap. 9 by Amsler and Chap. 11 by Potin). Although endophytes do not necessarily harm their hosts, serious pathogenic effects of filamentous algal endophytes on their seaweed hosts are well known (e.g., Apt 1988; Correa and Sánchez 1996; Peters and Schaffelke 1996). Whether or not, and if so how frequently, the abundant algal endophytes in Antarctic seaweeds are deleterious to their hosts and whether or not Antarctic seaweeds use chemical defenses against endophytes are important areas for future study.

Although seaweeds in Arctic Spitsbergen support a diversity of macrofauna (Lippert et al. 2001; Włodarska-Kowalczuk et al. 2009), only 2 of 19 such species examined were shown to consume the seaweeds (Wessels et al. 2006). However, it is likely that, as in Antarctica, seaweed carbon is important in this system via the detrital food web (Hop et al. 2002). In the Alaskan Beaufort Sea, stable isotope techniques have shown that kelps are an important carbon source to many invertebrate species including direct herbivores and animals deriving the carbon via detritus (Dunton and Schell 1987).

13.5 Impact of Global Climate Changes on Seaweeds and Their Communities

As stratospheric ozone depletion is highest in polar regions, numerous studies were performed on the effect of enhanced UV radiation (UVR) on seaweeds from the Arctic and Antarctic (see Chap. 20 by Bischof and Steinhoff; Karsten et al. 2011). The most UV-susceptible stages in the life history of seaweeds are their spores (Roleda et al. 2007). Spores of shallow water species exhibit a high UV tolerance, whereas spores of mid- and lower sublittoral species are increasingly UV susceptible (Wiencke et al. 2006). Tolerance of spores to UVR is a major if not one of the most important factors determining the upper depth distribution limit of kelps and kelp-like species. Enhanced UVR due to stratospheric ozone depletion may lead to changes in the depth distribution of seaweeds which may cause significant snowballing effects.

As temperature is one of the most important factors controlling biogeographic distribution of seaweeds, distributional shifts are an inevitable effect of global warming, especially in polar and cold-temperate regions (see Chap. 18 by Bartsch et al.). Modeled temperature changes through the end of twenty-first century indicate that North Atlantic-polar to cold-temperate seaweeds will extend their range into the high Arctic, but retreat along the northeastern Atlantic coasts. In contrast, many Antarctic seaweeds will presumably not strongly alter latitudinal distributions due to changes in temperature (Müller et al. 2011). Clearly, the distributional changes of key species as so-called ecological engineers will provoke substantial and cascading effects in polar and cold-temperate transition areas with strong consequences for biodiversity and ecosystem functioning.

Although increasing temperatures due to climate change may not affect the latitudinal distribution of Antarctic seaweeds directly, it very likely is doing so indirectly via changes in sea-ice extent and duration, particularly along the southern portion of the western Antarctic Peninsula. Sea-ice cover in this region has been dramatically changed by increases in air temperatures over the past 30 years, advancing nearly 2 months later in winter and retreating approximately 1 month earlier in spring (Smith and Stammerjohn 2001; Stammerjohn et al. 2008). These changes are likely to continue and are thought to be significantly impacting marine communities (Clarke et al. 2007; Ducklow et al. 2007; McClintock et al. 2008). The marked decrease in seaweed biomass and species richness historically observed along the southern portion of the Antarctic Peninsula is thought to be a result of the increasing sea-ice coverage as one moves south. It seems likely that seaweed communities typical of the northern portion of the western Antarctic Peninsula are expanding southward. Unfortunately, seaweed floras in the area between Anvers and Adelaide Islands (64°S–67°35′S) are very poorly studied, both historically and currently, so the extent to which such changes have and are occurring is unknown.

Polar regions are not only affected by stratospheric ozone depletion and global warming, but certainly also threatened by ocean acidification as a result of human use of fossil fuels (see Chap. 19 by Roleda and Hurd). When CO_2 enters the ocean surface, CO_2 and bicarbonate concentrations will increase, and in contrast, pH and carbon ion concentrations will decrease (Feely et al. 2004). As the solubility of gases is higher in colder waters, polar regions will be most strongly affected. In general, effects of ocean acidification on seaweeds and their communities are poorly understood. Elevated CO_2 levels may support primary production, although species with a carbon concentrating mechanism (Giordano et al. 2005) may show less strong responses. Calcification of calcareous algae will be impaired (Leclerc et al. 2000). For polar waters, there is only one publication addressing this important question (McClintock et al. 2009). Therefore, we need to study the responses of polar seaweeds to elevated CO_2 levels and ocean acidification in more detail and also investigate possible acclimation processes.

13.6 Conclusion

In the above summary of our present knowledge about polar seaweeds, gaps are clearly apparent. The biodiversity of polar algae must be studied more intensively to better understand their evolutionary history, biogeographical relationships, and their physiological performance. With respect to their phenology, an open question is the perception of daylength, which controls seasonal growth and reproduction in endemic Antarctic and Arctic seaweeds. Storage of photosynthetic products, their remobilization and translocation to the meristems, and prerequisites for growth in darkness and low light are not fully explored. Moreover, it is necessary to identify the thresholds of tolerance to low and high light, UV radiation, temperature, freezing, and other stresses and to quantify the related physiological processes by genomic, proteomic, and metabolomic approaches in order to better understand the ecology of individual species.

In seaweed communities from the Antarctic, the function of secondary metabolites in chemical defense and functioning of these systems need to be studied more intensively, not only with respect to herbivory, but also to biofouling. The putative photoprotective role of phlorotannins needs to be studied in greater detail. As the molecular structure of phlorotannins is still obscure, more studies are necessary also in this field. An improved knowledge of these compounds should allow us to investigate the effects of these brown algal secondary metabolites on grazers in a better way. In particular, the effects on the digestive system of consumers should come into the focus of science. Compared to Antarctic seaweeds, much less is known about seaweed–herbivore relationships in the Arctic, which therefore need to be studied much more intensively. Finally, the effect of global climate changes, especially of global warming and ocean acidification on seaweeds and their communities, must urgently be studied in order to develop scenarios about changes in the functioning of polar seaweed systems. In this respect, special emphasis should be laid on interactive effects between temperature, CO_2 level, pH, and radiation conditions to obtain a more plastic picture of future trends.

Acknowledgments We thank M. Amsler for editorial suggestions. Manuscript preparation was supported in part by US National Science Foundation award ANT-0838773 from the Antarctic Organisms and Ecosystems program.

References

Adey WH, Lindstrom SC, Hommersand MH, Müller KM (2008) The biogeographic origin of Arctic endemic seaweeds: a thermogeographic view. J Phycol 44:1384–1394

Aguilera J, Bischof K, Karsten U, Hanelt D, Wiencke C (2002) Seasonal variation in ecological patterns in macroalgae from an Arctic fjord II. Pigment accumulation and biochemical defence systems against high light stress. Mar Biol 140:1087–1095

Alongi G, Cormaci M, Furnari G (2002) The Corallinaceae (Rhodophyta) from the Ross Sea (Antarctica): a taxonomic revision rejects all records except *Phymatolithon foecundum*. Phycologia 41:140–146

Amsler CD, Rowley RJ, Laur DR, Quetin LB, Ross RM (1995) Vertical distribution of Antarctic Peninsular macroalgae: cover, biomass, and species composition. Phycologia 34:424–430

Amsler CD, McClintock JB, Baker BJ (1998) Chemical defense against herbivory in the Antarctic marine macroalgae *Iridaea cordata* and *Phyllophora antarctica* (Rhodophyceae). J Phycol 34:53–59

Amsler CD, McClintock JB, Baker BJ (1999) An antarctic feeding triangle: defensive interactions between macroalgae, sea urchins, and sea anemones. Mar Ecol Prog Ser 183:105–114

Amsler CD, Iken K, McClintock JB, Amsler MO, Peters KJ, Hubbard JM, Furrow FB, Baker BJ (2005a) Comprehensive evaluation of the palatability and chemical defenses of subtidal macroalgae from the Antarctic Peninsula. Mar Ecol Prog Ser 294:141–159

Amsler CD, Okogbue IN, Landry DM, Amsler MO, McClintock JB, Baker BJ (2005b) Potential chemical defenses against diatom fouling in Antarctic macroalgae. Bot Mar 48:318–322

Amsler CD, McClintock JB, Baker BJ (2008) Macroalgal chemical defenses in polar marine communities. In: Amsler CD (ed) Algal chemical ecology. Springer, Berlin, pp 91–103

Amsler CD, Amsler MO, McClintock JB, Baker BJ (2009) Filamentous algal endophytes in macrophytic Antarctic algae: prevalence in hosts and palatability to mesoherbivores. Phycologia 48:324–334

Amsler CD, Iken K, McClintock JB, Baker BJ (2011) Defenses of polar macroalgae against herbivores and biofoulers. In: Wiencke C (ed) Biology of polar benthic algae. De Gruyter, Berlin, pp 101–120

Andersson B, Salter AH, Virgin I, Vass I, Styring S (1992) Photodamage to photosystem II-primary and secondary events. J Photochem Photobiol B: Biol 15:15–31

Apt KE (1988) Ethiology and development of hyperplasia induced by *Streblonema* sp. (Phaeophyta) on members of the Laminariales (Phaeophyta). J Phycol 24:28–34

Aro EM, Virgin I, Andersson B (1993) Photoinhibition of photosystem II. Inactivation, protein damage and turnover. Biochim Biophys Acta 1143:113–134

Aumack CF, Amsler CD, McClintock JB, Baker BJ (2010) Chemically mediated resistance to mesoherbivory in finely branched macroalgae along the western Antarctic Peninsula. Eur J Phycol 45:19–26

Aumack CF, Amsler CD, McClintock JB, Baker BJ (2011) Impacts of mesograzers on epiphyte and endophyte growth associated with chemically defended macroalge from the western Antarctic Peninsula: a mesocosm experiment. J Phycol 47:36–41

Barnes D, Brockington S (2003) Zoobenthic biodiversity, biomass and abundance at Adelaide Island. Antarctica Mar Ecol Prog Ser 249:145–155

Becker EW (1982) Physiological studies on Antarctic *Prasiola crispa* and *Nostoc commune* at low temperatures. Polar Biol 1:99–104

Becker S, Walter B, Bischof K (2011) Freezing tolerance and photosynthetic performance of polar seaweeds at low temperatures. In: Wiencke C (ed) Biology of Polar Benthic Algae. De Gruyter, Berlin, pp 221–236

Beuchel F, Gulliksen B (2008) Temporal patterns of benthic community development in an Arctic fjord (Kongsfjorden, Svalbard): results of a 24-year manipulation study. Polar Biol 31:913–924

Bischoff B, Wiencke C (1993) Temperature requirements for growth and survival of macroalgae from Disko-Island (Greenland). Helgol Mar Res 47:167–191

Bischoff B, Wiencke C (1995) Temperature ecotypes and biogeography of Acrosiphoniales (Chlorophyta) with Arctic-Antarctic disjunct and Arctic/cold-temperate distributions. Eur J Phycol 30:19–27

Bischoff-Bäsmann B, Wiencke C (1996) Temperature requirements for growth and survival of Antarctic Rhodophyta. J Phycol 32:525–535

Bolton JJ, Lüning K (1983) Optimal growth and maximal survival temperatures of Atlantic *Laminaria* species (Phaeophyta) in culture. Mar Biol 66:89–94

Bowden DA (2005) Quantitative characterization of shallow marine benthic assemblages at Ryder Bay, Adelaide Island. Antarctica. Mar Biol 146:1235–1249

Brouwer PEM (1996a) In situ photosynthesis and estimated annual production of the red macroalga *Myriogramme mangini* in relation to underwater irradiance at Signy Island (Antarctica). Antarctic Sci 8:245–252

Brouwer PEM (1996b) Decomposition in situ of the sublittoral Antarctic macroalga *Desmarestia anceps* Montagne. Polar Biol 16:129–137

Brouwer PEM, Geilen EFM, Gremmen NJM, van Lent F (1995) Biomass, cover and zonation pattern of sublittoral macroalgae at Signy Island, South Orkney Islands. Antarctica. Bot Mar 38:259–270

Busdosh M, Beehler CL, Robillard GA, Tarbox KR (1983) Distribution and abundance of kelp in the Alaskan Beaufort Sea near Prudhoe Bay. Arctic 38:18–22

Campana GL, Zacher K, Fricke F, Molis M, Wulff A, Quartino ML, Wiencke C (2011) Drivers of colonization and succession in polar benthic macro- and microalgal communities. In: Wiencke C (ed) Biology of polar benthic algae. De Gruyter, Berlin, pp 299–320

Cattaneo-Vietti R, Chiantore M, Gambi MC, Albertelli G, Cormaci M, Di Geronimo I (2000) Spatial and vertical distribution of benthic littoral communities in Terra Nova Bay. In: Faranda FM, Guglielmo L, Ianora A (eds) Ross sea ecology. Italiantartide expeditions (1985–1995). Springer, Berlin, pp 503–514

Chapman ARO, Lindley JE (1980) Seasonal growth of *Laminaria solidungula* in the Canadian high Arctic in relation to irradiance and dissolved nutrient concentration. Mar Biol 57:1–5

Chung H, Oh YS, Lee IK, Kim D-Y (1994) Macroalgal vegetation of Maxwell Bay in King George Island. Antarctica Korean J Phycol 9:47–58

Clarke DL (1990) Arctic Ocean ice cover; Geologic history and climatic significance. In: Grantz A, Johnson L, Sweeney JL (eds) The arctic ocean region. Geol Soc, America, Boulder Colorado, pp 53–62

Clarke A, Murphy EJ, Meredith MP, King JC, Peck LS, Barnes DKA, Smith RC (2007) Climate change and the marine ecosystem of the western Antarctic Peninsula. Philos Trans R Soc B-Biol Sci 362:149–166

Corbisier TN, Petti MAV, Skowronski RSP, Brito TAS (2004) Trophic relationships in the nearshore zone of Martel Inlet (King George Island, Antarctica): $\delta^{13}C$ stable-isotope analysis. Polar Biol 27:75–82

Cormaci M, Furnari G, Scammacca B (2000) The macrophytobenthos of Terra Nova Bay. In: Faranda FM, Guglielmo L, Ianora A (eds) Ross Sea Ecology. Italiantartide Expeditions (1987-1995). Springer, Berlin, pp 493-502

Correa JA, Sánchez PA (1996) Ecological aspects of algal infectious diseases. Hydrobiologia 326–327:89–95

Crame JA (1993) Latitudinal range fluctuations in the marine realm through geological time. Trends Ecol Evol 8:162–266

Dawson R, Schramm W, Bolter M (1985) Factors influencing the production, decomposition and distribution of organic matter in Admirality Bay, King George Island. In: Siegfried WR, Condy PR, Laws RM (eds) Antarctic nutrient cycles and food webs. Springer, Berlin, pp 109–114

DeLaca TE, Lipps JH (1976) Shallow water marine associations. Antarctic Peninsula. Antarctic J US 11:12–20

Delépine R, Lamb IM, Zimmerman MH (1966) Preliminary report on the marine vegetation of the Antarctic Peninsula. Proc Int Seaweed Symp 5:107–116

Dhargalkar V, Reddy C, Deshmukhe G, Unatawale A (1987) Biochemical composition of some benthic marine algae of the Vestfold Hills, Antarctica. Indian J Mar Sci 16:269–271

Dhargalkar V, Burton H, Kirkwood J (1988) Animal associations with the dominant species of shallow water macrophytes along the coastline of the Vestfold Hills, Antarctica. Hydrobiologia 165:141–150

Dieckmann G, Reichardt W, Zielinski D (1985) Growth and production of the seaweed, *Himantothallus grandifolius*, at King George Island. In: Siegfried WR, Condy P, Laws RM (eds) Antarctic nutrient cycles and food webs. Springer, Berlin, pp 104–108

Drew EA, Hastings RM (1992) A year-round ecophysiological study of *Himantothallus grandifolius* (Desmarestiales, Phaeophyta) at Signy Island, Antarctica. Phycologia 31:262–277

Dring MJ (2006) Stress resistance and disease resistance in seaweeds: The role of reactive oxygen metabolism. Adv Bot Res 43:175–207

Ducklow HW, Baker K, Martinson DG, Quetin LB, Ross RM, Smith RC, Stammerjohn SE, Vernet M, Fraser W (2007) Marine pelagic ecosystems: the west Antarctic Peninsula. Philos Trans R Soc B 362:67–94

Dummermuth AL, Wiencke C (2003) Experimental investigation of seasonal development in six Antarctic red macroalgae. Antarct Sci 15:449–457

Dunton KH (1985) Growth of dark-exposed *Laminaria saccharina* (L.) Lamour and *Laminaria solidungula* J. Ag (Laminariales: Phaeophyta) in the Alaskan Beaufort Sea. J Exp Mar Biol Ecol 94:181–189

Dunton KH (1990) Growth and production in *Laminaria solidungula*: relation to continuous underwater light levels in the Alaskan high Arctic. Mar Biol 106:297–304

Dunton KH (1992) Arctic biogeography: The paradox of the marine benthic fauna and flora. Trends Ecol Evol 7:183–189

Dunton K (2001) $d^{15}N$ and $d^{13}C$ measurements of Antarctic Peninsula fauna: trophic relationships and assimilation of benthic seaweeds. Am Zool 41:99–112

Dunton KH, Schell DM (1986) Seasonal carbon budget and growth of *Laminaria solidungula* in the Alaskan High Arctic Mar. Ecol Prog Ser 31:57–66

Dunton KH, Schell DM (1987) Dependence of consumers on macroalgal (*Laminaria solidungula*) carbon in an arctic kelp community. $d^{13}C$ evidence. Mar Biol 93:615–625

Dunton KH, Schonberg SV (2000) The benthic faunal assemblage of the Boulder Patch kelp community. In: Johnson SR, Truett JC (eds) The natural history of an arctic oil field. Academic, San Diego, pp 372–397

Dunton KH, Reimnitz E, Schonberg S (1982) An Arctic kelp community in the Alaskan Beaufort Sea. Arctic 35:465–484

Elner R, Vadas RJ (1990) Inference in ecology: the sea urchin phenomenon in the northwest Atlantic. Am Nat 136:108–125

Estes J, Duggins D (1995) Sea otter and kelp forests in Alaska: generality and variation in a community ecological paradigm. Ecol Monogr 65:75–100

Feely RA, Sabine CL, Lee K, Berelson W, Kleypas J, Fabry VJ, Millero FJ (2004) Impact of anthropogenic CO_2 on the $CaCO_3$ system in the oceans. Science 305:362–366

Fischer G, Wiencke C (1992) Stable carbon isotope composition, depth distribution and fate of macroalgae from the Antarctic Peninsula region. Polar Biol 12:341–348

Giordano MJ, Beardall J, Raven JA (2005) CO_2 concentrating mechanisms in algae: mechanisms, environmental modulation, and evolution. Annu Rev Plant Biol 56:99–131

Gómez I, Wiencke C (1997) Seasonal growth and photosynthetic performance of the Antarctic macroalga *Desmarestia menziesii* (Phaeophyceae) cultivated under fluctuating Antarctic daylengths. Bot Acta 110:25–31

Gómez I, Wiencke C, Weykam G (1995) Seasonal photosynthetic characteristics of the brown alga *Ascoseira mirabilis* from King George Island (Antarctica). Mar Biol 123:167–172

Gómez I, Wiencke C, Thomas DN (1996) Variations in photosynthetic characteristics of the Antarctic marine brown alga *Ascoseira mirabilis* in relation to thallus age and size. Eur J Phycol 31:167–172

Gómez I, Weykam G, Klöser H, Wiencke C (1997) Photosynthetic light requirements, daily carbon balance and zonation of sublittoral macroalgae from King George Island (Antarctica). Mar Ecol Progr Ser 148:281–293

Gómez I, Wulff A, Roleda MY, Huovinen P, Karsten U, Quartino ML, Dunton K, Wiencke C (2011) Light and temperature demands of marine benthic microalgae and seaweeds in polar regions. In: Wiencke C (ed) Biology of polar benthic algae. De Gruyter, Berlin, pp 195–220

Hanelt D (1998) Capability of dynamic photoinhibition in Arctic macroalgae is related to their depth distribution. Mar Biol 131:361–369

Hanelt D, Jaramillo J, Nultsch W, Senger S, Westermeier R (1994) Photoinhibitioin as a regulative mechanism of photosynthesis in marine algae of Antarctica. Ser Cient INACH 44:67–77

Hanelt D, Melchersmann B, Wiencke C, Nultsch W (1997) Effects of high light stress on photosynthesis of polar macroalgae in relation to depth distribution. Mar Ecol Prog Ser 149:255–266

Hanelt D, Tüg H, Bischof K, Gross C, Lippert H, Sawall T, Karsten U, Wiencke C (2001) Light regime in an Arctic fjord: a study related to stratospheric ozone depletion as a basis for determination of UV effects on algal growth. Mar Biol 138:649–658

Henley WJ, Dunton KH (1995) A seasonal comparison of carbon, nitrogen, and pigment content in *Laminaria solidungula* and *L. saccharina* (Phaeophyta) in the Alaskan Arctic. J Phycol 31:325–331

Hommersand MH, Moe RL, Amsler CD, Fredericq S (2011) Notes on the systematics and biogeographical relationships of Antarctic and sub-Antarctic Rhodophyta with descriptions of four new genera and five new species. In: Wiencke C (ed) Biology of Polar Benthic Algae. De Gruyter, Berlin, pp 53–100

Hop H, Pearson T, Hegseth EN, Kovacs KM, Wiencke C, Kwasniewski S, Eiane K, Mehlum F, Gulliksen B, Wlodarska-Kowalczuk M, Lydersen C, Weslawski JM, Cochrane S, Gabrielsen GW, Leakey RJG, Lønne OJ, Zajaczkowski M, Falk-Petersen S, Kendall M, Wängberg S-Å, Bischof K, Voronkov AY, Kovaltchouk NA, Wiktor J, Poltermann M, di Prisco G, Papucci C, Gerland S (2002) The marine ecosystem of Kongsfjorden. Svalbard Polar Res 21:167–208

Horn M, Neighbors M (1984) Protein and nitrogen assimilation as a factor in predicting the seasonal macroalgal diet of the monkeyface prickleback. Trans Amer Fish Soc 113:388–396

Huang YM, Amsler MO, McClintock JB, Amsler CD, Baker BJ (2007) Patterns of gammarid amphipod abundance and species composition associated with dominant subtidal macroalgae along the western Antarctic Peninsula. Polar Biol 30:1417–1430

Iken K (1999) Feeding ecology of the Antarctic herbivorous gastropod *Laevilacunaria antarctica* Martens. J Exp Mar Biol Ecol 236:133–148

Iken K, Amsler CD, Amsler MO, McClintock JB, Baker BJ (2011) Field studies on deterrent properties of phlorotannins in Antarctic brown algae. In: Wiencke C (ed) Biology of Polar Benthic Algae. De Gruyter, Berlin, pp 121–140

Irving A, Connell S, Johnston E, Pile A, Gillanders B (2005) The response of encrusting coralline algae to canopy loss: an independent test of predictions on an Antarctic coast. Mar Biol 147:1075–1083

Jacob A, Kirst GO, Wiencke C, Lehmann H (1991) Physiological responses of the Antarctic green alga *Prasiola crispa* ssp. *antarctica* to salinity stress. J. Plant Physiol 139:57–62

Jacob A, Wiencke C, Lehmann H, Kirst GO (1992) Physiology and ultrastructure of desiccation in the green alga *Prasiola crispa* from Antarctica. Bot Mar 35:297–303

Johnston E, Connell S, Irving A, Pile A, Gillanders B (2007) Antarctic patterns of shallow subtidal habitat and inhabitants in Wilke's Land. Polar Biol 30:781–788

Kain JM (1989) The seasons in the subtidal. Br Phycol J 24:203–215

Karsten U, Wiencke C, Kirst GO (1991a) The effect of salinity changes upon physiology of eulittoral green macroalgae from Antarctica and Southern Chile. I. Cell viability, growth, photosynthesis and dark respiration. J. Plant Physiol 138:667–673

Karsten U, Wiencke C, Kirst GO (1991b) The effect of salinity changes upon physiology of eulittoral green macroalgae from Antarctica and Southern Chile II. Inorganic ions and organic compounds. J Exp Bot 42:1533–1539

Karsten U, Kück K, Vogt C, Kirst GO (1996) Dimethylsulfoniopropionate production in phototrophic organisms and its physiological function as a cryoprotectant. In: Kiene RP (ed) Biological and environmental chemistry of DMSP and related sulfonium compounds. Plenum, New York, pp 143–153

Karsten U, Bischof K, Wiencke C (2001) Photosynthetic performance of Arctic macroalgae after transplantation from deep to shallow waters. Oecologia 127:11–20

Karsten U, Wulff A, Roleda MY, Müller R, Steinhoff FS, Fredersdorf J, Wiencke C (2011) Physiological responses of polar benthic algae to ultraviolet radiation. In: Wiencke C (ed) Biology of polar benthic algae. De Gruyter, Berlin, pp 271–298

Kirkwood J, Burton H (1988) Macrobenthic assemblages in Ellis Fjord, Vestfold Hills. Antarctica Mar Biol 97:445–457

Kirst GO (1990) Salinity tolerance of eukaryotic marine algae. Annu Rev Plant Physiol Plant Mol Biol 41:21–53

Klöser H, Mercuri G, Laturnus F, Quartino ML, Wiencke C (1994) On the competive balance of macroalgae at Potter Cove (King George Island, South Shetlands). Polar Biol 14:11–16

Klöser H, Quartino ML, Wiencke C (1996) Distribution of macroalgae and macroalgal communities in gradients of physical conditions in Potter Cove, King George Island, Antarctica. Hydrobiologia 333:1–17

Lamb IM, Zimmerman MH (1977) Benthic marine algae of the Antarctic Peninsula. Antarctic Research Series 5:130–229

Lawver LA, Sclater JG, Meinke M (1985) Mesozoic and Cenozoic Reconstructions of the South Atlantic. Tectonophysics 114:233–254

Leclerc N, Gattuso JP, Jaubert J (2000) CO_2 partial pressure controls the calcification rate of a coral community. Global Change Biol 6:329–334

Liebezeit G, von Bodungen B (1987) Biogenic fluxes in the Bransfield Strait: Planktonic *versus* macroalgal sources. Mar Ecol Prog Ser 36:23–32

Lippert H, Iken K, Rachor E, Wiencke C (2001) Macrofauna associated with macroalgae in the Kongsfjord (Spitsbergen). Polar Biol 24:512–522

Lowell RB, Markham JH, Mann KH (1991) Herbivore-like damage induces increased strength and toughness in a seaweed. Proc R Soc London: Biol Sci 243:31–38

Lüder UH, Knoetzel J, Wiencke C (2001a) Acclimation of photosynthesis and pigments to seasonally changing light conditions in the endemic Antarctic red macroalga *Palmaria decipiens*. Polar Biol 24:598–603

Lüder UH, Knoetzel J, Wiencke C (2001b) Two forms of phycobilisomes in the Antarctic red macroalga *Palmaria decipiens* (Palmariales, Florideophyceae). Physiol Plant 112:572–581

Lüder UH, Knoetzel J, Wiencke C (2002) Acclimation of photosynthesis and pigments during and after six months of darkness in *Palmaria decipiens* (Rhodophyta)—a study to simulate Antarctic winter sea ice cover. J Phycol 38:904–913

Lüning K (1988) Photoperiodic control of sorus formation in the brown alga *Laminaria saccharina*. Mar Ecol Prog Ser 45:137–144

Lüning K (1990) Seaweeds—their environment, biogeography and ecophysiology. Wiley, New York

Lüning K (1991) Circannual growth rhythm in a brown alga. *Pterygophora californica*. Bot Acta 104:157–162

Lüning K, Dring MJ (1979) Continuous underwater light measurement near Helgoland (North Sea) and its significance for characteristic light limits in the sublittoral region. Helgol Meeresunters 32:403–424

Lüning K, Kadel P (1993) Daylength range for circannual rhythmicity in *Pterygophora californica* (Alariaceae, Phaeophyta) and synchronisation of seasonal growth by daylength cycles in several other brown algae. Phycologia 32:379–387

McClintock JB, Ducklow H, Fraser W (2008) Ecological responses to climate change on the Antarctic Peninsula. Am Sci 96:302–310

McClintock JB, Angus RA, McDonald MR, Amsler CD, Catledge SA, Vohra YK (2009) Rapid dissolution of shells of weakly calcified Antarctic benthic macroorganisms indicates high vulnerability to ocean acidification. Antarct Sci 21:449–456

Miller KA, Pearse JS (1991) Ecological studies of seaweeds in McMurdo Sound. Antarctica. Am Zool 31:35–48

Moe RL, DeLaca TE (1976) Occurrence of macroscopic algae along the Antarctic Peninsula. Antarctic J US 11:20–24

Molis M, Wessels H, Hagen W, Karsten U, Wiencke C (2008) Do sulphuric acid and the brown alga *Desmarestia viridis* support community structure in Arctic kelp patches by altering grazing impact, distribution patterns, and behaviour of sea urchins? Polar Biol 32:71–82

Müller R, Laepple T, Bartsch I, Wiencke C (2011) Impact of oceanic warming on the distribution of seaweeds in polar and cold-temperate waters. In: Wiencke C (ed) Biology of polar benthic algae. de Gruyter, New York, pp 237–270

Neushul M (1965) Diving observation of sub-tidal Antarctic marine vegetation. Bot Mar 8:234–243

Norkko A, Thrush SF, Cummings VJ, Funnell GA, Schwarz AM, Andrew NL, Hawes I (2004) Ecological role of *Phyllophora antarctica* drift accumulations in coastal soft-sediment communities of McMurdo Sound. Antarctica Polar Biol 27:482–494

Novaczek I, Lubbers GW, Breeman AM (1990) Thermal ecotypes in amphi-Atlantic algae. I. Algae of Arctic to cold-temperate distribution (*Chaetomorpha melagonium*, *Devaleraea ramentacea* and *Phycodrys rubens*). Helgoländer Meeresunters 44:459–474

Pedersen PM (1976) Marine, benthic algae from southernmost Greenland. Meddr Grønland 3:1–80, 7 plates

Peters AF (2003) Molecular identification, taxonomy and distribution of brown algal endophytes, with emphasis on species from Antarctica. Proc Int Seaweed Symp 17:293–302

Peters AF, Breeman AM (1993) Temperature tolerance and latitudinal range of brown algae from temperate Pacific South America. Mar Biol 115:143–150

Peters AF, Schaffelke B (1996) *Streblonema* (Ectocarpales, Phaeophyceae) infection in the kelp *Laminaria saccharina* (Laminariales, Phaeophyceae) in the western Baltic. Hydrobiologia 326–327:111–116

Peters AF, van Oppen MJH, Wiencke C, Stam WT, Olsen JL (1997) Phylogeny and historical ecology of the Desmarestiaceae (Phaeophyceae) support a Southern Hemisphere origin. J Phycol 33:294–309

Peters KJ, Amsler CD, Amsler MO, McClintock JB, Dunbar RB, Baker BJ (2005) A comparative analysis of the nutritional and elemental composition of macroalgae from the western Antarctic Peninsula. Phycologia 44:453–463

Quartino ML, Boraso de Zaixso AL (2008) Summer macroalgal biomass in Potter Cove, South Shetland Islands. Antarctica: its production and flux to the ecosystem. Polar Biol 31:281–294

Quartino ML, Klöser H, Schloss IR, Wiencke C (2001) Biomass and associations of benthic marine macroalgae from the inner Potter Cove (King George Island, Antarctica) related to depth and substrate. Polar Biol 24:349–355

Rakusa-Suszczewski S, Zielinski K (1993) Macrophytobenthos. In: Rakusa-Suszczewski S (ed) The maritime Antarctic coastal ecosystem of Admiralty Bay. Polish Academy of Sciences, Warsaw, pp 57–60

Raymond JA, Fritsen CH (2001) Semipurification and ice recrystallization inhibition activity of ice-active substances associated with Antarctic photosynthetic organisms. Cryobiology 43:63–70

Raymond JA, Knight CA (2003) Ice binding, recrystallization inhibition, and cryoprotective properties of ice-active substances associated with Antarctic sea ice diatoms. Cryobiology 46:174–181

Reichardt W (1987) Burial of antarctic macroalgal debris in bioturbated deep-sea sediments. Deep-Sea Res 34:1761–1770

Richardson MG (1977) The ecology including physiological aspects of selected Antarctic marine invertebrates associated with inshore macrophytes. PhD Dissertation. Department of Zoology, University of Durham

Richter A, Wuttke S, Zacher K (2008) Two years of in situ UV measurements at the Dallmann Laboratory/Jubany Station. Ber Polarforsch Meeresforsch 571:12–19

Roleda MY (2009) Photosynthetic response of Arctic kelp zoospores exposed to radiation and thermal stress. Photochem Photobiol Sci 8:1302–1312

Roleda MY, Wiencke C, Hanelt D, Bischof K (2007) Sensitivity of the early life stages of macroalgae from the northern hemisphere to ultraviolet radiation. Photochem Photobiol 83:1–12

Schaffelke B, Lüning K (1994) A circannual rhythm controls seasonal growth in the kelp *Laminaria hyperborea* and *L. digitata* from Helgoland (North Sea). Eur J Phycol 29:49–56

Schwarz AM, Hawes I, Andrew N, Norkko A, Cummings V, Thrush S (2003) Macroalgal photosynthesis near the southern global limit for growth; Cape Evans, Ross Sea, Antarctica. Polar Biol 26:789–799

Sevak HP (2010) Potential chemical defenses against diatom fouling in macroalgae from the Antarctic Peninsula: insights from bioassay guided fractionation. MS Thesis. Department of Biology, University of Alabama at Birmingham

Smith RC, Stammerjohn SE (2001) Variations of surface air temperature and sea-ice extent in the western Antarctic Peninsula region. Ann Glaciol 33:493–500

Stammerjohn S E, Martinson DG, Smith RC, Yuan X, Rind D (2008) Trends in Antarctic annual sea ice retreat and advance and their relation to El Niño–Southern Oscillation and Southern Annular Mode variability, J Geophys Res 113:C03S90, doi:10.1029/2007JC004269.

Svendsen H, Beszczynska-Møller A, Hagen JO, Lefauconnier B, Tverberg V, Gerland S, Bischof K, Papucci C, Ørbæk JB, Zajaczkowski M, Azzolini R, Bruland O, Wiencke C, Winther JG, Dallmann W (2002) The physical environment of Kongsfjorden-Krossfjorden, an Arctic fjord system in Svalbard. Polar Res 21:133–166

tom Dieck I (1992) North Pacific and North Atlantic digitate *Laminaria* species (Phaeophyta): hybridisation experiments and temperature responses. Phycologia 31:147–163

tom Dieck I (1993) Temperature tolerance and survival in darkness of kelp gametophytes (Laminariales, Phaeophyta): ecological and biogeographical implications. Mar Ecol Prog Ser 100:253–264

van den Hoek C (1982a) Phytogeographic distribution groups of benthic marine algae in the North Atlantic Ocean. A review of experimental evidence from life history studies. Helgoländer Meeresunters 35:153–214

van den Hoek C (1982b) The distribution of benthic marine algae in relation to the temperature regulation of their life histories. Biol J Linn Soc 18:1–144

van Oppen MJH, Olsen JL, Stam W, van den Hoek C, Wiencke C (1993) Arctic-Antarctic disjunctions in the benthic seaweeds *Acrosiphonia arcta* (Chlorophyta) and *Desmarestia viridis/willii* (Phaeophyta) are of recent origin. Mar Biol 115:381–386

van Oppen MJH, Diekmann OE, Wiencke C, Stam WT, Olsen JL (1994) Tracking dispersal routes: Phylogeography of Arctic-Antarctic disjunct seaweed *Acrosiphonia arcta* (Chlorophyta). J Phycol 30:67–80

Wessels H, Hagen W, Molis M, Wiencke C, Karsten U (2006) Intra- and interspecific differences in palatability of Arctic macroalgae from Kongsfjorden (Spitsbergen) for two benthic sympatric invertebrates. J Exp Mar Biol Ecol 329:20–33

Weykam G, Gómez I, Wiencke C, Iken K, Kloser H (1996) Photosynthetic characteristics and C:N ratios of macroalgae from King George Island (Antarctica). J Exp Mar Biol Ecol 204:1–22

Weykam G, Thomas DN, Wiencke C (1997) Growth and photosynthesis of the Antarctic red algae *Palmaria decipiens* (Palmariales) and *Iridaea cordata* (Gigartinales) during and following extended periods of darkness. Phycologia 36:395–405

Wiencke C (1990a) Seasonality of brown macroalgae from Antarctica - a long-term culture study under fluctuating Antarctic daylengths. Polar Biol 10:589–600

Wiencke C (1990b) Seasonality of red and green macroalgae from Antarctica - a long-term culture study under fluctuating Antarctic daylengths. Polar Biol 10:601–607

Wiencke C, Clayton MN (2002) Antarctic Seaweeds. ARG Gantner Verlag, KG Ruggell

Wiencke C, Fischer G (1990) Growth and stable carbon isotope composition of cold-water macroalgae in relation to light and temperature. Mar Ecol Prog Ser 65:283–292

Wiencke C, Bartsch I, Bischoff B, Peters AF, Breeman AM (1994) Temperature requirements and biogeography of Antarctic, Arctic and amphiequatorial seaweeds. Bot Mar 37:247–259

Wiencke C, Clayton MN, Langreder C (1996) Life history and seasonal morphogenesis of the endemic Antarctic brown alga *Desmarestia anceps* Montagne. Bot Mar 39:435–444

Wiencke C, Vögele B, Kovaltchouk NA, Hop H (2004) Species composition and zonation of marine benthic macroalgae at Hansneset in Kongsfjorden, Svalbard. Ber Polarforsch Meeresforsch 492:55–62

Wiencke C, Roleda MY, Gruber A, Clayton MN, Bischof K (2006) Susceptibility of zoospores to UV radiation determines upper depth distribution limit of Arctic kelps: evidence through field experiments. J Ecol 94:455–463

Wiencke C, Clayton MN, Gómez I, Iken K, Lüder U, Amsler CD, Karsten U, Hanelt D, Bischof K, Dunton K (2007) Life strategy, ecophysiology and ecology of seaweeds in polar water. Rev Environ Sci Biotech 6:95–126

Wiencke C, Gómez I, Dunton K (2011) Phenology and seasonal physiological performance of polar seaweeds. In: Wiencke C (ed) Biology of Polar Benthic Algae. De Gruyter, Berlin, pp 181–194

Wilce RT (1990) Role of the Arctic Ocean as a bridge between the Atlantic and Pacific Oceans: fact and hypothesis. In: Garbary DJ, South GR (eds) Evolutionary biogeography of the marine algae of the North Atlantic. Springer, Berlin, pp 323–348

Wilce RT (1994) The Arctic subtidal as habitat for macrophytes. In: Lobban CS, Harrison PJ (eds) Seaweed ecology and physiology. Cambridge University Press, Cambridge, pp 89–92

Wilce RT, Pedersen PM, Sekida S (2009) *Chukchia pedicellata* gen. et sp. nov. and *C. endophytica* nov. comb., Arctic endemic brown algae (Phaeophyceae). J Phycol 45:272–286

Włodarska-Kowalczuk M, Kukliński P, Ronowicz M, Legeżyńska J, Gromisz S (2009) Assessing species richness of macrofauna associated with macroalgae in Arctic kelp forests (Hornsund, Svalbard). Polar Biol 32:897–905

Wulff A, Iken K, Quartino ML, Al-Handal A, Wiencke C, Clayton MN (2011) Biodiversity, biogeography and zonation of marine benthic micro- and macroalgae in the Arctic and Antarctic. In: Wiencke C (ed) Biology of polar benthic algae. De Gruyter, Berlin, pp 23–52

Zacher K, Wulff A, Molis M, Hanelt D, Wiencke C (2007) Ultraviolet radiation and consumer effects on a field-grown intertidal macroalgal assemblage in Antarctica. Global Change Biol 13:1201–1215

Zacher K, Rautenberger R, Hanelt D, Wulff A, Wiencke C (2011) The abiotic environment of polar benthic algae. In: Wiencke C (ed) Biology of polar benthic algae. De Gruyter, Berlin, pp 9–22

Zachos J, Pagani M, Sloan L, Thomas E, Billups K (2001) Trends, rhythms, and aberrations in global climate 65 Ma to present. Science 292:686–693

Zamzow JP, Amsler CD, McClintock JB, Baker BJ (2010) Habitat choice and predator avoidance by Antarctic amphipods: the roles of algal chemistry and morphology. Mar Ecol Prog Ser 400:155–163

Zamzow JP, Aumack CF, Amsler CD, McClintock JB, Amsler MO, Baker BJ (2011) Gut contents and stable isotope analyses of the Antarctic fish, *Notothenia coriiceps* Richardson, from two macroalgal communities. Antarctic Sci 23:107–116

Zaneveld JS (1966) The occurrence of benthic marine algae under shore fast-ice in the western Ross Sea, Antarctic. Proc Int Seaweed Symp 5:217–231

Zielinski K (1981) Benthic macroalgae of Admirality Bay (King George Island, South Shetland Islands) and circulation of algal matter between water and the shore. Pol Polar Res 2:71–94

Zielinski K (1990) Bottom macroalgae of the Admiralty Bay (King George Island, South Shetlands, Antarctica). Pol Polar Res 11:95–131

Chapter 14
Cold-Temperate Seaweed Communities of the Southern Hemisphere

Pirjo Huovinen and Iván Gómez

Please note the Erratum to this chapter at the end of the book

14.1 The Regions and Their Basic Abiotic Conditions

In the southern hemisphere, two circumpolar fronts roughly define the boundaries of the cold-temperate region (Fig. 14.1). The southern limit, the Antarctic Polar Front (APF; also known as Antarctic Convergence), is characterized by cold (ranging 3–5°C) surface waters. The northern limit, the Subtropical Front (STF; also known as Subtropical Convergence), separates subtropical water in the north from sub-Antarctic water, and follows approximately the 10°C surface isotherm in winter and the 15°C surface isotherm in summer (Heath 1985; Parsons 1985; Lüning 1990; see also Chap. 3 by Eggert and Chap. 18 by Bartsch). The STF is a complex, irregular frontal zone with seasonal migration of up to 2° latitude (Nelson and Broom 2010). The large eastward-moving Antarctic Circumpolar Current (ACC; called also West Wind Drift) strongly determines the oceanography in the Southern Hemisphere (Orsi et al. 1995) (Fig. 14.1). The circulation of the ACC and circumpolar frontal zones strongly define the diversity, biogeography, and the ecology of seaweeds in all the adjacent coastal regions.

14.1.1 Southwestern South America Region

In the Southeastern Pacific coast of Chile, there are two major current systems, the southward Cape Horn Current and the northward Humboldt Current, derived from the ACC around 40°S (Fig. 14.1). In the south, the Cape Horn Current flows through the Drake Passage. Three major regions along the coast can be identified: the

P. Huovinen (✉) • I. Gómez
Instituto de Ciencias Marinas y Limnológicas, Universidad Austral de Chile, Campus Isla Teja s/n, Casilla 567, Valdivia, Chile
e-mail: pirjo.huovinen@uach.cl

C. Wiencke and K. Bischof (eds.), *Seaweed Biology*, Ecological Studies 219,
DOI 10.1007/978-3-642-28451-9_14,

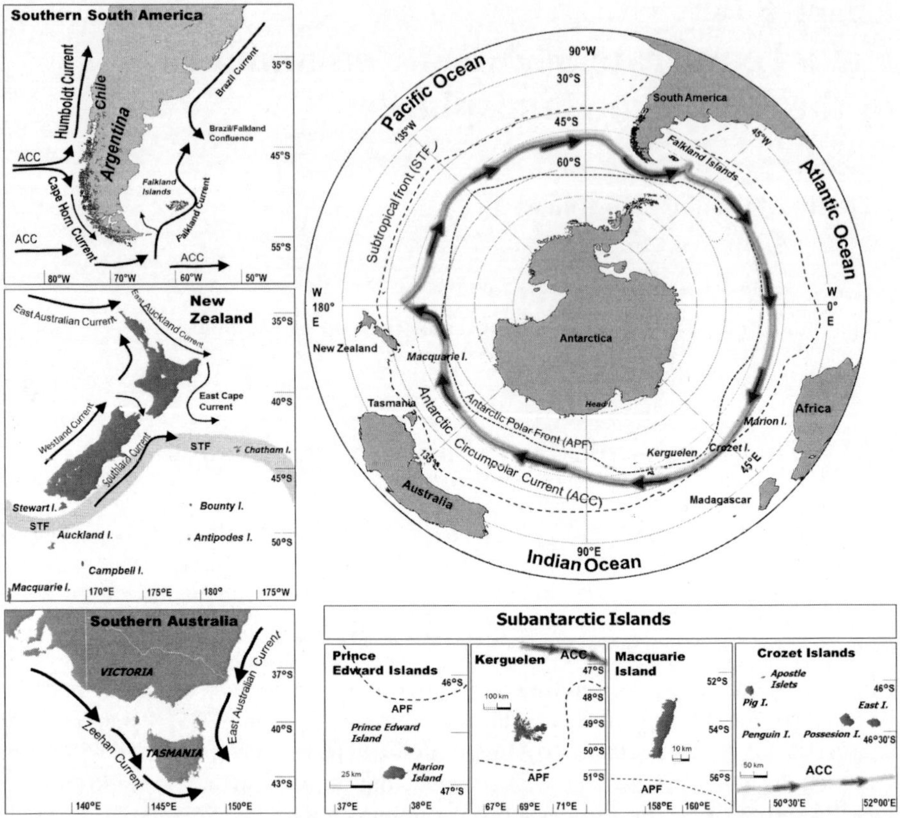

Fig. 14.1 The five cold-temperate regions of the southern hemisphere: southwestern South America, southeastern South America, Victoria-Tasmania, southern New Zealand, and the sub-Antarctic islands. Major oceanographic fronts, the Subtropical Front (STF; also known as Subtropical Convergence), the Antarctic Polar Front (APF; also known as Antarctic Convergence), and the Antarctic Circumpolar Current (ACC; called also West Wind Drift), are presented

southern Austral Fjords Region (sub-Antarctic, surface temperature 5–12°C), the Central/South Upwelling Region (Mediterranean characteristics, surface temperature 8–15°C), and the Northern Upwelling Region (subtropical, temperature 15–21°C) (Strub et al. 1998; Thiel et al. 2007; see also Chap. 18 by Bartsch). Habitats for seaweeds are characterized by exposed rocky shores between 18 and 41°S, while southwards the littoral systems exhibit higher variability with many fjords, archipelagos, and estuaries as well as glaciers (Santelices 1989).

14.1.2 Southeastern South America Region

In the coast of Argentina, the northward, cold-water (sub-Antarctic) Falkland current, derived from the ACC, meets the southward Brazil (subtropical) current around 36°S in an area called the Brazil/Falkland Confluence (Acha et al. 2004; Fig. 14.1). The cold-temperate section of the southeastern South America (approximately 41–55°S) is characterized by large bays and harbors, which provide heterogeneous habitats for seaweeds. Northwards of the Gulf San Matías (41°S), temperature increase and substrate become less suitable for seaweeds due to the massive runoff and sedimentation from Rio de la Plata (Acha et al. 2004). The Falkland Islands (Islas Malvinas), with sub-Antarctic characteristics, are situated in the path of the Falkland Current (Fig. 14.1) with water temperatures around 3–14°C (see also Chap. 18 by Bartsch). The coastal oceanography is characterized by the two branches of the Falkland Current, the weaker western branch (the Patagonian Current) rejoining the eastern branch north of the islands (Upton and Shaw 2002).

14.1.3 Victoria-Tasmania Region

The warm East Australian Current is the largest and the warm Leeuwin Current the second largest ocean current in Australia. The Leeuwin Current flows from the west Australian coast towards south-east, becoming the South Australia Current as it flows across the Great Australian Bight, reaching the west Tasmanian coast as the Zeehan Current (Fig. 14.1). The importance of these currents is evident from the tropical elements present in southern Australian waters at temperate latitudes. The southern coast of Tasmania is situated a few degrees north of the STF, the limit of sub-Antarctic waters (summarized by Poloczanska et al. 2007). However, in the coasts of Tasmania and Victoria, the surface water temperature remains cool, generally not exceeding 15°C in summer (Lüning 1990; see also Chap. 18 by Bartsch). The southern Australian coastline has large sections of rocky substrata, providing suitable habitats for seaweeds (Phillips 2001).

14.1.4 Southern New Zealand Region

Subtropical waters with relatively high salinity and temperature around New Zealand are mainly derived from the East Australian Current, while sub-Antarctic waters, with relatively low salinity and temperature, are driven north by the ACC. These two water masses meet in the STF (following roughly the 10°C winter and the 15°C summer surface isotherm; see also Chap. 18 by Bartsch). Along the east coast of the South Island, the Southland Current brings cool water towards north. It turns towards east and back to south around 40°S (Cape Turnagain), the northern limit for many species, combining with the warmer East Cape Current (Heath 1985; Parsons 1985) (Fig. 14.1). In the Fiordland region of the South Island, the exposed coastline is interrupted by channels, inlets, fjords, and many estuaries. Thus, the littoral systems in these locations are strongly influenced by the inflow of freshwater from rivers and glaciers (Nelson et al. 2002).

14.1.5 Sub-Antarctic Islands Region

The sub-Antarctic islands region consists of the island groups of Prince Edward (Prince Edward and Marion Islands), Crozet, Kerguélen, Heard (included also in the Antarctic region), and Macquarie (sometimes regarded in the Southern New Zealand region) Islands. The sub-Antarctic islands near New Zealand include Snares, Auckland, Campbell, Bounties, Antipodes, and Chatham Islands (Fig. 14.1). The location of the Southern Ocean islands in relation to the APF influences strongly their climatic conditions. Overall, these islands exhibit particularly cool, wet, and windy climates (Bergstrom and Chown 1999). The surface water temperatures along the coasts of the sub-Antarctic islands range 3–11°C in winter and 5–14°C in summer, depending on the geographical location (Lüning 1990).

14.2 Structure and Function of Seaweed Communities

14.2.1 Southwestern South America Region (Chile)

The southeastern Pacific coast can be divided into three biogeographic provinces: (1) the Peruvian province (from Peru to 30°S) with high presence of warm-temperate biota, (2) the Magellan province (from 40–42°S to 56°S) with dominance of sub-Antarctic cold-temperate elements, and (3) the intermediate area (from 30°S to 40–42°S) with mixed components from both provinces. According to Briggs (1974, cited in Lüning 1990) the cold-temperate region in the south-eastern Pacific is limited to the north at around 40°S (see also Chap. 18 by Bartsch et al.). Recent evidence (Strub et al. 1998; Camus 2001; Thiel et al. 2007) suggests that this region extends to around 30°S due to the influence of the Humboldt Current. In general, the seaweed flora in this region presents a remarkable degree of endemism (32%) (Santelices 1980), although recent accounts indicate increases in subtropical and alien species (Meneses and Santelices 2000; Castilla et al. 2005). Due to the higher occurrence of cold-temperate species, the total number of species increases southwards (Santelices 1980; Santelices and Marquet 1998). According to a recent update, the Magellan and Tierra del Fuego regions in the southern part of South America comprise 234 seaweed species (red 60%, brown 22%, green 18%) (Ramírez 2010). The species composition shows a break point around the latitude 42°S (Meneses and Santelices 2000).

In the wave-exposed coast at 30–55°S latitudes, the zonation and the structure of seaweed communities are characterized by the presence of *Porphyra columbina* and filamentous and foliose green algae such as *Ulva intestinalis* at upper littoral zones (Santelices 1989; Gómez and Huovinen 2011) (Fig. 14.2). In the mid-littoral, coarsely branched red algae such as *Mazzaella laminarioides and Nothogenia fastigiata* dominate. The red alga *Gelidium* sp. and the brown alga *Petalonia fascia*

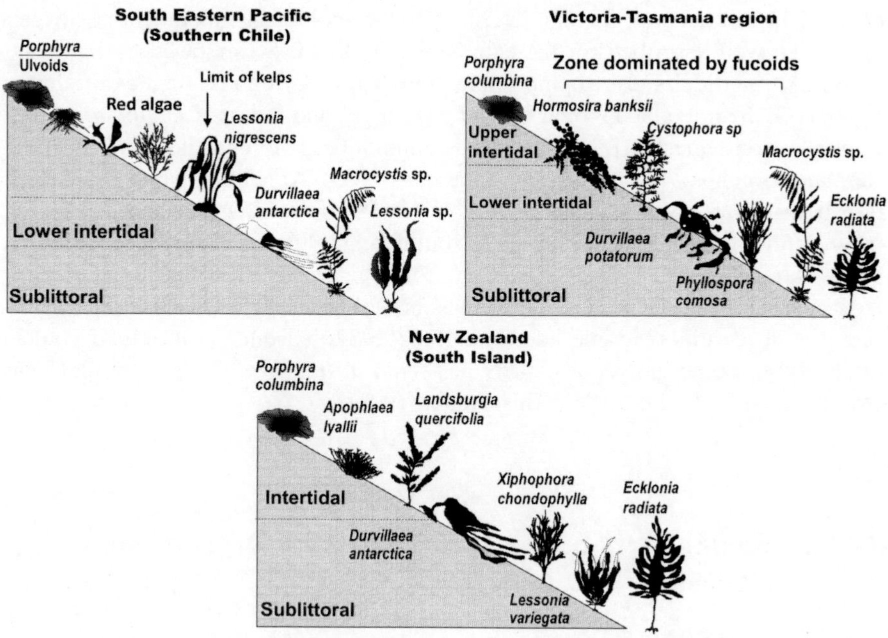

Fig. 14.2 General patterns of seaweed zonation, with dominant species in each littoral zone, in southern Chile, Victoria-Tasmania, and New Zealand (South Island) (the schemes represent a compilation of the literature revised in the text)

can colonize shaded pools and rocky crevices. The large kelp *Lessonia nigrescens* and the fucoid *Durvillaea antarctica* define the structure in the wave-battered infralittoral zone together with crustose/articulate species (*Corallina*) and red algae adapted to episodic erosion–accretion of sand such as *Ahnfeltiopsis durvillaei*. The sublittoral zone (up to 10 m) is characterized by a heterogeneous group with a dominance of kelps such as *Macrocystis pyrifera*, *Lessonia trabeculata*, *L. vadosa,* and *L. flavicans* (Villouta and Santelices 1986), mixed with thin leathery red algae (e.g., *Rhodymenia* sp. and *Callophyllis* sp.). Giant kelp forests are also found in protected bays and fjords, while the rhodophyte *Gracilaria chilensis* dominates in estuarine zones (Santelices 1980; Buschmann et al. 2004). At midlittoral and subtidal zones of the Magellan coast and some sub-Antarctic islands south of Tierra del Fuego (e.g., Diego Ramirez Islands), some species common to the Antarctic region (e.g., *Adenocystis utricularis*, *Iridaea cordata*, *Gigartina skottsbergii*, *Desmarestia* sp.) can be abundant (Santelices and Ojeda 1984; Westermeier and Rivera 1986).

The coexistence of *L. nigrescens* and *D. antarctica* at infralittoral zones conforms one of the most remarkable ecological associations. The spatial distribution and abundance of these two habitat-forming species result from complex interactions and morphofunctional adaptations, both species exhibiting different strategies to persist in areas with strong surf (Santelices et al. 1980; Westermeier et al. 1994). In general, factors determining the ecological functions of seaweeds have been well described for the ecosystems at 30–40°S (reviewed by Santelices 1989). At intertidal sites, trophic interactions with gastropods, such as the limpet *Fissurella*, the polyplacophoran *Chiton granossus,* and the snail *Tegula atra*, have important consequences for the seaweed communities (Moreno and Jaramillo 1983; Jara and Moreno 1984; Otaíza and Santelices 1985). At subtidal sites, the interaction between *M. pyrifera* and the sea urchins *Loxechinus albus* and *Tetrapygus niger* defines the ecological process of southern kelp forests (Santelices and Ojeda 1984; Dayton 1985). Unlike the situation for kelp forests of California, sea urchins from southern region preferentially consume algal drifts and act regulating the biomass of recruits (Moreno and Sutherland 1982; Vásquez et al. 1984; Graham et al. 2007). Some gastropods, such as *Tegula atra*, graze also on subtidal kelps (Moreno and Sutherland 1982; Buschmann 1992).

14.2.2 Southeastern South America Region (Argentinean Patagonia)

Approximately 400 seaweed species have been reported for the Atlantic coast of Argentina (reviewed by Lüning 1990). Two main biogeographical provinces have been proposed: (1) the Magellan province with cold water covering the southern Patagonia, and (2) the Argentinean province with warmer waters extending from North Patagonia (around Peninsula Valdés at 42–43°S) northwards (Lutz et al. 2003). In this region, species richness has been characterized as being relatively poor (Bolton 1994). Intertidal rocky shores here are characterized by a depauperate seaweed flora, mainly due to harsh physical conditions (Paruelo et al. 1998; Bertness et al. 2006), calcareous algae covering low tidal height in wave-protected sites (Bertness et al. 2006). Northwards from the Gulf San Matías (41°S), the massive runoff and sedimentation from Rio de la Plata limit the availability of hard substrates for seaweeds. In this region, temperate chlorophytes adapted to estuarine conditions dominate (Acha et al. 2004). Southwards, in the limit between the Argentinean and the Magellan biogeographical regions, subtropical species, such as *Dictyota*, are mixed with sub-Antarctic (e.g., *Bryopsis australis*) and endemic (e.g., *Myrogloia major*) components. The kelps *Macrocystis pyrifera*, *Lessonia flavicans,* and *L. vadosa* occur continuously from the Pacific to the Atlantic coast (up to 42°S for *Macrocystis*, 47–48°S for *Lessonia*) (Barrales and Lobban 1975; Searles 1978). The fucoid *Durvillaea antarctica* has its northern limit around 51°S in the Argentinean side of Tierra del Fuego (Boraso and Zaixso 2011). The invasive kelp *Undaria pinnatifida* has been associated with a decrease of species richness and diversity of native seaweeds in Patagonia (Casas et al. 2004; see also Chap. 12 by Andreakis and Schaffelke).

In the Falkland Islands, a total of 169 seaweed species (55% Rhodophyta, 29% Phaeophyceae, 16% Chlorophyta) were recently listed (Ramírez 2010). The lower intertidal zone is dominated by *Durvillaea antarctica*, while the subtidal zone is characterized by extensive *Macrocystis pyrifera* forests (van Tussenbroek 1993). Other large kelps, such as *Lessonia flavicans*, are also present (Ramírez 2010). Based on phylogenetic analyses, a highly diverse flora of bladed Bangiales (at least nine species) has been reported for the Falkland Islands (Broom et al. 2010). A high floristic similarity (60%) is found with the flora from Magellan-Tierra del Fuego region, principally due to species with a sub-Antarctic origin. The Falkland Islands and Magellan-Tierra del Fuego region form a group that separates from the Antarctic region (Hommersand et al. 2009; Ramírez 2010). These studies do not support the previously proposed biogeographic province with continuity of the marine flora in the Antarctic region and the sub-Antarctic South American region (John et al. 1994).

Ecological studies on seaweeds from the southeastern South American region are scarce, and thus the understanding of the distribution and abundance patterns as well as the functional roles of the seaweeds in these ecosystems is poorly known. Apparently, intertidal seaweed communities are strongly regulated by physical factors, as grazing pressure is weak due to the absence of common predaceous crabs and snails and the limited impact by limpets (Bertness et al. 2006). In subtidal environments, evidence based on population dynamics from Argentinean mainland and the Falkland Islands indicates that giant kelp beds are regulated by storms and nutrient availability (Barrales and Lobban 1975; van Tussenbroek 1989). In the kelp (*Macrocystis*) forests no consistent grazers, such as sea urchins, were reported either (Barrales and Lobban 1975). However, growth patterns and demography of these forests show considerable local variation (van Tussenbroek 1993).

14.2.3 Victoria-Tasmania Region

The seaweed flora of the southern Australia is characterized by high species richness and endemism (Bolton 1994; Phillips 2001; Kerswell 2006), resulting from a complex interaction of biogeographical, phylogenetic, and ecological processes (reviewed by Phillips 2001). Over 1,100 species (Bolton 1994) and four major floristic elements, i.e., endemic, widely distributed temperate, tropical, and polar elements, have been identified (Phillips 2001). Endemic (40–77%) and widely distributed temperate (17–45%) species account for the majority of the seaweeds, whereas species with tropical (4.5–9.7%) and polar (1.5–5.1%) affinities are much less represented. In Rhodophyta and Phaeophyceae, endemic elements dominate (77% and 59%, respectively), while in Chlorophyta widely distributed temperate (45%) flora account for the majority (Phillips 2001). Recent quantitative seaweed analyses confirmed the traditionally recognized Peronian (eastern), Flindersian (western), and Maugean (south-eastern) marine biogeographic provinces of the southern coast of Australia (Waters 2010). Furthermore, differentiation of marine communities in eastern and western elements has been associated with the Bassian Isthmus as a historical vicariant barrier (Waters 2008b; Fraser et al. 2009a).

The cold-temperate region of southern Australia (Victoria) and Tasmania is characterized by the dominance of fucoids, such as *Durvillaea potatorum* in the upper sublittoral zone in sites with high wave exposure (Cheshire and Hallam 1988) (Fig. 14.2). At medium-exposed sites the fucalean *Phyllosphora comosa* and the kelps *Macrocystis pyrifera* (former *M. angustifolia*) and *Ecklonia radiata* are common. In the intertidal zone, *Porphyra* dominate at higher locations, while at lower intertidal zones, the fucoid *Hormosira banksii* is abundant coexisting with green algae (e.g., *Codium*) and the Rhodophyta *Corallina*. Below the fringe of *H. banksii*, in the eulittoral zone *Cystophora torulosa* is the dominant fucoid. In southern Tasmania, *M. pyrifera* can form dense sublittoral forests in some locations, the upper sublittoral zone being dominated by *D. potatorum* and the endemic *Lessonia corrugata*. In general, many of the dominant fucoids are restricted to cold-temperate waters, and only *P. comosa* and *H. banksii* have distribution along the warm-temperate eastern coast of Australia (to a latitude 32°S). The northern limit of *D. potatorum* is at latitude 36°S, *M. pyrifera* (former *M. angustifolia*) extending to the northern border of Victoria (Lüning 1990).

Large brown algae, due to their size and foundational character, have been a special topic in many ecological studies (Sanderson and Thomas 1987). In the east coast of Tasmania, the sea urchins *Heliocidaris erythrogramma* and *Centrostephanus rodgersii* and the abalone *Haliotis rubra* regulate the population dynamics of various canopy-forming algae such as *Phyllospora comosa* and *Ecklonia radiata* (Valentine and Johnson 2005; Ling and Johnson 2009). Interestingly, the sea urchin barren zone is rapidly colonized by the introduced kelp *Undaria pinnatifida* (Valentine and Johnson 2004, 2005). In Tasmania, overgrazing of seaweeds by introduced sea urchin *Centrostephanus rodgersii* together with oceanographic changes is causing dramatic decline in the giant kelp beds (Johnson et al. 2011).

14.2.4 Southern New Zealand Region

Coasts of New Zealand represent areas of high seaweed diversity (Norton et al. 1996; Kerswell 2006). A recent account indicated that of the 770 currently known seaweed species, 265 are endemic (34%) and 22 alien (2.9%) (Hurd et al. 2004). According to the revision of Parsons (1985), the South Island hosts around 497 species, while for the North Island 555 species were reported. Rhodophyta accounts for around 60% of the total taxa, Chlorophyta and Phaeophyceae each contributing with approximately 20% (Parsons 1985). Within the cold-temperate southern New Zealand region are included the South Island, the southern part of the North Island, and the Stewart Island. Also the sub-Antarctic islands Auckland, Campbell, Antipodes, and Bounty, as well as the Chatham Island, are in this region (Fig. 14.2).

Although lower in number compared to green and red algae, brown algae, particularly fucoids, dominate the rocky shores of Southern New Zealand, while only few kelps can be regarded as dominant organisms (Schiel 1990) (Fig. 14.2). In the middle and lower intertidal zone, *Cystophora torulosa* and *Landsburgia quercifolia* are abundant, while *Durvillaea antarctica* is the most conspicuous brown alga in the lower littoral fringe at exposed sites, accompanied by red algae such as *Gigartina*, *Laurencia*, *Polysiphonia*, *Caulacanthus*, *Gelidium*, and *Nemalion*, the brown algae such as *Hormosira banksii* and *Colpomenia sinusosa*, and the green algae *Codium* and *Caulerpa*. In the southern part of the North Island (Wellington), the fucalean species *L. quercifolia*, *Carpophyllum flexuosum*, *C. maschalocarpum*, and *C. angustifolium* and the laminarian *Lessonia variegata* are common in shallow waters. *Macrocystis pyrifera* and *Ecklonia radiata* generally reach their highest abundance in deeper waters (Schiel 1990).

In Fiordland in the South Island, the lower intertidal is dominated by *Durvillaea antarctica*, particularly in the east coast. At around a depth of 2 m, the fucalean species *Xiphophora chondrophylla* is abundant. *Ecklonia radiata* is common at 3–6 m and at 12–18 m depth zones (Schiel 1990; Villouta et al. 2001). The upper sublittoral zone is inhabited by *Lessonia variegata* as well as *D. willana*, which is endemic to New Zealand (Cheshire et al. 1995). In sheltered rocky shores within the fjords, diversity and structure of seaweed assemblages have been less studied. Surveys carried out in Doubtful Sound (45°S) indicated the presence of, e.g., red alga *Pachymenia lusoria* and diverse *Ulva* species in the most estuarine locations. The lower intertidal zones are characterized by *D. antarctica*, *Xiphophora chondrophylla*, and sparse populations of *M. pyrifera* (Boyle et al. 2001).

The structure and function of seaweed communities from the New Zealand cold-temperate region, especially of fucalean and kelps, have been well addressed in some locations. However, many aspects of population dynamics and underlying processes such as phenology, dispersal, and recruitment have not been intensively addressed (reviewed by Hurd et al. 2004; Schiel 1990, 2004). On the other hand, biotic control of seaweed assemblages is different between regions from the main islands, which reflect their intrinsic differences in seaweed structure and composition (Schiel and Hickford 2001). For example, the impact of the sea urchin *Evechinus chloroticus* on different large brown algae is higher in northern communities dominated by *Ecklonia*, compared to communities dominated by cold-temperate seaweeds, e.g., in Fiordland and Chatham Island, where the *Ecklonia* stands are replaced by fucoids and other kelps such as *Lessonia* spp. (Schiel 1990; Villouta et al. 2001). The presence of large brown algae has different impact on understory species, especially red algae (Lilley and Schiel 2006), and in systems where the control by herbivores is less intense, the interspecific competence between large canopy-forming seaweeds can become a relevant factor (Schiel and Hickford 2001).

14.2.5 Sub-Antarctic Islands Region

According to a catalog of Papenfuss (1964), the seaweed flora of the sub-Antarctic islands and the Antarctic region contains 550 species, approximately half being endemic to these two regions, while approximately 75% of these species are distributed in the sub-Antarctic islands (Clayton 1994). These regions share a considerable number of cold-temperate species, e.g., the brown alga *Adenocystis utricularis*, the red alga *Iridaea cordata* and *Gigartina skottsbergii*, however, some endemic Antarctic species such as *Himantothallus grandifolius* and *Ascoseira mirabilis* do not inhabit the sub-Antarctic islands (Lüning 1990; Clayton 1994). The distribution of the endemic Antarctic rhodophyte *Palmaria decipiens* just extends to Macquarie Island (Wiencke and Clayton 2002). On the other hand, the circumpolar kelps *Macrocystis pyrifera* and *Lessonia* spp. as well as the large intertidal fucoid *Durvillaea antarctica* are absent from the Antarctic (John et al. 1994; Clayton et al. 1997).

The general zonation pattern for the shores of Macquarie Island has been described to comprise an upper littoral *Porphyra* zone, a "bare" zone (dominated by siphonariid mollusc *Kerguelenella lateralis*), an upper red algal zone (*Rhodymenia* sp., *Chaetangium fastigiatum*, *Palmaria georgica*, *Acrosiphonia pacifica*, *Porphyra columbina*), a kelp zone (*Durvillaea antarctica*), and lower red algal zone below the infralittoral with understory species such as *Delesseria* spp., *Iridaea* sp., and *Desmarestia* sp. (Kenny and Haysom 1962; Smith and Simpson 2002). Similarly, in Marion Island, *D. antarctica* is the dominant species at the infralittoral fringe, while *Desmarestia rossii* and *Durvillaea* sp. occur in a zone of 3–6 m. Encrusting coralline algae are particularly abundant in shallow areas (Beckley and Branch 1992). At subtidal locations, beds of the endemic *Macrocystis laevis* are abundant (Perissinotto and McQuaid 1992; Beckley and Branch 1992). In general, infralittoral *D. antarctica*, shallow-water encrusting coralline algae, *Desmarestia* spp., and subtidal *Macrocystis* beds seem to overall characterize the benthic seaweed communities in many sub-Antarctic islands (Beckley and Branch 1992).

For the sub-Antarctic islands near New Zealand, 14–148 seaweed species have been reported, with the highest species richness being recorded for Chatham Islands and the lowest for the Bounty Island (Parsons 1985). Although notable differences in species composition and dominance patterns can exist among islands (Freeman et al. 2011), *Durvillaea antarctica* and *Macrocystis pyrifera* occur in all these island groups, *Lessonia flavicans* reported for Auckland, Campbell, the Antipodes, and Bounty. The fucalean *Xiphophora chondrophylla* is present in all the island groups, except for the Bounty (Lüning 1990). Endemic *Durvillaea* species have been proposed for the Antipodes (undescribed species) and Chatham Island (*D. chathamensis*) (Cheshire et al. 1995; Fraser et al. 2010b; Fig. 14.3). In the light of a re-examination of the description of *Porphyra columbina* using molecular techniques, this species is distributed in the Auckland, Campbell, Antipodes, Chatham, and Falkland Islands, with a very restricted distribution in mainland New Zealand (Nelson and Broom 2010).

In the Chatham Island, the shallowest subtidal areas are dominated by *Durvillaea*. Several fucalean species (e.g., *Xiphophora chondrophylla*, *Carpophyllum plumosum* and *C. maschalocarpum*) are abundant to a depth of around 10 m, *C. flexuosum* occurring mostly in deeper waters (9–20 m). The endemic *Lessonia tholiformis* and *L. variegata* are abundant at 2.5–15 m depth. In some sites, *Macrocystis pyrifera* forms dense beds at 15–20 m depth. In the Auckland Islands,

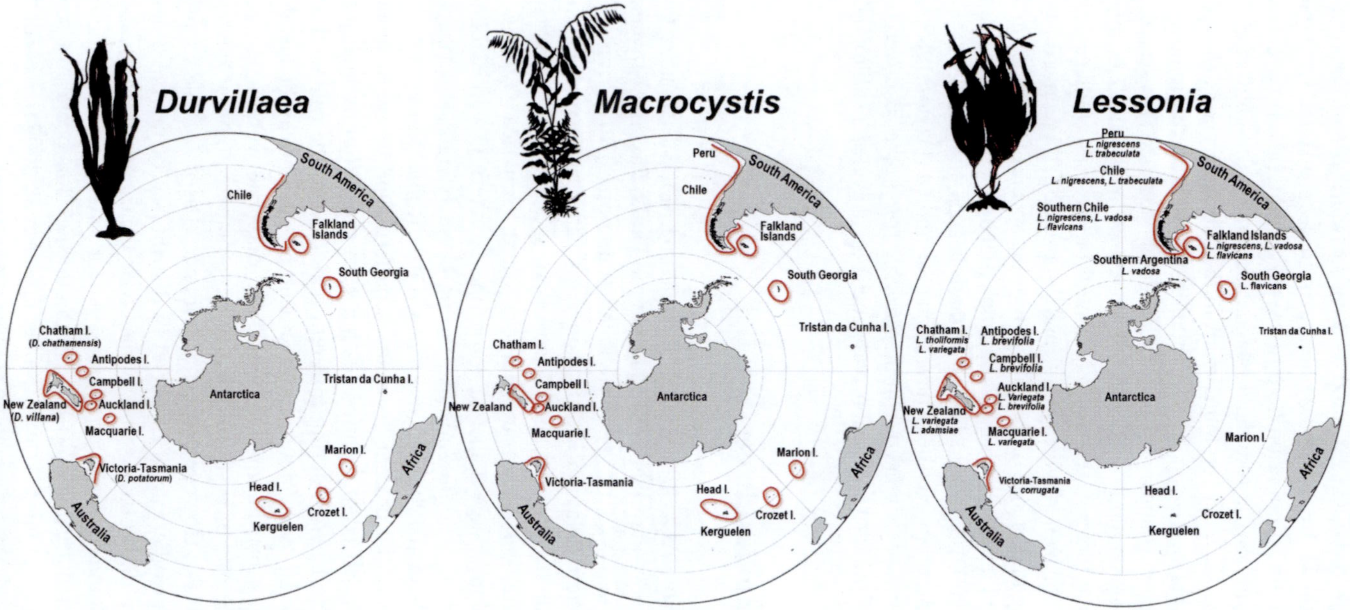

Fig. 14.3 Distribution of the large brown algal genera *Durvillaea* and laminariales *Macrocystis* and *Lessonia* in the southern hemisphere. Different species of *Durvillaea* and *Lessonia* are presented. As the current evidence indicates the existence of only one species of *Macrocystis* (*M. pyrifera*) (see the text), this information is presented as *Macrocystis* sp. Note: *Macrocystis* is also present in the northern hemisphere in the Pacific coast from Baja California, Mexico, to Alaska (the schemes represent a compilation of the literature revised in the text)

shallowest areas are also dominated by *D. antarctica*. *Xiphophora* is common at 2 m depth and *Halopteris funicularis* around 2–5 m. *Lessonia brevifolia* forms beds around 4–12 m depth, *M. pyrifera* being abundant in deep waters. The laminarian *Ecklonia radiata,* the most common kelp of New Zealand, has not been recorded neither at the Chatham nor the Auckland Islands (Schiel 1990; Schiel et al. 1995).

The ecology of the seaweed communities from the sub-Antarctic islands region is poorly known. The harsh environmental conditions as well as the remoteness and inaccessibility of many islands impede experimental studies. The scarce studies characterizing the structure and functions of seaweeds of intertidal zones reveal that physical environmental factors exert a strong control on these assemblages, probably regulating also the action by herbivores (Smith and Simpson 2002; Freeman et al. 2011). On the other hand, the importance of the habitat-forming species *Durvillaea antarctica* on the ecology of the littoral system has been emphasized (Klemm and Hallam 1988; Smith and Simpson 2002). In subtidal areas of these islands, the biological processes occurring within the *Macrocystis* forests have been also studied (Dayton 1985; Graham et al. 2007), including aspects related with their trophic role in the food web (Kaehler et al. 2006), aspects of primary production, and geochemical cycles (Delille et al. 2009).

14.3 Biogeographical Processes

The development of the ACC as a result of the opening of the Tasmanian Gateway between Australia and Antarctica (~33 Ma) and the Drake Passage between South America and Antarctica (~23 Ma) permitted the circumpolar distribution of biota in the southern hemisphere (Clayton 1994). Many cool water taxa evolved in the early Tertiary at high latitudes (the so-called Weddellian species) and colonized lower latitudes, e.g., Tierra del Fuego in southern South America (Lüning 1990). A relatively recent long-range dispersal across the Southern Ocean is reflected in many species reported in South America, the Falkland Islands, and various sub-Antarctic islands (Clayton 1994; see also Chap. 17 by Rothäusler et al.).

In order to interpret the present distribution patterns and origin of biota in the southern hemisphere, different theories dealing with separation through plate tectonics, long-distance dispersal during different periods, and the role of the formation of the ACC linking the land masses have been proposed over the last decades. Models of the species distribution involving ACC dispersal were widely replaced by vicariance and the role of past geological events (plate tectonic) (Cheshire et al. 1995; Waters 2008a). However, recent evidence based on new molecular techniques emphasizes the role of dispersal in distribution of marine biota in the southern hemisphere and relates some evidence of vicariance in southern waters with more localized and recent vicariant models, e.g., the isolation by the Bassian Isthmus that connected Tasmania with Australia during Pleistocene low sea-level, rather than with plate tectonics (Waters 2008a; Fraser et al. 2009a). Recent ecological, phylogenetic, and phylogeographic studies suggest also that rafting cannot be ignored as an important mechanism of long-distance dispersal in southern waters (Waters 2008a; see also Chap. 17 by Rothäusler et al.).

14.3.1 Cold-Temperate Kelps as Biogeographical Models

The large "bull kelp" *Durvillaea* has proved to be an interesting model for seaweed biogeography. This genus has five recognized species: one (*D. antarctica*) is buoyant and with circumpolar distribution, while other four are nonbuoyant and with restricted distribution (*D. willana* present in southern New Zealand, *D. potatorum* in southern Australia and Tasmania, *D. chathamensis* in Chatham Islands, an undescribed species *Durvillaea* sp. in Antipodes Islands) (Cheshire et al. 1995) (Fig. 14.3). Recently, a phylogeographic circum-sub-Antarctic study provided evidence of recent (postglacial) recolonization by *D. antarctica*, most probably by transoceanic rafting of adult specimens (see also Chap. 17 by Rothäusler et al.). This is supported by high genetic homogeneity over vast geographic distances in sub-Antarctic latitudes, while along the coasts of New Zealand and Chile, substantial genetic differentiation was detected over relatively small geographic distances (e.g., tens of kilometers) (Fraser et al. 2009b).

In the coast of Chile, two different lineages of *D. antarctica* with a biogeographic break at 44–49°S have recently been revealed. Interestingly, populations from southern Patagonia (49–56°S) were found to be genetically homogenous and identical to a haplotype found throughout the sub-Antarctic region, pointing to the transoceanic ancestry and a relative recent recolonization (since the Last Glacial Maximum) (Fraser et al. 2010a). Similarly, genetically distinct species have been postulated for two populations of *D. antarctica* in New Zealand (Fraser et al. 2009c). Under a phylogenetic species concept, the genus *Durvillaea* has been recently interpreted as a complex of nine distinct evolutionary lineages, only one of which has an intercontinental distribution (Fraser et al. 2010b).

The most widely distributed kelp, *Macrocystis* (Fig. 14.3), displays an amphi-equatorial distribution. Four species (*M. pyrifera*, *M.*, *integrifolia*, *M. angustifolia,* and *M. laevis*) are currently recognized, however, increasing evidence strongly suggests the existence of a single morphologically plastic species (reviewed by Graham et al. 2007; Demes et al. 2009). Phylogenetic analyses by Coyer et al. (2001) point to a very recent (0.01–3 Ma) dispersal of *Macrocystis* from the northern to the southern hemisphere. This direction of the dispersal was supported by the greater genetic diversity of the individuals of the northern hemisphere, displaying paraphyletic clades, while the southern individuals formed monophyletic clades and showed little genetic differentiation across a wide southern geographic range, which refers to a recent dispersal.

Unlike the floating *D. antarctica*, the species of the genus *Lessonia* show a disperse distribution along the southern hemisphere (Fig. 14.3) characterized by a remarkable regional endemism (Searles 1978). The only intertidal species, *L. nigrescens*, extends from 17°S to 56°S and is the dominant organism at the infralittoral zone of the wave exposed locations of Chile and Peru (Santelices 1989). The subtidal *L. trabeculata* has a distribution range from 12 to 41°S in the eastern Pacific coast (from Peru to Chiloé Island) (Villouta and Santelices 1986). Populations of *L. vadosa* are found in the southern South America (47–56°S) in the coasts of Chile and Argentina as well as in the Falkland Islands (Searles 1978; Scrosati 1991). *L. flavicans* has also been described for South Georgia (John et al. 1994). *L. variegata* is endemic to New Zealand region including the mainland (Schwarz et al. 2006) and some adjacent islands (Chatham, Auckland, and Macquarie) (Kenny and Haysom 1962; Hay 1981). Other species show restricted distribution in this region, *L. tholiformis* being endemic to Chatham Island (Hay 1989), *L. brevifolia* present in Auckland, Campbell, Antipodes, and Bounty Islands (Hay 1981, 1987), *L. adamsiae* in Snares Islands (Hay 1987), and *L. corrugata* in Tasmania (Lane et al. 2006) (Fig. 14.3).

Recent phylogeographic analyses indicated the presence of two main divergent lineages, possibly two cryptic species, within *L. nigrescens* of the coast of Chile (Tellier et al. 2009), which matches the biogeographical transition zone around 30°S. In this species complex, a very limited dispersal has been reported (Faugeron et al. 2005; Tellier et al. 2009), confirming the marked endemism in many regions. Apparently, environmental factors (e.g., temperature) and anthropogenic activities can reduce gene flow between populations of *L. nigrescens* (Faugeron et al. 2005; Oppliger et al. 2011). Despite some molecular analyses (Lane et al. 2006), the origin of the genus *Lessonia* still remains unclear. Interestingly, within Lessoniaceae, it is the only genus not present in the northern hemisphere, and furthermore, it does not appear to be closely related to the other genera of this family (Bolton 2010).

14.4 Major Differences with Cold Temperate Regions of the Northern Hemisphere

Cold-adapted seaweeds are believed to have developed during the Tertiary decline of seawater temperature in temperate and polar regions. A permanent separation of the southern and northern hemisphere cold-water flora was produced by the formation of the pan-tropical Tethys Sea during the Mezozoic era (251–65 Ma). The presence of amphi-equatorial species with cold-temperate characteristics might be the result of processes such as paleoclimatic vicariance during the Miocene when intrusion of glacial water masses allowed many species crossing the equator (Lüning 1990; see also Chap. 18 by Bartsch et al.). However, some evidence indicates that disjunctions for some species are of a recent origin, e.g., during the last Pleistocene glacial maximum (18,000 years ago) (van Oppen et al. 1994). In the genus *Macrocystis*, phylogenetic analyses point to a very recent (0.01–3 Ma) dispersal from the northern to the southern hemisphere (Coyer et al. 2001). These transequatorial migrations took place at the east side of the oceans where the tropical regions are compressed. Crossing the equator was facilitated by "stepping stones," as well as by further constriction of the tropical regions during the Pleistocene and by glacial temperature drops as high as 8°C (reviewed by Lüning 1990; see also Chap. 18 by Bartsch et al.).

Compared to the cold-temperate regions in the northern hemisphere, the southern hemisphere has been characterized by the greater importance of the prevailing ocean currents for the seaweed distribution as well as by a more pronounced geographical isolation of the seaweed flora (Lüning 1990). In the southern hemisphere, the cold temperate regions have more island-like or peninsular character, thus resulting in a higher endemism. In the northern hemisphere, migrations at higher latitudes are possible mainly along coastlines. Since its formation the ACC has been the dominating biogeographical element at cold-temperate latitudes of the southern hemisphere and has served as a migration route for floating seaweed species (e.g., *Macrocystis pyrifera* and *Durvillaea antarctica*) creating similarities in floras between these regions (Lüning 1990; Fraser et al. 2009a; see also Chap. 17 by Rothäusler et al.).

One of the striking differences between the northern and southern cold-temperate regions is the almost exclusive dominance of species of the order Laminariales in littoral systems in the northern hemisphere, whereas in the southern hemisphere the only genera of the Laminariales, i.e., *Macrocystis*, *Lessonia*, *Ecklonia*, *Eisenia,* and *Laminaria*, grow intermixed with various dominant fucalean species. Nearshore reef assemblages in temperate New Zealand and Australia differ from those in other parts of the world: fucalean seaweeds dominate in the shallow subtidal zone, the kelp *E. radiata* forms dense beds at middle depths, and the giant kelp *M. pyrifera* occurs in the southern regions, but is abundant at only a few sites (Schiel et al. 1995). In the mid-intertidal zone, *Chondrus crispus*, a common dominant rhodophyte in the northern hemisphere, is replaced in regions of southern South America and many sub-Antarctic islands by species of *Nothogenia* and by the brown alga *Adenocystis utricularis* (Westermeier and Rivera 1986; Ingólfsson 2005). Among the main similarities between the northern and southern hemisphere cold-temperate floras is the presence of various members of the order Desmarestiales. In the coast of southern Chile, South Africa, and many sub-Antarctic islands several species of *Desmarestia* are regional endemics, and some species such as *D. ligulata* and *D. confervoides*/*D. viridis* exhibit a disjunct amphi-equatorial distribution (Wiencke et al. 1994; Peters et al. 1997). Species such as *D. anceps*, *D. menziesii,* and *Himantothallus grandifolius* are dominant in the Antarctic, but absent from the cold-temperate region. These species are regarded as ecological equivalent of Laminariales in the northern hemisphere (Moe and Silva 1977; Clayton 1994; Wiencke et al. 1996).

14.5 Concluding Remarks

Seaweed flora in the cold-temperate regions of the southern hemisphere has been modified not only by large-scale geological and paleoclimatic processes but also by long-distance dispersal events and an inherent capacity of the species to adapt to new habitats. Diverse present and future threats arising from anthropogenic activities will impose new challenges for these seaweeds communities. Many related aspects, such as temperature rise (see also Chap. 18 by Bartsch et al.), ocean acidification (see also Chap. 19 by Roleda and Hurd.), stratospheric ozone depletion and solar UV-B radiation (see also Chap. 20 by Bischof and Steinhoff), marine eutrophication and pollution (see also Chap. 21 by Teichberg.), as well as invasive seaweed species (see also Chap. 12 by Andreakis and Schaffelke.) and aquaculture (see also Chap. 22 by Buchholz et al.), are addressed in more detail in separate chapters throughout this book.

One of the most important issues is the scarce basic knowledge on the seaweeds communities over large areas. Although advances have been made in some regions, such as central/southern Chile, southern Australia, and New Zealand, seaweed communities of, e.g., the fjord region of southern Chile, the Argentinean region, and many sub-Antarctic islands are poorly known, partly due to their remoteness. The lack of studies on species diversity, community structure, and function of seaweeds in these locations impede suitable estimations on their contribution to the global geochemical fluxes, the potential genetic loss due to environmental shifts, and, in general, their conservation status. These are basic elements to understand the potential for tolerance and resilience in scenarios of global change (Harley et al. 2006).

Due to the closeness of the Antarctic ozone hole, the increase of solar UV-B radiation and resulting adverse effects on the biota of the cold-temperate region in the southern hemisphere is of a concern. Considering the latitudinal gradients of UV radiation and episodes of stratospheric ozone depletion reaching latitudes as low as 40°S (Orce and Helblin 1997), it is reasonable to argue that this factor can impose geographical barriers to species exhibiting low stress tolerance capacity. Recently, the potential effects of enhanced ultraviolet radiation on seaweed physiology have been examined in the context of adaptations to the intertidal life both in the Pacific and Atlantic coasts of Patagonia (Häder et al. 2003; Gómez et al. 2004; Huovinen et al. 2006; Rautenberger et al. 2009; Huovinen and Gómez 2011).

In some islands of the Southern Ocean, especially in Marion, Kerguelen, and Macquarie Island, a marked climate change, expressed as increases in temperature and declines of precipitation, has begun to be examined in the context of global warming. Because of the closeness of these islands to the APF, a relatively mobile climatic boundary, the impact of these climate shifts and its relation with, e.g., invasion of species in these islands has drawn attention, hitherto for terrestrial ecosystems (Bergstrom and Chown 1999). In Tasmania, intensification of southward intrusions of warmer, nutrient-poor East Australian Current waters has been reported to cause ocean warming and transport of biota to eastern Tasmania. The shift in large-scale oceanography has been related with a decline in the extent of the giant kelp beds. Furthermore, overgrazing of seaweeds by a recently established sea urchin species is affecting fundamentally the rocky reef systems in Tasmania (Johnson et al. 2011).

As seaweed communities of the cold-temperate region in the southern hemisphere are relatively isolated and thus, their structure, function, and biogeographic processes are strongly dependent on the environmental context, the direct impact of human activities requires urgent attention. Important questions related with impact of fish aquaculture and industrial effluents, introduction of alien species, and the overexploitation of commercially important seaweeds have not been addressed in detail. This information clearly is essential in order to understand the ecological processes of these organisms and their adaptive potential to large-scale physical forcings.

Acknowledgements This review has been written in the frame of the Project Fondecyt 1090494.

References

Acha EM, Mianzan HW, Guerrero RA, Favero M, Bava J (2004) Marine fronts at the continental shelves of austral South America physical and ecological processes. J Mar Syst 44:83–105

Barrales H, Lobban CS (1975) The comparative ecology of *Macrocystis pyrifera*, with emphasis on the forests of Chubut, Argentina. Ecology 63:657–677

Beckley LE, Branch GM (1992) A quantitative scuba-diving survey of the sublittoral macrobenthos at subantarctic Marion Island. Polar Biol 11:553–563

Bergstrom DM, Chown SL (1999) Life at the front: history, ecology and change on southern ocean islands. Trends Ecol Evol 14:472–477

Bertness MD, Crain CM, Silliman BR, Bazterrica MC, Reyna MV, Hildago F, Farina JK (2006) The community structure of western Atlantic Patagonian rocky shores. Ecol Monogr 76:439–460

Bolton JJ (1994) Global seaweed diversity: patterns and anomalies. Bot Mar 37:241–245

Bolton JJ (2010) The biogeography of kelps (Laminariales, Phaeophyceae): a global analysis with new insights from recent advances in molecular phylogenetics. Helgol Mar Res 64:263–279

Boraso A, Zaixso JM (2011). Atlas de sensibilidad ambiental de la costa y el Mar Argentino. Algas marinas bentónicas. pp 1–28. http://atlas.ambiente.gov.ar/tematicas/mt_02/pdfs/AL_01_Introduccion.pdf

Boyle MC, Jillett JB, Mladenov PV (2001) Intertidal communities in Doubtful Sound, New Zealand: changes over time. NZ J Mar Freshwat Res 35:663–673

Broom JES, Nelson WA, Farr TJ, Phillips LE, Clayton M (2010) Relationships of the *Porphyra* (Bangiales, Rhodophyta) flora of the Falkland Islands: a molecular survey using *rbc*L and nSSU sequence data. Aust Syst Bot 23:27–37

Buschmann AH (1992) Algal communities of a wave-protected intertidal rocky shore in southern Chile. In: Seeliger U (ed) Coastal plant communities of Latin America. Academic, Orlando, pp 91–104

Buschmann AH, Vásquez JA, Osorio P, Reyes E, Filún L, Hernández-González MC, Vega A (2004) The effect of water movement, temperature and salinity on abundance and reproductive patterns of *Macrocystis* spp. (Phaeophyta) at different latitudes in Chile. Mar Biol 145:849–862

Camus PA (2001) Biogeografía marina de Chile continental. Rev Chi Hist Nat 74:587–617

Casas G, Scrosati R, Piriz ML (2004) The invasive kelp *Undaria pinnatifida* (Phaeophyceae, Laminariales) reduces native seaweed diversity in Nuevo Gulf (Patagonia, Argentina). Biol Inv 6:411–416

Castilla JC, Uribe M, Bahamonde N, Clarke M, Desqueyroux-Faúndez R, Kong I, Moyano H, Rozbaczylo N, Santelices B, Valdovinos C, Zavala P (2005) Down under the southeastern Pacific: marine non-indigenous species in Chile. Biol Inv 7:213–232

Cheshire AC, Hallam ND (1988) Biomass and density of native stands of *Durvillaea potatorum* (southern bull-kelp) in south eastern Australia. Mar Ecol Prog Ser 48:277–283

Cheshire AC, Conran JG, Hallam ND (1995) A cladistic analysis of the evolution and biogeography of *Durvillaea* (Phaeophyta). J Phycol 1995:644–655

Clayton MN (1994) Evolution of the Antarctic marine benthic algal flora. Rev J Phycol 30:897–904

Clayton MN, Wiencke C, Klöser H (1997) New records of temperate and sub-Antarctic marine benthic macroalgae from Antarctica. Polar Biol 17:141–149

Coyer JA, Smith GJ, Andersen RA (2001) Evolution of *Macrocystis* spp. (Phaeophyceae) as determined by ITS1 and ITS2 sequences. J Phycol 37:574–585

Dayton PK (1985) Ecology of kelp communities. Annu Rev Ecol Syst 16:215–245

Delille B, Borges AV, Delille D (2009) Influence of giant kelp beds (*Macrocystis pyrifera*) on diel cycles of pCO_2 and DIC in the Sub-Antarctic coastal area. Est Coast Shelf Sci 81:114–122

Demes KW, Graham MH, Suskiewicz TS (2009) Phenotypic plasticity reconciles incongruous molecular and morphological taxonomies: the giant kelp, *Macrocystis* (Laminariales, Phaeophyceae), is a monospecific genus. J Phycol 45:1266–1269

Faugeron S, Martínez EA, Correa JA, Billot C (2005) Long-term copper mine waste disposal in northern Chile associated with gene flow disruption of the intertidal kelp *Lessonia nigrescens*. Mar Ecol Prog Ser 288:129–140

Fraser CI, Spencer HG, Waters JM (2009a) Glacial oceanographic contrasts explain phylogeography of Australian bull kelp. Mol Ecol 18:2287–2296

Fraser CI, Nikula R, Spencer HG, Waters JM (2009b) Kelp genes reveal effects of subantarctic sea ice during the Last Glacial Maximum. Proc Nat Acad Sci USA 106:3249–3253

Fraser CI, Hay CH, Spencer HG, Waters JM (2009c) Genetic and morphological analyses of the southern bull kelp *Durvillaea antarctica* (Phaeophyceae: Durvillaeales) in New Zealand reveal cryptic species. J Phycol 45:436–443

Fraser CI, Thiel M, Spencer HG, Waters JM (2010a) Contemporary habitat discontinuity and historic glacial ice drive genetic divergence in Chilean kelp. BMC Evol Biol 10:203b

Fraser CI, Winter DJ, Hamish GS, Waters JM (2010b) Multigene phylogeny of the southern bull-kelp genus *Durvillaea* (Phaeophyceae: Fucales). Mol Phylogenet Evol 57:1301–1311

Freeman D, Cooper S, Funnell G, Neale D (2011) Nearshore benthic community structure at the Bounty and Antipodes Islands subantarctic New Zealand. Polar Biol. doi:10.1007/s00300-011-1006-1

Gómez I, Huovinen P (2011) Morpho-functional patterns and zonation of South Chilean seaweeds: the importance of photosynthetic and bio-optical traits. Mar Ecol Prog Ser 422:77–91

Gómez I, López-Figueroa F, Ulloa N, Morales V, Lovengreen C, Huovinen P, Hess S (2004) Patterns of photosynthesis in 18 species of intertidal macroalgae from southern Chile. Mar Ecol Prog Ser 270:103–116

Graham MH, Vásquez JA, Buschmann AH (2007) Global ecology of the giant kelp *Macrocystis*: from ecotypes to ecosystems. Oceanogr Mar Biol Annu Rev 45:39–88

Häder D-P, Lebert M, Helbling EW (2003) Effects of solar radiation on the Patagonian Rhodophyte, *Corallina officinalis* (L.). Photosynth Res 78:119–132

Harley CDG, Hughes RA, Hultgren KM, Miner BG, Sorte CJB, Thornber CS, Rodriguez LF, Tomanek L, Williams SL (2006) The impacts of climate change in coastal marine systems. Ecol Lett 9:228–241

Hay ME (1981) The functional morphology of turf-forming seaweeds: persistence in stressful marine habitats. Ecology 62:739–750

Hay CH (1987) *Lessonia adamsiae* sp. nov. (Phaeophyta: Laminariales) from the Snares Islands, New Zealand. NZ J Bot 25:295–308

Hay CH (1989) *Lessonia tholiformis* sp. nov. (Phaeophyta: Laminariales) from the Chatham Islands, New Zealand. NZ J Bot 27:461–469

Heath RA (1985) A review of the physical oceanography of the seas around New Zealand–1982. NZ J Mar Freshwat Res 19:79–124

Hommersand MH, Moe RL, Amsler CD, Fredericq S (2009) Notes on the systematics and biogeographical relationships of Antarctic and sub-Antarctic Rhodophyta with descriptions of four new genera and five new species. Bor Mar 52:509–534

Huovinen P, Gómez I (2011) Spectral attenuation of solar radiation in Patagonian fjord and coastal waters and implications for algal photobiology. Cont Shelf Res 31:254–259

Huovinen P, Gómez I, Lovengreen C (2006) A Five-year study of solar ultraviolet radiation in southern Chile (39°S): Potential impact on physiology of coastal marine algae? Photochem Photobiol 82:515–522

Hurd CL, Nelson WA, Falshaw R, Neill KF (2004) History, current status and future of marine macroalgal research in New Zealand: Taxonomy, ecology, physiology and human uses. Phycol Res 52:80–106

Ingólfsson A (2005) Community structure and zonation patterns of rocky shores at high latitudes: an interocean comparison. J Biogeogr 32:169–182

Jara HF, Moreno CA (1984) Hervibory and structure in a mid-littoral rocky community: a case in southern Chile. Ecology 65:28–38

John DM, Tittley I, Lawson GW, Pugh PJA (1994) Distribution of seaweed floras in the Southern Ocean. Bot Mar 37:235–239

Johnson CR, Banks SC, Barrett NS, Cazassus F, Dunstan PK, Edgar GJ, Frusher SD, Gardner C, Haddon M, Helidoniotis F, Hill KL, Holbrook NJ, Hosie GW, Last PR, Ling SD, Melbourne-Thomas J, Miller K, Pecl GT, Richardson AJ, Ridgway KR, Rintoul SR, Ritz DA, Ross DJ, Sanderson JC, Shepherd SA, Slotwinski A, Swadling KM, Taw N (2011) Climate change cascades: Shifts in oceanography, species ranges and subtidal marine community dynamics in eastern Tasmania. J Exp Mar Biol Ecol 400:17–32

Kaehler S, Pakhomov EA, Kalin RM, Davis S (2006) Trophic importance of kelp-derived suspended particulate matter in a through-flow sub-Antarctic system. Mar Ecol Prog Ser 316:17–22

Kenny R, Haysom N (1962) Ecology of rocky shore organisms at Macquarie Island. Pacific Sci 16:245–263

Kerswell AP (2006) Global biodiversity patterns of benthic marine algae. Ecology 87:2479–2488

Klemm MF, Hallam ND (1988) Standing crop of *Durvillaea antarctica* (Chamisso) Hariot (Phaeophyta) on the Australian sub-Antarctic Macquarie and Heard Islands. Phycologia 27:505–509

Lane CE, Mayes C, Druehl LD, Saunders GW (2006) A multi-gene molecular investigation of the kelp (Laminariales, Phaeophyta) supports substantial taxonomic re-organization. J Phycol 42:493–512

Lilley SA, Schiel DR (2006) Community effects following the deletion of a habitat forming alga from rocky marine shores. Oecologia 148:672–681

Ling SD, Johnson CR (2009) Population dynamics of an ecologically important range-extender: kelp beds versus sea urchin barrens. Mar Ecol Prog Ser 374:113–125

Lüning K (1990) Seaweeds, Their environment, biogeography and ecophysiology. Wiley, New York

Lutz VA, Boschi EE, Bremec CS, Cousseau MB, Figueroa DE et al (2003) Perspectives of marine biodiversity studies in Argentina. Gayana 67:371–382

Meneses I, Santelices B (2000) Patterns and breaking points in the distribution of benthic algae along the temperate Pacific coast of South America. Rev Chi Hist Nat 73:615–623

Moe RC, Silva PC (1977) Antarctic marine flora: uniquely devoid of kelps. Science 196:1206–1208

Moreno CA, Jaramillo E (1983) The role of grazers in the zonation of intertidal macroalgae of the Chilean coast. Oikos 41:73–76

Moreno CA, Sutherland JP (1982) Physical and biological processes in a *Macrocystis pyrifera* community near Valdivia, Chile. Oecologia 55:1–6

Nelson WA, Broom JES (2010) The identity of *Porphyra columbina* (Bangiales, Rhodophyta) originally described from the New Zealand subantarctic islands. Aust Syst Bot 23:16–26

Nelson WA, Villouta E, Neill KF, Williams GC, Adams NM, Slivsgaard R (2002) Marine macroalgae of Fiordland, New Zealand. Tuhinga 13:117–152

Norton TA, Melkonian M, Andersen RA (1996) Algal biodiversity. Phycologia 35:308–326

Oppliger LV, Correa JA, Faugeron S, Beltrán J, Tellier F, Valero M, Destombe C (2011) Sex ratio variation in the *Lessonia nigrescens* complex (Laminariales, Phaeophyceae): Effect of latitude, temperature, and marginality. J Phycol 47:5–12

Orce VL, Helblin EW (1997) Latitudinal UVR-PAR measurements in Argentina: Extent of the "Ozone Hole". Global Planet Change 15:113–121

Orsi AH, Whitworth T, Nowlin WD (1995) On the meridional extent and fronts of the Antarctic Circumpolar Current. Deep-Sea Res I 42:641–673

Otaíza RD, Santelices B (1985) Vertical distribution of chitons (Mollusca: Polyplacophora) in the rocky intertidal zone of central Chile. J Exp Mar Biol Ecol 86:229–240

Papenfuss G (1964) Catalogue and bibliography of Antarctic and subantarctic benthic marine algae. Am Geophys Union, Antarctic Res Ser 1:1–76

Parsons MJ (1985) New Zealand seaweed flora and its relationship. NZ J Mar Freshw Res 19:131–138

Paruelo JM, Beltrán A, Jobbágy E, Sala OE, Golluscio RA (1998) The climate of Patagonia: general patterns and controls on biotic processes. Ecol Aust 8:85–101

Perissinotto R, McQuaid CD (1992) Deep occurrence of the giant kelp *Macrocystis laevis* in the Southern Ocean. Mar Ecol Prog Ser 81:89–95

Peters AF, van Oppen MJH, Wiencke C, Stam WT, Olsen JL (1997) Phylogeny and historical ecology of the Desmarestiaceae (Phaeophyceae) support a southern hemisphere origin. J Phycol 33:294–309

Phillips JA (2001) Marine macroalgal biodiversity hotspots: why is there high species richness and endemism in southern Australian marine benthic flora? Biodiv Conserv 10:1555–1577

Poloczanska ES, Babcock RC, Butler A, Hobday AJ, Hoegh-Guldberg O, Kunz TJ, Matear R, Milton DA, Okey TA, Richardson AJ (2007) Climate change and Australian marine life. Oceanogr Mar Biol Annu Rev 45:407–478

Ramírez ME (2010) Flora marina bentónica de la región austral de Sudamérica y la Antártica. An Inst Patagonia 38:57–71

Rautenberger R, Mansilla A, Gómez I, Wiencke C, Bischof K (2009) Photosynthetic acclimation to UV-radiation of intertidal macroalgae from the Strait of Magellan (Chile). Rev Chil Hist Nat 82:43–61

Sanderson JC, Thomas DP (1987) Subtidal macroalgal communities in the D'Entrecasteaux Channel, Tasmania. Aust J Ecol 12:41–51

Santelices B (1980) Phytogeographic characterization of the temperate coast of Pacific South America. Phycologia 19:1–12

Santelices B (1989) Algas marinas de Chile. Distribución, ecología, utilización y diversidad. Ediciones Universidad Católica de Chile, Santiago, Chile

Santelices B, Marquet P (1998) Seaweeds, latitudinal diversity patterns, and Rapaport's rule. Div Dist 4:71–75

Santelices B, Ojeda FP (1984) Population dynamics of coastal forests of *Macrocystis pyrifera* in Puerto Toro, Isla Navarino, Southern Chile. Mar Ecol Prog Ser 14:175–183

Santelices B, Castilla J, Cancino J, Schmiede P (1980) Comparative ecology of *Lessonia nigrescens* and *Durvillaea antarctica* (Phaeophyta) in central Chile. Mar Biol 59:119–132

Schiel DR (1990) Macroalgal assemblages in New Zealand: structure, interactions and demography. Hydrobiologia 192:59–76

Schiel DR (2004) The structure and replenishment of rocky shore intertidal communities and biogeographic comparisons. J Exp Mar Biol Ecol 300:309–342

Schiel DR, Hickford MJH (2001) Biological structure of nearshore rocky subtidal habitats in southern New Zealand, vol 182, Conserv Ser. N.Z. Department of Conservation Science, New Zealand, p 55

Schiel DR, Andrew NL, Foster MS (1995) The structure of subtidal algal and invertebrate assemblages at the Chatham Islands, New Zealand. Mar Biol 123:355–367

Schwarz AM, Hawes I, Wendy N, Neil A (2006) Growth and reproductive phenology of the kelp *Lessonia variegata* in central New Zealand. NZ J Mar Freshw Res 40:273–284

Scrosati RA (1991) Estudios anatómicos en *Lessonia vadosa* (Phaeophyta, Laminariales) de la Argentina. Boletín de la Sociedad Argentina de Botánica 27:165–171

Searles RB (1978) The genus *Lessonia* Bory (Phaeophyta, Laminariales) in southern Chile and Argentina. Br Phycol J 13:361–381

Smith SDA, Simpson RD (2002) Spatial variation in the community structure of intertidal habitats at Macquarie island (sub-Antarctic). Antarctic Sci 14:374–384

Strub PT, Mesías JM, Montecino V, Rutllant J, Salinas S (1998) Coastal ocean circulation off western South America. In: Robinson AR, Brink KH (eds) The sea. Wiley, New York, pp 273–314

Tellier F, Meynard AP, Correa JA, Faugeron S, Valero M (2009) Phylogeographic analyses of the 30°S south-east Pacific biogeographic transition zone establish the occurrence of a sharp genetic discontinuity in the kelp *Lessonia nigrescens*: Vicariance or parapatry? Mol Phylogenet Evol 53:679–693

Thiel M, Macaya EC, Acuña E, Arntz WE, Bastias H et al (2007) The Humboldt current system of northern and central Chile. Oceanographic processes, ecological interactions and socioeconomic feedback. Oceanogr Mar Biol Annu Rev 45:195–344

Upton J, Shaw CJ (2002) An overview of the oceanography and meteorology of the Falkland Islands. Aquatic Conserv Mar Freshwat Ecosyst 12:15–25

Valentine JP, Johnson CR (2004) Establishment of the introduced kelp *Undaria pinnatifida* following dieback of the native macroalga *Phyllospora comosa* in Tasmania, Australia. Mar Freshw Res 55:223–230

Valentine JP, Johnson CR (2005) Persistence of sea urchin (*Heliocidaris erythrogramma*) barrens on the east coast of Tasmania: inhibition of macroalgal recovery in the absence of high densities of sea urchins. Bot Mar 48:106–115

van Oppen MJH, Diekmann OE, Wiencke C, Stam WT, Olsen JL (1994) Tracking dispersal routes: phylogeography of the Arctic-Antarctic disjunct seaweed *Acrosiphonia arcta* (Chlorophyta). J Phycol 30:67–80

van Tussenbroek BI (1989) Seasonal growth and composition of fronds of *Macrocystis pyrifera* in the Falkland Islands. Mar Biol 100:419–430

van Tussenbroek BI (1993) Plant and frond dynamics of the giant kelp, *Macrocystis pyrifera*, forming a fringing zone in the Falkland Islands. Eur J Phycol 28:161–165

Vásquez JA, Castilla JC, Santelices B (1984) Distributional patterns and diets of four species of sea urchins in giant kelp forest (*Macrocystis pyrifera*) of Puerto Toro, Navarino Island, Chile. Mar Ecol Prog Ser 19:55–63

Villouta E, Santelices B (1986) *Lessonia trabeculata* sp. nov. (Laminariales, Phaeophyta), a new kelp from Chile. Phycologia 25:81–86

Villouta E, Chadderton WL, Pugsley CW, Hay CH (2001) Effects of sea urchin (*Evechinus chloroticus*) grazing in Dusky Sound, Fiordland, New Zealand. NZ J Mar Freshw Res 35:1007–1024

Waters JM (2008a) Driven by the West Wind Drift? A synthesis of southern temperate marine biogeography, with new directions for dispersal. J Biogeogr 35:417–427

Waters JM (2008b) Marine biogeochemical disjunction in temperate Australia: historical landbridge, contemporary currents, or both? Div Dist 14:692–700

Waters JM (2010) Australia's marine biogeography revisited: Back to the future? Aust Ecol 35:988–992

Westermeier R, Rivera PJ (1986) Caracterización ficológica del intermareal rocoso de la Xa. Región (Valdivia, Osorno, Llanquihue y Chiloé) y de la XII región (Islas Diego Ramírez), Chile. In: Westermeier R (ed), Actas Segundo Congreso Nacional sobre Algas Marinas Chilenas. pp 125–144

Westermeier R, Müller DG, Gómez I, Rivera P, Wenzel H (1994) Population biology of *Durvillaea antarctica* and *Lessonia nigrescens* (Phaeophyta) on the rocky shores of southern Chile. Mar Ecol Prog Ser 110:187–194

Wiencke C, Clayton MN (2002) Antarctic Seaweeds. ARG Gantner Verlag KG, Ruggell

Wiencke C, Bartsch I, Bischoff B, Peters AF, Breeman AM (1994) Temperature requirements and biogeography of Antarctic, Arctic and amphiequatorial seaweeds. Bot Mar 37:247–259

Wiencke C, Clayton MN, Langreder C (1996) Life history and seasonal morphogenesis of the endemic Antarctic brown alga *Desmarestia anceps* Montagne. Bot Mar 39:435–444

Chapter 15
Warm Temperate Seaweed Communities: A Case Study of Deep Water Kelp Forests from the Alboran Sea (SW Mediterranean Sea) and the Strait of Gibraltar

Antonio Flores-Moya

15.1 Introduction: Kelp Forests and Kelp Beds in the Alboran Sea and the Strait of Gibraltar

Kelp forests dominate the shallower sublittoral of temperate oceans worldwide (Mann 1973; Dayton 1985, 1994; Steneck et al. 2002) although they are also present in the Arctic Ocean (Dunton et al. 1982; Steneck et al. 2002; see also Chap. 13 by Wiencke and Amsler) and in deep waters near Ecuador (Graham et al. 2007). There are singular kelp forest communities inhabiting deep waters (20–100 m depth) in some localities at the Strait of Gibraltar and in the Mediterranean Sea, e.g., the Alboran Sea (the westernmost portion of the Mediterranean Sea; Flores-Moya 2004; Templado et al. 2006) and the Strait of Messina (SW Italy; Huvé 1958; Fredj and Giermann 1969; Giaccone 1969, 1972; Drew 1972, 1974).

Curiously, the first record of the presence of kelps in the areas close to the Strait of Gibraltar may be due to the "Father of Botany," Theophrastus of Eresus (371–286 BC): "Again in the ocean, about the Pillars of Hercules, there is a [seaweed] kind of marvelous size, they say, which is larger, about a palmsbreadth. This is carried into the inner sea along with the current from the outer sea, and they call it "sea-leek"; and in this [outer] sea in some parts it grows higher than a man's waist" (translation of Hort 1926). The "Pillars of Hercules" are peaks at either side of the Strait of Gibraltar, while the Atlantic Ocean and the Alboran Sea can be recognized as the "outer sea" and "inner sea," respectively. Drift thalli of *Saccorhiza polyschides* are very frequent in the autumn and winter on the beaches of the area; thus, this species is possibly the best candidate for Theophrastus

A. Flores-Moya (✉)
Department of Plant Biology (Botany), University of Málaga, Campus de Teatinos s/n, 29071 Málaga, Spain
e-mail: floresa@uma.es

C. Wiencke and K. Bischof (eds.), *Seaweed Biology*, Ecological Studies 219,
DOI 10.1007/978-3-642-28451-9_15, © Springer-Verlag Berlin Heidelberg 2012

Fig. 15.1 Kelp forest of *Laminaria ochroleuca* (**a**) at 46 m depth at Alboran Island and *Saccorhiza polyschides* (**b**) at 24 m depth at Tarifa (Strait of Gibraltar). *Phyllariopsis brevipes* (**c**) at 42 m depth at Tarifa (Strait of Gibraltar). *Phyllariopsis purpurascens* (**d**) at 45 m depth at Tarifa (Strait of Gibraltar). Photographs courtesy from OCEANA (**a**) and from JC Moreno (**b–d**)

"sea-leek" (Amigues 1989). However, the first evidence of the presence of kelp forests in the Mediterranean Sea (in particular at Alboran Island, located at the center of the Alboran Sea) was found in 1958 in the course of dives by an expedition of the vessel *Calypso* (Cousteau and Dugan 1963). Before this discovery, it was thought that kelp forests were not present in the Mediterranean Sea. These forests remain poorly studied in comparison to similar communities in other oceans due to their locations in deep waters and the few phycologists working on this area.

The species occurring in the deep water kelp forests are *Laminaria ochroleuca* Bachelot de la Pylaie, *S. polyschides* (Lightfoot) Batters, and the small kelps *Phyllariopsis brevipes* (C. Agardh) E. C. Henry & G. R. South and *Phyllariopsis purpurascens* (C. Agardh) E. C. Henry & G. R. South, which also form open kelp bed communities in very deep waters. While *L. ochroleuca* is a member of Laminariales, the other three species are members of Phyllariaceae, a family that has been transferred from Laminariales to Tilopteridales (Sasaki et al. 2001; Adl et al. 2005).

L. ochroleuca (Fig. 15.1a) inhabits the NW Atlantic Ocean (from Brittany and South England in the North to the Atlantic shores of Morocco in the South); it is also present in the Azores, the Alboran Sea, and the Strait of Messina and, in the southern hemisphere, in Namibia (detailed references on distribution of all of the

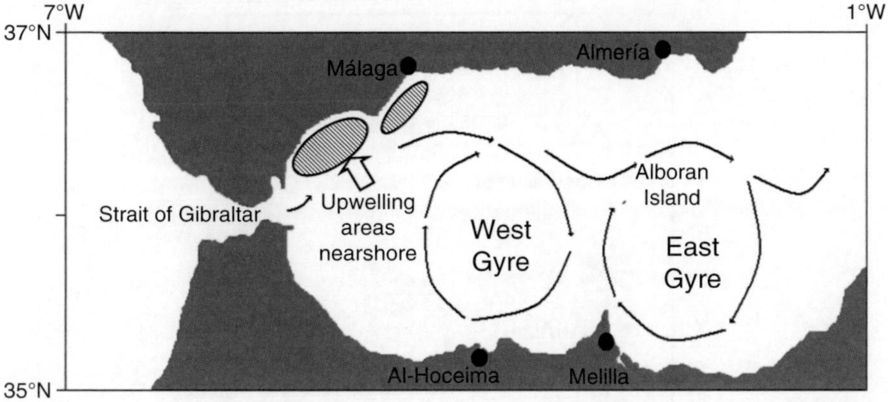

Fig. 15.2 Main surface currents and upwelling zones in the Strait of Gibraltar and the Alboran Sea

cited species are available in Guiry and Guiry 2011). The perennial thalli have a holdfast with branched haptera, a stipe (in the longest individuals reaching 3 m length) and a blade (up to 3–4 m length) divided into laciniae.

S. polyschides (Fig. 15.1b) is a North Atlantic species (from the Norwegian shores at the North to Mauritania at the South) that can also be found in the Canary Islands and in deep waters from the Alboran Sea and the Strait of Messina. It has a holdfast with a large bulb and small unbranched haptera, a stipe with lateral frills, and a blade divided into laciniae. Thalli are smaller than those of *L. ochroleuca*, but they can reach 3–4 m length. This species is annual or biennial.

Both species of *Phyllariopsis* are annuals and inhabit the NW and W Iberian Peninsula, NW Morocco, the western Mediterranean Sea, and the Strait of Messina, but *P. purpurascens* has also been found in the Canary Islands. *P. brevipes* (Fig. 15.1c) has an entire, cordate blade of <0.5 m length and a short stipe; the holdfasts have unbranched haptera. *P. purpurascens* (Fig. 15.1d) has a single blade that, in the longest individuals, reaches 2 m in length. It holds to the substratum by a disc of around 1 cm diameter, and it has a stipe 10–12 cm in length. The Mediterranean endemic *Laminaria rodriguezii* Bornet inhabits very deep waters (50–120 m depths) in the Western Mediterranean Sea and in the Adriatic Sea (Feldmann 1934; Giaccone 1969), but this species has not been found in the Alboran Sea (Pérez-Ruzafa et al. 2003).

The presence of kelp forests and beds in the Alboran Sea and the Strait of Gibraltar is linked to the particular oceanographic conditions (Fig. 15.2). The negative water balance in the Mediterranean Sea (i.e., the loss of water by evaporation is higher than the inputs by precipitation and rivers) is compensated by a permanent current flowing from the Atlantic Ocean into the Mediterranean Sea across the Strait of Gibraltar. Thus, the salinity in the Mediterranean Sea is higher than that at the same latitude in the neighbor Atlantic Ocean, but it does not increase further because high-salinity deep water current flows from the Mediterranean to

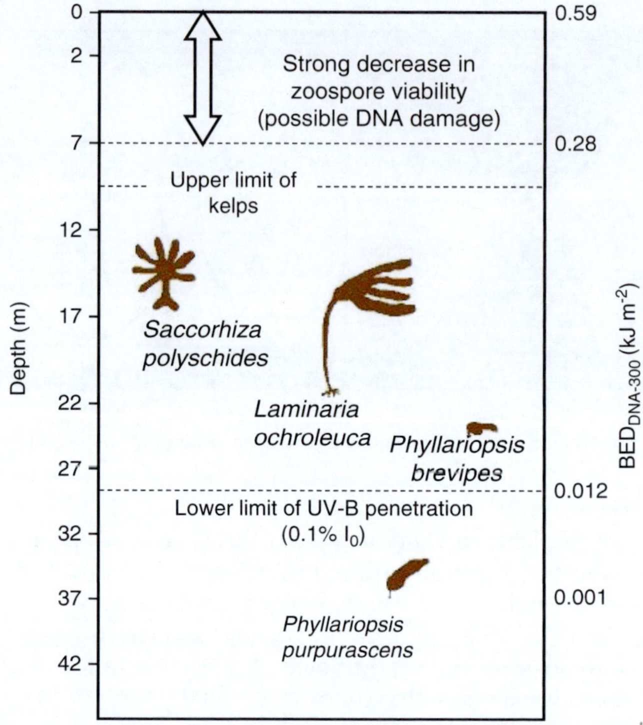

Fig. 15.3 Zonation of the four kelps at Tarifa (Strait of Gibraltar) and its correlation to the prevailing vertical ultraviolet-B radiation conditions and major biological and bio-optical events, expressed as biologically effective ultraviolet-B doses for DNA damage ($BED_{DNA\text{-}300}$) assuming constant photon fluence rates during 8 h (Wiencke et al. 2000)

the Atlantic. The surface current of Atlantic water flows close to the shores of the Iberian Peninsula and then it turns around, like an anticyclone, in the Western Alboran Sea. Simultaneously, upwelling of cold, deep water close to the shores compensates for the accumulation of water by the action of the anticyclone (Lacombe and Tchernia 1972; Rodríguez 1982; Sarhan et al. 2000).

The kelp forests are located in the Strait of Gibraltar, in the cold and nutrient-rich upwelling zones close to the South Iberian Peninsula, similar as in other kelp regions worldwide (Druehl 1981; Bolton 2010; Steneck et al. 2002) and at Alboran Island. Because of the high number of sunny days in the Mediterranean and the high light transmittance of the clear water, kelp forests are confined to deep water as a consequence of the photoinhibitory effects of the strong irradiance in shallow water (see Fig. 15.3). Moreover, surface temperatures >22°C in summer also contribute to the restriction in depth because these high temperatures are at the edge of thermal limits for the kelp species inhabiting Alboran Island (see Table 15.1).

Table 15.1 Growth and reproduction thermal limits for the kelps inhabiting the Strait of Gibraltar and the Alboran Sea

Species	Growth/reproduction (sporophyte)		Growth/reproduction (gametophyte)		References
	Upper	Lower	Upper	Lower	
Laminaria ochroleuca	22–23°C/*	*/*	*/21°C	*/5°C	Lüning (1990), Izquierdo et al. (2002)
Saccorhiza polyschides	24°C/*	3°C/*	25°C/<17°C	+/5°C	Norton (1970)
Phyllariopsis brevipes	*/*	*/*	20°C/20°C	10°C/10°C	Henry (1987)
Phyllariopsis purpurascens	*/*	*/*	20°C/20°C	10°C/10°C	Unpublished

Lethal limits can be estimated 1–2°C beyond the growth thermal limits
Asterisks: no data are available

15.2 Abiotic Factors Affecting Distribution of Kelps in the Alboran Sea and the Strait of Gibraltar

15.2.1 Substratum

The four species of kelps inhabit hard substrata. A singular case is *P. brevipes*, which always grows on living thalli of the crustose coralline *Mesophyllum alternans* (Foslie) Cabioch & Mendoza (Henry 1987; Pérez-Cirera et al. 1989). The specificity for the substratum in *P. brevipes* is possibly due to the requirements of the microscopic gametophyte (Sauvageau 1918) rather than to those of the sporophyte.

15.2.2 Hydrodynamics

The bottoms where kelps grow in the Strait of Gibraltar are exposed to very strong currents linked to the semidiurnal tidal cycle, with peaks of 2.5 m s^{-1} at low or high tide at spring tides (Lacombe and Richez 1982). The current may be a factor controlling the largest blade area that kelps can reach. Blades of 0.1 m^2 of *P. purpurascens* (the largest blade area found) can be detached from the stipe with current velocities ranging from 2.4 to 3.5 m s^{-1} and an inverse correlation was detected between blade area and current exposure on different bottoms (Flores-Moya et al. 1993; Flores-Moya 1997). A similar control by peak currents decapitating older thalli was suggested for *L. ochroleuca* in the Strait of Messina (Drew et al. 1982).

15.2.3 Irradiance

It is accepted that the lower limit of distribution of kelps corresponds to approximately 1% of the surface solar irradiance (Lüning and Dring 1979) and this assessment agrees with the maximum depths found for the kelps from the Alboran Sea and the Strait of Gibraltar. The lower limit of *L. ochroleuca* in the Strait of Gibraltar (Jerlov's optical water oceanic II–III and coastal 1–2 types) is close to 25 m depth, but at Alboran Island (Jerlov's optical water oceanic I–II type) it reaches 70–80 m depth (Flores-Moya 2004; Templado et al. 2006; Aguilar et al. 2010). There is a report of possible heterotrophic growth of *L. ochroleuca* in deep water from the Strait of Messina (Drew et al. 1982), but data are inconclusive. The most complete set of data about irradiance requirements was derived from a study of the sporophyte of *P. purpurascens*. The compensation of dark respiration by photosynthesis is achieved at 6.5 μmol photons m^{-2} s^{-1}. This irradiance level occurs at 35–40 m depth in summer, the lowest limit of the deepest populations in the Strait of Gibraltar (Flores-Moya et al. 1995). Similarly, the compensation irradiance occurs at 85–100 m depth at Alboran Island (Templado et al. 2006; Aguilar et al. 2010). This is the reason why this kelp and, in general, the four species cited in this chapter must be recognized as "shade plants" (see also Chap. 1 by Hanelt and Figueroa). Moreover, it can be hypothesized that the irradiance requirements of the gametophytes are lower than for the sporophytes, although no data are available.

Although the high PAR levels are usually considered to be responsible for determination of the upper depth limit of sublittoral seaweeds (Lüning and Dring 1979), the ultraviolet-B radiation (UV-B) also affects the germination of zoospores and the growth of microscopic gametophytes of kelps. Thus, the capacity to tolerate UV-B by zoospores and gametophytes of the four kelps in the Strait of Gibraltar is correlated with those levels of solar UV-B that penetrate seawater column (Fig. 15.3; Wiencke et al. 2000; Roleda et al. 2004; see also Chap. 20 by Bischof and Steinhoff). This could also be the reason for a deeper upper limit of kelps at Alboran Island, because the water there is more transparent than in the Strait of Gibraltar (see above).

15.2.4 Temperature

In the deep waters where kelps occur in the Alboran Sea and the Strait of Gibraltar (below 20 m depth), temperature never exceeds 15°C in summer. Thus, the thermal regime is closer to those of cold temperate oceans than to warm temperate ones. Moreover, this habitat is very constant in temperature. For instance, the temperature in the Strait of Gibraltar at 35–45 m depth (where *P. purpurascens* form kelp beds) ranges from 12.5 to 14.5°C during the year (Flores-Moya et al. 1993; Flores-Moya 1997).

The thermal limits for growth and reproduction of the four species are compiled in Table 15.1, but no data are available on the lethal limits. Moreover, data from *L. ochroleuca* and *S. polyschides* must be considered with caution since they correspond to seaweed populations very distant from the Alboran Sea and the occurrence of different thermal ecotypes in their distribution areas is probable.

15.2.5 Inorganic Nutrients

There are no time series of inorganic nutrients in the area. However, they must be high due to the permanent cold, rich-nutrient upwelling in the kelp forests. It can be hypothesized that the perennial *L. ochroleuca* stores nutrients as other perennial kelps do (Mann 1973; Gagné and Mann 1981; see Chap. 4 by Gordillo). The nutrient storage strategy is very different in the annual *P. purpurascens* from the Strait of Gibraltar. The C:N and N:P ratios increased from spring (when recruitment of sporophytes took place) to summer (when the thalli reached the greatest size and the production of sori began), and they remained constant until the first storm period in the autumn when the sporophytes were detached. The ratios suggested that the growth of *P. purpurascens* could be limited by both N and P, but the time course of the C:N:P Redfield ratio (Redfield 1934; Atkinsons and Smith 1984) showed that the growth was P-limited (Flores-Moya et al. 1995).

P. purpurascens shows external carbonic anhydrase (CA) activity (see also Chap. 19 by Roleda and Hurd). This enzyme catalyzes the dehydration of HCO_3^- (the most abundant inorganic carbon species in seawater) to CO_2 which diffuses into the cells. However, under fast flowing water due to tidal currents the extracellular CA may be physically removed from the thallus surface (Flores-Moya and Fernández 1998).

15.3 Growth and Reproduction

An estimate of the mean age of *L. ochroleuca* individuals of about 4–6 m length collected at 45 m depth at Alboran Island (assuming one annual cycle per pair of bands formed by the meristoderm; Klinger and DeWreede 1988) resulted in an age of 8–11 years (Flores-Moya, unpublished data). Similar age structure was found in the populations growing at 60 m depth in the Strait of Messina (Drew 1972). *S. polyschides* and both species of *Phyllariopsis* can be considered as "monocarpic" (the term used for flowering plants which die after setting fruit) since the thalli become senescent after sorus formation. The basic difference between the two genera is that the *S. polyschides* thalli sometimes do not become fertile during the first year (i.e., they could be considered biennial) whereas growth and reproduction in *Phyllariopsis* occurs in less than 1 year (Flores-Moya et al. 1993). The thalli are recruited in spring and are detached from the substratum in autumn–winter.

L. ochroleuca bears sori all year long but the highest percentage of fertile blades occurs in the spring and summer. The optimal temperature range for both zoospore germination and growth of gametophytes is 15–18°C (Table 15.1); gametogenesis is inhibited at temperatures below 10°C. Growth of young sporophytes is maximal at irradiances of 20–40 μmol photons $m^{-2}\ s^{-1}$ (Izquierdo et al. 2002). *S. polyschides* and both species in *Phyllariopsis* become fertile in summer (Flores-Moya et al. 1993). Gametophytes of both species of *Phyllariopsis* are dioecious but unusually monomorphic (Henry 1987; Flores-Moya and Henry 1998). Thermal limits for growth and reproduction for *Phyllariopsis* gametophytes range from 10 to 20°C (Table 15.1).

15.4 Productivity

There are no data on net primary productivity of *L. ochroleuca* or *S. polyschides* in the Alboran Sea or the Strait of Gibraltar. The measurements for populations of *L. ochroleuca* from the shallower sublittoral of NW Spain ranged from 1 to 2 kg C year^{-1} m^{-2} sea bottom (Fernández and Niell 1981). These figures were similar to those found in other species of *Laminaria* (Mann 1982; Sjøtun et al. 1995) but lower than those found in *Laminaria hyperborea* growing at depths lower than 30 m in Norway (3 kg C year^{-1} m^{-2} sea bottom; Abdullah and Fredriksen 2004). The net primary productivity of *S. polyschides* from North Spain ranged from 0.5 to 1 kg C year^{-1} m^{-2} sea bottom (Fernández and Niell 1981). The net annual primary productivity measured in *P. purpurascens* growing at 30–35 m depth in the Strait of Gibraltar was approximately 50 g C year^{-1} m^{-2} sea bottom, a very low value in comparison to the values found in other kelps (Flores-Moya et al. 1993). This result was later interpreted as a consequence of the P-limited (Flores-Moya et al. 1995) and irradiance-limited (Flores-Moya et al. 1996, 1997; Flores-Moya 1997) growth.

15.5 The Communities of Kelp Forests in the Alboran Sea and the Strait of Gibraltar

The best-known kelp forests considered here are located on the sea bottom at Alboran Island (Templado et al. 2006). *S. polyschides* forests range from 15 to 35 m depth, with a maximum density of 5–8 thalli m^{-2} sea bottom. Between 25 and 50–60 m depth, the kelp forest is formed by *L. ochroleuca*, with densities of 2–4 thalli m^{-2}. Both species form a continuous canopy due to the more or less horizontal position of the blades. In the case of *L. ochroleuca* forests, new individuals are recruited after a catastrophic event causing the loss of old thalli; in fact, young thalli only appear in cleared areas. The intraspecific competition is much more intense in the very deep water kelp beds of *P. purpurascens*. The irradiance level below one blade at 30 m depth was lower than the photosynthetic irradiance-compensation point. This could explain two facts. On the one hand, the overdispersed distribution of thalli on the sea bottom which avoided the overlapping of blades (Flores-Moya 1997). On the other hand, the growth of sporophytes which occurred only after the detaching of some individuals (Flores-Moya et al. 1996, 1997) following a defined self-thinning pattern (Fig. 15.4).

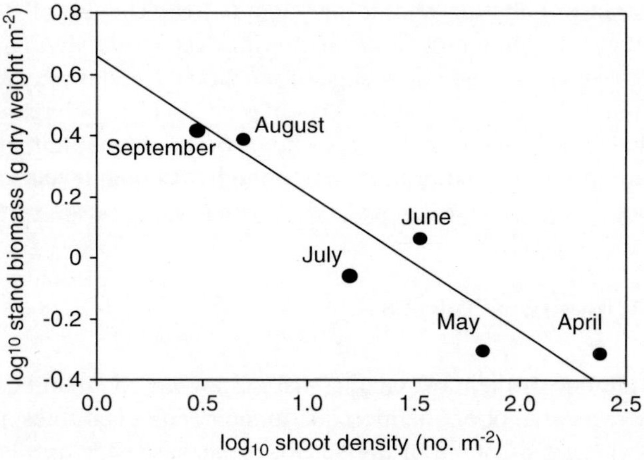

Fig. 15.4 Relationship between $\log_{10}$ stand biomass and $\log_{10}$ shoot density along the life span of the sporophytes of *Phyllariopsis purpurascens* (from April to September) growing at 30 m depth in Tarifa (Strait of Gibraltar). A significant negative slope suggested the occurrence of self-thinning in this population (Flores-Moya et al. 1997)

Kelp forests are considered as one of the most productive and dynamic ecosystems on Earth (Mann 1973). Moreover, they have very high species richness. About 1,800 different species from different phyla find refuge or breed in the kelp forests in the United Kingdom (Birkett et al. 1998) and a total of 238 faunal species were recorded in the Norwegian *Laminaria hyperborea* forests (Christie et al. 2003, 2009). The species richness associated with the *L. ochroleuca* forests at Alboran Island (Templado et al. 2006) has been estimated to be around 1,800–2,000 species, although this island is minuscule (614 m maximum length and 300 m maximum width) in comparison to the British Isles or to the shoreline of Norway, and its shelf is not extensive due to the steep slopes. Thus, it could be hypothesized that the Alboran Island kelp forests may be the ecosystem with the highest biodiversity in Europe.

15.6 Conservation and Prospects

The conservation status of kelp forest in the Strait of Gibraltar and the Alboran Sea is satisfactory, principally due to the deep locations of these communities. The best conserved kelp forests are located on the sea bottoms of Alboran Island. The Natural Park of the Strait and the Marine and Fishing Reserve of Alboran Island are protected areas that help to preserve these singular deep water kelp forests. Recent observations with unmanned ROVs from the organization OCEANA (Aguilar et al. 2010) have contributed more knowledge about the extent, depth distribution, and organisms associated with the deep water kelp forests and kelp beds in the Strait of Gibraltar and in the Alboran Sea. The use of ROVs makes it possible to avoid the problems associated with deep water diving in areas with strong currents. However, our knowledge of functional ecology and trophic ecology of these communities is very sparse. This gap needs to be addressed in order to understand the contribution of kelp forests in the Strait of Gibraltar and the Alboran Sea to the nutrient cycles and productivity of higher trophic levels.

15.7 Concluding Remarks

Kelp forests are one of the most characteristic seaweed communities in temperate oceans. They receive a lot of attention due to their ecological roles, their commercial interests (i.e., as a source of alginates), or because they are the nursery and habitat of fishery species such as lobsters, crabs, and mollusks. However, the singular deep water kelp forests from the Strait of Gibraltar and the Alboran Sea long remained almost unknown to the phycological community. The structure of kelp forests and kelp beds, the species richness, and some autoecological topics have been addressed. However, ecological processes such as the role of nutrient cycles or their contribution to higher trophic levels remain unstudied.

Acknowledgments Agustín Barrajón was my excellent dive partner for more than 2,000 deep water dives in different locations in the Alboran Sea and the Strait of Gibraltar. The research work was enriched by collaboration (and sometimes diving) with the following colleagues worldwide: María Altamirano, Francisco Conde, José A. Fernández, Félix L. Figueroa, Eric C. Henry (who also revised the English style and usage of this chapter), Hiroshi Kawai, Ángel A. Luque, F. Xavier Niell, José Templado, and Christian Wiencke. José C. Moreno and the organization OCEANA were the authors of the submarine photographs. The writing of the chapter was financially supported by the Junta de Andalucía Research Group RNM-115.

References

Abdullah MI, Fredriksen S (2004) Production, respiration and exudation of dissolved matter by the kelp *Laminaria hyperborea* along the west coast of Norway. J Mar Biol Ass UK 84:887–894

Adl SM, Simpson AGB, Farmer MA, Andersen RA, Anderson OR, Barta JR, Bowser SS, Brugerolle G, Fensome RA, Fredericq S, James TY, Karpov S, Kugrens P, Krug J, Lane CE, Lewis LA, Lodge J, Lynn DH, Mann DG, McCourt RM, Mendoza L, Moestrup Ø, Mozley-Standridge SE, Nerad TA, Shearer CA, Smirnov AV, Speigel FW, Taylor MFJR (2005) The new higher level classification of eukaryotes with emphasis on the taxonomy of protists. J Eukaryot Microbiol 52:399–451

Aguilar R, García S, Ubero J (2010) Distribution of deep-sea laminarians around three Spanish marine protected areas. http://eu.oceana.org/sites/default/files/reports/OCEANA_Poster_Laminarias_2010.pdf Cited 14 March 2011

Amigues S (1989) Théophraste. Recherches sur les plantes, vol II. Les Belles Lettres, Paris

Atkinsons MJ, Smith SV (1984) C:N:P ratios of benthic marine plants. Limnol Oceanogr 28:568–574

Birkett DA, Maggs CA, Dring MJ, Boaden PJS (1998) Infralittoral reef biotopes with kelp species. An overview of dynamic and sensitivity characteristics for conservation management of marine SACs. Scottish Association of Marine Science (UK Marine SACs Project), Scotland

Bolton JJ (2010) The biogeography of kelps (Laminariales, Phaeophyceae): a global analysis with new insights from recent advances in molecular phylogenetics. Helgoland Mar Res 64:263–279

Christie H, Jørgesen NM, Norderhaug KM, Waage-Nielsen E (2003) Species distribution and habitat exploitation of fauna associated with kelp (*Laminaria hyperborea*) along the Norwegian coast. J Mar Biol Ass UK 83:687–699

Christie H, Norderhaug KM, Fredriksen S (2009) Macrophytes as habitat for fauna. Mar Ecol Prog Ser 396:221–233

Cousteau J-Y, Dugan J (1963) The living sea. Hamish Hamilton, London

Dayton PK (1985) Ecology of kelp communities. Annu Rev Ecol Syst 16:215–245

Dayton PK (1994) Essay: kelp forest. In: Lobban CS, Harrison PJ (eds) Seaweed ecology and physiology. Cambridge University Press, New York, pp 75–77

Drew EA (1972) Growth of a kelp forest at 60 metres in the Straits of Messina. Mem Biol Mar Oceanogr Univ Messina 2:135–157

Drew EA (1974) An ecological study of *Laminaria ochroleuca* Pyl. Growing below 50 metres in the Straits of Messina. J Exp Mar Biol Ecol 15:11–24

Drew EA, Ireland JF, Muir C, Robertson WA, Robinson JD (1982) Photosynthesis, respiration, and other factors influencing the growth of *Laminaria ochroleuca* Pyl. below 50 metres in the Straits of Messina. PSZNI Mar Ecol 3:335–355

Druehl LD (1981) The distribution of Laminariales in the North Pacific with reference to environmental influences. Proc Int Congr Syst Evol Biol 2:248–256

Dunton KH, Reimnitz E, Schonberg S (1982) An Arctic kelp community in the Alaskan Beaufort Sea. Arctic 35:465–484

Feldmann J (1934) Les Laminaricées de la Méditerranée et leur répartition géographique. Bull Trav Stat d'Aquicult Pêche de Castiglione 1932:141–184

Fernández C, Niell FX (1981) Discusión sobre los métodos usados en la estimación de la producción de macrófitos intermareales. Oecol Aquat 5:43–52

Flores-Moya A (1997) Changes in reproductive effort, lamina-area index, and standing crop with depth in the deep water alga *Phyllariopsis purpurascens* (Laminariales, Phaeophyta). Phycologia 36:32–37

Flores-Moya A (2004) Los bosques de laminariales. In: Luque AA, Templado J (eds) Praderas y bosques marinos de Andalucía. Consejería de Medio Ambiente de la Junta de Andalucía, Sevilla, pp 183–198

Flores-Moya A, Fernández JA (1998) The role of external carbonic anhydrase in the photosynthetic use of inorganic carbon in the deep-water alga *Phyllariopsis purpurascens* (Laminariales, Phaeophyta). Planta 207:115–119

Flores-Moya A, Henry EC (1998) Gametophyte and first stages of the sporophyte of *Phyllariopsis purpurascens* (Laminariales, Phaeophyta). Phycologia 37:398–401

Flores-Moya A, Fernández JA, Niell FX (1993) Reproductive phenology, growth and primary production of *Phyllariopsis purpurascens* (Phyllariaceae, Phaeophyta) from the Straits of Gibraltar area. Eur J Phycol 28:223–230

Flores-Moya A, Fernández JA, Niell FX (1995) Seasonal variations of photosynthetic pigments, total C, N, and P content, and photosynthesis in *Phyllariopsis purpurascens* (Phaeophyta) from the Strait of Gibraltar. J Phycol 31:867–874

Flores-Moya A, Fernández JA, Niell FX (1996) Growth pattern, reproduction and self-thinning in seaweeds. J Phycol 32:767–769

Flores-Moya A, Fernández JA, Niell FX (1997) Growth pattern, reproduction and self-thinning in seaweeds: a re-evaluation in reply to Scrosati. J Phycol 33:1080–1081

Fredj G, Giermann G (1969) Observations en SP300 de peuplements de Laminariales dans le detroit de Messine. Rapp et PV du CIESM, Monaco 20:259–261

Gagné JA, Mann KH (1981) Comparison of growth strategy in *Laminaria* populations living under differing seasonal patterns of nutrient availability. Proc Int Seaweed Symp 10:297–302

Giaccone G (1969) Note sistematiche ed osservazioni fitosociologiche sulle Laminariales del Mediterraneo occidentale. G Bot Ital 193:457–474

Giaccone G (1972) Struttura, ecologia e corologia dei popolameti a laminarie dello Stretto di Messina e del mare di Alboran. Mem Biol Mar Oceanogr NS 2:37–59

Graham MH, Kinlan BP, Druehl LD, Garske LE, Banks S (2007) Deep-water kelp refugia as potential hotspots of tropical marine diversity and productivity. Proc Natl Acad Sci USA 104:16576–16580

Guiry MD, Guiry GM (2011) AlgaeBase. World-wide electronic publication, National University of Ireland, Galway. http://www.algaebase.org. Cited 14 March 2011

Henry EC (1987) The life history of *Phyllariopsis brevipes* (=*Phyllaria reniformis*) (Phyllariaceae, Laminariales, Phaeophyceae), a kelp with dioecious but sexually monomorphic gametophytes. Phycologia 26:17–22

Hort AF (1926) Theophrastus, enquiry into plants Vol. II. William Heinemann, Harvard University Press, London

Huvé H (1958) Contribution à l'étude des peuplements des Phyllariacées du détroit de Messine. Réun Commiss internation Explor sci Médit 14(NS):525–533

Izquierdo JL, Pérez-Ruzafa IM, Gallardo T (2002) Effect of temperature and photon fluence rate on gametophytes and young sporophytes of *Laminaria ochroleuca* Pylaie. Helgol Mar Res 55:285–292

Klinger T, DeWreede RE (1988) Stipe rings, age, and size in populations of *Laminaria setchellii* Silva (Laminariales, Phaeophyta) in British Columbia, Canada. Phycologia 27:234–240

Lacombe H, Richez P (1982) The regime of the Strait of Gibraltar. In: Nihoul JCJ (de) Hydrodynamics of the semi-enclosed seas, 13th International Liège Colloquium on Ocean Hydrodynamics. Elsevier Oceanography Series 34, Elsevier, Amsterdam, pp 13–73

Lacombe H, Tchernia P (1972) Caràcteres hydrologiques et circulation des eaux en Méditerranée. In: Stanley DJ (ed) The Mediterranean Sea: a natural sedimentation laboratory. Dowden Hutchinson and Ross, Stroudsburg, pp 26–36

Lüning K (1990) Seaweeds. Their environment, biogeography and ecophysiology. Wiley, New York

Lüning K, Dring M (1979) Continuous underwater light measurement near Helgoland (North Sea) and its significance for characteristics light limits in the sublittoral region. Helgoländer wiss Meeresunters 32:403–424

Mann KH (1973) Seaweeds: their productivity and strategy for growth. Science 182:975–981

Mann KH (1982) Ecology of coastal waters. A system approach. Blackwell, Oxford

Norton TA (1970) Synopsis of biological data on *Saccorhiza polyschides*. FAO Fisheries Synopsis 83:1–35

Pérez-Cirera JL, Cremades J, Bárbara I (1989) Precisiones sistemáticas y sinecológicas sobre algunas algas nuevas para Galicia o para las costas atlánticas de la Península Ibérica. Anales Jard Bot Madrid 46:35–45

Pérez-Ruzafa I, Izquierdo JL, Araújo R, Sousa-Pinto I, Pereira L, Bárbara I (2003) Mapas de distribución de algas marinas de la Península Ibérica e Islas Baleares. XVII. *Laminaria rodriguezii* Bornet y adiciones a los mapas de *L. hyperborea* (Gunner.) Foslie, *L. ochroleuca* Bach. Pyl. y *L. saccharina* (L.) Lamour. (*Laminariales, Fucophyceae*). Bot Complut 27:155–164

Redfield AC (1934) On the proportions of organic derivations in sea water and their relation to the composition of plankton. In: Daniel RJ (ed) James Johnstone memorial volume. University Press of Liverpool, Liverpool, England, pp 177–192

Rodríguez J (1982) Oceanografía del mar Mediterráneo. Pirámide SA, Madrid

Roleda MY, Hanelt D, Kräbs G, Wiencke C (2004) Morphology, growth, photosynthesis and pigments in *Laminaria ochroleuca* (Laminariales, Phaeophyta) under ultraviolet radiation. Phycologia 43:603–613

Sarhan T, García Lafuente J, Vargas M, Vargas JM, Plaza F (2000) Upwelling mechanisms in the northwestern Alboran Sea. J Mar Syst 23:317–331

Sasaki H, Flores-Moya A, Henry EC, Müller D, Kawai H (2001) Molecular phylogeny of Phyllariaceae, Halosiphonaceae and Tilopteridales (Phaeophyceae). Phycologia 40:123–134

Sauvageau C (1918) Sur les plantules d'une Laminaire à prothalle parasite (*Phyllaria reniformis* Rostanf.). C R Acad Sci Paris 166:787–789

Sjøtun K, Fredriksen S, Rueness J, Lein TE (1995) Ecological studies of the kelp *Laminaria hyperborea* (Gunnerus) Foslie in Norway. In: Skjoldal HR, Hopkins C, Erikstad KE, Leinaas HP (eds) Ecology of fjords and coastal waters. Elsevier, Amsterdam, pp 525–536

Steneck RS, Graham MH, Bourque BJ, Corbett D, Erlandson JM, Estes JA, Tegner MJ (2002) Kelp forest ecosystems: biodiversity, stability, resilience and future. Environ Conserv 29:436–459

Templado J, Calvo M, Moreno D, Flores-Moya A, Conde F , Abad R, Rubio J, López-Fé CM, Ortiz M (2006) Flora y fauna de la reserva marina y reserva de Pesca de la isla de Alborán. Ministerio de Agricultura, Pesca y Alimentación, Secretaría General de Pesca Marítima, Madrid

Wiencke C, Gómez I, Pakker H, Flores-Moya A, Altamirano M, Hanelt D, Bischof K, Figueroa FL (2000) Impact of UV radiation on viability, photosynthetic characteristics and DNA of brown algal zoospores: implications for depth zonation. Mar Ecol Prog Ser 197:217–229

Chapter 16
Macroalgae in Tropical Marine Coastal Systems

Astrid Y. Mejia, Gregory N. Puncher, and Aschwin H. Engelen

16.1 Introduction

Tropical marine environments are oligotrophic, warm, and expose primary producers to relatively high levels of photosynthetically active radiation (PAR) that penetrates as far as several hundred meters into the water column. Many tropical marine coastal communities are structurally dependent upon ecosystem engineering organisms such as seagrasses, mangroves, and corals in which the role of macroalgae is fundamentally different from temperate marine systems. This chapter will begin with a brief overview of the abiotic conditions, such as temperature, light, and nutrients, which have a defining influence on algal diversity, distribution, and dynamics with particular emphasis given to coral reef habitats because of the extensive body of research focused on this ecosystem. This will be followed by more detailed information on the role of macroalgae in the three main tropical marine ecosystems. Although they are not the main habitat providers in

A.Y. Mejia (✉)
Laboratory of Experimental Ecology and Aquaculture, University of Rome Tor Vergata, Via Cracovia No. 1, 00133 Rome, Italy

Bay Islands Conservation Association (BICA-Roatan), Bay Islands, Honduras
e-mail: aymejia@gmail.com

G.N. Puncher
Molecular Genetics for Environmental and Fishery Resources Laboratory (GenMAP), Interdepartment Center for Research in Environmental Sciences, via Sant'Alberto, 163-48100 Ravenna, Italy
e-mail: greg@respectnature.com

A.H. Engelen
Centre for Marine Sciences (CCMAR), University of the Algarve, Campus de Gambelas 8005-139, Faro, Portugal
e-mail: aengelen@ualg.pt

C. Wiencke and K. Bischof (eds.), *Seaweed Biology*, Ecological Studies 219,
DOI 10.1007/978-3-642-28451-9_16, © Springer-Verlag Berlin Heidelberg 2012

tropical coastal areas, macroalgae are essential to the productivity, structure, and function of each of these communities. Throughout this chapter, the role of macroalgae in regime shifts from systems dominated by the aforementioned habitat providers to ones dominated by algae as well as the factors that bring about these changes will be discussed.

16.2 Abiotic Conditions

Compared to the extreme seasonal changes in temperate regions, the variability of abiotic conditions in the tropical zone is limited, aside from the influence of the rainy season which can reduce salinity, increase water velocity, and decrease PAR.

16.2.1 Temperature

Three types of physiological responses to temperatures determine a species' biogeographical boundaries (Breeman 1988): growth, reproduction, and mortality (see Chap. 3 by Eggert). Pakker et al. (1995) suggested that tropical macroalgae are the most stenothermal macroalgae of any biogeographical region; their growth generally reaching a peak between 25 and 30°C, with those restricted to the tropical western Atlantic surviving a total variation of only 10/13°C. In comparison, macroalgae from colder regions are tolerant of a much wider range of temperatures (see Chap. 3 by Eggert and Chap. 18 by Bartsch et al.). Macroalgae of the tropical to warm-temperate Eastern Atlantic Ocean and Mediterranean Sea show the same temperature limitations as those of the Western tropical Atlantic Ocean and Caribbean Sea. Unsurprisingly, species restricted to the subtidal regions, and thus rarely exposed to extreme variations in abiotic conditions, are least tolerant of high temperatures, generally succumbing to temperatures exceeding 33°C. Furthermore, both subtidal and intertidal species appear to be limited by temperatures lower than 18/20°C.

16.2.2 Nutrients

Seawater in tropical regions is generally low in nutrients due to strong stratification of the water column that prevents vertical mixing of nutrients. Tropical marine habitats that are not exposed to terrestrial effluents are particularly nutrient poor, thereby preventing the proliferation of phytoplankton and macroalgae and increasing the light availability for benthic organisms. Occasionally nutrients become temporarily available through terrestrial runoff during the rainy season or upwelling events, encouraging strong seasonal responses (see also Chap. 21 by Teichberg et al.). After several decades of scientific research it has become clear that nutrients are cycled tightly within tropical ecosystems to prevent losses and enable a high level of productivity. The higher light levels and temperatures of the tropics make continuous productivity possible despite lower nutrient concentrations.

The increased supply of nutrients to the marine environment, resulting from human activities (agriculture, development, sewage, etc.), has been linked to increased algal growth in environments where they were previously a limiting factor (see below). However, the concentration of nutrients in seawater is not always directly related to macroalgal growth rates. For example, under low ambient nutrient conditions, high water motion can also stimulate increased growth rates of macroalgae (McCook 1999). Tropical macroalgae are very efficient nutrient users that are capable of taking advantage of nutrient pulses through surge uptake mechanisms. Additionally, they can store and recycle nutrients in their tissues for extended periods of time (Fong et al. 2003). In tropical macroalgae, small increases in nutrient concentrations can trigger an uptake response comparable to that of bloom-forming species from temperate estuaries responding to much higher nutrient concentrations (Kennison 2008; Fong et al. 2001, 2003). The concept that nitrogen and phosphorus are limiting nutrients in the tropical environment remains a source of controversy (Bell 1992; Lapointe 1997; Bell et al. 2007). Results of nutrient-enrichment experiments using macroalgae taken from coral reef and seagrass habitats indicate that nitrogen limitation is common (e.g., Lapointe et al. 1987; Littler et al. 1991; McGlathery et al. 1992; Delgado and Lapointe 1994; Collado-Vides et al. 2007). Several investigations have reached the same conclusion that phosphorus availability limits the productivity of fleshy macroalgae in oligotrophic reef waters (Lapointe et al. 1987, 1992; Littler et al. 1991). However, alkaline phosphatase activity (APA) enables some macroalgae to fulfill their phosphorus requirements by facilitating the utilization of organic phosphorus. These mechanisms have been well documented in macroalgae growing in the inshore regions of the Great Barrier Reef (GBR) where nutrient supply can be extremely variable and dependent on inshore human activities (Schaffelke 2001).

16.2.3 Light

Light is the most important physical factor affecting primary productivity in the marine realm. At low latitudes, light levels and water transparency are high due to limited suspended sediments, particulate and dissolved organic matter, and phytoplankton. Additionally, very little PAR is lost due to limited scattering and reflection of light on the sea surface. Although light penetrates deeper in the water column in the tropics than in temperate regions, light attenuation with depth still creates a gradient of light quantity and quality (see also Chap. 1 by Hanelt and Figueroa). Light in this way plays a structuring role in the distribution of tropical macroalgae. Rhodophytes, like crustose coralline algae, are common throughout all reef zones, from the intertidal to 80–90°m deep, where the light intensity has been estimated at ~0.2% of the surface radiation (Van den Hoek et al. 1978), and even as deep as 268 m in the Bahamas (Littler et al. 1985, 1986). On Curaçao sparse algal turfs reach their lower limit around 65 m, where PAR levels are as low as 1% of that found at the surface (Van den Hoek et al. 1978). Green macroalgae generally are most common in shallow waters, but have been found forming entire communities consisting of *Udotea, Caulerpa*, and *Halimeda* species at depths of 25 m on Saba Bank (Lesser Netherlands Antilles) (Littler et al. 2010a) as well as *Johnson-sealinkia profunda* colonies found at the extreme depth of 200 m in the Bahamas (Littler and Littler 1988). In addition, the lack of a conspicuous seasonal cycle in photoperiod in the tropics compared to mid and high latitudes provides a relatively constant supply of PAR throughout the year.

Although ultraviolet radiation (UVR) levels are high in the tropics, studies focused on its impact on macroalgae have for the most part been limited to polar and temperate seaweeds due to ozone depletion in these regions (Bischof et al. 2006; see also Chap. 20 by Bischof and Steinhoff). Recently, Figueroa et al. (2009) showed that seaweeds from the Red Sea show dynamic photoinhibition as an efficient mechanism of acclimation to high PAR as well as UV irradiance levels. This is not unique to tropical seaweeds but has been shown for other seaweeds in high light exposed environments (Figueroa et al. 1997). A recent investigation of tropical marine macroalgae collected from Belize at depths of 0.3–26 m depths showed that UVR caused an additional decrease of photosynthetic performance during high light stress which varied according to species, depth of growth, and UV penetration at the site of collection (Hanelt and Roleda 2009; see also Chap. 1 by Hanelt and Figueroa). However, this same study also provided evidence that UVB radiation may not only cause negative effects on cellular processes of seaweeds but may also support recovery processes in high UVR adapted seaweeds (Hanelt and Roleda 2009).

16.3 Vegetation Structure

16.3.1 Succession

Succession patterns depend on the characteristics of vacant space, bottom-up availability of nutrients, and top-down grazing characteristics. Many studies have documented the recovery of coral reef systems after storms and hurricanes and the succession of algae during this process as well as the experimental succession on settlement plates. In general, there is first a dominance of simple turf like green and brown filamentous seaweeds, a midsuccessional stage with thin and finely branched red filamentous seaweeds, and a late stage dominated by coarsely branched and thick canopy-forming seaweeds. The 37-year history of the Tiahura Outer Reef Sector (Moorea) constitutes one of the longest records of coral reef dynamics (Adjeroud et al. 2009). Between 1991 and 1994, the decline in coral cover was accompanied by colonization by turf algae (16.2 ± 5.5% in 1991 to 48.5 ± 2.5%

in 1994). However, instead of a successional sequence of algal growth towards the proliferation of canopy-forming macroalgae, the cover of algal turf decreased after 1994, and returned to the predisturbance levels within a decade (Adjeroud et al. 2009). These results suggest that the availability of vacant space was not sufficient to cause a persistent increase in algal cover, and that other factors, such as a reduction in grazing pressure or an increase in nutrients, may be necessary for a regime shift to macroalgal dominance (McManus and Polsenberg 2006; Mumby 2009; see also Chap. 21 by Teichberg et al.). A similar story emerged from the protracted loss of coral cover in Jamaica following two hurricanes, three bleaching events, a reduction of grazing pressure (overfishing and disease), and a potential pulse of nutrients which all contributed to a dramatic increase of algal cover, from ~4% in 1977 to ~92% in 1993 (Hughes 1994). Experimental studies, like that of Hixon and Brostoff (1996), demonstrate the importance of herbivores on the successional trajectory in tropical marine ecosystems. In conclusion, the impacts that macroalgae can have on tropical marine communities are complex and are dependent upon the identity of the herbivores present as well as the developmental stage of the seaweed community (see Coral reefs below).

16.3.2 Diversity

Tropical marine ecosystems are among the most diverse ecosystems on our planet. While invertebrates have received a great deal of attention during biodiversity surveys and studies in the past, macroalgae have received much less consideration. Some early studies that have assessed macroalgal diversity patterns across latitudinal gradients have arrived at conflicting conclusions, finding high species diversity at low and midlatitudes as well as decreasing diversity towards lower latitudes (e.g., Pielou 1977; Santelices and Marquet 1998). A more recent literature review by Kerswell (2006) revealed distinct gradients in species and genus diversity worldwide. When all macroalgal genera were considered, tropical regions had lower diversity than temperate regions. Only when a more reef associated order like the Bryopsidales was examined, diversity peaked at low latitudes. Furthermore, Konar et al. (2010) confirmed the common trend of higher taxa diversity at midlatitudes compared to low latitudes in the northern hemisphere, particularly in the intertidal.

Until recently, the coral reef seaweed hotspots of the Caribbean Sea were considered to be at Diamond Rock in Martinique and the Pelican Cays in Belize. However, on a recent expedition to the Saba Bank, Littler et al. (2010a) collected between 150 and 200 seaweed species from 17 different dive sites, demonstrating that there is still much to discover. Their checklist of 98 taxa including 43 Rhodophyta, 26 Chlorophyta, 26 Phaeophyceae (Heterokontophyta), and three Cyanophyta contained several unknown species as well as some previously unrecorded seaweed communities. These communities were located at depths of 25–30 m and were dominated by green, brown, or fleshy red macroalgae. In contrast to the predominant idea that abundant macroalgae indicate compromised reef health, no indications were found that conflicted with the existing definition of "pristine" oligotrophic reef systems (Littler et al. 2010a).

Turf algal assemblages can provide the majority of the primary productivity on a coral reef (Adey and Steneck 1985; Fricke et al. 2011). These assemblages are multispecies associations, characterized by undifferentiated upright axes and a rapid turnover (Carpenter et al. 1985). Although inconspicuous, they occur almost everywhere on reefs where space is available (Littler and Littler 1984; Steneck and Dethier 1994). Although they are fast growing, fleshy macroalgae often outcompete with them for limited light and space (Hay 1981). Algal turfs are able to persist under intense herbivory and physical stress in areas where other algae are continuously excluded. Their high productivity rather than their standing crop maintains the high standing stocks of grazers characteristic of coral reefs (Hatcher 1988; Carpenter 1986; McCook et al. 2001). In deep and shallow reef zones, where grazing pressure is low, frondose algae rather than turfs commonly dominate (Van den Hoek et al. 1978; Vuki and Price 1994).

Crustose coralline algae (CCA) play three important roles on coral reefs: (1) limestone formation, (2) the consolidation of loose substrates (Fabricius and Dea'th 2001), and (3) primary production (Littler and Littler 1984). In shallow waters they can build massive carbonate structures on reef crests despite the typically turbulent conditions (Adey and Vassar 1975). Due to their unpalatable nature, coralline algae are more resistant to grazing and therefore survive well in areas where grazing intensity is high.

Fleshy frondose macroalgae are mainly composed of tropical fucoids like *Sargassum*, *Turbinaria*, and *Cystoseira*. They often dominate on shallow reef flats and crests of inner reefs or in deeper fore reef zones and are generally absent from mid and outer shelf reefs (Vuki and Price 1994; McCook 1996). *Sargassum* species are often dominant in terms of biomass and canopy cover, due in part to their high productivity (Vuki and Price 1994; McCook 1996; Schaffelke and Klumpp 1997). Foliose macroalgae are seasonal opportunists and can be found in areas less dominated by canopy-forming macroalgae (Schaffelke and Klumpp 1997). Due to high biomass turnover, reefs with a high cover of large Phaeophyta are often regarded as detritus-driven ecosystems (Schaffelke and Klumpp 1997).

16.3.3 Zonation

The zonation and distribution of macroalgae is highly variable, which complicates extrapolations between similar habitats and across habitat types within regions and between different parts of the world. The distribution and zonation of macroalgae is strongly dependent on the structure of the habitats which is formed by historical (geology: reef types), abiotic (nutrients, temperature, and light), and biotic (grazing and competition) conditions (Berner 1990). We therefore describe the zonation using examples from a Caribbean fringing reef, a Pacific Atoll, and a Barrier reef.

On Curaçao in the Caribbean, van den Hoek et al. (1978) described the vertical distribution of reef macroalgae and distinguished seven different algal communities. The narrow eulittoral zone is inhabited by cyanophytes, turfs, and smaller macroalgae. In the top part of the surf area, the seaweed community is composed of *Ulva* and the brown seaweeds *Giffordia duchassaigniana* and *Sphacelaria tribuloides*, whereas in the deeper part the brown seaweed *Sargassum polyceratium* and the red algae *Hypnea musciformis* and *Laurencia* spp. dominate. In the sublittoral zone on the platform, from the surf area down to 3 m deep is a shallow reef occupied by coral rubble fragments aggregated by the crustose red alga *Porolithon pachydermum* and macroalgal turfs. The foliose brown seaweed *Lobophora variegata* forms a girdle-like vegetation at 30–38 m deep and the erect *Sargassum polyceratium* occurs from 15 to more than 65 m depth. Many shade adapted species are present in this zone reaching to depths of 55–65 m, including the green macroalgae *Udotea* and *Caulerpa* species and numerous red algae. Crustose red algae reach into the deep where the reef terminates and a sandy plateau forms at a depth of 75–90 m.

Womersley and Bailey (1969) described the zonation of macroalgae in the Solomon Islands on four types of coral reefs, with differences in hydrodynamic activity. The seaward rim consists of crustose coralline algae, mainly *Lithophyllum* and corals, and behind this the coralline alga *Porolithon onkodes* dominates. On shallow intertidal rubble near the island shore, crustose and endolithic green algae and cyanobacteria dominate.

Studies of the algal assemblages on the GBR have shown marked latitudinal, cross-shelf, and within-reef variations in composition and abundance (Wismer et al. 2009). In contrast to midshelf and outer-shelf reefs, inshore or coastal reefs typically have abundant and conspicuous macroalgal communities (Diaz-Pulido et al. 2007). The reef flat zone in particular is often dominated by dense and highly productive beds of large fleshy brown macroalgae, predominantly *Sargassum* (e.g., McCook 1997; Schaffelke and Klumpp 1997).

16.3.4 Seasonal Dynamics

Varying weather conditions throughout the year have a strong influence on the abundance of macroalgae in tropical regions. In particular, lower temperatures and the influx of nutrient-rich freshwater during the rainy season can cause a significant increase of algal biomass. For example, in Kenya epiphyte coverage on seagrasses can double or triple during the shift from the NE monsoon (warm temperatures, less rain, and light winds) to the SE monsoon (cool temperatures, heavy rains, and strong winds) (Uku and Björk 2005). In Florida, researchers observed an increase of epiphyte biomass during the rainy season so extreme that it became three times greater than seagrass blade biomass (Lapointe et al. 2004). Of course, the hurricane/cyclone season presents potential for extreme events in the tropical marine environment each year as well. It is widely recognized that these events characterized by intense water velocities, higher freshwater input, and mechanical disturbance can both disrupt and maintain the balance between macroalgae and seagrass, mangrove, and coral reef communities.

Macroalgae within coral reef habitats display a great amount of seasonality as well (Coles and Fadlalah 1991). A very comprehensive overview of seasonality of four major functional groups: canopy, foliose, turf, and CCA was accomplished by Aterweberhan et al. (2006). They monitored biomass of all these groups in four reef zones at two localities in the southern Red Sea. All functional groups showed seasonal variation in biomass, although seasonal variation differed among groups and reef zones. Canopy and foliose macroalgae were highly seasonal, whereas turf algae and crustose corallines showed much less seasonal variability. The biomass variation of *Sargassum* spp. greatly determined the seasonal variation of standing biomass of all algal types on the shallow inner and middle reef flats. On the deeper, outer reef flat, *Turbinaria triquetra* contributed most to the seasonal variation. Foliose algae contribute most to the seasonal variation only where canopy-forming macroalgae were lacking, like on the middle reef flat. Crustose corallines and turf algae only contribute considerably to the total biomass during the hot season when macroalgae are strongly reduced. On the fore reef, crustose corallines contribute most to the biomass throughout the year. In general, turf algae contributed least to the standing biomass during all seasons, at all sites, and all reef zones.

16.4 Macroalgae and Their Roles and Interactions in the Three Main Tropical Ecosystems

16.4.1 Macroalgae and Coral Reef Interactions

16.4.1.1 Herbivory

Herbivory is widely recognized as the main driver affecting the distribution and abundance of macroalgae on tropical coral reefs (Carpenter 1986; Hay 1997; Hughes et al. 2007; Burkepile and Hay 2010). Several experimental studies have demonstrated that coral reef herbivores can remove up to 100% of daily algal production in some coral reef habitats (Duffy and Hay 1990; Hughes et al. 2007). In tropical marine habitats, herbivorous fish and sea urchins are the dominant grazers of macroalgae (Williams et al. 2001; Mumby et al. 2006; Burkepile and Hay 2010). Sparisomid parrotfishes represent more than 80% of the herbivorous fish biomass that is dominant in Caribbean reefs, along with the acanthurids (surgeonfishes, tangs, and unicornfish), kyphosids (sea chubs), and pomacentrids (damselfishes) (Mumby 2009). Herbivorous fish feed on an array of algal types, including fleshy macroalgae, algal turfs, and encrusting coralline algae (Hay 1997; Mumby 2009). Algal turfs are easily digested and have a high energetic and protein value relative to other macroalgae and are thus the preferred food of many herbivorous fish, in both the Caribbean Sea and other tropical habitats (Bruggemann et al. 1994; Kopp et al. 2010). Based on their experiments in the Caribbean (Guadaloupe), Kopp et al. (2010) determined that *Acanthurus coeruleus* (Blue tangs) and *Sparisoma aurofrenatum* (Redband parrotfish) graze preferentially on algal turfs, while *Sparisoma rubripinne (Redfin parrotfish)* and *Sparisoma viride* (Stoplight parrotfish) are more inclined towards *Halimeda* spp. while *Acanthurus bahianus* (Ocean surgeon) prefer Phaeophytes. Their results also revealed the aversion of herbivorous fish towards *Dictyota* spp., a brown macroalgae that uses chemical compounds as a grazing deterrent (Hay 1997; Paul et al. 1990). On the GBR, Mantyka and Bellwood (2007a) report that grazers of macroalgae on the reef crest of Pioneer Bay are composed of six reef fish species (two rabbitfishes, three parrotfishes, and one damselfish) which display species-specific feeding behavior. For example, the rabbitfishes, *Siganus doliatus* feed heavily on *Hypnea* spp. while *Siganus canaliculatus* feed intensively on *Sargassum,* and the three parrotfishes *Chlorurus microrhinos*, *Hipposcarus longiceps,* and *Scarus rivulatus* were the dominant grazers of calcified *Halimeda* spp. (*H. cylindracea*, *H. discoidea*, *H. opuntia)*, and *Amphiroa* spp. During their experiment, *Chlorodesmis fastigiata* and *Galaxaura* spp. were least affected by herbivory, suggesting a reduced palatability owing to their chemical deterrence and calcareous structures (Paul et al. 1990). More recently, Cvitanovic and Bellwood (2009) have reported that the dominance of herbivorous fish over macroalgae species and specific reef areas can show local variability. In an experiment conducted on the Orpheus Islands, they found similar rates of macroalgae removal in three separate bays; however, three different species of herbivores were responsible for the grazing control. They also found that *Kyphosus vaigiensis* (Brassy chub) were responsible for high rates of removal of *Sargassum* spp., although past studies had identified *Siganus canaliculatus* (White spotted spinefoot) and *Platax pinnatus* (Dusky batfish) as the dominant grazers of *Sargassum* spp. in Pioneers Bay *(*Mantyka and Bellwood 2007b). Most recent studies highlight the importance of herbivore richness and diversity (Burkepile and Hay 2010) as well as abundance and biomass (Mumby et al. 2006; Kopp et al. 2010) in maintaining reef macroalgae community structure, reducing fleshy macroalgal blooms and enhancing coral reef recovery and resilience. For example, Burkepile and Hay (2010) determined that complementarity feeding among fish species not only impacted structure of macroalgal communities, but also enhanced coral survivorship and growth on reefs in the Caribbean Sea (Florida Keys, USA). The authors observed that fast grazing by ocean surgeonfish (*Acanthurus bahianus*) and princess parrotfish (*Scarus taeniopterus*) kept macroalgae communities dominated by short, filamentous algae and crustose coralline algae, both algal states that do not suppress coral growth. Conversely, within the established benthic communities, the authors observed that the redband parrotfish (*Sparisoma aurofrenatum*) played a vital role in the removal of upright macroalgae cover which can be detrimental for coral growth and survival. To further investigate the role of herbivory on regulation of macroalgae community structure, numerous herbivory exclusion experiments have been conducted in tropical coral reefs (Lirman 2001; Hughes et al. 2007; Smith et al. 2010a; Hoey and Bellwood 2010). These studies provide evidence of how exclusion of large herbivores is followed by undesirable blooms of upright macroalgal species, which are considered less susceptible to herbivory and capable of causing phase shifts from coral to macroalgae dominated systems. Mork et al. (2009) conducted a 6-week herbivore exclusion study on a moderately disturbed coral reef on the Kenyan coast. Their results show a 77% increase of algal biomass, largely attributed to a 1,000% increase in corticated forms of algae, dominated by *Dictyota* spp. Sotka and Hay (2009) reported similar results from an experiment conducted in the Florida Keys. After 142 days of monitoring, they observed that coral slabs exposed to natural densities of large herbivorous fishes were dominated by crustose coralline algae, short (<0.5 cm) filamentous turf algae, and upright macrophytes, which never exceeded 15% cover. In contrast, herbivore exclusion treatments were subject to 80–100% upright macroalgae cover.

The role of herbivory in regulating macroalgal communities can be severely compromised in overfished and degraded reefs and may result in phase shifts from coral to algae dominated systems (Mumby et al. 2006). In areas suffering from moderate levels of overfishing, a lack of top predators can stimulate an increase of herbivore biomass. However, in severely overfished areas, herbivorous fish are not able to control macroalgae because they too become subject to overfishing and are unable to become sufficiently large in size to regulate algal production, even if they are numerically abundant (Kopp et al. 2010). A significant reduction of fishing pressure of herbivorous fish (Mumby and Harborne 2010) as well as a reduction of confounding disturbances such as nutrient pollution (Smith et al. 2010a) has been suggested as a means of reversing algal phase shifts.

16.4.1.2 Chemical Defenses

Macroalgae defend against herbivory using a number of strategies, including morphological, structural, and chemical defenses as well as associations with other algae or benthic organisms (Hay 1997; Paul and Puglisi 2004; Amsler 2008; Smith et al. 2010b; Fong and Paul 2011; see Chap. 9 by Amsler). Morphological and structural defenses include calcification and toughness, which are characteristics of the *Halimeda, Dictyota,* and *Lobophora* genera (Paul et al. 2001). These species are coarsely branched, leathery, or rubbery and have hard calcium carbonate tissues which are of little interest or nutritional value for herbivores (Schupp and Paul 1994; Hay 1997). Some macroalgae utilize secondary metabolites to help protect against pathogens, fouling organisms, and herbivores (Fong and Paul 2011). A significant quantity of natural products has been extracted and isolated from tropical marine green, red, and brown macroalgae, of which halides like bromine and chloride are very common as well as compounds like terpenoids (Fong and Paul 2011). Chemical deterrents in calcified green algae from the genera *Caulerpa* and *Halimeda* have been well studied and experiments have shown that their chemical composition deters both parrotfishes and sea urchins alike (Schupp and Paul 1994). Brown algae are known to utilize polyphetanolic compounds as deterrents along with terpenoids, which are commonly present in the order Dictyotales (Pereira and Da Gama 2008). Red macroalgae possess a great variety of secondary metabolites and the genus *Laurencia* has been widely studied for this reason (Blunt et al. 2007).

The study of chemical defenses of macroalgae has been widely investigated for over two decades and most of the research has been focused on their utilization against grazers (fishes, sea urchins, and gastropods) and their effect on coral species. For example, from their experiments conducted in the Caribbean and tropical Pacific, Rasher and Hay (2010) provide evidence that five of seven seaweeds (71%) caused bleaching of the coral species *Porites porites* in Panama. While in Fiji, three of eight species (38%) caused bleaching of *Porites cylindrica*. Seaweeds were observed to damage corals via abrasion, shading, or lipid-soluble allelopathic compounds transferred through direct contact. The effects of a lipid-soluble extract from various species of macroalgae (*Ochtodes secundaramea, Dictyota bartayresiana, Lobophora variegata, Halimeda opuntia, and Amphiroa fragillisima* in Panama) caused significant coral bleaching and suppression of photosynthetic efficiency in assays using both intact seaweeds and chemical extracts. The two species that did not cause bleaching in any of the two assays were *Padina perindusiata* and *Sargassum* spp. These results illustrate how overfishing of herbivorous fish, capable of suppressing undesirable macroalgae, will inevitably result in an increase of direct coral–algae contacts and further impair corals through allelopathic interactions. Major gaps still persist concerning the ecological role that macroalgal secondary metabolites play apart from feeding deterrence (Pereira and Da Gama 2008).

16.4.1.3 Providers of Spatial Refuge

Persistence of unpalatable macroalgae on coral reefs can be facilitated by spatial refuges and associational assemblages. An associational refuge is one in which a host species provides protection that enhances the survival of associate species (Bittick et al. 2010). These interactions are becoming increasingly recognized as important drivers of ecosystem function. Smith et al. (2010b) have found that patches of the green algae *Caulerpa sertularioides* found in reef areas of Uva Island in Panama can persist due to their association with an epiphytic cyanobacterium (*Lyngbya majuscula*). The cyanobacterial epiphytes on *C. sertularioides* thalli provide protection from herbivory for both upright assimilators and stolons. Similarly, associational defenses provided by cyanobacteria have been reported to be responsible for the 5-year persistence of the highly palatable alga *Acanthophora spicifera* on the Uva Island reef (Fong et al. 2006). On the coral reefs of Moorea, French Polynesia, Bittick et al. (2010) observed that aggregations of the brown macroalga *Turbinaria ornata* provide mechanical and chemical refuge from herbivory to associated macroalgae. The study showed that aggregations of *Turbinaria* significantly enhanced species richness, especially of fleshy species like *Caulerpa serrulata*, *Dictyota bartayresiana*, *Sargassum mangarevense*, *Valonia utricularis*, and *Ventricaria ventricosa*. On the contrary, in coral heads without *Turbinaria*, there was less algal density and these were dominated by crustose coralline algae, filamentous algal turfs, and cyanobacteria. As a result, unprotected coral heads without *Turbinaria* had the lowest diversity of all treatments.

16.4.1.4 Nutrient Enrichment

The influence of excess nutrient supply on the development of reef macroalgae is a topic of major debate between marine researchers (Littler and Littler 2006; Mork et al. 2009; Smith et al. 2010a). Experimental studies have often investigated the effect of nutrient enrichment in the presence and absence of herbivory to determine its role in regulation of macroalgal growth and cover in tropical habitats, especially coral reefs. The available literature is divided between studies that report a significant effect of nutrient enrichment on enhancing growth and abundance of fleshy macroalgae and cyanobacteria in coral reef habitats and others that report no significant effect. For example, Littler et al. (2010b) conducted a 12-month study in a sedimentary lagoon at Carrie Bay Cay in Belize's barrier reef to compare the enrichment effects of dissolved inorganic nitrogen (DIN) and soluble reactive phosphorus (SRP) additions on algal community development. They found that SRP and SRP + DIN treatments stimulated the growth of cyanobacteria (*Spirulina* spp.) and inhibited settlement and colonization of CCA. Meanwhile, the DIN enrichments increased CCA cover, decreased cyanobacteria cover, and had no influence on fleshy macroalgae. Sotka and Hay (2009) reported similar findings related to fleshy macroalgae from a 20-week nutrient-enrichment experiment conducted in the shallow fore reefs of the Florida Keys. They concluded that herbivory is stronger than nutrient abundance in regulating macroalgae abundance and composition on coral reefs. Their results revealed that large herbivorous fish selectively grazed on nutrient-enriched macroalgae. On the contrary, experiments conducted by Smith et al. (2010a) on the reef slope at Puako Reef, Hawai'i found that both fertilization and herbivory exclusion impacted macroalgae community structure at different temporal scales. They observed that reducing herbivory led to rapid changes in benthic communities (1 month), while the effects of nutrient enrichment were only visible after 3–4 months. This can explain the lack of strong nutrient-enrichment effects noted in other studies, as they may be a result of short-term experiments. Vermeij et al. (2010) conducted experiments in Curaçao, Netherlands Antilles, to examine the effects of Caribbean turf algae on neighboring corals and interactions of these algae with nutrient enrichment and grazing. They found that nutrient enrichment significantly enhanced the capacity of turf algae to overgrow the coral *Montrastrea annularis*, since turf algae is also able to overgrow coral at ambient nutrient concentrations when herbivory is excluded or reduced.

16.4.1.5 Competition

Competition is an important force influencing the structure of coral reef communities and its effects can impact all stages of a coral colony's life cycle (Lirman 2001; Nugues and Bak 2006; Foster et al. 2008; Ritson-Williams et al. 2010; Vermeij et al. 2011). Macroalgae are known to compete with corals using different strategies which include the occupation of settlement substrate (Birrell et al. 2005; Vermeij 2006), physical disturbance (e.g., abrasion, shading, smothering) (McCook et al. 2001; Box and Mumby 2007), allelopathy (Gross et al. 2003; Kuffner et al. 2006; Foster et al. 2008), and direct transmission of pathogenic microbes (Vermeij et al. 2009). This competitive interaction is often species specific and capable of causing different levels of stress on coral species, potentially leading to coral mortality (Jompa and McCook 2002, 2003). In the Caribbean, Box and Mumby (2007) determined that *Lobophora variegata* and *Dictyota pulchella* inhibited coral growth on juvenile *Agaricia* spp. and *L. variegata* through shading and abrasion. In Hawaii, Vermeij et al. (2009) found that algal-induced microbes can cause mortality of the planular larvae of *Montipora capitata*. This effect on corals may be related to the ability of some algae to exude organic compounds that enhance rapid microbial growth, causing hypoxia and the reduction of larval resistance via the hampering of the coral's defense mechanisms against opportunistic microbes.

Although competition between macroalgae and corals has received significant attention from marine researchers, more evidence is still required in order to determine if algal growth is a cause rather than a consequence of coral mortality (McCook et al. 2001). More information is also needed concerning the interaction between coral larvae, macroalgae, and microbes in order to improve reef restoration techniques (Vermeij et al. 2009).

16.4.1.6 Facilitation of Coral Recruitment

CCA have been identified as facilitators of settlement and metamorphosis of a variety of marine invertebrate larvae as well as coral larvae (Ritson-Williams et al. 2010; Vermeij et al. 2011). CCA and coral interactions are known to be species specific, since corals have displayed different degrees of specificity in their requirement for CCA to induce successful settlement and metamorphosis. For example, an experiment conducted in the Caribbean (Belize) revealed that both *Acropora palmata* and *Acropora cervicornis* have higher rates of metamorphosis on the top surfaces of the CCA species *Hydrolithon boergesenii* and/or *Titanoderma prototypum* than on *Porolithon pachydermum (*Ritson-Williams et al. 2010). Further observations in the field (after 6 weeks) showed 15% survival of transplanted *A. palmata* recruits on both *T. prototypum* and *H. boergesenii*, and only 13% of A. *cervicornis* recruits on *T. prototypum*. These results suggest that CCA and settling coral polyp interactions can be species specific and also highlight the importance of benthic community composition for successful coral recruitment.

Aside from the positive role of CCA in coral recruitment, Vermeij et al. (2011) have reported that mixed communities of CCA (*Porolithon* spp. and *Hydrolithon* spp.) are capable of suppressing macroalgal growth and recruitment of the green macroalgae *Ulva fasciata,* based on studies conducted on a coral reef area in Hawai'i. Likewise, Birrell et al. (2005) have highlighted the negative impact of the association of macroalgal turfs and sediments have on coral recruitment of *Acropora millepora* found in the Orpheus Islands of the GBR. The results of that study indicated that sediment addition and algal turfs are capable of reducing coral settlement, although the effects of different algal turfs varied. In some cases, algal turfs inhibited coral settlement while other turfs inhibited settlement only when combined with sediments. Their results are suggestive of the need to study the impacts of similar associations to further understand the role of CCA in securing long-term resilience and recovery of coral reefs.

16.4.2 Macroalgae–Seagrass Interactions

In low nutrient environments, seagrasses and slow-growing macroalgae constitute a dominant regime in stable coastal waters. Seagrass habitats are important nurseries for commercially important fish species found within seagrasses and adjacent habitats such as coral reefs and mangroves (Dorenbosch et al. 2005). Within seagrass meadows macroalgae contribute to substrate complexity, act as a vital food resource for grazers, and have a significant influence on the cycling of nutrients. In nutrient-rich locations, macroalgal epiphytes are capable of achieving the same level of productivity as their seagrass hosts (Uku 2005). Some calcareous algae generate such immense quantities of $CaCO_3$ that entire beaches are constructed from the sand that they produce (Muzuka et al. 2005). However, enrichment of coastal waters can encourage the proliferation of algal blooms and initiate a breakdown in the balance of the system, resulting in decreased seagrass coverage (McGlathery 2001; Hauxwell et al. 2003; Lapointe et al. 2004). Macroalgae may provide short-term spatial heterogeneity and an enriched trophic resource, thereby appearing to be beneficial for invertebrate community diversity; however, they are often ephemeral and thus represent a threat to long-term biodiversity, relative to more permanent substrates, such as seagrasses (Cardosoa et al. 2004). Three distinct types of macroalgae (Epiphytic, Drift, and Calcareous Algae) have an important influence on the seagrass environment and thus dictate the structure of the following section.

16.4.2.1 Seagrass Epiphytes

In the tropics, epiphytic macroalgae are abundant and play a major ecological role in several habitats (see also Chap. 11 by Potin). In the Western Indian Ocean, macroalgal epiphytes constitute the most abundant group of epiphytes and can account for as much as 40% of the fresh weight of seagrass shoots (de la Torre-Castro et al. 2008; Uku and Björk 2001). In a Bahamas lagoon, the rates of net primary production (NPP) for seagrass epiphytes (5.2 ± 1.4 gC $kg^{-1}day^{-1}$) have been measured at approximately 40% of the NPP of the seagrasses themselves (Koch and Madden 2001). In highly enriched waters, it is not uncommon for seagrass-associated macroalgae to reach abundances higher than 0.5 kg m^{-2} and obtain canopy heights greater than 0.5 m (McGlathery 2001).

In Kenya, encrusting red coralline algae are the first to colonize seagrass leaves, establishing a purchase upon which subsequent epiphytes may settle, such as the green algae *Cladophora* spp. and red algae *Ceramium* spp. (Uku 2005). This succession of algal groups follows a consistent and predictable community gradient along the vertical length of seagrass leaves, which is likely determined by light exposure (Uku 2005). Of course, leaf colonization by epiphytes inevitably exacts a toll on the host. Dixon (2000) determined that epiphytes covering seagrass leaves can be responsible for reducing as much as 33.1% of available PAR. However, studies in the tropical waters of Eastern Africa provide evidence that shading by epiphytes has no effect on overall photosynthetic output of *Thalassodendron ciliatum* shoots (Uku 2005). It seems that *T. ciliatum* compensate for the lost PAR by maintaining photosynthetic activity in older epiphyte covered leaves while increasing the production of new leaves capable of higher photosynthetic activity.

In the presence of excessive nutrient concentrations, seagrasses may respond to smothering epiphytes by increasing their growth rate (Ferdie and Fourqurean 2004). Nutrient-enrichment studies have revealed decreased carbon reserves in the rhizomes of seagrasses in the presence of excessive concentrations of nitrate and ammonium (Invers et al. 2004). The unbalanced metabolism caused by this type of enrichment can result in a significant loss of nonstructural carbohydrates which allows seagrasses to persist during unfavorable conditions. Koch and Madden (2001) observed surge growth and *N*-storage in macroalgal seagrass epiphytes during enrichment studies, indicating an adaptive trait that allows for a competitive advantage when nutrients are limiting.

Epiphytes constitute an important food resource for seagrass herbivores and omnivores. In Zanzibar, the majority of commercially important fish species found within seagrass habitats, ranging from herbivores, piscivores, and invertebrate feeders, were found to graze directly on seagrasses and epiphytes (de la Torre-Castro et al. 2008). It seems that the higher nitrogen content of epiphytes and their resident invertebrates is an important supplement to nitrogen-poor seagrass diets. Various species of herbivorous parrotfish and surgeonfish collected from the seagrass beds of Curaçao were found to have diets consisting primarily of seagrass epiphytes (filamentous algae) and leaves (Nagelkerken and van der Velde 2004). From their work with manipulative studies in Florida, Baggett et al. (2010) suggest that increased grazing of epiphytes by herbivores (isopods, amphipods, gastropods, and caridean shrimps) compensates for increased coverage during nutrient-enrichment treatments. Recent studies in temperate seagrasses suggest that herbivorous invertebrates exert a strong top-down control of epiphytes on seagrasses (Moksnes et al. 2008). In fact, Spivak et al. (2009) found that the strongest effect during their nutrient-enrichment experiments was an increase in grazer biomass, not epiphytes, indicating an efficient transfer of nutrients through trophic levels. This evidence suggests that removal of large predators (tertiary consumers) is likely to cause trophic cascades, resulting in the increased severity and frequency of algal blooms in the seagrass environment.

16.4.3 Drift Algae Within Seagrass Meadows

In the tropics, drift algae can have a significant impact on the well-being of seagrass communities. Drift algae originate as epiphytes growing within seagrass meadows and neighboring habitats, having been liberated from the substrate during heightened hydrological disturbances. Accumulation of drift macroalgae within seagrass meadows is a function of water circulation characteristics determined by coastal physiography (Kopecky and Dunton 2006) and rates of entanglement in the seagrass canopy (Biber 2007). Transport of drift algae across the seabed is dependent upon substrate complexity, and in the case of seagrasses, it is determined by a meadow's patchiness, blade length, and shoot density. In shallow waters where water velocity is reduced drift algae can become entrained in bare patches. Algae also have a tendency to accumulate where wind or tidal generated water currents are sufficiently reduced to allow deposition out of the water column, be it on the leeward side of a bank or within a protected depression (Kopecky and Dunton 2006). Once entangled, drift algae can rapidly proliferate under favorable conditions generated by freshwater inputs, ideal temperatures, and nutrient loading (Biber 2007). In the Gulf of Mexico, drift macroalgae can reach a biomass exceeding that of the dominant seagrass, *Thalassia testudinum* (Kopecky and Dunton 2006). The adverse weather conditions typical of the tropical hurricane season generate current velocities sufficient for the removal of accumulated algal biomass (10 cm s^{-1}) (Biber 2007). Drift algae biomass varies by season as well as from 1 year to the next (Houk and Camacho 2010).

It has long been recognized that excessive quantities of drift macroalgae can have serious impacts on underlying seagrasses via competition for nutrients and PAR (Hauxwell et al. 2003). Excessive macroalgal loads can become so dense that they restrict water flow, thereby depriving seagrasses of oxygen and driving up concentrations of sulfides in the subsurface sediments (Holmer and Nielsen 2007). In addition to its impact on seagrasses, light attenuation by blooms of drift algae off the coast of Texas has caused significant reductions in seagrass epiphyte biomass (Kopecky and Dunton 2006). Given that the integrity of seagrasses can be severely compromised during these short-term seasonal blooms, the long-term repercussions of persistent blooms that may arise as a result of climate change demand serious consideration. Climate experts suspect that a changing climate will increase the number of intense storms and decrease those of intermediate intensity (Webster et al. 2005). The consequences of these changing weather patterns on the distribution of drift algae and their persistence within the coastal environment require close monitoring.

16.4.3.1 Calcareous Algae Within the Seagrass Habitat

High densities of seagrasses have the capacity to increase the pH of entire water bodies due to their uptake of CO_2 and release of OH^- (Beer et al. 2006). For example, *Cymodocea rotunda, Halophila ovalis,* and *Thalassia hemprichii* have the capacity to raise water pH from 8.1 to 8.6, 8.8, and 9.2, respectively (Beer et al. 2006). The pH compensation point of each species as well as the inability of seagrasses like *H. ovalis* to tolerate the high pH values and low inorganic carbon concentrations generated by the other two species forces these seagrasses into monospecific tide pools. However, Semesi et al. (2009) determined that calcifying macroalgae, such as *Hydrolithon* spp., *Mesophyllum* spp., and *Halimeda renschii*, can buffer against this increase in pH via intensified calcification. In fact calcification rates of *Hydrolithon* spp. can be five to six times higher in the presence of seagrasses. Similarly, photosynthetic activity of *Mesophyllum* spp. increased by 15% in the presence of seagrasses. Thus, when coexisting these two groups are capable of maintaining high levels of productivity while encouraging higher levels of biodiversity. Monitoring of the balance between these two groups is important as the dominance of one over the other could lead to dramatic shifts in water chemistry and a regime shift. Particularly important are the consequences that ocean acidification may have on this balance.

16.4.3.2 Regime Shifts and Controls Within the Seagrass Habitat

Human development of the coastal zone has often been identified as the source of nutrient pollution that disrupts the balance between seagrasses and macroalgae. In 2008, Houk and van Woesik collected evidence suggesting that seagrass species respond differently to environmental pressures. In a Saipan lagoon they found that increased watershed development stimulated macroalgal proliferation within *Halodule uninervis* seagrass meadows, severely impacting the integrity of the meadows. However, the same study revealed that the growth of a different seagrass, *Enhalus acoroides,* was positively correlated with human development and watershed size. The authors suggest that the physical and/or morphological characteristics of seagrass species determine their susceptibility to macroalgal overgrowth. Similarly, McGlathery (2001) suggests that *Thalassia testudinum* may be more resistant to the pressures of algal blooms because of a proportionately greater allocation of resources to below-ground biomass, allowing them to persist during temporary algal blooms.

The consequences of nutrient enrichment are not always easily predicted. Nutrient addition experiments in Florida resulted in increased seagrass biomass and decreased macroalgal epiphyte coverage offshore. Conversely, the inshore treatment had no effect on two species of seagrasses while macroalgae biomass increased significantly (Ferdie and Fourqurean 2004). In a similar Florida-based study, Armitage et al. (2005) found that enrichment of sediments stimulated highly variable group and site-specific responses by macroalgae. The resulting macroalgal blooms (calcareous green and filamentous red) were ephemeral and had no long-lasting impact on the system. Their results also showed that areas containing a high abundance of macroalgae were exposed to an array of additional chronic pressures including nutrient enrichment, current and erosion pressures, and heavy boat traffic. During nutrient treatments, phosphate (not nitrate) stimulated growth, prompting the researchers to conclude that Florida Bay is phosphate limited, a consequence of the high carbonate content of the sediments, perhaps owing to the calcareous algae contained therein. Results from a similar study in Puerto Rico show that several species of nutrient-depleted nuisance bloom (*Hypnea musciformis*) and mat-forming (*Acanthophora spicifera, Dictyota cervicornis*) macroalgae increased growth rates significantly when exposed to nutrient enrichment. However, enriched algae of the same species had no response to increased nutrient concentrations. Calcareous green algae (CGA) (e.g., *Halimeda incrassata*) did not respond to treatments within the 3-day experiment, suggesting limitation by some other metabolic process, such as calcification (Fong et al. 2003). Therefore, in order to predict the consequences of nutrient enrichment, coastal managers must consider the existing concentration of each relevant nutrient within the system as well as the primary producers contained therein.

Evidence suggests that storms arriving in the late summer cyclone/hurricane season play an important role in maintaining a balance between seagrasses and algae. Frequent disturbances such as large swells and intense currents prevent epiphyte growth and thus maintain seagrasses as the dominant flora (Houk and Camacho 2010). However, the impacts of massive natural disturbances on coastal systems (erosion, mechanical thinning, and burial) can completely transform benthic communities, often irreversibly. Following Hurricane Georges in 1998, Fourqurean and Rutten (2004) observed that while only 3% of *T. testudinum* was removed, 24% of CGAs were lost. CGAs are much more susceptible to hurricane damage because they lack the extensive rhizomes of seagrasses and are thus easily liberated from the substrate during sediment displacement. As the system recovered, CGAs were the first primary producers to recolonize bare patches, although its density did not return to prehurricane levels for another 3 years. This storm-dependent balance within the seagrass–algae complex and the biodiversity that it maintains can be considered within the context of the Intermediate Disturbance Hypothesis (Connell 1978).

16.4.4 Macroalgae–Mangrove Interactions

The vast majority of studies focused on mangroves come out of the Indian and Southwest Pacific Ocean in which water transparency can be low due to high tidal amplitude and intense terrestrial runoff typical of estuaries and deltas (Kristensen 2008). This limits subsurface primary productivity and promotes the development of detrital communities (Kristensen 2008). Studies coming out of Southeast Asia (Thailand, Malaysia) and the Indus Delta describe mangrove forests as nutrient exporting systems that have low epifaunal biodiversity, low macroalgal abundance, and organically rich sediments (Saifullah and Ahmed 2007). In this environment, macrophytes and filter feeding assemblages are extremely limited. As a result, many of the inhabitants are generalists and opportunists with microalgae and allochthonous phytoplankton providing an additional and essential source of nitrogen (Kristensen 2008). However, in the Atlantic Ocean and Caribbean Sea tidal amplitudes are minimal and terrestrial runoff is limited, leading to high water clarity, allowing for the proliferation of macroalgal and filter feeding communities within fringe regions (Kieckbusch et al. 2004). Therefore, the disproportionate amount of research coming out of Asia and Australia has led to a skewed perception of the role of macroalgae within mangrove systems.

16.4.4.1 Abundance and Productivity

The NPP (NPP) of mangrove prop root epiphytes in the Bahamas (8.5 ± 6.4 gC $kg^{-1}day^{-1}$) has been measured to rival that of nearby seagrass epiphytes (Koch and Madden 2001). In the Karachi mangroves of the Indus Delta, Saifullah and Ahmed (2007) have estimated that the algal productivity is much higher than that of the mangroves themselves. However, the actual biomass of algae growing within some mangrove systems may be quite low due to light limitation. Nevertheless, their role in coastal food webs is likely to be consistently significant, owing to their nitrogen content, which is $10\times$ higher than mangrove tissues (Alongi 1998). Since decomposition of mangrove tissues is a slow process, due to its high lignin, cellulose, and tannin content, it is likely that algae and bacteria provide a significant amount of nitrogen to the mangrove food web. Therefore, it is probable that any algal tissue appearing within algae-deficient mangrove habitats is likely to be quickly utilized by resident fauna (Alfaro 2008).

In 2001, Koch and Madden attempted to determine the source of nutrients within oligotrophic lagoons in the Bahamas. They found that total nitrogen and phosphorus concentrations were highest within fringe mangrove and seagrass meadow sediment samples directly adjacent to mangroves, relative to sand flats and open lagoon seagrass meadows. On a biomass basis, the gross primary production of prop root macroalgae was twice as high as nearby seagrass epiphytes and their NPP was even comparable to that of seagrasses. At that time they were unable to identify the cause of this nutrient gradient, but they suggested remineralization by bacteria, nutrient mobilization, and even delivery of metabolic wastes by fishes that migrate between habitats during the day.

Recently, a great deal of debate has been focused on whether or not mangrove systems are a source of nutrients for adjacent benthic and pelagic food webs (Kieckbusch et al. 2004; Bouillon et al. 2008). Of course, the retention of organic material within mangroves is completely dependent upon the tidal regimes of the area. For example, from their studies in Thailand, Kristensen (2008) suggests that the main source of organic matter in mangrove systems is derived from the mangroves themselves and that macroalgae contribute very little to the organic pool. Conversely, researchers in the Indus Delta argue that epiphytic algae are an important source of energy within the mangrove environment (Saifullah and Ahmed 2007). However, Bouillon et al. (2002) have provided evidence from studies in India, suggesting that there is very little export of this material to adjacent habitats. According to stable isotope analysis of benthic invertebrates in neighboring habitats, the labile organic matter that is exported from the mangroves is not consumed in significant quantities by faunal assemblages. Instead, grazers outside of the mangrove system prefer to consume local sources of algae which are likely to have a higher content of nitrogen. In fact, they argue that the primary production of mangrove epiphytes in Karachi is higher than that of mangrove trees. Therefore, caution must be taken when reviewing the literature and an understanding that each mangrove system is likely to be unique based upon hydrodynamics and algal communities.

16.4.5 Distribution of Macroalgae Within Mangroves

The mangrove systems of the Atlantic Ocean and Caribbean Sea are quite distinct from those of Asia and Australia, especially in the context of the importance of macroalgae. A recent study revealed that the most abundant epiphytes growing on mangrove prop roots in Panama were sponges and macroalgae (*Caulerpa verticillata* and *Halimeda* spp.) (Diaz and Rützler 2009). Sponges and macroalgae clearly compete for space on prop roots, with algae occasionally dominating where elevated sun exposure tips the balance in favor of algal dominance. In Belize, macroalgae only dominate where deforestation and dredging have recently occurred (Diaz and Rützler 2009). These two disturbances promote the suspension of fine sediments that negatively impact all filter feeders (sponges and ascidians), presenting the resulting macroalgal blooms with a competitive advantage.

Within mangrove habitats, algae growing on vertical surfaces are subject to an array of extremely variable conditions such as submersion and exposure, light, nutrients, temperature, and salinity. These challenging conditions have brought about the evolution of mangrove-specific algae. A survey of macroalgae growing on the pneumatophores of the mangroves of Karachi in the Indus Delta revealed a collection of 14 species belonging to Chlorophyta, Phaeophyta, and Xanthophyta phyla (Saifullah and Ahmed 2007). Within this habitat, algae and cyanobacteria covered 70–81% of the vertical length of mangrove pneumatophores. In Australia, species of macroalgae show a pattern of zonation along the vertical extent of the pneumatophores, likely arising as a result of competition, light exposure, and/or submersion/desiccation tolerance (Melville and Pulkownik 2007). Most algae are positioned above the low water mark where they are able to avoid extended periods of desiccation during the changing tides as well as maximize exposure to sunlight. Melville and Pulkownik (2007) suggest that species are arranged into three distinct ranges within the intertidal region according to their tolerance to desiccation. Low water mark algae are tolerant to submersion and have been suggested to be more aggressive competitors. High water mark algae are tolerant to desiccation and salinity changes (see Chap. 5 by Karsten) and the algae that are found in between these two groups are adaptable to both but have been so arranged as a consequence of competition.

16.4.6 Herbivory Within Mangroves

Studies have shown that many mangrove invertebrates rely on algae as a primary resource and some of them are likely to share symbiotic relationships with algae and/or their photosynthetic components (Bouillon et al. 2008). For example, an opisthobranch found in Indian mangroves feeds exclusively on red and green algae, sequestering the algal chloroplasts into its intercellular spaces in order to yield the products of continued photosynthesis (Bouillon et al. 2008). Grazers of the primary producers within the mangrove habitat (microphytobenthos, macroalgae, and mangroves themselves) are essential for channeling organic matter to higher trophic levels. Littorinids have been suggested as the dominant grazers of macroalgae growing on the vertical surfaces of mangrove prop roots, trunks, and branches and have even been said to have a "strong top-down force" in these coastal ecosystems (Alfaro 2008). Primary and secondary consumers, such as juvenile fishes and invertebrates, which inhabit the fringe of mangroves depend upon the roots of mangroves as a refuge alone and acquire most of their nutritional requirements from algal sources within the mangroves and seagrasses, as well as the seagrasses themselves (Kieckbusch et al. 2004). In 2004, Boyer et al. measured herbivory in three separate habitats dominated by seagrasses (*T. testudinum*), corals, and mangroves (*Rhizophora mangle*) off the northern coast of Honduras. Their results revealed that herbivores in all three habitats are capable of differentiating between food sources based upon nutritive value and thus preferentially grazed on nutrient-rich food items when given the opportunity. Compared to seagrass and coral habitats, overall macroalgae herbivory was lowest in the mangroves (31%, 33%, and 15%, respectively). However, the herbivores within the mangroves displayed the highest capacity for increased grazing (91%) following the nutrient enrichment.

Algae swept into the mangroves from adjacent rocky shore and estuarine habitats can become entangled in pneumatophores, persist, and at times reproduce to establish persistent free living colonies. It is suspected that these drift algae increase the likelihood of colonization by mollusks in mangroves by increasing substrate complexity and possibly serve as an additional food source (Bishop et al. 2009). The complex surface of fucalean macroalgae also represents an additional surface area for the growth of microfilms. Bishop et al. (2009) determined that algae originating from estuaries were capable of attracting more mollusks than algae coming from rocky shores. This is likely due to the typically higher nutrient content of estuarine systems which is capable of enriching algae and increasing its palatability for mollusks.

Some herbivorous fishes living within and adjacent to mangroves depend upon macroalgae growing as epiphytes on mangrove prop roots as a primary food source. For example, in Curaçao, the doctorfish, *Acanthurus chirurgus*, acquires approximately 55% of its macroalgal diet directly from mangrove surfaces (Nagelkerken and van der Velde 2004). The largest herbivorous reef fish in the Atlantic Ocean, *Scarus guacamaia*, is exclusively dependent on the mangrove habitat during its development and has suffered local extinctions where mangroves have been removed (Mumby et al. 2004). Of particular concern is the recent appearance of the invasive predatory lionfish, *Pterois volitans*, in Caribbean mangrove systems and the impacts that it will have on the recruitment of vital herbivorous fish species to adjacent coral reef habitats (Barbour et al. 2010; Albins and Hixon 2008).

16.5 Conclusion

Tropical macroalgae are essential elements of the complex coastal ecosystems they inhabit. They provide a wide range of vital ecological functions, like the stabilization of reefs, retention of nutrients, and provision of habitat and food. Macroalgae are able to provide all of these different functions due to their enormous diversity and habitat-specific adaptations. Over the last few decades, macroalgae have become more abundant across the globe, in coral reef, seagrass, and mangrove habitats, often as a response to human activities. At times, the proliferation of algae has led to regime shifts in these already threatened habitats. These ecosystem shifts have had profound consequences on tropical systems worldwide. In most cases a combination of coral mortality, eutrophication, reductions in grazing pressure due to disease, and overfishing form the basis for these shifts. In the near future, these shifts are likely to be exacerbated and prolonged, resulting in reduced overall habitat complexity and biodiversity.

References

Adey WH, Steneck R (1985) Highly productive Eastern Caribbean reefs: synergistic effects of biological, chemical, physical, and geological factors. In: Reaka M (ed) The ecology of coral reefs, vol 3. NOAA Symposia Series for Undersea Research, Rockville, pp 163–187

Adey WH, Vassar JM (1975) Colonization, succession and growth rates of tropical crustose coralline algae (Rhodophyta, Cryptonemiales). Phycologia 14:55–69

Adjeroud M, Michonneau F, Edmunds PJ, Chancerelle Y, Lison de Loma T, Penin L, Thibaut L, Vidal-Dupiol J, Salvat B, Galzin R (2009) Recurrent disturbances, recovery trajectories, and resilience of coral assemblages on a South Central Pacific reef. Coral Reefs 28:775–780

Albins MA, Hixon MA (2008) Invasive Indo-Pacific lionfish *Pterois volitans* reduce recruitment of Atlantic coral-reef fishes. Mar Ecol Prog Ser 367:233–238

Alfaro AC (2008) Diet of *Littoraria scabra*, while vertically migrating on mangrove trees: gut content, fatty acid, and stable isotope analyses. Estuar Coast Shelf Sci 79:718–726

Alongi DM (1998) Coastal ecosystem processes. CRC, New York, p 419

Amsler CD (2008) Algal chemical ecology. Springer, Berlin, p 313

Armitage AR, Frankovich TA, Heck KL Jr, Fourqurean JW (2005) Experimental nutrient enrichment causes complex changes in seagrass, microalgae, and macroalgae community structure in Florida Bay. Estuaries 28:422–434

Aterweberhan M, Bruggemann JH, Breeman AM (2006) Effects of extreme seasonality on community structure and functional group dynamics of coral reef algae in the southern Red Sea (Eritrea). Coral Reefs 25:391–406

Baggett LP, Heck KL Jr, Frankovich TA, Armitage AR, Fourqurean JW (2010) Nutrient enrichment, grazer identity, and their effects on epiphytic algal assemblages: field experiments in subtropical turtlegrass *Thalassia testudinum* meadows. Mar Ecol Prog Ser 406:33–45

Barbour AB, Montgomery ML, Adamson AA, Díaz-Ferguson E, Silliman BR (2010) Mangrove use by the invasive lionfish *Pterois volitans*. Mar Ecol Prog Ser 401:291–294

Beer S, Mtolera M, Lyimo T, Bjork M (2006) The photosynthetic performance of the tropical seagrass *Halophila ovalis* in the upper intertidal. Aquat Bot 84:367–371

Bell PRF (1992) Eutrophication and coral reefs: some examples in the Great Barrier Reef lagoon. Water Res 26:553–568

Bell PRF, Lapointe BE, Elmetri I (2007) Reevaluation of ENCORE: Support for the eutrophication threshold model for coral reefs. Ambio 36:416–424

Berner T (1990) Coral reef algae. In: Dubinsky Z (ed) Ecosystems of the world, vol 25. Elsevier, Amsterdam, pp 253–264

Biber PD (2007) Transport and persistence of drifting macroalgae (Rhodophyta) are strongly influenced by flow velocity and substratum complexity in tropical seagrass habitats. Mar Ecol Prog Ser 343:115–122

Birrell CL, McCook LJ, Willis BL (2005) Effects of algal turfs and sediment on coral settlement. Mar Pollut Bull 51:408–414

Bischof K, Gómez I, Molis M, Hanelt D, Karsten U, Lüder U, Roleda MY, Zacher K, Wiencke C (2006) Ultraviolet radiation shapes seaweed communities. Rev Environ Sci Biotechnol 5:141–166

Bishop MJ, Morgan T, Coleman MA, Kelaher BP, Hardstaff LK, Evenden RW (2009) Facilitation of molluscan assemblages in mangroves by the fucalean alga *Hormosira banksii*. Mar Ecol Prog Ser 392:111–122

Bittick SJ, Bilotti ND, Peterson HA, Stewart HL (2010) *Turbinaria ornata* as an herbivory refuge for associate algae. Mar Biol 157:317–323

Blunt JW, Copp BR, Hu W-P, Munro MHG, Northcote PT, Prinsep MR (2007) Marine natural products. Nat Prod Rep 24:31–86

Bouillon S, Raman AV, Dauby P, Dehairs F (2002) Carbon and Nitrogen stable isotope ratios of subtidal benthic invertebrates in an estuarine mangrove ecosystem (Andhra Pradesh, India). Estuar Coast Shelf Sci 54:901–913

Bouillon S, Connolly RM, Lee SY (2008) Organic matter exchange and cycling in mangrove ecosystems: Recent insights from stable isotope studies. J Sea Res 59:44–58

Box SJ, Mumby PJ (2007) Effects of macroalgal competition on growth and survival of juvenile Caribbean corals. Mar Ecol Prog Ser 342:139–149

Boyer KE, Fong P, Armitage AR, Cohen RA (2004) Elevated nutrient content of tropical macroalgae increases rates of herbivory in coral, seagrass, and mangrove habitats. Coral Reefs 23:530–538

Breeman AM (1988) Relative importance of temperature and other factors in determining geographic boundaries of seaweeds: experimental and phenological evidence. Helgol Mar Res 42:199–241

Bruggemann JH, van Oppen MJH, Breeman AM (1994) Foraging by the stoplight parrotfish *Sparisoma viride* I. Food selection in different socially determined habitats. Mar Ecol Prog Ser 106:41–55

Burkepile DE, Hay ME (2010) Impact of herbivore Identity on algal succession and coral growth on a Caribbean Reef. PLoS One 5:e8963. doi:10.1371/journal.pone.0008963

Cardosoa PG, Pardal MA, Lillebø AI, Ferreira SM, Raffaelli D, Marques JC (2004) Dynamic changes in seagrass assemblages under eutrophication and implications for recovery. J Exp Mar Biol Ecol 302:233–248

Carpenter RC (1986) Partitioning herbivory and its effects on coral reef algae communities. Ecol Monogr 56:345–363

Carpenter RC, Hackney JM, Adey WH (1985) Measurements of primary productivity and nitrogenase activity of coral reef algae in a chamber incorporating oscillatory flow. Limnol Oceanogr 36:40–49

Coles SL, Fadlalah YH (1991) Reef coral survival and mortality at low temperatures in the Arabian Gulf: new species-specific lower temperature limits. Coral Reefs 9:231–237

Collado-Vides L, Caccia VG, Boyer JN, Fourqurean JW (2007) Tropical seagrass-associated macroalgae distributions and trends relative to water quality. Estuar Coast Shelf Sci 73:680–694

Connell JH (1978) Diversity in tropical rain forests and coral reefs. Science 199:1302–1310

Cvitanovic C, Bellwood DR (2009) Local variation in herbivore feeding activity on an inshore reef of the Great Barrier Reef. Coral Reefs 28:127–133

de la Torre-Castro M, Eklöf JS, Rönnbäck P, Björk M (2008) Seagrass importance in food provisioning services: fish stomach content as a link between seagrass meadows and local fisheries. West Indian Ocean J Mar Sci 7:95–110

Delgado O, Lapointe BE (1994) Nutrient-limited production of calcareous versus fleshy macroalgae in a eutrophic, carbonate- rich tropical marine environment. Coral Reefs 13:151–159

Diaz MC, Rützler K (2009) Biodiversity and abundance of sponges in Caribbean Mangrove: indicators of environmental quality. Smithsonian Contrib Zool 38:151–172

Diaz-Pulido G, McCook LJ, Larkum AWD, Lotze HK, Raven JA, Schaffelke B, Smith JE, Steneck RS (2007) Vulnerability of macroalgae of the great barrier reef to climate change. In: Marshall PA, Johnson J (eds) Climate change and the great barrier reef. Great Barrier Reef Marine Park Authority, Townsville, pp 153–192

Dixon LK (2000) Establishing light requirements for the seagrass *Thalassia testudinum*: an example from Tampa Bay, Florida. In: Bortone SA (ed) Seagrasses: monitoring, ecology, physiology and management. CRC, Boca Raton, pp 9–31

Dorenbosch M, Grol MGG, Christianen MJA, Nagelkerken I, van der Velde G (2005) Indo-Pacific seagrass beds and mangroves contribute to fish density and diversity on adjacent coral reefs. Mar Ecol Prog Ser 302:63–76

Duffy JE, Hay ME (1990) Seaweed adaptations to herbivory. Bioscience 40:368–375

Fabricius K, Dea'th G (2001) Environmental factors associated with the spatial distribution of crustose coralline algae on the Great Barrier Reef. Coral Reefs 19:303–309

Ferdie M, Fourqurean JW (2004) Responses of seagrass communities to fertilization along a gradient of relative availability of nitrogen and phosphorus in a carbonate environment. Limnol Oceanogr 49:2082–2094

Figueroa FL, Salles S, Aguilera J, Jiménez C, Mercado J, Viñegla B, Flores-Moya A, Altamirano M (1997) Effects of solar radiation on photoinhibition and pigmentation in the red alga *Porphyra leucosticta*. Mar Ecol Prog Ser 151:81–90

Figueroa FL, Martínez B, Israel A, Neori A, Malta E, Ang P Jr, Inken S, Marquardt R, Rachamim T, Arazi U, Frenk S, Korbee N (2009) Acclimation of Red Sea macroalgae to solar radiation: photosynthesis and thallus absorptance. Aquat Biol 7:159–172

Fong P, Paul VJ (2011) Coral reef algae. In: Dubinsky Z, Stambler N (eds) Coral reefs: an ecosystem in transition. Springer, New York, pp 241–272

Fong P, Kamer K, Boyer KE, Boyle KA (2001) Nutrient content of macroalgae with differing morphologies may indicate sources of nutrients to tropical marine systems. Mar Ecol Prog Ser 220:137–152

Fong P, Boyer KE, Kamer K, Boyle KA (2003) Influence of initial tissue nutrient status of tropical marine algae on response to nitrogen and phosphorus additions. Mar Ecol Prog Ser 262:111–123

Fong P, Smith TB, Wartian MJ (2006) Epiphytic cyanobacteria maintain shifts to macroalgal dominance on coral reefs following enso disturbance. Ecology 87:1162–1168

Foster NL, Box SJ, Mumby PJ (2008) Competitive effects of macroalgae on the fecundity of the reef-building coral *Montastraea annularis*. Mar Ecol Prog Ser 367:143–152

Fourqurean JW, Rutten LM (2004) The impact of Hurricane Georges on soft-bottom, back reef communities: site and species-specific effects in South Florida seagrass beds. Bull Mar Sci 75(2):239–257

Fricke A, Teichberg M, Beilfuss S, Bischof K (2011) Succession patterns of turf algal communities in a Caribbean coral reef. Bot Mar 54:111–126

Gross EM, Erhard D, Ivanyi E (2003) Allelopathic activity of *Ceratophyllum demersum* L. and *Najas marina* sspp. Intermedia (Wolgang) Casper. Hydrobiologia 506–509:583–589

Hanelt D, Roleda MY (2009) UVB radiation may ameliorate photoinhibition in specific shallow-water tropical marine macrophytes. Aquat Bot 91:6–12

Hatcher BG (1988) Coral reef primary productivity: a beggar's banquet. Trends Ecol Evol 3:106–111

Hauxwell J, Cebrián J, Valiela I (2003) Eelgrass Zostera marina loss in temperate estuaries: relationship to land-derived nitrogen loads and effect of light limitation imposed by algae. Mar Ecol Prog Ser 247:59–73

Hay ME (1981) The functional morphology of turf-forming seaweeds: persistence in stressful marine habitats. Ecology 62(3):739–750

Hay ME (1997) The ecology and evolution of seaweed-herbivore interactions on coral reefs. Coral Reefs 16:67–76

Hixon MA, Brostoff WN (1996) Succession and herbivory: effects of differential fish grazing on Hawaiian coral-reef algae. Ecol Monogr 66:67–90

Hoey AS, Bellwood DR (2010) Cross-shelf variation in browsing intensity on the Great Barrier Reef. Coral Reefs 29:499–508

Holmer M, Nielsen RM (2007) Effects of filamentous algal mats on sulfide invasion in eelgrass (*Zostera marina*). J Exp Mar Biol Ecol 353:245–252

Houk P, Camacho R (2010) Dynamics of seagrass and macroalgal assemblages in Saipan Lagoon, Western Pacific Ocean: disturbances, pollution, and seasonal cycles. Bot Mar 53:205–212

Houk P, van Woesik R (2008) Dynamics of shallow-water assemblages in the Saipan Lagoon. Mar Ecol Prog Ser 356:39–50

Hughes TP (1994) Catastrophes, phase shifts, and large-scale degradation of a Caribbean coral reef. Science 265:1547–1551

Hughes TP, Rodrigues MJ, Bellwood DR, Ceccarelli D, Hoegh-Guldberg O, McCook LJ, Moltschaniwskyj N, Pratchett MS, Steneck RS, Willis B (2007) Phase shifts, herbivory, and the resilience of coral reefs to climate change. Curr Biol 17:360–365

Invers O, Kraemer GP, Pérez M, Romero J (2004) Effects of nitrogen addition on nitrogen metabolism and carbon reserves in the temperate seagrass *Posidonia oceanica*. J Exp Mar Biol Ecol 303:97–114

Jompa J, McCook LJ (2002) Effects of competition and herbivory on interactions between a hard coral and a brown alga. J Exp Mar Biol Ecol 271:25–39

Jompa J, McCook LJ (2003) Coral-algal competition: macroalgae with different properties have different effects on corals. Mar Ecol Prog Ser 258:87–95

Kennison RL (2008) Evaluating ecosystem function of nutrient retention and recycling in excessively eutrophic estuaries. PhD Dissertation. University of California, Los Angeles

Kerswell AP (2006) Global biodiversity patterns of benthic marine algae. Ecology 87:2479–2488

Kieckbusch DK, Koch MS, Serafy JE, Anderson WT (2004) Trophic linkages among primary producers and consumers in fringing mangroves of subtropical lagoons. Bull Mar Sci 74:271–285

Koch MS, Madden CJ (2001) Patterns of primary production and nutrient availability in a Bahamas lagoon with fringing mangroves. Mar Ecol Prog Ser 219:109–119

Konar B, Iken K, Cruz-Motta JJ et al (2010) Current patterns of macroalgal diversity and biomass in northern hemisphere rocky shores. PLoS One 5:e13195. doi:10.1371/journal.pone.0013195

Kopecky AL, Dunton KH (2006) Variability in drift macroalgal abundance in relation to biotic and abiotic factors in two seagrass dominated estuaries in the Western Gulf of Mexico. Estuar Coasts 29(4):617–629

Kopp D, Bouchon-Navarro Y, Cordonnier S, Haouisee ML, Bouchon C (2010) Evaluation of algal regulation by herbivorous fishes on Caribbean coral reefs. Helgol Mar Res 64:181–190

Kristensen E (2008) Mangrove crabs as ecosystem engineers; with emphasis on sediment processes. J Sea Res 59:30–43

Kuffner IB, Walters LJ, Becerro MA, Paul VJ, Ritson-Williams R, Beach K (2006) Inhibition of coral recruitment by macroalgae and cyanobacteria. Mar Ecol Prog Ser 323:107–117

Lapointe BE (1997) Nutrient thresholds for bottom-up control of macroalgal blooms on coral reefs in Jamaica and southeast Florida. Limnol Oceanogr 42:1119–1131

LaPointe BE, Littler MM, Littler DS (1987) A comparison of nutrient-limited productivity in macroalgae from a Caribbean barrier reef and from a mangrove ecosystem. Aquat Bot 28:243–255

Lapointe BE, Littler MM, Littler DS (1992) Nutrient availability to marine macroalgae in siliciclastic versus carbonate-rich coastal waters. Estuaries 15:75–82

Lapointe BE, Barile PJ, Matzie WR (2004) Anthropogenic nutrient enrichment of seagrass and coral reef communities in the Lower Florida Keys: discrimination of local versus regional nitrogen sources. J Exp Mar Biol Ecol 308:23–58

Lirman D (2001) Competition between macroalgae and corals: effects of herbivore exclusion and increased algal biomass on coral survivorship and growth. Coral Reefs 19:392–399

Littler MM, Littler DS (1984) A relative-dominance model for biotic reefs. In: Proceedings of the joint meeting of the Atlantic reef committee society of reef studies, Miami, 1984

Littler MM, Littler DS (1988) Structure and role of algae in tropical reef communities. In: Lembi CA, Waaland JR (eds) Algae and human affairs. Cambridge University Press, Cambridge, pp 29–56

Littler MM, Littler DS (2006) Review: assessment of coral reefs using herbivory/nutrient assays and indicator groups of benthic primary producers: a critical synthesis, proposed protocols, and critique of management strategies. Aquat Conserv Mar Freshwater Ecosystem 17:195–215. doi:10.1002/aqc.790

Littler MM, Littler DS, Blair SM, Norris JN (1985) Deepest known plant life discovered on an uncharted seamount. Science 227:57–59

Littler MM, Littler DS, Blair SM, Norris JN (1986) Deep-water plant communities from an uncharted seamount off San Salvador Island, Bahamas: distribution, abundance, and primary productivity. Deep Sea Res 33:881–892

Littler MM, Littler DS, Titlyanov EA (1991) Comparisons of N- and P-limited productivity between high Granitic Islands vs low carbonate atolls in the seychelles archipelago: a test of the relative-dominance paradigm. Coral Reefs 10:199–209

Littler MM, Littler DS, Brooks BL (2010a) Marine macroalgal diversity assessment of saba bank Netherlands antilles. PLoS One 5:e10677. doi:10.1371/journal.pone.0010677

Littler MM, Littler DS, Brooks BL (2010b) The effects of nitrogen and phosphorus enrichment on algal community development: artificial mini reefs on the Belize Barrier Reef sedimentary lagoon. Harmful Algae 9:255–263

Mantyka CS, Bellwood DR (2007a) Direct evaluation of macroalgal removal by herbivorous coral reef fishes. Coral Reefs 26:435–442

Mantyka CS, Bellwood DR (2007b) Macroalgal grazing selectivity among herbivorous coral reef fishes. Mar Ecol Prog Ser 352:177–185

McCook LJ (1996) Effects of herbivores and water quality on the distribution of Sargassum on the central Great Barrier Reef: cross-shelf transplants. Mar Ecol Prog Ser 139:179–192

McCook LJ (1997) Effects of herbivory on zonation of Sargassum spp. within fringing reefs of the central Great Barrier Reef. Mar Biol 129:713–722

McCook LJ (1999) Macroalgae, nutrients and phase shifts on coral reefs: scientific issues and management consequences for the Great Barrier Reef. Coral Reefs 18:357–367

McCook L, Jompa J, Diaz-Palido G (2001) Competition between corals and algae on coral reefs: a review of evidence and mechanisms. Coral Reefs 19:400–417

McGlathery KJ (2001) Macroalgal blooms contribute to the decline of seagrass in nutrient-enriched coastal waters. J Phycol 37:453–456

McGlathery KJ, Howarth RW, Marino R (1992) Nutrient limitation of the macroalga, Penicillus capitatus, associated with subtropical seagrass meadows in Bermuda. Estuaries 15:18–25

McManus JW, Polsenberg JF (2006) Coral-algal phase shifts on coral reefs: ecological and environmental aspects. Prog Oceanogr 60:263–279

Melville F, Pulkownik A (2007) Seasonal and spatial variation in the distribution of mangrove macroalgae in the Clyde River, Australia. Estuar Coast Shelf Sci 71:683–690

Moksnes PO, Gullström M, Tryman K, Baden S (2008) Trophic cascades in a temperate seagrass community. Oikos 117:763–777

Mork E, Sjoo GL, Kautsky N, McClanahan TR (2009) Top-down and bottom-up regulation of macroalgal community structure on a Kenyan reef. Estuar Coast Shelf Sci 84:331–336

Mumby PJ (2009) Phase shifts and the stability of macroalgal communities on Caribbean coral reefs. Coral Reefs 28:761–773

Mumby PJ, Harborne AR (2010) Marine reserves enhance the recovery of corals on Caribbean reefs. PLoS One 5:e8657

Mumby PJ, Edwards AJ, Arias-González JE, Lindeman KC, Blackwell PG, Gall A, Gorczynska MI, Harborne AR, Pescod CL, Renken H, Wabnitz CCC, Llewellyn G (2004) Mangroves enhance the biomass of coral reef fish communities in the Caribbean. Nature 427:533–536

Mumby PJ, Dahlgren CP, Harborne AR, Kappel CV et al (2006) Fishing, trophic cascades, and the process of grazing on coral reefs. Science 311:98

Muzuka ANN, Shaghude YW, Wannas KO, Nyandwi N (2005) Sediment sources and their distribution in Chwaka Bay, Zanzibar Island. West Indian Ocean J Mar Sci 4:185–197

Nagelkerken I, van der Velde G (2004) Are Caribbean mangroves important feeding grounds for juvenile reef fish from adjacent seagrass beds? Mar Ecol Prog Ser 274:143–151

Nugues M, Bak RP (2006) Differential competitive abilities between Caribbean coral species and a brown alga: a year of experiments and a long-term perspective. Mar Ecol Prog Ser 315:75–86

Pakker H, Breeman AM, Prud'homme van Reine WF, Van den Hoek C (1995) A comparative study of temperature responses of Caribbean seaweeds from different biogeographic groups. J Phycol 31:499–507

Paul VJ, Puglisi MP (2004) Chemical mediation of interactions among marine organisms. Nat Prod Rep 21:189–209

Paul VJ, Nelson SG, Sanger HR (1990) Feeding preferences of adult juvenile rabbitfish Siganus argenteus in relation to chemical defenses of tropical seaweeds. Mar Ecol Prog Ser 60:23–24

Paul VJ, Cruz-Rivera E, Thacker RW (2001) Chemical mediation of macroalgal-herbivore interactions: ecological and evolutionary perspectives. In: McClintock J, Baker B (eds) Marine chemical ecology. CRC Press, LLC, Boca Raton, pp 227–265

Pereira RC, Da Gama BAP (2008) Macroalgal chemical defenses and their roles in structuring tropical marine communities. In: Amsler DC (ed) Algal chemical ecology. Springer, Germany, pp 25–49

Pielou EC (1977) The latitudinal spans of seaweed species and their patterns of overlap. J Biogeogr 4:299–311

Rasher DB, Hay ME (2010) Chemically rich seaweeds poison corals when not controlled by herbivores. Proc Natl Acad Sci 107:9683–9688

Ritson-Williams R, Paul VJ, Arnold SN, Steneck R (2010) Larval settlementpreferences and post-settlement survival of the threatened Caribbean corals *Acropora palmata* and *A. cervicornis*. Coral Reefs 29:71–81

Saifullah SM, Ahmed W (2007) Epiphytic algal biomass on pneumatophores of mangroves of Karachi, Indus Delta. Pakistan J Bot 39:2097–2102

Santelices B, Marquet PA (1998) Seaweeds, latitudinal diversity patterns and Rapoport's rule. Div Dist 4:71–75

Schaffelke B (2001) Surface alkaline phosphatase activities of macroalgae on coral reefs of the central Great Barrier Reef, Australia. Coral Reefs 19:310–317

Schaffelke B, Klumpp DW (1997) Growth of germlings of the macroalga *Sargassum baccularia* (Phaeophyta) is stimulated by enhanced nutrients. Proc 8th Int Coral Reef Symp 2:1839–1842

Schupp PJ, Paul VJ (1994) Calcification and secondary metabolites in tropical seaweeds: variable effects on herbivorous fishes. Ecology 75:1172–1185

Semesi IS, Beer S, Björk M (2009) Seagrass photosynthesis controls rates of calcification and photosynthesis of calcareous macroalgae in a tropical seagrass meadow. Mar Ecol Prog Ser 382:41–47

Smith JE, Hunter CL, Smith CM (2010a) The effects of top-down versus bottom-up control on benthic coral reef community structure. Oecologia 163:497–507

Smith TB, Fong P, Kennison R, Smith J (2010b) Spatial refuges and associational defenses promote harmful blooms of the alga Caulerpa sertularioides onto coral reefs. Oecologia 164:1039–1048

Sotka EE, Hay ME (2009) Effects of herbivores, nutrient enrichment, and their interactions on macroalgal proliferation and coral growth. Coral Reefs 28:555–568

Spivak AC, Canuel EA, Duffy JE, Richardson JP (2009) Nutrient enrichment and food web composition affect ecosystem metabolism in an experimental seagrass habitat. PLoS One 4:e7473. doi:10.1371/journal.pone.0007473

Steneck RS, Dethier MN (1994) A functional group approach to the structure of algal-dominated communities. Oikos 69:476–498

Uku J (2005) Seagrasses and their epiphytes: characterization of abundance and productivity in tropical seagrass beds. Doctoral dissertation. Stockholm University. Stockholm, Sweden

Uku J, Björk M (2001) The distribution of epiphytic algae on three Kenyan seagrass species. S Afr J Bot 67:475–482

Uku J, Björk M (2005) Productivity aspects of three tropical seagrass species in areas of different nutrient levels in Kenya. Estuar Coast Shelf Sci 63:407–420

van den Hoek C, Breeman AM, Bak RPM, Van Buurt G (1978) The distribution of algae, corals and gorgonians in relation to depth, light attenuation, water movement and grazing pressure in the fringing coral reef of Curaçao, Netherlands Antilles. Aquat Bot 5:1–46

Vermeij MJA (2006) Early life-history dynamics of Caribbean coral species on artificial substratum: the importance of competition, growth, and variation in life-history strategy. Coral Reefs 25:59–71

Vermeij MJA, Smith JE, Smith CM, Thurber RV, Sandin SA (2009) Survival and settlement success of coral planulae: independent and synergistic effects of macroalgae and microbes. Oecologia 159:325–336

Vermeij MJA, van Moorselaar I, Engelhard S, Hörnlein C, Vonk SM, Visser PM (2010) The effects of nutrient enrichment and herbivore abundance on the ability of turf algae to overgrow coral in the Caribbean. PLoS One 5:e14312. doi:10.1371/journal.pone.0014312

Vermeij MJA, Dailer ML, Smith CM (2011) Crustose coralline algae can suppress macroalgal growth and recruitment on Hawaiian coral reefs. Mar Ecol Prog Ser 422:1–7

Vuki VC, Price IR (1994) Seasonal changes in the *Sargassum* populations on a fringing coral reef, magnetic Island, Great barrier reef region, Australia. Aquat Bot 48:153–166

Webster PJ, Holland GJ, Curry JA, Chang HR (2005) Changes in tropical cyclone number, duration, and intensity in a warming environment. Science 309:1844–1846

Williams ID, Polunin VC, Hendrick VJ (2001) Limits to grazing by herbivorous fishes and the impact of low coral cover on macroalgal abundance on a coral reef in Belize. Mar Ecol Prog Ser 222:187–196

Wismer S, Hoey A, Bellwood D (2009) Cross-shelf benthic community structure on the Great barrier reef: relationships between macroalgal cover and herbivore biomass. Mar Ecol Prog Ser 376:45–54

Womersley HBS, Bailey A (1969) The marine algae of the Solomon Islands and their place in biotic reefs. Philos Trans R Soc B 255:432–433

Chapter 17
Floating Seaweeds and Their Communities

Eva Rothäusler, Lars Gutow, and Martin Thiel

17.1 Introduction

Floating seaweeds have fascinated sailors and naturalists since the first ships sailed across the oceans, where they discovered immense rafts of seaweeds far from shore. Christopher Columbus was one of the first who documented dense aggregations of floating algae. In September 1492, his crew encountered masses of floating seaweed as they were approaching the Bahamas: "They saw so much weed that the sea appeared to be covered with it. . . .and there was a great deal of weed and they found crabs in it" (Farlow 1914).

Floating *Sargassum natans* and *S. fluitans* from the Sargasso Sea and especially their specifically adapted fauna are remarkable biological communities. Full-grown algae that form dense masses without ever having been attached to the seafloor pass their entire floating life cycle via the propagation of vegetative fragments (Parr 1939). While floating algal populations in other seas appear to originate from benthic source populations, the size of algal rafts and the length of their ocean voyages equal or even surpass that of *S. natans* and *S. fluitans*.

E. Rothäusler
Universidad Católica del Norte, Larrondo 1281, 1781421 Coquimbo, Chile

Atmosphere and Ocean Research Institute, University of Tokyo, Kashiwanoha 5-1-5, Kashiwa-shi, Chiba-ken, 277-8564 Japan
e-mail: Rothaeusler@gmail.com

L. Gutow
Alfred Wegener Institute for Polar and Marine Research, Box 12 01 61, 27515 Bremerhaven, Germany
e-mail: Lars.Gutow@awi.de

M. Thiel (✉)
Universidad Católica del Norte, Larrondo 1281, 1781421 Coquimbo, Chile

Centro de Estudios Avanzados en Zonas Áridas (CEAZA), Coquimbo, Chile
e-mail: thiel@ucn.cl

C. Wiencke and K. Bischof (eds.), *Seaweed Biology*, Ecological Studies 219,
DOI 10.1007/978-3-642-28451-9_17, © Springer-Verlag Berlin Heidelberg 2012

Centuries after Columbus' voyage, contemporary marine scientists have reported floating seaweeds from all major oceans (Ohno 1984; Kingsford 1992; Helmuth et al. 1994; Ingólfsson 1998; Hobday 2000a; Hirata et al. 2001; Macaya et al. 2005; Vandendriessche et al. 2006; Komatsu et al. 2008; Hinojosa et al. 2010). Recent studies confirm that floating algae can cross large ocean basins, facilitating the colonization of new habitats on remote shores by the algae themselves but also by organisms living associated with floating algae (Fraser et al. 2009, 2011; Macaya and Zuccarello 2010; Nikula et al. 2010). Accordingly, these authors emphasized the importance of dispersal of organisms via these floating islands.

There is a wide variety of positively buoyant seaweeds in the oceans that become detached by breakage of stipes, thallus fragmentation, erosion of holdfasts, storms, or grazing activities. Once detached, floating seaweeds continue to grow if environmental conditions are optimal, and they form habitat for many associated organisms. Also these algae are known to travel considerable distances at the mercy of winds and currents, having the potential to release spores or gametes at a new site and thereby connecting distant populations (Fraser et al. 2009). In order to understand the fate of floating seaweeds it is crucial to address a set of questions. Where can we regularly find floating algae and where do they come from? How do abiotic and biotic factors affect the persistence of floating algae at the sea surface? How can floating algae acclimate or withstand sea surface conditions? Which organisms are traveling with them and how can they affect raft viability? Answering these questions will help to understand the role of floating algae as dispersal agents in the world's oceans.

17.2 Spatial and Seasonal Distribution of Floating Seaweeds

A wide diversity of seaweeds can be found floating in the world's oceans, where they persist at the sea surface thanks to the presence of gas-filled tissues or pneumatocysts (Fig. 17.1). The dominant floating seaweeds in the world's oceans are brown algae belonging to the genera *Macrocystis*, *Fucus*, *Sargassum*, *Ascophyllum, Durvillaea, Carpophyllum, Phyllospora,* and *Cystophora*. Some red and green algae have also been reported floating, albeit with a more limited floating potential than brown algae (Thiel and Gutow 2005a; Gagnon et al. 2011). Besides a few entirely pelagic species such as *Sargassum natans* and *S. fluitans*, most floating algae grow in benthic habitats during their earlier life stages and can vary widely in size after detachment, ranging from a few centimeters to tens of meters. Floating seaweeds can aggregate at the sea surface and form large rafts of many individuals with biomasses of up to 300 kg wet weight per seaweed patch (e.g., Senta 1962 for the Japanese coast) or more (Mitchell and Hunter 1970 for the southern Californian coast and Baja California).

Fig. 17.1 (**a**) Cross section through a thallus of *Durvillaea antarctica* showing the gas-filled honeycomb structure; (**b**) vesicles of *Sargassum horneri*, providing positive buoyancy to the algae; and (**c**) raft of *S. horneri* floating in the eastern East China Sea

17.2.1 Global Distribution Patterns of Floating Seaweeds

The presence of floating algae at the sea surface strongly depends on their supply from benthic source populations. Most reports of floating macroalgae come from mid and high latitudes in both hemispheres (Kingsford 1992; Helmuth et al. 1994; Hobday 2000a; Macaya et al. 2005; Komatsu et al. 2008; Hinojosa et al. 2010; Thiel et al. 2011) (Fig. 17.2), where extensive benthic populations provide abundant supply (Lüning 1990; Graham et al. 2007). For instance, in the northern hemisphere algal rafts can be found in the NE Pacific where extensive subtidal kelp forests along the Pacific coast of North America nourish abundant floating populations of *Macrocystis pyrifera* (Kingsford 1995; Hobday 2000a). In the NW Pacific a diverse *Sargassum* community is floating in coastal waters around Japan (Ohno 1984; Hirata et al. 2001). In the N Atlantic pelagic *S. natans* and *S. fluitans* occur in the Gulf of Mexico, the Gulf Stream, and in the Sargasso Sea (Parr 1939) while floating fronds of *Ascophyllum nodosum* and *Fucus vesiculosus* are common along the northern N Atlantic >45°N (Tully and Ó Céidigh 1986; Ingólfsson 1998; Vandendriessche et al. 2006; Muhlin et al. 2008; Thiel et al. 2011).

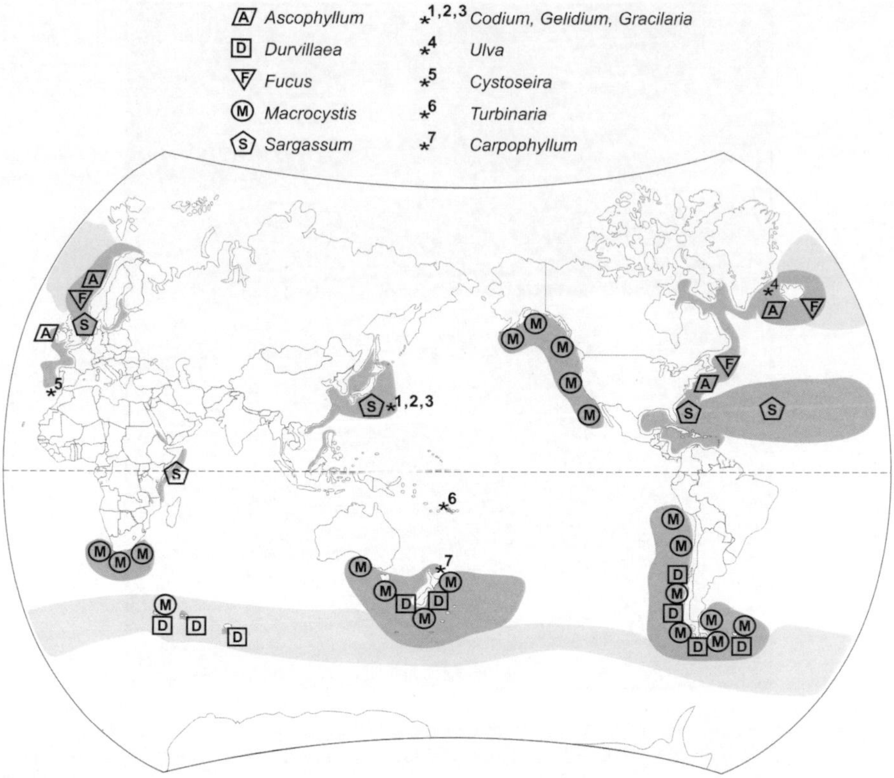

Fig. 17.2 Distribution of the most common genera of benthic algae (*indicated with symbols and letters*) that are positively buoyant, as well as examples of a few benthic genera (*indicated with asterisks and numbers*) that are less commonly found floating in the world's oceans. *Shadings* indicate areas where floating algae have been reported abundantly

M. pyrifera and *Durvillaea antarctica* are the most common floating seaweeds in the southern hemisphere. They occur in the SE Pacific along the Chilean coast (Macaya et al. 2005; Hinojosa et al. 2007, 2011), around Tasmania and New Zealand (Edgar 1987; Kingsford 1992) and throughout the entire West Wind Drift where floating sporophytes connect the algal populations of the sub-Antarctic islands (Smith 2002; Fraser et al. 2009). Also the Agulhas Bank region off South Africa "literally swarms with *Macrocystis*" (Hooker 1847). Whereas *Macrocystis* and *Durvillaea* are widespread throughout cold temperate waters of the southern hemisphere, other floating seaweeds have a more restricted distribution but may become locally very abundant. For example, high densities of floating *Carpophyllum* spp. and *Cystophora* spp. can be found in waters around New Zealand (Kingsford 1993) and buoyant *Turbinaria ornata* are common among the Polynesian islands (Stewart 2008). Although floating seaweeds can bridge large oceanic distances in subpolar regions, the warm waters of the tropics and the subtropics appear to be an impenetrable dispersal barrier mostly preventing the exchange of floating algae between the two hemispheres.

17.2.2 Regional Aggregations of Floating Seaweeds

Complex interactions between winds and mesoscale oceanographic features such as fronts, eddies, and currents determine the spatial distribution of algae at the sea surface. Accordingly, abundances of floating seaweeds vary substantially on a regional scale and are often highly unpredictable (Kingsford 1992; Ingólfsson 1998). For example, Kingsford and Choat (1986) observed consistently higher densities of floating algae inside than outside transient surface slicks, which were generated by tidally forced internal waves. In other regions, floating algae accumulate along fronts (Kingsford 1995) where single sporophytes may entangle at the sea surface to form large floating mats. Franke et al. (1999) reported that during calm summer conditions dense aggregations of floating algae form at boundary zones between tidal currents of different speeds and directions around the island of Helgoland (North Sea). Hinojosa et al. (2010) could show for the fjords of southern Chile that floating seaweeds are aggregated in an estuarine front in the large channel openings by an interaction of wind and the surface outflow of estuarine waters (Fig. 17.3). These regions are suggested to act as retention zones, which receive and accumulate floating seaweeds. Due to the aggregation of floating algae, other pelagic organisms, and organic matter, these retention areas are considered centers

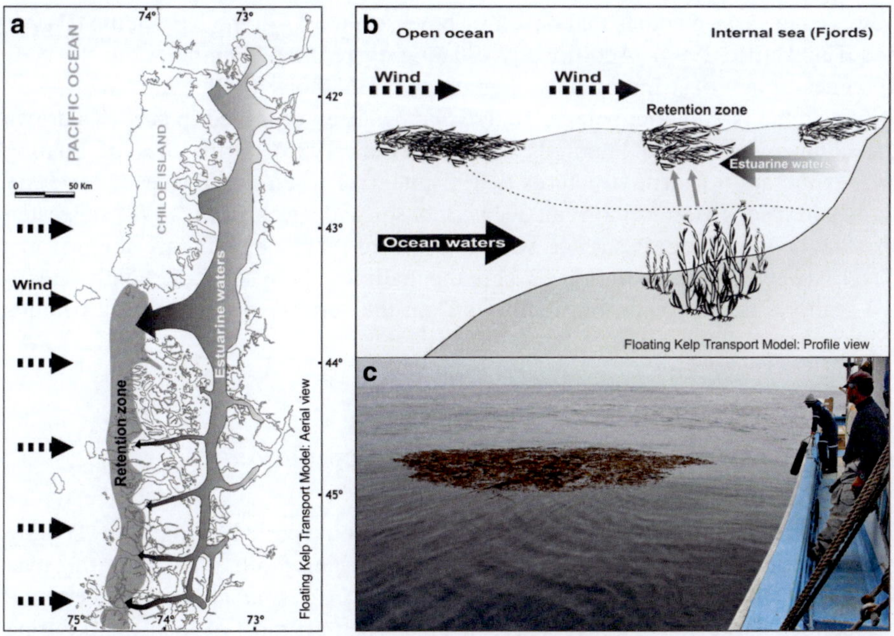

Fig. 17.3 Retention zones for floating seaweeds formed by winds, currents, and estuarine influence in the fjord regions of southern Chile, in (**a**) aerial view and (**b**) profile view; figure modified after Hinojosa et al. (2010). (**c**) Floating patch of *Macrocystis pyrifera*; image courtesy of Iván Hinojosa

of high biological activity, where decomposers that are associated with the floating algae recycle organic remains into nutrients and other compounds (Thiel and Gutow 2005b).

Floating seaweeds are often more abundant in coastal waters than in the open ocean (Segawa et al. 1961; Kingsford 1992; Ingólfsson 1995; Hobday 2000b). High nearshore densities of floating algae are primarily due to the proximity to coastal source populations. Additionally, floating algae are known to accumulate in coastal areas with complex current conditions and frontal zones (Segawa et al. 1961). Harrold and Lisin (1989) tracked floating *M. pyrifera* equipped with radio transmitters in coastal waters of California. They could show that the majority of algal rafts are deposited on beaches soon after detachment while others become entangled with benthic sporophytes in coastal kelp forests (see also Dayton et al. 1984; Graham et al. 1997). The onshore transport of floating *M. pyrifera* in coastal waters of California was confirmed by Hernández-Carmona et al. (2006) (Fig. 17.4). Results from experiments with artificial drifters revealed that directional net transport of floating algae in coastal systems is mainly the result of prevailing winds that interact with surface currents (Tapia et al. 2004). During periods of low wind, no substantial transport of the drifters by currents alone could be detected (Fig. 17.4). Apparently, only a relatively small fraction of detached seaweeds escapes to offshore waters where they might enter larger oceanic currents.

In offshore regions conspicuous amounts of seaweeds have repeatedly been observed in surface eddies that develop above seamounts in the NE Pacific (Parker and Tunnycliffe 1994). Additionally, floating algae often accumulate in the convergence zones of wind-induced Langmuir circulations (Faller and Woodcock 1964). In the N Atlantic, pelagic *Sargassum* has been observed in such windrows extending over tens to hundreds of miles (Carr 1986). Drift lines of floating *Sargassum* are transient structures that disintegrate when wind exceeds a certain threshold speed (Marmorino et al. 2011). Johnson and Richardson (1977) suggested that the downwelling of surface waters in the zones of convergence might carry *Sargassum* below a critical depth where the thalli lose buoyancy and finally sink to the seafloor, thereby exporting biomass from the surface to the deep sea benthos (Schoener and Rowe 1970).

17.2.3 Seasonal Patterns of Floating Populations

While large quantities of seaweeds appear during some periods of the year, floating algae can virtually disappear at other times. For instance, along the coast of Japan high densities of floating individuals of the annual *Sargassum horneri* can be found throughout spring and summer when the algae shed off most of their thalli during their reproductive season (Yoshida 1963; Ohno 1984; Hirata et al. 2001). Similar observations have been made by Kingsford (1992) for floating *S. sinclairii* from coastal waters of New Zealand, indicating that the appearance of large quantities of *Sargassum* is driven by their seasonal growth cycles. Additionally, the invasive

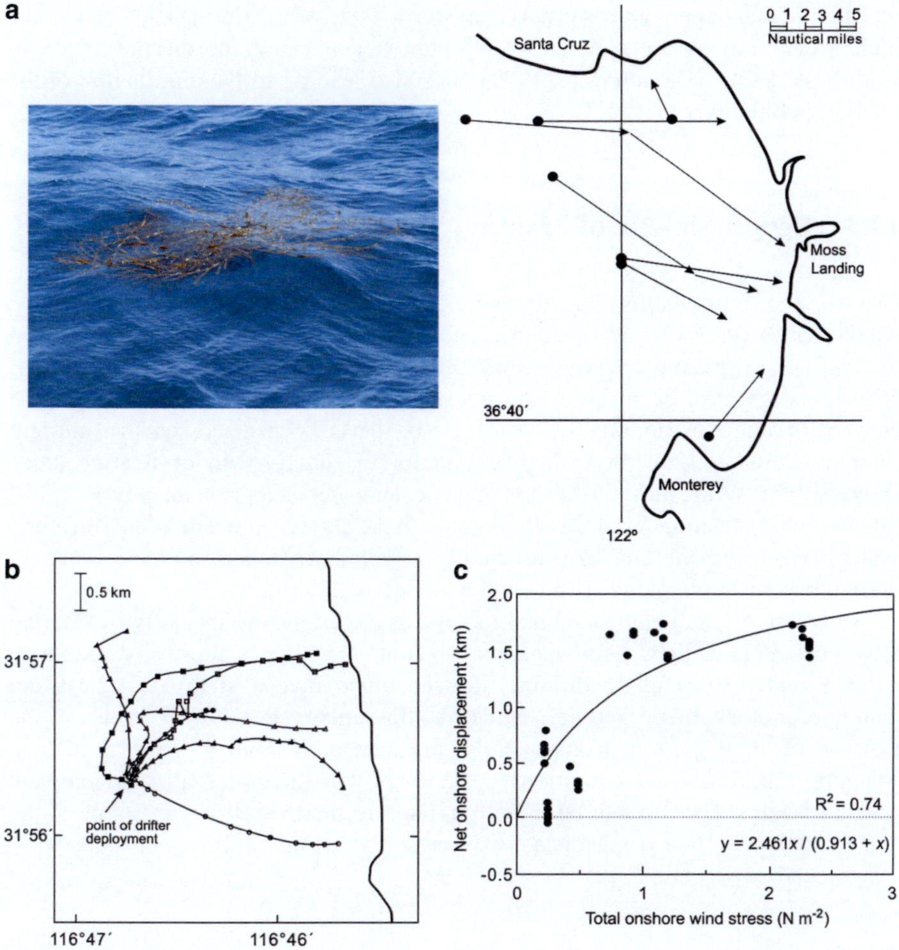

Fig. 17.4 Similar trajectories of (**a**) floating *Macrocystis pyrifera* and (**b**) artificial drifters towards the shore in coastal waters of California and Baja California, respectively. (**c**) Drift direction and velocity of the artificial drifters correlated with prevailing wind conditions. Figures modified after Hernández-Carmona et al. (2006) and Tapia et al. (2004)

species *S. muticum,* which originates from Japan, has been reported floating abundantly in May in the North Sea (Vandendriessche et al. 2006); thallus fragmentation and rafting dispersal may have facilitated the rapid spread of this species along the coasts of NW Europe (Rueness 1989; Harries et al. 2007). The holopelagic *Sargassum* species from the Gulf of Mexico seemed to have similar seasonal growth cycles, with highest amounts during March and June (Gower and King 2008). Seasonal variations of floating populations have also been revealed for perennial algae such as *M. pyrifera* and *D. antarctica* in southern Chile, with highest densities being found during austral spring (Hinojosa et al. 2010). Floating seaweeds can also reach high abundances after storms (Kingsford 1992), while during El Niño events, high water temperature and nutrient limitation can cause the disappearance of benthic seaweeds (Dayton et al. 1999), thus also leading to the elimination of the floating populations.

17.3 Ecophysiology of Floating Seaweeds

Growth and reproduction of floating seaweeds depend (as for their benthic counterparts) on a variety of abiotic and biotic conditions. At the sea surface, factors such as grazing activity, epibiont overgrowth, high water temperature, and solar radiation have been repeatedly inferred to have a negative effect on growth and health status of floating seaweeds. Also, it has been discussed that nutrient limitation might affect physiological functioning and growth of floating algae (Edgar 1987). While nutrient-limited open ocean waters suppress the physiological functioning of floating *S. natans* (Lapointe 1995), algae that accumulate in frontal systems where organic matter is efficiently recycled are thought to have sufficient nutrients to sustain algal growth (Thiel and Gutow 2005b).

Abrupt changes in environmental factors, as experienced especially by floating algae, can impact their photosynthetic apparatus, which is most susceptible to damage under stressful conditions. Stressed algae invest energy to adjust and maintain photosynthetic activity. However, this energy investment comes at the expense of algal growth. Consequently, algae respond with variable growth to changing environmental conditions because growth integrates all physiological costs and gains. Growth can be directly related to the overall health status of the algae and thus to their persistence at the sea surface (Fig. 17.5).

17.3.1 Light

At the sea surface floating algae are often exposed to intense visible (400–700 nm) and ultraviolet radiation (280–400 nm), which may induce photoprotective processes. In large outdoor mesocosm studies conducted along the Chilean coast, *M. pyrifera* reacted to high solar irradiance by lowering their pigment contents and by energy dissipation via heat (Rothäusler et al. 2011a, b). Similar physiological responses were observed for the holopelagic *S. natans*, floating in tropical waters of the Gulf of Mexico (Schofield et al. 1998). These processes are relevant because they permit algae to tolerate the stressful conditions of extreme irradiance and even to continue growing at the sea surface.

While the positively buoyant *M. pyrifera*, of which the attached sporophytes grow throughout the entire water column (up to 40 m in length), can efficiently acclimate to a broad range of irradiance prevailing along the Chilean Pacific coast (Rothäusler et al. 2011a, b), negatively buoyant seaweeds showed strong

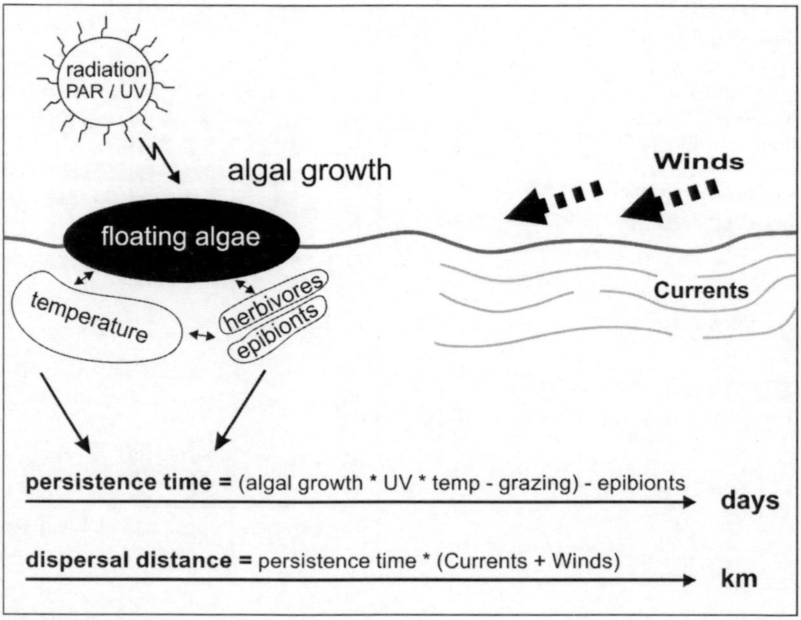

Fig. 17.5 Conceptual figure of the main abiotic and biotic factors affecting the growth of floating algae and thus their persistence at the sea surface

photoinhibition when transplanted close to the sea surface (Karsten et al. 2001). At low latitudes these photoacclimation responses were costly for *M. pyrifera*, which was reflected in their overall diminished growth responses, implying lower persistence due to combined effects of high water temperatures and high light conditions (Rothäusler et al. 2011a). While short-term exposure to UVA and UVB radiation induces the production of UV-absorbing phlorotannins in blades of benthic *M. pyrifera* (Swanson and Druehl 2002), at present it is not known whether floating *M. pyrifera* can also increase the phlorotannin content in response to elevated UV exposure as a protection of their photosynthetic tissues.

17.3.2 Temperature

Temperature is one of the most important factors controlling growth and persistence of floating seaweeds (Hobday 2000a; Rothäusler et al. 2009). For instance, algae from temperate waters lose biomass (Fig. 17.6), and thus floating capacity, when experiencing thermal stress on their floating voyages into warmer oceanic regions. Hobday (2000a) showed for the temperate *M. pyrifera* from southern California that the aging rate of kelp blades was low at surrounding water temperatures <20°C but substantially increased above this threshold. Recent

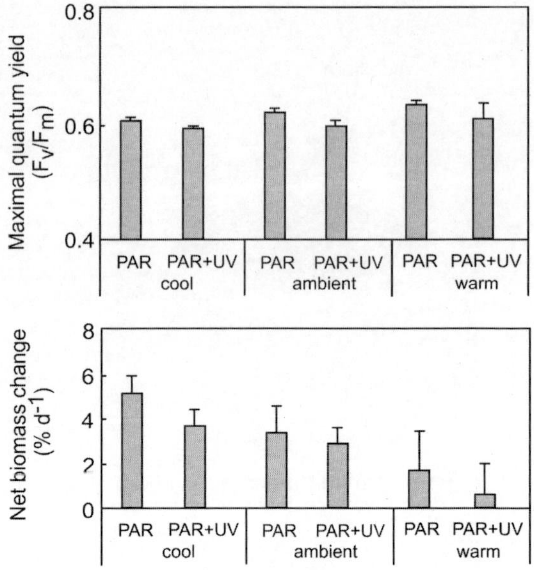

Fig. 17.6 Photosynthetic performance and growth of floating *Macrocystis pyrifera* maintained under different temperature and light conditions in outdoor mesocosm experiments. Figure modified after Rothäusler et al. (2011c)

studies in outdoor mesocosms confirmed that floating individuals of *M. pyrifera* lost biomass at water temperatures >20°C. A rapid disintegration of the algal thalli was observed at 24°C, where all algae sank after 5 days, even in treatments without grazers (Rothäusler et al. 2009). Similarly, floating persistence of temperate *A. nodosum* and *F. vesiculosus* from the southwestern North Sea was reduced at water temperatures >15°C due to high losses of biomass (Vandendriessche et al. 2007a). At equatorial latitudes in the Atlantic, John (1974) documented floating individuals of *A. nodosum* that were small (<0.5 m long) and not particularly robust. Possibly, high water temperatures are responsible for the rapid demise of floating algae in these oceanic regions. Besides temperature, also desiccation may affect the physiological functioning and thus the disintegration of algae, but this effect probably depends on the degree of buoyancy of each algal species (Thiel et al. unpublished).

Suboptimal water temperatures can also affect the reproductive activity of floating algae (Macaya et al. 2005; Rothäusler et al. 2009) (Fig. 17.7). Along the Chilean Pacific coast, in situ sampling and mesocosm experiments with floating *M. pyrifera* confirmed the absence of reproductive tissues (sporophylls) at low latitudes, while at higher latitudes many rafts had fertile sporophylls (Macaya et al. 2005). The authors inferred that unfavorable sea surface conditions, such as water temperatures >20°C, at low latitudes are responsible for the observed pattern. Overall, results imply that low water temperatures (and possibly also winter conditions) favor the persistence and reproductive success of temperate species at the sea surface, which consequently have higher potential for long-distance dispersal.

Contrary to meropelagic rafts, the tropical holopelagic *S. natans* and *S. fluitans* in the Sargasso Sea have growth optima and thus an optimal persistence at

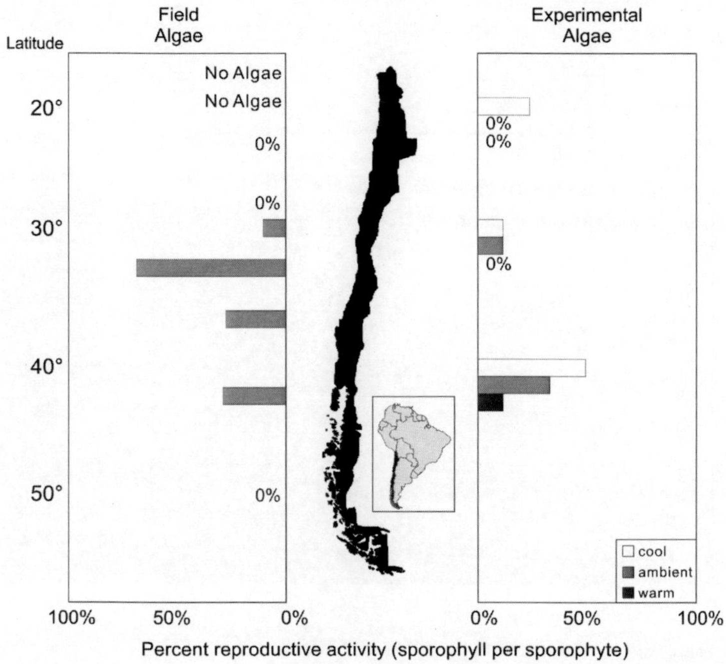

Fig. 17.7 Percent reproductively active sporophytes of freely floating *M. pyrifera* and in outdoor mesocosm experiments simulating the temperature gradient (ambient, cool, warm) along the Chilean Pacific coast. Figures modified after Macaya et al. (2005) and Rothäusler et al. (2009), respectively

temperatures between 18°C and 30°C (Hanisak and Samuel 1987). These algae, commonly known as gulfweed, appear to have evolved efficient protection mechanisms to endure surface levels of high temperature and irradiance (Schofield et al. 1998 for *S. natans*), which is in accordance with their distributional ranges in the tropical and subtropical N Atlantic. No growth was revealed for both species at water temperatures around 12°C (Hanisak and Samuel 1987).

17.3.3 Herbivory

Floating seaweeds are commonly inhabited by dense populations of mobile mesoherbivores such as amphipods and isopods, which can affect algal persistence by removing photosynthetic tissue and thereby weakening the ability to fix carbon (Gutow 2003; Thiel and Gutow 2005b; Vandendriessche et al. 2007a; Rothäusler et al. 2009). These grazers mainly feed on vegetative blades and it was shown for floating individuals of *M. pyrifera* in an outdoor mesocosm study that high densities of a herbivorous amphipod led to a reduction in pigment contents (e.g., by the

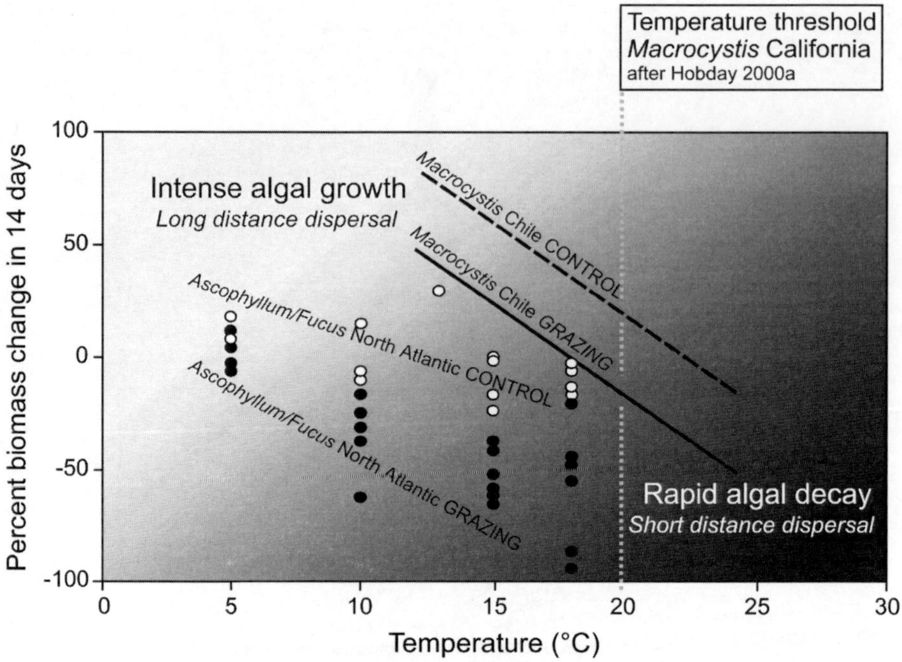

Fig. 17.8 Conceptual model of the influence of the combined effects of temperature and grazing (CONTROL versus GRAZING) on percent biomass change of floating algae during 14 days; data for *Ascophyllum nodosum* and *Fucus vesiculosus* were taken from Vandendriessche et al. (2007a), and data for *Macrocystis pyrifera* from Rothäusler et al. (2009)

allocation of N to the synthesis of other compounds) and algal growth at mid latitudes (Rothäusler et al. 2009, 2011b), while under moderate grazing pressure algae could compensate tissue loss by enhanced growth (Cerda et al. 2009). Grazing by isopods also provoked fast sinking of *A. nodosum* and *F. vesiculosus* but it seemed that the destructive effects of the grazers depended on water temperature and the availability of alternative food resources (Gutow 2003; Vandendriessche et al. 2007a). The negative effect of isopod grazing on the floating potential of the algae was enhanced by high water temperatures, which resulted in a reduced persistence of the brown algae at the sea surface (Vandendriessche et al. 2007a). Contrarily, in cooler waters (<15°C) grazing seemed to play only a minor role for the survival of floating algae because they can compensate grazer-induced tissue losses via growth (Vandendriessche et al. 2007a; Rothäusler et al. 2009) (Fig. 17.8). This might also explain why Ingólfsson (1998) observed no signs of decay in *A. nodosum* after >40 days of floating in cold Icelandic waters.

17.3.4 Epibiosis

Floating seaweeds are commonly colonized by larvae of sessile epibionts such as bryozoans and lepadid barnacles (Thiel and Gutow 2005b), which can influence algal growth by covering photosynthetic tissues. At mid latitudes in the Humboldt Current System, it was observed that bryozoan cover of *M. pyrifera* increased with distance from potential source populations and that physiological parameters (e.g., maximal quantum yield and overall photosynthetic efficiency) declined with increasing bryozoan size (Rothäusler et al. 2011d). Small benthic algae can maintain high photosynthetic activity under bryozoan cover by increasing the pigment content (Muñoz et al. 1998). However, such shade adaptation was not observed in benthic and floating *M. pyrifera* (Hurd et al. 2000; Hepburn et al. 2006; Rothäusler et al. 2011d). While dense epibiont cover can increase the specific density of floating algae and even cause sinking, uncalcified young bryozoan colonies might have a positive effect on algal tissue by shielding them against high solar radiation. Algae may also benefit from encrusting bryozoans through the provision of carbon dioxide and ammonium released directly from epibionts onto the algal tissue (Muñoz et al. 1998; Hurd et al. 1994, 2000), but these advantageous effects probably depend on the degree of colonization. In general, over long floating times, continuous epibiont growth and grazing pressure negatively affect algal persistence at the sea surface. However, depending on the presence of alternative food sources and the colonization progress, not all organisms associated with floating algae (e.g., detritus feeders, scavengers, and predators) may have the same destructive effects on their rafts.

17.4 Rafting Communities

Floating seaweeds carry with them a wide diversity of organisms (Thiel and Gutow 2005b and references therein). The composition of the rafting community and the abundance of individual species can be highly variable in time and space, and successional changes are probably related to travel time of the floating algae, their distance from the shore, competition, and predation.

Not all organisms are equally adapted to the rafting life style because not all of them can efficiently hold onto floating algae. Large invertebrates such as sea urchins, crustaceans, and gastropods immediately drop off when detached algae float to the sea surface. Also small organisms such as the boring peracarid isopods *Limnoria* spp. quickly left their burrows after kelp holdfasts were detached from benthic substrata (Miranda and Thiel 2008). Similarly, Gutow et al. (2009) showed that the densities of some peracarid crustaceans dropped immediately after algae became dislodged. Some species, such as ampithoid amphipods that build nest-like domiciles on algal blades (e.g., *Peramphithoe femorata* Cerda et al. 2010), may persist and even reproduce after detachment, resulting in a population increase during rafting journeys (see also Thiel 2003). Even though these motile organisms can stay and proliferate on algal rafts, other effective rafters are sessile organisms such as hydrozoans and bryozoans, which can form dense colonies on algal blades where they overgrow each other when competing for space (Thiel and Gutow 2005b and references therein). Some of these sessile epibionts colonize the rafting substrata at a later stage via the propagules from the water column.

The most common marine invertebrates on floating algae are peracarid crustaceans, which incubate their embryos in a marsupium, from which fully developed juveniles emerge. Direct development has been considered an important life history trait for rafters, because offspring can immediately recruit onto the algae (Ingólfsson 1998; Gutow 2003; Thiel and Haye 2006; Vandendriessche et al. 2006). Thus, their populations might not only persist but also increase over time. This mode of local recruitment is advantageous for rafters, since local populations on the algae can be maintained, even when the duration of the rafting journey exceeds the lifetime of the species (Thiel and Gutow 2005b). Direct development also favors successful colonization of new habitats.

17.4.1 Successional Changes of Epibiont Communities on Algal Rafts

Rafting communities can experience substantial changes during their journey. These changes are immediately initiated with the detachment of the algae from the benthic substratum (Kingsford and Choat 1985; Miranda and Thiel 2008; Gutow et al. 2009) and continue throughout the entire floating period. For instance, Helmuth et al. (1994) reported that with distance from their benthic source populations, kelp rafts had more sessile epibionts. Vásquez (1993), who tethered kelp holdfasts to a main line, demonstrated that densities of associated peracarid crustaceans increased towards the end of the experiment. In natural algal rafts around Iceland, Ingólfsson (1995) observed that the diversity of associated organisms decreased with increasing distance from source regions.

These temporal changes of rafting communities can be influenced by species interactions. Floating seaweeds attract fish (Kingsford 1992, 1995) and seabirds (Vandendriessche et al. 2007b), which prey upon associated organisms, thereby suppressing the populations of these organisms (including small grazers) on the rafts. In addition, rafters may compete with each other for food and space, which also affects species succession during a long voyage (Tsikhon-Lukanina et al. 2001).

17.5 Rafting Dispersal of Seaweeds and Associated Organisms

The dispersal of algae and their associated organisms strongly depends on the persistence of the rafts at the sea surface and on the capability of the organisms to withstand the rafting conditions. Under favorable conditions, algae can continuously grow and thus provide a long-lasting food source for herbivorous passengers. The floating thalli of, e.g., *M. pyrifera* even continue to reproduce while afloat, thereby acting as successful spore carriers, dispersing over hundreds of kilometers (Macaya et al. 2005; Hernández-Carmona et al. 2006). Thus, floating rafts facilitate population connectivity and expansion of geographic ranges of the floating algae themselves and of associated organisms. Winds and ocean currents mainly determine the floating directions, velocities, and distances that algae can cover.

17.5.1 Floating Velocities and Trajectories

Driven by the major oceanic currents, floating algae are transported across the world's oceans. While ocean currents determine large-scale algal movements, strong winds largely influence algal dispersal on a regional scale. Overall both forces in combination determine algal trajectories and velocities (Thiel and Gutow 2005a and references therein).

Following patches of *M. pyrifera*, Harrold and Lisin (1989) showed that rafts within Monterey Bay (California) occasionally move very slowly and at other times can advance very fast (0.2–2.3 km h^{-1}). Within the southern California Bight a similar trend was observed by Hobday (2000b) for satellite-tracked drifters. Very high velocities of up to 0.8 km h^{-1} have also been reported for algae floating in southern Japan (Segawa et al. 1962; Yoshida 1963), which are probably related to the prevailing current velocities (Thiel and Gutow 2005a).

Trajectories of floating algae and their associated organisms closely follow the directions of the major current systems (Thiel and Gutow 2005a). For instance, Helmuth et al. (1994), who sampled *M. pyrifera* rafts, underlined the importance of the West Wind Drift in the Southern Ocean in transporting algae in eastward direction between South America and sub-Antarctic islands (see also Chap. 14 by Huovinen and Gómez). A similar route in the West Wind Drift had been reported for *D. antarctica* (Smith 2002). Along the Chilean Pacific coast, within the Humboldt Current, floating seaweeds are generally displaced towards the north (Macaya et al. 2005) but southward movements are also observed occasionally. In the N Pacific, Hobday (2000b) underlined that floating *M. pyrifera* are transported in southward direction within the Californian Current. These observations confirm that algal rafts together with their passengers are transported with the main oceanic currents, both across vast areas of open ocean and over extensive distances in alongshore direction.

17.5.2 Connectivity of Populations

There is increasing evidence from molecular studies confirming population connectivity of floating seaweeds and their associated fauna (e.g., Thiel and Haye 2006 and references therein; Muhlin et al. 2008; Fraser et al. 2009; Coleman and Kelaher 2009; Nikula et al. 2010; Coyer et al. 2011; Buchanan and Zuccarello 2012). Bull kelp *D. antarctica* showed high genetic homogeneity among populations from different sub-Antarctic islands (Fraser et al. 2009) (Fig. 17.9). Similarly, results by Macaya and Zuccarello (2010) showed that a single haplotype of *M. pyrifera* dominated the populations from sub-Antarctic islands that are separated by thousands of kilometers (see green dots in Fig. 17.9). Results of both studies show strong evidence that *D. antarctica* and *M. pyrifera* can occasionally be transported across some of the world's largest gaps of open ocean (Fig. 17.9). Efficient dispersal of these floating kelps and of associated organisms is facilitated by the West Wind Drift, the strong surface current moving steadily in eastward direction (Waters 2008). Rafting on seaweeds can thus transport epifaunal organisms to distant shores as confirmed by, e.g., Nikula et al. (2010) for two peracarid crustaceans inhabiting the holdfasts of *D. antarctica*.

Algal rafting also contributes strongly to population connectivity over smaller spatial scales. The fucoid alga *Phyllospora comosa* from eastern Australia showed gene flow between populations that were separated by a 70-km distributional gap (Coleman and Kelaher 2009). Similarly, Muhlin et al. (2008) confirmed connectivity between *Fucus vesiculosus* populations from two adjacent peninsulas in the Gulf of Maine. Results suggest that floating of fertile thalli, which can release gametes when reaching a new site, is responsible for population connectivity. Floating dispersal also seems to affect the genetic population structure of some species of *Sargassum* along the coasts of SE Asia (Cheang et al. 2010, but see also Uwai et al. 2009). Along the European Atlantic coast, there is evidence of high genetic homogeneity between populations of *A. nodosum* (Olsen et al. 2010) and similar

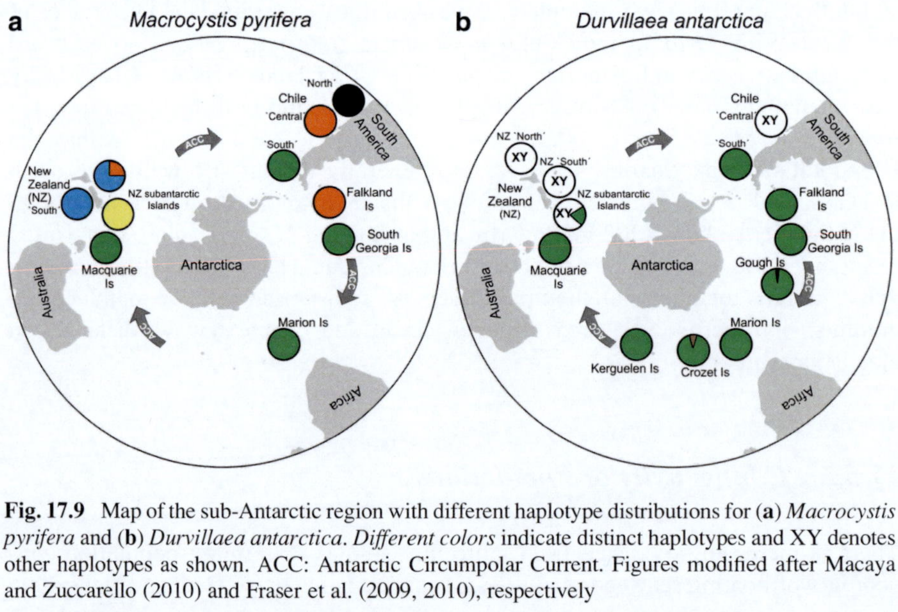

Fig. 17.9 Map of the sub-Antarctic region with different haplotype distributions for (**a**) *Macrocystis pyrifera* and (**b**) *Durvillaea antarctica*. *Different colors* indicate distinct haplotypes and XY denotes other haplotypes as shown. ACC: Antarctic Circumpolar Current. Figures modified after Macaya and Zuccarello (2010) and Fraser et al. (2009, 2010), respectively

results were shown for *Cystoseira amentacea* var. *stricta* from the Mediterranean Sea (Susini et al. 2007). While *A. nodosum* is highly buoyant, individuals of *C. amentacea* var. *stricta* lack vesicles but thallus fragments mix with other floating algae at the sea surface (Susini et al. 2007). Overall, the dispersal of gametes and spores via floating reproductive fragments or adults may also facilitate the reestablishment of algal populations after large-scale disturbances (such as El Niño events) (e.g., Dayton et al. 1999).

17.6 Outlook

Floating seaweeds can be found from the tropics to the Arctic and the sub-Antarctic islands. Since most studies on the physiology and ecology of floating algae were conducted at mid and high latitudes (e.g., Hobday 2000a, b; Macaya et al. 2005; Rothäusler et al. 2009), our knowledge about floating algae from tropical and subtropical regions is scarce. We propose that future research should examine the floating persistence of algae traveling in warm waters and under high radiation conditions, which will help to understand the role of low latitude algae as potential dispersal agents. In the face of global climate change, which predicts to drive species ranges towards the poles (see also Chap. 18 by Bartsch et al.; Parmesan and Yohe 2003), it can be expected that tropical and subtropical algae with positive buoyancy might expand farther towards mid latitudes, while the dispersal potential of temperate floating algae will be suppressed at mid latitudes (see above).

While afloat, algae can be concentrated in frontal systems (convergence zones) and eddies, which can result in the formation of huge biomass accumulations (e.g., Gower et al. 2006). Particulate organic matter such as algal detritus can also aggregate within these zones, contributing to biogeochemical cycles (Thiel and Gutow 2005b). Decomposers may use the organic matter as a substrate and nutrients released can be assimilated by the algae. Efforts should be made to better understand these processes within zones where floating algae and other floating items accumulate (Hinojosa et al. 2011).

So far, molecular studies have helped to reveal population connectivity of seaweeds and associated organisms over a broad range of spatial scales. However, they should also include the temporal scale. Rafting frequencies mainly depend on the availability and the persistence of the floating algae at the sea surface. Depending on latitude and local conditions, supply of floating algae can be spatially and temporally variable. Furthermore, the benthic distribution of floating algae might have changed substantially over evolutionary and ecological timescales. Future studies should therefore not only examine the relevance of rafting for recent population connectivity, but also how episodic rafting in the past has facilitated the establishment of isolated populations and subsequently led to allopatric speciation.

17.7 Conclusions

Floating seaweeds are an important component of the pelagic environment, predominantly in cold temperate and subpolar regions. Although most of the common floating algae naturally grow in benthic environments, these species show specific adaptations to the floating lifestyle which allow for persistence at the sea surface over prolonged time periods. These adaptations indicate selection for traits that facilitate floating persistence, suggesting that floating is a common and integral part of the species' life cycle (see e.g., Rothäusler et al. 2011a). Long-term floating supports genetic exchange between distant populations, natural range expansion, and regional population persistence through re-colonization after local extinctions.

Floating seaweeds significantly contribute to the complexity of the pelagic environment providing habitat for a rich community of associated rafters. Large patches of floating algae can be considered biodiversity hotspots, which catalyze the accumulation of species and the development of interspecific interactions among rafters and with their algal raft. For the associated rafters the patches of floating algae provide a highly dynamic habitat, which is subject to continuous changes depending on fluctuations of the biotic and abiotic environment. Considerable changes of the algal habitat during the pelagic journey are likely to exert strong selective pressure on the associated rafters.

Genetic homogeneity among distant populations of seaweeds and associated rafters, especially in the southern ocean, is the result of continuous exchange of floating individuals across extensive areas of open ocean. Continuous large-scale dispersal via floating algae adds a cosmopolitan dimension to marine biogeography that might not have been achieved through dispersal of larvae and spores alone.

Acknowledgements ER and MT were supported through FONDECYT 1100749 during the preparation of this contribution. ER is grateful for a JSPS Research Fellowship from the Japan Society for the Promotion of Science for providing financial support.

References

Buchanan J, Zuccarello GC (2012) Decoupling of short and long distance dispersal pathways in the endemic New Zealand seaweed *Carpophyllum maschalocarpum* (Phaeophyceae, Fucales). J Phycol. doi:10.1111/j.1529-8817.2012.01167.x

Carr AF (1986) Rips, FADs, and little loggerheads. Bioscience 36:92–100

Cerda O, Karsten U, Rothäusler E, Tala F, Thiel M (2009) Compensatory growth of the kelp *Macrocystis integrifolia* (Phaeophyceae, Laminariales) against grazing of *Peramphithoe femorata* (Amphipoda, Ampithoidae) in northern-central Chile. J Exp Mar Biol Ecol 377:61–67

Cerda O, Hinojosa IA, Thiel M (2010) Nest-building behavior by the amphipod *Peramphithoe femorata* (Krøyer) on the kelp *Macrocystis pyrifera* (Linnaeus) C Agardh from northern-central Chile. Biol Bull 218:248–258

Cheang CC, Chu KH, Ang PO (2010) Phylogeography of the marine macroalga *Sargassum hemiphyllum* (Phaeophyceae, Heterokontophyta) in northwestern Pacific. Mol Ecol 19:2933–2948

Coleman MA, Kelaher BP (2009) Connectivity among fragmented populations of a habitat-forming alga, *Phyllospora comosa* (Phaeophyceae, Fucales) on an urbanised coast. Mar Ecol Prog Ser 381:63–70

Coyer JA, Hoarau G, Costa JF, Hogerdijk B, Serrao EA, Billard E, Valero M, Pearson GA, Olsen JL (2011) Evolution and diversification within the intertidal brown macroalgae *Fucus spiralis*/*F. vesiculosus* species complex in the North Atlantic. Mol Phylogenet Evol 58:283–296

Dayton PK, Currie V, Gerrodette T, Keller BD, Rosenthal RJ, Van Tresca D (1984) Patch dynamics and stability of some California kelp communities. Ecol Monogr 54:253–289

Dayton PK, Tegner MJ, Edwards PB, Riser KL (1999) Temporal and spatial scales of kelp demography: the role of oceanographic climate. Ecol Monogr 69:219–250

Edgar GJ (1987) Dispersal of fauna and floral propagules associated with drifting *Macrocystis pyrifera* plants. Mar Biol 95:599–610

Faller AJ, Woodcock AH (1964) The spacing of windrows of *Sargassum* in the ocean. J Mar Res 22:22–29

Farlow WG (1914) The vegetation of the Sargasso Sea. Proc Am Philos Soc 53:257–262

Franke HD, Gutow L, Janke M (1999) The recent arrival of the oceanic isopod *Idotea metallica* Bosc off Helgoland (German Bight, North Sea): an indication of a warming trend in the North Sea? Helgoländer Meeresun 52:347–357

Fraser CI, Nikula R, Spencer HG, Waters JM (2009) Kelp genes reveal effects of subantarctic sea ice during the Last Glacial Maximum. Proc Natl Acad Sci USA 106:3249–3253

Fraser CI, Thiel M, Spencer H, Waters J (2010) Contemporary habitat discontinuity and historic glacial ice drive genetic divergence in Chilean kelp. BMC Evol Biol 10:203

Fraser CI, Nikula R, Waters JM (2011) Oceanic rafting by a coastal community. Proc R Soc B 278:649–655

Gagnon K, McKindsey CW, Johnson LE (2011) Dispersal potential of invasive algae: the determinants of buoyancy in *Codium fragile* ssp. *fragile*. Mar Biol 158:2449–2458

Gower J, King S (2008) Satellite images show the movement of floating *Sargassum* in the Gulf of Mexico and Atlantic Ocean. Nat Proc (hdl:10101/npre.2008.1894.1)

Gower J, Hu C, Borstard G, King S (2006) Ocean color satellites show extensive lines of floating *Sargassum* in the Gulf of Mexico. IEEE T Geosci Remote 44:3619–3625

Graham MH, Harrold C, Lisin S, Light K, Watanabe JM, Foster MS (1997) Population dynamics of giant kelp *Macrocystis pyrifera* along a wave exposure gradient. Mar Ecol Prog Ser 148:269–279

Graham MH, Vásquez JA, Buschmann AH (2007) Global ecology of the giant kelp *Macrocystis*: from ecotypes to ecosystems. Oceanogr Mar Biol Annu Rev 45:39–88

Gutow L (2003) Local population persistence as a pre-condition for large-scale dispersal of *Idotea metallica* (Crustacea, Isopoda) on drifting habitat patches. Hydrobiologia 503:45–48

Gutow L, Gimenez L, Boos K, Saborowski R (2009) Rapid changes in the epifaunal community after detachment of buoyant benthic macroalgae. J Mar Biol Assoc UK 89:323–328

Hanisak MD, Samuel MA (1987) Growth rates in culture of several species of *Sargassum* from Florida, USA. Hydrobiologia 151–152:399–404

Harries DB, Cook E, Donnan DW, Mair JM, Harrow S, Wilson JR (2007) The establishment of the invasive alga *Sargassum muticum* on the west coast of Scotland: Rapid northwards spread and identification of potential new areas for colonization. Aquat Invas 2:367–377

Harrold C, Lisin S (1989) Radio-tracking rafts of giant kelp: local production and regional transport. J Exp Mar Biol Ecol 130:237–251

Helmuth B, Veit RR, Holberton R (1994) Long-distance dispersal of a subantarctic brooding bivalve (*Gaimardia trapesina*) by kelp-rafting. Mar Biol 120:421–442

Hepburn CD, Hurd CL, Frew RD (2006) Colony structure and seasonal differences in light and nitrogen modify the impact of sessile epifauna on the giant kelp *Macrocystis pyrifera* (L.) C Agardh. Hydrobiologia 560:373–384

Hernández-Carmona G, Hughes B, Graham M (2006) Reproductive longevity of drifting kelp *Macrocystis pyrifera* (Phaeophyceae) in Monterey Bay, USA. J Phycol 42:1199–1207

Hinojosa I, González E, Ugalde P, Valdivia N, Macaya E, Thiel M (2007) Distribution and abundance of floating seaweeds and their associated peracarid fauna in the fjords and channels of the XI. Region, Chile. Cienc Tecnol Mar (Chile) 30:37–50

Hinojosa IA, Pizarro M, Ramos M, Thiel M (2010) Spatial and temporal distribution of floating kelp in the channels and fjords of southern Chile. Est Coast Shelf Sci 87:367–377

Hinojosa IA, Rivadeneira MM, Thiel M (2011) Temporal and spatial distribution of floating objects in coastal waters of central-southern Chile and Patagonian fjords. Cont Shelf Res 31:172–186

Hirata T, Tanaka J, Iwami T, Ohmi T, Dazai A, Aoki M, Ueda H, Tsuchiya Y, Sato S, Yokohama Y (2001) Ecological studies on the community of drifting seaweeds in the south-eastern coastal waters of Izu Peninsula, central Japan. I: Seasonal changes of plants in species composition, appearance, number of species and size. Phycol Res 49:215–229

Hobday AJ (2000a) Age of drifting *Macrocystis pyrifera* (L.) C. Agardh rafts in the Southern California Bight. J Exp Mar Biol Ecol 253:97–114

Hobday AJ (2000b) Abundance and dispersal of drifting kelp *Macrocystis pyrifera* rafts in the Southern California Bight. Mar Ecol Prog Ser 195:101–116

Hooker JD (1847) Algae. In: The botany of the Antarctic voyage of H.M. Discovery Ships Erebus and Terror in the years 1839–1843. I. Flora Antarctica, Part 2. Reeve, London, pp 454–502

Hurd CL, Durante KM, Chia FS, Harrison PJ (1994) Effect of bryozoan colonization on inorganic nitrogen acquisition by the kelps *Agarum fimbriatum* and *Macrocystis integrifolia*. Mar Biol 121:167–173

Hurd CL, Durante KM, Harrison PJ (2000) Influence of bryozoan colonization on the physiology of the kelp *Macrocystis integrifolia* (Laminariales, Phaeophyta) from nitrogen-rich and -poor sites in Barkley Sound, British Columbia, Canada. Phycologia 39:435–440

Ingólfsson A (1995) Floating clumps of seaweed around Iceland: natural microcosms and a means of dispersal for shore fauna. Mar Biol 122:13–21

Ingólfsson A (1998) Dynamics of macrofaunal communities of floating seaweed clumps off western Iceland: a study patches on the surface of the sea. J Exp Mar Biol Ecol 231:119–137

John DM (1974) New records of *Ascophyllum nodosum* (L.) Le Jol. from the warmer parts of the Atlantic Ocean. J Phycol 10:243–244

Johnson DL, Richardson PL (1977) On the wind-induced sinking of *Sargassum*. J Exp Mar Biol Ecol 28:255–267

Karsten U, Bischof K, Wiencke C (2001) Photosynthetic performance of Arctic macroalgae after transplantation from deep to shallow waters. Oecologia 127:11–20

Kingsford MJ (1992) Drift algae and small fish in coastal waters of northeastern New Zealand. Mar Ecol Prog Ser 80:41–55

Kingsford MJ (1993) Biotic and abiotic structure in the pelagic environment: importance to small fishes. Bull Mar Sci 53:393–415

Kingsford MJ (1995) Drift algae—a contribution to near-shore habitat complexity in the pelagic environment and an attractant for fish. Mar Ecol Prog Ser 116:297–301

Kingsford MJ, Choat JH (1985) The fauna associated with drift algae captured with a plankton-mesh purse seine net. Limnol Oceanogr 30:618–630

Kingsford MJ, Choat JH (1986) Influence of surface slicks on the distribution and onshore movement of small fish. Mar Biol 91:161–171

Komatsu T, Daisuke Matsunaga D, Mikami A, Sagawa T, Boisnier E, Tatsukawa K, Aoki M, Ajisaka T, Uwai S, Tanaka K, Ishida K, Tanoue H, Sugimoto T (2008) Abundance of drifting seaweeds in eastern East China Sea. J Appl Phycol 20:801–809

Lapointe BE (1995) A comparison of nutrient-limited productivity in *Sargassum natans* from neritic vs. oceanic waters of the western North Atlantic Ocean. Limnol Oceanogr 40:625–633

Lüning K (1990) Seaweeds: their environment. Biogeography and ecophysiology. Wiley, New York

Macaya EC, Zuccarello GC (2010) Genetic structure of the giant kelp *Macrocystis pyrifera* along the southeastern Pacific. Mar Ecol Prog Ser 420:103–112

Macaya EC, Boltaña S, Hinojosa IA, Macchiavello JE, Valdivia NA, Vásquez N, Buschmann AH, Vásquez J, Vega JMA, Thiel M (2005) Presence of sporophylls in floating kelp rafts of *Macrocystis* spp. (Phaeophyceae) along the Chilean Pacific coast. J Phycol 41:915–922

Marmorino GO, Miller WD, Smith GB, Bowles JH (2011) Airborne imagery of a disintegrating *Sargassum* drift line. Deep Sea Res 58:316–321

Miranda L, Thiel M (2008) Active and passive migration in boring isopods *Limnoria* spp. (Crustacea, Peracarida) from kelp holdfasts. J Sea Res 60:176–183

Mitchell CT, Hunter JR (1970) Fishes associated with drifting kelp, *Macrocystis pyrifera*, off the coast of southern California and northern Baja California. Calif Fish Game 56:288–297

Muhlin JF, Engel CR, Stessel R, Weatherbee RA, Brawley SH (2008) The influence of coastal topography, circulation patterns, and rafting in structuring populations of an intertidal alga. Mol Ecol 17:1198–1210

Muñoz J, Cancino JM, Molina MX (1998) Effect of encrusting bryozoans on the physiology of their algal substratum. J Mar Biol Assoc UK 71:877–882

Nikula R, Fraser CI, Spencer HG, Waters JM (2010) Circumpolar dispersal by rafting in two subantarctic kelpdwelling crustaceans. Mar Ecol Prog Ser 405:221–230

Ohno M (1984) Observations on the floating seaweeds of near-shore waters of southern Japan. Hydrobiologia 116(117):408–412

Olsen JL, Zechman FW, Hoarau G, Coyer JA, Stam WT, Valero M, Aberg P (2010) The phylogeographic architecture of the fucoid seaweed *Ascophyllum nodosum*: an intertidal 'marine tree' and survivor of more than one glacial-interglacial cycle. J Biogeogr 37:842–856

Parker T, Tunnycliffe V (1994) Dispersal strategies of the biota of an oceanic seamount: implications for ecology and biogeography. Biol Bull 187:336–345

Parmesan C, Yohe G (2003) A globally coherent fingerprint of climate change impacts across natural systems. Nature 421:37–42

Parr AE (1939) Quantitative observations on the pelagic *Sargassum* vegetation of the western North Atlantic. Bull Bingham Oceanogr Coll 6:1–94

Rothäusler E, Gómez I, Hinojosa IA, Karsten U, Tala F, Thiel M (2009) Effect of temperature and grazing on growth and reproduction of floating *Macrocystis* spp. (Phaephyceae) along a latitudinal gradient. J Phycol 45:547–559

Rothäusler E, Gómez I, Karsten U, Tala F, Thiel M (2011a) UV-radiation versus grazing pressure: long-term floating of kelp rafts (*Macrocystis pyrifera*) is facilitated by efficient photoacclimation but undermined by grazing losses. Mar Biol 158:127–141

Rothäusler E, Gómez I, Hinojosa IA, Karsten U, Tala F, Thiel M (2011b) Physiological performance of floating giant kelp *Macrocystis pyrifera* (Phaeophyceae): Latitudinal variability in the effects of temperature and grazing. J Phycol 47:269–281

Rothäusler E, Gómez I, Karsten U, Tala F, Thiel M (2011c) Physiological acclimation of floating *Macrocystis pyrifera* to temperature and irradiance ensures long-term persistence at the sea surface at mid-latitudes. J Exp Mar Biol Ecol 405:33–41

Rothäusler E, Gómez I, Hinojosa IA, Karsten U, Miranda L, Tala F, Thiel M (2011d) Kelp rafts in the Humboldt Current: Interplay of abiotic and biotic factors limit their floating persistence and dispersal potential. Limnol Oceanogr 56:1751–1763

Rueness J (1989) *Sargassum muticum* and other introduced Japanese macroalgae. Mar Poll Bull 20:173–176

Schoener A, Rowe GT (1970) Pelagic *Sargassum* and its presence among the deep sea benthos. Deep Sea Res 17:923–925

Schofield O, Evens TJ, Millie DF (1998) Photosystem II quantum yields and xanthophyll-cycle pigments of the macroalga *Sargassum natans* (Phaeophyceae): responses under natural sunlight. J Phycol 34:104–112

Segawa S, Sawada T, Higaki M, Yoshida T, Kamura S (1961) The floating seaweeds of the sea west of Kyushu. Rec Oceanogr Work Jpn 5:179–186

Segawa S, Sawada T, Higaki M, Yoshida T, Ohshiro H, Hayashida F (1962) Some comments on the movement of the floating seaweeds. Rec Oceanogr Work Jpn 6:153–159

Senta T (1962) Studies on floating seaweeds in early summer around Oki Islands and larvae and juveniles of fish accompanying them. Physiol Ecol 10:68–78

Smith SDA (2002) Kelp rafts in the Southern Ocean. Global Ecol Biogeogr 2:67–69

Stewart HL (2008) The role of spatial and ontogenetic morphological variation in the expansion of the geographic range of the tropical brown alga, *Turbinaria ornata*. Integr Comp Biol 48:713–719

Susini ML, Thibaut T, Meinesz A, Forcioli D (2007) A preliminary study of genetic diversity in *Cystoseira amentacea* (C. Agardh) Bory var. *stricta* Montagne (Fucales, Phaeophyceae) using random amplified polymorphic DNA. Phycologia 46:605–611

Swanson AK, Druehl LD (2002) Induction, exudation and the UV protective role of kelp phlorotannins. Aquat Bot 73:241–253

Tapia FJ, Pineda J, Ocampo-Torres FJ, Fuchs HL, Parnell PE, Montero P, Ramos S (2004) High-frequency observations of wind-forced onshore transport at a coastal site in Baja California. Cont Shelf Res 24:1573–1585

Thiel M (2003) Rafting of benthic macrofauna: important factors determining the temporal succession of the assemblage on detached macroalgae. Hydrobiologia 503:49–57

Thiel M, Gutow L (2005a) The ecology of rafting in the marine environment I. The floating substrata. Oceanogr Mar Biol Annu Rev 42:181–263

Thiel M, Gutow L (2005b) The ecology of rafting in the marine environment II. The rafting organisms and community. Oceanogr Mar Biol Annu Rev 43:279–418

Thiel M, Haye PA (2006) The ecology of rafting in the marine environment III. Biogeographical and evolutionary consequences. Oceanogr Mar Biol Annu Rev 44:323–429

Thiel M, Hinojosa IA, Joschko T, Gutow L (2011) Spatio-temporal distribution of floating objects in the German Bight (North Sea). J Sea Res 65:368–379

Tsikhon-Lukanina EA, Reznichenko OG, Nikolaeva GG (2001) Ecology of invertebrates on the oceanic floating substrata in the Northwest Pacific Ocean. Oceanology 41:525–530

Tully O, Ó Céidigh P (1986) The ecology of *Idotea* species (Isopoda) and *Gammarus locusta* (Amphipoda) on surface driftweed in Galway Bay (west of Ireland). J Mar Biol Assoc UK 66:931–942

Uwai S, Kogame K, Yoshida G, Kawai H, Ajisaka T (2009) Geographical genetic structure and phylogeography of the *Sargassum horneri/filicinum* complex in Japan, based on the mitochondrial *cox3* haplotype. Mar Biol 156:901–911

Vandendriessche S, De Keersmaecker G, Vincx M, Degraer S (2006) Food and habitat choice in floating seaweed clumps: the obligate opportunistic nature of the associated macrofauna. Mar Biol 149:1499–1507

Vandendriessche S, Vincx M, Degraer S (2007a) Floating seaweed and the influences of temperature, grazing and clump size on raft longevity—A microcosm study. J Exp Mar Biol Ecol 343:64–73

Vandendriessche S, Stienen EWM, Vincx M, Degraer S (2007b) Seabirds foraging at floating seaweeds in the Northeast Atlantic. Ardea 95:289–298

Vásquez JA (1993) Effects on the animal community of dislodgement of holdfasts of *Macrocystis pyrifera*. Pac Sci 47:180–184

Waters JM (2008) Driven by the West Wind Drift? A synthesis of southern temperate marine biogeography, with new directions for dispersalism. J Biogeogr 35:417–427

Yoshida T (1963) Studies on the distribution and drift of the floating seaweed. Bull Tohoku Reg Fish Res Lab 23:141–186

Part IV
Effects of Global and Local Changes on the Performance of Seaweeds

Chapter 18
Global Seaweed Biogeography Under a Changing Climate: The Prospected Effects of Temperature

Inka Bartsch, Christian Wiencke, and Thomas Laepple

18.1 Introduction

Climate changes and concurrent changes in temperatures, atmospheric CO_2 concentration, and other greenhouse gases have often occurred during the earth's history (Zachos et al. 2008) and it is well known that biogeographical distribution patterns of species are directly controlled by climate (Pearson and Dawson 2003 and references therein). Considering only the last 740,000 years, we know of about eight glaciations associated with strong variations in temperature, precipitation, and thermohaline circulation worldwide (Augustin et al. 2004; Knutti et al. 2004). In the past century, overall global warming was around 0.8°C (Hansen et al. 2006). Changes are particularly strong in polar and cold-temperate regions of the northern hemisphere (Levitus et al. 2000; Hansen et al. 2006). At the western Antarctic Peninsula a recent rapid regional warming with a sea surface temperature (SST) increase of 3.7°C century^{-1} has been recorded (Vaughan et al. 2003). Massive retreat of icefronts and glaciers and an increase of melt water production has been documented since 1956 (Braun and Gossmann 2002). For the Arctic, a similar warming trend has been observed over recent decades. Monthly temperature anomalies can be as large as 3–4°C (Turner et al. 2007). In cold- and warm-temperate localities, the reported mean annual temperature increase was as high

I. Bartsch (✉) • C. Wiencke
Department Seaweed Biology, Section Functional Ecology, Alfred Wegener Institute for Polar and Marine Research, Am Handelshafen 12, 27570 Bremerhaven, Germany
e-mail: inka.bartsch@awi.de; christian.wiencke@awi.de

T. Laepple
Section Palaeoclimate Dynamics, Alfred Wegener Institute for Polar and Marine Research, Bussestraße 24, 27570 Bremerhaven, Germany
e-mail: thomas.laepple@awi.de

C. Wiencke and K. Bischof (eds.), *Seaweed Biology*, Ecological Studies 219,
DOI 10.1007/978-3-642-28451-9_18, © Springer-Verlag Berlin Heidelberg 2012

as >2°C depending on location and duration of time series within the last decades (e.g., California: Sagarin et al. 1999, Portugal: Lima et al. 2007, North Sea: Wiltshire et al. 2008, Baltic: Andersen et al. 2011, Australia: Poloczanska et al. 2011). Current warming is larger over landmasses than over the open ocean and is larger in higher latitudes than in the tropical region. In the Equatorial Tropical Atlantic and the Western Equatorial Pacific (WEP), the SST increase between 2001 and 2005 relative to 1870–1900 was 0.5–1°C, while the Eastern Equatorial Pacific (EEP) did not show any temperature increase (Hansen et al. 2006). Whether the increased temperature difference between the near-equatorial WEP and EEP may be responsible for dampening or enhancing the frequency of El Niño-Southern Oscillation (ENSO) activity is still in debate (e.g., Hansen et al. 2006; Collins et al. 2010). The two most pronounced ENSO events in the last 100 years took place in 1983 and 1998 accompanied by an unprecedented warming in the EEP (Hansen et al. 2006).

Within the last two decades, there have been many attempts to develop predictive modeling approaches to project present-day biogeographical distribution patterns into the future. Principally, two different directions have been followed: (1) the "bioclimate envelope" models correlate species distributions with climate variables including the knowledge about the physiological responses of species to climate change (Pearson and Dawson 2003 and references therein). A special form of these models is niche modeling which correlates the macroecological preferences of a species at sample locations (e.g., demands for temperature, substrate, light, etc.) with their distributional records. Niche models predict potentially suitable habitats and the fundamental biogeographical niche (Guisan and Thuiller 2005; Graham et al. 2007; Verbruggen et al. 2009). (2) On the other hand, marine ecologists stressed the importance to also consider the variety of biotic interactions between species which are mostly responsible for shaping the realized niche. A recent review extensively summarizes possible consequences of both the abiotic and biotic environment and their interactions in coastal marine environments with respect to climate change (Harley et al. 2006). Although biotic interactions locally shape communities, they do not explain global biogeographical patterns (Santelices et al. 2009). Recently, Müller et al. (2009, 2011) presented a new bioclimate envelope model comparing observed winter and summer SSTs of 1980–1999 to model SSTs of 2080–2099 based on a moderate greenhouse gas emission scenario in order to predict future distributional range shifts of selected polar and cold-temperate seaweed species of both hemispheres. Here, we use the same approach providing a macroecological view on seaweed distribution based on a worldwide model of present and future oceanic isotherms, carving out the resulting changes in spatial extent of major biogeographical coastal regions (after Briggs 1995) due to temperature change and discussing expected changes of seaweed floras on a worldwide scale for the end of the century.

18.2 Coastal Marine Biogeographical Regions and Their Relation to Oceanic Isotherms and Temperature Responses of Seaweeds

Especially temperature changes had and will have strong effects in the geographical distribution of seaweeds as temperature is the main abiotic factor directly controlling geographic boundaries of seaweed species (van den Hoek 1982a; Lüning 1990; Chap. 3 by Eggert) and also has recently been identified in shaping global marine biodiversity pattern (Tittensor et al. 2010). Principally, the presence of seaweeds in these regions is determined by the physiological tolerance of their life cycle stages to temperature. There are two fundamentally different types of temperature boundaries in seaweeds, lethal boundaries and growth or reproduction boundaries (Breeman 1988; van den Hoek and Breeman 1989). Lethal boundaries are determined by the capacity of the species to survive during the unfavorable season, e.g., a cold-water alga in the summer season. Growth and/or reproduction boundaries are determined by the ability of the species to grow and reproduce during the favorable season, e.g., a cold-water alga in the winter/spring season. Moreover, seaweed distribution is affected by temperature–daylength interactions (Dring 1984; Molenaar 1996) affecting the coordinated timing between life cycle stages.

Seven broad biogeographical regions have been recognized in coastal marine zoogeography and phytogeography (Briggs 1974, 1995; Lüning 1990): the tropical region of the Atlantic and Indo-Pacific, the cold- and warm-temperate regions of both hemispheres, and the two Polar regions (for details see Chap. 3 by Eggert). Although Spalding et al. (2007) recently proposed a new and more sophisticated global system of marine ecoregions than Briggs (1995), the major biogeographical boundaries along the continents are similar in both approaches. We therefore chose the Briggs (1995) system as it is better suited for demonstrating the broad global pattern of modeled future change. The boundaries between biogeographical regions after Briggs (1995) are determined by fundamental changes in the composition of coastal biota and have been defined by their degree of endemism (>10%; Spalding et al. 2007). The transition between regions may be characterized by SSTs as temperature is a major factor in shaping marine phytogeographical regions (e.g., van den Hoek 1982a, b; Lüning 1990). Lüning (1990) described the boundaries of phytogeographical regions by intermediate mean summer and winter isotherms to show the general broad pattern. The specific mean summer and winter isotherms prevailing at the eastern and western coasts of the Oceans and in the northern and southern hemisphere at the phytogeographic boundaries deviate, however, considerably from these averages and are presented in Table 18.1 based on isotherm data from 1980 to 1999 provided by Müller et al. (2009).

The current tropical coastal biogeographical regions are broad and characterized by summer isotherms between 24 and 29°C and winter isotherms between 18 and 23°C (Table 18.1, Figs. 18.1a and 18.2a). The adjacent warm-temperate regions are much smaller and generally slightly compressed on the western sides of the Pacific

Table 18.1 Biogeographic regions after Briggs (1995) with the respective northern and southern mean oceanic sea surface isotherms limiting the respective region derived from global sea-surface isotherm maps 1980–1999 (Müller et al. 2009)

	Biogeographic region		Atlantic				Pacific				Indic			
			Western coastlines		Eastern coastlines		Western coastlines[a]		Eastern coastlines		Western coastlines[c]		Eastern coastlines	
			Winter	Summer	Winter	Summer	Winter	Summer	Winter	Summer	Winter	Summer	Winter	Summer
Northern hemisphere	Polar-Arctic	SB	<0°C	<10°C	<0/2°C[b]	<9°C	<0°C	<9°C	<0°C	<9°C	–	–	–	–
	Cold-temperate	NB	0°C	10°C	0/2°C[b]	9°C	0°C	9°C	0°C	9°C	–	–	–	–
		SB	10°C	27°C	10°C	17°C	12/14°C[c]	26°C	14°C	18°C	–	–	–	–
	Warm-temperate	NB	10°C	27°C	10°C	17°C	12/14°C[c]	26°C	14°C	18°C	–	–	–	–
		SB	23°C	29°C	21°C	27°C	22°C	29°C	20°C	26°C	–	–	–	–
	Tropical	NB	>23°C	>29°C	>21°C	>27°C	>22°C	>29°C	>20°C	>26°C	–	–	–	–
Southern hemisphere		SB	>22°C	>27°C	>18°C	>24°C	>21°C	>27°C	>21°C	>26°C	20°C	24°C	21°C	25°C
	Warm-temperate	NB	22°C	27°C	18°C	24°C	21°C	27°C	21°C	26°C	20°C	24°C	21°C	25°C
		SB	10°C	19°C	–	–	14°C	20°C	11°C	16°C	–	–	14°C	18°C
	Cold-temperate	NB	10°C	19°C	–	–	14°C	20°C	11°C	16°C	–	–	14°C	18°C
		SB	–	–	–	–	–	–	–	–	–	–	–	–
	Polar-Antarctic	NB	Winter: <1°C / Summer: <4°C											

SB southern boundary, NB northern boundary, – no exact data available as there are no further land masses south of the continents in the southern hemisphere or in the case of the Indian Ocean no ocean-land boundaries further north

[a] The border of the W-Pacific to the Indian Ocean was considered to be located in central Asia (northern hemisphere) and in southeastern Australia (southern hemisphere)

[b] Temperature boundaries for Barents Sea/North Atlantic

[c] Temperature boundaries for Korea/Japan

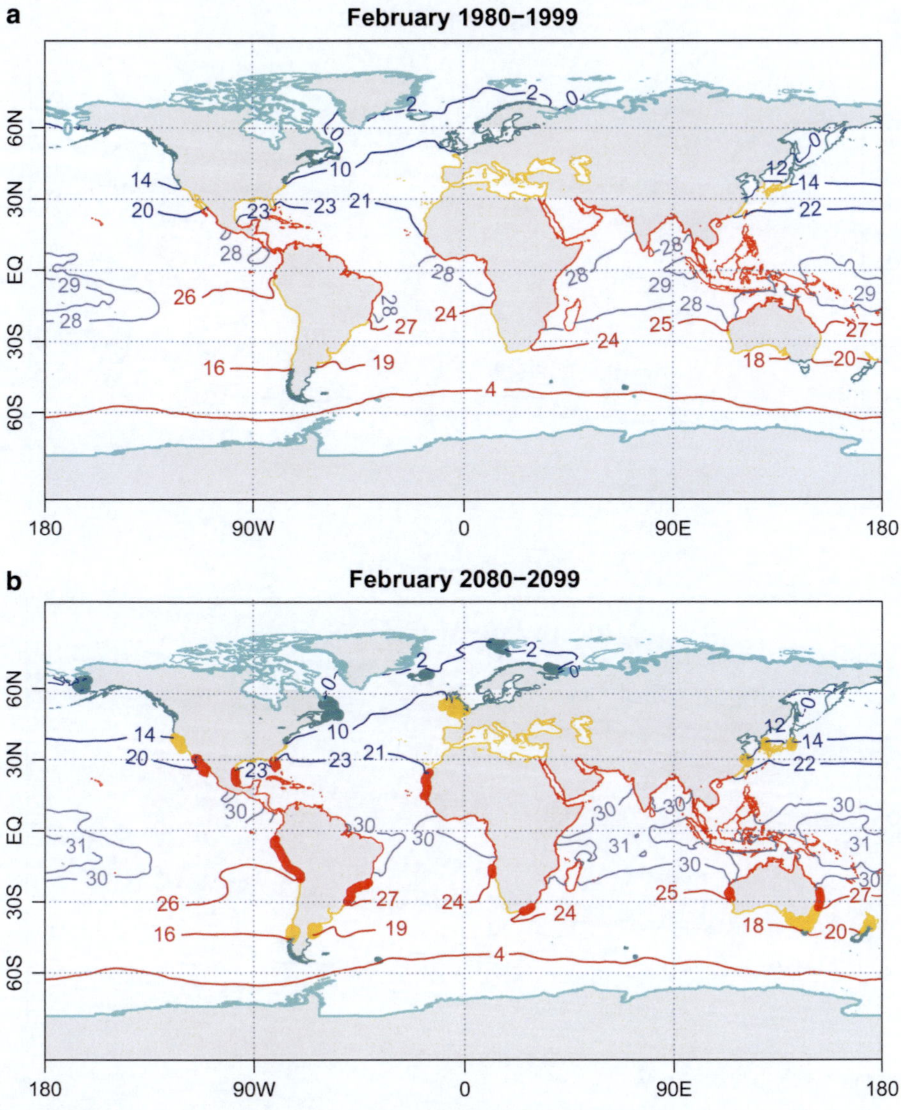

Fig. 18.1 Changes in the extent of biogeographical regions after Briggs (1995) due to global warming at the end of the twentieth century inferred from current (1980–1999) and future (2080–2099) mean February sea surface isotherms. The mean February sea-surface isotherms delimiting current biogeographical regions are depicted in figure **a**. The future extension of the biogeographical regions (**b**) is based on the modeled change of the mean February sea surface isotherms at the present boundaries of the respective regions. Projected changes are indicated by bold coastal lines. Color code of biogeographical regions: *turquoise:* Polar regions, *blue-green*: cold-temperate regions, *yellow*: warm-temperate regions, *red*: tropical regions. Color code of isotherms: *blue*: winter isotherms, *red*: summer isotherms, *gray*: isotherms characterizing the central tropics

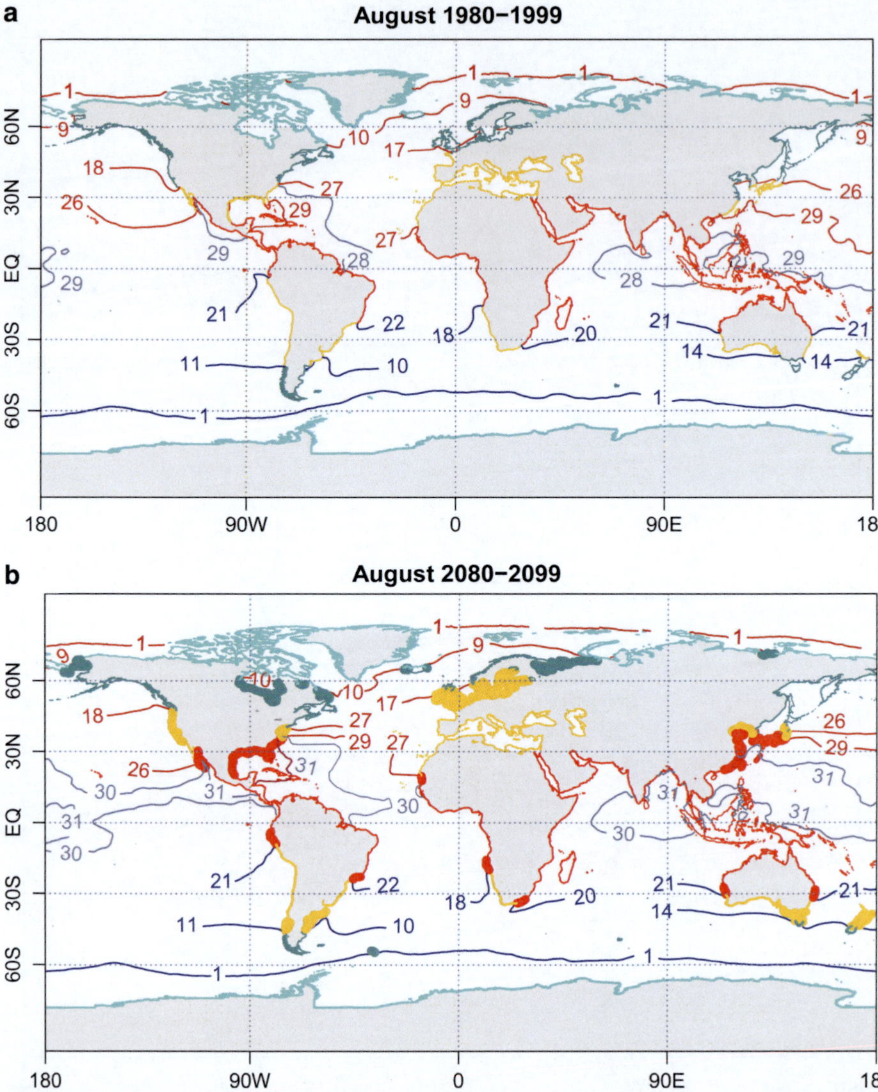

Fig. 18.2 Changes in the extent of biogeographical regions after Briggs (1995) due to global warming at the end of the twentieth century inferred from current (1980–1999) and future (2080–2099) mean August sea surface isotherms. The mean August sea-surface isotherms delimiting current biogeographical regions are depicted in figure **a**. The future extension of the biogeographical regions (**b**) is based on the modeled change of the mean August sea surface isotherms at the present boundaries of the respective regions. Projected changes are indicated by bold coastal lines. Color code of biogeographical regions: *turquoise:* Polar regions, *blue-green*: cold-temperate regions, *yellow*: warm-temperate regions, *red*: tropical regions. Color code of isotherms: *blue*: winter isotherms, *red*: summer isotherms, *gray*: isotherms characterizing the central tropics

and Atlantic compared to the eastern sides of the oceans. The boundary between the warm- and cold-temperate regions is characterized by 16–29°C summer SSTs and by 10–20°C winter SSTs. These extreme differences are attributed to the pronounced compression of isotherms along the western part of the Atlantic and Pacific in contrast to the eastern part of the oceans (see Figs 12.2, 12.3 in Müller et al. 2011). When comparing both hemispheres pronounced differences become apparent as well. The southern boundary of the Arctic region, for example, is described by the 9–10°C summer isotherm, whereas the northern boundary of the Antarctic region is the 4°C summer isotherm (Table 18.1, Figs. 18.1a and 18.2b). The situation in the southern hemisphere might be partially attributed to missing continuous land masses in the circumpolar Southern Ocean. Thus, the limit between the polar and cold-temperate coastal regions in the southern hemisphere is not clearly defined. The largest differences between northern and southern hemispherical biogeographic regions become apparent at the boundary of the warm- and cold-temperate regions at the western coasts of both the Atlantic and the Pacific. In the N-Atlantic, this boundary is situated at the 27°C summer isotherm. In contrast, in the S-Atlantic it is located at the 19°C summer isotherm (Table 18.1, Figs. 18.1a and 18.2a). At the west coasts of the Pacific the differences are smaller (26°C August isotherm in the north to 20°C February isotherm in the south).

18.3 Responses of Seaweeds to Temperature Changes

18.3.1 General Responses

Seaweeds can principally respond to environmental changes in four ways which all may result in distributional and diversity changes: on short timescales, they can acclimate. On medium to long timescales they either adapt to new conditions or are able to slowly migrate keeping pace with the changing environmental pressure. Species unable to acclimate, adapt, or disperse may be captured in isolated refugia or become extinct. Acclimation to temperature stress in seaweeds has been mostly studied on the level of photosynthesis or growth (e.g., Davison et al. 1991; Kübler and Davison 1995; Eggert et al. 2006; see Chap. 3 by Eggert). Locally, the acclimation potential of thermal traits can shape the vertical or seasonal distribution pattern of species (Davison and Pearson 1996; Ateweberhan et al. 2005; Zardi et al. 2011) and on a broader scale eurythermal species have a broader distribution range than stenothermal species (see also Chap. 3 by Eggert). Hitherto, it is unknown whether species with a broad acclimation potential also had or will have a faster genetic adaptation potential—a fact which would help to explain historical or future biogeographical processes.

Adaptive processes on the physiological and molecular level are explained in detail in Chap. 3 by Eggert. On the organism level, the adaptation processes to temperature changes have been identified by comparing the temperature requirements of tropical and cold-water seaweeds from both hemispheres in relation to the climatic history of the various regions (Wiencke et al. 1994). Since Mesozoic times [251–65.5 million years before present (My)] there was a continuous warm water girdle around the earth so that cold-water seaweed species probably only evolved after the glaciation events in the Tertiary (65 My) (Lüning 1990). An upper survival temperature of 33–35°C may still be found in representatives of temperate and tropical seaweed species, indicating that this thermal trait is rather deeply entrenched and not subject to fast adaptation (Lüning 1990). While strictly tropical seaweed species are stenothermal and may survive 30–37°C and grow best between 25 and 30°C, they do not have the ability to live below temperatures of 7–12°C (Pakker et al. 1995; Bischoff-Bäsmann et al. 1997). The first steps in the adaptation of seaweeds to lower temperatures are an increase in cold tolerance and an increase of growth and reproduction rates at lower temperatures leading to eurythermal species. This temperature trait is apparent in tropical to warm-temperate species which acquired a better lower temperature tolerance with survival temperatures between 6 and ≤1°C without losing the upper temperature tolerance (Yarish et al. 1984; Pakker et al. 1995; Bischoff-Bäsmann et al. 1997). Later, the ability to survive temperatures ≥20°C and to grow and reproduce at ≥15–20°C was lost. This type of temperature response is typical for endemic Arctic and Arctic to cold-temperate distributed seaweeds exposed to low temperatures for about 3 My (Briggs 1995). The last steps in the adaptation to low temperatures are the loss of the ability to grow and reproduce at ≥5–10°C and to survive temperatures ≥6–13°C (Wiencke et al. 1994; Bischoff-Bäsmann and Wiencke 1996). This type of temperature response is exemplified in endemic Antarctic species exposed to cold water for at least 14 My (Crame 1993; Briggs 1995). So, the climatic history during species evolution determines the temperature requirements of seaweeds in all biogeographical regions. All these cases give moreover an insight into the time periods needed for adaptation to changing temperatures. Physical barriers and differential environmental gradients along coastlines also may produce ecotypic adaptation. There has been a wealth of studies tackling this question. It became evident that upper tolerance limits seemingly are quite stable within several seaweed species (Lüning 1990 and references therein). True temperature ecotypes have only been found in a few species yet (e.g., *Ectocarpus siliculosus*: Bolton 1983, *Saccharina latissima*: Gerard and Du Bois 1988).

18.3.2 Changes of Seaweed Distribution and Oceanic Temperature in the Geological Past

Interesting examples for migration in the geological past are present-day amphiequatorial species such as *Acrosiphonia arcta* (Chlorophyta) or the species pair *Desmarestia confervoides/D. viridis* (Phaeophyceae), which are absent in the tropics. The fast evolving internal transcribed spacer region of the rDNA of northern and southern hemisphere populations of both entities only exhibit minor variation, suggesting that a migration across the equator took place at the maximum of the Würm/Wisconsin glaciation 18,000 years ago, the so-called last glacial maximum (LGM; van Oppen et al. 1993). The hypothesized migrationist transit of the tropics is supported by the temperature tolerance of both species which is high enough (25°C) to survive a passage through the tropics at the LGM (Peters and Breeman 1992; van Oppen et al. 1994; Bischoff and Wiencke 1995). Similarly it is assumed that a common ancestor of the warm-temperate NE-Atlantic kelp species *Laminaria ochroleuca* and its S-Atlantic sister species *L. abyssalis* and *L. pallida* was able to survive the passage through the tropics during glacial lowering of seawater temperatures by its gametophytic microstages being able to survive at least 25°C (tom Dieck 1992; tom Dieck and de Oliveira 1993).

During the last glaciations, the Arctic ice cap extended south to ~45–55°N and there was a considerable southward shift of the Gulf Stream (CLIMAP Project members 1981; Bradley 1985) inducing major latitudinal dislocations of marine biota in the N-Atlantic between glacial and interglacial periods (van den Hoek and Breeman 1989; Breeman 1990). Some seaweed species were faced with an extreme reduction in their distribution area. The conditions were particularly severe in the NW-Atlantic, where distributions became excessively reduced. A pertinent example is the cold-temperate green alga *Cladophora sericea*. By comparing temperature demands of the species with modeled glacial sea surface isotherms (van den Hoek and Breeman 1989; Cambridge et al. 1990), it became evident that the distribution area of the species was probably very strongly reduced during the LGM in the NW-Atlantic. In the NE-Atlantic, the distribution was shifted from the coasts of Scandinavia, Great Britain, and France to the Iberian Peninsula, Northwest Africa, and even the Mediterranean became hospitable. These proposed migrational shifts during the LGM within the N-Atlantic have been corroborated by recent phylogeographic studies of diverse seaweed species, indicating the English Channel Region as a primary refugium (e.g., Provan et al. 2005) from which species redispersed to their current distributional range.

An example for a probable extinction during the LGM is the current restriction of the kelp species *Laminaria hyperborea* to the NE-Atlantic. The much greater compression of the distribution areas on the NW-Atlantic during the LGM—if compared to the NE-Atlantic coasts—suggests that *L. hyperborea* was not able to survive the inhospitable conditions in the NW-Atlantic during the LGM and became locally extinct (van den Hoek and Breeman 1989). Other kelp species possibly became isolated in cold-water pockets after the LGM. Examples are the deep-water species *L. rodriguezii* (Huvé 1955; Žuljević et al. 2011) or the isolated population of the southern European kelp *L. ochroleuca* in southern Italy (Giaccone 1972). Nowadays both species are only found below the thermocline in selected Mediterranean habitats.

18.3.3 Changes of Oceanic Temperature and Seaweed Distribution in the Future

Following Müller et al. 2009, we analyzed the observed modern (1980–1999; Rayner et al. 2003) and projected end of century (2080–2099) SST distribution for February and August as representatives of summer and winter extreme temperatures. The SST changes are based on the multimodel ensemble mean of coupled climate model simulations prepared for phase 3 of the Coupled Model Intercomparison Project (CMIP3; http://www-pcmdi.llnl.gov) using an emission scenario which assumes a moderate increase of greenhouse gases (SRESA1B) (for details, see Müller et al. 2009). Both the global mean February and August SSTs are projected to warm by 1.9°C until the end of the century (Fig. 18.3). The maximum

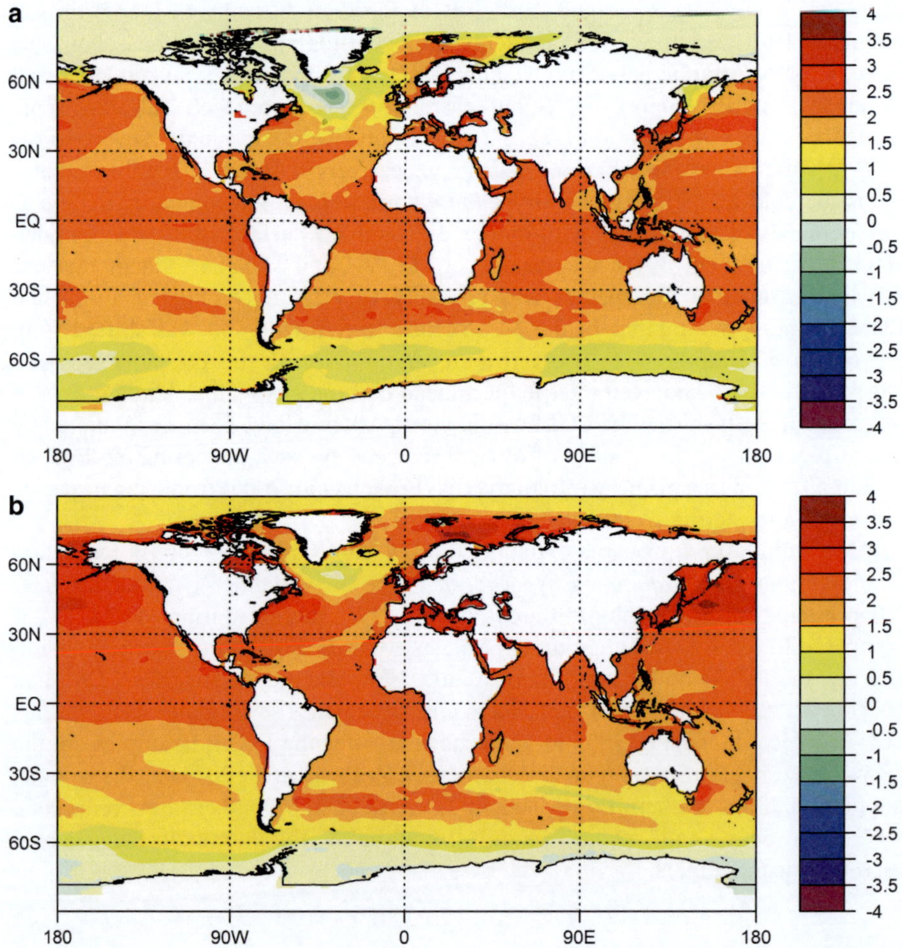

Fig. 18.3 Simulated future changes in sea surface temperature. The ensemble mean anomaly for 2080–2099 relative to 1980–1999 is shown for February (**a**) and August (**b**)

warming of around 4°C is predicted for high latitudes of the northern hemisphere in summer. The larger land area in the northern hemisphere leads to a stronger response to the radiative forcing than in the southern hemisphere. Additionally, the decrease in summer sea-ice cover amplifies the summer warming. In contrast, minimum warming is predicted in the Southern ocean due to strong ocean heat uptake in this region. In addition to this global scale pattern, some regional scale patterns in the temperature changes become visible. As most climate models predict a reduction in the thermohaline circulation as a response to the increasing greenhouse gas concentration, the heat transport from the tropics to the Polar regions will be affected in the northern hemisphere (Schmittner et al. 2005). This will result in a local cooling in the Labrador Sea in winter and a reduced warming in summer according to the model simulations (Fig. 18.3).

The simulated future changes in sea surface temperatures shown in Fig. 18.3 will result in a general poleward movement of biogeographical regions (see Sect. 18.2) until the end of the twentieth century. The extent of the present-day biogeographical regions and the resulting future changes are shown in Figs. 18.1 and 18.2. This projection is based on the location of modeled mean winter and summer sea surface isotherms for the end of the twentieth century (Müller et al. 2009, Appendix Figs. 1 and 2) characteristic for the current boundaries of the respective regions (see Table 18.1, Figs. 18.1a and 18.2a). Despite the proven recent regional warming in the western Antarctic Peninsula (Vaughan et al. 2003) there will be almost no changes in the northern delimitation of the Antarctic region until the end of this century based on our model. The Arctic region, in contrast, will shrink considerably at its southern border, especially due to warmer winter temperatures while new ice-free coastal Arctic habitats will become available in the north following the retreating pack-ice border (1°C August isotherm) in summer (Fig. 18.2b; Müller et al. 2009). The current cold-temperate regions of both hemispheres will become compressed as the warm-temperate regions are shifting polewards. The cold-temperate regions will gain much new area at the expense of the Polar region only in the N-Atlantic while in the S-Atlantic only the small sub-Antarctic islands will become cold-temperate. The poleward expansion of the warm-temperate regions at the expense of the cold-temperate regions is especially evident in western and eastern S-America, in southeastern Australia and New Zealand, western N-America, and along the European coastline. The warm-temperate regions themselves will lose habitats at the expense of the projected widening of the tropical regions (Figs. 18.1 and 18.2). Recently, it was shown that this process has already started and five different types of climatological measurements revealed a widening of the tropical region of several degrees latitude since 1979 (Seidel et al. 2008). According to our results, this future widening will be especially pronounced along the west and east coast of S-America, along the northern and eastern coastlines of S-Africa and Australia, in SE-Asia, W-Africa, and the Gulf of Mexico (Figs. 18.1 and 18.2).

It becomes apparent that the location of the boundary isotherms characterizing present-day biogeographical regions (Table 18.1) will not move polewards at the same pace according to the model simulations (Figs. 18.1 and 18.2). Thereby, the annual mean temperature minima and maxima and the resulting temperature gradient will change along vast coastlines compared to present day. As a consequence, some biogeographical regions will not be extended or reduced as a whole

(Figs. 18.1 and 18.2). Two examples shall demonstrate resulting consequences. Along the coast of China and Japan tropical winter temperatures will not change in the future so that the 22°C February isotherm will stay at approx. 23°N (Fig. 18.1b). In contrast, the corresponding 29°C summer isotherm characterizing the warm-temperate/tropical boundary nowadays will move from approx. 23°N up to the Yellow Sea and Strait of Korea to approx. 36°N (Fig. 18.2b). Hence, this whole area will become a transitional region where biota will experience a reorganization according to the individual growth, reproduction, and lethal limits of the species present (van den Hoek 1982a, b; see Chap. 3 by Eggert). These temperature changes will probably also have consequences for the intensive seaweed aquaculture industry along the Chinese and Japanese coastlines (see Chap. 22 by Buchholz et al.). Another prominent example of a future transitional region is the west coast of S-America. The current austral 21°C winter and 26°C summer SST delimit the biogeographical boundary between the warm-temperate and tropical regions approx. at a latitude of 5°S (northern Peru). According to our simulations, both isotherms will move southwards to approx. 10°S in future austral winter or to approx. 20°S (northern Chile) in future austral summer. The difference between the current and future boundary region thereby spans either 5° or 15° latitude (Figs. 18.1 and 18.2). Thus, a coastline of more than 1,000 km length will become a biogeographical transition region and composition of biota will probably change. It should be noted, however, that due to the influence of the cold Humboldt current and coastal upwelling regions in general, biogeographical regions may not always correspond to the system developed by Briggs (1995) and the model assumptions. Recent evidence, for example, suggests that the northern boundary of the cold-temperate region along the western coastline of S-America set to approx. 40°S by Briggs (1995) actually extends further north to approx. 30°S (Camus 2001; see also Chap. 14 by Huovinen and Gómez).

In addition, the model data identify areas where winter and summer isotherms and thereby possibly the whole biogeographical region will shift homogenously. Examples are the warm-temperate region in eastern and western S-America or the tropical region along eastern and western southern Africa (Figs. 18.1b and 18.2b). Although it is not yet predictable whether minimal and maximal or mean annual SST or altogether will be most responsible for shaping future phytogeographical regions, our data obviously show that new correlations of biogeographical regions with SSTs will establish.

18.3.4 Specific Effects of Oceanic Warming on Seaweed Distribution and Ecology

18.3.4.1 Polar Regions

Since the late 1970s the glacial ice sheets have retreated by up to 2% per decade (Serreze et al. 2007) and the Arctic will probably be ice free by the end of this century (Johannessen et al. 2004). In the Antarctic, 87% of the glaciers of the West Antarctic Peninsula are retreating (Cook et al. 2005), the ice season has shortened by about 90 days, and perennial ice does not occur any more at this location (Martinson et al. 2008; Stammerjohn et al. 2008). In the Arctic, oceanic warming leads to a retreat of the pack-ice border coinciding with the 1°C summer isotherm. This will provide new habitats for algal colonization in the Arctic and Antarctic along rocky coastlines (Fig. 18.1; Müller et al. 2009, 2011). As ice-related pressures on shallow water biota of the Arctic and Antarctic will be reduced, perennial macroalgae, which are so far restricted mainly to the sublittoral, will be able to colonize the eulittoral, resulting in an increase in biomass and diversity (Weslawski et al. 2010, 2011). On the other hand, prolonged inflow of glacial melt water will reduce salinity and increase turbidity of the water due to a higher sediment impact (Campana et al. 2011). The concomitant reduction of the euphotic region will change production rates (Pivovarov et al. 2003; Deregibus et al. personal communication; Spurkland and Iken 2011) and probably will cause an upward shift of the depth limit of seaweeds. Biomass and seaweed cover already increased between 1988 and 2008 in the rocky littoral of Sorkappland (Svalbard; Weslawski et al. 2010) in the Arctic accompanied by an increase in air temperature and SST and a marked decrease in the duration and extent of sea-ice cover. However, no "new" species were detected but are expected in future (Müller et al. 2009). The described upward shift of seaweeds might though be counteracted by high levels of ultraviolet-B radiation (UVBR) due to stratospheric ozone depletion (Weatherhead and Andersen 2006; Zacher et al. 2011) which still prevails in the Arctic and Antarctica. UVBR is one of the most important factors controlling the upper depth distribution of seaweeds. Effects have been demonstrated from the cellular to the ecosystem level, affecting community structure and diversity in the Arctic and Antarctic (Bischof et al. 2006; Zacher et al. 2007; Campana et al. 2011; Karsten et al. 2011; Fricke et al. 2011; see Chap. 20 by Bischof and Steinhoff). UVBR, turbidity, water temperature, and sea-ice conditions are interdependent factors but multifactorial interactive effects on polar biota have scarcely been investigated (Müller et al. 2008; Fredersdorf et al. 2009). Bifactorial experiments on Arctic kelp species indicated that negative effects of UVBR can be mitigated by the interaction with increased temperature. For example, germination of zoospores of the kelp *Laminaria digitata* was inhibited almost completely by UVBR at 2°C, but not at 7°C (Müller et al. 2008).

Compared to changes in the Arctic, the distributional changes of seaweeds in the Antarctic will probably be minor (Müller et al. 2009, 2011) as the model data predict an SST increase of only 1°C throughout the year in the Antarctic region. Moreover, only few cold-temperate species will be able to colonize present-day Antarctic coasts. One example might be the brown alga *Chordaria linearis*, which has been found already on two locations in West Antarctica (Müller et al. 2009, 2011). The estimation of minor changes for seaweed richness along the coastal West Antarctica Peninsula under climate change conditions contrasts to demonstrated changes in the respective pelagic ecosystem in response to rapid climate changes, which include a shift in phytoplankton biomass, in zooplankton community structure, and expected effects on higher trophic levels (Schofield et al. 2010).

18.3.4.2 Temperate Regions

Within the last decades, an increase in SSTs of >2°C has been documented in many cold- and warm-temperate regions (see Sect. 18.1). At many temperate European coastlines, migrational shifts of benthic and pelagial species have already taken place (e.g., benthos overview: Mieszkowska et al. 2006; Hawkins et al. 2008; plankton: Beaugrand and Reid 2003; fish: Rijnsdorp et al. 2009). Climate driven biomass changes or loss of kelp vegetation has recently been reported from many local sites worldwide (e.g., Japan: Kirihara et al. 2006, Tasmania: Johnson et al. 2011, Norway: Andersen et al. 2011, Spain: Díez et al. 2012). In contrast, no evidence for broadscale latitudinal shifts of kelps since 1850 was found in the transition region between the boreal and sub-arctic region in the NW-Atlantic (Merzouk and Johnson 2011). A recent investigation into the decline of the sugar kelp *Saccharina latissima* along southern Norwegian shorelines (Andersen et al. 2011) provides a good example for the complex interactions in the field. After transplantation of *S. latissima* from healthy to impacted sites, normal growth and maturation took place in winter and spring, but heavy fouling of epiphytes occurred over summer followed by mortality. Although duration of periods with summer temperatures >20°C increased in recent years, a temperature which is sublethal for *S. latissima* (Bolton and Lüning 1982), mortality could not unequivocally be correlated to high summer temperatures alone. Instead, Andersen et al. (2011) assumed a cascade of reduced growth at sublethal temperatures, followed by heavy epiphytism at locations with low wave activity leading to shading, thereby causing a negative carbon balance and brittleness of thalli and finally mortality—all factors together possibly preventing recruitment and recovery of the species at the impacted sites. Successful recruitment is crucial for the continuous recovery of boundary populations which becomes impacted if the environmental pressure surpasses critical limits. Within a few years of unfavorable abiotic conditions, the reproductive capacity was dramatically reduced in southern European marginal populations of the intertidal brown alga *Fucus serratus* (Viejo et al. 2011). Similarly, along the SST gradient in western Australia density of kelp recruits was inversely related to increasing mean ocean temperature, suggesting an effect of temperatures on either reproduction or recruits (Wernberg et al. 2010).

There are a few long-term case studies from temperate coastal regions investigating possible ecological consequences of environmental warming for rocky shore communities. Three examples are given here. A 10-year thermal outfall of a power station which induced a long-lasting SST increase by up to 3.5°C resulted in a complete change of the intertidal rocky shore community structure at Diablo Potter Cove in California. The initial dense cover of foliose algae was replaced by bare rocks, algal crusts, or turf algae and there was a major replacement of species and decrease of algal richness (Schiel et al. 2004). A mean SST increase of 0.79°C in the intertidal of Monterey Bay California between 1931 and 1996 led to a significant increase in southern species and a decrease of northern species (Sagarin et al. 1999). In the northern Baltic, another monitoring study of long-lasting warming in the sea (up to 10°C) diminished ice cover and thereby increased light availability in winter. The situation also caused major changes in the quantitative composition of species over the seasonal cycle. While growth of cyanobacteria was promoted and red and brown algae decreased in abundance or disappeared over summer, the latter had a prolonged growth season in autumn and winter due to "better" winter temperatures. Generally, a species-specific response was evident (Snoeijs and Prentice 1989). In future, we expect similar transitional changes in rocky shore communities along all warm- and cold-temperate shorelines possibly subjected to change according to our model results (Figs. 18.1 and 18.2).

18.3.4.3 Tropical Regions

Many coastal hard-bottom tropical and subtropical regions are characterized by coral reefs which also inherit a high seaweed species richness (Diaz-Pulido et al. 2007). The abundance of macroalgae in reefs has been thought to be generally low and controlled by grazing pressure of herbivorous fish (e.g., Wanders 1977; Carpenter 1986; Hay 1997, see Chap. 16 by Mejia et al.). Only in recent years, it was realized that tropical reefs are also algal reefs and a high coverage of macroalgae among corals and natural variability of seaweed abundance on coral reefs is not necessarily indicative of environmental degradation (Vroom et al. 2006, 2010; Vroom and Timmers 2009). Coral–algal interactions are manifold and it is known that algae may inhibit or kill corals (e.g., Titlyanov et al. 2007; Rasher et al. 2011) and vice versa dead corals may negatively influence macroalgal growth (Liu et al. 2009).

As tropical corals and seaweeds are currently living near to their lethal limit, a slight temperature increase of 1–2°C above the mean summer temperatures as predicted for the end of the twentieth century (Fig. 18.3) may already lead to catastrophic events. Coral reefs worldwide have faced severe damage by periodic heat waves especially through extreme ENSO activities since the 1980s inducing so-called coral bleaching events which involve the loss of the symbiotic zooxanthellae after thermal stress (Jokiel and Coles 1990). Baker et al. (2008) describe in their extensive review all facets of this phenomenon. There is a correlation between coral bleaching with maximum monthly SSTs (Manzello et al. 2007). Temperature thresholds for coral bleaching are not uniform but site specific and range from 27.5 to 32°C (Baker et al. 2008). Thereby, they are generally above current mean tropical summer SST of 27–29°C (Fig. 18.2a; Müller et al. 2009, Appendix Figs. 1 and 2), but this will change in future when this region will experience an unprecedented warming (Solomon et al. 2007) with annual mean SSTs of 30–31°C over wide areas (Fig. 18.2b). A possible acclimation of corals to increased temperatures has been observed in the Great Barrier Reef as threshold temperatures increased over time (Berkelmans 2009) and up to now no coral species became extinct. Currently, reefs are still able to recuperate with highly variable rates, indicating a differential recovery capacity. But full recovery of reefs after thermal stress probably needs decades to centuries (Baker et al. 2008). As growth of macroalgae is much faster and their temperature tolerance is several degree Celsius higher than those of corals (Pakker et al. 1995; Bischoff-Bäsmann et al. 1997), a shift from coral to seaweed dominated coastal ecosystems has been proposed for future tropical areas (Hoegh-Guldberg et al. 2007) and has already been observed during recent decades (Rasher et al. 2011 and references therein).

Current warm-temperate coastlines which will become tropical in future (Figs. 18.1 and 18.2) will not be able to compensate for the prospected loss of reefs as rates needed to establish coral reefs are slow (Baker et al. 2008). Similarly as corals, subtidal reef macroalgae potentially face local extinction if temperature exceeds algal tolerance limits which are firmly set to 30–33°C (Pakker et al. 1995). In contrast, eulittoral tropical macroalgae with their higher lethal temperature limits of up 32–37°C (Bischoff-Bäsmann et al. 1997) will be better able to withstand future temperature increase in the central tropics. First local extinctions after warming events have been recorded: In the Galapagos Archipelago six tropical macroalgal species disappeared after the ENSO warming event in 1982/1983. Here a transition of the macroalgal and coral habitats to heavily grazed reefs dominated by crustose coralline "urchin barrens" was observed (Edgar et al. 2010). Additionally to ocean warming, ocean acidification may decrease coralline abundance in future which will enhance the cascade effect of decreasing coral recruitment, opening space for turf algal species, and further inhibition of coral recruitment, coral fecundity, and coral growth (Hoegh-Guldberg et al. 2007). Unfortunately, the functional ecology and thermal traits of coralline red algae which are important contributors to reef structure and facilitate settlement of corals are virtually unknown (Nelson 2009).

As the tropical region will considerably extend polewards at the expense of the current warm-temperate region (Figs. 18.1 and 18.2), substantial new transitional areas will develop along rocky shore coastlines characterized, for example, by assemblages dominated by tropical to subtropical members of the brown algal order Dictyotales and not corals such as described for tropical to warm-temperate transitional areas of the Canary Islands (Sangil et al. 2011). Locally, other factors such as wave exposure, local currents, and physical barriers may be more important than temperature for biogeographic distribution as has been observed along a tropical to warm-temperate coastal transition region in E-Australia where species distribution is still stable despite recent rapid warming (Poloczanska et al. 2011).

18.4 Assumptions for Global Seaweed Biodiversity

Global patterns of marine diversity differ from terrestrial habitats where species numbers decrease moving away from the equator in both hemispheres (Pianka 1966; Willig et al. 2003). Marine coastal taxa also show clear latitudinal trends, but overall seaweed genus diversity peaks in mid latitudes and not the tropical region (Kerswell 2006; Santelices et al. 2009; Tittensor et al. 2010). SST was identified as one significant predictor of overall coastal species richness (Tittensor et al. 2010) and high interannual temperature variability was proposed as explanation for the depauperate brown algal flora along the southwest African and north-central Chilean coastline (Bolton 1996). Thus, the predicted changes in SST for the end of the twentieth century will probably have a substantial effect not only on distribution of seaweeds and marine communities in general, but also on global marine diversity. Kerswell (2006) analyzed the latitudinal pattern of global seaweed genus diversity and seaweed endemism in detail. As in other coastal taxonomic groups (Tittensor et al. 2010) there is a band of high diversity in the northwestern Pacific surrounding the Japanese archipelago and in the southwestern Pacific along the southern Australian coastline (Kerswell 2006). In the Atlantic, there is a major hotspot of diversity along the European coastline (Kerswell 2006). All these regions inherit a high seaweed genus diversity and have been identified by our model data to be impacted by future warming. Thus, major changes in seaweed species richness, and the functionality of assemblages through species extinctions, species invasions, and changes in trophic relationships (Sala and Knowlton 2006) are expected especially in these regions. Within the last decade, a wealth of cryptic seaweed species has been described with the help of modern molecular biological tools (e.g., van der Strate et al. 2002; Brodie et al. 2007; Verbruggen et al. 2009). Thus, species richness of marine algae and genetic diversity of their populations probably is strongly underestimated (Zuccarello et al. 2011) and thereby possibly also their adaptive potential to change. But it is not yet clear whether genetic differentiation always coincides with ecological differentiation (Tronholm et al. 2010).

18.5 Synopsis

Clearly, the prospected worldwide changes in SSTs will exert a differential pressure on seaweed species and assemblages along biogeographical regions. As the ice-free Arctic coastlines and all other regions will expand polewards, the tropical biogeographical region will widen considerably. In contrast, there will be almost no change in the northern limit of the Antarctic region. Most effects on biota will be expected in biogeographical transition regions which have been identified here, for example along the warm- to cold-temperate European coastline or along the warm-temperate to tropical coasts of SW and SE-America, Japan, or China. In these areas, summer and winter SST will move polewards with a differential magnitude so that annual temperature gradients will become more pronounced in some areas. Hence, severe biotic changes are expected as the assemblages characterizing the biogeographic regions will not be able to shift as a whole. Rather, we predict differential species-specific shifts depending on the respective temperature-dependent life cycle characteristics of species which additionally will be shaped by other abiotic and biotic factors. Comparison of coastal areas comprising present-day highest seaweed genus diversity with future biogeographical transition areas revealed a correlation in some cases, for example along western European, Japanese, or southern Australian shorelines. In the future, it will be particularly interesting to investigate the changing structure and function in these transitional biogeographical regions to gain a better understanding of fast acclimation and adaptation rates on an ecosystem level.

Acknowledgements We acknowledge the modeling group WCRP's Working Group on Coupled Modelling (WGCM) for making available the WCRP CMIP3 multi-model dataset. Support of this dataset is provided by the Office of Science, U.S. Department of Energy. The authors thank R. Müller and C. Buchholz for helpful advice on an earlier draft of this chapter.

References

Andersen GS, Stehen H, Christie H, Fredriksen S, Moy FE (2011) Seasonal patterns of sporophyte growth, fertility, fouling, and mortality of *Saccharina latissima* in Skagerrak, Norway: Implications for forest recovery. J Mar Biol 2011:Article ID 690375

Ateweberhan M, Bruggemann JH, Breeman AM (2005) Seasonal patterns of biomass, growth and reproduction in *Dictyota cervicornis* and *Stoechospermum polypodioides* (Dictyotales, Phaeophyta) on a shallow reef flat in the southern Red Sea (Eritrea). Bot Mar 48:8–17

Augustin L, Barbante C, Barnes PRF, Barnola JM, Bigler M, Castellano E, Cattani O, Chappellaz J, Dahl-Jensen D, Delmonte B, Dreyfus G, Durand G, Falourd S, Fischer H, Flückiger J, Hansson ME, Huybrechts P, Jugie G, Johnsen SJ, Jouzel J, Kaufmann P, Kipfstuhl J, Lambert F, Lipenkov VY, Littot GC, Longinelli A, Lorrain R, Maggi V, Masson-Delmotte V, Miller H, Mulvaney R, Oerlemans J, Oerter H, Orombelli G, Parrenin F, Peel DA, Petit J-R, Raynaud D, Ritz C, Ruth U, Schwander J, Siegenthaler U, Souchez R, Stauffer B, Steffensen JP, Stenni B, Stocker TF, Tabacco IE, Udisti R, van de Wal RSW, van den Broeke M, Weiss J, Wilhelms F, Winther JG, Wolff EW, Zucchelli M (2004) Eight glacial cycles from an Antarctic ice core. Nature 429:623–628

Baker AC, Glynn PW, Riegl B (2008) Climate change and coral reef bleaching: An ecological assessment of long-term impacts, recovery trends and future outlook. Estuar Coastal Shelf Sci 80:435–471

Beaugrand G, Reid PC (2003) Long-term changes in phytoplankton, zooplankton and salmon related to climate. Global Change Biol 9:801–817

Berkelmans R (2009) Bleaching and mortality thresholds: How much is too much? In: van Oppen MJH, Lough JM (eds) Coral bleaching: patterns and processes, causes and consequences, Ecol Studies, vol 205. Springer, New York, pp 103–120

Bischof K, Gómez I, Molis M, Hanelt D, Karsten U, Lüder U, Roleda MY, Zacher K, Wiencke C (2006) Ultraviolet radiation shapes seaweed communities. Rev Environ Sci Biotechnol 5:141–166

Bischoff B, Wiencke C (1995) Temperature ecotypes and biogeography of Acrosiphoniales (Chlorophyta) with Arctic-Antarctic disjunct and Arctic/cold-temperate distributions. Eur J Phycol 30:19–27

Bischoff-Bäsmann B, Wiencke C (1996) Temperature requirements for growth and survival of Antarctic Rhodophyta. J Phycol 32:525–535

Bischoff-Bäsmann B, Bartsch I, Xia B, Wiencke C (1997) Temperature responses of macroalgae from the tropical island Hainan (P.R. China). Phycol Res 45:91–104

Bolton JJ (1983) Ecoclinal variation in *Ectocarpus siliculosus* (Phaeophyceae) with respect to temperature growth optima and survival limits. Mar Biol 73:131–138

Bolton JJ (1996) Patterns of species diversity and endemism in comparable temperate brown algal floras. Hydrobiologia 327:173–178

Bolton JJ, Lüning K (1982) Optimal growth and maximal survival temperatures of Atlantic *Laminaria* species (Phaeophyta) in culture. Mar Biol 66:89–94

Bradley RS (1985) Quaternary paleoclimatology. Allan and Unwin, London

Braun M, Gossmann H (2002) Glacial changes in the areas of Admiralty Bay and Potter Cove, King George Island, Maritime Antarctica. In: Beyer L, Bölter M (eds) Geoecology of Antarctic ice-free coastal landscapes, Ecol Studies, vol 154. Springer, Heidelberg, pp 75–90

Breeman AM (1988) Relative importance of temperature and other factors in determining geographic boundaries of seaweeds: experimental and phenological evidence. Helgoländer Meeresunters 42:199–241

Breeman AM (1990) Expected effects of changing seawater temperatures on the geographic distribution of seaweed species. In: Beukema JJ et al (eds) Expected effects of climatic change on marine coastal ecosystems. Kluwer Academic, Netherlands, pp 69–76

Briggs JC (1974) Marine zoogeography. McGraw-Hill, New York

Briggs JC (1995) Global biogeography. Elsevier, Amsterdam

Brodie J, Bartsch I, Neefus C, Orfanidis S, Bray T, Mathieson AC (2007) New insights into the cryptic diversity of the North Atlantic–Mediterranean '*Porphyra leucosticta*' complex: *P. olivii* sp. nov. and *P. rosengurttii* (Bangiales, Rhodophyta). Eur J Phycol 42:3–28

Cambridge ML, Breeman AM, van den Hoek C (1990) Temperature responses limiting the geographical distribution of two temperate species of *Cladophora* (Cladophorales; Chlorophyta) in the North Atlantic. Phycologia 29:74–85

Campana G, Zacher K, Fricke A, Molis M, Wulff A, Wiencke C (2011) Drivers of colonization and succession in polar benthic macro- and microalgal communities. In: Wiencke C (ed) Biology of polar benthic algae. de Gruyter, Berlin, pp 299–320

Camus PA (2001) Biogeografía marina de Chile continental. Rev Chi Hist Nat 74:587–617

Carpenter RC (1986) Partitioning herbivory and its effects on coral reefs algal communities. Ecol Monogr 56:345–363

CLIMAP Project Members (1981) Seasonal reconstructions of the earth's surface at the last glacial maximum. The Geological Society of America, Map and Chart Service MC-36, Washington DC

Collins M, An S-I, Cai W, Ganachaud A, Guilyardi JF-F, Jochum M, Lengaigne M, Power S, Timmermann A, Vecchi G, Wittenberg A (2010) The impact of global warming on the tropical Pacific ocean and El Niño. Nat Geosci 3:391–397

Cook AJ, Fox AJ, Vaughan DG, Ferrigno JG (2005) Retreating glacier fronts on the Antarctic Peninsula over the past half-century. Science 308:541–544

Crame JA (1993) Latitudinal range fluctuations in the marine realm through geological time. Trends Ecol Evol 10:1096–1111

Davison IR, Pearson GA (1996) Stress tolerance in intertidal seaweeds. J Phycol 32:197–211

Davison IR, Greene RM, Podolak EJ (1991) Temperature acclimation of respiration and photosynthesis in the brown alga *Laminaria saccharina*. Mar Biol 110:449–454

Diaz-Pulido G, McCook LJ, Larkum AWD, Lotze HK, Raven JA, Schaffelke B, Smith JE, Steneck RS S (2007) Vulnerability of macroalgae of the Great Barrier Reef to climate change. In: Johnson JE, Marshall PA (eds) Climate change and the Great Barrier Reef. Great Barrier Reef Marine Park Authority and Australian Greenhouse Office, Townsville, Queensland, pp 153–192

Díez I, Muguerza N, Santolaria A, Ganzedo U, Gorostiaga JM (2012) Seaweed assemblage changes in the Eastern Cantabrian Sea and their potential relationship to climate change. Estuar, Coastal Shelf Sci 99:108–120

Dring MJ (1984) Photoperiodism and phycology. In: Round F, Chapman DJ (eds) Progress in phycological research, vol 3. Biopress, Bristol, pp 159–192

Edgar GJ, Banks SA, Brandt M, Bustamante RH, Chiriboga A, Earle SA, Garske LE, Glynn PW, Grove JS, Henderson S, Hickmam CP, Miller KA, Rivera F, Wellington GM (2010) El Niño, grazers and fisheries interact to greatly elevate extinction risk for Galapagos marine species. Global Change Biol 10:2876–2890

Eggert A, Visser RJW, van Hasselt PR, Breeman AM (2006) Differences in acclimation potential of photosynthesis in seven isolates of the tropical to warm temperate macrophyte *Valonia utricularis* (Chlorophyta). Phycologia 45:546–556

Fredersdorf J, Müller R, Becker S, Wiencke W, Bischof K (2009) Interactive effects of radiation, temperature and salinity on different life history stages of the Arctic kelp *Alaria esculenta* (Phaeophyceae). Oecologia 160:483–492

Fricke A, Molis M, Wiencke C, Valdivia N, Chapman AS (2011) Effects of UV radiation on the structure of Arctic macrobenthic communities. Polar Biol 34:995–1009

Gerard VA, Du Bois KR (1988) Temperature ecotypes near the southern boundary of the kelp *Laminaria saccharina*. Mar Biol 97:575–580

Giaccone G (1972) Struttura, ecologia e corologia dei popolamenti a Laminarie dell stretto di Messina e del mare di Alboran. Mem Biol Mar Ocean NS 2:37–49

Graham MH, Kinlan BP, Druehl LD, Garske LE, Banks S (2007) Deep-water kelp refugia as potential hotspots of tropical marine diversity and productivity. Proc Nat Acad Sci USA 104:16576–16580

Guisan A, Thuiller W (2005) Predicting species distribution: offering more than simple habitat models. Ecol Lett 8:993–1009

Hansen J, Sato M, Ruedy R, Lo K, Lea DW (2006) Global temperature change. Proc Natl Acad Sci 103:14288–14293

Harley CDG, Hughes AR, Hultgren KM, Miner BG, Sorte CJB, Thornber CS, Rodriguez LF, Tomanek L, Williams SL (2006) The impacts of climate change in coastal marine systems. Ecol Lett 9:228–241

Hawkins SJ, Moore PJ, Burrows MT, Poloczanska E, Mieszkowska N, Herbert RJH, Jenkins SR, Thompson RC, Genner MJ, Southward AJ (2008) Complex interactions in a changing world: Responses of rocky shore communities to recent climate change. Climate Res 37:123–133

Hay ME (1997) The ecology and evolution of seaweed-herbivore interactions on coral reefs. Coral Reefs 16:S67–S76

Hoegh-Guldberg O, Mumby PJ, Hooten AJ, Steneck RS, Greenfield P, Gomez E, Harvell CD, Sale PF, Edwards AJ, Caldeira K, Knowlton N, Eakin CM, Iglesias-Prieto R, Muthiga N, Bradbury RH, Dubi A, Hatziolos ME (2007) Coral reefs under rapid climate change and ocean acidification. Science 318:1737–1742

Huvé H (1955) Présence de *Laminaria rodriguezii* Bornet sur les côtes de Mediterraneée. Rec Trav Stat Mar Endoume 15:73, 89 + 11 plates

Johannessen O, Bengtsson L, Miles M, Kuzmina S, Semenov V, Alekseev G, Nagurnyi A, Zakharov V, Bobylev L, Pettersson L (2004) Arctic climate change: observed and modeled temperature and sea-ice variability. Tellus 56(A):328–341

Johnson CR, Banks SC, Barrett NS, Cazassus F, Dunstan PK, Edgar GJ, Frusher SD, Gardner C, Haddon M, Helidoniotis F, Hill KL, Holbrook NJ, Hosie GW, Last PR, Ling SD, Melbourne-Thomas J, Miller K, Pecl GT, Richardson AJ, Ridgway KR, Rintoul SR, Ritz DA, Ross DJ, Sanderson JC, Shepherd SA, Slotwinski A, Swadling KM, Taw N (2011) Climate change cascades: Shifts in oceanography, species' ranges and subtidal marine community dynamics in eastern Tasmania. J Exp Mar Biol Ecol 400:17–32

Jokiel PL, Coles SL (1990) Response of Hawaiian and other Indo-Pacific reef corals to elevated temperature. Coral Reefs 8:155–162

Karsten U, Wulff A, Roleda M, Müller R, Steinhoff FS, Fredersdorf J, Wiencke C (2011) Physiological responses of polar benthic algae to ultraviolet radiation. In: Wiencke C (ed) Biology of polar benthic algae. De Gruyter, Berlin, pp 271–298

Kerswell AP (2006) Global biodiversity patterns of benthic marine algae. Ecology 87:2479–2488

Kirihara S, Nakamura T, Kon N, Fujita D, Notoya M (2006) Recent fluctuations in distribution and biomass of cold and warm temperature species of Laminarialean algae at Cape Ohma, northern Honshu, Japan. J Appl Phycol 18:521–527

Knutti R, Flückinger J, Stocker TF, Timmermann A (2004) Strong hemispheric coupling of glacial climate through freshwater discharge and ocean circulation. Nature 430:851–856

Kübler JE, Davison IR (1995) Thermal acclimation of light-use characteristics of *Chondrus crispus* (Rhodophyta). Eur J Phycol 30:189–195

Levitus S, Antonov JI, Boyer TP, Stephens C (2000) Warming of the world ocean. Science 287:2225–2229

Lima FP, Ribeiro PA, Queiroz N, Hawkins SJ, Santos AM (2007) Do distributional shifts of northern and southern species of algae match the warming pattern? Glob Change Biol 13:2592–2604

Liu PJ, Lin SM, Fan TY, Meng PJ, Shao KT, Lin HJ (2009) Rates of overgrowth by macroalgae and attack by sea anemones are greater for live coral than dead coral under conditions of nutrient enrichment. Limnol Oceanogr 54:1167–1175

Lüning K (1990) Seaweeds: Their environment, biogeography and ecophysiology. Wiley, New York

Manzello DP, Berkelmans R, Hendee JC (2007) Coral bleaching indices and thresholds for the Florida reef tract, Bahamas, and St. Croix, US Virgin Islands. Mar Pollut Bull 54:1923–1931

Martinson DG, Stammerjohn SE, Iannuzzi RA, Smith RC, Vernet M (2008) Western Antarctic Peninsula physical oceanography and spatio-temporal variability. Deep Sea Res II 55:1964–1987

Merzouk A, Johnson LE (2011) Kelp distribution in the northwest Atlantic Ocean under a changing climate. J Exp Mar Biol Ecol 400:90–98

Mieszkowska N, Kendall MA, Hawkins SJ, Leaper R, Williamson P, Hardman-Mountfort NJ, Southward AJ (2006) Changes in the range of some common rocky shore species in Britain—a response to climate change? Hydrobiologia 555:241–251

Molenaar F (1996) Seasonal growth and reproduction of North Atlantic red seaweeds. PhD dissertation, University of Groningen, Netherlands, p 111

Müller R, Wiencke C, Bischof K (2008) Interactive effects of UV radiation and temperature on microstages of Laminariales (Phaeophyceae) from the Arctic and North Sea. Clim Res 37:203–213

Müller R, Bartsch I, Laepple T, Wiencke C (2009) Impact of oceanic warming on the distribution of seaweeds in polar and cold-temperate waters. Bot Mar 52:617–638

Müller R, Bartsch I, Laepple T, Wiencke C (2011) Impact of oceanic warming on the distribution of seaweeds in polar and cold-temperate waters. In: Wiencke C (ed) Biology of polar benthic algae. De Gruyter, Berlin, pp 237–270

Nelson WA (2009) Calcified macroalgae—critical to coastal ecosystems and vulnerable to change: a review. Mar Freshwater Res 60:787–801

Pakker H, Breeman AM, Prud'homme van Reine WF, van den Hoek C (1995) A comparative study of temperature responses of Carribean seaweeds from different biogeographic groups. J Phycol 31:499–507

Pearson RG, Dawson TP (2003) Predicting the impacts of climate change on the distribution of species: are bioclimate envelope models useful? Global Ecol Biogeogr 12:361–371

Peters AF, Breeman AM (1992) Temperature responses of disjunct temperate brown algae indicate long-distance dispersal of microthalli across the tropics. J Phycol 28:428–438

Pianka ER (1966) Latitudinal gradients in species diversity: a review of concepts. Am Natural 100:33–46

Pivovarov S, Schlitzer R, Novikhin A (2003) River runoff influence on the water mass formation in the Kara Sea. In: Stein R, Fahl K, Fütterer DK, Galimov EM, Stepanets OV (eds) Siberian river run-off in the Kara Sea: characterisation, quantification, variability, and environmental significance. Elsevier, Amsterdam, pp 9–26

Poloczanska ES, Smith S, Fauconnet L, Healy J, Tibbetts IR, Burrows MT, Richardson AJ (2011) Little change in the distribution of rocky shore faunal communities on the Australian east coast after 50 years of rapid warming. J Exp Mar Biol Ecol 400:145–154

Provan J, Wattier RA, Maggs CA (2005) Phylogeographic analysis of the red seaweed *Palmaria palmata* reveals a Pleistocene marine glacial refugium in the English Channel. Mol Ecol 14:793–803

Rasher DB, Stout EP, Engel S, Kubanek J, Hay ME (2011) Macroalgal terpernes function as allelopathic agents against reef corals. Proc Natl Acad Sci USA 43:17726–17731

Rayner NA, Parker DE, Horton EB, Folland CK, Alexander LV, Rowell DP, Kent EC, Kaplan A (2003) Global analyses of sea surface temperature, sea ice, and night marine air temperature since the late nineteenth century. J Geophys Res 108(D14):4407

Rijnsdorp AD, Peck MA, Engelhard GH, Mollmann C, Pinnegar JK (2009) Resolving the effect of climate change on fish populations. ICES J Mar Sci doi:10.1093/icesjms/fsp056

Sagarin RD, Barry JB, Gilman SE, Baxter CH (1999) Climate-related change in an intertidal community over short and long time scales. Ecol Monogr 69:465–490

Sala E, Knowlton N (2006) Global marine biodiversity trends. Annu Rev Environ Resour 31:93–122

Sangil C, Sanson M, Afonso-Carillo J (2011) Spatial variation patterns of subtidal assemblages along a subtropical oceanic archipelago: Thermal gradient vs herbivore pressure. Estuar Coast Shelf Sci 94:322–333

Santelices B, Bolton JJ, Meneses I (2009) 6. Marine algal communities. In: Whitman JD, Roy K (eds) Marine macroecology. Chicago University Press, Chicago, pp 153–192

Schiel DR, Steinbeck JR, Foster MS (2004) Ten years of induced ocean warming causes comprehensive changes in marine benthic communities. Ecology 85:1833–1839

Schmittner A, Latif M, Schneider B (2005) Model projections of the North Atlantic thermohaline circulation for the 21st century assessed by observations. Geophys Res Lett 32:L23710. doi:10.1029/2005GL024368

Schofield O, Ducklow HW, Martinson DG, Meredith MP, Moline MA, Fraser WR (2010) How do polar marine ecosystems respond to rapid climate change? Science 328:1529–1523

Seidel DJ, Fu Q, Randel WJ, Reichler TJ (2008) Widening of the tropical belt in a changing climate. Nat Geosci 1:21–24

Serreze MC, Holland MM, Stroeve J (2007) Perspectives on the Arctic's shrinking sea-ice cover. Science 315:1533–1536

Snoeijs PJM, Prentice IC (1989) Effects of cooling water discharge on the structure and dynamics of epilithic algal communities in the northern Baltic. Hydrobiologia 184:99–123

Solomon S, Qin D, Manning M, Chen Z, Marquis M, Averyt KB, Tignor M, Miller HL (eds 2007) Contribution of Working Group I to the Fourth Assessment Report of the Intergovernmental Panel on Climate Change. Cambridge University Press, Cambridge, New York

Spalding MD, Fox HE, Allen GR, Davidson N, Ferdaña ZA, Finlayson M, Halpern BS, Jorge MA, Lombana A, Lourie SA, Martin KD, McManus E, Molnar J, Recchia CA, Robertson J (2007) Marine ecoregions of the world: a bioregionalization of coastal and shelf areas. BioScience 57:573–583

Spurkland T, Iken K (2011) Salinity and irradiance effects on growth and maximum photosynthetic quantum yield in subarctic *Saccharina latissima* (Laminariales, Laminariaceae). Bot Mar 54:355–365

Stammerjohn SE, Martinson DG, Smith RC, Yuan X, Rind D (2008) Trends in Antarctic annual sea ice retreat and advance and their relation to El Niño-southern oscillation and southern annual mode variability. J Geophys Res 113:C03S90

Titlyanov EA, Yakovleva IM, Titlyanova TV (2007) Interaction between benthic algae (*Lyngbya bouillonii, Dictyota dichotoma*) and scleractinian coral *Porites lutea* in direct contact. J Exp Mar Biol Ecol 342:282–291

Tittensor DP, Mora C, Jetz W, Lotze HK, Ricard D, van den Berghe E, Worm B (2010) Global patterns and predictors of marine biodiversity across taxa. Nature 466:1098–1101

tom Dieck I (1992) North Pacific and North Atlantic digitate *Laminaria* species (Phaeophyta): hybridization experiments and temperature responses. Phycologia 31:147–163 tom Dieck I, de Oliveira EC (1993) The section Digitatae of the genus *Laminaria* (Phaeophyta) in the northern and southern Atlantic: crossing experiments and temperature responses. Mar Biol 115:151–160

Tronholm A, Steen F, Tyberghein L, Leliaert F, Verbruggen H, Siguan MAR, De Clerk O (2010) Species delimitation, taxonomy, and biogeography of *Dictyota* in Europe (Dictyotales, Phaeophyceae). J Phycol 46:1301–1321

Turner J, Overland JE, Walsh JE (2007) An Arctic and Antarctic perspective on recent climate change. Int J Climatol 27:277–293

van den Hoek C (1982a) The distribution of benthic marine algae in relation to the temperature regulation of their life histories. Biol J Linn Soc 18:81–144

van den Hoek C (1982b) Phytogeographic distribution groups of benthic marine algae in the North Atlantic Ocean. A review of experimental evidence from life history studies. Helgoländer Meeresunters 35:53–214

van den Hoek C, Breeman AM (1989) Seaweed biogeography in the North Atlantic: Where are we now? In: Garbary DJ, South GR (eds) Evolutionary biogeography of the marine algae of the North Atlantic. Springer, Berlin, pp 55–86

van der Strate HJ, Boele-Bos SA, Olsen JL, van den Zande L, Stam WT (2002) Phylogeographic studies in the tropical seaweed *Cladophoropsis membranacea* (Chlorophyta, Ulvophyceae) reveal a cryptic species complex. J Phycol 38:572–582

van Oppen MJH, Olsen JL, Stam WT, van den Hoek C, Wiencke C (1993) Arctic-Antarctic disjunctions in the benthic seaweeds *Acrosiphonia arcta* (Chlorophyta) and *Desmarestia viridis/willii* (Phaeophyta) are of recent origin. Mar Biol 115:381–386

van Oppen MJH, Diekmann OE, Wiencke C, Stam WT, Olsen JL (1994) Tracking dispersal routes: Phylogeography of the Arctic-Antarctic disjunct seaweed *Acrosiphonia arcta* (Chlorophyta). J Phycol 30:67–80

Vaughan DG, Marshall GJ, Connolley WM, Parkinson C, Mulvaney R, Hodgson DA, King JC, Pudsey CJ, Turner J (2003) Recent rapid regional climate warming on the Antarctic Peninsula. Clim Change 60:243–274

Verbruggen H, Tyberghein L, Pauly K, Vlaeminck C, van Nieuwenhuyze K, Kooistra WHCF, Leliaert F, De Clerk O (2009) Macroecology meets macroevolution: evolutionary niche dynamics in the seaweed *Halimeda*. Global Ecol Biogeogr 18:393–405

Viejo RM, Martínez B, Arrontes J, Astudillo C, Hernández L (2011) Reproductive patterns in central and marginal populations of a large brown seaweed: drastic changes at the southern range limit. Ecography 34:75–84

Vroom PS, Timmers MAV (2009) Spatial and temporal comparison of algal biodiversity and benthic cover at Gardner Pinnacles, northwestern Hawai'ian islands. J Phycol 45:337–347

Vroom PS, Page KN, Kenyon JC, Brainard RE (2006) Algae-dominated reefs. Am Scientist 94:430–437

Vroom PS, Musburger CA, Cooper SW, Maragos JE, Page-Albins KN, Timmers MAV (2010) Marine biological community baselines in unimpacted tropical ecosystems: spatial and temporal analysis of reefs at Howland and Baker Islands. Biodivers Conserv 19:797–812

Wanders JBW (1977) The role of benthic algae in the shallow reef of Curaçao (Netherlands Antilles). 3. The significance of grazing. Aquat Bot 3:357–390

Weatherhead EC, Andersen SB (2006) The search for signs of recovery of the oregion layer. Nature 441:39–45

Wernberg T, Thomsen MS, Tuya F, Kendrick GA, Staehr PA, Toohey BD (2010) Decreasing resilience of kelp beds along a latitudinal temperature gradient: potential implications for a warmer future. Ecol Lett 13:685–694

Weslawski JM, Wiktor J, Kotwicki L (2010) Increase in biodiversity in the arctic rocky littoral, Sorkappland, Svalbard, after 20 years of climate warming. Mar Biodiv 40:123–130

Weslawski JM, Kendall MA, Wlodarska-Kowalczuk M, Iken K, Kedra M, Legezynska J, Sejr MK (2011) Climate change effects on Arctic fjord and coastal macrobenthic diversity—observations and predictions. Mar Biodiv 41:71–85

Wiencke C, Bartsch I, Bischoff B, Peters AF, Breeman AM (1994) Temperature requirements and biogeography of Antarctic, Arctic and amphiequatorial seaweeds. Bot Mar 37:247–259

Willig MR, Kaufman DM, Stevens RD (2003) Latitudinal gradients of biodiversity: pattern, process, scale, and synthesis. Annu Rev Ecol Evol Syst 34:273–309

Wiltshire KH, Malzahn M, Wirtz K, Greve W, Janisch S, Mangelsdorf P, Manly BFJ, Boersma M (2008) Resilience of North Sea phytoplankton spring bloom dynamics: an analysis of long-term data at Helgoland Roads. Limnol Oceanogr 53:1294–1302

Yarish C, Breeman AM, van den Hoek C (1984) Temperature, light and photoperiod responses of some Northeast American and west European endemic rhodophytes in relation to their geographic distribution. Helgoländer Meeresunters 38:273–304

Zacher K, Wulff A, Molis M, Hanelt D, Wiencke C (2007) Ultraviolet radiation and consumer effects on a field-grown intertidal macroalgal assemblage in Antarctica. Global Change Biol 13:1201–1215

Zacher K, Rautenberger R, Hanelt D, Wulff A, Wiencke C (2011) The abiotic environment of polar marine benthic algae. In: Wiencke C (ed) Biology of polar benthic algae. De Gruyter, Berlin, pp 9–21

Zachos JC, Dickens GR, Zeebe RE (2008) An early Cenozoic perspective on green house warming and carbon-cycle dynamics. Nature 451:279–283

Zardi GI, Nicastro KR, Canovas F, Costa JF, Serrão EA, Pearson GA (2011) Adaptive traits are maintained on steep selective gradients despite gene flow and hybridization in the intertidal region. PLoS One 6(e19402):1–12

Zuccarello GC, Buchanan J, West JA, Pedroche FF (2011) Genetic diversity of the mangrove-associated alga *Bostrychia radicans*/*Bostrychia moritziana* (Ceramiales, Rhodophyta) from southern Central America. Phycol Res 59:98–104

Žuljević A, Antolić B, Nikolić V, Isajlović I (2011) Review of *Laminaria rodriguezii* records in the Adriatic Sea. 5th European Phycological Congress, Abstract Book, p 194

Chapter 19
Seaweed Responses to Ocean Acidification

Michael Y. Roleda and Catriona L. Hurd

19.1 Introduction

Since the start of the industrial revolution (~1850), burning fossil fuels, cement manufacturing, and changes to land use have increased atmospheric carbon dioxide (CO_2) concentrations from ~280 to 385 ppm. For the coming decades, the 2007 IPCC report on climate change predicts a continued rapid rise in atmospheric CO_2 leading to significant temperature increases in the atmosphere and ocean (IPCC 2007). Another consequence of increasing anthropogenic CO_2 is a reduction in the pH of the world's ocean and concomitant changes in seawater carbonate chemistry, termed ocean acidification. Ocean acidification (OA) may have profound impact on biogeochemical cycles, marine organisms, ecosystems, and the services they provide. In combination with warming seas and other stress factors (e.g., eutrophication, UVR), this will create multiple threats to the marine environment. This chapter will discuss seawater chemistry and algal carbon physiology, and concentrate on seaweed responses to OA by themselves and in synergy with other environmental stresses. Reference to phytoplankton, seagrasses, and corals is made where appropriate.

19.2 Anthropogenic CO_2 and Seawater Carbonate Chemistry

The world's oceans have taken up ~30% of anthropogenic CO_2 and have buffered climate change in terrestrial systems by reducing the atmospheric concentration of this greenhouse gas (Feely et al. 2004). The air–sea gas exchange equilibrates surface seawater CO_2 to atmospheric levels with a timescale of ~1 year (Doney

M.Y. Roleda (✉) • C.L. Hurd
Department of Botany, University of Otago, P.O. Box 56, Dunedin 9054, New Zealand
e-mail: michael.roleda@botany.otago.ac.nz; catriona.hurd@botany.otago.ac.nz

C. Wiencke and K. Bischof (eds.), *Seaweed Biology*, Ecological Studies 219,
DOI 10.1007/978-3-642-28451-9_19,

Table 19.1 Average changes to ocean chemistry and pH based on ocean carbon models and surface-ocean measurements (modified from The Royal Society 2005)

	Preindustrial	Today	2× preindustrial	3× preindustrial	4× preindustrial
CO_{2atm} (ppm)	280	380	560	840	1,120
H_2CO_3 (μmol kg^{-1})	9	13	19	28	38
HCO_3^- (μmol kg^{-1})	1,768	1,867	1,976	2,070	2,123
CO_3^{2-} (μmol kg^{-1})	225	185	141	103	81
Total DIC (μmol kg^{-1})	2,003	2,065	2,136	2,201	2,242
Mean surface pH	8.18	8.07	7.92	7.77	7.65
$\Omega_{calcite}$	5.3	4.4	3.3	2.4	1.9
$\Omega_{aragonite}$	3.4	2.8	2.1	1.6	1.2

et al. 2009), and the seawater carbonate chemistry is governed by a series of chemical reactions that are in equilibrium (see also Chap. 2 by Gómez and Huovinen and Chap. 4 by Gordillo):

$$CO_{2(atoms)} \leftrightarrow CO_{2(aq)} + H_2O \leftrightarrow H_2CO_3 \leftrightarrow H^+ + HCO_3^- \leftrightarrow 2H^+ + CO_3^{2-}$$

CO_2 reacts with H_2O causing the equilibrium to shift such that the concentration of the bicarbonate ion (HCO_3^-) and carbonic acid (H_2CO_3) increases, while that of carbonate ions (CO_3^{2-}) declines. The increase in hydrogen ion concentration lowers pH because pH $= -\log_{10}[H^+]$. The mean surface ocean pH has already decreased by 0.1 unit (equivalent to a 30% increase in H^+) from ~8.18 at the beginning of the industrial revolution. For surface seawater with pH of ~8.07, approximately 91% of the inorganic carbon is HCO_3^-, 8% CO_3^{2-}, and only 1% $CO_{2(aq)}$. By 2100, concentrations of $CO_{2(aq)}$ and HCO_3^- are predicted to increase by 192% and 14%, respectively, and CO_3^{2-} decrease by 56%, with a concomitant decline in pH to 7.65 (The Royal Society 2005). Aside from the decline in carbonate concentration, a reduction in the saturation state of aragonite and calcite is also predicted (Table 19.1). As the seawater saturation state for a mineral measures its thermodynamic potential to form or to dissolve, a reduction will make calcified structures more vulnerable to erosion.

Carbonate ions play an important role in the formation of calcareous "skeletons" for a large number of phototrophic marine organisms including corals, phytoplankton, and macroalgae. If mitigation of emissions does not occur, CO_2 concentrations could reach >800 ppm by 2100 that could result in a decrease of ~0.4 pH units. Under a worst-case scenario of continued and unabated usage of known fossil fuel reserves, modeling indicates that average surface ocean pH could fall by a maximum of 0.77 pH units by 2300 (Caldeira and Wickett 2003). Marine organisms would then experience the lowest pH level for ~20 million years and the most rapid rate of decline in pH ever (Turley 2008). While the OA event experienced 55 million years ago during the Paleocene–Eocene thermal maximum probably occurred over several thousand years, the present-day anthropogenically induced event is happening over decades to centuries (Zachos et al. 2005).

In 2005, a Royal Society Report highlighted the potential effects of OA on marine organisms. Initial research has focused on how the lowered availability of carbonate ions might impact marine calcifiers, particularly corals, calcareous micro- and macroalgae, and other shell-forming invertebrates. However, because the relative availability of CO_2 and HCO_3^- will also change and algae use one or both of these as a CO_2 source for photosynthesis, OA has the potential to affect the rates of photosynthesis, and hence growth and productivity, of all algae.

19.3 Carbon Sources and Acquisition

In seawater, there are two sources of inorganic carbon (C_i: also known as Dissolved Inorganic Carbon, DIC, and Total Inorganic Carbon, C_T) to seaweeds, HCO_3^-, which is available in high concentrations (2,200 μM), and CO_2 (14 μM). Most seaweeds are able to take up HCO_3^- and/or CO_2 by active transport (termed a carbon concentrating mechanism, CCM), while some seaweeds (mostly reds) rely on CO_2 uptake by passive diffusion (Hurd et al. 2009; see also Chap. 2 by Gómez and Huovinen and Chap. 4 by Gordillo).

The enzyme carbonic anhydrase (carbonate dehydrase) EC 4.2.1.1 (CA) catalyses the reversible hydration of CO_2 and is widely distributed not only in plants but also in animals and prokaryotes. CA is synthesized within the cytoplasm but may also be found associated with the chloroplast. They are also found on the plasmalemma, where they have access to the external medium (Giordano and Maberly 1989). Although the significance of CA is not yet fully understood, it is generally believed to play a role in carbon acquisition and photosynthesis, where high CA activity is correlated to aquatic plant's ability to concentrate CO_2 internally in response to low CO_2 concentrations in the growth medium.

Inorganic carbon concentrating mechanisms (CCMs) cause the accumulation of CO_2 around RUBISCO (ribulose-1,5-bisphosphate carboxylase oxygenase) in organisms capable of oxygenic photosynthesis (e.g., all cyanobacteria, most algae and aquatic plants, and in C_4 and crassulacean acid metabolism (CAM) of vascular plants). In algae, CCMs are polyphyletic (more than one evolutionary origin) and involve active transport of HCO_3^-, CO_2, and/or H^+, or an energized biochemical mechanism similar to C_4 and CAM plants (Raven et al. 2008). CCMs involve both the active transport of HCO_3^- and/or CO_2 and the catalytic conversion of HCO_3^- to CO_2 with subsequent passive diffusion of CO_2. CCMs effectively alleviate the low CO_2 affinity of the enzyme RUBISCO which is less than half saturated under current dissolved CO_2 concentrations [$CO_{2(aq)}$]. Different types of CCM are present in different phototrophs (Table 1 in Giordano et al. 2005) and different models for inorganic carbon transport and CCM are schematized in cyanobacteria (Fig. 1 in Giordano et al. 2005) and eukaryotic algae (Figs. 1 and 2 in Thoms et al. 2001, and Moroney and Ynalvez 2007; Fig. 1 in Giordano et al. 2005). While pyrenoids in eukaryotic algae play an important role in CCMs (Badger et al. 1998), their absence does not necessarily imply the absence of a CCM (e.g., *Caulerpa scalpelliformis*, Kevekordes et al. 2006).

A number of seaweeds lack a CCM and rely solely on CO_2 diffusion from the external environment (Raven et al. 2005). In order to carry out net C assimilation under current CO_2 and O_2 levels, RUBISCO of the organism should have high selectivity factor (S_{rel}) that defines the relative rates of carboxylase and oxygenase reactions to diffusive CO_2 supply (Giordano et al. 2005). Macroalgae lacking CCMs are often found inhabiting fast-flow environments, to minimize the diffusion boundary-layer thickness (see Sect. 19.7.1), or in deep water with low light levels where a smaller CO_2 flux is required to satisfy cellular assimilatory requirements (Sherlock and Raven 2001; Raven et al. 2002a, b; Hepburn et al. 2011). Oxygenic photolithotrophs have a diversity of incompletely understood mechanisms of inorganic carbon acquisition that require further studies. Raven (2010) recommended four areas of research where more studies are needed: (1) Diffusive CO_2 entry, where a number of algae are, in various respects, intermediate between diffusive CO_2 entry and occurrence of a CCM, (2) the nature and role of cyanelles (plastids) in organic carbon assimilation, (3) the energetics of CCM in *Chlamydomonas reinhardtii*, the eukaryotic alga with the best understood CCM, and (4) the occurrence of C_4-like metabolism in the CCMs of marine diatoms.

Methods used to determine whether an alga is a HCO_3^- or CO_2 user include: (1) pH drift experiments, with HCO_3^--using algae being capable of raising pH of the surrounding medium to values in excess of those attained by species using only CO_2, (2) pH dependence of half saturation constants, $K_{0.5}$ (HCO_3^-) or $K_{0.5}$ (CO_2), (3) HCO_3^- utilization pathway inhibitors, (4) isotope disequilibrium, and (5) membrane-inlet mass spectroscopy (Giordano et al. 2005). Aspects of inorganic carbon acquisition, metabolism, and carbon isotope discrimination by marine macroalgae are discussed by Kremer (1981), Johnston (1991), Maberly et al. (1992), and Raven et al. (2002b).

Applying one or a combination of these various techniques shows that most algae examined can take up both HCO_3^- and CO_2. However, there is some evidence that some algae can take up only CO_2 while other species can only actively transport HCO_3^-. Using the natural abundance of $^{13}C/^{12}C$ ratios (as $\delta^{13}C$) as a proxy, marine macroalgae with $\delta^{13}C$ values lower than –30‰ rely on diffusive CO_2 supply to RUBISCO (Maberly et al. 1992) while organisms with $\delta^{13}C$ higher than –10‰ (a value more positive than $\delta^{13}C$ of CO_2 in seawater) must involve HCO_3^- use (Raven et al. 2002b). In combination with pH drift experiments, HCO_3^--using macroalgae can shift seawater pH to 9.0 or higher (Maberly 1990) and have corresponding $\delta^{13}C$ values less negative than −30‰. Conversely, CO_2-using species have $\delta^{13}C$ values more negative than −30‰ and were unable to raise pH above a critical value of pH 9 (Maberly et al. 1992).

The expression Δ which represents the $\delta^{13}C$ value of the algal sample ($\delta^{13}C_{alga}$) relative to the $\delta^{13}C$ of the source of inorganic carbon used in photosynthesis ($\delta^{13}C_{CO_2}$), is also indicative of certain photosynthetic pathways. Previous work strongly suggests that Δ values in excess of 20‰ are associated with diffusive CO_2 entry and C_3 pathway of CO_2 fixation while Δ values below 20‰ are consistent with HCO_3^- use and CO_2 uptake by a nondiffusive mechanism (Raven et al. 1995 and references therein). A correlation was found between the two proxies for

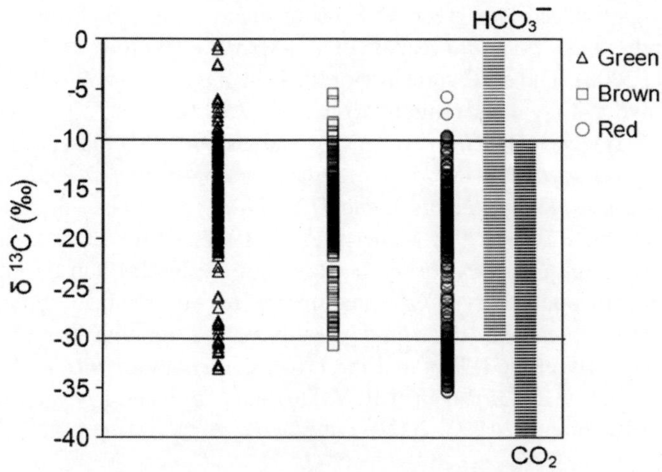

Fig. 19.1 $\delta^{13}C$ signatures of different macroalgae as a proxy of their carbon use strategy ($n = 857$, representing 370 species sampled from ten thermographic regions described by Adey and Steneck 2001). *Horizontal lines and bars* to the right delimit the range of possible values for species restricted wholly to CO_2 or HCO_3^- (see text and cf Maberly et al. 1992). The sources of data used here are Black and Bender (1976), Bode et al. (2006), Corbisier et al. (2006), Dunton (2001), Fischer and Wiecke (1992), Kang et al. (2008), Kevekordes et al. (2006), Kübler and Raven (1994), Laurand and Riera (2006), Maberly et al. (1992), Pinnegar and Polunin (2000), Raven et al. (1995, 2002b), Runcie et al. (2008; unidentified Corallinales not included), Vizzini and Mazzola (2006), Wang and Yeh (2003), and Wozniak et al. (2006)

diffusive CO_2 uptake where macroalgae with $\delta^{13}C$ more negative than –30‰ also have Δ above 20‰ (Raven et al. 1995; Runcie et al. 2008). Several conditions of using stable isotopes to describe the carbon physiology of macroalgae are discussed by Raven et al. (2002b).

Using pH drift experiments and carbonic anhydrase activity measurements, the preference of different macroalgae for either CO_2 or HCO_3^- or both as source of inorganic carbon has previously been compiled and reviewed (Table 3 in Gao and McKinley 1994 and references therein). For example, pH drift measurements showed that the brown alga *Carpophyllum* sp. most likely uses CO_2 rather than HCO_3^- (Dromgoole 1978). On the other hand, using $\delta^{13}C$ data in the literature as a proxy for carbon use, the values for seaweeds (Fig. 19.1) vary from –0.64‰ for *Halimeda distorta* (Chlorophyta: Bryopsidophyceae) collected from 99 m depth at Penguin Bank off the island of Moloka'i, Hawaii archipelago (Runcie et al. 2008) to −35.3‰ for *Georgiella confluens* (Rhodophyta: Floridiophyceae) from Anvers Island off the Antarctic Peninsula (Dunton 2001). Most seaweeds have $\delta^{13}C$ between −30 and −10‰, within the range in which both CO_2 and HCO_3^- use can occur (Maberly et al. 1992; Raven et al. 1995). Macroalgae that rely solely on HCO_3^- as carbon source ($\delta^{13}C > -10$‰) are 10 greens, 5 browns, and 6 reds while macroalgae that depend on CO_2 diffusion ($\delta^{13}C < -30$‰) consist of 6 greens, 1 brown, and 51 reds. Most of these CO_2-using deep water red macroalgae belongs to the Class Florideophyceae (94%), with 51% from the Order Ceramiales, 18% Plocamiales, and 14% Gigartinales. The green CO_2-using macroalgae are different species of *Caulerpa* and *Udotea petiolata* (Raven et al 2002b; Kevekordes et al. 2006; Vizzini and Mazzola 2006). The $\delta^{13}C$ signature of the brown macroalga *Desmarestia anceps* is inconclusive with values ranging from −25.3‰ (Dunton 2001) to −30.68‰ (Fischer and Wiecke 1992). Inconclusive values are also observed in six red macroalgae, *Lomentaria articulata, Membranoptera alata, Odonthalia dentata, Palmaria decipiens, Ptilota gunneri*, and *Trailliella intricata* (Fischer and Wiencke 1992; Kübler and Raven 1994; Maberly et al. 1992; Raven et al. 2002b) and in the green macroalga *Caulerpa obscura* (Raven et al. 2002b). On the other side of the spectrum, inconclusive $\delta^{13}C$ signatures were observed in the green *Ulva intestinalis* (*Enteromorpha intestinalis*), with values ranging from −8.81‰ to −20.30‰ (Maberly et al. 1992), and the browns *Adenocystis utricularis* (−8.81‰ to −20.30‰, Dunton 2001; Raven et al. 2002b) and *Colpomenia peregrina* (−6.14‰ to −12.09‰, Raven et al. 1995, 2002b).

19.4 Calcification

The biogenic formation of calcium carbonate from Ca^{2+} and CO_3^{2-} is termed calcification. Three polymorphs of calcium carbonate are made by seaweeds: calcite, aragonite, and high-magnesium calcite, and in the present-day surface ocean, these are supersaturated so that biogenic precipitation by algae is "thermodynamically favored" (Raven and Giordano 2009). However, the lower pH predicted for future oceans causes a reduction in the saturation state of the carbonate ion, making it more difficult for calcifying organisms to calcify (Cao and Caldeira 2008). High-magnesium calcite, synthesized by coralline seaweeds, is the most soluble polymorph, potentially making this group most susceptible to OA. The site of calcification varies among seaweeds. For example, coralline seaweeds deposit high magnesium calcite into their cell walls, the brown genus *Padina* deposits a fine layer of aragonite onto its cell surface, and for the tropical genus *Halimeda*, aragonite mineralizes in the intercellular spaces between the utricles (Stanley 2008; Raven and Giordano 2009; Nelson 2009; Ries 2009, 2010).

There have been few studies (e.g., Borowitzka and Larkum 1976; Borowitzka 1987) on the underlying physiological mechanisms of calcification in seaweeds, but the recent interest in OA has stimulated research in this important area (see Ries 2009, 2010). Photosynthesis (and primary production) and calcification are thought to be intimately linked because rates of calcification are enhanced in the light, but the mechanisms are not fully understood (Ries 2009, 2010). Also affecting calcification rate is the ratio of magnesium to calcite (Mg/Ca ratio) and the carbonate saturation state, both of which are influenced by pH (Ries 2009, 2010). Over geological timescales, the world's ocean has switched between "aragonite seas" (today's ocean) and "calcite seas". When the green seaweeds *Halimeda*, *Penicillus*, and *Udotea* were grown in seawater with a Mg/Ca ratio > 2 (aragonite sea), rates of growth, calcification, and primary production were greater than when grown in

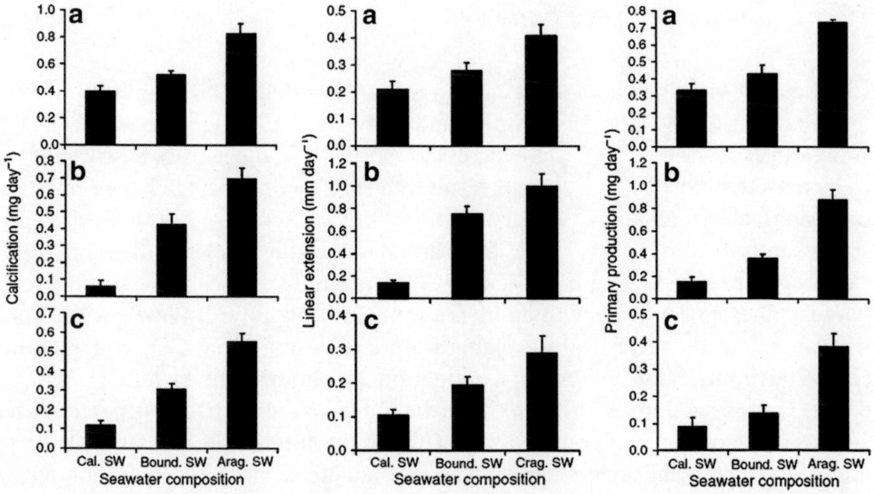

Fig. 19.2 Calcification, linear extension, and primary production in (**a**) *Halimeda*, (**b**) *Penicillus*, and (**c**) *Udotea* grown in experimental calcite, boundary, and aragonite seawaters. Rates of calcification, linear extension, and production significantly increase ($P < 0.05$) with increasing seawater Mg/Ca. *Vertical bars* are standard errors (From Ries 2009, with permission of Blackwell Publishing Ltd)

calcite seas (Fig. 19.2). In the calcite sea treatments, however, they were still able to precipitate aragonite, demonstrating some level of internal control over the polymorph of calcite produced. For coccolithophores (calcifying phytoplankton), the physiological basis of calcification, and its relationship to photosynthesis, is being elucidated using various physiological and molecular tools (e.g., Brownlee and Taylor 2004); similar approaches need to be applied to calcifying red, green, and brown seaweeds if we are to better understand how they will be affected by OA.

19.5 Physiological Performance in a High CO_2 World

The sensitivity of seaweeds to OA can be species specific. Their response may vary depending on their carbon physiology, mode of calcification, morphology (functional growth forms), and life history. For example, filamentous turf algae with shorter generation times may be able to acclimate and adapt to the changing environment faster than large canopy-forming kelp species with a longer generation time.

19.5.1 Noncalcifying Seaweeds

CCM activity in algae is in most cases associated with the ability to use, indirectly or directly, HCO_3^- ions (e.g., Smith and Bidwell 1989; Björk et al. 1992, 1993; Axelsson et al. 1999, 2000; Mercado et al. 2006). Since the shift in DIC species due to increased atmospheric CO_2 will result in a small proportional change in HCO_3^- compared to CO_2 and CO_3^{2-} concentrations, this will confer no/little advantage on species relying on HCO_3^- use for carbon acquisition (Beardall et al. 1998). Consequently, species with an active CCM are unlikely to show stimulation of photosynthesis or growth with an increase in atmospheric CO_2 levels. Conversely, when modern day terrestrial C_3 plants were grown at Glacial CO_2 concentrations (180–200 ppm), they exhibited a reduction in photosynthesis and growth, and delayed reproduction (Ward 2005; Gerhart and Ward 2010). Further long-term studies on seaweed's response to high CO_2 of the future and low CO_2 of the past may provide crucial understanding on their adaptive capability to changing CO_2 over geological and evolutionary timescales.

Macroalgae grown at elevated CO_2 showed downregulation of CCM activity (e.g., *Fucus serratus*, Johnston and Raven 1991; *Ulva lactuca*, Magnusson et al. 1996) and the energy savings from downregulating CCM are hypothesized to increase growth, but no increase in growth rate was observed in brown seaweed species with known active CCMs, e.g., *Laminaria digitata* and *Saccharina latissima*, grown under Glacial, Preindustrial, and 750 ppm CO_2 concentrations (Roleda MY, Stecher A, Gutow L, Bartsch I and Wiencke C, unpublished data). Among red seaweeds, photosynthesis and growth rates of *Porphyra yezoensis* in culture were enhanced under high (1,000 and 1,600 ppm) CO_2 concentrations (Gao et al. 1991). Enhancement of growth with increasing CO_2 is also reported in *Gracilaria* species (Gao et al. 1993b). For the CO_2-user *Lomentaria articulata*, higher relative carbon growth (52%) was measured under 2× of the current ambient CO_2 compared to its growth (23%) under 5× CO_2 (Kübler et al. 1999).

At a high photon flux density (PFD), the seawater concentration of DIC becomes insufficient to saturate photosynthesis of intertidal algae (Mercado et al. 2001). Therefore, when photosynthetic rate exceeds the rate at which CO_2 is supplied, photosynthesis becomes inhibited (Murru and Sandgren 2004). Under such carbon-limiting conditions, employing the energetically expensive CCM becomes ecologically advantageous making most intertidal macroalgal species DIC saturated (e.g., Surif and Raven 1989, 1990, Beer 1994). Carbon acquisition in 38 species of red macroalgae showed that intertidal species used both dissolved CO_2 and HCO_3^- while subtidal algae are typically restricted to the use of DIC in the form of dissolved CO_2 (Murru and Sandgren 2004). Another study showed that three species of intertidal Gelidiales have a low affinity for HCO_3^- in their natural habitat (Mercado et al. 2001).

Because DIC availability is relatively high and replenished from carbonates in shells and rocks, DIC-limited aquatic photosynthesis is considered rare; however, there is evidence that some subtidal species are DIC limited, e.g., *Dilophus guineensis*, *Lobophora variegata*, and *Laurencia papillosa* collected at 3–5 m depths, and *Turbinaria turbinata* collected at 10 m (Holbrook et al. 1988). But if in situ photosynthetic rates of deep macroalgal species are light limited, they may essentially be DIC saturated under present-day CO_2 levels (Johnston et al. 1992, Surif and Raven 1989). Under elevated CO_2, downregulation of CCMs among sublittoral macroalgal species growing under light limited conditions will require less light energy for HCO_3^- transport; this energy saving may stimulate growth rate (Hepburn et al. 2011).

19.5.2 Calcifying Seaweeds

Calcified macroalgae, represented by members of phylogenetically diverse brown, green, and red species, are distributed from polar to tropical latitudes and inhabit shores ranging from the mid-high intertidal to the deepest reaches of the euphotic zone. These macroalgae produce marine sediments for reef accretion, provide three-dimensional habitat structure functioning as autogenic ecosystem engineers, contribute structural frameworks for coral reef ecosystems, and play critical ecological roles for invertebrate recruitment processes (reviewed by Nelson 2009, Ries 2009, 2010).

Regardless of the increased DIC availability, the severe consequences of the predicted lower pH and lower carbonate concentration to calcifying macroalgae are relatively well documented. The most studied calcified macroalgal group in relation to OA belongs to the red algal order Corallinales. Calcification in *Corallina pilulifera* is inhibited at pH 7.6 compared to 8.2 (Gao et al. 1993a). A 20% decrease in calcification was also observed in *Hydrolithon* sp. when incubated at 900 ppm CO_2 (pH = 7.8), while the photosynthetic rate was enhanced by 13% (Semesi et al. 2009b). For the green seaweed *Halimeda tuna* from the Great Barrier Reef, calcification declined when exposed to a pH of 7.5 compared to 8.0 (Borowitzka and Larkum 1986).

The fundamental changes in carbonate chemistry of seawater due to increasing atmospheric CO_2 levels may enhance the competitive advantage of noncalcifying over calcifying species (e.g., Fig. 19.5, Hall-Spencer et al. 2008; Kuffner et al. 2008). Moreover, its interaction with increasing sea surface temperatures may have serious impacts on calcifying macroalgae, associated biota, and coastal ecosystems as a whole.

19.6 Natural Fluctuations in Seawater Carbonate Chemistry

Most projections on how seawater carbonate chemistry and pH will change in the coming decades are based on the open ocean, where pH is relatively stable and only tends to vary on annual cycles. In coastal oceans, however, local-scale processes

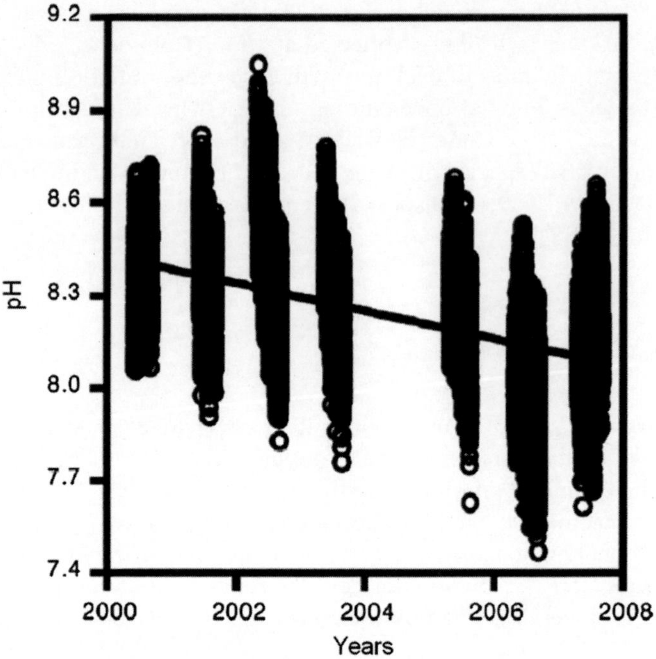

Fig. 19.3 Interannual variation in ocean pH at Tatoosh Island ($n = 24{,}519$). pH measurements as a function of date and time between 2000 and 2007, with significant decline ($P < 0.05$). (From Wootton et al. 2008, with permission of National Academy of Sciences)

including upwelling (Feely et al. 2010), river outflows (Mathis et al. 2011), and metabolic activity of the primary producers themselves (Semesi et al. 2009a) also alter pH. Predicting the effects of OA on coastal communities thus requires field information on the natural daily cycles of pH encountered by organisms in the field.

Coastal ocean pH measurements spanning 8 years around Tatoosh Island in the eastern Pacific revealed several patterns contrary to the historical perspective that ocean is well buffered (Wootton et al. 2008). A diurnal pH oscillation of 0.24 units is due to daily variation in photosynthesis (CO_2 uptake) which causes pH to increase during the light and decrease at night due to respiration and diffusion from the atmosphere which replenish CO_2 (Bensoussan and Gattuso 2007) causing a pH decline. A complex seasonal pattern in pH variation was also observed that is linked to key physical and biological drivers with known mechanistic ties to pH (cf Wootton et al. 2008). Looking at the entire temporal span of the data from year 2000 to 2008, however, a general trend of declining pH is apparent which can be attributed to the effect of increasing atmospheric CO_2 (Fig. 19.3; Wootton et al. 2008).

Diurnal variation in pH, total alkalinity (TA), and DIC within a seagrass bed is related to tidal fluctuations (Semesi et al. 2009a). During low tide, restricted water motion, higher irradiance, and temperature caused enhanced rates of seagrass photosynthesis which resulted in pH increasing from 7.9 to >8.9 and a decrease in TA and DIC. On the incoming tide, the pH was lowered and DIC and TA increased, effectively buffering the system. Such an upward shift in seawater pH due to photosynthesis also affects the physiology of local organisms. For example, higher pH (10.1), lower inorganic carbon (DIC = 0.6 mM), and possibly a supersaturated O_2 condition generated by *Ulva intestinalis* in temperate rockpools inhibit the growth of *Fucus vesiculosus* and *Chondrus crispus*; these species normally coexist in the eulittoral where the prevailing natural open water chemistry lies at pH 8.2 and 2.2 mM DIC (Björk et al. 2004). Conversely, biologically induced increases in pH (>1 pH unit) in a tropical seagrass meadow caused higher calcification rates of three seaweeds species in the community, *Hydrolithon* sp. and *Mesophyllum* sp., and *Halimeda renschii* (Semesi et al. 2009a).

Spatial and seasonal variations in $p\mathrm{CO_2}$ and DIC are also observed in sub-Antarctic kelp beds (Delille et al. 2000, 2009). Outside the kelp bed, the buffer factor (β) indicates that DIC dynamics are mainly influenced by air–sea exchange and photosynthesis, but inside the kelp bed, β suggests that DIC dynamics are controlled by calcification by the epiphytic community (Delille et al. 2000). Diel variations in $p\mathrm{CO_2}$ and DIC were prominent in spring and summer but absent in winter. The seasonal drivers responsible for variations in $p\mathrm{CO_2}$ and DIC are the higher photosynthetic rates in spring and summer, and the decay of organic matter in autumn which leads to a strong oversaturation of $p\mathrm{CO_2}$ (Delille et al. 2009).

The above examples illustrate that seaweeds themselves modify the carbonate chemistry of the surrounding seawater (in terms of total alkalinity, $p\mathrm{CO_2}$, DIC, and pH) due to their metabolic activities. This ability of seaweeds to modify their local environment needs to be taken into account in future experiments to test the effects of OA on individual seaweeds and communities.

19.7 Physicochemical Coupling

19.7.1 Water Motion and Diffusion Boundary Layers

At the surface of all aquatic organisms that have metabolic exchange across their outer surface (i.e., plankton, seaweed, and some invertebrates) is a diffusion boundary layer (DBL), a region of stagnant water across which molecules and ions move by molecular diffusion (Hurd et al. 2011). In seaweeds, the thickness of the DBL depends on thallus morphology, seawater velocity, and the type of water motion (waves vs. currents) (Hurd 2000; Hurd and Pilditch 2011). The traditional view is that thick DBLs reduce seaweed growth rates because the flux of essential nutrients (including inorganic carbon) is hampered, but there is little empirical data to support this for temperate seaweeds (reviewed by Hurd and Pilditch 2011). Metabolic processes at the seaweed surface cause the local pH to increase (photosynthesis) and decrease (respiration and calcification). Laboratory experiments demonstrate that under slow flows that are typical of many subtidal habitats (<5 cm s^{-1}), the pH at the surface of the seaweed could be considerably higher

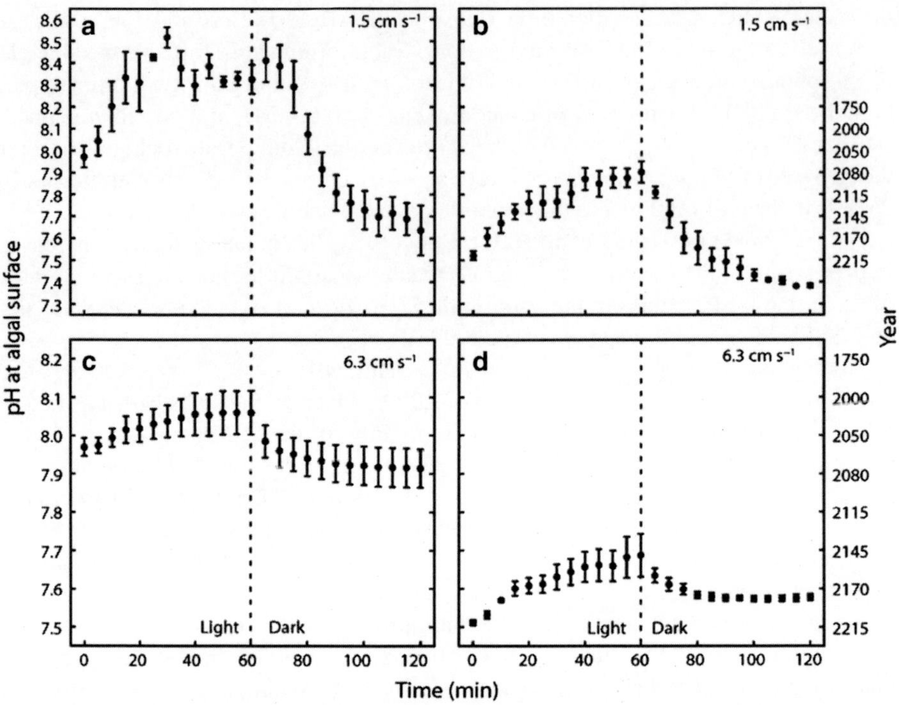

Fig. 19.4 pH fluctuations at the surface of the coralline seaweed *Sporolithon durum* measured for 1 h in the light and 1 h in the dark at an initial seawater pH of ~7.9 (**a**, **c**) and ~7.5 (**b**, **d**) at mainstream velocities of 1.5 cm s^{-1} (**a**, **b**) and 6.3 cm s^{-1} (**c**, **d**). Data are also plotted against modeled pH of surface waters (right-hand side) projected on centennial timescales, i.e., for the period 1750–2215 (Caldeira and Wickett 2003). *Symbols* represent the mean of three replicates (±1 s.e.m.). (Modified from Hurd et al. 2011, with permission of Blackwell Publishing Ltd)

or lower than that of the bulk/mainstream seawater (De Beer and Larkum 2001; Hurd et al. 2011). For example, in the light, at a mainstream flow of 1.5 cm s^{-1}, pH at the surface of *Sporolithon durum* increases to ~8.4 and declines to ~7.65 in the dark (Fig. 19.4a, Hurd et al. 2011). As seawater velocity increases, the range of these surface pH fluctuations declines because the DBL is thinner (Fig. 19.4c). When the mainstream seawater pH is decreased to 7.5 (worst-case scenario for 2215), *S. durum* is still able to raise the surface pH by 0.2 units in the light at flows of 6.3 cm s^{-1} (Fig. 19.4d) illustrating an ability to biologically modify the local pH environment, at least in the short term.

The ubiquitous nature of DBLs has implications for how seaweed communities might respond to OA. Seaweeds growing on wave-impacted shores or in fast flowing currents will experience a pH at their surface that will be similar to that of the mainstream seawater because DBLs will be thin: there are exceptions to this, however, because some seaweeds such as the giant kelp *Macrocystis pyrifera* have small (<1 mm) morphological features that trap seawater and cause thick DBLs (0.67 mm) even under fast flows. Similarly, at flows of 1.5 cm s^{-1}, the coralline red

Sporolithon durum has DBLs that are up to 1.7 mm thick, and although DBL thickness declines as seawater velocity increases, average DBL thickness at a mainstream velocity of 10 cm s^{-1} ranges from ~0.2 to 0.4 mm. In wave-sheltered sites, or for small, foliose or turf-forming algae in understory canopies, the seawater velocity at the seaweed surface will be much slower than that of mainstream flows, and they may encounter a wide range of pHs at their surface; this range is likely to vary on daily and tidal cycles. Therefore, the possibility exists that at their surface, many seaweeds naturally experience pH at night that are as low as those predicted for OA, and during the day pH at their surface, will increase, perhaps ameliorating any detrimental low-pH effects. Essentially, seaweeds in slow flows may be adapted to a wider range of pH fluctuations than those in fast flows which has implications for differential susceptibility of organisms to OA (Hurd et al. 2011). However, experiments examining the effects of water motion on the susceptibility to OA have been over short timescales (hours) and the question is whether over the long term (i.e., generations) seaweeds will have the physiological flexibility to adapt to lower pH seawater.

19.7.2 Temperature and Regional Vulnerability

Due to the greater CO_2 solubility in cold waters, seawater is expected to acidify more strongly making high latitude and polar regions more vulnerable to OA compared to tropical regions (McNeil and Matear 2008). Likewise, carbonate ion concentrations are highest in the tropics and decrease poleward in concert with decreasing temperature (see also Chap. 18 by Bartsch et al.). Carbonate ion concentrations will therefore remain higher in warmer parts of the oceans, so calcification rates in these regions should remain higher than in cooler regions. However, tropical coral reef ecosystems will be affected in multiple ways by rising CO_2, but most directly by the simultaneous changes in temperature and in seawater chemistry. Differential responses among populations of cosmopolitan species, or species with broad latitudinal distribution, to OA and greenhouse conditions (high CO_2 and high temperature) will be an important line of investigation. For example, comparisons could be made between tropical, temperate, and arctic populations of cosmopolitan macroalgae (e.g., *Acrosiphonia, Ulva, Plocamium*), and different *Macrocystis* populations; this species is widely distributed throughout the continental margins of the northern (Alaska to Mexico) and southern (Peru to southernmost Chile) Pacific Ocean, and is also found in isolated regions of South Africa and around most of the sub-Antarctic islands to 60°S (Graham et al. 2007). A poleward shift in the geographic range of major subtidal kelp species has been suggested for future climate scenarios in marine coastal ecosystems (Merzouk and Johnson 2011).

19.7.3 pH, Chemical Speciation, and Nutrient Availability

The effect of OA on seawater chemistry goes beyond the carbonate system (see also Chap. 4 by Gordillo). Chemical species that may undergo speciation under lower pH include major (phosphorus, silicon, and nitrogen) and minor (boron) elements as well as trace elements such as iron and zinc (Doney et al. 2009). Moreover, the concentration of chemical species such as phosphate, silicate, and ammonia decreases with decreasing pH (Fig. 1.2.11 in Zeebe and Wolf-Gladrow 2001). Recently, a decline in oceanic nitrification rates is reported as a consequence of OA (Beman et al. 2011). The estimated 3–44% reduction in the next decades can affect oceanic nitrous oxide production, reducing oxidized nitrogen (NO_2^-, NO_3^-) supplies in the upper layers of the ocean, and fundamentally altering nitrogen cycling in the ocean (Beman et al. 2011). The consequence of nutrient speciation and their bioavailability to macroalgae is largely unknown; fractional changes may alter photosynthesis, growth, and nutritional value which could affect the rest of the food web (see Sect. 19.8.1). Likewise, dissolved organic matter (DOM) which undergoes hydrolytic reactions in seawater, e.g., organic acids, amino acids, nucleic acids, proteins, and humic materials, will also be effectively altered by changing pH (Doney et al. 2009). Therefore, the overall impact of decreasing pH on these biologically important organic compounds requires further in-depth investigation.

19.8 Trophic Dynamics and Coastal Ecosystem Response

19.8.1 Stoichiometric Ratio of Aquatic Ecosystems

Interactions between global climate stressors (e.g., CO_2 and temperature) and local perturbation (e.g., eutrophication) will not only cause regime shifts in coastal ecosystems favoring spatial expansion of ephemeral turfs and persistence beyond their normal seasonal limits (Russell and Connell 2009) but also stoichiometry of aquatic ecosystems.

When the ocean becomes enriched with carbon due to rising levels of atmospheric CO_2, the associated warming and vertical stratification of water suppresses nutrient supply from deep water into the surface layers. The consequent increase in carbon and decrease in nutrient concentration results in an increase in the cellular carbon:nutrient ratio. Phytoplankton with high carbon-to-nutrient ratios are of low nutritional value for zooplankton; such change in stoichiometry may cascade throughout the entire aquatic food web (van de Waal et al. 2010).

Conversely, if a coastal ecosystem receives a concurrent high supply of nutrients of anthropogenic sources, the probable consequence of altering the carbon:nutrient ratios of seaweeds, and their nutritional values to herbivorous fishes and invertebrates, is largely underappreciated. For example, the molar C:N ratio of *Ulva rigida* under nitrogen-limited conditions increased by ~10% when the CO_2 concentration was increased from 350 to 10,000 ppm. However, when the elevated CO_2 condition was coupled with a 20× increase in nitrogen, a >20% decrease in C:N ratio was observed (Gordillo et al. 2001). On the other hand, significant increases in the C:N ratio in a marine angiosperm, *Thalassia hemprichii*, is reported under increasing CO_2; the increase in tissue carbon content was seen as a positive

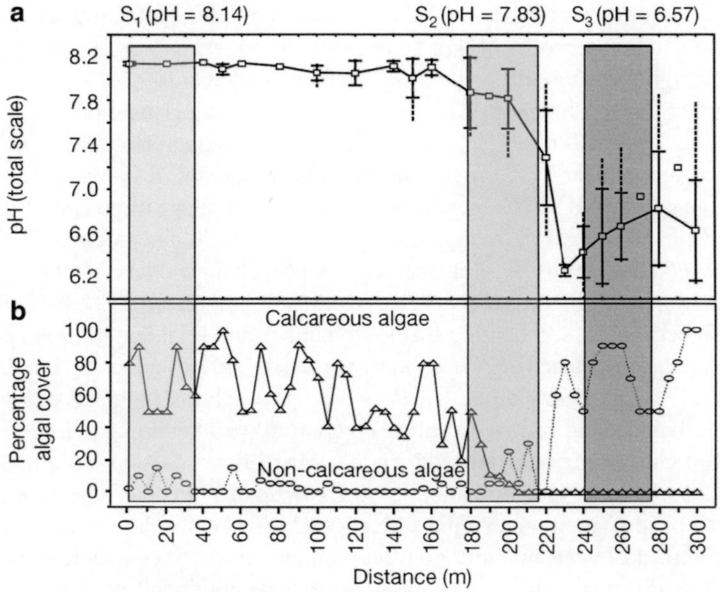

Fig. 19.5 Variation in pH at CO_2 vents south of Castello d' Aragonese (Ischia Island, Italy) and abundance of macroalgae. (**a**) Mean pH ± s.d. data from stations S1–S3. (**b**) Percent cover of calcareous (*triangles*) and noncalcareous macroalgae (*circle*). (Modified from Hall-Spencer et al. 2008, with permission of Macmillan Publishers Limited)

response to OA in coastal seas (Jiang et al. 2010). However, neither study considered the consequence of the shift in C:N ratio on ecosystem trophic dynamics.

19.8.2 Community Structure

Natural volcanic CO_2 vents have conditions comparable to the high CO_2/low pH conditions progressively developing in oceans through the uptake of fossil-fuel CO_2. Hall-Spencer et al. (2008) used this natural ecosystem to demonstrate the effect of OA on biodiversity and ecosystem structure and function. They showed that organisms such as sea urchins, coralline algae, and stony corals decline in abundance or completely disappear with decreasing pH close to the vents, while seagrasses and brown algae benefit from elevated CO_2 availability close to the vent by increasing their biomass. Percentage cover of calcareous algae dropped from >60% outside the vent area (S1, pH = 8.14) to zero cover within it (S2–S3, pH = 7.83–6.57), while the reverse was observed with noncalcareous algal cover (Fig. 19.5). Macroalgae observed to be resilient to naturally high amounts of pCO_2 include *Caulerpa*, *Cladophora*, *Asparagopsis*, *Dictyota*, and *Sargassum*. This study showed that lowered pH due to OA, even without global warming that is associated with rise in atmospheric CO_2, can bring about loss in biodiversity and shift in ecosystem structure and function (Hall-Spencer et al. 2008). Recently, Porzio et al. (2011) showed that reproduction of some resilient seaweeds was not affected under extremely high CO_2 conditions near the volcanic CO_2 vents.

Manipulative experiments (using the HCl method of pH manipulation) showed that OA, at 2× present day (pH = 7.91), is responsible for a 79% reduction in crustose coralline algae (CCA) recruitment, 40% reduction in growth rate and 92% reduction in cover, and a 52% increase of a mixed assemblage of noncalcifying algae composed of macroalgal germlings, diatoms, and small filamentous algae (Kuffner et al. 2008). Likewise, using multispecies Markov chain models, a method used to link environmental change to species dynamics, Wootton et al. (2008) showed a strong link between in situ benthic species dynamics and variation in ocean pH. They showed that the abundance of calcareous invertebrate species generally declined with declining pH while fleshy algae (*Halosaccion glandiforme*, ephemeral algae, filamentous red algae, and foliose red algae) increased with declining pH.

Elevated CO_2 may potentially influence competitive exclusion of certain species within a community. Quantification of carbon uptake mechanism (using carbon isotope and pH drift methods) within a kelp forest community showed that the canopy (Laminarians and Fucaleans) and dominant understory species (calcifying coralline algae) all had functional CCMs, while the less dominant understory species, consisting of mostly red noncalcifying species, relied on diffusive CO_2 uptake and had no functional CCMs (Hepburn et al. 2011). Results of the above study suggest that in the low-pH coastal waters of the future, the growth rates and competitive abilities of noncalcifying and CO_2-only seaweeds will increase, while those of calcifying CCM seaweeds will decrease. This could cause major shifts in community structure, altering functional diversity and near shore ecosystem functioning (Hepburn et al. 2011). However, it is difficult to predict how ecosystems might change because some HCO_3^--using species can switch to CO_2 use at lower pH 7.5, thereby increasing the energy available for growth and reproduction (Cornwall et al. 2012).

19.8.3 Interactive Effects of Ocean Acidification and Other Stressors

The results of single-factor experiments may be different, or the physiological response be of a lower magnitude, compared to the interactive effects of multiple stressors that most seaweeds are exposed to in their natural habitats. For example, calcification rate of CCA *Lithophyllum cabiochae* was not affected when independently exposed to elevated CO_2 or temperature. However, when exposed to a combination of elevated CO_2 and temperature, calcification rates decreased by 50% and an accelerated and extensive (~60%) tissue necrosis was observed (Martin and Gattuso 2009). Higher sensitivity to OA under higher temperature is also reported in the CCA *Hydrolithon onkodes* which sustained higher skeletal dissolution and concurrent increase in the abundance of their co-habiting endolithic algae (Diaz-Pulido et al. 2012).

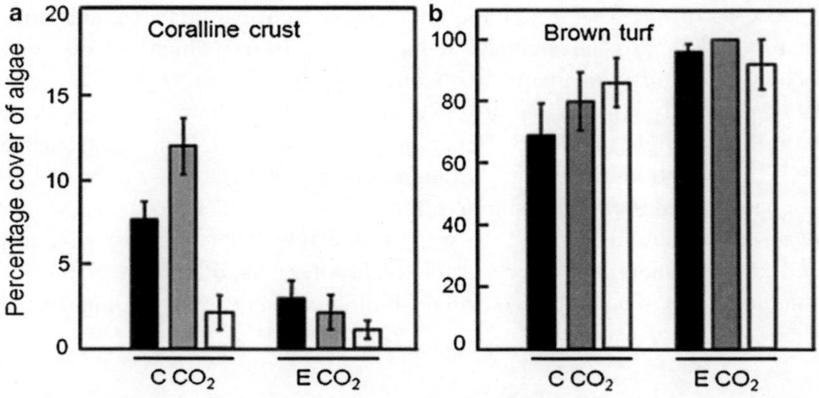

Fig. 19.6 Percentage cover (mean ± SE) of (**a**) coralline crusts and (**b**) turfs recruited to unoccupied substrates when exposed to crossed combination of CO_2 (control [CCO_2] vs. elevated [ECO_2]) and light (shade vs. low UV vs. full sunlight; in *black, gray, and white bars*, respectively). (From Russell et al. 2011, with permission of Blackwell Publishing Ltd)

On the other hand, elevated CO_2 concentration (1,000 ppm) in synergy with ultraviolet radiation is reported to systemically negatively impact the physiology of an articulate coralline alga *Corallina sessilis* in terms of photosynthesis, photosynthetic and accessory pigments, growth, and calcification rates (Gao and Zheng 2010). Positive effects of elevated CO_2 in synergy with other environmental stressors are also reported. Emersed net photosynthesis of *Porphyra haitanensis* increased with higher CO_2 (700 ppm), temperature, and level of desiccation (Zou and Gao 2002).

At the community level, the synergistic effect of global increase in CO_2 and a local stressor such as eutrophication can select for resilient species and cause phase shifts in coastal ecosystem. For example, the combined effect of high CO_2 and nutrients accelerates the expansion of filamentous turfs at the expense of calcifying algae (Russell et al. 2009). Likewise, the synergistic effect of high CO_2 and high temperature favors the filamentous turfs over the kelps (Connell and Russell 2010). Conversely, increased photosynthesis and growth of seagrasses under nutrient replete and elevated CO_2 may benefit some calcifiers, as shown in the higher calcification rate of *Hydrolithon* sp. in the proximity of seagrasses, due to the drawdown of CO_2 from the water column (Semesi et al. 2009a).

Light (including ultraviolet radiation) is known for its role in shaping the subtidal ecology of photosynthetic organism (Bischof et al. 2006). However, Russell et al. (2011) found that the forecasted increase in CO_2 levels can alter how light influences community structure. For example, under control CO_2 (pH = 8.06–8.16) the effects of light on the percent cover of the two groups of algae studied, the calcifying red crustose *Lithophyllum* sp. and the turf-forming brown *Feldmannia* spp., corresponded to their growth habitat's light environments: the subcanopy red coralline grew better under low UV while the highest cover of the brown turf was observed under full sunlight (Fig. 19.6). However, under elevated CO_2 (pH = 7.84–8.12), the influence of light on the percentage cover of understory crusts was substantially reduced (Fig. 19.6a) which caused crusts to grow less under all light treatments. Conversely, elevated CO_2 had the opposite effect of positively influencing turf cover (Fig. 19.6b), thereby reducing the structuring effects of light and UVB. Hence, to predict the ecological consequences of future CO_2 conditions, the role of contemporary processes (e.g., light and UVB) cannot be assumed to produce similar effects relative to other processes, which will change with a changing climate (Russell et al. 2011). Therefore, more studies are needed on the synergistic effects of different stress factors at the organism and community levels to better understand the biological and ecosystem impacts of OA.

19.9 Experimental Considerations

The possible effects of OA have been studied for ~10 years and an enormous volume of research has been produced during this time. Most studies have focused on reducing pH and measuring response variables of individual algae or communities. There are three methods available to lower SW pH: 1. addition of HCl, 2. bubbling with CO_2 gas, or 3. HCl/NaOH additions (Hurd et al. 2009; Riebesell et al. 2010). Methods 2 and 3 alter seawater carbonate chemistry in a way that mimics OA (TA remains constant, DIC increases), while using HCl, alkalinity decreases but there is no change in DIC. Importantly, two seawater carbonate parameters must be measured during experiments so that the carbonate speciation within experiments can be followed. Some first-generation OA experiments on seaweeds suffered from pseudoreplication or lack of independent replication (Hurd et al. 2009). However, with the design of new systems for controlling and monitoring pH, many of these experimental problems have been resolved (e.g., McGraw et al. 2010). Because seawater carbonate chemistry is complex, and because algae themselves modify their local pH environment, it is important that biologists and carbonate chemists collaborate on OA studies to avoid the many pitfalls that will result in a nonrigorous experimental design.

19.10 Conclusions and Future Works

The responses of ecosystems to OA are difficult to predict because all algae, calcifying and noncalcifying, could be affected. Most macroalgae are HCO_3^- users, and the predicted increases in DIC may have little effect on growth rates compared to seagrasses and seaweeds that rely on CO_2. For calcifiers, several studies have shown that calcification rates are depressed by increased levels of CO_2. This is one of the many indications of a negative biological response in the marine environment. Predicting the responses of ocean chemistry to rising CO_2 is reasonably straightforward (e.g., Caldeira and Wickett 2003), but with fairly modest biological data available, we cannot easily predict long-term species and community responses to climate and other environmental changes, and their feedback on climate in the future. Our present understanding of the potential effects of OA on marine organisms stems largely from short-term laboratory and mesocosm experiments. But chronic exposure to increased pCO_2 may have complex effects that are different to, and/or not observed, in short-term experiments. For example, under long-term exposure to elevated CO_2, certain organisms may employ a range of mechanisms for adaptation. Moreover, the response of individual organisms, populations, and communities to a more realistic gradual increase in CO_2 is largely unknown (cf Boyd et al. 2008). Nevertheless, highly controlled short-term single or multifactorial laboratory experiments are also important in identifying the species' preadapted sensitivities to increasing CO_2. Considering that the CO_2 level expected in 2100 will be the highest for at least the past 24 million years, physiological acclimation and genetic adaptation might, at least in part, counteract adverse effects. Therefore, studies at molecular and genetic levels can contribute to understanding the genetic bases of resistance or resilience in marine macrophytes to changes in atmospheric CO_2 concentration.

The susceptibility of the early life history stages (reproductive cells, e.g., spores and gametes) as well as reproductive success (e.g., sporogenesis and gametogenesis) in the complex life histories of different seaweeds, e.g., bi- to triphasic, and heteromorphic involving microstages, under elevated CO_2/low pH needs to be addressed (e.g., Roleda et al. 2012). The microstages and juveniles are known to be more susceptible to environmental stressors compared to adult macrostages, e.g., life-stage-dependent UVR susceptibility (Roleda et al. 2007, 2009). Survival of the early life history stages to global climate change related stressors is essential for recruitment success to maintain species diversity and standing stock to sustain community metabolism.

Acknowledgments The authors were funded by the Royal Society of New Zealand Marsden Fund (UOO0914). We thank the two anonymous reviewers for their helpful comments.

References

Adey WH, Steneck RS (2001) Thermography over time creates biogeographic regions: a temperature/space/time-integrated model and an abundance-weighted test for benthic marine algae. J Phycol 37:677–698

Axelsson L, Larsson C, Ryberg H (1999) Affinity, capacity and oxygen sensitivity of two different mechanisms for bicarbonate utilization in *Ulva lactuca* L. (Chlorophyta). Plant Cell Environ 22:969–978

Axelsson L, Mercado JM, Figueroa FL (2000) Utilization of HCO_3^- at high pH by the brown macroalgae *Laminaria saccharina*. Eur J Phycol 35:53–59

Badger MR, Andrews TJ, Whitney SM, Ludwig M, Yellowlees DC, Leggat W, Price GD (1998) The diversity and coevolution of Rubisco, plastids, pyrenoids, and chloroplast-based CO_2-concentrating mechanisms in algae. Can J Bot 76:1052–1071

Beardall J, Beer S, Raven JA (1998) Biodiversity of marine plants in an era of climate change: some predictions based on physiological performance. Bot Mar 41:113–123

Beer S (1994) Mechanisms of inorganic carbon acquisition in marine macroalgae (with special reference to the Chlorophyta). Prog Phycol Res 10:179–207

Beman JM, Chow CE, King AL, Feng YY, Fuhrman JA, Andersson A, Bates NR, Popp BN, Hutchins DA (2011) Global declines in oceanic nitrification rates as a consequence of ocean acidification. Proc Natl Acad Sci USA 108:208–213. doi:10.1073/pnas.1011053108

Bensoussan N, Gattuso J-P (2007) Community primary production and calcification in a NW Mediterranean ecosystem dominated by calcareous macroalgae. Mar Ecol Prog Ser 334:37–45

Bischof K, Gómez I, Molis M, Hanelt D, Karsten U, Lüder U, Roleda MY, Zacher K, Wiencke C (2006) Ultraviolet radiation shapes seaweed communities. Rev Environ Sci Biotechnol 5:141–166. doi:10.1007/s11157-006-002-3

Björk M, Haglund K, Ramazanov Z, Garcia-Reina G, Pedersen M (1992) Inorganic-carbon assimilation in the green seaweed *Ulva rigida* C. Ag. (Chlorophyta). Planta 187:152–156

Björk M, Haglund K, Ramazanov Z, Pedersen M (1993) Inducible mechanisms for HCO_3^- utilization and repression of photorespiration in protoplasts and thalli of three species of *Ulva* (Chlorophyta). J Phycol 29:166–173

Björk M, Axelsson L, Beer S (2004) Why is *Ulva intestinalis* the only macroalga inhabiting isolated rock pools along the Swedish Atlantic coast? Mar Ecol Prog Ser 284:109–116

Black CC, Bender MM (1976) δ^{13}C values in marine organisms from the Great Barrier Reef. Aust J Plant Physiol 3:25–32

Bode A, Alvarez-Ossorio MT, Varela M (2006) Phytoplankton and macrophyte contributions to littoral food webs in the Galician upwelling estimated from stable isotopes. Mar Ecol Prog Ser 318:89–102

Borowitzka MA (1987) Calcification in algae: mechanisms and the role of metabolism. CRC Cr Rev Plant Sci 6:1–45

Borowitzka MA, Larkum AWD (1976) Calcification in the green alga *Halimeda*. 3. sources of inorganic carbon for photosynthesis and calcification and a model of mechanism of calcification. J Exp Bot 27:879–893

Borowitzka MA, Larkum AWD (1986) Reef algae. Oceanus 29:49–54

Boyd P, Doney SC, Strzepek R, Dusenberry J, Lindsay K, Fung I (2008) Climate-mediated changes to mixed-layer properties in the Southern Ocean: assessing the phytoplankton response. Biogeosciences 5:847–864

Brownlee C, Taylor A (2004) Calcification in coccolithophores: A cellular perspective. In: Thierstein HR, Young JR (eds) Coccolithophores: from cellular processes to global impact. Springer, Berlin, pp 31–49

Caldeira K, Wickett ME (2003) Anthropogenic carbon and ocean pH. Science 425:365–365. doi:10.1038/425365a

Cao L, Caldeira K (2008) Atmospheric CO_2 stabilization and ocean acidification. Geophys Res Lett. doi:10.1029/2008GL035072

Connell SD, Russell BD (2010) The direct effects of increasing CO_2 and temperature on non-calcifying organisms: increasing the potential for phase shifts in kelp forest. Proc R Soc B-Biol Sci 227:1409–1415. doi:10.1098/rspb.2009.2069

Corbisier TN, Soares LSH, Petti MAV, Muto EY, Silva MHC, McClelland J, Valiela I (2006) Use of isotopic signatures to assess the food web in a tropical shallow marine ecosystem of Southeastern Brazil. Aquat Ecol 40:381–390. doi:10.1007/s10452-006-9033-7

Cornwall CE, Hepburn CD, Pritchard D, Currie KI, McGraw CM, Hunter KA, Hurd CL (2012) Carbon use-strategies in macroalgae: differential responses to lowered pH and implications for ocean acidification. J Phycol 48:137–144. doi:10.1111/j.1529-8817.2011.01085.x

De Beer D, Larkum AWD (2001) Photosynthesis and calcification in the calcifying algae *Halimeda discoidea* studied with microsensors. Plant Cell Environ 24:1209–1217

Delille B, Delille D, Fiala M, Prevost C, Frankignoulle M (2000) Seasonal changes of pCO_2 over a subantarctic *Macrocystis* kelp bed. Polar Biol 23:706–716

Delille B, Borges AV, Delille D (2009) Influence of giant kelp beds (*Macrocystis pyrifera*) on diel cycles of pCO_2 and DIC in the Sub-Antarctic coastal area. Estuar Coast Shelf S 81:114–122. doi:10.1016/j.ecss.2008.10.004

Diaz-Pulido G, Anthony KRN, Kline DI, Dove S, Hoegh-Guldberg O (2012) Interactions between ocean acidification and warming on the mortality and dissolution of coralline algae. J Phycol 48:32–39. doi:10.1111/j.1529-8817.2011.01084.x

Doney SC, Fabry VJ, Feely RA, Kleypas JA (2009) Ocean acidification: the other CO_2 problem. Annu Rev Mar Sci 1:169–192. doi:10.1146/annurev.marine.010908.163834

Dromgoole FI (1978) The effects of pH and inorganic carbon on photosynthesis and dark respiration of *Carpophyllum* (Fucales, Phaeophyceae). Aquat Bot 4:11–22. doi:10.1016/0304-3770(78)90003-7

Dunton KH (2001) $\delta^{15}N$ and $\delta^{13}C$ measurements of Antarctic peninsula fauna: trophic relationships and assimilation of benthic seaweeds. Am Zool 41:99–112

Feely RA, Sabine CL, Lee K, Berelson W, Kleypas JA, Fabry VJ, Millero FJ (2004) Impact of anthropogenic CO_2 on the $CaCO_3$ system in the oceans. Science 305:362–366. doi:10.1126/science.1097329

Feely RA, Alin SR, Newton J, Sabine CL, Warner M, Devol A, Krembs C, Maloy C (2010) The combined effects of ocean acidification, mixing, and respiration on pH and carbonate saturation in an urbanized estuary. Estuar Coast Shelf Sci 88:442–449. doi:10.1016/j.ecss.2010.05.004

Fischer G, Wiecke C (1992) Stable carbon isotope composition, depth distribution and fate of macroalgae from the Antarctic Peninsula region. Polar Biol 12:341–348

Gao K, McKinley KR (1994) Use of macroalgae for marine biomass production and CO_2 remediation: a review. J Appl Phycol 6:45–60

Gao KS, Zheng YQ (2010) Combined effects of ocean acidification and solar UV radiation on photosynthesis, growth, pigmentation and calcification of the coralline alga *Corallina sessilis* (Rhodophyta). Glob Change Biol 16:2388–2398. doi:10.1111/j.1365-2486.2009.02113.x

Gao K, Aruga Y, Asada K, Ishihara T, Akano T, Kiyohara M (1991) Enhanced growth of the red alga *Porphyra yezoensis* Ueda in high CO_2 concentration. J Appl Phycol 3:355–362

Gao K, Aruga Y, Asada K, Ishihara T, Akano T, Kiyohara M (1993a) Calcification in the articulated coralline alga *Corallina pulifera* with special reference to the effect of elevated CO_2. Mar Biol 117:129–132

Gao K, Aruga Y, Asada K, Kiyohara M (1993b) Influence of enhanced CO_2 on growth and photosynthesis of the red algae *Gracilaria* sp. & *G. chilensis*. J Appl Phycol 5:563–571

Gerhart LM, Ward JK (2010) Plant responses to low $[CO_2]$ of the past. New Phytol 188:674–695. doi:10.1111/j.1469-8137.2010.03441.x

Giordano M, Maberly SC (1989) Distribution of carbonic anhydrase in British marine macroalgae. Oecologia 81:534–539

Giordano M, Beardall J, Raven JA (2005) CO_2 concentrating mechanisms in algae: mechanisms, environmental modulation, and evolution. Annu Rev Plant Biol 56:99–131. doi:10.1146/annurev.arplant.56.032604.144052

Gordillo FJL, Niell FX, Figueroa FL (2001) Non-photosynthetic enhancement of growth by high CO_2 level in the nitrophilic seaweed *Ulva rigida* C. Agardh (Chlorophyta). Planta 213:64–70

Graham MH, Vásquez JA, Buschmann AH (2007) Global ecology of the giant kelp *Macrocystis*: from ecotypes to ecosystems. Oceanogr Mar Biol 45:39–88

Hall-Spencer J, Rodolfo-Metalpa R, Martin S, Ransome E, Fine M, Turner SM, Rowley SJ, Tedesco D, Buia M-C (2008) Volcanic carbon dioxide vents show ecosystem effects of ocean acidification. Nature 454:96–99. doi:10.1038/nature07051

Hepburn CD, Pritchard DW, Cornwall CE, McLeod RJ, Beardall J, Raven JA, Hurd CL (2011) Diversity of carbon use strategies in a kelp forest community: implications for a high CO_2 ocean. Global Change Biol 17:2488–2497. doi:10.1111/j.1365-2486.2011.02411.x

Holbrook GP, Beer S, Spenser WE, Reiskind J, Davis JS, Bowes G (1988) Photosynthesis in marine macroalgae: evidence for carbon limitation. Can J Bot 66:577–582

Hurd CL (2000) Water motion, marine macroalgal physiology, and production. J Phycol 36:453–472

Hurd CL, Pilditch CA (2011) Flow-induced morphological variations affect diffusion boundary-layer thickness of *Macrocystis pyrifera* (Heterokontophyta, Laminariales). J Phycol 47:341–351. doi:10.1111/j.1529-8817.2011.00958.x

Hurd CL, Hepburn CD, Currie KI, Raven JA, Hunter KA (2009) Testing the effects of ocean acidification on algal metabolism: considerations for experimental designs. J Phycol 45:1236–1251

Hurd CL, Cornwall CE, Currie K, Hepburn CD, McGraw CM, Hunter KA, Boyd PW (2011) Metabolically-induced pH fluctuations by some coastal calcifiers exceed projected 22nd century ocean acidification: a mechanism for differential susceptibility? Glob Change Biol doi:10.1111/j.1365-2486.2011.02473.x

IPCC (2007) Climate Change 2007: The physical science basis. Summary for policymakers. Contribution of working group I to the fourth assessment report. The International Panel on Climate Change, http://www.ipcc.ch/SPM2feb07.pdf

Jiang ZJ, Huang X-P, Zhang J-P (2010) Effects of CO_2 enrichment on photosynthesis, growth, and biochemical composition of seagrass *Thalassia hemprichii* (Ehrenb.) Aschers. J Integr Plant Biol 52:904–913. doi:10.1111/j.1744-7909.2010.00991.x

Johnston AM (1991) The acquisition of inorganic carbon by marine macroalgae. Can J Bot 69:1123–1132

Johnston AM, Raven JA (1991) Effects of culture in high CO_2 on the photosynthetic physiology of *Fucus serratus*. Br Phycol J 25:75–82

Johnston AM, Maberly SC, Raven JA (1992) The acquisition of inorganic carbon by four red macroalgae. Oecologia 92:317–326

Kang C-K, Choy EJ, Son Y, Lee J-K, Kim JK, Kim Y, Lee K-S (2008) Food web structure of a restored macroalgal bed in the eastern Korean peninsula determined by C and N stable isotope analyses. Mar Biol 153:1181–1198. doi:10.1007/s00227-007-0890-y

Kevekordes K, Holland D, Häubner N, Jenkin S, Koss R, Roberts S, Raven JA, Scrimgeour CM, Shelly K, Stojkovic S, Beardall J (2006) Inorganic carbon acquisition by eight species of *Caulerpa* (Caulerpaceae, Chlorophyta). Phycologia 45:442–449. doi:10.2216/05-55.1

Kremer BP (1981) Aspects of carbon metabolism in marine macroalgae. In: Barnes HB, Barnes M (eds) Oceanography and marine biology: an annual review, 19. Aberdeen University Press, Aberdeen, UK, pp 41–94

Kübler JE, Raven JA (1994) Consequences of light limitation for carbon acquisition in three rhodophytes. Mar Ecol Prog Ser 110:203–209

Kübler JE, Johnston AM, Raven JA (1999) The effects of reduced and elevated CO2 and O2 on the seaweed *Lomentaria articulata*. Plant Cell Environ 22:1303–1310

Kuffner IB, Andersson AJ, Jokiel PL, Rodgers KS, Mackenzie FT (2008) Decreased abundance of crustose coralline algae due to ocean acidification. Nat Geosci 1:114–117. doi:10.1038/ngeo100

Laurand S, Riera P (2006) Trophic ecology of the supralittoral rocky shore (Roscoff, France): a dual stable isotope ($\delta^{13}C$, $\delta^{15}N$) and experimental approach. J Sea Res 56:27–36. doi:10.1016/j.seares.2006.03.002

Maberly SC (1990) Exogenous sources of inorganic carbon for photosynthesis by marine macroalgae. J Phycol 26:439–449

Maberly SC, Raven JA, Johnston AM (1992) Discrimination between ^{12}C and ^{13}C by marine plants. Oecologia 91:481–492

Magnusson G, Larsson C, Axelsson L (1996) Effects of high CO_2 treatment on nitrate and ammonium uptake by *Ulva lactuca* grown in different nutrient regimes. Sci Mar 60(Supplement 1):179–189

Martin S, Gattuso JP (2009) Response of Mediterranean coralline algae to ocean acidification and elevated temperature. Glob Change Biol 15:2089–2100. doi:10.1111/j.1365-2486.2009.01874.x

Mathis JT, Cross JN, Bates NR (2011) Coupling primary production and terrestrial runoff to ocean acidification and carbonate mineral suppression in the eastern Bering Sea. J Geophys Res-Oceans. doi:10.1029/2010JC006453

McGraw CM, Cornwall CE, Reid MR, Currie KI, Hepburn CD, Boyd P, Hurd CL, Hunter KA (2010) An automated pH-controlled culture system for laboratory-based ocean acidification experiments. Limnol Oceanogr Meth 8:686–694. doi:10:4319/lom.2010.8.686

McNeil BI, Matear RJ (2008) Southern Ocean acidification: a tipping point at 450-ppm atmospheric CO_2. Proc Natl Acad Sci USA 105:18860–18864. doi:10.1073/pnas.0806318105

Mercado JM, Niell FX, Gil-Rodríquez MC (2001) Photosynthesis might be limited by light, not inorganic carbon availability, in three intertidal Gelidiales species. New Phytol 149:431–439

Mercado JM, Andria JR, Pérez-Llorens JL, Vergara JJ, Axelsson L (2006) Evidence for a plasmalemma-based CO_2 concentrating mechanism in *Laminaria saccharina*. Photosynth Res 88:259–268. doi:10.1007/s11120-006-9039-y

Merzouk A, Johnson LE (2011) Kelp distribution in the northwest Atlantic Ocean under a changing climate. J Exp Mar Biol Ecol 400:90–98. doi:10.1016/j.jembe.2011.02.020

Moroney JV, Ynalvez RA (2007) Proposed carbon dioxide concentrating mechanism in *Chlamydomonas reinhardtii*. Eukaryot Cell 6:1251–1259. doi:10.1128/EC.00064-07

Murru M, Sandgren CD (2004) Habitat matters for inorganic carbon acquisition in 38 species of red macroalgae (Rhodophyta) from Puget Sound, Washington, USA. J Phycol 40:837–845. doi:10.1111/j.1529-8817.2004.03182.x

Nelson WA (2009) Calcified macroalgae- critical to coastal ecosystems and vulnerable to change: a review. Mar Freshwater Res 60:787–801. doi:10.1071/MF08335

Pinnegar JK, Polunin NVC (2000) Contributions of stable-isotope data to elucidating food webs of Mediterranean rocky littoral fishes. Oecologia 122:399–409

Porzio L, Buia MC, Hall-Spencer JM (2011) Effects of ocean acidification on macroalgal communities. J Exp Mar Biol Ecol 400:278–287. doi:10.1016/j.jembe.2011.02.011

Raven JA (2010) Inorganic carbon acquisition by eukaryotic algae: four current questions. Photosynth Res 106:123–134. doi:10.1007/s11120-010-9563-7

Raven JA, Giordano M (2009) Biomineralization by photosynthetic organisms: evidence of coevolution of the organisms and their environment? Geobiology 7:140–154. doi:10.1111/j.1472-4669.2008.00181.x

Raven JA, Walker DI, Johnston AM, Handley LL, Kübler JE (1995) Implications of ^{13}C natural abundance measurements for photosynthetic performance by marine macrophytes in their natural environment. Mar Ecol Prog Ser 123:193–205

Raven JA, Johnston AM, Kübler JE, Korb R, McInroy SG, Handley LL, Scrimgeour CM, Walker DI, Beardall J, Clayton MN, Vanderklift M, Fredriksen S, Dunton KH (2002a) Seaweeds in cold seas: evolution and carbon acquisition. Ann Bot 90:525–536. doi:10.1093/aob/mcf171

Raven JA, Johnston AM, Kübler JE, Korb R, McInroy SG, Handley LL, Scrimgeour CM, Walker DI, Beardall J, Vanderklift M, Fredriksen S, Dunton KH (2002b) Mechanistic interpretation of carbon isotope discrimination by marine macroalgae and seagrasses. Funct Plant Biol 29:355–378. doi:10.1071/PP01201

Raven JA, Ball LA, Beardall J, Giordana M, Maberly SC (2005) Algae lacking carbon-concentrating mechanisms. Can J Bot 83:879–890. doi:10.1139/B05-074

Raven JA, Cockell CS, De La Rocha CL (2008) The evolution of inorganic carbon concentrating mechanisms in photosynthesis. Philos Trans R Soc B 363:2641–2650. doi:10.1098/rstb.2008.0020

Riebesell U, Fabry VJ, Hansson L, Gattuso J-P (2010) Guide to best practices for ocean acidification research and data reporting. Publications Office of the European Union, Luxembourg

Ries JB (2009) Effects of secular variation in seawater Mg/Ca ratio (calcite-aragonite seas) on $CaCO_3$ sediment production by the calcareous algae *Halimeda*, *Penicillus* and *Udotea* – evidence from recent experiments and the geological record. Terra Nova 21:323–339. doi:10.1111/j.1365-3121.2009.00899.x

Ries JB (2010) Review: geological and experimental evidence for secular variation in seawater Mg/Ca (calcite-aragonite seas) and its effects of marine biological calcification. Biogeosciences 7:2795–2849. doi:10.5194/bg-7-2795-2010

Roleda MY, Wiencke C, Hanelt D, Bischof K (2007) Sensitivity of the early life history stages of macroalgae from the Northern Hemisphere to ultraviolet radiation. Photochem Photobiol 83:851–862. doi:10.1562/2006-08-17-IR-1005

Roleda MY, Campana GL, Wiencke C, Hanelt D, Quartino ML, Wulff A (2009) Sensitivity of Antarctic *Urospora penicilliformis* (Ulotrichales, Chlorophyta) to ultraviolet radiation is life-stage dependent. J Phycol 45:600–609. doi:10.1111/j.1529-8817.2009.00691.x

Roleda MY, Morris JN, McGraw CM, Hurd CL (2012) Ocean acidification and seaweed reproduction: increased CO_2 ameliorates the negative effect of lowered pH on meiospore germination in the giant kelp *Macrocystis pyrifera* (Laminariales, Phaeophyceae). Global Change Biol 18:854–864. doi:10.1111/j.1365-2486.2011.02594.x

Royal Society (2005) Ocean acidification due to increasing atmospheric carbon dioxide. Policy document 12/05 Royal Society, London. The Clyvedon press Ltd, Cardiff

Runcie JW, Gurgel CFD, McDermid KJ (2008) In situ photosynthetic rates of tropical marine macroalgae at their lower depth limit. Eur J Phycol 43:377–388. doi:10.1080/09670260801979303

Russell BD, Connell SD (2009) Eutrophication science: moving into the future. Trends Ecol Evol 24:527–528. doi:10.1016/j.tree.2009.06.001

Russell BD, Thompson JAI, Falkenberg LJ, Connell SD (2009) Synergistic effects of climate change and local stressors: CO_2 and nutrient-driven change in subtidal rocky habitats. Glob Change Biol 15:2153–2162. doi:10.1111/j.1365-2486.2009.01886.x

Russell BD, Passarelli CA, Connell SD (2011) Forecasted CO2 modifies the influence of light in shaping subtidal habitat. J Phycol doi:10.1111/j.1529-8817.2011.01002.x

Semesi IS, Beer S, Bjork M (2009a) Seagrass photosynthesis controls rates of calcification and photosynthesis of calcareous macroalgae in a tropical seagrass meadow. Mar Ecol Prog Ser 382:41–47. doi:10.3354/meps07973

Semesi IS, Kangwe J, Björk M (2009b) Alterations in seawater pH and CO_2 affect calcification and photosynthesis in the tropical coralline alga, *Hydrolithon* sp. (Rhodophyta). Estuar Coast Shelf Sci 84:337–341. doi:10.1016/j.ecss.2009.03.038

Sherlock DJ, Raven JA (2001) Interactions between carbon dioxide and oxygen in the photosynthesis of three species of marine red algae. Bot J Scotl 53:33–43

Smith RG, Bidwell RGS (1989) Inorganic carbon uptake by photosynthetically active protoplast of the red macroalga *Chondrus crispus*. Mar Biol 102:1–4

Stanley SM (2008) Effects of global seawater chemistry on biomineralization: past, present and future. Chem Rev 108:4483–4498. doi:10.1021/cr800233u

Surif MB, Raven JA (1989) Exogenous inorganic carbon sources for photosynthesis in seawater by members of the Fucales and the Laminariales (Phaeophyta): ecological and taxonomic implications. Oecologia 78:97–105

Surif MB, Raven JA (1990) Photosynthetic gas exchange under emersed conditions in intertidal and normally submersed members of the Fucales and Laminariales: interpretation in relation to C isotope ratio and N and water use efficiency. Oecologia 82:68–80

Thoms S, Pahlow M, Wolf-Gladrow DA (2001) Model of the carbon concentrating mechanism in chloroplasts of eukaryotic algae. J Theor Biol 208:295–313. doi:10.1006/jtbi.2000.2219

Turley C (2008) Impacts of changing chemistry in a high-CO_2 world. Mineral Mag 72:359–362. doi:10.1180/minmag.2008.072.1.359

van de Waal D, Verschoor AM, Verspagen JMH, van Donk E, Huisman J (2010) Climate-driven changes in the ecological stoichiometry of aquatic ecosystem. Front Ecol Environ 8:145–152. doi:10.1890/080178

Vizzini S, Mazzola A (2006) The effects of anthropogenic organic matter inputs on stable carbon and nitrogen isotopes in organisms from different trophic levels in a southern Mediterranean coastal area. Sci Total Environ 368:723–731. doi:10.1016/j.scitotenv.2006.02.001

Wang W-L, Yeh H-W (2003) $\delta^{13}C$ values of marine macroalgae from Taiwan. Bot Bull Acad Sin 44:107–112

Ward JK (2005) Evolution and growth of plants in a low CO_2 world. In: Ehleringer JR, Cerling TE, Dearing MD (eds) A history of atmospheric CO_2 and its effects on plants, animals, and ecosystem, vol 177, Ecological Studies. Springer, New York, pp 232–257

Wootton JT, Pfister CA, Forester JD (2008) Dynamic patterns and ecological impacts of declining ocean pH in a high-resolution multi-year dataset. Proc Natl Acad Sci USA 105:18848–18853. doi:10.1073/pnas0810079105

Wozniak AS, Roman CT, Wainright SC, McKinney RA, James-Pirri M-J (2006) Monitoring food web changes in tide-restored salt marshes: a carbon stable isotope approach. Estuar Coast 29:568–578

Zachos JC, Rohl U, Schellenberg SA, Sluijs A, Hodell DA, Kelly DC, Thomas E, Nicolo M, Raffi I, Lourens LJ, McCarren H, Kroon D (2005) Rapid acidification of the ocean during the Paleocene-Eocene thermal maximum. Science 308:1611–1615. doi:10.1126/science.1109004

Zeebe RE, Wolf-Gladrow D (2001) CO_2 in seawater: equilibrium, kinetics. Isotopes. Elsevier Science B.V, Amsterdam

Zou D, Gao K (2002) Effects of desiccation and CO_2 concentrations on emersed photosynthesis in *Porphyra haitanensis* (Bangiales, Rhodophyta), a species farmed in China. Eur J Phycol 37:587–592. doi:10.1017/S0967026202003876

Chapter 20
Impacts of Stratospheric Ozone Depletion and Solar UVB Radiation on Seaweeds

Kai Bischof and Franciska S. Steinhoff

20.1 Introduction

Reports on the thinning of the stratospheric ozone (O_3) layer in the early 1980s resulted in intense research activities aiming to assess and predict effects of an increased ultraviolet-B (UVB) radiation on biological systems, human skin, physiological key processes, e.g., photosynthesis, plant development, animal performance as well as terrestrial and marine ecosystem structure. Despite recent reports on the recovery of the Antarctic ozone hole (Newman et al. 2009; Kerr 2011), the impact of UVB on humans is still regarded as one of the most striking health issues in areas close to the Antarctic convergence (e.g., New Zealand, Southern Chile, Southern Argentina). Over the Arctic, however, the less stable atmospheric conditions result in extreme interannual variation in O_3 concentration, with a record loss in stratospheric O_3 being reported in spring 2011 (see press release by the European Space Agency (ESA) http://www.esa.int/esaCP/SEMIF24SZLG_index_0.html). Therefore, serious concern about the impacts of high UVB radiation on the biosphere still drives significant research efforts in the Polar regions of both hemispheres, based on the calculation that a 10% decline in column ozone would result in an approximately 5% increase of surface irradiance at 320 nm while the same decline would be accompanied by a 100% increase at 300 nm (Frederick et al. 1989), thus in the harmful range of UVB radiation. Among scientists there is large consensus about the success of the Montreal protocol, which

K. Bischof (✉)
Department of Marine Botany, Bremen Marine Ecology, University of Bremen,
Leobener Str. NW2, 28359 Bremen, Germany
e-mail: kbischof@uni-bremen.de

F.S. Steinhoff
Department of Biological and Environmental Sciences, University of Gothenburg, P.O. Box 461
40530 Göteborg, Sweden
e-mail: Franciska.Steinhoff@ntnu.no

C. Wiencke and K. Bischof (eds.), *Seaweed Biology*, Ecological Studies 219,
DOI 10.1007/978-3-642-28451-9_20,

was signed in 1987 with the intention to minimize emission of O_3-depleting CFCs (chlorofluorocarbons), and thus ozone depletion might be further slowed down in the future. However, due to the inertia of the atmosphere, recovery of the ozone layer may still not be observed for the next two decades. Furthermore, there is large uncertainty on how ozone depletion will interact with other aspects of climate change, e.g., the greenhouse effect (UNEP 2010). These interactions will be crucial also to the future radiation climate on the earth's surface.

In general, the effects of UVB exposure on biological systems are manifold, primarily acting on the molecular level, but with the potential to reach out to changes in ecosystem structure and function. The prerequisite for UVB-induced damage is its absorption by any kind of biomolecule. UV-chromophores in plants include nucleic acids (such as DNA, RNA) and proteins (Vass 1997). The results of UVB absorption by DNA are structural alterations like the formation of cyclobutane dimers, single-strand breaks, and pyrimidine (6–4)-pyrimidone (6–4)-photoproducts (Lois and Buchanan 1994). Furthermore, DNA damage may indirectly be induced by free oxygen radicals, generated by the electron transfer from chromophore molecules, excited by UVR absorption (Mitchell and Karentz 1993). UVB-mediated damage to proteins includes the splitting of disulfide bonds, which are crucial for protein folding, and thus, are essential for its proper functioning (Vass 1997). In algae, pigments of the photosynthetic apparatus are critical targets of UVR (Strid et al. 1990), with the phycobilins being highly sensitive, and carotenoids being less affected than chlorophylls (Häder and Häder 1989). The UVB-mediated damage to molecular targets may directly perpetuate with central physiological processes like, e.g., photosynthesis (Strid et al. 1990; Bischof et al. 2006a; see Chap. 1 by Hanelt and Figueroa) and nutrient uptake (Flores-Moya et al. 1998; Gómez et al. 1998). These physiological effects may also be reflected by changes on the ultrastructural level like, e.g., by the formation of "inside-out" vesicles from thylakoids, which was demonstrated in some red algal species (Poppe et al. 2002, 2003). Evidently, the impairment of physiological processes may further result in impairment of growth, production, and reproduction and may thus also alter ecosystem function (Han 1996a, b; Makarov 1999; Aguilera et al. 1999; Altamirano et al. 2000a, b).

Due to their crucial importance to coastal ecosystem function, seaweeds became a prominent group of organisms in UVB research. This chapter focuses on ecophysiological/autecological UV responses found in seaweeds, for further information on synecological implications like, e.g., interspecific competition and trophic interactions the reader is referred to the review by Bischof et al. (2006a). From the number of studies on UVB effects on seaweeds conducted so far, some general patterns can be concluded:

1. The species-dependent susceptibility of the alga involved is strongly correlated with the vertical zonation pattern the alga displays in the field. Moreover, even algal material, which is in culture for several years revealed that there is a genetically fixed adaptation and differential UVB susceptibility reflecting the UVB climate of the original distribution patterns (both vertical and latitudinal) of the species (Bischof et al. 1998a, b).

2. Within one species, juvenile and smaller specimens are generally more vulnerable than adult and larger algae. The reproductive microstages are more susceptible than adult sporophytes and acclimation to UVB exposure is possible, to at least some extent, for most of the species investigated so far (Dring et al. 1996; Roleda et al. 2007).
3. Phylogenetically old organismic groups, like algae, have evolved under conditions with much higher UVB irradiances in the atmosphere than nowadays, even under conditions of severe ozone depletion. Thus, most organisms have developed some kind of strategies to cope with the adverse effects of UVB exposure, which can be summarized by the terms: avoidance, protection, and repair (see Bischof et al. 2006a).
4. UVR may also confer promoting effects, like, e.g., the induction of photorepair processes (Hanelt and Roleda 2009).

20.2 Species-Dependent UV Susceptibility Based on Latitudinal and Depth Distribution: The Frame Set by Adaptation

The frame of species-dependent reactions of seaweeds to UVB exposure is set by genetically fixed adaptation. Early laboratory studies on Antarctic seaweeds, which were kept in stock cultures for more than 10 years still revealed distinct species-specific reaction patterns toward UV exposure once they were grown to macrothalli and exposed to identical UV conditions (Bischof et al. 1998a). Thus, it was confirmed that shallow water species, like the two green algae *Enteromorpha bulbosa* (now termed *Ulva bulbosa,* also referred to as *U. hookeriana*) and *Acrosiphonia arcta,* were still extremely UV tolerant even after long-term exposure to dim light conditions in stock cultures. In contrast, photosynthesis of the red algal species, *Phycodrys austrogeorgica* and *Delesseria lancifolia,* responded extremely sensitive toward UV exposure. Overall, the general zonation patterns seaweeds display, e.g., at the islands off the Antarctic Peninsula are reflected by their UV sensitivity even decades after their isolation from the field (Bischof et al. 1998a).

The same adaptational setting has been observed in six red algal species from cold-temperate regions and with different zonation patterns with respect to their species-specific growth rates and ability to cope with UVB-mediated DNA damage (van de Poll et al. 2001). Again, these algae were cultivated under identical culture conditions and exposed to similar irradiance of UVB and, evidently, the respective inhibition of growth was in line with the species position on the shore, with the two shallow water species (*Palmaria palmata* and *Chondrus crispus*) hardly exhibiting inhibition in growth, whereas in deep-water algae (*Phycodrys rubens* and *Polyneura hilliae)* growth was inhibited almost completely. Furthermore, accumulation of damaged DNA, expressed as thymine dimer formation, was found in the two deep-water species, which apparently lack the ability for sufficient DNA repair. This confirms the strong adaptation of these species to their low irradiance environment at their natural growth site. Furthermore, the accumulation of DNA damage in seaweeds can be regarded as a very good measure for the degree of adaptation of a given species towards UVB exposure (van de Poll et al. 2001).

These findings were also confirmed in a transplantation experiment on another deep-water red algal species *Delesseria sanguinea*, in which growth was also impaired when the alga was exposed to surface solar radiation (Pang et al. 2001). Results obtained indicated the absence of any protecting mechanisms against excessive radiation in this algal species, which are, however, not relevant to the species as it inhabits the low light environments in the shade of canopy algae or in great water depths. Thus, sublittoral seaweeds may avoid being exposed to UVB simply by growing in water depths, in which most of the UVB irradiance is already attenuated by the water column above. A prominent example of this avoidance strategy is also the Arctic endemic deep-water kelp *Laminaria solidungula* (Brey 2009). Other algal species following this strategy may also be growing in the understory being protected by the algae above or following a cryptic life cycle.

Generally, all modulation of ecophysiological reactions toward variation in abiotic factors is conditioned by genetic adaptation. This is also visible in two red algal species from Spitsbergen (Arctic) with slightly different vertical zonation preferences (Karsten et al. 1999): *Devaleraea ramentacea* as a species from shallow waters is permanently equipped with high activities of reactive oxygen scavenging superoxide dismutase (SOD, see Chap. 6 by Bischof and Rautenberger). This high but static activity is reasonable for a species from shallow waters, where usually strong variation in abiotic conditions and, thus, the onset of stressful conditions to photosynthesis potentially resulting in increased ROS production is more likely than in more stable deeper waters. However, maintaining a protective system on such a high level throughout the year is probably energetically cost intense. Thus, species which are not permanently exposed to stressful conditions, e.g., in deeper waters, may favor the strategy to rather respond to abiotic stress and to increase protective strategies, like SOD activity, only when they are needed during times of, e.g., high UV irradiance. *Palmaria palmata* inhabiting slightly deeper waters than *D. ramentacea* is applying this strategy (Karsten et al. 1999; see also Bischof et al. 2006a).

20.3 Acclimation to UV Exposure in Space and Time: Depth Distribution, Small-Scale Gradients, and Season

Seaweeds populating a flexible environment, i.e., the shallow water zones, coastlines with strong seasonality, etc., require mechanisms of acclimation in order to set physiological performance to the variation of environmental requirements. Thus, also acclimation toward UV exposure has been widely observed in seaweeds. Acclimation of photosynthesis to UV exposure has been demonstrated in the Arctic/cold-temperate kelp *Alaria esculenta* from Spitsbergen

(Bischof et al. 1999; see also Chap. 1 by Hanelt and Figueroa). The potential for acclimation is not only the precondition to endure stress caused by exposure to harmful radiation, but is a prerequisite to establish over wide depth ranges and to endure the seasonal variation of radiation conditions (Bischof et al. 1998b, 1999). However, the process of acclimation of photosynthetic activity in Arctic brown seaweeds to changing radiation conditions showed a distinct sequence of events, which may be indicative for the different molecular mechanisms involved: in *A. esculenta* under repeated UV exposure, the competence of recovery from UV-induced photoinhibition increased after just a few exposure/recovery cycles. This might indicate an activation of different repair mechanisms, counteracting the impact of UV exposure by a faster replacement of damaged molecules. Moreover, the degree of inhibition became smaller (Bischof et al. 1998b, 1999), which might also be related to an activated ROS defense system counteracting UVB-mediated oxidative stress (see Chap. 6 by Bischof and Rautenberger) or to the formation of UV-screening compounds, like phlorotannins (Schoenwaelder 2002a, b).

In fact, biosynthesis and accumulation of UV-screening substances has been described as one of the most important physiochemical acclimation mechanisms against biologically harmful UV radiation. In red algae, mycosporine-like amino acids (MAAs) have been extensively studied as potent UV-screening substances (Karsten et al. 1998; Conde et al. 2000). While MAAs have been mainly observed in the Rhodophyta (Hoyer et al. 2001; Huovinen et al. 2004), Phaeophyta and most Chlorophyta typically lack these compounds, with the green alga *Prasiola crispa* ssp. *antarctica* being an interesting exception and containing high concentrations of a unique MAA with an absorption maximum at 324 nm (Hoyer et al. 2001; Karsten et al. 2005). The role of MAA accumulation as an acclimatory response toward the respective radiation environment becomes apparent from the observed decrease in cellular concentration with increasing depth (Hoyer et al. 2001). In general, cellular MAA concentrations in red algae have been shown to be positively correlated with the natural UV doses (Karsten et al. 1998; Huovinen et al. 2004). The flexibility of acclimation with respect to the synthesis of UV-screening compounds is reflected by strong seasonal as well as microscale variation in MAA contents: thus, specimens of the red alga *Palmaria decipiens* collected in Antarctic winter contained low concentrations of UV-absorbing compounds and exhibited significantly higher values in summer (Post and Larkum 1993). Furthermore, in individual red algal specimens cross-sectional and longitudinal MAA concentration gradients have been described depending on the respective microenvironment of radiation, the age, or the tissue type (Karsten and Wiencke 1999; Hoyer et al. 2001). Based on their ability and flexibility of MAA synthesis, red algae have been classified into three categories (Hoyer et al. 2001): Type I—species which completely lack the ability to synthesize MAAs, as e.g., most of the deep-water algae; Type II—species which synthesize MAAs in variable concentrations in response to the variation of the respective environmental conditions, and Type III—species which always contain high concentrations of MAAs, typically these are algal species populating very shallow waters or even the intertidal zone and are, thus, exposed to strong irradiances and large amplitudes of environmental variation.

Similar traits of optical protection against UVB exposure can also be found in brown seaweeds, however, based on another class of chemical compounds, the so-called phlorotannins (Ragan and Glombitza 1986). Phlorotannins have been invoked as multifunctional compounds acting in deterring herbivores and microbes, in adhesion and in strengthening algal cell walls (Schoenwaelder 2002b), and in absorbing UV radiation (Ragan and Glombitza 1986; Pavia et al. 1997; Henry and Van Alstyne 2004). Phlorotannins do also possess a high antioxidant activity (e.g., Ahn et al. 2007) and are, thus, important for scavenging ROS. With respect to the acclimation of brown seaweeds toward exposure to detrimental UV radiation several strategies have been described in which phlorotannins do play a vital role: (1) a generally high tissue concentration of phlorotannins acting as UV screen (Schoenwaelder 2002a, b), (2) an induction of phlorotannins in response to the environmental radiation conditions (Halm et al. 2010), (3) an exudation of phlorotannins and other phenolic compounds into the surrounding medium creating an UV-absorbing microenvironment (Pérez-Rodríguez et al. 1998; Roleda et al. 2010; Steinhoff et al. 2011b), or (4) an excess inclusion of phlorotannins in cell walls shielding harmful radiation (Schoenwaelder 2002b). An induction of phlorotannins after exposure to UVB radiation was first described in the brown alga *Ascophyllum nodosum* (Pavia et al. 1997). An induction of phlorotannins due to UVB and UVA radiation was also described for *Macrocycstis integrifolia* (Swanson and Druehl 2002). An increase in the size of phlorotannin containing vesicles, the so-called physodes, was observed in various Laminariales from Spitsbergen after UVB exposure indicating an induction of phlorotannin synthesis (Wiencke et al. 2006). This has recently been verified in the UV-tolerant species *Alaria esculenta* and *Saccorhiza dermatodea* (Steinhoff unpublished data, Steinhoff et al. 2011b). An exudation of phlorotannins as response to artificial UVB radiation was observed in *Macrocystis integrifolia* (Swanson and Druehl 2002) creating so-called UV-refugia. Again, high phlorotannin contents and high exudation rates might reflect an adaptation of seaweeds to the radiation climate and overall habitat requirements. In contrast, induction and variable exudation rates of phlorotannins reflect the acclimation potential to environmental changes. Interestingly, the precursors of MAAs and phlorotannins are both synthesized by the Shikimate pathway, with phenylalanine being an important intermediate. The reason why finally different specific UV screens are applied in different algal divisions represents an interesting aspect to study in terms of algal evolution.

In contrast to the specific UV-absorbing compounds found in red and brown algae, there is inconsistency about the role of optical UV protection in the group of green algae. The occurrence of MAAs in supralittoral green algae such as *Prasiola* sp. (Karsten et al. 2005) seems to be quite exceptional. Similar as phlorotannin exudation in brown algae, some green seaweeds have also been found to generate microenvironments depleted from UVB, e.g., by excreting UVB-absorbing substances. The green *Dasycladus vermicularis* is commonly found in the shallow waters of the Mediterranean as well as in rock pools. Under conditions of high irradiance and increased water temperature and salinity, the alga excretes high concentration of hydroxycumarine, a phenolic compound which may provide a yellowish color to the water body around and thus may increase UVB absorption in the water, before it may reach algal specimens (Pérez-Rodríguez et al. 1998, 2001). However, an extended screening of 71 green macroalgal species revealed that in fact optical UV screening in green seaweeds is rather exceptional (Pescheck et al. 2010), and thus green algae largely need to rely on alternative mechanisms to resist the harmful effects of UV exposure. An alternative strategy to achieve optical shielding on a cellular base is the arrangement of specimens in dense assemblages, e.g., macroalgal mats. In these systems which are usually formed by highly opportunistic species (e.g., *Ulva, Chaetomorpha*) under high nutrient loads and low wave exposure, physiologically healthy and productive algal material is shielded by the algal layers above, generating a steep small-scale gradient of UVB and PAR, but also chlorophyll content and photosynthetic performance in those assemblages (Bischof et al. 2002, 2006b). A prominent example for the effectiveness of this strategy is found in mats of *Ulva rotundata* from southern Spain (Bischof et al. 2002, 2003). Thus, this protection by canopy arrangement does not require the synthesis of specific UV-screening compounds; however, it is at the expense of the top algal layers.

20.4 Interaction with Other Abiotic Factors

Seaweeds and particularly those inhabiting the intertidal and shallow subtidal are exposed to (inter-) dependently changing abiotic parameters; thus the interactive effects of UV radiation in concert with the change of other abiotic parameters are important to address. In the early studies on UV effects on seaweed physiology conducted under laboratory conditions, the proper ratio of radiative power of the different wavelength ranges was largely neglected and the significance to the effects observed was largely underestimated. Among others, this was shown in a study conducted on the green macroalga *Ulva lactuca* isolated from Greenland (Fredersdorf and Bischof 2007). Here algal material was exposed to a constant irradiance of UVB radiation, however, at different irradiances of background photosynthetically active radiation (PAR). The different effects on algal photosynthesis of the respective wavelengths ranges (PAR, UVA, UVB) were identified by the combination of various cutoff filters. In this study, the importance of background PAR for the impact of UVB effects observed became evident: The extent of UVB-induced inhibition of photosynthetic quantum yield was highly dependent on the irradiance of PAR in the setup and was diminishing with increasing PAR irradiance. However, due to the different mechanisms of PAR- and UVB-induced inhibition of photosynthesis, the additional effect of UVB rather became apparent in the recovery kinetics from photoinhibition, resulting in a delay of recovery under UVB exposure, presumably due to the increased defragmentation of photosynthetic units, i.e., the D1 reaction centre protein. In contrast, increased levels of PAR as well as UVA may provide an additional energy input to fuel simultaneously operating repair processes counteracting the damaging impact of UV exposure, e.g., by an increased activity of DNA photolyase, or elevated expression of psbA-gene and consequently enhanced turnover of D1. Therefore, the ecological implications of UVB exposure can only be validated in approaches mimicking natural radiation conditions, particularly reflecting the naturally occurring relative ratios of the different wavelength ranges. In particular, algae and aquatic plants from high irradiance environments (e.g., tropical shallow water ecosystems) show distinct (sometimes even promoting) interactions of UVB- and PAR-induced photoinhibition (Hanelt et al. 2006; Hanelt and Roleda 2009).

The importance of temperature for the respective UV susceptibility observed in seaweeds has also been demonstrated; however, the level of knowledge is still far behind compared to that existing for higher plants, in which, e.g., a chilling-induced increase in optical UV screening has been observed, even in the absence of UVB (Bilger et al. 2007). This observation formed the base for intense studies on the interactive effects of temperature and UVB exposure. Low temperatures generally pose the problem of slowed down enzyme reactions and consequently generally reduced metabolic activity. *In concreto*, this also applies to enzymatic repair processes in response to UVB exposure. At low temperatures, the activity of repair enzymes (DNA-photolyase, excision repair), as well as synthetic pathways (e.g., D1-turnover) might be not operative at the velocity required to keep up with rate of damage occurring at a certain irradiance (of UVB and/or PAR); thus, accumulation of damage will be the consequence. In turn, under conditions of reduced activity of repair enzymes, organisms may rely more strongly on optical protection strategies. However, studies on the photoprotective potential of phlorotannins in Arctic brown algal zoospores and juvenile gametophytes revealed that changes in phlorotannin content were neither affected by low nor by high temperatures (Müller et al. 2009; Steinhoff et al. unpublished). Taking the high antioxidative potential of phlorotannins (Connan et al. 2006; Zubia et al. 2007) into consideration, phlorotannins might be produced and rapidly oxidized again (Steinhoff et al. 2012). Therefore, photoprotective substances might also help to protect brown macroalgal cells by scavenging ROS within a wide temperature range. Nevertheless, several studies investigating the interactive effects of temperature and UV exposure on macroalgal spores (Wiencke et al. 2006; Müller et al. 2008, 2009; Steinhoff et al. 2011a, b) lead to the conclusion that at low temperatures, solar radiation effects might be better compensated than at increased water temperatures indicating the potential impact of rising water temperatures on early macroalgal life stages.

Still, a moderately increased temperature has been found to compensate for UVB-induced damage due to the higher activity of repair pathways. The impact of temperature increase on UV susceptibility of photosynthetic activity was studied in two *Ulva* species from Antarctic and subantarctic regions (Rautenberger and Bischof 2006). An isolate of the Antarctic/cold-temperate *Ulva bulbosa* (now also referred to as *U. hookeriana*) was compared to the cosmopolitan *U. clathrata* by exposing them to identical conditions of UV radiation at 0° and 10°C. In both species, exposure to 10°C almost completely compensated for the UV-induced inhibition of photosynthetic quantum yield observed at 0°C (see Fig. 20.1). Observed results were striking for two reasons: (1) in *U. bulbosa* UV-induced

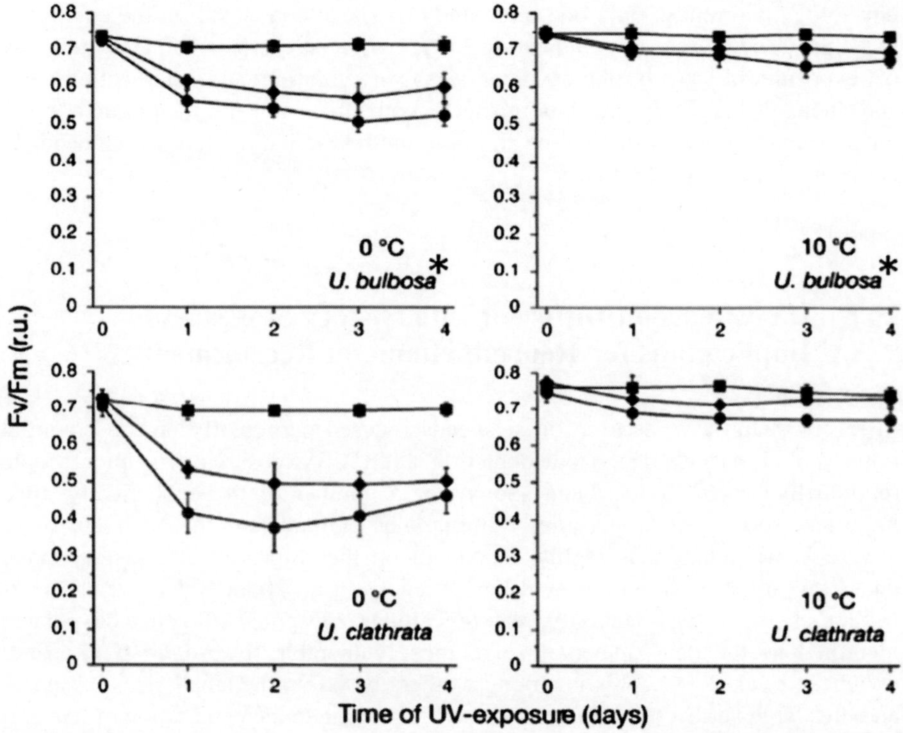

Fig. 20.1 Interactive effects of UV exposure and temperature on optimum quantum yield of photosynthesis (F_v/F_m): *Ulva bulbosa,* (*now also referred to as *U. hookeriana*) (isolated from King George Island, Antarctica) and *U. clathrata* (isolated from Puerto Williams, Tierra del Fuego, South America) have been exposed to identical irradiances of PAR, UVA, and UVB at 0° and 10°C; (*filled square*) PAR alone, (*filled diamond*) PAR + UVA, (*filled circle*) PAR + UVA + UVB (Rautenberger and Bischof 2006)

inhibition was compensated by temperatures, which are higher than those usually encountered at its natural growth sites; (2) in the cosmopolitan species *U. clathrata* exposure to lower temperatures resulted in strongly pronounced UV effects, even stronger than in the more southern species *U. bulbosa*, which might be based on higher SOD activities measured in the latter species, presumably as a result of higher cellular SOD concentrations. It has been described as a common adaptive feature in Polar organisms to increase respective enzyme concentrations in order to compensate for activity loss at low environmental temperatures. However, temperature compensation of UV effects is evidently only possible within the temperature tolerance range of the species under investigation.

A further aspect of environmental change has raised new concerns on marine ecosystem functions and resulted in further research activities with respect to the potential impacts of ocean acidification (see Chap. 19 by Roleda and Hurd). Hitherto, there is very little information available on how the impacts of UV radiation and changing pH levels might interact with respect to seaweed physiology. Currently, only one pilot study has been conducted on the calcifying red alga *Corallina sessilis* indicating some synergistic effects of UV- and low pH-exposure on growth, photosynthetic oxygen evolution, and calcification (Gao and Zheng 2010). Revealing synergisms of both environmental factors on a wider range of marine organisms will be an important research direction, which needs to come up in the very near future.

20.5 UV Effects on Different Life History Stages and Implications for Reproduction and Recruitment

In recent years, UV research on seaweeds focused particularly on the potential impacts of climate change, ozone depletion, and UVB exposure on the microscopic reproductive stage. It has been shown for a number of the ecologically most important species of kelp (order Laminariales and related) that the sensitivity toward UVB radiation is highly dependent on the respective life history stage under investigation (Dring et al. 1996; Wiencke et al. 2006; Roleda et al. 2007; Müller et al. 2008). Generally, the unicellular zoospores of kelps have been identified as the developmental stage most vulnerable toward UVB exposure (Wiencke et al. 2000). This is amongst others based on the small sizes of spores and thus large surface-to-volume ratios, which facilitate the penetration of harmful radiation into the cell and thus the damage to cellular components. Another explanation for the high UVB susceptibility of spores is provided by the level of ploidy. With respect to zoospores and gametes being haploid stages, UVB-mediated damage to the DNA bears even more severe consequences, as single copy genes might be damaged without a potential substitute by a second allele, as e.g., in the diploid sporophytes (Roleda et al. 2008). Evidently, the impact of UVB exposure on the reproductive success of seaweeds is determined by the most UVB susceptible stage in the life history of a species. There is common sense that if there will be an adverse impact of UVB radiation on seaweed communities it will be mediated by a disruption of the developmental cycle due to pronounced damage to the haploid, unicellular life history stages (zoospores and gametes), which will ultimately preclude reproduction and recruitment (Roleda et al. 2007; Müller et al. 2008, 2009; Steinhoff et al. 2011b).

Via its significance to the reproductive microstages, impacts of solar UV radiation on the cellular level are closely related to effects on a community level as it may result in, e.g., impaired release of zoospores (Makarov and Voskoboinikov 2001) and recruitment (Wiencke et al. 2000, 2006; Roleda et al. 2007). At present, only a small number of publications describe UVR effects on cellular ultrastructure of seaweed spores (Steinhoff et al. 2008). UV effects observed on the ultrastructural level include wrinkled thylakoid membranes, dilatations of the thylakoid lumen, and disrupted mitochondria with plastoglobuli formation as well as changes of the nucleoplasm structure (Poppe et al. 2002, 2003; Holzinger et al. 2004; Holzinger and Lütz 2006; Steinhoff et al. 2008; see also Karsten et al. 2009).

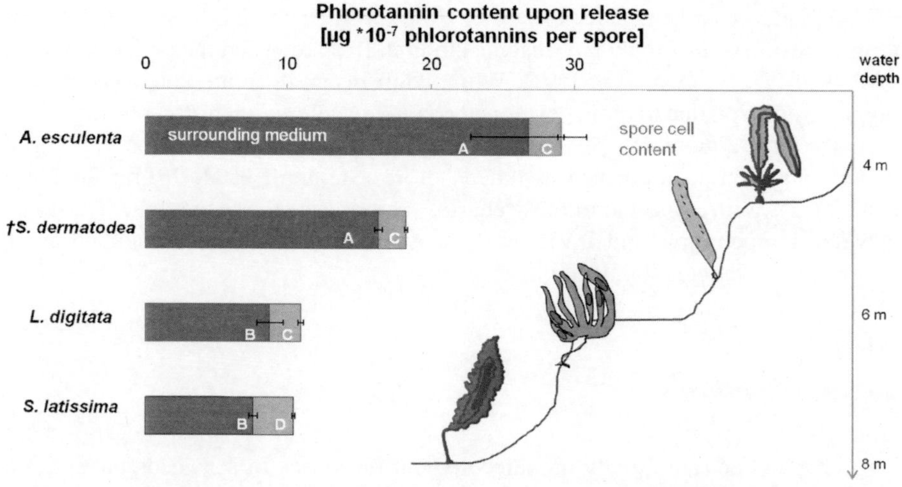

Fig. 20.2 Collection depth of *Alaria esculenta, Saccorhiza dermatodea* [†data from Steinhoff et al. (2011b)], *Laminaria digitata,* and *Saccharina latissima* from Kongsfjorden, Spitsbergen in correlation with the sum of phlorotannin content [$\mu g \times 10^{-7}$ phlorotannin spore^{-1}] within the surrounding medium and the zoospores. Vertical zonation of sublittoral Arctic brown algae species reflects parental phlorotannin content liberated upon spore release. *Capital letters* indicate statistical differences ($p < 0.05$) and vertical bars standard deviation ($n = 3$). Steinhoff et al. previously unpublished

Consequently, cellular protection against harmful UVB is particularly vital for the microscopic reproductive stages of seaweeds. Studies on brown algal spores under elevated radiation scenarios suggest that phlorotannins are highly variable due to inter- and intraspecific variations (Toth et al. 2005; Connan et al. 2006), developmental stage (van Alstyne et al. 2001), habitat (Hemmi and Jormalainen 2004), and season (Plouguerné et al. 2006). Likewise are the multiple roles and tasks of phlorotannins within different species and their developmental stages very likely to be determined by interaction of a variety of biotic and abiotic factors (Pavia et al. 1997). The vertical zonation pattern of the parental algae seems to be coupled with the amount of exuded phlorotannins, with high amounts of released phlorotannins in species and specimens from the upper sublittoral compared to individuals from the lower sublittoral (Fig. 20.2, Steinhoff et al. unpublished, Wiencke et al. 2006). The induction of phlorotannins (Pavia et al. 1997) is most likely time dependent; a significant phlorotannin formation at mid-term scales (i.e., 10 days) within macroalgal juveniles of *S. latissima* and *A. esculenta* (Steinhoff et al. 2011a) could be detected while it is lacking on short-term (i.e., 8 h) scales (Müller et al. 2009; Steinhoff et al. 2011b). Parental efforts and the respective conditioning of parental seaweeds toward the respective UV exposure at the natural growth site seem to play a major role to increase viability of the offspring by "transferring" the acquired UVR tolerances to their reproductive stages (Steinhoff et al. 2011b).

Again, not only the isolated effects of UV exposure should be considered: Short-term exposure of macroalgal propagules to high PAR revealed the underestimation of PAR effects compared to UVR. While adult macroalgae are able to cope with higher irradiation due to more developed adaptive/acclimation processes, high PAR seems to have a more detrimental impact on spore germination and total fatty acid content and composition than assumed before (Steinhoff et al. 2011a; Steinhoff et al. 2012). With respect to future scenarios for seaweed ecosystem functions under elevated temperatures and UVB irradiances, high PAR should be studied as an interactive parameter for all life history stages.

20.5.1 Synthesis

This chapter focused largely on autecological responses of seaweeds under UVB exposure. However, species-specific responses ultimately result in significant ecological implications, which include changes in distributional patterns (latitude vs. depth distribution), succession patterns, and species diversity, and may further result in altered biotic relationships, e.g., competition and trophic interactions (see Bischof et al. 2006a for review), which are, however, beyond the scope of this chapter. Taking into account the multitude of effects UVB exerts on the physiology and ecology of a species, it is apparent that UVB exposure plays indeed an important role in shaping seaweed-dominated coastal ecosystems. Based on the knowledge acquired so far it might be concluded that in line with the respective growth site, physiological activity of most seaweeds is sufficiently buffered and thus protected against increasing UVB. However, this assumption still neglects the importance of the interaction with other abiotic and biotic factors to be expected to change under future environmental perturbations. Still most data on UV impacts on seaweeds available so far have been obtained from laboratory experiments: Studies under more realistic environmental conditions are still needed. In particular, mesocosm approaches under natural irradiation and experimental variation of temperature and pH should be designed to obtain a better understanding of organismic interaction (e.g., changes in benthic/shallow water food web structure) in an era of climate change (Zacher et al. 2007).

References

Aguilera J, Karsten U, Lippert H, Vögele B, Philipp E, Hanelt D, Wiencke C (1999) Effects of solar radiation on growth, photosynthesis and respiration of macroalgae from the Arctic. Mar Ecol Prog Ser 19:109–119

Ahn GN, Kim KN, Cha SH, Song CB, Lee J, Heo MS, Yeo IK, Lee NH, Jee YH, Kim JS, Heu MS, Jeon YJ (2007) Antioxidant activities of phlorotannins purified from *Ecklonia cava* on free radical scavenging using ESR and H_2O_2-mediated DNA damage. Eur Food Res Technol 226:71–79

Altamirano M, Flores-Moya A, Figueroa FL (2000a) Long-term effects of natural sunlight under various ultraviolet radiation conditions on growth and photosynthesis of intertidal *Ulva rigida* (Chlorophyceae) cultivated in situ. Bot Mar 43:119–126

Altamirano M, Flores-Moya A, Figueroa FL (2000b) Growth seasonality, photosynthetic pigments, and C and N content in relation to environmental factors: a field study on *Ulva olivascens* (Ulvales, Chlorophyta). Phycologia 39:50–58

Bilger W, Rolland M, Nybakken L (2007) UV screening in higher plants induced by low temperature in the absence of UV-B radiation. Photochem Photobiol Sci 6:190–195

Bischof K, Hanelt D, Wiencke C (1998a) UV-radiation can affect depth-zonation of Antarctic macroalgae. Mar Biol 131:597–605

Bischof K, Hanelt D, Tüg H, Karsten U, Brouwer PEM, Wiencke C (1998b) Acclimation of brown algal photosynthesis to ultraviolet radiation in Arctic coastal waters (Spitsbergen, Norway). Polar Biol 20:388–395

Bischof K, Hanelt D, Wiencke C (1999) Acclimation of maximal quantum yield of photosynthesis in the brown alga *Alaria esculenta* under high light and UV radiation. Plant Biol 1:435–444

Bischof K, Peralta G, Kräbs G, van de Poll WH, Perez-Llorens JL, Breeman AM (2002) Effects of solar UV-B radiation on canopy structure of *Ulva* communities from Southern Spain. J Exp Bot 53:2411–2421

Bischof K, Janknegt PJ, Buma AGJ, Rijstenbil JW, Peralta G, Breeman AM (2003) Oxidative stress and enzymatic scavenging of superoxide radicals induced by solar UV-B radiation in *Ulva* canopies from southern Spain. Sci Mar 67:353–359

Bischof K, Gomez I, Molis M, Hanelt D, Karsten U, Lüder U, Roleda MY, Zacher K, Wiencke C (2006a) Ultraviolet radiation shapes seaweed communities. Rev Environ Sci Biotechnol 5:141–166

Bischof K, Rautenberger R, Brey L, Perez-Llorens JL (2006b) Physiological acclimation along gradients of solar irradiance within mats of the filamentous green macroalga *Chaetomorpha linum* from southern Spain. Mar Ecol Prog Ser 306:165–175

Brey L (2009) Acclimation of kelp photosynthesis to seasonal changes in the underwater radiation regime of an Arctic fjord system. PhD thesis, Universität Bremen, pp 98

Conde FR, Churio MS, Previtali CM (2000) The photoprotector mechanism of mycosporine-like amino acids. Excited-state properties and photostability of porphyra-334 in aqueous solution. J Photochem Photobiol 56B:139–144

Connan S, Delisle F, Deslandes E, Gall EA (2006) Intra-thallus phlorotannin content and antioxidant activity in Phaeophyceae of temperate waters. Bot Mar 49:39–46

Dring MJ, Makarov V, Schoschina E, Lorenz M, Lüning K (1996) Influence of ultraviolet-radiation on chlorophyll fluorescence and growth in different life history stages of three species of *Laminaria*. Mar Biol 126:183–191

Flores-Moya A, Gomez I, Vinegla B, Altamirano M, Perez-Rodriguez E, Maestr C, Caballero RM, Figerueroa FL (1998) Effects of solar radiation on the endemic Mediterranean red alga *Rissoella verruculosa*: photosynthetic performance, pigment content and the activities of enzymes related to nutrient uptake. New Phytol 139:673–683

Frederick JE, Snell HE, Haywood EK (1989) Solar ultraviolet radiation at the earth's surface. Photochem Photobiol 50:443–450

Fredersdorf J, Bischof K (2007) Irradiance of photosynthetically active radiation determines UV-susceptibility of photosynthesis in *Ulva lactuca* L. (Chlorophyta). Phycol Res 55:295–301

Gao K, Zheng Y (2010) Combined effects of ocean acidification and solar UV radiation on photosynthesis, growth, pigmentation and calcification of the coralline alga *Corallina sessilis* (Rhodophyta). Global Change Biol 16:2388–2398

Gómez I, Pérez-Rodríguez E, Viñegla B, Figueroa FL, Karsten U (1998) Effects of solar radiation on photosynthesis, UV-absorbing compounds and enzyme activities of the green alga *Dasycladus vermicularis* from southern Spain. J Photochem Photobiol B Biol 47:46–57

Häder DP, Häder M (1989) Effects of solar and artificial radiation on motility and pigmentation in *Cyanophora paradoxa*. Arch Microbiol 152:453–457

Halm H, Lüder UH, Wiencke C (2010) Induction of phlorotannins through mechanical wounding and radiation conditions in the brown macroalga *Laminaria hyperborea*. Eur J Phycol 46:16–26

Han T (1996a) Effect of ultraviolet-B radiation on *Ulva pertusa* Kjellman (Chlorophyta) I Growth and pigment content. Algae (Kor J Phycol) 11:155–159

Han T (1996b) Far-UV action on growth, pigmentation and photosynthesis of *Ulva lactuca* (Chlorophyta). Algae (Kor J Phycol) 11:203–206

Hanelt D, Roleda MY (2009) UVB radiation may ameliorate photoinhibition in specific shallow-water tropical marine macrophytes. Aquat Bot 91:6–12

Hanelt D, Hawes I, Rae R (2006) Reduction of UV-B radiation causes an enhancement of photoinhibition in high light stressed aquatic plants from New Zealand lakes. J Photochem Photobiol B Biol 84:89–102

Hemmi A, Jormalainen V (2004) Geographic covariation of chemical quality of the host alga *Fucus vesiculosus* with fitness of the herbivorous isopod *Idotea baltica*. Mar Biol 145:759–768

Henry BE, van Alstyne KL (2004) Effects of UV radiation on growth and phlorotannins in *Fucus gardneri* (Phaeophyceae) juveniles and embryos. J Phycol 40:527–533

Holzinger A, Lütz C (2006) Algae and UV irradiation: effects on ultrastructure and related metabolic functions. Micron 37:190–207

Holzinger A, Lütz C, Karsten U, Wiencke C (2004) The effect of ultraviolet radiation on ultrastructure and photosynthesis in the red macroalgae *Palmaria palmata* and *Odonthalia dentata* from Arctic waters. Plant Biol 6:568–577

Hoyer K, Karsten U, Sawall T, Wiencke C (2001) Photoprotective substances in Antarctic macroalgae and their variation with respect to depth distribution, different tissues and developmental stages. Mar Ecol Prog Ser 211:117–129

Huovinen P, Gómez I, Figueroa FL, Ulloa N, Morales V, Lovengreen C (2004) Ultraviolet-absorbing mycosporine-like amino acids in red macroalgae from Chile. Bot Mar 47:21–29

Karsten U, Wiencke C (1999) Factors controlling the formation of UV-absorbing mycosporine-like amino acids in the marine red alga Palmaria palmata from Spitsbergen (Norway). J Plant Physiol 155:407–415

Karsten U, Sawall T, Hanelt D, Bischof K, Figueroa FL, Flores-Moya A, Wiencke C (1998) An inventory of UV-absorbing mycosporine-like amino acids in macroalgae from polar to warm-temperate regions. Bot Mar 41:443–453

Karsten U, Bischof K, Hanelt D, Tüg H, Wiencke C (1999) The effect of UV-radiation on photosynthesis and UV-absorbing substances in the endemic Arctic macroalga *Devaleraea ramentacea* (Rhodophyta). Physiol Plant 105:58–66

Karsten U, Friedl T, Schumann R, Hoyer K, Lembcke S (2005) Mycosporine-like amino acids (MAAs) and phylogenies in green algae: *Prasiola* and its relatives from the Trebouxiophyceae (Chlorophyta). J Phycol 41:557–566

Karsten U, Wulff A, Roleda MY, Müller R, Steinhoff FS, Fredersdorf J, Wiencke C (2009) Physiological responses of polar benthic algae to ultraviolet radiation. Bot Mar 52:639–654

Kerr RA (2011) First detection of ozone hole recovery claimed. Science 332(6026):160

Lois R, Buchanan BBN (1994) Severe sensitivity to ulraviolet radiation in an *Arabidopsis* mutant deficient in flavonoid accumulation: II. Mechanisms of UV-resistance in Arabidopsis. Planta 194:504–509

Makarov M (1999) Influence of ultraviolet radiation on the growth of the dominant macroalgae of the Barents Sea. Chemosphere Global Change Sci 1:461–467

Makarov MV, Voskoboinikov GM (2001) The influence of ultraviolet-B radiation on spore release and growth of the kelp *Laminaria saccharina*. Bot Mar 44:89–94

Mitchell DL, Karentz D (1993) The induction and repair of DNA photodamage in the environment. In: Young AR, Björn LO, Moan J, Nultsch W (eds) Environmental UV photobiology. Plenum, New York, pp 345–377

Müller R, Wiencke C, Bischof K (2008) Interactive effects of UV radiation and temperature on microstages of Laminariales (Phaeophyceae) from the Arctic and North Sea. Clim Res 37:203–213

Müller R, Wiencke C, Bischof K, Krock B (2009) Zoospores of three Laminariales under different UV radiation and temperature conditions: Exceptional spectral absorbance properties and lack of phlorotannin induction. Photobiol Photochem 85:970–977

Newman PA, Nash ER, Douglass AR, Nielsen JE, Stolarski RS (2009) Estimating when the Antarctic ozone hole will recover. In: Zerefos C, Contopoulos G, Skalkeas G (eds) Twenty years of ozone decline. Springer, Berlin, pp 191–200

Pang S, Gómez I, Lüning K (2001) The red macroalga *Delesseria sanguinea* as a UVB-sensitive model organism: selective growth reduction by UVB in outdoor experiments and rapid recording of growth rate during and after pulses. Eur J Phycol 36:216–297

Pavia H, Cervin G, Lindgren A, Åberg P (1997) Effects of UV-B radiation and simulated herbivory on phlorotannins in the brown alga *Ascophyllum nodosum*. Mar Ecol Prog Ser 157:139–146

Pérez-Rodríguez E, Gómez I, Karsten U, Figueroa FL (1998) Effects of UV radiation on photosynthesis and excretion of UV-absorbing compounds of *Dasycladus vermicularis* (Dasycladales, Chlorophyta) from southern Spain. Phycologia 37:379–387

Pérez-Rodríguez E, Aguilera J, Gómez I, Figueroa FL (2001) Excretion of coumarins by the Mediterranean green alga Dasycladus vermicularis in response to environmental stress. Mar Biol 139:633–639

Pescheck F, Bischof K, Bilger W (2010) Screening of UV-A and UV-B radiation in marine green macroalgae (Chlorophyta). J Phycol 46:444–455

Plouguerné E, Le Lann K, Connan S, Jechoux G, Deslandes E, Stiger-Pouvreau V (2006) Spatial and seasonal variation in density, reproductive status, length and phenolic content of the invasive brown macroalga *Sargassum muticum* (Yendo) Fensholt along the coast of Western Brittany (France). Aquat Bot 85:344–377

Poppe F, Hanelt D, Wiencke C (2002) Changes in ultrastructure, photosynthetic activity and pigments in the Antarctic red alga *Palmaria decipiens* during acclimation to UV radiation. Bot Mar 45:253–261

Poppe F, Schmidt RAM, Hanelt D, Wiencke C (2003) Effects of UV radiation on the ultrastructure of several red algae. Phycol Res 51:11–19

Post A, Larkum AW (1993) UV-absorbing pigments, photosynthesis and UV exposure in Antarctica: comparision of terrestrial and marine algae. Aquat Bot 45:231–243

Ragan MA and Glombitza KW (1986) In: Round FE and Chapman DJ (eds) Phlorotannins, brown algal polyphenols. Prog Phycol Res 4:129–241

Rautenberger R, Bischof K (2006) Impact of temperature on UV susceptibility of two species of *Ulva* (Chlorophyta) from Antarctic and Subantarctic regions. Polar Biol 29:988–996

Roleda MY, Wiencke C, Hanelt D, Bischof K (2007) Sensitivity of the early life stages of macroalgae to ultraviolet radiation. Photochem Photobiol 83:851–862

Roleda MY, Zacher K, Wulff A, Hanelt D, Wiencke C (2008) Susceptibility of spores of different ploidy levels from Antarctic Gigartina skottsbergii (Gigartinales, Rhodophyta) to ultraviolet radiation. Phycologia 47:361–370

Roleda MY, Lüder UH, Wiencke C (2010) UV-susceptibility of zoospores of the brown macroalga *Laminaria digitata* from Spitsbergen. Polar Biol 33:577–588

Schoenwaelder MEA (2002a) Physode distribution and the effects of "Thallus Sunburn"in *Hormosira banksii* (Fucales, Phaeophyceae). Bot Mar 45:262–266

Schoenwaelder MEA (2002b) The occurence and cellular significance of physodes in brown algae. Phycologia 41:125–139

Steinhoff FS, Wiencke C, Müller R, Bischof K (2008) Effects of ultraviolet radiation and temperature on the ultrastructure of zoospores of the brown macroalga *Laminaria hyperborea*. Plant Biol 10:388–397

Steinhoff FS, Graeve M, Wiencke C, Wulff A, Bischof K (2011a) Lipid content and fatty acid consumption in zoospores/developing gametophytes of *Saccharina latissima* (Laminariales, Phaeophyceae) as potential precursors for secondary metabolites as phlorotannins. Polar Biol 34(7):1011–1018

Steinhoff FS, Wiencke C, Wuttke S, Bischof K (2011b) Effects of enhanced water temperatures, UV radiation and low versus high PAR on phlorotannin content and germination in zoospores of *Saccorhiza dermatodea* (Tilopteridales, Phaeophyceae). Phycologia 50(3):256–263

Steinhoff FS, Graeve M, Bischof K, Wiencke C (2012) Phlorotannin production and lipid oxidation as a potential protective function against high photosynthetically active and UV radiation in gametophytes of *Alaria esculenta* (Alariales, Phaeophyceae). Photochem Photobiol 88(1): 46–57

Strid A, Chow WS, Anderson JM (1990) Effects of supplementary ultraviolet-B radiation on photosynthesis in *Pisum sativum*. Biochim Biophys Acta 1020:260–268

Swanson AK, Druehl LD (2002) Induction, exudation and the UV protective role of kelp phlorotannins. Aquat Bot 73:241–253

Toth GB, Langhamer O, Pavia H (2005) Inducible and constitutive defenses of valuable seaweed tissues: consequences for herbivore fitness. Ecology 86:612–618

UNEP 2010 Annual Report, United Nations Environment Programme, pp 124

van Alstyne KL, Whitman SL, Ehlig JM (2001) Differences in herbivore preferences, phlorotannin production, and nutritional quality between juvenile and adult tissues from marine brown algae. Mar Biol 139:201–210

van de Poll WH, Eggert A, Buma AGJ, Breeman AM (2001) Effects of UV-B-induced DNA damage and photoinhibition on growth of temperate marine red macrophytes: habitat-related differences in UV-B tolerance. J Phycol 37:30–37

Vass I (1997) Adverse effects of UV-B light on the structure and function of the photosynthetic apparatus. In: Pessarakli M (ed) Handbook of photosynthesis. Marcel Dekker, New York, pp 931–949

Wiencke C, Gómez I, Pakker H, Flores-Moya A, Altamirano M, Hanelt D, Bischof K, Figueroa FL (2000) Impact of UV-radiation on viability, photosynthetic characteristics and DNA of brown algal zoospores: implications for depth zonation. Mar Ecol Prog Ser 197:217–229

Wiencke C, Roleda M, Gruber A, Clayton M, Bischof K (2006) Susceptibility of zoospores to UV radiation determines upper depth distribution limit of Arctic kelps: evidence through field experiments. J Ecol 94:455–463

Zacher K, Wulff A, Molis M, Hanelt D, Wiencke C (2007) Ultraviolet radiation and consumer effects on a field-grown intertidal macroalgal assemblage in Antarctica. Global Change Biol 13:1201–1215

Zubia M, Robledo D, Freile-Pelegrin Y (2007) Antioxidant activities in tropical marine macroalgae from the Yucatan Peninsula, Mexico. J Appl Phycol 19:449–458

Chapter 21
Bottom-Up Versus Top-Down Control of Macroalgal Blooms

Mirta Teichberg, Paulina Martinetto, and Sophia E. Fox

21.1 Introduction

Macroalgal blooms are increasing worldwide and have many detrimental effects on ecosystems, including more frequent occurrences of anoxia, and loss of seagrasses, coral cover, and ecologically and commercially important species. The increased occurrences of macroalgal blooms have been attributed primarily to increases in the supply of the limiting nutrients, referred to as "bottom-up" control, and decreased grazer abundance due to overfishing and habitat degradation, referred to as "top-down" control. This chapter begins with an overview of the occurrence of macroalgal blooms worldwide and their ecological consequences. It discusses the role of coastal eutrophication and overfishing as the two main global drivers of bloom formations. The relative role of bottom-up versus top-down control in structuring ecological communities is controversial and has been heavily debated within the scientific community. This chapter will provide an overview of the main arguments of this debate in marine coastal waters. Specifically, scientific evidence supporting bottom-up versus top-down controls in shallow estuaries and coastal bays, coral reefs, and intertidal or subtidal rocky shores will be reviewed to understand how systems differ in response to these controls. We will provide an outlook on the future

M. Teichberg (✉)
Leibniz-Center for Tropical Marine Ecology, Bremen, Germany
e-mail: mirta.teichberg@zmt-bremen.de

P. Martinetto
Laboratorio de Ecología, Departamento de Biología, Facultad de Ciencias Exactas y Naturales, Universidad Nacional de Mar del Plata, Instituto de Investigaciones Marinas y Costeras (IIMyC), CONICET-UNMDP, Mar del Plata, Argentina
e-mail: pmartin@mdp.edu.ar

S.E. Fox
Cape Cod National Seashore, National Park Service, Wellfleet, MA, USA
e-mail: Sophia_Fox@nps.gov

C. Wiencke and K. Bischof (eds.), *Seaweed Biology*, Ecological Studies 219,
DOI 10.1007/978-3-642-28451-9_21,

trends of eutrophication and overfishing and their impact on algal blooms in combination with other global changes. We will conclude with a discussion on management options to help reduce the escalation of bloom occurrences worldwide.

21.2 Occurrence of Macroalgal Blooms and Their Ecological Consequences

Macroalgal blooms are widespread along the coasts of the world (Table 21.1; Morand and Merceron 2005; Ye et al. 2011), forming dense mats or canopies in shallow estuaries and coastal lagoons (Valiela et al. 1992; Sfriso et al. 1992), accumulating on beaches as wrack (Morand and Merceron 2005), overgrowing corals (Smith 1981; Hughes 1994; Lapointe et al. 2005a, b), and affecting seagrass, soft-bottom, and rocky shore benthic communities (Díaz et al. 2002; Fox et al. 2009). Blooms of green, red, and brown macroalgae are common, although among these there are a few cosmopolitan species belonging to the genera *Ulva, Chaetomorpha, Cladophora*, *Gracilaria*, and *Pylaiella* (Table 21.1). Recent reviews on green algal tides showed bloom occurrences in most marine water bodies adjacent to highly urbanized or agricultural areas throughout Europe, Australia, New Zealand, Asia, and North and Central America (Morand and Merceron 2005; Teichberg et al. 2010; Ye et al. 2011). Some of the most famous examples of macroalgal blooms include the drift mats of *Ulva* spp. in Brittany, France (Morand and Merceron 2005) and Venice Lagoon (Sfriso et al. 1992; Sfriso and Marcomini 1997), *Cladophora* spp. in Peel Harvey Estuary, Australia (Lavery et al. 1991), and the more recent widely publicized *Ulva prolifera* blooms in the Yellow Sea, China during the 2008 summer Olympics with an estimated biomass of approximately 20 million wet tons (Liu et al. 2009; Gao et al. 2010; Ye et al. 2011).

Blooms of macroalgae have become a major ecological and economic problem in coastal habitats (Table 21.1, Fig. 21.1). Seaweed wrack accumulating on beaches can produce foul odors and release toxic hydrogen sulfide gas (Wilce et al. 1982), mats and deep canopies of seaweeds physically obliterate other primary producers including seagrasses (Valiela et al. 1997; Hauxwell et al. 2001; McGlathery 2001) and corals (Hughes 1994; Lapointe 1997; Lapointe et al. 2005a, b), and their decay causes anoxic conditions that lead to fish and shellfish kills (D'Avanzo et al. 1996; Valiela et al. 1997; Diaz 2001). Macroalgal blooms not only affect natural coastal communities and ecosystem function (Duarte 1995), but have high economic consequences by making coastal environments increasingly undesirable for recreational uses and threatening commercially important harvests (Valiela et al. 1992; Raffaelli et al. 1998; Oesterling and Pihl 2001). Often, the macroalgal species that form blooms are nonnative, such as *Gracilaria vermiculophylla*, *Codium* spp., and *Undaria pinnatifida,* and have the ability to quickly adapt to new environments and outcompete native species due to high stress tolerances, rapid nutrient uptake and growth rates, and varying reproductive strategies (Smith et al. 2002; Piriz et al. 2003;

Table 21.1 Examples of macroalgal blooms reported in different parts of the world's coastlines and some of their ecological and economic consequences (adapted from Teichberg et al. 2010)

Site	Seaweed taxa	Some effects	References
North America			
Gulf of California, Mexico	*Ulva, Gracilaria, Cladophora*	Anoxia, loss of species diversity	Ochoa-Izaguirre et al. (2002), Piñon-Gimate et al. (2008)
Nahant Bay, USA	*Pilayella*	Noxious odor, accumulated on beaches, nuisance to swimming and fishing	Wilce et al. (1982), Pregnall and Miller (1988)
Waquoit Bay, USA	*Cladophora, Gracilaria, Ulva*	Replaced seagrasses, anoxia, shell- and fin-fish kills	Valiela et al. (1997), Hauxwell et al. (2001), Fox et al. (2008)
Hog Island Bay, USA	*Ulva, Gracilaria, Codium*	Loss of species diversity	Thomsen et al. (2006)
San Francisco Bay, USA	*Ulva*	Anoxia, replaced benthic fauna	Fong et al. (1996)
Kaneohe Bay, Hawaii	*Dictyosphaeria*	Replaced corals	Smith (1981)
Southeast Florida, USA	*Codium*	Impact coral reefs	Lapointe et al. (2005a, b)
Bermuda	*Cladophora, Laurencia, Codium*	Anoxia, reduced benthic diversity and commercial fisheries	Lapointe and O'Connell (1989)
Southern California lagoons, USA	*Ulva*	Shifts in primary producers, anoxic conditions, cascading effects up food web	Kamer et al. (2001), Fong and Kennison (2010)
Europe			
Laholm Bay, Sweden NW Baltic Sea	*Ulva, Cladophora*	Replaced seagrasses, nuisance to swimming fishing and boating	Baden et al. (1990), Rosenberg et al. (1990)
Maasholm Bay, Germany	*Ulva, Pilayella*	Replaced native macroalgae, lowered benthic diversity and fishery yield, nuisance to swimming, fishing, and boating	Lotze et al. (2000), Worm et al. (1999)
Mondego Estuary, Portugal	*Ulva*	Replaced seagrasses, reduced benthic diversity	Martins et al. (2001), Cardoso et al. (2004)
Venice Lagoon, Italy	*Ulva, Gracilaria, Dictyota,*	Anoxia, fish kills, nutrient re-release, phytoplankton blooms	Sfriso et al. (1992), Sfriso and Marcomini (1997)

(continued)

Table 21.1 (continued)

Site	Seaweed taxa	Some effects	References
South America			
Nuevo Gulf and San Antonio Bay, Patagonia, Argentina	*Ulva, Undaria*	Accumulated on beaches, interferes with recreational uses	Díaz et al. (2002), Piriz et al. (2003), Teichberg et al. (2010), Martinetto et al. (2011)
Asia			
Quingdao, China	*Ulva*	Loss of species diversity, accumulated on beaches and nuisance for recreational activities	Liu et al. (2007, 2009)
Seto Inland Sea, Japan	*Ulva*	Replaced seagrasses	Sugimoto et al. (2007)
Australia			
Peel-Harvey Estuary, Western Australia	*Cladophora, Ulva, Chaetomorpha*	Accumulated on beaches	Lavery et al. (1991)
Tuggerah Lakes Estuary, New South Wales	*Ulva*	Replaced seagrasses, reduced benthic diversity	Cummins et al. (2004)
Africa			
Saldanha Bay, South Africa	*Ulva*	Competed with other commercial use algae	Anderson et al. (1996), Monteiro et al. (1997)

Lapointe et al. 2005a; Thomsen et al. 2006). Invasive macroalgae have often become nuisance species causing major damage to coastal benthic communities, reducing marine biodiversity, and altering food webs (Schaffelke and Hewitt 2007, see Chap. 12 by Andreakis and Schaffelke).

21.3 Coastal Eutrophication

One of the main triggers of macroalgal blooms is increased nutrient loading due to rapid urbanization and development of the coastal zone (Howarth 2008). Coastal eutrophication can be described as the increase in organic matter, including extensive blooms of phytoplankton and ephemeral macroalgae, due to nutrient loading along the coast (Nixon 1995; Cloern 2001). Cloern (2001) provides a detailed review and conceptual model of the rapidly changing problem of coastal eutrophication. This conceptual model describes both direct and indirect responses to increased nutrient loading along the coast, including increases in phytoplankton and macroalgal biomass, and changes in biogeochemistry, water transparency, and

Fig. 21.1 Macroalgal blooms in estuaries and coral reefs habitats: (**a**) blooms of *Ulva lactuca* (photo by Ylva Olsen) and (**b**) *Cladophora variegata* (photo by Gabrielle Tomasky Holmes) along the subestuaries of Waquoit Bay, Cape Cod, USA; (**c**) *Ulva* spp. blooms attached to the walls in Venice Lagoon, Italy (photo by Mirta Teichberg) and (**d**) in the intertidal channel of San Antonio Bay, Argentina (photo by Paulina Martinetto); (**e**) *Lobophora variegata,* overgrowing coral reefs in Curaçao, Netherland Antilles (photo by Anna Fricke); (**f**) blooms of the green and brown macroalgae *Chaetomorpha* and *Dictoya* spp. on reef flats of Spermonde Archipelago, Indonesia (photo by Mirta Teichberg)

habitat quality. Most large occurrences of algal blooms have been associated with increased nutrient enrichment, primarily nitrogen, although phosphorus may also play a role (Howarth 2008; Teichberg et al. 2010). Here, we focus primarily on occurrences of macroalgal blooms reported worldwide that are linked to coastal eutrophication. Ecophysiological uptake and utilization mechanisms of nutrients, nitrogen and phosphorus limitation, assimilation of carbon, and the role of nutrient availability in stimulating macroalgal growth rates are reviewed in the Chap. 4 by Gordillo.

In the USA, approximately 65% of estuaries surveyed showed moderate to high levels of eutrophic conditions, influenced primarily by land-based nutrient loads from human activities (Bricker et al. 2008). In many of these estuaries, extensive chlorophyte and rhodophyte blooms are prevalent and have caused anoxic conditions (Bricker et al. 2008) and changes in the benthic community (Table 21.1). Waquoit Bay, in Cape Cod, Massachusetts, serves as a good example of how increased urbanization has led to drastic changes in the marine community structure over the past six decades (Valiela et al. 1992, 1997; Hauxwell et al. 1998, 2001; Fox et al. 2008). Macroalgal blooms of *Cladophora*, *Gracilaria*, and *Ulva* spp. have been directly linked to increased nitrogen loading rates in the subestuaries of Waquoit Bay predominantly due to increases in wastewater nitrate from septic systems that enters the estuary via groundwater (Valiela 2006; Fox et al. 2008; Fig. 21.1). Along the Florida coast, blooms of green macroalgae, including invasive species of *Codium* and *Caulerpa*, and drift red macroalgae have also been linked to increasing in nutrient inputs from sewage, shown through their high tissue N and P content and heavier nitrogen isotopic signatures that reflect that of wastewater nitrogen (Lapointe et al. 2005a, b; Lapointe and Bedford 2007, 2010).

Outside of North America, frequency of macroalgal blooms driven by high nutrient loads has also been reported (Table 21.1). Most sites where macroalgal blooms reach a remarkably high standing crop are locations where waters are enriched with nutrients as a result of urban growth in surrounding areas. A 20-year study in Peel Harvey Estuary, Australia, showed how large blooms of *Cladophora*, *Ulva*, and *Chaetomorpha* spp. have been caused by high nutrient loading, with interannual differences in biomass associated with nutrient and light conditions (Lavery et al. 1991). In San Antonio Bay, Argentina, high wastewater nitrogen and phosphorus inputs have been linked to high growth rates and biomass of *Ulva lactuca* that forms extensive blooms in the main channel of the bay (Teichberg et al. 2010; Martinetto et al. 2011; Fig. 21.1). Other sites known for macroalgal blooms that have been linked to coastal pollution include Venice Lagoon, Italy and Mondego Estuary, Portugal (Table 21.1, Fig. 21.1). In these sites, *Ulva* blooms persist under high nutrient loads (Sfriso et al. 1992; Martins et al. 2001; Teichberg et al. 2010).

21.4 Overfishing

Drastic changes in consumer abundance due to the overharvesting of commercially and recreationally important fish, mammal, and invertebrate species have occurred in many marine ecosystems throughout history (Jackson et al. 2001). A recent study shows approximately 90% of predatory fish stocks have been removed by fishing fleets worldwide (Myers and Worm 2003) with cascading effects down to lower trophic levels (Hughes 1994; Jackson et al. 2001; Mumby et al. 2006). Here, we focus on the role overfishing plays in the occurrence of macroalgal blooms in marine coastal waters. A classification of the different type of seaweed grazers and an overview of algal–herbivore interactions can be found in detail in Chap. 8 by Iken.

Most cases where overfishing has directly impacted macroalgal growth occur in coral reef and kelp ecosystems. This is due primarily to the number of relatively larger and ecologically important consumers that have been removed from these ecosystems compared to others (Jackson et al. 2001). Overharvesting of herbivorous fish, marine mammals, and invertebrates that directly graze on macroalgae have led to increased occurrence of macroalgal blooms in coral reefs (Hughes 1994; Jackson et al. 2001; Mumby et al. 2006). The opposite trend has occurred in kelp forests, where the removal of large consumers has often led to increases in invertebrate grazer abundance which then reduce macroalgal cover in these systems (Estes and Duggins 1995; Jackson et al. 2001).

It has been argued that shifts in benthic and pelagic communities due to overfishing long predated coastal eutrophication problems in marine systems (Jackson et al. 2001; Lotze and Milewski 2004; Heck and Valentine 2007). A review paper by Heck and Valentine (2007) stressed the importance of overfishing in regulating ecosystem structure by indirectly leading to altered food webs. They support that many of the negative effects on coastal ecosystems attributed to coastal eutrophication may actually be confounded by indirect effects of historical changes in consumer abundances. They state that most manipulative experiments that test the importance of nutrients and consumers on benthic ecosystems today are doing so in ecosystems that have already been long devoid of apex predators and many other consumers lower in the food web. Another study (Eriksson et al. 2009) showed that declines in predatory fish were promoting blooms of macroalgae in the Baltic Sea through cascading effects down the food web. Through surveys they showed that macroalgal cover was highest when abundances of large piscivorous fish were low, which led to higher abundances of smaller fish that feed on invertebrate grazers. Further manipulative experiments supported the notion that these predators when abundant exerted a strong top-down forcing on the food web (Eriksson et al. 2009).

21.5 Bottom-Up Versus Top-Down Control in Marine Coastal Ecosystems

There are an overwhelming number of studies focusing on bottom-up versus top-down controls of primary producer biomass and community structure in marine coastal waters (Hauxwell et al. 1998; Lapointe 1999; Balducci et al. 2001; Lotze et al. 2001; Smith et al. 2010; Martinetto et al. 2011; Fox et al. 2012). Marine communities are, in general, naturally regulated by bottom-up and top-down forces; however, the increase in anthropogenic activities in the coastal zone has influenced the relative strength of bottom-up and top-down controls on these communities. In many instances, increased nutrient inputs seem to be the main driver of macroalgal blooms (Valiela et al. 1997; Hauxwell et al. 1998; Lapointe 1999), while in others changes in grazing pressure are thought to be responsible (Hughes 1994; Smith et al. 2010).

In order to determine the relative influence of these two factors, many experimental manipulations of nutrients and grazing have been carried out in marine habitats under a range of different in situ conditions. Recent metaanalyses by Burkepile and Hay (2006) and Gruner et al. (2008) have compiled the results of these experimental studies to see whether any general patterns can be deduced regarding the relative influence of top-down and bottom-up controls on these ecosystems. They found that in most cases nutrients and grazers independently affected producer biomass, but that the relative role of top-down versus bottom-up control is context dependent, that is, it may differ by latitude, by type of marine habitat, by the functional groups of algae or type of grazers in the system, by the productivity of the ecosystem, or by the sampling method of producer response (Burkepile and Hay 2006; Gruner et al. 2008). They also found that in some cases there were interactive or synergistic effects, in which nutrient enrichment in the absence of herbivores showed the highest productivity and percent cover of macroalgae (Gruner et al. 2008; Burkepile and Hay 2006).

21.5.1 Latitudinal Patterns

Most reports from temperate latitudes addressing the forces behind macroalgal blooms appear to suggest that nutrients were involved (Baden et al. 1990; Valiela et al. 1992; Sfriso et al. 1992; Raffaelli et al. 1998), with the exception of rocky shores (Lubchenco 1978; Lotze and Worm 2000). Similar studies from tropical latitudes place less emphasis on nutrients and suggest large impacts of grazing on macroalgal blooms (Lewis 1986; Hughes 1994), but opinions differ (Lapointe 1997; Littler et al. 2006). The metaanalysis of Burkepile and Hay (2006) showed that in tropical marine habitats, macroalgae responded independently to both nutrient enrichment and herbivore removal by increasing biomass or cover. Nutrient effects were, however, smaller than the grazing effects, and only important when herbivores were absent. This was not the case for temperate marine habitats, which tended to vary in the relative effects of bottom-up and top-down control based on the nutrient status of the system. In high nutrient environments bottom-up effects were stronger, while in low nutrient environments top-down effects were more significant (Burkepile and Hay 2006).

One of the main explanations for these latitudinal differences in the control of macroalgal blooms is the types and sizes of grazers and their grazing rates. Other possible explanations may be the extent of nutrient loading, the macroalgal taxa involved in different systems, the length of the growing season, and the range in light intensity in temperate versus tropical coastal waters.

21.5.2 *Dominant Control in Different Habitats*

In this section, we provide some key examples from the literature defining the relative roles of bottom-up and top-down controls within different marine habitats, including estuaries and coastal bays, coral reefs, and intertidal and subtidal rocky shores. Through these examples, we show the variation in macroalgal responses to their controls, and also try to include some of the studies, which have paved the way or stimulated the debate in this field.

21.5.2.1 Estuaries and coastal bays

The relative roles of bottom-up and top-down controls in estuarine systems have been thoroughly studied due to the high degree of disturbance found in these ecosystems as a consequence of coastal urbanization and development. This over-development leads to higher nutrient inputs from point and nonpoint sources of wastewater and fertilizers into coastal waters, while wetlands and forests that generally buffer and act as a filter along the coastline are being removed at alarming rates. Most sites receiving high nutrient enrichment generally are found to be bottom-up controlled, whereas undisturbed low nutrient environments are often found to be top-down controlled (Hauxwell et al. 1998; Burkepile and Hay 2006; Fox et al. 2012).

This general model of the relative importance of bottom-up and top-down controls on macroalgal blooms in estuaries is again exemplified by the Waquoit Bay example. Waquoit Bay macroalgal blooms are stimulated by high nitrogen loading to its receiving subestuaries (Valiela et al. 1997; Hauxwell et al. 1998; Fox et al. 2008). In the high nitrogen loaded sites of Waquoit Bay, macroalgal growth rates and biomass were highest (Hauxwell et al. 1998). Additionally, more frequent hypoxic events lowered the abundance of small crustaceans, the dominant grazers, and therefore, grazing rates were too low to compensate for the increase in algal biomass (Hauxwell et al. 1998; Fox et al. 2009, 2012). In contrast, in the low nitrogen loaded site macroalgal growth rates and biomass were lower, grazers were very abundant, and macroalgae were easily controlled by consumption rates (Hauxwell et al. 1998; Fox et al. 2009, 2012).

Another well-studied case of bottom-up and top-down controls of macroalgal blooms is that of Venice Lagoon. In the 1970s and 1980s, the lagoon received inputs of nutrients from urbanized areas in and around Venice from agricultural, industrial, and treated and untreated sewage effluent sources (Sfriso et al. 1992). With high nutrient loading came noticeable changes in benthic community structure driven by large blooms of *Ulva rigida* and other green macroalgae (Sfriso et al. 1987, 1992). Between 1987 and 1998, macroalgal standing crop in Venice Lagoon declined to only 1.6% of what was present in 1987 (Sfriso et al. 2003). This dramatic reduction was initially thought to be due to a combination of changes in climate, sedimentation fluxes, and management of nutrient loading entering the lagoon. Additionally, as macroalgal growth declined, fewer anoxic events allowed for the recovery of invertebrate grazers, which were able to help control macroalgae blooms from the top-down (Balducci et al. 2001).

Other biotic and abiotic factors, however, may affect the relative roles of bottom-up and top-down controls of macroalgal communities in estuarine systems. For example, reproduction and recruitment of early life history stages of macroalgae may respond differently under nutrient enrichment and grazing pressure than adult life stages (Lotze et al. 1999, 2000, 2001; Lotze and Worm 2000). Lotze et al. (1999) found that the bottom-up and top-down controls on early life stages may act as a bottleneck for bloom-forming species of macroalgae in some cases. Lotze et al. (2001), however, showed that total recruit density of ephemeral bloom-forming macroalgae *Ulva* and *Pilayella* spp. in the Baltic Sea was positively influenced by nutrient enrichment, while grazing only limited recruitment and growth of the more palatable of the two species without changing the total recruitment (Lotze et al. 2001).

Macroalgal community structure may also influence the strength of bottom-up and top-down controls. In the presence of canopy-forming macroalgal species, such as *Fucus vesiculosus*, the response of ephemeral algae to nutrients was found to be limited by as much as 90% compared to those without canopies due to a reduction in light availability (Eriksson et al. 2007). In contrast, the presence of epiphytes growing on macroalgae may actually stimulate macroalgal growth if epiphytes are preferentially consumed by grazers (Kamermans et al. 2002; see also Chap. 11 by Potin). Furthermore, on a smaller spatial scale, the presence of grazers within the macroalgal canopy may also be an additional source of nutrients through their excretion (Taylor and Rees 1998, see also Chap. 4 by Gordillo).

Hydrodynamics can also alter the strength of bottom-up and top-down controls on a system. For example, in Mondego Estuary, Portugal, mitigation measures to improve the hydrodynamics of the estuary have been found to alleviate macroalgal blooms occurrences caused by high nutrient loading by increasing the circulation and diverting inflow of nutrient-rich waters (Lillebø et al. 2005). In San Antonio Bay, Argentina, high nutrient loads enter the bay exposing macroalgae to elevated nutrient concentrations during low tide and supporting a large macroalgal biomass (Teichberg et al. 2010; Martinetto et al. 2010, 2011). Additionally, large tidal flushing helps to remove anoxic waters and nutrients from the system, minimizing hypoxia-related stress on grazers and negative cascading effects up the food web (Martinetto et al. 2010, 2011). Therefore, the high biomass of nutrient-rich macroalgae can provide a large quantity of food with higher nutritional quality to grazers and support higher grazer abundances (Martinetto et al. 2010, 2011).

21.5.2.2 Coral reefs

In the last four to five decades, there has been an increasing trend of macroalgal blooms in coral reefs coinciding with a decrease in coral cover (Hughes 1994; McManus and Polsenberg 2004; Nugues and Bak 2008). This shift from coral to macroalgal-dominated reefs, known as coral–algal phase shifts, has stimulated much of the debate in the relative role of top-down and bottom-up controls in marine habitats. Some studies have shown that the removal of the dominant herbivores, through overfishing or natural causes, has been more important as a control on macroalgal growth than nutrient enrichment (Hughes 1994; Hughes et al. 1999; Burkepile and Hay 2006; Sotka and Hay 2009). Others state that recent increases in nutrient inputs from land to coastal reefs have triggered macroalgal blooms (Lapointe 1997; Lapointe et al. 2005a, b). And then there are those studies that support both top-down and bottom-up controls as important influencing factors (McClanahan et al. 2003; Littler et al. 2006; Smith et al. 2001). In this chapter, we do not attempt to provide a comprehensive review of the current extensive literature on coral–algal phase shifts in coral reefs, but rather focus on a few specific studies that examine the effects of top-down and bottom-up controls on macroalgal growth.

In Hughes (1994), one of the first long-term studies of coral reef decline in the Caribbean reported large-scale natural and human disturbances were linked to coral–algal phase shifts. In most of the reefs around Jamaica, the driving factors of these phase shifts were thought to be the increase in human population, which drove an increase in overfishing, followed by the mass mortality of the sea urchin *Diadema antillarum*. Without herbivorous fish and *D. antillarum*, the dominant grazers of macroalgae on the reefs, blooms of filamentous macroalgae began to overgrow corals, which were then replaced by late successional stage species, such as *Dictyota*, *Lobophora*, *Halimeda*, and *Sargassum* spp. These macroalgae inhibited coral reef recovery by effectively competing for open space needed for recruitment of coral larvae (Hughes 1994). Lapointe (1997) introduced another potential trigger of macroalgal blooms on these Jamaican reefs by showing that bottom-up control also played a role. It was demonstrated that dissolved inorganic nitrogen from wastewater entered the reefs through groundwater discharge and elevated reef nutrient water concentrations sufficiently to sustain macroalgal blooms in these habitats (Lapointe 1997). These studies spurred a debate as to the relative role of bottom-up and top-down controls on coral reefs (Hughes et al. 1999; Lapointe 1999).

Since this debate began, a few studies have demonstrated how both top-down and bottom-up controls can be important. Littler et al. (2006) developed a conceptual model, the Relative Dominance Model, to describe the process of bloom formation on reefs under independent and combined factors of nutrients and grazing pressure. Results of manipulative experiments, bioassays, and surveys of communities showed that: (1) under reduced nutrients alone fleshy algae grow when herbivory is low, (2) under high herbivory alone fleshy algae grow when nutrients are high, (3) reduced nutrients and high herbivory prevent blooms of macroalgae (Littler et al. 2006). It has also been found that different controls are important for different groups of algae; for example, McClanahan et al. (2003) found that turf algae were positively affected by nutrients and negatively affected by herbivory, while frondose brown algae grew better under low nutrients and low herbivory. Vermeij et al. (2010) also found that turf algae can overgrow corals under high nutrients, but that herbivores are not able to control their growth under these conditions.

Despite the evidence that bottom-up control does play a role in triggering macroalgal bloom formations on reefs, the majority of studies support that top-down control is the primary driving factor (Thacker et al. 2001; Burkepile and Hay 2006; Sotka and Hay 2009). A recent study in a pristine fisheries management area along a Hawaiian reef revealed that in the absence of herbivores, fleshy macroalgal abundance was higher than in any other treatment, while in the presence of herbivores corals and crustose coralline algae were more abundant (Smith et al. 2010). The effects of nutrient enrichment supported slightly higher macroalgal percent cover than in nonenriched treatments, but the species composition depended on the herbivore treatment. After reexposure to natural conditions, the effects of enrichment and herbivore absence disappeared over a relatively short time frame, supporting the notion that increasing herbivore populations is critical to reduce macroalgal blooms on reefs (Smith et al. 2010).

Based on the variable results of studies examining top-down and bottom-up controls of macroalgal blooms on coral reefs, the relative role of these controls is still not fully resolved. Only through larger scale changes in fishing regulations and better management of nutrient inputs to coastal reefs will we be able to better distinguish which of these or the combination of both is the causal factor.

21.5.2.3 Rocky Intertidal and Subtidal Ecosystems

Much of the knowledge defining the role of top-down controls on macroalgal community dynamics began with the well-known classical ecological studies in rocky intertidal and subtidal communities (Paine and Vadas 1969; Lubchenco 1978, 1983; Estes et al. 1978; Lubchenco and Menge 1978). These studies showed how removal of the dominant grazers or predators influenced macroalgal diversity and community structure directly or indirectly through cascading effects from the top-down. In the experimental study by Paine and Vadas (1969), the removal of the sea urchin *Strongylocentrotus purpuratus* resulted in an increase in the biomass of brown algae (*Hedophyllum sessile* and *Laminaria complanata*) in intertidal pools and subtidally on the rocky coast of Washington, USA. Lubchenco (1978) showed that when snails were removed from tide pools where *Chondrus crispus* dominated, green ephemeral species settled quickly and outgrew *C. crispus*.

Only more recently has the role of bottom-up control, in terms of nutrient supply, been explored in rocky shores (Menge 1992; Nielsen 2003; Bokn et al. 2003; Korpinen et al. 2007a, b). A review by Menge (1992) discusses some of the reasons the role of bottom-up controls on rocky shores has been neglected for so long, including the overemphasis of physical stress in these environments, logistical and methodological constraints of experiments, and different training backgrounds of marine benthic ecologists. Possible explanations of how nutrient enrichment cascades up the food web and affects the community structure of rocky shores were also discussed using a case study along the Oregon Coast. In this environment, nutrient loading from upwelling was thought to increase primary production and lead to higher abundances of filter-feeding prey that then supported higher abundances and feeding rates of predators (Menge 1992). Further studies have also shown that increased phytoplankton due to upwelling events and nearshore hydrography drives this bottom-up regulation of the benthic community (Menge et al. 1997).

The relative effects of bottom-up and top-down control on the rocky shore macrophyte community along the same rocky shoreline were later experimentally tested in intertidal pools by Nielson (2003). Small increases in nutrient supply strongly increased the total percent cover of macroalgae and the diversity of benthic macrophytes under low herbivore abundance, but not under high herbivore abundance. This response in the macrophyte community was primarily due to an increase in the functional group of corticated algae. The results of this study suggest that open-coast marine intertidal communities are more sensitive to fluctuations in nutrient regimes than previously thought, and that nutrient effects are likely to be amplified where consumers are overexploited or naturally low in abundance (Nielsen 2003).

In both the above cases, bottom-up control of intertidal communities was driven by natural sources of nutrients from upwelling events rather than anthropogenic sources. However, other scientists have linked changes in the macroalgal community structure of rocky shores to coastal eutrophication, where macroalgal assemblages nearer to urbanized areas consisted of higher abundances of the opportunistic chlorophyte *Ulva* spp. than in control sites (Díaz et al. 2002; Worm and Lotze 2006). Overall the consensus is that top-down controls are dominant on rocky shores but that bottom-up controls can interact with top-down controls to change patterns in macrophyte community structure and food web interactions.

21.6 Conclusion

There may be no general solution to the bottom-up top-down debate in macroalgal communities, but rather different responses based on species or functional groups, physical and chemical environments, and types of grazers. The fact that so many rigorous experimental studies have found conflicting results shows that both factors are important under different circumstances and often interact simultaneously.

The human population is projected to steadily increase, which will result in further anthropogenic pressure on coastal marine systems. It is likely that nutrient loading rates and fishing pressure will intensify, despite management efforts. The impacts of nutrient loading and overfishing will also be exacerbated by other global changes, such as rising seawater temperatures, sea level rise, and ocean acidification. Nutrients in combination with warmer waters may promote more occurrences of macroalgal blooms, as photosynthesis, growth rates of macroalgae, and recruitment are thought to be positively affected by an interaction of these factors (Longstaff et al. 2002; Lotze and Worm 2002; Bintz et al. 2003; Tsai et al. 2005). Additionally, enhanced CO_2 conditions have been shown to benefit some ephemeral bloom-forming macroalgal species over other slower growing species (Gordillo et al. 2001; Wootton et al. 2008), while negatively impacting many benthic invertebrates (Fabry et al. 2008). Additionally, the increase in invasions by nonnative macroalgal species to new coastal habitats may promote blooms of macroalgae that are better adapted to changing environmental conditions (Thomsen et al. 2006). Cascading effects up or down the food web and, thus, the strength of bottom-up and top-down controls acting on macroalgal communities will likely be altered due to these shifts in species composition and environmental change (Fox et al. 2010).

Management of all aspects (nutrients, fishing, habitat protection, reduction of CO_2 emissions) is important, and choosing one management practice over another will not be successful at controlling macroalgal blooms. The management and removal of nitrogen and phosphorus is necessary as a first step. Wastewater treatment plants and improvements in septic systems have shown significant positive effects on water quality through the removal of the bulk of nutrients prior to entering the coast or groundwater. Reduction of fertilizer use and phosphate-free detergents can also reduce the amount of nutrients entering these systems. Preservation and restoration of wetlands will help to filter nutrient inputs before they enter rivers and coastal waters. Additionally, improving hydrodynamics of low-flow systems by increasing tidal flushing could help to mitigate nutrient effects. Increases in fishing regulations and protected areas with no take zones may help to increase fish and invertebrate populations. This will impact food web communities in multiple ways and will strengthen top-down controls on macroalgal growth.

In many cases, these management practices may not be enough to facilitate recovery of marine habitats to their natural state prior to disturbances, and much is still unknown as to how bottom-up and top-down controls will interact with future global change projections. It is very likely that macroalgal blooms will continue to become more frequent and severe in coastal systems before conditions improve. A better understanding of how these controls will interact under predicted future conditions is critical to develop tools to mitigate these changes.

References

Anderson RJ, Monteiro PMS, Levitt GJ (1996) The effect of localized eutrophication on competition between *Ulva lactuca* (Ulvaceae, Chlorophyta) and a commercial resource of *Gracilaria verrucosa* (Gracilariaceae, Rhodophyta). Hydrobiologia 326(327):291–296

Baden SP, Loo LO, Pihl L, Rosenberg R (1990) Effects of eutrophication on benthic communities including fish: Swedish west coast. Ambio 19:113–122

Balducci C, Sfriso A, Pavoni B (2001) Macrofauna impact on *Ulva rigida* C Ag. Production and relationship with environmental variables in the lagoon of Venice. Mar Environ Res 52:27–49

Bintz JC, Nixon SW, Buckley BA, Granger SL (2003) Impacts of temperature and nutrients on coastal lagoon plant communities. Estuaries 76:765–776

Bokn TL, Duarte CM, Pedersen MF, Marba N, Moy FE, Barrón C, Bjerkeng B, Borum J, Christie H, Engelbert S, Fotel FL, Hoell EE, Karez R, Kersting K, Kraufvelin P, Lindblad C, Olsen M, Sanderud KA, Sommer U, Sørensen K (2003) The response of experimental rocky shore communities to nutrient additions. Ecosystems 6:577–594

Bricker SB, Longstaff B, Dennison W, Jones A, Boicourt K, Wicks C, Woerner J (2008) Effects of nutrient enrichment in the nation's estuaries: a decade of change. Harmful Algae 8:21–32

Burkepile DE, Hay ME (2006) Herbivore vs. nutrient control of marine primary producers: context-dependent effects. Ecology 87:3128–3139

Cardoso PG, Pardal MA, Lillebø AI, Ferreira SM, Raffaelli D, Marques JC (2004) Dynamic changes in seagrasses assemblages under eutrophication and implications for recovery. J Exp Mar Biol Ecol 302:233–248

Cloern JE (2001) Our evolving conceptual model of the coastal eutrophication problem. Mar Ecol Prog Ser 210:223–253

Cummins SP, Roberts DE, Zimmerman KD (2004) Effects of the green macroalga *Enteromorpha intestinalis* on macrobenthic and seagrass assemblages in a shallow coastal estuary. Mar Ecol Prog Ser 266:77–87

D'Avanzo C, Kremer JN, Wainright SC (1996) Ecosystem production and respiration in response to eutrophication in shallow temperate estuaries. Mar Ecol Prog Ser 141:263–274

Diaz RJ (2001) Overview of hypoxia around the world. J Environ Qual 30:275–281

Díaz P, López Gappa JJ, Piriz ML (2002) Symptoms of eutrophication in intertidal macroalgal assemblages of Nuevo Gulf (Patagonia, Argentina). Bot Mar 45:267–273

Duarte CM (1995) Submerged aquatic vegetation in relation to different nutrient regimes. Ophelia 41:87–112

Eriksson BK, Rubach A, Helmut H (2007) Dominance by a canopy forming seaweed modifies resource and consumer control of bloom-forming macroalgae. Oikos 116:1211–1219

Eriksson BK, Ljunggren L, Sandström A, Johansson G, Mattila J, Rubach A, Råberg S, Snickars M (2009) Declines in predatory fish promote bloom-forming macroalgae. Ecol Appl 19:1975–1988

Estes JA, Duggins DO (1995) Sea otters and kelp forests in Alaska: generality and variation in a community ecological paradigm. Ecol Monogr 65:75–100

Estes JA, Smith NS, Palmisano JF (1978) Sea otter predation and community organization in the western Aleutian Islands, Alaska. Ecology 59:822–833

Fabry VJ, Seibel BA, Feely RA, Orr JC (2008) Impacts of ocean acidification on marine fauna and ecosystem processes. ICES J Mar Sci 65:414–432

Fong P, Kennison RL (2010) Phase shifts, alternative stable states, and the status of Southern California lagoons. In: Kennish MJ, Pearl HW (eds) Coastal Lagoons: critical habitats of environmental change, vol 164. CRC, New York, pp 227–251

Fong P, Boyer KE, Desmond JS, Zedler JB (1996) Salinity stress, nitrogen competition, and facilitation: what controls seasonal succession of two opportunistic green macroalgae? J Exp Mar Biol Ecol 206:203–221

Fox SE, Stieve E, Valiela I, Hauxwell J, McClelland J (2008) Macrophyte abundance in Waquoit Bay: effects of land-derived nitrogen loads on seasonal and multi-year biomass patterns. Estuar Coast 31:532–541

Fox SE, Teichberg M, Olsen YS, Heffner L, Valiela I (2009) Restructuring of benthic communities in eutrophic estuaries: lower abundance of prey leads to trophic shifts from omnivory to grazing. Mar Ecol Prog Ser 380:43–57

Fox SE, Olsen YS, Spivak AC (2010) Effects of bottom-up and top-down controls and climate change on estuarine macrophyte communities and the ecosystem services they provide. In: P.F. Kemp (ed.) Eco-DAS VIII Symposium Proceedings, ASLO, pp 129–145

Fox SE, Teichberg M, Heffner L, Valiela I (2012) The relative role of nutrients, grazing, and predation as controls on macroalgal growth in a temperate estuary. Estuar Coasts (in press)

Gao S, Chen X, Yi Q, Wang G, Pan G, Lin A, Peng G (2010) A strategy for the proliferation of *Ulva prolifera*, main causative species of green tides, with formation of sporangia by fragmentation. PLoS One 5:e8571. doi:10.1371/journal.pone.0008571

Gordillo FJL, Niell FX, Figueroa FL (2001) Non-photosynthetic enhancement of growth by high CO_2 level in the nitrophilic seaweed *Ulva rigida* Agardh (Chlorophyta). Planta 213:64–70

Gruner DS, Smith JE, Seabloom EW, Ngai JT, Hillebrand H, Harpole WS, Elser JJ, Cleland EE, Bracken MES, Borer ET, Bolker BM (2008) A cross-system synthesis of consumer and nutrient resource control on producer biomass. Ecol Lett 11:740–755

Hauxwell J, McClelland J, Behr P, Valiela I (1998) Relative importance of grazing and nutrient controls of macroalgal biomass in three temperate shallow estuaries. Estuaries 21:347–360

Hauxwell J, Cebrian J, Furlong C, Valiela I (2001) Macroalgal canopies contribute to eelgrass (*Zostera marina*) decline in temperate estuarine ecosystems. Ecology 82:1007–1022

Heck KL, Valentine JF (2007) The primacy of top-down effects in shallow benthic ecosystems. Estuar Coast 30:371–381

Howarth RW (2008) Coastal nitrogen pollution: a review of sources and trends globally and regionally. Harmful Algae 8:14–20

Hughes TP (1994) Catastrophes, phase-shifts, and large-scale degradation of a Caribbean coral reef. Science 265:1547–1551

Hughes T, Szmant AM, Steneck R, Carpenter R, Miller S (1999) Algal blooms on coral reefs: what are the causes? Limnol Oceanogr 44:1583–1586

Jackson JBC, Kirby MX, Berger WH, Bjorndal KA, Botsford LW, Bourque BJ, Bradbury RH, Cooke R, Erlandson J, Estes JA, Hugues TP, Kidwell S, Lange CB, Lenihan HS, Pandolfi JM, Peterson CH, Steneck RS, Tegner MJ, Warner RR (2001) Historical overfishing and the recent collapse of coastal ecosystems. Science 293:629–638

Kamer K, Boyle KA, Fong P (2001) Macroalgal bloom dynamics in a highly eutrophic Southern California estuary. Estuaries 24:623–635

Kamermans P, Malta E, Verschuure JM, Schrijvers L, Lentz LF, Lien ATA (2002) Effect of grazing by isopods and amphipods on growth of *Ulva* spp. (Chlorophyta). Aquat Ecol 36:425–433

Korpinen S, Jormalainen V, Honkanen T (2007a) Bottom-up and cascading top-down control of macroalgae along a depth gradient. J Exp Mar Biol Ecol 343:52–63

Korpinen S, Jormalainen V, Honkanen T (2007b) Effects of nutrients, herbivory, and depth on the macroalgal community in the rocky sublittoral. Ecology 88:839–852

Lapointe BE (1997) Nutrient thresholds for bottom-up control of macroalgal blooms on coral reefs in Jamaica and southeast Florida. Limnol Oceanogr 42:1119–1131

Lapointe B (1999) Simultaneous top-down and bottom-up forces control macroalgal blooms on coral reefs (Reply to the comment by Hughes et al.). Limnol Oceanogr 44:1586–1592

Lapointe BE, Bedford BJ (2007) Drift rhodophyte blooms emerge in Lee County, Florida, USA: evidence of escalating coastal eutrophication. Harmful Algae 6:421–437

Lapointe B, Bedford B (2010) Ecology and nutrition of invasive *Caulerpa brachypus* f. *parvifolia* blooms on coral reefs off southeast Florida, USA. Harmful Algae 9:1–12

Lapointe BE, O'Connell J (1989) Nutrient-enhanced growth of *Cladophora prolifera* in Harrington Sound, Bermuda: eutrophication of a confined, phosphorus-limited marine ecosystem. Estuar Coast Shelf Sci 28:347–360

Lapointe BE, Barile PJ, Littler MM, Littler DS (2005a) Macroalgal blooms on southeast Florida coral reefs II cross-shelf discrimination of nitrogen sources indicates widespread assimilation of sewage nitrogen. Harmful Algae 4:1106–1122

Lapointe BE, Barile PJ, Littler MM, Littler DS, Bedford BJ, Gasque C (2005b) Macroalgal blooms on southeast Florida coral reefs I nutrient stoichiometry of the invasive green alga Codium isthmocladum in the wider Caribbean indicates nutrient enrichment. Harmful Algae 4:1092–1105

Lavery PS, Lukatelich RJ, McComb AJ (1991) Changes in the biomass and species composition of macroalgae in a eutrophic estuary. Estuar Coast Shelf Sci 33:1–22

Lewis SM (1986) The role of herbivorous fishes in the organization of a Caribbean reef community. Ecol Monogr 56:183–200

Lillebø AI, Neto JM, Martins I, Verdelhos T, Leston S, Cardoso PG, Ferreira SM, Marques JC, Pardal MA (2005) Management of a shallow temperate estuary to control eutrophication: the effect of hydrodynamics on the system's nutrient loading. Estuar Coast Shelf Sci 65:697–707

Littler MM, Littler DS, Brooks BL (2006) Harmful algae on tropical coral reefs: bottom-up eutrophication and top-down herbivory. Harmful Algae 5:565–585

Liu D, Bai J, Song S, Zhang J, Sun P, Li Y, Han G (2007) The impact of sewage discharge on the macroalgal community in the Yellow Sea coastal area around Quingdao, China. Water Air Soil Pollut Focus 7:683–692

Liu D, Keesing JK, Xing Q, Shi P (2009) World's largest macroalgal bloom caused by expansion of seaweed aquaculture in China. Mar Pollut Bull 58:888–895

Longstaff BJ, Kildea T, Runcie JW, Cheshire A, Dennison WC, Hurd C, Kana T, Raven JA, Larkum AW (2002) An in situ study of photosynthetic oxygen exchange and electron transport rate in the marine macroalga *Ulva lactuca* (Chlorophyta). Photosynth Res 74:281–293

Lotze HK, Milewski I (2004) Two centuries of multiple human impacts and successive changes in a North Atlantic food web. Ecol Appl 14:1428–1447

Lotze HK, Worm B (2000) Variable and complementary effects of herbivores on different life stages of bloom forming macroalgae. Mar Ecol Prog Ser 200:167–175

Lotze HK, Worm B (2002) Complex interactions of climatic and ecological controls on macroalgal recruitment. Limnol Oceanogr 47:1734–1741

Lotze HK, Schramm W, Schories D, Worm B (1999) Control of macroalgal blooms at early development stages: *Pilayella littoralis* versus *Enteromorpha* spp. Oecologia 119:46–54

Lotze HK, Worm B, Sommer U (2000) Propagule banks, herbivory and nutrient supply control population development and dominance patterns in macroalgal blooms. Oikos 89:46–58

Lotze HK, Worm B, Sommer U (2001) Strong bottom-up and top-down control of early life stages of macroalgae. Limnol Oceanogr 46:749–757

Lubchenco J (1978) Plant species diversity in a marine intertidal community: importance of herbivore food preference and algal competitive ability. Am Nat 112:23–39

Lubchenco J (1983) *Littorina* and *Fucus*: effects of herbivores, substratum, heterogeneity, and plant escapes during succession. Ecology 64:1116–1123

Lubchenco J, Menge BA (1978) Community development and persistence in a low rocky intertidal zone. Ecol Monogr 48:67–94

Martinetto P, Daleo P, Escapa M, Alberti J et al (2010) High abundance and diversity of consumers associated with eutrophic areas in a semi-desert macrotidal coastal ecosystem in Patagonia, Argentina. Estuar Coast Shelf Sci 88:357–364

Martinetto P, Teichberg M, Valiela I, Montemayor D, Iribarne O (2011) Top-down and bottom-up regulation in a high nutrient-high herbivory coastal ecosystem. Mar Ecol Prog Ser 432:69–82

Martins I, Pardal M, Lillebo AI, Flindt MR, Marques JC (2001) Hydrodynamics as a major factor controlling the occurrence of green macroalgal blooms in a eutrophic estuary: a case study on the influence of precipitation and river management. Estuar Coast Shelf Sci 52:165–177

McClanahan TR, Sala E, Stickel PA, Cokos BA, Baker AC, Starger CJ, Jones SH IV (2003) Interaction between nutrients and herbivory in controlling algal communities and coral condition on Glover's Reef, Belize. Mar Ecol Prog Ser 261:135–147

McGlathery KJ (2001) Macroalgal blooms contribute to the decline of seagrass in nutrient-enriched coastal waters. J Phycol 37:453–456

McManus JW, Polsenberg JF (2004) Coral-algal phase shifts on coral reefs: ecological and environmental aspects. Prog Oceanogr 60:263–279

Menge BA (1992) Community regulation: under what conditions are bottom-up factors important on rocky shores? Ecology 73:755–765

Menge BA, Daley BA, Wheeler PA, Dahlhoff E, Sanford E, Strub PT (1997) Benthic-pelagic links and rocky intertidal communities: bottom-up effects on top-down control? Proc Natl Acad Sci 94:14530–14535

Monteiro PMS, Anderson RJ, Woodbourne S (1997) $d^{15}N$ as a tool to demonstrate the contribution of fish-waste-derived nitrogen to an *Ulva* bloom in Saldanha Bay, South Africa. S Afr J Mar Sci 18:1–9

Morand P, Merceron M (2005) Macroalgal population and sustainability. J Coast Res 21:1009–1020

Mumby PJ, Dahlgren CP, Harborne AR, Kappel CV, Micheli F, Brumbaugh DR, Holmes KE, Mendes JM, Broad K, Sanchirico JN, Buch K, Box S, Stoffle RW, Gill AB (2006) Fishing, trophic cascades, and the process of grazing on coral reefs. Science 311:98–101

Myers RA, Worm B (2003) Rapid worldwide depletion of predatory fish communities. Nature 423:280–283

Nielsen KJ (2003) Nutrient loading and consumers: agents of change in open-coast macrophyte assemblages. Proc Natl Acad Sci 100:7660–7665

Nixon SW (1995) Coastal eutrophication: a definition, social causes, and future concerns. Ophelia 41:199–220

Nugues MM, Bak RPM (2008) Long-term dynamics of the brown macroalga *Lobophora variegata* on deep reefs in Curacao. Coral Reefs 27:389–393

Ochoa-Izaguirre MJ, Carballo JL, Paez-Osuna F (2002) Qualitative changes in macroalgal assemblages under two contrasting climatic conditions in a subtropical estuary. Bot Mar 45:130–138

Oesterling M, Pihl L (2001) Effects of filamentous green algal mats on benthic macrofauna functional feeding groups. J Exp Mar Biol Ecol 263:159–183

Paine RT, Vadas RL (1969) The effects of grazing by sea urchins, *Strongylocentrotus* spp., on benthic algal populations. Limnol Oceanogr 14:710–719

Piñon-Gimate A, Serviere-Zaragoza E, Ochoa-Izaguirre MJ, Páez-Osuna F (2008) Species composition and seasonal changes in macroalgal blooms in lagoons along the southeastern Gulf of California. Bot Mar 51:112–123

Piriz ML, Eyras MC, Rostagno CM (2003) Changes in biomass and botanical composition of beach-cast seaweeds in a disturbed coastal area from Argentine Patagonia. J Appl Phycol 15:67–74

Pregnall AM, Miller SL (1988) Flux of ammonium from surf-zone and nearshore sediments in Nahant Bay, Massachusetts, USA in relation to free-living *Pilayella littoralis*. Mar Ecol Prog Ser 50:161–167

Raffaelli DG, Raven JA, Poole LJ (1998) Ecological impact of green macroalgal blooms. Oceanogr Mar Biol 36:97–125

Rosenberg R, Elmgren R, Fleischer S, Jonsson P, Persson G, Dahlin H (1990) Marine eutrophication case studies in Sweden. Ambio 19:102–108

Schaffelke B, Hewitt CL (2007) Impacts of introduced seaweeds. Bot Mar 50:397–417

Sfriso A, Marcomini A (1997) Macrophyte production in a shallow coastal lagoon: part II coupling with chemico-physical parameters and nutrient concentrations in waters. Mar Environ Res 44:351–375

Sfriso A, Marcomini A, Pavoni B (1987) Relationships between macroalgal biomass and nutrient concentrations in a hypertrophic area of the Venice Lagoon. Mar Environ Res 22:297–312

Sfriso A, Pavoni B, Marcomini A, Orio AA (1992) Macroalgae, nutrient cycles, and pollutants in the Lagoon of Venice. Estuaries 15:517–528

Sfriso A, Facca C, Ghetti PF (2003) Temporal and spatial changes of macroalgae and phytoplankton in a Mediterranean coastal area: the Venice lagoon as a case study. Mar Environ Res 56:617–636

Smith SV (1981) Kanehoe Bay sewage diversion experiment: perspectives on ecosystem responses to nutritional perturbation. Pac Sci 35:279–402

Smith JE, Smith CM, Hunter CL (2001) An experimental analysis of the effects of herbivory and nutrient enrichment on benthic community dynamics on a Hawaiian reef. Coral Reefs 19:332–342

Smith JE, Hunter CL, Smith CM (2002) Distribution and reproductive characteristics of nonindigenous and invasive marine algae in the Hawaiian islands. Pac Sci 56:299–315

Smith JE, Hunter CL, Smith CM (2010) The effects of top-down versus bottom-up control on benthic coral reef community structure. Oecologia 163:497–507

Sotka EE, Hay ME (2009) Effects of herbivores, nutrient enrichment, and their interactions on macroalgal proliferation and coral growth. Coral Reefs 28:555–568

Sugimoto K, Hiraoka K, Ohta S, Niimura Y, Terawaki T, Okada M (2007) Effects of ulvoid (*Ulva* spp.) accumulation on the structure and function of eelgrass (*Zostera marina* L.) bed. Mar Poll Bull 54:1582–1585

Taylor RB, Rees TAV (1998) Excretory products of mobile epifauna as a nitrogen source for seaweeds. Limnol Oceanogr 43:600–606

Teichberg M, Fox SE, Olsen YS, Valiela I, Martinetto P, Iribarne O, Muto EY, Petti MAV, Corbisier TN, Soto-Jimenez M, Páez-Osuna F, Castro P, Freitas H, Zitelli A, Cardinaletti M, Tagliapietra D (2010) Eutrophication and macroalgal blooms in temperate and tropical coastal waters: nutrient enrichment experiments with *Ulva* spp. Glob Change Biol 16:2624–2637

Thacker RW, Ginsburg DW, Paul VJ (2001) Effects of herbivore exclusion and nutrient enrichment on coral reef macroalgae and cyanobacteria. Coral Reefs 19:318–329

Thomsen MS, McGlathery KJ, Tyler AC (2006) Macroalgal distribution patterns in a shallow, soft-bottom lagoon, with emphasis on the nonnative *Gracilaria vermiculophylla* and *Codium fragile*. Estuar Coasts 29:465–473

Tsai CC, Chang JS, Sheu F, Shyu YT, Yu AY, Wong SL, Dai CF, Lee TM (2005) Seasonal growh dynamics of *Laurencia papillosa* and *Gracilaria coronipifolia* from a highly eutrophic reef in southern Taiwan: temperature limitation and nutrient availability. J Exp Mar Biol Ecol 315:49–69

Valiela I (2006) Global coastal change. Blackwell, Massachusetts, p 368

Valiela I, Foreman K, LaMontagne M, Hersh D, Costa J, Peckol P, DeMeo-Anderson B, D'Avanzo C, Babione M, Sham C, Brawley J, Lajtha K (1992) Couplings of watersheds and coastal waters: sources and consequences of nutrient enrichment in Waquoit Bay, Massachusetts. Estuaries 15:443–457

Valiela I, McClelland J, Hauxwell J, Behr PJ, Hersh D, Foreman K (1997) Macroalgal blooms in shallow estuaries: controls and ecophysiological and ecosystem consequences. Limnol Oceanogr 42:1105–1118

Vermeij MJA, Moorselaar I, Engelhard S, Hörnlein C, Vonk SM, Visser PM (2010) The effects of nutrient enrichment and herbivore abundance on the ability of turf algae to overgrow coral in the Caribbean. PLoS One 5:1–8

Wilce RT, Schneider CW, Quinlan AV, Vanden Bosch K (1982) The life history and morphology of free-living *Pilayella littoralis* (L.) Kjellm. (Ectocarpaceae, Ectocarpales) in Nahant Bay, Massachusetts. Phycologia 21:336–345

Wootton JT, Pfister CA, Forester JD (2008) Dynamic patterns and ecological impacts of declining ocean pH in a high-resolution multi-year dataset. Proc Natl Acad Sci 10:18848–18853

Worm B, Lotze HK (2006) Effects of eutrophication, grazing, and algal blooms on rocky shores. Limnol Oceanogr 51:569–579

Worm B, Lotze HK, Bostrom C, Engkvist R, Labanauskas V, Sommer U (1999) Marine diversity shift linked to interactions among grazers, nutrients, and propagule banks. Mar Ecol Prog Ser 185:309–314

Ye N, Zhang X, Mao Y, Liang C, Xu D, Zou J, Zhuang Z, Wang Q (2011) 'Green tides' are overwhelming the coastline of our blue planet: taking the world's largest example. Ecol Res 26:477–485

Part V
Aquaculture and Economical Importance of Seaweeds

Chapter 22
Seaweed and Man

Cornelia M. Buchholz, Gesche Krause, and Bela H. Buck

22.1 Aquacultural Production of Seaweeds and Its Economic Relevance

22.1.1 Introduction

Despite an Asian aquaculture tradition of many centuries, aquatic farming on the global scale is still a young sector of food production that has grown rapidly in the last 50 years. Seaweeds, a colloquial but widely used term for macroalgae, play an important role in this business which remains a growing, vibrant, and important production sector for healthy human food.

As archaeological investigations in Chile testified, seaweeds have been used by humans for about 14,000 years (Dillehay et al. 2008). According to earliest written records, they were consumed in Japan during the Asukaand Nara Era approx. 1,500 years ago. Even food products directly made from seaweeds have a long

C.M. Buchholz (✉)
Alfred Wegener Institute for Polar and Marine Research (AWI), Am Handelshafen 12, 27550 Bremerhaven, Germany
e-mail: Cornelia.Buchholz@awi.de

G. Krause
Center for Tropical Marine Ecology (ZMT), Fahrenheitstrasse 6, 28359 Bremen, Germany
e-mail: Gesche.Krause@zmt-bremen.de

B.H. Buck
Alfred Wegener Institute for Polar and Marine Research (AWI), Am Handelshafen 12, 27550 Bremerhaven, Germany

Institute for Marine Resources (IMARE), Bussestrasse 27, 27570 Bremerhaven, Germany

Bremerhaven University of Applied Sciences, Applied Marine Biology, An der Karlstadt 8, 27568 Bremerhaven, Germany
e-mail: Bela.H.Buck@awi.de

C. Wiencke and K. Bischof (eds.), *Seaweed Biology*, Ecological Studies 219,
DOI 10.1007/978-3-642-28451-9_22, © Springer-Verlag Berlin Heidelberg 2012

tradition and can be traced back to the fourth century in Japan and the sixth century in China (Tseng 1987; Mc Hugh 2003). Exclusively wild seaweed was used, which limited it as a food source up to the Middle Ages. Later, during the Shogun regime in the Tokugawa Era (1600–1800 AD) fishermen constructed artificial substrates for fish farming which also allowed various seaweed species to grow upon. Ever since, seaweeds have been cultivated in the sea (Tamura 1966). Increasing demand over the last 50 years has outstripped the ability to supply the required biomass from natural (wild) stocks which triggered a dramatic growth of seaweed production from aquaculture sources.

Following Bartsch et al. (2008) farmed seaweeds are used for various applications, as food as well as in the textile, pharmaceutical, cosmetic, and biotechnological industry. As a source of food for human consumption seaweeds can be used in different forms – for instance in salads, sushi recipes, or as various food additives. Other purposes are the use on the health market advertising its minerals and enzymes. Industrial macroalgal use includes the extraction of phycocolloids and biochemicals (Sahoo and Yarish 2005; Pereira and Yarish 2008). A wide range of potential utilizations of seaweeds and/or algal compounds are referred to in Sect. 22.3.

The accessibility and reliability of data on "aquatic plants" (FAO classification) concerning collection from the wild as well as aquaculture production is still not sufficiently consolidated and spread. Acknowledging the shortcomings the Food and Agriculture Organisation of the United Nations (FAO) has continually improved its assessment of the available sources of information, evaluated and updated them, and with addition of some educated estimates published annual statistics that may well serve as a useful guide to world seaweed production and marketing. The latest report by the FAO on "The State of World Fisheries and Aquaculture 2010" contains data up to 2008 (FAO 2010a). Including the latest available data from 2009 (FAO 2011b), we can show the state and development of this industry up to that year and present the new numbers adjusted by the FAO for the period 1997–2005 (Fig. 22.1) after China revised its production statistics based on its Second National Agricultural Census 2007 (FAO 2010–2011). Since China by far runs the most intensive aquaculture business worldwide, numbers on global aquaculture production had to be decreased by about 8% (FAO 2011b).

22.1.2 Aquaculture Production of Seaweeds

Farm production of "aquatic plants" has permanently been expanded since 1970 with an average annual growth rate of 7.7%. It is overwhelmingly dominated by macroalgae (seaweeds) while cultivation of microalgae on a large commercial scale is still in its infancy.

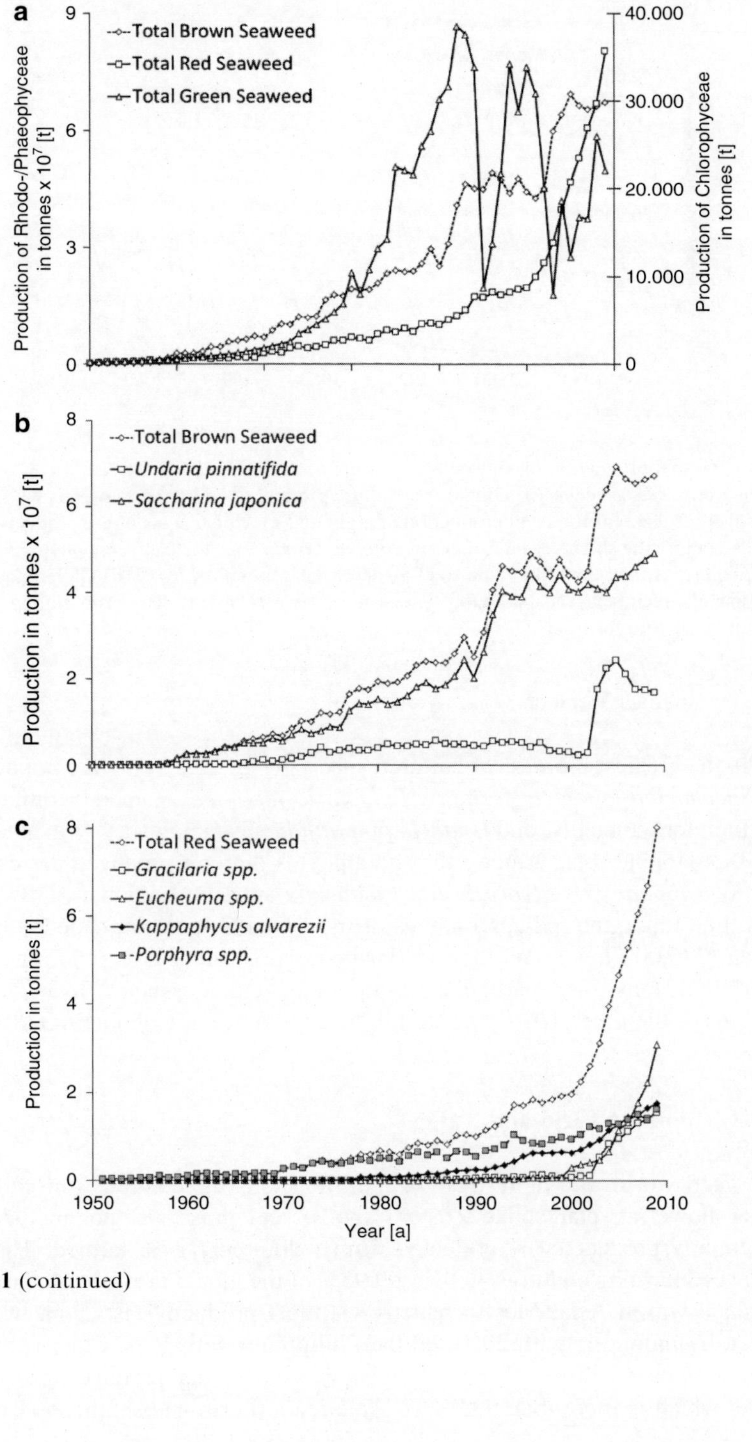

Fig. 22.1 (continued)

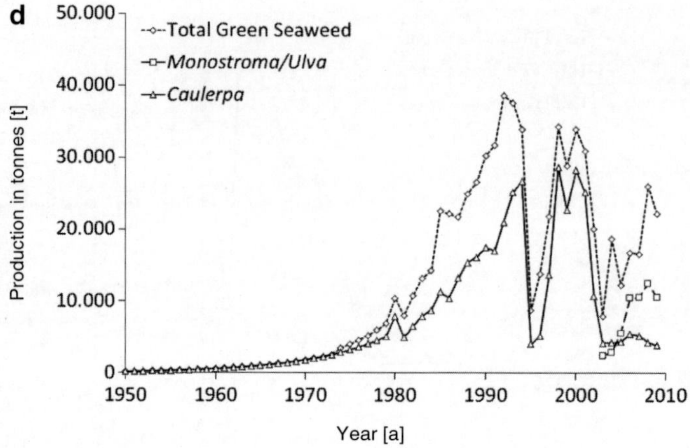

Fig. 22.1 Global production of seaweeds over time according to FAO (2011a). (**a**) Overview of the three groups of seaweeds. (**b**) Global aquaculture production of brown seaweeds with the most important crops used for food and production of alginate. (**c**) Global aquaculture production of red seaweeds comprising the most important agarophyte *Gracilaria*, the carageenophytes *Eucheuma* and *Kappaphycus*, and the high value food algae of the genus *Porphyra*. (**d**) Global aquaculture production of green seaweeds showing "green nori" as a relatively new crop among the most important algae sold for food

22.1.2.1 Species Variety

In 2009, the greatest biomass of cultured species (Fig. 22.2) were the popular food kelps *Saccharina japonica* (formerly *Laminaria japonica*), named "Kombu," with 4.9 million tons annually, and *Undaria pinnatifida,* also known as "Wakame," with 1.7 million annual tons. Among the rhodophytes that are produced the carrageenophytes *Kappaphycus alvarezii* and *Eucheuma* spp., both known as "Cottonii," (4.8 million tons), the red agarophytes of the genus *Gracilaria*, called "Ogonori" in Japan (296,000 tons), and the red *Porphyra* spp. valuable as food alga "Nori" (1.6 million tons) are particularly important. Other species like *Palmaria*, *Chondrus*, or the green *Ulva*, etc. are produced to a minor extent (FAO 2011a).

22.1.2.2 Biomass Yield and Value

To date and worldwide more than 14.7 million tons of seaweeds (miscellaneous vascular flowering plants like Zostera spp. or eel grass etc. not included) are commercially produced, 6% collected from wild stock, 94% farmed. The seven top seaweed farming countries deliver 99.95% of the global farmed volume and are all situated within Asia: Most productive is most productive is China with 54% followed by Indonesia with 20% and the Philippines with 12%. Chile is the most important seaweed farming country outside Asia having produced 88,147 tons in 2009, which is more than 99.9% of America's (north and south) total volume.

Fig. 22.2 Examples of macroalgae grown in aquaculture. (**a**) *Solieria*, a carrageenophyte; (**b**) *Chondrus crispus* "Irish Moss" carrageenophyte; (**c**) *Palmaria palmata*, used as feed for abalone; (**d**) *Ulva lactuca*, used for bioremediation (uptake of excess nutrients) and feed; (**e**, **f**) *Saccharina latissima*, after 9 months in tank culture and (**f**) phylloids drying for storage demonstrating how evenly they are grown. (Photographs **a**, **b**, **c**, **e**, **f** by K. Lüning with permission)

Next in the ranking according to production volume are countries mostly from Africa (e.g., Tanzania, Madagascar, South Africa, and Namibia) with 108,400 t in 2009 and Western Europe (e.g., Spain, France, Italy, and the Russian Federation), which are responsible for the remaining biomass production volumes. Finally, the Pacific Ocean Islands grow just a small amount of seaweed and produced 2,377 tons in 2009 (Fiji, Kiribati, and Solomon Islands) (FAO 2011a, 2010b).

22.1.3 Methods of Production and Technical Design

Quite a great amount of preliminary observations and experimental setups are necessary for a commercially successful cultivation of seaweeds. To pick the right choice of species thorough knowledge of the alga's often complicated life cycle and a good control of the different life stages are crucial. Likewise local weather conditions, the temperature range, wave action, currents, tidal amplitude, and salinity levels must be appropriate for the respective target species. Further factors to consider are nutrient supply in the water, water depth, and transparency to maintain beneficial irradiance levels, which may, in the case of shallow water farming, also be influenced by the color and composition of the bottom sediments. Moreover, ideally, grazers should not be found in the vicinity of a farm and the presence of epiphytes or other unwanted macroalgal species competing for light and nutrients should have been tested before a commercial farm is ventured (see also Chap. 11 by Potin).

Since the worldwide demand for seaweeds and their products could not be met by simple collection from natural populations, several decades of effort have gone into farming (Tseng 1984, 1987). A considerable number of technical variations in cultivating seaweeds are presently used depending on species, local conditions, and experience (Pereira and Yarish 2008). Meanwhile cultivation methods comprise not only single species cultures but also integrated multi-trophic aquaculture (IMTA; Chopin et al. 2008; Buschmann et al. 2008).

22.1.3.1 Monocultures of Seaweeds

Basically, seaweeds are either "seeded" on ropes or nets (e.g., *Porphyra)* or thallus fragments are fastened on or pinched into ropes, which are subsequently fixed to various suspended or floating culture structures (e.g., *Gracilaria*). From Chile a system to anchor *Gracilaria* cuttings in the sand is known and has also been successful (Trono 1990; Pereira and Yarish 2008; FAO 2011c: National Aquaculture Sector Overview—NASO). Paddle wheel ponds that keep algae floating and moving are a suitable device to grow the green *Ulva* to large quantities (Chopin et al. 2008; Butterworth 2010).

"Nori" production comprising several species of the genus *Porphyra* is a big business worldwide valued at US$ 1,400 million in 2008 according to the FAO statistical yearbook 2010 (FAO 2010b). After about 300 years of culture efforts dependent on natural seeding, Baker (1949) discovered the conchocelis phase of this genus. Only then could the present-day effective multistage culture system be established (Pereira and Yarish 2008, 2010). *Porphyra* is mainly cultivated in China, Japan, and Korea: Mollusk shells, mostly of oysters, are seeded with diploid carpospores from preselected thalli and kept in large shallow indoor tanks for approx. 5 months until under nutrient, temperature, and light control conchospores are released by the conchocelis filaments. Appropriate spore density and agitation of the suspension facilitate even settlement on collecting nets. After germination, grow-out of the thalli takes place in relatively shallow bays, nets fixed, semi-floating, or floating (Sahoo and Yarish 2005; Pereira and Yarish 2010). Fixed or semi-floating, the alternation of immersion and desiccation with the tide is guaranteed helping to avoid diseases caused by fungi or bacteria, reducing epiphytic diatoms and improving the taste. Floating nets can be kept over slightly deeper water (10–20 m), thereby extending the farming area. However, a nursery system must be included that periodically dries the nets and hardens them (Pereira and Yarish 2008, 2010). It is also possible and even improves the quality of the final product to carefully freeze the nets with young thalli and store them at −20°C for later grow-out (FAO 2005–2011a). Depending on the species it takes 40–50 days at sea, for *P. haitanensis* and *P. yezoensis*, respectively, before the first crop can be attained. Six to eight harvests are possible during 5 months of cultivation (Pereira and Yarish 2010).

The species with the highest production is the brown alga *Saccharina japonica* (formerly *Laminaria japonica*), "kombu." 4.9 million tons of kombu were produced in 2009, 84% of it grown in China, where the species is not endemic, but was introduced in 1927 (Tseng 1987; Lowther 2006). Conventional "2 year cultivation" of *S. japonica* took a period of 18 months at sea with at least another 2 months for "seeding," which resulted in relatively high prices for the product (Ohno 1993). "Cultivation by transplanting" uses natural *Saccharina* sporophytes either washed ashore or manually thinned out. As the activity of the meristem increases in late winter to early spring, new haptera are easily formed and allow a new attachment on ropes during this time. Time from transplantation to harvest lasts 12–18 months (Ohno 1993). Only 12 months are needed for the widely applied so-called forced-cultivation technique (Hasegawa 1971; Ohno 1993; Critchley and Ohno 1997; Sahoo and Yarish 2005). This became possible due to scientific control of the entire biphasic life cycle, where indoor facilities are necessary to manage the labor and cost-intensive "seedling phase" (Tseng 1989; Mc Hugh 2003; Chen 2006). A large independence of naturally available seedstock could be attained by detaching *Laminaria* frond fractions from the meristem (Buchholz and Lüning 1999; see also Lüning et al. 2000; Pang and Lüning 2004). Meiospores are artificially released from sporogenous thalli, germinate to microscopic gametophytes, form zygotes, and eventually produce young sporophytes that stick to ropes. In "the grow-out phase" (Tseng 1989), culture ropes with juvenile sporophytes are transferred to the open sea where they grow to a frond length of approx. 1–2.5 m, depending on the species. If the predicted shift of biogeographic areas becomes true (see also Chap. 18 by Bartsch et al.), aquaculture of *S. japonica,* as an example, may be strongly impaired in that the space and the period for grow-out in coastal waters are reduced. Young seedlings do not tolerate more than 20°C and fronds have to be harvested at ≤21°C, because they start to rot at higher temperatures (FAO 2005–2011b) .

The most common design for grow-out of Laminariales in the open water is a longline system of horizontal ropes parallel to the sea surface with anchoring weights to stabilize the entire system and with buoys to provide flotation. Combining vertical arrangement of seeded culture lines as the first step and later suspension and lifting of these lines into a horizontal position overcomes first overexposure, then shading problems in growing sporophytes, while reducing effects of cross currents and storms on only horizontally attached lines (FAO 2005–2011b). From the various methods (e.g., Holt and Kain 1983; Kawashima 1984; Kain and Dawes 1987; Dawes 1988; Kain 1991; Merrill and Gillingham 1991; Critchley and Ohno 1997; Buck and Buchholz 2004), the locally appropriate one has to be chosen for the special conditions of a given farming site.

Longline systems installed in harsh offshore conditions, to where farms could be expanded, were not robust enough as there is a considerable stress on support material and algae (Buck 2004; Buck and Buchholz 2004). Among the various suggestions for technical structures that have been made (Polk 1996; Hesley 1997; Stickney 1998; Bridger and Costa-Pierce 2003) a ring design did withstand strong currents and wind waves and is still the most promising (Buck and Buchholz 2004). The idea of utilizing the grounding structures of offshore wind generators for the fixation of aquaculture systems is intriguing (e.g., Buck 2002; Krause et al. 2003; Buck et al. 2004) and the first experiments on *Laminaria* species show that adapted to strong currents as young individuals, they grow well at exposed sites (Buck and Buchholz 2005).

The cultivation of seaweeds at sea or in ponds flushed by incoming seawater has lately been supplemented or completely exchanged for land-based tank cultivation for smaller rhodophytes and chlorophytes (see Sect. 22.1.3.2) as well as large kelps. Lüning and Pang (2003) kept free floating sporophytes of laminarians or *Palmaria palmata* circulating in the water agitated by air. The tanks allowed a high cultivation density of 10 kg m^{-2}, since shading was amended by the continuous turnover of fronds toward the light. Uniform exposure to nutrients was likewise facilitated and the infestation with epiphytes was kept very low (Ryther et al. 1979; Bidwell et al. 1985). The Lüning and Pang (2003) system was additionally supported by a continuous short day treatment. The same short day treatment (8 h light) using outdoor tanks with automatic blinds resulted in prolonged growth activity of *Laminaria digitata* that also seemed to deter epiphyte settlement (Gómez and Lüning 2001). The experience gained in tank cultivation is a valuable basis for some of the internationally developing integrated mariculture systems (see citations in Neori et al. 2004, p. 376; Abreu et al. 2011; Pereira et al. 2011).

The world population recently reached the seven billion mark and a sustained or rather growing supply of protein from aquatic animals is highly desirable. Concomitantly there is an increasing concern about the negative consequences of intensive and constantly spreading aquaculture of fish, shrimps, and mollusks. Therefore, the remediation of negative consequences has been a field of intensive research during the last decade. Two ecologically sensible strategies to meet the requirements for more space allotted to aquaculture have been and will continue to be tested: One is the offshore aquaculture that to date seems very expensive and technically demanding, but will allow considerable mass production. The other is the very promising but likewise complicated Integrated Multi-Trophic Aquaculture (IMTA) approach.

22.1.3.2 Integrated Multi-Trophic Aquaculture (IMTA)

Extensive polyculture pond systems with organisms of several species in the same water body have traditionally been applied in Asian countries and were based on trial and error. Only since the 1970s, a more systematic approach has resulted in the development of integrated intensive land-based mariculture systems (Ryther et al. 1979). The aim of IMTA is the creation of a manageable small ecosystem with several species of different trophic levels combined in one system in the right proportions, each utilizing waste products or the biomass generated by another member of the system. All of the individual components must be marketable since the commercial viability is an important factor of any such IMTA design (Chopin et al. 2008). If the benefits for the environment were accounted for, the value of IMTA production systems would be highly increased and political support for the development of these structures would mirror this.

Neori et al. (2004, 2007) argued that seaweed-based integrated aquaculture systems will most probably facilitate the expansion and sustainability of the worldwide aquaculture industry. Nevertheless, the major aim of global aquaculture enterprises is the production of fish, shrimp, or shellfish protein. With the dwindling of wild resources through overfishing there is an ever growing demand for those products. At the same time the demand for certain seaweeds for human consumption or animal feed or else for algal ingredients (mainly phycocolloids) has to be met and is already a large market, worth 22.4 billion US$ in the year 2008 (FAO 2010b).

Since the culture of fed species like fish or shrimp inevitably results in eutrophication of the adjacent waters (Stead and Laird 2002; Fei 2004; Sanderson et al. 2006; Troell et al. 1999, 2003), a bioremedial complementary culture design makes sense for environmentally protective reasons. Simultaneously the biomass of carefully chosen extractive and marketable species could at least partly counterbalance the considerable costs for fish or shrimp feed (Neori et al. 2004; Abreu et al. 2009). Moreover, there are oligotrophic seawater conditions, like in Israel or Australia, that do not allow the growth of algae and it makes sense to try intelligent new aquaculture approaches under these circumstances (Schuenhoff et al. 2003; Butterworth 2010; Neori et al. 2004).

An encompassing review of the multiple IMTA activities is found in Barrington et al. (2009) for temperate and Troell (2009) for tropical regions of the world. Potential candidates for integrated systems are not only the hitherto monocultured and expensively fed fish and crustaceans, filter-feeding bivalve or herbivore mollusks (*Haliotis*), and more than 20 species of seaweeds but also echinoderms and polychaetes. Presently existing IMTA systems usually contain no more than three components at different trophic levels. One is fed fish or shrimp, one extracting organic bound nutrient particles, either feed leftovers or feces, and one, seaweeds, extracting the effluents utilizing inorganic nitrogen and phosphate for growth. In a Sustainable Ecological Aquaculture effort Cascadia SEAfood even integrates sablefish with two species of bivalves, kelp, and sea urchins that feed on the fouling organisms on cages, their gonads being offered to seafood gourmets (Cross 2010; Cook and Kelly 2007). In the case of an abalone farm in South Africa, the resulting seaweed crop of *Ulva lactuca* can partly be used for feeding the abalone (Nobre et al. 2010; Robertson-Andersson et al. 2008). Since it is known that *Ulva* synthesizes more protein (>40% dw^{-1}) with higher ammonia-N in the water, it turns out to be a valuable feed allowing *Haliotis* to grow significantly faster than with Ulva kept in low nitrogen concentrations and containing only 12% protein (Shpigel et al. 1999).

While *Ulva* spp. with their thin thalli and large surface-to-volume ratio are perfect inorganic extractors their biomass does not realize a high price. Therefore, commercially valuable red algae like *Porphyra* (nori), *Gracilaria* (as an agarophyte), and *Kappaphycus* (as a carrageenophyte) are being tested in IMTA systems (Chopin et al. 1999; Abreu et al. 2009, 2011; McVey et al. 2002; Pereira and Yarish 2010; Rawson et al. 2002; Robertson-Andersson et al. 2008; Lombardi et al. 2006). Depending on the region, particularly those with pronounced seasons, seaweeds in IMTAs may have to be exchanged for other species in the course of the year, which requires additional engineering efforts to suit each species' requirements. There are for example different demands of red algae versus brown kelps concerning surface area, water flow rates, and nutrient exchange, etc.

Owners of profitable finfish or shrimp cultures are not necessarily concerned about eutrophication of the environment and have to be convinced that no diseases are introduced by co-cultured organisms (Troell et al. 2003). It needs the owners' consent to use existing fish or shrimp cultures to establish an IMTA system: In open sea, but nearshore systems, different components of an IMTA can simply be placed in each other's vicinity with mutual beneficial effects provided local currents and other water dynamics allow sufficient exchange between them (Abreu et al. 2009; Sanderson 2009; McVey et al. 2002; Chopin et al. 2001; Troell et al. 1999; Petrell and Alie 1996). The alternatives are land-based tank systems or ponds that can be separated from sea water inflow at periods of toxic algal blooms or oil spills, etc. In these systems, fish farm effluents are redirected through a series of tanks containing complementing organisms. The Sea Or Marine Enterprises in Israel integrated gilthead seabream with *Ulva* or *Gracilaria*, with the algae serving as a feed for commercially valuable abalone (Neori et al. 2004). With abalone (*Haliotis midae*) and seaweeds alone, a partial recirculation of the culturing water is feasible (Robertson-Andersson et al. 2008).

To establish a managed small ecosystem, that is adapted to site and region and that in addition sustains stakeholders and environment, is a very complex task, impossible to be tackled by individual farmers. Therefore, Canada has launched a new network CIMTAN (2011). CIMTAN is directed to collect additional knowledge and intensify interdisciplinary as well as multi-institutional information exchange between the experts on aquaculture on both coasts of Canada. Research and development of IMTAs has been conducted in Canada since 2001. The joint goal is to establish an easily adaptable system of IMTA with a set of nutritionally interacting species that mimics the natural condition of a diverse ecosystem, thereby being less vulnerable to e.g., microbial infections or parasite infestation.

At the same time most/all of the systems components should be of commercial value supplementing each other in reaching an acceptable profit for stakeholders and environment. Between January 2010 and December 2014, about 250 people will be involved in 14 projects systematically investigating the various aspects of IMTA from nutrient plumes to microbial interactions, detritivores (such as polychaetes and echinoderms), bivalves, fish parasites, and seaweeds (including *Saccharina latissima,* formerly *Laminaria saccharina*, *Alaria esculenta*, *Palmaria palmata*, and *Ulva* spp.) to infrastructure components, ecosystem modeling, and social implications for coastal communities. With the insights gained it will be easier to promote a sustainable and vibrant aquaculture industry in Canada and probably other temperate regions of the world.

22.1.3.3 Offshore Aquaculture

Aquaculture is continuously expanding in coastal seas and ashore, comprising farming in marine and brackish water environments (FAO 2010a, b). However, coastal waters host a highly competitive group of uses such as commercial shipping, areas exclusively reserved for the navy, extraction or disposal of sand, oil exploration and exploitation, as well as pipelines, cables, wind farms, nature reserves, and other marine and coastal protected areas. Recreational activities and fisheries are additional interests that deserve attention. This massive utilization of marine areas leads to stakeholder conflicts (Buck et al. 2004; Langan et al. 2006; Rensel et al. 2006). Additionally, farming activities may also generate negative environmental impacts on coastal ecosystems at local up to regional scales (e.g., Buck and Krause 2012), thus leaving little room for further expansion of modern coastal aquaculture systems. Locating aquaculture activities further offshore appears as a viable option to avoid stakeholder conflicts and to reduce environmental impacts to the coast (Corbin 2007). The term "offshore" within the context of aquaculture was defined by Ryan (2005) and is based on the moving of farm installations from nearshore sheltered environments to more exposed environments, which are commonly described as "high energy environment".

Following Troell et al. (2012) and North (1987) considerable controversy has emerged over the proper development of offshore aquaculture, and its actual advantages over existing nearshore aquaculture. In general, many of the challenges for offshore aquaculture engineering involve adaptations of farm installation designs and operation protocols to a variety of physical factors, such as currents and wave actions: The robustness of the aquaculture systems to withstand harsh oceanographic conditions is one challenge, while the difficulties in anchoring and/or submerging structures in deep water is another. Major shipping routes have to be considered as well as migration routes of marine mammals. Logistic difficulties of transport and the operation and maintenance of offshore platforms of any farming enterprise must be evaluated.

Due to the scarcity of space even in the open ocean island territories or countries with relatively short coastlines, the concept of "multiple use" needs to be addressed. Germany is an example where the plans for the massive expansion of wind farms in offshore areas of the North Sea triggered the idea of a combination of wind turbines with installations for extensive shellfish and seaweed aquaculture (Buck 2002, 2004). A combined design of fish cages in the foundation of the turbines in addition to the extractive components of IMTA systems was discussed (McVey and Buck 2008). Offshore wind farms provide an appropriately sized area for farming that is free of shipping traffic. At the same time the infrastructure for regular service support is readily available. Such sites provide an ideal opportunity for devising and implementing a multiple-use concept (Buck et al. 2004; Michler-Cieluch 2009).

Some experimental-scale operations have shown the feasibility of offshore macroalgal farming (for review, see e.g., Buck et al. 2008). The focus of those systems was placed upon the technical design needed to withstand hydrodynamic forces and investigations on cultivation techniques to avoid dislodgement of laminarians (Buck and Buchholz 2004, 2005). Ebeling, Griffin, and Buck (unpublished data) were the first to calculate the economic potential of a seaweed farm (*Saccharina latissima*) within a planned wind farm off the coast of Woods Hole (Massachusetts, USA) in Nantucket Sound and found it being beneficial on a large scale.

22.2 Socioeconomic Aspects

Traditionally, the academic community has tended to approach aquaculture primarily from technological and environmental perspectives (Marra 2005). However, it has been recognized that aquaculture increasingly generates direct socioeconomic benefits through the supply of highly nutritious foods and other commercially valuable products, providing jobs and creating incomes. For example, the FAO reports for the Philippines that seaweed farming is currently the largest and most productive form of livelihood among the coastal population of the Philippines. In 2004, more than 116,000 families consisting of more than one million individuals were farming more than 58,000 ha of seaweed (FAO 2005–2011c). Enough and affordable manpower to maintain the farms is an indispensable prerequisite. Personnel on all levels of skills are required. The benefits for the well-being of coastal communities are reflected in the finding of a recent case study on a South African IMTA farm of abalone and seaweed presented by Nobre et al. (2010). In this study, the impact of direct permanent employment within the South African aquaculture industry on local communities was exemplified: The selected communities were characterized by high unemployment (85.7%), with more than 50% of the labor force being unskilled and semiskilled. It could be shown that employment of a high number of unskilled and semiskilled personnel in the aquaculture sector had a large local impact in previously disadvantaged coastal communities, where any increase in employment is valuable (Nobre et al. 2010). This is particularly relevant where unemployment is not only an economic issue but also a sociopolitical concern.

Thus, in addition to its own economic contribution, aquaculture can also induce, as a spin-off, economic contribution to other sectors that supply materials to aquaculture or use aquaculture products as inputs (ICES 2011). The numbers of people engaged in other ancillary activities, such as processing, farm construction, manufacturing of processing equipment, packaging, marketing, and distribution can be substantial. Indeed, estimates indicate that, for each person employed in aquaculture production, about three other jobs can be produced in secondary activities. The total aquaculture sector, encompassing finfish, shellfish, and seaweed aquaculture, and those supplying services and goods to them, provides employment and livelihoods to a total of about 20 million people (compiled from FAO 2011c).

Despite these positive effects, decisions about aquaculture development are often based on incomplete information, particularly in relation to the socioeconomic dimensions. As a consequence, inadequate accounts for trade-offs associated with different development options are made. Therefore, there is a risk that anticipated and much needed socioeconomic benefits from aquaculture expansion may come at the expense of increased and possible unsustainable pressure on ecosystem goods and services (Naylor et al. 2000), ultimately jeopardizing people's food security and livelihoods.

In contrast to many finfish aquaculture enterprises, there are, however, encouraging experiences made with seaweed aquaculture. An example on the important role of seaweed cultivation for local livelihoods and sustainable development is the introduction of seaweed farming on Zanzibar, Tanzania in 1989. In that year, the seaweed *Eucheuma* was imported from the Philippines and successfully grown on the East Coast of Unguja Island. Today, more than 90% of the farmers are women, which have changed the life in the villages. Not only did the women gain independent economic power, but the number of children suffering from malnutrition has also decreased, which indicates that the health of their mothers has improved. As daily income is secured, children are able to attend schools regularly. Furthermore, seaweed farming has also reversed the trend of rural depopulation, since it fostered self-employment of the village youths (Msuya 1997, 2006; Msuya et al. 2007).

The question remains though, how negative effects on the environment and positive socioeconomic consequences from aquaculture development can be balanced. For instance, although methods of cultivation can be adapted and vary being equally successful, the careful choice of the farming site seems to be essential for any aquaculture success (Trono 1990; Buck and Buchholz 2005). Notwithstanding, the seascapes are increasingly managed for multiple functions and services in addition to provision of food, and this requires the integration of ecological and socioeconomic research, policy innovation, and public education (ICES 2011). The multiuse dilemma has driven many researchers, experts, and policy makers to try and address issues related to the sustainability of aquaculture development from disciplinary/sectoral perspectives. However, aquaculture development raises questions that cannot be addressed in isolation. If it is to bring about the expected benefits, such as in the case of the seaweed farmers on Zanzibar, seaweed farming must address the interactions and functioning within wider ecosystem, social, economic, and political contexts.

A critical question is how to best guide the development of aquaculture that has the potential to support a portfolio of sustainable livelihoods and assist in poverty alleviation and food security (ICES 2011). Broader systematic perspectives on aquaculture, such as the "Ecosystem Approach to Aquaculture" (Soto et al. 2008), may enable analysis of trade-offs and sustainability aspects, especially with respect to net benefits for poorer resource users. Furthermore, local knowledge generated through active bottom-up participation and the application of transparent decision-making processes are some of the building blocks behind improved coordination of all the sector's stakeholders. Strengthening of institutional capacity and resources (including human capacity), both at national and international levels, is needed to enable development of aquaculture for poverty reduction and improved human well-being.

22.3 Direct Seaweed Applications and Bioactive Compounds

22.3.1 Introduction

Following the twentieth International Seaweed Symposium in Ensenada, Mexico, in 2010, several quite encompassing reviews have lately been published comprising various potential uses of seaweeds as functional food, feed supplement, or manure and soil conditioner with biological or pharmaceutical activities. Adding some more the following paragraphs provide an overview of the currently available information in published literature guiding specific interests in seaweed applications to the great number of detailed references collated already.

22.3.2 Seaweed for Food and Medication

Seaweeds have been used as food and for medical purposes since the late Pleistocene as Dillehay et al. (2008) reported from an excavation site at Monte Verde in southern Chile. Nine species of marine algae were recovered, among them edible species (*Durvillaea antarctica*, *Porphyra columbina*, *Sarcothalia crispata*, and *Macrocystis pyrifera*) and two nonedible ones (in the genus *Gigartina* and *Sargassum*). Some are nowadays being used as medical plants by indigenous people of that area and may have served the same purpose 14,000 years ago.

To date the food sector is still the most important field of application for the various species of seaweeds farmed or collected from the wild. While direct consumption is most common in the Asia-Pacific region, algal hydrocolloids are used worldwide in a great variety of food items as emulsifying, gelling, or water retention agents (Indergaard and Østgaard 1991; Murata and Nakazoe 2001; Bartsch et al. 2008). 86,100 tons of hydrocolloids were traded in 2009 comprising

58% of carrageen, ~31% alginates, and ~11% agar (Bixler and Porse 2011). The demand particularly for carrageen could not be met lately mainly due to the increased demands of the Chinese hydrocolloid industry. Moreover, collective quantity does not always suffice, since species as well as geographical location and climate where the seaweeds are grown and the season of harvest determine the chemical characteristics of the hydrocolloids and their quality. Quality is also influenced by the extraction methodology (Bixler and Porse 2011).

Aware of the fact that there is a great variety of chemical compositions and therefore bioactive properties in the different species of seaweeds Løvstad Holdt and Kraan (2011) supply a wealth of current knowledge on bioactive compounds of the most important species in 21 tables comprising the various polysaccharides, proteins, peptides, and amino acids as well as lipids and fatty acids, pigments, vitamins, iodine, phenolic components, and undesirable substances like heavy metals: The vast range of biological activities they listed originated from in vitro investigations up to clinical studies. Most spectacular are the antibacterial and antiviral activities that may partly be responsible for the records on positive effects against tumors and HIV. Important beneficial effects for human health lie also in the reduction of blood cholesterol levels and anti-diabetes and anti-hypertension effects. In addition to direct pharmaceutical uses of algal ingredients, a high-tech medical use of alginate as part of a matrix that can carry protein drugs is being developed. It utilizes the mucoadhesive property of alginate helping to retain the drugs in the gastrointestinal tract for a longer period, thereby improving drug bioavailability and effectiveness in the intestine (George and Abraham 2006). While the list of beneficial effects of seaweeds and their ingredients on humans is long, the process of getting them authorized as food or medical items can be as well (Løvstad Holdt and Kraan 2011).

22.3.3 Seaweed in Agriculture and Animal Diets

The large amounts of minerals, trace elements, vitamins, and iodine among other components render seaweeds, particularly the brown ones, a valuable addition not only to food but also to livestock feed and soil fertilization (lit. in Bartsch et al. 2008; Craigie 2011). Some direct or indirect beneficial effects hold for plants as well as animals of very different classes. An example is antihelmintic properties of seaweed extracts that can help not only mammals but also plants like tomatoes which suffer from nematode infestations in their roots (Løvstad Holdt and Kraan 2011; Craigie 2011). While commercial seaweed extracts have been available for 60 years, only 1% of the current seaweed industrial production goes into agricultural use, even though according to Craigie (2011) "seaweed extracts can modify plant and animal responses at a fundamental level." However, the appropriate utilization of seaweed meal or extracts in agriculture and as feed addition has to be experimentally secured in advance of extensive use. Craigie (2011) reviews the history of seaweed utilization and the development of extracts and the responses of the soil and crops to various applications, e.g., different concentrations of extracts that can decide between inhibition and promotion of germination and growth. Table 12 (in Craigie 2011) summarizes the bioactive properties reported for seaweed extracts in plants and animals. As for plants the choice of algal species and extract concerning its nature and proportion in the feed of the respective livestock has to be carefully tested to avoid detrimental effects and improve overall health and benefit reproduction (Craigie 2011). Addition of 4% *Sargassum* meal to the feed of shrimp cultures reduced cholesterol contents of their muscle tissue by 29%, quite desirable for shrimp grown for human consumption (Casas-Valdez et al. 2006). In the case of cattle there can even be an advantage elicited in the shelf life of steaks, since a short-term feed addition of 2% *Ascophyllum* meal (Tasco) prior to slaughter not only results in a better marbling of steaks but also retains the red color over a longer period. The latter is due to a higher proportion of oxymyoglobin compared to conventionally fed cattle (Braden et al. 2007).

Of the 10 larger and 17 smaller producers of commercial seaweed extracts for agriculture that Craigie mentions, only three are presently located in Asia, that is in China, even though Asia is producing >98% of all seaweeds. The indisputable benefits of seaweed utilization in plant and animal farming excite expectations toward a more extended production and use in the future.

22.3.4 Other Applications of Seaweeds

The phycocolloids agar, carrageen and alginates have long been used for their water binding and thickening properties: While agar is not only known as a neutrally flavored thickener of stews, sauces, desserts etc., it is also indispensable as a solid culture medium in medical bacteriology and microbiological research. Alginates from brown seaweeds are used in printing dyes and for better adsorptivity of textiles. They are together with other seaweed components valued ingredients of cosmetics. Extracts typically found in cosmetics are made from *Ulva lactuca, Ascophyllum nodosum, Laminaria longicruris, Saccharina latissima, Laminaria digitata, Alaria esculenta,* various *Porphyra species, Chondrus crispus*, and *Mastocarpus stellatus* (Cosmetic Ingredient Dictionary 2002–2011).

The latest innovation is a textile fiber named SeaCell®. It is a cellulose-based fiber produced from seaweeds like *Ascophyllum nodosum* and used as a yarn for clothing or for filling duvets (Smartfiber 2010).

Another field of high interest and great demand is the partial replacement of fish meal by seaweeds. Only a small selection of seaweed species will probably be suited. Experiments by Walker et al. (2009) showed positive results with up to 30% *Porphyra* spp. in the diet of juvenile Atlantic cod. If *Porphyra* ("nori") was introduced into the commercial aquaculture of fish, demand and price of the already valuable seaweed would probably rise enormously.

Due to their high carbohydrate content seaweeds can be fermented to methane (biogas) and in many places (mainly beach cast) seaweeds are considered a potential CO_2-neutral and renewable energy supply (lit. in Bartsch et al. 2008; Roesijadi et al. 2008; Chung et al. 2011).

Bioremediation of eutrophic waters has been mentioned above in relation to aquaculture. It would also work for waters supplied with an excess of nutrients from other sources. If the resulting quantity of algal biomass was not good enough to be introduced into a high quality production line, it may still serve as a good feedstock for biofuels.

22.4 Conclusion

Seaweeds have accompanied human history for about 14,000 years. Seaweed research from simple observation to organized experiments has helped to install an extensive aquaculture industry that produced close to 14.8 million tons in 2009, while it was just a few kg in the 1950s. Large and many small enterprises worldwide secure thousands of family incomes and are therefore of high socioeconomic importance. The global "hot spot" of seaweed farming as well as direct use of these macroalgae as food items is Asia. Algal ingredients like the phycocolloids agar, alginate, and carrageenan are in demand by the food industry, medical, and technical applications. The use of seaweed meal or extracts for agricultural applications bears a great potential for expansion. To satisfy the market demand not only for seaweeds but also highly requested protein sources like finfish, shellfish, and crustacea, the relocation of culturing sites to offshore areas is suggested for wind farms at sea as a multiuse concept. Another promising and intensively developing field of aquaculture is the expansion of Integrated Multi-Trophic Aquaculture (IMTA) systems where commercially valuable organisms from different trophic levels, some fed and some extractive, are combined in a culturing system, ideally sustaining each other and with the help of seaweeds even bioremedial for the environment.

References

Abreu MA, Varela DA, Henríquez L, Villarroel A, Yarish C, Sousa-Pinto I, Buschmann AJ (2009) Traditional vs. Integrated Multi-Trophic Aquaculture of *Gracilaria chilensis* Bird CJ, McLachlan J & Oliveira EC: Productivity and physiological performance. Aquaculture 293:211–220

Abreu MH, Pereira R, Yarish C, Buschmann AH, Sousa-Pinto I (2011) IMTA with *Gracilaria vermiculophylla*: Productivity and nutrient removal performance of the seaweed in a land-based pilot scale system. Aquaculture 312:77–87

Baker KD (1949) Conchocelis-phase in the life-history of *Porphyra umbilicalis* (L.) Kütz. Nature 164:748–749

Barrington K, Chopin T, Robinson S (2009) Integrated multi-trophic aquaculture (IMTA) in marine temperate waters. In: Soto D (ed) Integrated mariculture: a global review. FAO Fisheries and Aquaculture Technical Paper. No. 529, Rome. FAO pp 7–46

Bartsch I, Wiencke C, Bischof K, Buchholz CM, Buck BH, Eggert A, Feuerpfeil P, Hanelt D, Jacobsen S, Karez R, Karsten U, Molis M, Roleda MY, Schumann R, Schubert H, Valentin K, Weinberger F, Wiese J (2008) The genus *Laminaria* sensu lato: recent insights and developments. Eu J Phycol 43(1):1–86

Bidwell RGS, McLachlan J, Lloyd NDH (1985) Tank cultivation of Irish Moss, *Chondrus crispus*. Bot Mar 28:87–97

Bixler HJ, Porse H (2011) A decade of change in seaweed hydrocolloids industry. J Appl Phycol 23:321–335

Braden KW, Blanton JR, Montgomery JL, van Santen E, Allen VG, Miller MF (2007) Tasco supplementation: Effects on carcass characteristics, sensory attributes, and retail display shelf-life. J Anim Sci 85:754–768

Bridger CJ, Costa-Pierce BA (2003) Open ocean aquaculture: from research to commercial reality. The World Aquaculture Society, Baton Rouge

Buchholz C, Lüning K (1999) Isolated, distal blade discs of the brown alga *Laminaria digitata* form sorus, but not discs, near to the meristematic transition zone. J Appl Phycol 16:579–584

Buck BH (2002) Open Ocean Aquaculture und Offshore Windparks. Eine Machbarkeitsstudie über die multifunktionale Nutzung von Offshore-Windparks und Offshore-Marikultur im Raum Nordsee, Berichte zur Polar- und Meeresforschung = Reports on polar and marine research, 412, p 252

Buck BH (2004). Farming in a High Energy Environment: Potentials and Constraints of Sustainable Offshore Aquaculture in the German Bight (North Sea), Dissertation, University of Bremen, p 258

Buck BH, Buchholz CM (2004) The offshore-ring: a new system design for the open ocean aquaculture of macroalgae. J Appl Phycol 16(5):355–368

Buck BH, Buchholz CM (2005) Response of offshore cultivated *Laminaria saccharina* to hydrodynamic forcing in the North Sea. Aquaculture 250:674–691

Buck BH, Krause G (2012) Integration of Aquaculture and Renewable Energy Systems. In: Meyers RA (ed) Encyclopedia of Sustainability Science and Technology, Springer Science + Business Media LLC. Chapter No. 180 http://www.springer.com/physics/book/978-0-387-89469-0. Cited 10 Oct 2011

Buck BH, Krause G, Rosenthal H (2004) Extensive open ocean aquaculture development within wind farms in Germany: the prospect of offshore co-management and legal constraints. Ocean Coast Manag 47(3–4):95–122

Buck B, Krause G, Michler-Cieluch T, Brenner M, Buchholz C, Busch J, Fisch R, Geisen M, Zielinski O (2008) Meeting the quest for spatial efficiency: progress and prospects of extensive aquaculture within offshore wind farms. Helg Mar Res 62:269–281

Buschmann AH, Hernández-González MC, Aranda C, Chopin T, Neori A, Halling C, Troell M (2008) Mariculture waste management. In: Jørgensen SE, Fath BD (eds) Encyclopedia of ecology, vol 3, Ecological engineering. Elsevier, Oxford, pp 2211–2217

Butterworth A (2010) Integrated Multi-Trophic Aquaculture systems incorporating abalone and seaweeds; Report for Nuffield Australia Project No 0914 http://www.nuffieldinternational.org/rep_pdf/1287395494Nuffield_Report-_Adam_Butterworth.pdf. Cited 10 Oct 2011

Casas-Valdez M, Portillo-Clark G, Aguila-Ramirez N, Rodriguez-Astudillo S, Sanchez-Rodriguez I, Carillo-Dominguez S (2006) Effect of the marine alga *Sargassum* spp. On the productive parameters and cholesterol content of the brown shrimp, *Farfantepenaeus californiensis* (Holmes, 1900). Revista Biol Mar Oceanogr 41:97–105, in Spanish, English abstract

Chen J (2006) Cultured aquatic species information programme—*Laminaria japonica*. Cultured Aquatic Species Fact Sheets. FAO Inland Water Resources and Aquaculture Service (FIRI)

Chopin T, Yarish C, Wilkes R, Belyea E, Lu S, Mathieson A (1999) Developing *Porphyra*/salmon integrated aquaculture for bioremediation and diversification of the aquaculture industry. J Appl Phycol 11:463–472

Chopin T, Buschmann AH, Halling C, Troell M, Kautsky N, Neori A, Kraemer GP, Zertuche-Gonzalez JA, Yarish C, Neefus C (2001) Integrating seaweeds into mariculture systems: a key towards sustainability. J Phycol 37:975–986

Chopin T, Robinson SMC, Troell M, Neori A, Buschmann AH, Fang J (2008) Multitrophic integration for sustainable marine aquaculture. In: Jørgensen SE, Fath BD (eds) Encyclopedia of ecology, vol 3, Ecological engineering. Elsevier, Oxford, pp 2463–2475

Chung IK, Beardall J, Mehta S, Sahoo D, Stojkovic S (2011) Using marine macroalgae for carbon sequestration:a critical appraisal. J Appl Phycol 23:877–886

CIMTAN, Aquaculture R&D Review (2011) Aquaculture Association of Canada Special Publication 16 (2011). ISBN: 978-0-9780943-5-5: 24–30 http://www.aquacultureassociation.ca/publications/special. Cited 20 Aug 2011

Cook EJ, Kelly MS (2007) Enhanced production of the sea urchin *Paracentrotus lividus* in integrated open-water cultivation with Atlantic salmon *Salmo salar*. Aquaculture 273:573–585

Corbin JS (2007) Hawaii aquaculture development: twenty-five years and counting, lessons learned. In: Leung P, Lee CS, O'Bryen PJ (eds) Species and system selection for sustainable aquaculture. Blackwell Publishing, Ames, pp 209–224

Cosmetic Ingredient Dictionary (2002–2011) Algae http://www.cosmeticscop.com/cosmetic-ingredient-dictionary/A.aspx. Cited 16 Dec 2011

Craigie JS (2011) Seaweed extract stimuli in plant science and agriculture. J Appl Phycol 23:371–393

Critchley AT, Ohno M (1997) Cultivation and Farming of Marine Plants. Biodiversity of Expert Centre for Taxonomic Identification (ETI). CD-ROM Version 1.0. Springer Electronic Media Dept, New York, USA

Cross (2010) Problem to Opportunity—Use of the sea urchin, *Strongylocentrotus droebachiensis*, to control biofouling in an Integrated Multi-Trophic Aquaculture System. The World Aquaculture Society Meeting 2010 San Diego, CA https://www.was.org/WasMeetings/Meetings/SessionAbstracts.aspx?Code=AQ2010. Cited 20 Aug 2011

Dawes CP (1988) Seaweed culture technology. In: Consultants M, Munro A (eds) Feasibility study on the technology of mariculture. Vol. II: Review of technologies and services. Aberdeen, University Marine Studies, pp 107–116

Dillehay TD, Ramírez C, Pino M, Collins MB, Rossen J, Pino-Navarro JD (2008) Monte Verde: seaweed, food, medicine, and the peopling of South America. Science 320:784–786. doi:10.1126/science.1156533

FAO (2005–2011a) Cultured Aquatic Species Information Programme. Porphyra spp. Text by Jiaxin Chen and Pu Xu. In: *FAO Fisheries and Aquaculture Department* [online]. Rome. Updated 18 February 2005. http://www.fao.org/fishery/culturedspecies/Porphyra_spp/en, Cited 11 Oct 2011

FAO (2005–2011b) Cultured Aquatic Species Information Programme. *Laminaria japonica*. Text by Chen J. In: *FAO Fisheries and Aquaculture Department* [online]. Rome. Updated 1 January 2004. http://www.fao.org/fishery/culturedspecies/Laminaria_japonica/en, Cited 13 Jul 2011

FAO (2005–2011c) Cultured Aquatic Species Information Programme. *Eucheuma* spp. Text by Gavino C. Trono Jr. In: *FAO Fisheries and Aquaculture Department* [online]. Rome. Updated 13 January 2005. http://www.fao.org/fishery/culturedspecies/Eucheuma_spp/en#tcNA0050, Cited 10 Oct 2011

FAO (2010–2011) Fisheries Global Information System (FAO-FIGIS) In: *FAO Fisheries and Aquaculture Department* [online]. Rome. http://www.fao.org/fishery/figis/en, Cited 13 Dec 2011

FAO (2010a) The State of World Fisheries and Aquaculture 2010 (SOFIA). FAO Fisheries and Aquaculture Department, Rome, p 197. http://www.fao.org/docrep/013/i1820e/i1820e00.htm, Cited 10 Oct 2011

FAO (2010b) 2008 FAO Yearbook of Fishery and Aquaculture Statistics. ftp://ftp.fao.org/FI/CDrom/CD_yearbook_2008/navigation/index_content_aquaculture_e.htm. Cited 10 Oct 2011

FAO (2011a) FAO Fisheries Department, Fishery Information, Data and Statistics Unit. FishStatPlus. Universal Software for fishery statistical time series. Version 2.3 in 2000. Last database update in April 2011

FAO (2011b) Fishery Statistical Collections Global Aquaculture Production, Status http://www.fao.org/fishery/statistics/global-aquaculture-production. Cited 10 Oct 2011

FAO (2011c) National Aquaculture Sector Overview (NASO) http://www.fao.org/fishery/naso/search/en Cited 10 Oct 2011

Fei X (2004) Solving the coastal eutrophication problem by large scale seaweed cultivation. Hydrobiologia 512:145–151

George M, Abraham TE (2006) Polyonic hydrocolloids for the intestinal delivery of protein drugs: Alginate and chitosan—a review. J Control Release 114:1–14

Gómez I, Lüning K (2001) Constant short–day treatment of outdoor–cultivated *Laminaria digitata* prevents summer drop growth rate. Eur J Phycol 36:391–395

Hasegawa Y (1971) Forced cultivation of *Laminaria*. Bull Hokkaido Reg Fish Res Lab 37:49–52

Hesley C (1997) Open Ocean Aquaculture: Chartering the Future of Ocean Farming. In: Proceedings of an International Conference, April 23–25, 1997, Maui, Hawaii. UNIHI-Seagrant-CP-98-08, Maui, University of Hawaii Sea Grant College Program p 353

Holt TJ and Kain (Jones) JM (1983) The cultivation of large brown algae as an energy crop. In: Strub A, Chartier and Schleser P (eds) Energy from biomass 2nd conference, Applied Science Publishers, London, pp 319–323

ICES (2011): Report of the Study Group on Social Dimensions of Aquaculture (SGSA). Bremen, Germany p 33

Indergaard M, Østgaard K (1991) Polysaccharides for food and pharmaceutical uses. In: Guiry MD, Blunden G (eds) Seaweed resources in Europe. Uses and potential. Wiley, Chichester, pp 169–183

Kain JM (1991) Cultivation of attached seaweeds. In: Guiry MD, Blunden G (eds) Seaweed resources in Europe: uses and potential. Wiley, Chichester, UK, pp 309–377

Kain JM, Dawes CP (1987) Useful European seaweeds: past hopes and present cultivation. Hydrobiologia 151(152):173–181

Kawashima S (1984) Kombu cultivation in Japan for human foodstuff. Jpn J Phycol 32:379–394

Krause G, Buck BH, Rosenthal H (2003) Multifunctional use and environmental regulations: potentials in the offshore aquaculture development in Germany, rights and duties in the coastal zone—multidisciplinary scientific Conference on sustainable coastal zone management, 12–14 June 2003, Stockholm (Sweden)

Langan R, Newell RIE, McVey JP, Newell C, Sowles JW, Rensel JE, Yarish C (2006) Country scenarios for ecosystem approaches for aquaculture: The United States. In: McVey JP, Lee C-S, O'Bryen PJ (eds) Aquaculture and ecosystems: an integrated coastal and ocean management approach. The World Aquaculture Society, Baton Rouge, Louisiana, pp 109–140

Lombardi JV, de Almeida Marques HL, Lima Pereira RT, Salée Barreto J, de Paula EJ (2006) Cage polyculture of the Pacific white shrimp *Litopenaeus vannamei* and the Philippines seaweed *Kappaphycus alvarezii*. Aquaculture 258:412–415

Løvstad Holdt S, Kraan S (2011) Bioactive compounds in seaweed: functional food applications and legislation. J Appl Phycol 23:543–597

Lowther A (2006) Highlights from the FAO database on Aquaculture Statistics. FAO Aquacult Newsletter 35:32–33

Lüning K, Pang S (2003) Mass cultivation of seaweeds: current aspects and approaches. J Appl Phycol 15:115–119

Lüning K, Wagner A, Buchholz C (2000) Evidence for inhibitors of sporangium formation in *Laminaria digitata* (Phaeophyceae) during the season of rapid growth. J Phycol 36:1129–1134

Marra J (2005) When will we tame the oceans? Nature 436:175–176

Mc Hugh, DJ (2003) A guide to the seaweed industry. FAO Fisheries Technical Papers T441

McVey JP, Buck BH (2008) IMTA-Design within an Offshore Wind Farm, "Aquaculture for Human Wellbeing—The Asian Perspective". The Annual Meeting of the World Aquaculture Society, 23rd May 2008, Busan (Korea)

McVey JP, Stickney R, Yarish C, Chopin T (2002) Aquatic polyculture and balanced ecosystem management: new paradigms for seafood production. In: Stickney RR, McVey JP (eds) Responsible aquaculture. CABI Wallingford, UK, pp 91–104

Merrill JE, Gillingham DM (1991) Bull kelp cultivation handbook. Publication No. NCRI-T-91-011. National Coastal Research and Development Institute, Portland, Oregon, USA

Michler-Cieluch T (2009) Co-Management processes in integrated coastal management : the case of integrating marine aquaculture in offshore wind farms. PhD Thesis, University of Hamburg

Msuya FE (1997) Women seaweed farmers in the Zanzibar Islands, Tanzania. InterCoast Network Narragansett, Rhode Island, USA, 29

Msuya FE (2006) The impact of seaweed farming on the social and economic structure of seaweed farming communities in Zanzibar, Tanzania. In: Critchley AT, Ohno M, Largo DB (eds), World Seaweed Resources, Version: 1.0. p 27

Msuya FE, Shalli MS, Sullivan K, Crawford B, Tobey J, Mmochi AJ (2007) A comparative economic analysis of two seaweed farming methods in Tanzania. The sustainable coastal communities and ecosystems program. Coastal Resources Center, University of Rhode Island and the Western Indian Ocean Marine Science Association, p 27

Murata M, Nakazoe J (2001) Production and use of marine algae in Japan. Jpn Agr Res Q 35:281–290

Naylor RL, Goldburg RJ, Primavera JH, Kautsky N, Beveridge MCM, Clay J, Folke C, Lubchenco J, Mooney H, Troell M (2000) Effect of aquaculture on world fish supplies. Nature 405:1017–1024

Neori A, Chopin T, Troell M, Buschmann AH, Kraemer GP, Halling C, Shpigel M, Yarish C (2004) Integrated aquaculture: rationale, evolution and state of the art emphasizing seaweed biofiltration in modern mariculture. Aquaculture 231:361–391

Neori A, Troell M, Chopin T, Yarish C, Critchley A, Buschmann AH (2007) The need for ecological balance in "blue revolution" aquaculture. Environment 49:36–42

Nobre AM, Robertson-Andersson D, Neori A, Sankar K (2010) Ecological–economic assessment of aquaculture options: comparison between abalone monoculture and integrated multi-trophic aquaculture of abalone and seaweeds. Aquaculture 306:116–126

North WJ (1987) Oceanic farming of Macrocystis. The problems and non-problems. In: Bird KT, Benson PH (eds) Seaweed Cultivation for Renewable Resources. Developments in Aquaculture and Fisheries Sciences, vol 16. Elsevier, Amsterdam, pp 39–67

Ohno M (1993) Cultivation methods and physiological aspect for edible seaweeds in Japan. Ser Ocasional 2:163–170

Pang SJ, Lüning K (2004) Breaking seasonal limitation: year-round sporogenesis in the brown alga *Laminaria saccharina* by blocking the transport of putative sporulation inhibitors. Aquaculture 240:531–541

Pereira R, Yarish C (2008) Mass production of marine macroalgae. In: Jørgensen SE, Fath BD (eds) Encyclopedia of ecology, vol 3, Ecological engineering. Elsevier, Oxford, pp 2236–2247

Pereira R, Yarish C (2010) The role of *Porphyra* in sustainable culture systems: physiology and applications. In: Israel A, Einav R (eds) Seaweeds and their role in a globally changing environment. Springer, Heidelberg, pp 339–354

Pereira R, Yarish C, Critchley A (2012) Seaweed aquaculture for human foods, land based. In: Costa-Pierce BA (ed) Ocean farming and sustainable aquaculture science and technology. Encyclopedia of sustainability science and technology. Springer Science, New York (in press)

Petrell RJ, Alie SY (1996) Integrated cultivation of salmonids and seaweeds in open systems. Hydrobiologia 326(327):67–73

Polk M (1996) Open Ocean Aquaculture. Proceedings of an International Conference, 8-10 May 1996, Portland, Maine. UNHMP-CP-SG-96-9, Portland, New Hamshire/Maine Sea Grant College Program, p 642

Rawson MV Jr, Chen C, Ji R, Zhu M, Wang D, Wang L, Yarish C, Sullivan JB, Chopin T, Carmona R (2002) Understanding the interaction of extractive and fed aquaculture using ecosystem modeling. In: Stickney RR, McVey JP (eds) Responsible aquaculture. CABI Wallingford, UK, pp 263–296

Rensel JE, Buschmann AH, Chopin T, Chung IK, Grant J, Helsley CE, Kiefer DA, Langan R, Newell RIE, Rawson M, Sowles JW, McVey JP, Yarish C (2006) Ecosystem based management: models and mariculture. In: McVey JP, Lee C-S, O'Bryen PJ (eds) Aquaculture and ecosystems: An integrated coastal and ocean management approach. The World Aquaculture Society, Baton Rouge, pp 207–220

Robertson-Andersson DV, Potgieter M, Hansen J, Bolton JJ, Troell M, Anderson RJ, Halling C, Probyn T (2008) Integrated seaweed cultivation on an abalone farm in South Africa. J Appl Phycol 20:579–595

Roesijadi G, Copping AE, Huesemann MH, Forster J, Benemann JR (2008) Techno-economic feasibility analysis of offshore seaweed farming for bioenergy and biobased products. Independent research and development report IR # PNWD-3931, Battelle Pacific Northwest Division, p 115

Ryan J (2005) Offshore aquaculture—do we need it, and why is it taking so long? International Salmon Farmers Association (Ireland). Expert workshop on "Sustainable Aquaculture. DG JRC European Commission, Institute for Prospective Technological Studies, 17–18 Jan 2005 Seville

Ryther JD, Deboer JA, Lapointe BE (1979) Cultivation of seaweeds for hydrocolloids, waste treatment and biomass for energy conversion. Proc Int Seaweed Symp 9:1–16

Sahoo D, Yarish C (2005) Mariculture of seaweeds. In: Andersen R (ed) Phycological methods: algal culturing techniques. Academic, Elsevier, Oxford, New York, pp 219–237

Sanderson JK (2009) Bioremediation using seaweed culture. Reducing the environmental impact of sea-cage fish farming through cultivation of seaweed. SAMS, UHI Milleneum Institute, Open University, 2006; VDM Verlag, Saarbrücken

Sanderson JC, Cromey CJ, Dring MJ, Kelly MS (2006) Distribution of nutrients for seaweed cultivation around salmon cages at farm sites in north-west Scotland. Aquaculture 278:60–68

Schuenhoff A, Shpigel M, Lupatsch I, Ashkenazi A, Msuya FE, Neori A (2003) A semi-recirculating, integrated system for the culture of fish and seaweed. Aquaculture 221 (1–4):167–181

Shpigel M, Ragg NL, Lupatsch I, Neori A (1999) Protein content determines the nutritional value of the seaweed *Ulva lactuca* L for the abalone *Haliotis tuberculata* L. and *H. discus hannai* Ino. J Shellfish Res 18(1):227–233

Smartfiber (2010) SeaCell® *pure*—SeaCell® *active*. http://www.smartfiber.de/index.php?option=com_content&view=article&id=9&Itemid=30&lang=en. Cited 16 Dec 2011

Soto D, Aguilar-Manjarrez J, Hishamunda N (2008) Building an Ecosystem Approach to Aquaculture. FAO/Universitat de les Illes Balears Expert Workshop, 7–11 May 2007, Palma de Mallorca, Spain. FAO Fisheries and Aquaculture Proceeding, No. 14. ftp://ftp.fao.org/docrep/fao/011/i0339e/i0339e.pdf. Cited 10 Oct 2011

Stead SM, Laird L (2002) Handbook of salmon farming. Springer, London, pp 163–166

Stickney RR (1998) Joining Forces With Industry—Open Ocean Aquaculture. Proceedings of the Third Annual International Conference, May 10-15, Corpus Christi, Texas. TAMU-SG-99-103, Corpus Christi, Texas Sea Grant College Program. p 152

Tamura T (1966) Marine aquaculture. 2nd edition (Translated from Japanese). Springfield, VA, USA

Troell M (2009) Integrated marine and brackishwater aquaculture in tropical regions: research, implementation and prospects. In D. Soto (ed) Integrated mariculture: a global review. FAO Fisheries and Aquaculture Technical Paper No. 529:47–131

Troell M, Rönnbäck P, Halling C, Kautsky N, Buschmann A (1999) Ecological engineering in aquaculture: use of seaweeds for removing nutrients from intensive mariculture. J Appl Phycol 11:89–97

Troell M, Halling C, Neori A, Buschmann AH, Chopin T, Yarish C, Kautsky N (2003) Integrated mariculture: asking the right questions. Aquaculture 226:69–90

Troell M, Buck BH, Angel D, Chopin T (2012) Possibilities for the development of IMTA, combined with other activities, in offshore environments. In: Chopin T, Neori A, Robinson S, Troell M (eds) Integrated multi-trophic aquaculture (IMTA) or the turquoise revolution: a greener ecosystem approach to the blue revolution. Springer, Berlin

Trono GC (1990) Technical resource papers—Regional workshop on the culture and utilization of seaweeds, vol. 2, 190 pp http://www.fao.org/docrep/field/003/AB728E/AB728E00.htm. Cited 10 Oct 2011

Tseng CK (1984) Phycological research in the development of the Chinese seaweed industry. Hydrobiologia 116(117):7–18

Tseng CK (1987) *Laminaria* mariculture in China. In: Doty MS, Caddy JF, Santelices B (eds) Case Studies of seven Commercial Seaweed Resources. FAO Fisheries Technical Paper 281, Food and Agriculture Organisation of the United Nations, Rome, Italy, pp 239–263

Tseng CK (1989) Farming and ranching of the sea in China. Proc 2nd Gen Conf Org Third World Acad Sci. In: Faruqui AM, Hassan MH, (eds), The Future of Science in China and the Third World, World Scientific, pp 92–106

Walker AB, Fournier HR, Neefus CD, Nardi GC, Berlinsky DL (2009) Partial replacement of fish meal with laver *Porphyra* spp. in diets for Atlantic cod. North Am J Aquacult 71:39–45

Erratum to: Cold-Temperate Seaweed Communities of the Southern Hemisphere

Pirjo Huovinen and Iván Gómez

Erratum to:
Chapter 14 in: C. Wiencke and K. Bischof (eds.), *Seaweed Biology*, Ecological Studies 219, DOI 10.1007/978-3-642-28451-9_14

Fig. 14.3: The bold line indicating the distribution of *Macrocystis* in South Africa should not be in black, but in red.

The online version of the original chapter can be found under
DOI 10.1007/978-3-642-28451-9_14

P. Huovinen (✉) • I. Gómez
Instituto de Ciencias Marinas y Limnoló gicas, Universidad Austral de Chile, Campus Isla Teja s/n, Casilla 567, Valdivia, Chile
e-mail: pirjo.huovinen@uach.cl

C. Wiencke and K. Bischof (eds.), *Seaweed Biology*, Ecological Studies 219,
DOI 10.1007/978-3-642-28451-9_23,

Index

C. Wiencke and K. Bischof (eds.), *Seaweed Biology*, Ecological Studies 219, DOI 10.1007/978-3-642-28451-9,

Ecological Studies

Further volumes can be found at springer.com

Volume 202
Ecosystem Organization of a Complex Landscape: Long-Term Research in the Bornhöved Lake District, Germany (2008)
O. Fränzle, L. Kappen, H.-P. Blume, and K. Dierssen (Eds.)

Volume 203
The Continental-Scale Greenhouse Gas Balance of Europe (2008)
H. Dolman, R. Valentini, and A. Freibauer (Eds.)

Volume 204
Biological Invasions in Marine Ecosystems: Ecological, Management, and Geographic Perspectives (2009)
G. Rilov and J.A. Crooks (Eds.)

Volume 205
Coral Bleaching: Patterns, Processes, Causes and Consequences
M.J.H van Oppen and J.M. Lough (Eds.)

Volume 206
Marine Hard Bottom Communities: Patterns, Dynamics, Diversity, and Change (2009)
M. Wahl (Ed.)

Volume 207
Old-Growth Forests: Function, Fate and Value (2009)
C. Wirth, G. Gleixner, and M. Heimann (Eds.)

Volume 208
Functioning and Management of European Beech Ecosystems (2009)
R. Brumme and P.K. Khanna (Eds.)

DOI 10.1007/978-3-642-28451-9,

Volume 209
Permafrost Ecosystems: Siberian Larch Forests (2010)
A. Osawa, O.A. Zyryanova, Y. Matsuura, T. Kajimoto, R.W. Wein (Eds.)

Volume 210
Amazonian Floodplain Forests: Ecophysiology, Biodiversity and Sustainable Management (2010)
W.J. Junk, M.T.F. Piedade, F. Wittmann, J. Schöngart, P. Parolin (Eds.)

Volume 211
Mangrove Dynamics and Management in North Brazil (2010)
U. Saint-Paul and H. Schneider (Eds.)

Volume 212
Forest Management and the Water Cycle: An Ecosystem-Based Approach (2011)
M. Bredemeier, S. Cohen, D.L. Godbold, E. Lode, V. Pichler, P. Schleppi (Eds.)

Volume 213
The Landscape Ecology of Fire (2011)
D. McKenzie, C.S. Miller, D.A. Donald (Eds.)

Volume 214
Human Population: Its Influences on Biological Diversity (2011)
R.P. Cincotta and L.J. Gorenflo (Eds.)

Volume 215
Plant Desiccation Tolerance (2011)
U. Lüttge, E. Beck, D. Bartels (Eds.)

Volume 216
Forest Hydrology and Biogeochemistry: Synthesis of Past Research and Future Directions (2011)
D.F. Levia, D. Carlyle-Moses, T. Tanaka (Eds.)

Volume 217
Peatlands of the Western Guayana Highlands, Venezuela: Properties and Paleogeographic Significance of Peats (2011)
J.A. Zinck and O. Huber (Eds.)

Volume 218
Aralkum—a Man-Made Desert: The Desiccated Floor of the Aral Sea (Central Asia) (2012)
S.-W. Breckle, W. Wucherer, L.A. Dimeyeva, N.P. Ogar (Eds.)

Volume 219
Seaweed Biology: Novel Insights into Ecophysiology, Ecology and Utilization (2012)
C. Wiencke and K. Bischof (Eds.)

MIX
Papier aus verantwortungsvollen Quellen
Paper from responsible sources
FSC® C105338

Printed by Books on Demand, Germany

Endoscopic Sinus Surgery
A Practical Approach

Springer
London
Berlin
Heidelberg
New York
Barcelona
Budapest
Hong Kong
Milan
Paris
Santa Clara
Singapore
Tokyo

S.K. Kaluskar
With a Contribution by Professor T. Ohinishi

Endoscopic Sinus Surgery

A Practical Approach

Foreword by Professor W. Draf

With 232 illustrations, 196 in colour

Springer

S.K. Kaluskar, MS, FRCS, DLO (Eng)
Consultant Otorhinolaryngologist, Tyrone County Hospital, Omagh,
Northern Ireland, UK

Contribution by
Toshio Ohinishi, MD
Department of Otolaryngology, St Luke's International Hospital,
Tokyo, Japan

ISBN 3-540-19989-6 Springer-Verlag Berlin Heidelberg New York

British Library Cataloguing in Publication Data
Kaluskar, S.K.
Endoscopic sinus surgery: a practical approach
1. Paranasal sinuses - Endoscopic surgery
I. Title
617.5′23
ISBN 3540199896

Library of Congress Cataloging-in-Publication Data
Kaluskar, S.K.
Endoscopic sinus surgery: a practical approach/S.K. Kaluskar.
p. cm.
Includes bibliographical references and index.
ISBN 3-540-19989-6 (hardback: alk. paper)
1. Paranasal sinuses - Endoscopic surgery. I. Title.
[DNLM: 1. Paranasal Sinus Diseases - surgery. 2. Surgery,
Endoscopic - methods. 3. Paranasal Sinuses - anatomy & histology.
WV 340 K14e 1996]
RF421.K35 1996
617.5′23 - dc20
DNLM/DLC
for Library of Congress 96-17383

Typeset by EXPO Holdings, Malaysia
Printed and bound at Art Printing Works, Malaysia
28/3830-543210 Printed on acid-free paper

Dedicated to
my wife, Hema, and children Komal, Anup and Soniya
who made it all worthwhile

Foreword

Ear, nose and throat surgeons all over the world remember well the evolution of ear surgery in the early fifties, when the concept of tympanoplasty was developed, providing cure for disease and its possible complications, and functional restoration of sound transmission of the middle ear. In a similar way, diagnosis and surgery of inflammatory diseases of paranasal sinuses had undergone basic changes within the last 20 years.

Endoscopy of the nose and paranasal sinuses was the first step, more extended surgery became safer and the first attempts at endoscopic surgery were made. Techniques for endo-nasal exploration of the paranasal sinuses had been described at the turn of the twentieth century but were soon abandoned because of frequent, severe complications. However, new visual aids such as the microscope and endoscope, a remarkable improvement in local and general anaesthesia, and an increasing reluctance of patients to accept the surgical sequelae of "radical" sinus procedures with almost total removal of mucosal lining all provided the impetus for a renaissance in endo-nasal practice.

The concept of functional endoscopic sinus surgery was first developed in Austria and Germany as a result of detailed studies of the physiology and pathology of the mucociliary transport mechanism. It facilitated the smallest possible intervention and the removal of severely diseased parts of mucosa, preservation of mucosa showing reversible pathology, and establishment of optimal drainage of all sinuses into the nose via an endo-nasal route. Functional endoscopic sinus surgery had gained overwhelming acceptance throughout the world by the mid 1980s.

Mr Kaluskar realised the importance of functional endoscopic sinus surgery at an early stage and the present monograph is the result of his diligent research and hard work in the field of this modern rhinosurgical technique, and his experience with a total of 532 operated cases and 1022 procedures. After intensive studies of the related anatomical structures, he started using mainly the technique of the Graz Messerklinger school, and restricting the more extended surgery of Wigand to severe cases and revisions. From the beginning he collected all the relevant data from his cases, which were processed by computer and analysed. He also developed modifications as, for example, combined approach middle meatal antrostomy (CAMMA).

This book is written for the beginner, describing step by step the route one has to follow until surgery may be started, the surgical technique itself, including some modifications and many "tricks". Chapter 14 (Tips for Beginners) makes clear why the old chestnut "practice makes perfect" is not good advice to the initiate in functional endoscopic sinus surgery. The discussion "Headaches and Facial Pains" is of utmost interest.

The final chapter, "Broadening the Horizons", gives a stimulating outlook for endonasal endoscopic sinus surgery. This publication is richly illustrated. The layout allows important points to be brought to the reader's particular attention. Appendix A shows the necessary equipment, and Appendix B supplies the necessary information for prospective documentation. With its meticulous review of results and complications, pitfalls and their prevention, this work will give significant support to the less experienced surgeon and cautions him or her of the possible dangers. For the advanced sinus surgeon seeking another experts opinion there is a rich mine of "tricks" and advice. We hope, this monograph will also stimulate head and neck and ear, nose and throat surgeons to expand their diagnostic as well as therapeutic abilities, and to improve the results of surgical management of patients with inflammatory paranasal sinus diseases.

Professor W. Draf
Director, University ENT Clinic, Fulda, Germany

Preface

The management of chronic inflammatory paranasal sinus disease has been transformed by an increasingly clear understanding of the pathophysiology of muco-ciliary clearance, the application of nasal endoscopy, both diagnostic and therapeutic and by recent developments in imaging techniques. Endoscopic sinus surgery (ESS) is a minimally invasive technique for the treatment of chronic sinus disease with significantly superior results.

This is a practical textbook on endoscopic sinus surgery with an emphasis on the basics of surgical anatomy, pathophysiology and, most importantly, endoscopic diagnosis; the surgeon must first diagnose the disease before he or she can treat, let alone operate! The book is intended not only for the practising ear, nose and throat and head and neck surgeons but also for trainees who will be able to enhance their understanding of the clinical presentation of chronic paranasal sinus disease and its management.

Information on the subject, surprisingly, is scarce, and falls mainly into two categories. Most standard ENT textbooks gloss over ESS, mentioning some historical aspects and basic techniques. The few books dealing specifically with the subject (they can almost be counted on the fingers of one hand!) assume a basic grasp of the topic on the part of the reader, and could pose difficulties for those who have limited or only recent access to endoscopy. There is, in my opinion, a need for a primer that provides a basic, visual overview of the why, the when and the how (*including the how not!*) of ESS.

This book takes the reader on a tour of the nose and sinuses in both health and disease, stopping to dwell on the intricacies of the ostio-meatal complex. Pre-operative evaluation using nasal endoscopy and computed tomography is discussed and illustrated. The surgical protocol is described step by step, as is post-operative care. Complications, and how to prevent and deal with them, are of particular importance in ESS. Since the surgeon is often operating millimetres away from the orbit and skull base, an intimate knowledge of anatomy gained in the dissection hall is of paramount value. Accordingly, numerous cadaveric and surgical illustrations are included.

I would like to emphasise my belief that nasal endoscopy is a very useful diagnostic procedure in its own right, and I would recommend it highly even to those not actively involved in endoscopic sinus surgery. I have enjoyed working on this venture tremendously, and would like to share my experience about nasal and sinus endoscopy. I hope this book guides the novice and stimulates the experienced endoscopist to greater heights.

Acknowledgements

I would like to thank the medical and nursing staff of the Tyrone County Hospital, Omagh for their help in my endeavor to mature this technique. A special mention must be made for the constant support and dedicated assistance that I have received from

Mr Hugh Mills, Chief Executive of Sperrin and Lakeland Management Unit, Mrs Bernie McCrory, General Services Manager and Mrs Martina Corrigan, Assistant General Services Manager and I should like to thank them with gratitude.

I acknowledge with thanks the commercial support received from Karl Storz, John Weiss of London and Richard Wolf (U.K.) Ltd.

I deeply appreciate the invaluable help of my son Anup in the preparation of this book especially with computer graphics and technical support and I would like to express my sincere thanks to him. I also would like to thank N. Patil for his technical assistance.

S.K. Kaluskar
Omagh, Northern Ireland, 1997

Contents

Abbreviations

A	adhesions
AN	agger nasi
BE	bulla ethmoidales
CG	crista galli
ET	Eustachian tube
ETh	ethmoids
FE	fovea ethmoidales
FR	frontal recess
HS	hiatus semilunaris
IM	inferior meatus
INA	inferior nasal antrostomy
IC	internal carotid
IT	inferior turbinate
LP	lamina papyracea
LW	lateral wall
MM	middle meatus
MMA	middle meatal antrostomy
MT	middle turbinate
MX	maxillary sinus
NP	nasopharynx
ON	optic nerve
OS	ostium
P	pus
PNS	post-natal space
PO	polyp
S	septum
SM	superior meatus
SP	sphenoid
ST	superior turbinate
UP	uncinate process

Chapter 1

A Historical Overview

The study of the internal nasal contours dates back to early mankind, when the nose was predominantly associated with the sense of smell. The earliest record of an anatomical observation is in the papyrus of Abers and Egyptian tomb inscriptions, dated well before 1500 BC. This ancient culture used the nose as a route for extracting the contents of the cranial vault as part of the mummification process, thereby avoiding any facial disfigurement. This implies an intimate knowledge of the intricate relationship between the roof of the ethmoids and the brain.

The first reference to a crude Arabian nasal speculum dates back to the twelfth century AD (As Sayzari). As medicine gradually advanced, more and more attempts where made to visualise this rather inaccessible region. The first modern-day nasal speculum was designed by Markusowsky. Gabriel Fallopius (1523–1562) used a nasal speculum to aid in the removal of posteriorly situated polyps. Other methods of illuminating the nasal cavity include Borel's (1620–1671) concave mirror which reflected sunlight into the nose. The sinuses were first demonstrated as a distinct anatomical feature by Johann Riolan in the seventeenth century. Czermak in Vienna coined the term "rhinoscopy" in 1859 when he examined the post-nasal space and choanae using a mirror. Even with better speculae such as Perret's valved instruments, the problem of illumination remained a major obstacle. Up to the middle of the last century it was usually the otologist who carried out nasal examination. In 1869 Wertheim designed a small tube with an apertured mirror set at 45° at one end ("conchoscope"); he used it routinely to visualise the anterior and mid-third of the nasal cavity. However, it was not until the introduction of the fenestrated head mirror in the latter half of the nineteenth century that the other regions of the nose were made more visible. The speculae we use today owe their origins to masters of the eighteenth century such as Sir St Clair Thomson (1859–1943), Gustav Killian (1862–1921) and Johann Thudicum (1829–1901).

Endoscopic nasal surgery has its roots both in intranasal procedures performed in the mid-nineteenth century as well as diagnostic sinoscopy which was relatively new at that time. The antral cavity was first accessed by von Mikulicz, who reported opening it through the middle meatus in 1886. Claoue published a 10-year survey of his antrostomies in 1912. The turbinates proved a major obstacle to adequate visualisation of the antrostomy site, but this was overcome by Dahmer (1909) who created a large inferior meatal antrostomy by resecting the anterior end of the inferior turbinate. He denuded the antral mucosa from the floor of the sinus, and claimed this resulted in a pain-free irrigation, often performed by the patient himself.

Claoue's survey suggests that other European rhinologists such as Boenninghaus and Hajek, who had accepted the method of wide inferior meatal antrostomy, with or without resection of the head of the inferior turbinate, were pleased with the almost immediate improvement in symptoms, especially the drying up of tenacious secretions. Lothrop in the USA (1897) carried out a wide inferior meatal antrostomy too, but he regarded this merely as an entry point for irrigation to control the infection, and inspection of the cavity.

As nasal surgeons became more aware of sinus pathophysiology, the concept of a middle meatal antrostomy was born. First described by Siebenmann in 1899, the procedure was once again aimed at giving the patient a method of irrigating his own sinuses over time. In addition, Siebenmann advocated resection of the head of the middle turbinate, and breaching the membranous fontanelle digitally. By this time, the delayed complications of inferior meatal windows were becoming apparent, as more and more surgeons reported closure and secondary problems. It is of interest that Zuckerkandl in 1882 stated that this window generally remained fully patent. Caldwell (1893) emphasised the importance of the anterior ethmoid cells as the primary focus of sinus disease, and not the frontal and maxillary sinuses. Kubo in 1912 preferred a middle meatal antrostomy, and Onodi designed a perforator for this purpose around the turn of the century.

Simultaneously with the advent of middle meatal antrostomies came various photographic and other studies on muco-ciliary clearance of the cavity itself. Hilding (1931) first exhibited that carbon particles bypassed an inferior meatal antrostomy, and could leave the antral cavity and re-enter it again via an accessory ostium before finally leaving through the natural os. Proetz (1941) added information about this phenomenon, while Messerklinger provided photographic documentation demonstrating this process in the 1980s, and elucidated the concepts of muco-ciliary clearance as we know them today.

Further procedures on the lateral wall of the nose included Killian's technique of resection of the uncinate process with widening of the neighbouring ostium in 1900. The foundations of endoscopic sinus clearance, such as creating a single ethmoid volume, problems of exenteration of the most anterior cells, the risks of blind dissection, the use of local decongestants and of special curved instruments, were all expounded by Halle in 1906. The risk of middle turbinate avulsion has been described by several authors such as Davison (1969), Freedman and Kern (1979) and Dixon (1983), who mention symptoms of dryness because the nasal cavity tends to become too large.

With the advent of nasal endoscopy, the intimate relations of the sinus openings assumed even greater importance; it is in this period that sinus surgery was elevated to "watchmaker" status. Today, it is as much an art as it is a science.

Chapter 2 Surgical Anatomy

Introduction

Being a focal facial feature, the nose inspired several quaint notions, it was even considered to be an exit for bad spirits from the body! The foundation for a scientific assessment of this organ was laid down by Emil Zuckerkandl of Austria who, not without a pinch of humour, in his treatise quips "you have to suffer and tolerate a lot on this earth, even if you are only a pneumatised cavity!". In 1882 his first book, a comparative anatomy, was published, and his illustrations were so accurate that they spawned interest all over the globe. He is credited with having given rhinology the pride of place it deserves. Along the road of discovery one finds monuments to masters such as Grunwald, Onodi and Hajek. Hand in hand with these advances came innovations in histology, local anaesthesia and radiology, leading to a burst of anatomical information, reflected today in our current state-of-the-art environment of multi-angled endoscopes, computed tomography (CT), magnetic resonance imaging (MRI) and even three-dimensional scanning.

Developmental Anatomy

The basic functions of the nose include respiration; humidification, filtration and temperature control of inspired air; protection via the mucociliary mechanism and lysozymes; and olfaction (the paranasal sinuses probably participate in some of these). Embryologically speaking, the nose begins as olfactory placodes in the fifth week of intra-uterine life on the ventral surfaces of the fronto-nasal process. These soon become depressed to form olfactory pits, which divide the fronto-nasal process into two parts: the medial and lateral nasal processes. These form the precursors of the nasal structure.

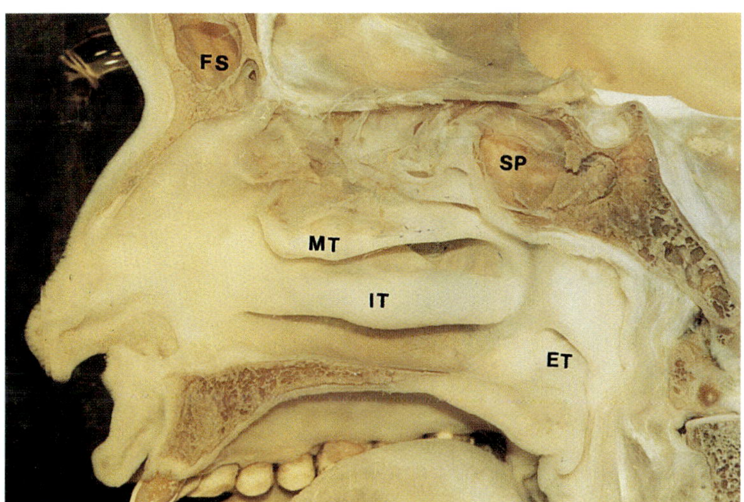

Fig. 2.1. The lateral wall of the nose showing the inferior turbinate (IT), middle turbinate (MT) and eustachian tube opening (ET). The sphenoid (SP) and frontal (FS) sinuses are also seen.

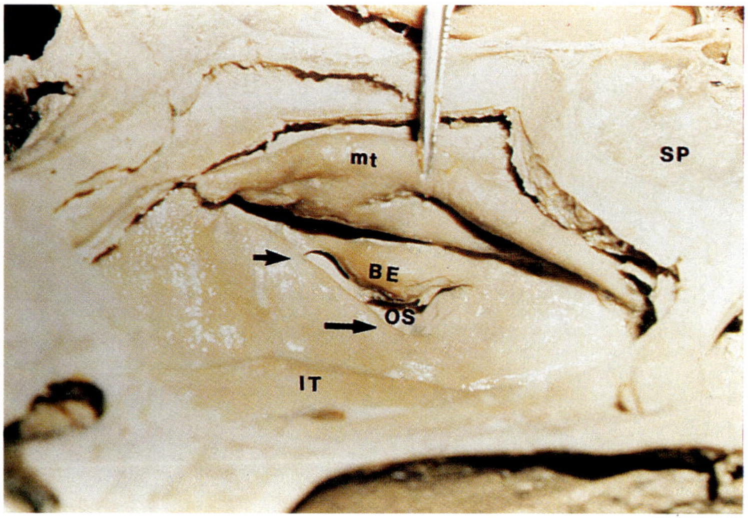

Fig. 2.2. On lifting the middle turbinate, the middle meatus comes into view. The short arrow points to the uncinate process, while the bulla ethmoidales (BE) and the natural os (OS) are also indicated.

Fig. 2.3. The middle turbinate has been resected to expose the hiatus semilunaris (HS), flanked by the uncinate process (UP) and the bulla ethmoidales (BE) The posterior ethmoidal cells (PE) and sphenoidal sinus (SP) are also seen. SB, skull base.

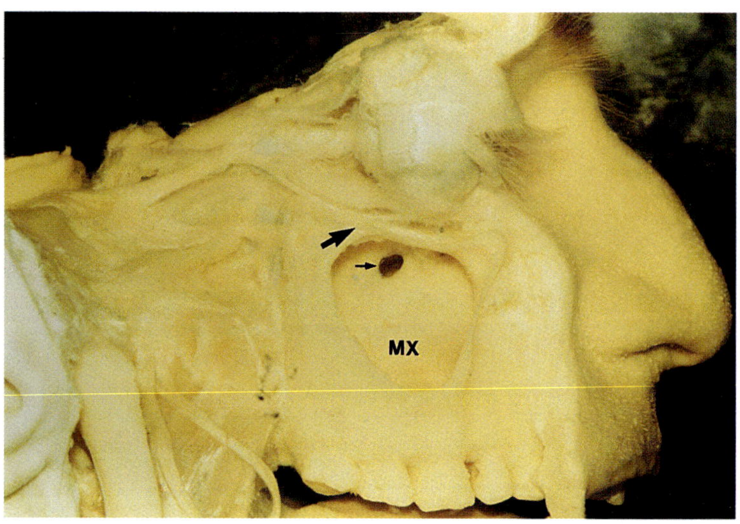

Fig. 2.4. Dissection of the maxillary sinus to show the natural os in relation to the inferior wall of the orbit. Small arrow, maxillary os; large arrow, infraorbital nerve.

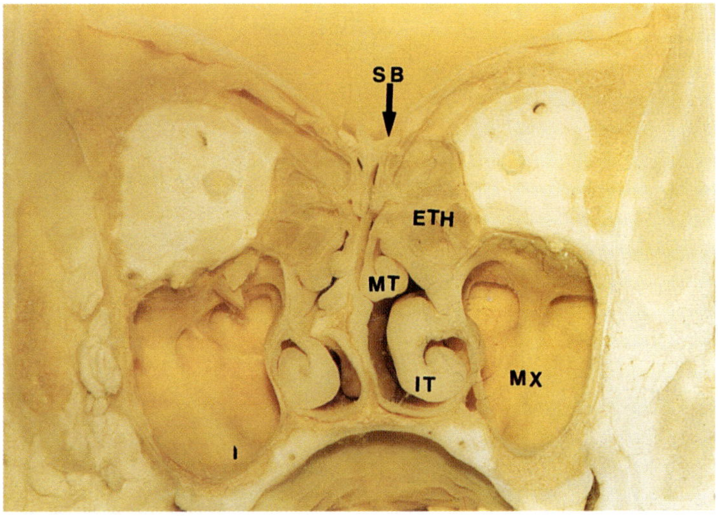

Fig. 2.5. Coronal section of the paranasal sinuses showing the maxillary sinus (MX), inferior turbinate (IT), middle turbinate (MT), ethmoid labyrinth (ETH) and the skull base (SB).

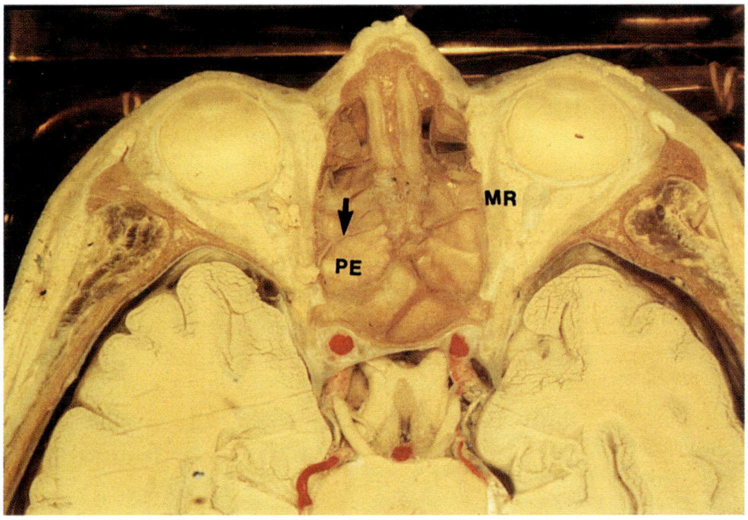

Fig. 2.6. Axial view of the ethmoid labyrinth, showing its relation to the medial rectus (MR). The internal carotid artery and optic nerves are seen to be in close proximity to the posterior-most ethmoidal cells. Arrow points to ground lamella. PE, posterior ethmoids.

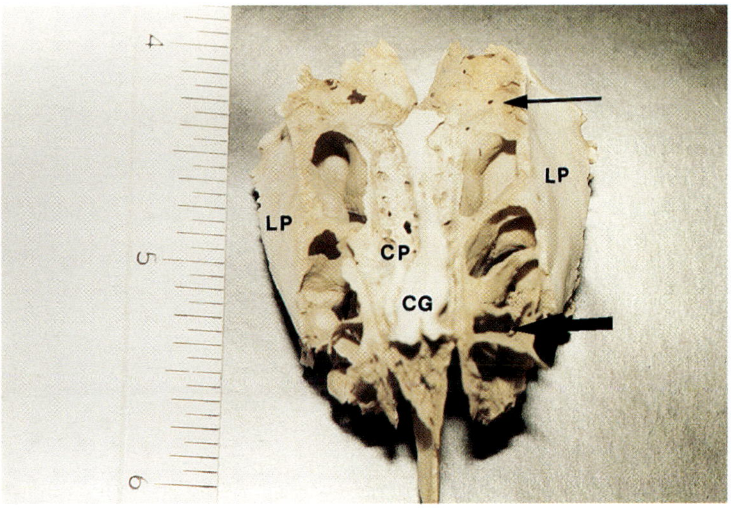

Fig. 2.7. The ethmoid bone showing the crista galli (CG), lamina papyracea (LP) and the cribriform plate (CP). Anterior (thick arrow) and posterior (thin arrow) ethmoidal cells are seen.

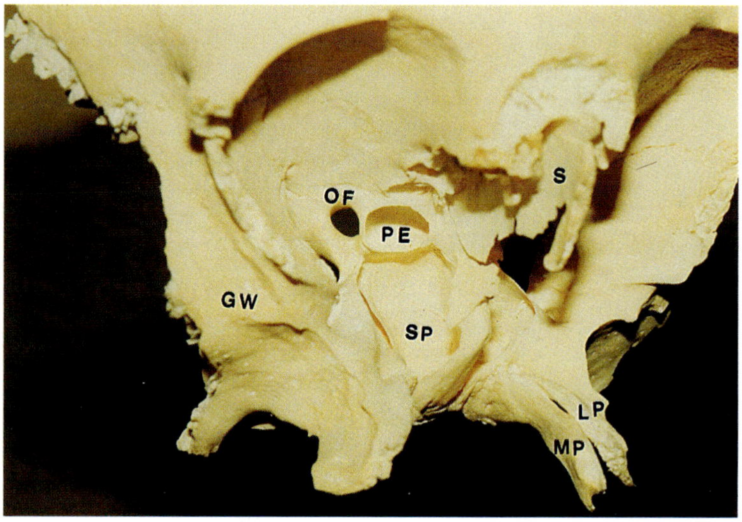

Fig. 2.8. Articluated frontal and sphenoid bones showing posterior ethmoidal cells (PE), their relation to the optic foramen (OF) and the sphenoid sinus (SP). Note that the sphenoid is not a continuation of the posterior ethmoid, but is infero-medial to it. GW, greater wing of sphenoid; S septum; MP, medial pterygoid plate; LP, lateral pterygoid plate

The maxillary sinus is seen as a shallow groove extending laterally from the infundibulum in the fourth month of intra-uterine life. The sinus begins to encroach into the maxilla at birth. Pneumatisation drives the floor of the sinus downwards until it approaches the nasal floor at the age of 8 years. The ethmoids originate from pre-formed furrows at around the same time, with the ethnoidal cells invaginating into the lateral ethmoid masses. The ethmoid labyrinth is seen to be well pneumatised at birth. Up to the age of 6 years the maxillary sinus shows a gradual growth curve, after which there is a spurt to reach its adult size by puberty.

The frontal sinus is absent at birth, and becomes obvious in the first year of life. It develops by direct extension of the whole frontal recess, from one or more anterior ethmoidal cells, or from the ethmoidal infundibulum. It reaches its adult size by puberty. The sphenoid sinus is recognisable from the third intra-uterine month as an invagination of the spheno-ethmoidal recess. It is well developed at the age of 8 years. The clinical application of this differential development is that children tend towards developing ethmoiditis more commonly, while adults show a preponderance of maxillary sinus involvement. Frontal sinusitis is rare before puberty. These facts have direct bearing on functional endoscopic sinus surgery (FESS) in children.

Macroscopic Anatomy

The anterior nares lead into the vestibule, which is lined by skin, sebaceous glands and hair follicles. The vestibule is limited above and behind by a curved ridge, the limen nasi.

The nasal passages, in contrast, are lined by pseudo-stratified ciliated columnar epithelium, each communicating with its own set of paranasal sinuses and with the nasopharynx via the posterior choana. The septum is made of both bony and cartilaginous segments (Fig. 2.9):

- perpendicular plate of ethmoid (above and behind, bony)
- vomer (below and behind, bony)
- quadrilateral cartilage (cartilaginous)

Fig. 2.9. C, cartilaginous septum; V, vomer; E, perpendicular plate of the ethmoid bone.

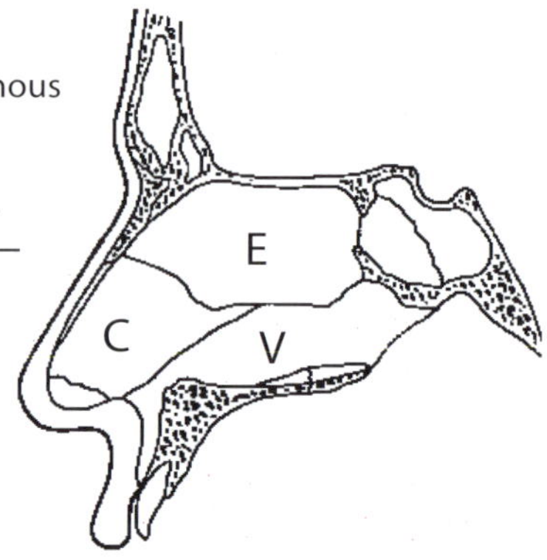

The lateral wall of the nose is its most important surface in terms of nasal dynamics. It consists of:

- medial wall of maxilla
- lateral portions of the ethmoid and lacrimal bones
- ascending process of maxilla/perpendicular plate of palatine/ medial pterygoid process

Average turbinal lengths in the adult	
Superior	7–25 mm, average 16.2 mms
Middle	30–50 mm, average 40 mms
Inferior	36–56 mm, average 46 mms

It bears three turbinates (superior, middle and inferior) overhanging their meati (occasionally a fourth "supreme" turbinate may be seen). Drainage of the paranasal sinuses occurs via ostia in the lateral wall of nose. The superior meatus bears the opening of the posterior ethmoidal cells, while the sphenoidal ostium opens into its own groove, the spheno-ethmoidal recess. The middle meatus is the most significant area of the lateral surface in this regard; it contains the openings of the anterior/middle ethmoids and the frontal sinuses. Into the anterior part of the middle meatus open the anterior ethmoid sinuses. In 50% of cases the frontal sinus opening is seen to coalesce with these openings. Into the posterior part opens the maxillary sinus. The middle ethmoidal cells have variable sites of opening, ranging from onto the bulla to above the bulla or near it. The inferior meatus receives the opening of the nasolacrimal duct.The paranasal sinus complex constitutes an intricate labyrinth of air spaces within the bones – four to each side. They are as follows:

- The *maxillary sinus* (antrum of Highmore): pyramidal in shape, its apex extends into the zygomatic process of the maxilla. It has a capacity of approximately 30 ml, thus making it the largest cavity. Its roof is the thin floor of the orbit, which is grooved by the infraorbital nerve. The base is the alveolar process and the hard palate. In the child, this lies at the level of, or just above, the nasal floor, while it drops to 1 cm below this level in the adult. The main ostium of the sinus is situated high up between the roof and medial wall. It opens into the hiatus semilunaris in its lower part.

The maxillary sinus (usually symmetrical bilaterally)	
Length	38.5 mm
Width	26.4 mm

- The *ethmoid cells* number 8-15. Predominantly occupying the mass of the ethmoid bone, they may "invade" the agger nasi and middle turbinate, and sometimes the surrounding bones such as the maxillary, frontal and sphenoid. They may be differentiated into anterior and posterior groups, separated by a ground lamella. The anterior cells are small, and open into the upper part of the hiatus semilunaris. The posterior cells are larger, and open into the superior meatus. The ethmoid labyrinth is related to the anterior cranial fossa, the orbit, lacrimal sac and optic nerve. The anterior and posterior ethmoidal vessels and nerves pass from the orbit along the roof into the nasal fossa.
- The frontal sinus is an extended anterior ethmoidal cell. It is of variable capacity, the right and left sinuses often being asymmetrically divided by one or many septae. Each sinus drains into the infundibulum of the ethmoid via the frontal recess, which opens into its antero-superior aspect. This area is identified as the frontal recess, and is in close anatomic proximity to the frontal sinus itself. The frontal recess is not a "duct" in the true sense of the word, and hence should not be referred to as such.
- The sphenoid sinus lies behind the upper part of the nasal fossa, and occupies the body, and sometimes the wings and pterygoid processes of the sphenoid. The sinus is related to the cavernous sinus, pituitary gland, optic chiasma, olfactory tract and frontal lobe. The vidian nerve runs below its floor. In its upper part, it measures approximately 14 mm from anterior to posterior wall.

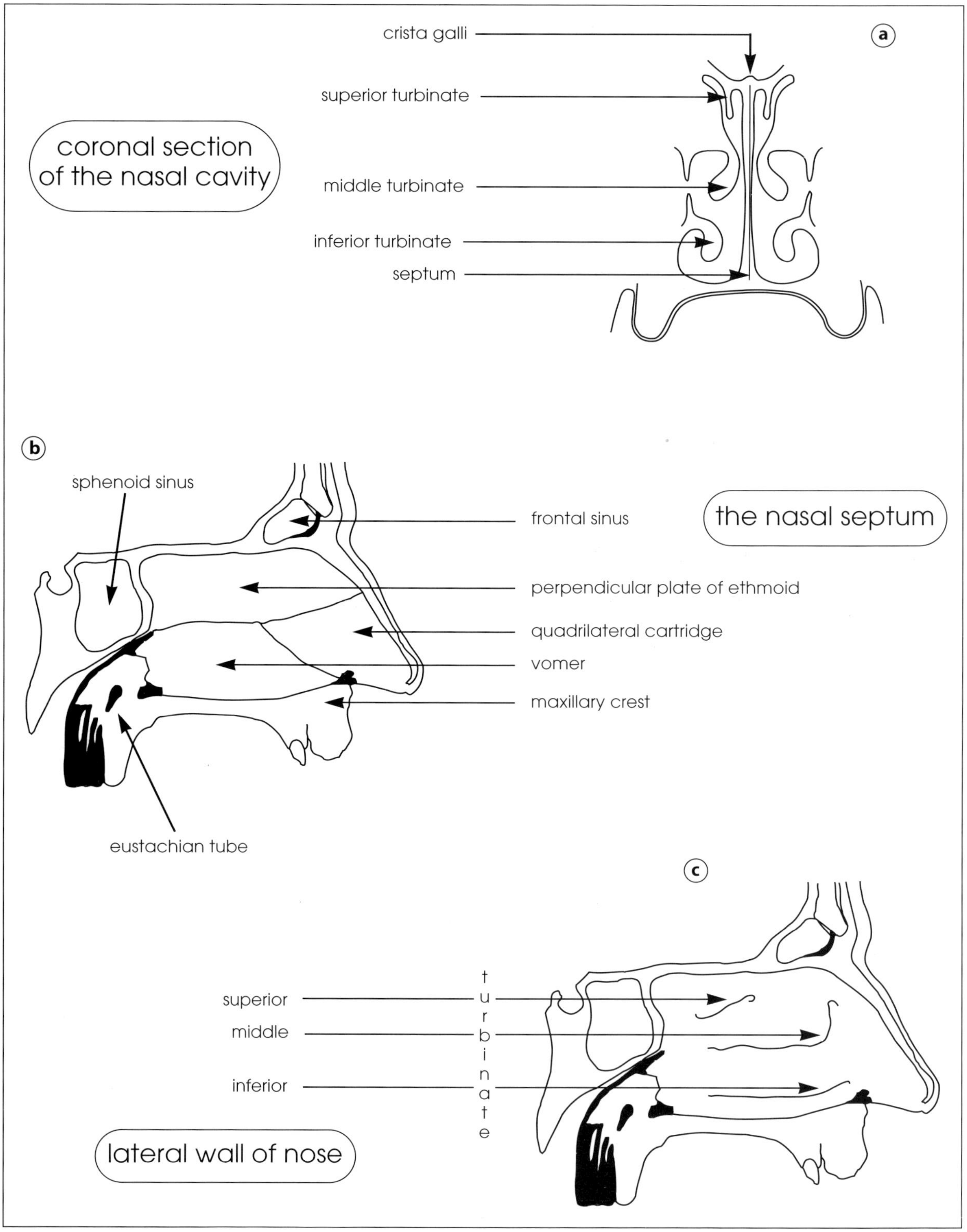

Fig. 2.10a–c. Various anatomical structures of the nasal cavity. **a** coronal section; **b** nasal septum; **c** lateral wall of the nose.

Chapter 3

Endoscopic Anatomy

Anatomy is often a matter of perspective, and this truism is apt as never before when applied to the nose and paranasal sinuses. Macroscopic descriptions found in most standard textbooks often do not correlate to the straight or angled endoscopic view the surgeon has to rely on when traversing these landscapes. He must develop a feel for volume as he looks at area, and must be able to build up a three-dimensional map allowing him to know exactly where he is, where the openings and the danger zones lie, and what structures are closely related. We present the anatomy of this complex region as a series of building blocks, to be put together and further enhanced by CT scanning. Extensive photographic documentation of endoscopic anatomy and pathological findings is to be found in chapter 5.

An endoscopic view of the lateral wall of nose shows:

- ostio-meatal complex
- turbinates
- fontanelles
- agger nasi cells

Ostio-meatal complex

The ostio-meatal complex (anterior to posterior) consists of:

- middle meatus
- uncinate process
- hiatus semilunaris
- infundibulum
- maxillary ostium
- bulla ethmoidales

Middle Meatus

The middle meatus is the space underneath the middle turbinate, in which lies the hiatus semilunaris, along with some of the paranasal sinus openings. It also contains parts of the fontanelles

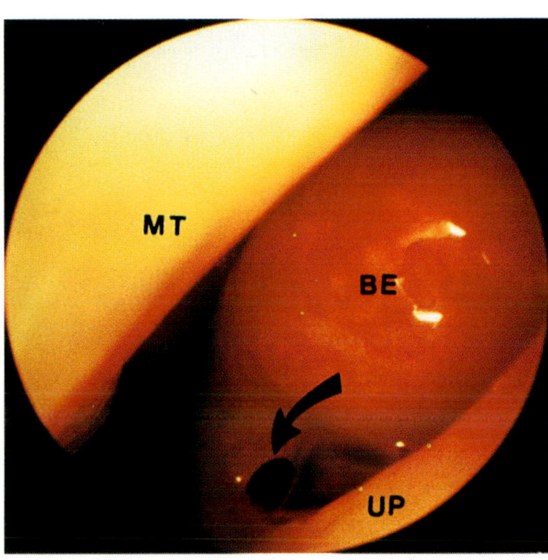

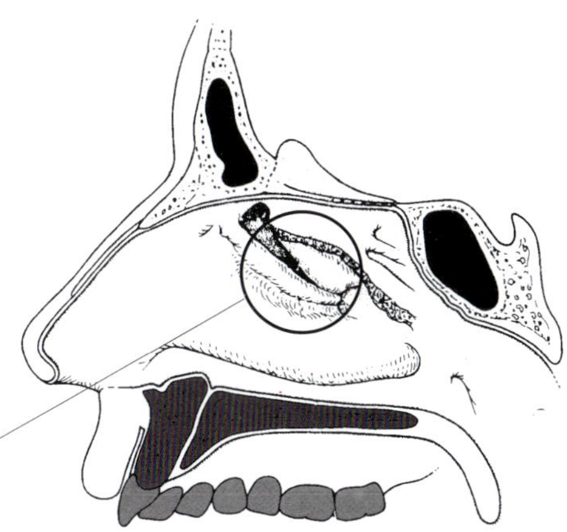

Fig. 3.1. Ostio-meatal complex (OMC). MT, middle turbinate; BE, bulla ethmoides; UP, uncinate process; arrow shows ostium.

and the bulla ethmoidales (middle ethmoidal cells). Anteriorly is the opening of the frontal sinus along with the anterior ethmoidal cells. The posterior portion bears the opening of the maxillary sinus, and may contain an accessory opening in 10%–40% of patients. It is important to note that the fronto-nasal "duct" is more a recess into which the frontal sinus opens, and should be called as such.

As Caldwell emphasised in 1893, "in the diagnosis ... beyond dispute the one absolute proof is the detection of pus escaping from the sinus. This may be secured in all the cells without resorting to exploratory operations ... the relative positions of the sinus openings is of diagnostic value".

Uncinate Process

The uncinate process (from the Latin *uncinatus* meaning hooked) is a part of the ethmoid bone, about 7–22 mm in length. It forms a constant curved landmark on the lateral wall of the nose, and runs from above downwards and from the front backwards. It is of crucial importance to the endoscopic surgeon as he prepares to access the infundibulum, of which it forms the anteromedial boundary. Its sharp, cephalic border forms the margin of the hiatus semilunaris.

Hiatus Semilunaris

The hiatus semilunaris is a 1–2 mm wide two-dimensional cleft, lined medially by the uncinate process and laterally by the ethmoidal infundibulum. In its anterior part lie the openings of the anterior ethmoidal cells and the frontal recess, while the lower and posterior part contains the natural opening of the maxillary sinus. The posterior relation is to the bulla ethmoidales.

Infundibulum

Mosher in 1929 described the infundibulum as the "uncinate groove" (a term which has not stood the test of time), while Myerson (1932) called it "a channel between the ethmoid bulla and the uncinate process". It is a three dimensional space, and is, in fact, a meeting point for drainage of the frontal, anterior/middle ethmoidal and maxillary sinuses. It belongs to the anterior ethmoids. The medial wall is constituted by the entire length of the uncinate process enveloped in its mucosa. The lateral wall is largely provided by the medial orbital wall, while the frontal process of the maxilla and, in some cases, the lacrimal bone makes up the remainder. At times, the infundibulum may end as a blind recess anteriorly, depending upon the position of the uncinate process in its upper part. The posterior boundary of the ethmoidal infundibulum is largely the anterior surface of the bulla ethmoidales, from where the infundibulum opens into the middle meatus, through the hiatus semilunaris.

From the surgical point of view, it is important to remember that the ethmoidal infundibulum may be quite shallow when the uncinate process is in close proximity to the lamina papyracea. Sometimes in the presence of paradoxical turbinate and/or concha bullosa the ethmoidal

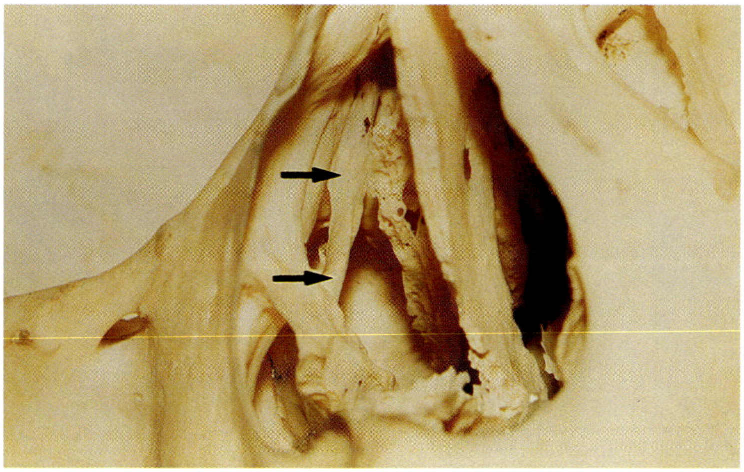

Fig. 3.2. A dissected bony specimen shows the uncinate process of the ethmoid bone (arrows) as a curved projection forming the medial wall of the ethmoidal infundibulum.

infundibulum may be completely obliterated, thus resulting in recurrent sinus infections.

Maxillary Ostium (antro-nasal channel)

This is not merely an opening into the lateral wall of the nose; rather it is a channel for communication between the infundibulum and the maxillary sinus. It is at the junction of the medial maxillary wall and the floor of the orbit, halfway between the anterior and posterior maxillary walls (approximately 2 cm from each). It is normally horizontal or oblique. It has two orifices, nasal and antral, which can differ in size, shape and direction. Looking out of the sinus at the ostium, one might see the uncinate process antero-superiorly, and the bulla ethmoidales postero-inferiorly. Accessory ostia may be seen in 10%–40% of patients.

Bulla Ethmoidales

The bulla ethmoidales is a large middle ethmoidal cell with thin delicate walls, often with its own opening. Sited just posteriorly to the hiatus semilunaris, it also forms the posterior boundary of the maxillary ostium. It opens into the ethmoid infundibulum or into the sinus lateralis. Sometimes there is a distinct furrow along its medial wall. Sometimes the bulla may extend to atttach to the vertical portion of the ground lamella, and/or the roof of the ethmoid. When it fuses with the roof of the ethmoid, the interveing space between it and the ground lamella is termed sinus lateralis. A very large bulla may obliterate this space, in which case the posterior wall of the bulla becomes the anterior surface of the ground lamella (see Fig. 3.3).

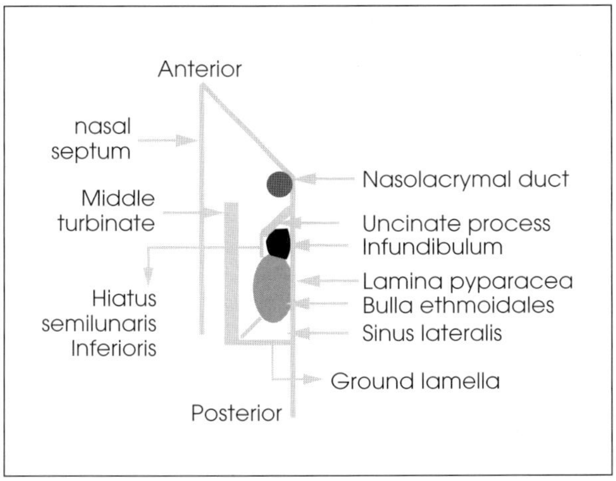

Fig. 3.3. Schematic diagram of the right ethmoidal infundibulum.

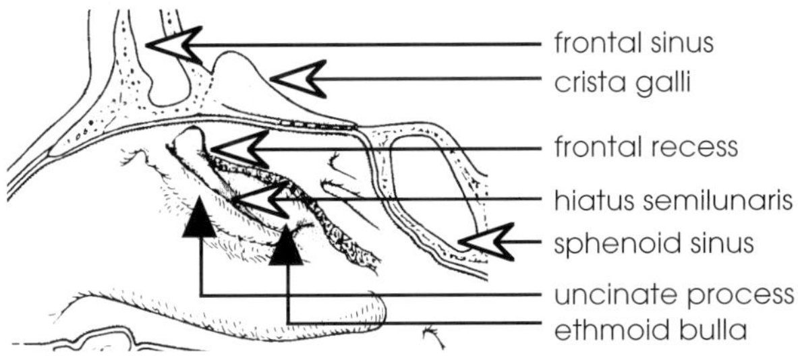

Fig. 3.4. Lateral wall of the nose.

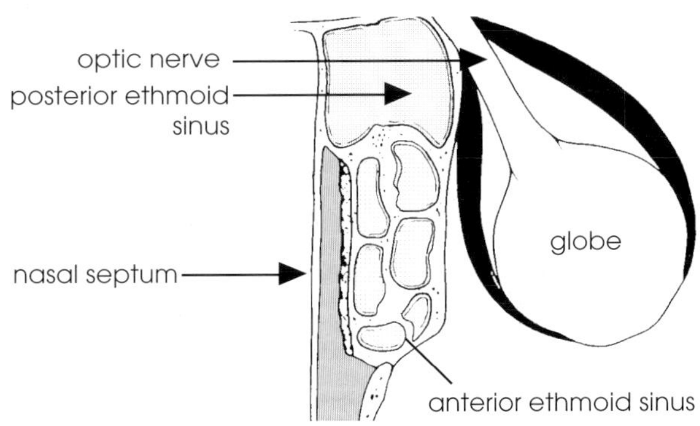

Fig. 3.5. Axial view of the ethmoids and orbit.

Turbinates

The three most important features on the lateral wall of the nose are the inferior, middle and superior turbinates with their meati (a fourth "supreme" turbinate may sometimes be seen). The supreme, superior and middle turbinates are an integral part of the ethmoid labyrinth, whereas the inferior turbinate is a separate bone. All the turbinates conceal their meati, into which open the paranasal sinuses.

A centimetre behind the anterior end of the inferior turbinate is the opening of the naso-lacrimal duct into the inferior meatus, often recognised by a surrounding crescent of mucosal fold. Gentle massage of the lacrimal sac often causes exudation of a tear-drop into the inferior meatus.

The middle turbinate is by far the most crucial landmark for the endoscopist. It has two attachments: an anterior one extending superiorly to articulate with the cribriform plate of the ethmoid bone; and the more important attachment to the lamina papyracea separating anterior and posterior ethmoids via several laminae, the most significant being the ground lamella. The attachment of the turbinate changes direction at its most posterior extent; instead of running in an antero-posterior direction, it curves laterally, the final lateral attachment of the middle turbinate being the lamina papyracea. The posterior termination of the middle turbinate lies along the frontal or coronal plane, and is called the basal or ground lamella. Behind this one finds a few large posterior ethmoidal cells, into which juts the sphenoid rostrum. Quite often, the anterior part of the middle turbinate contains an air space (conchal sinus) which may become infected in keeping with generalised inflammation of the ethmoid labyrinth.

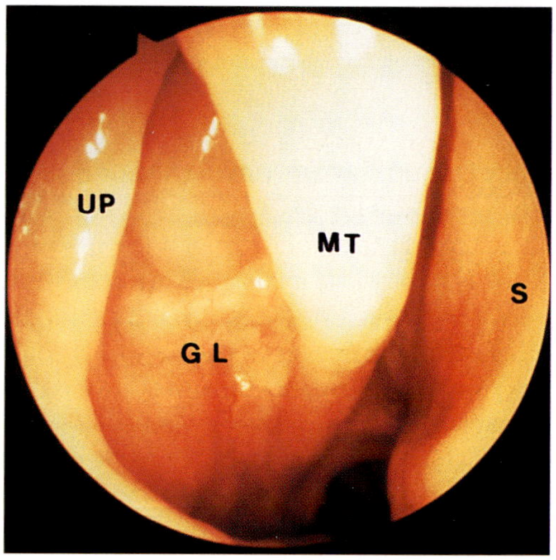

Fig. 3.6. Endoscopic view of the right middle meatus showing the ground lamella. UP, uncinate process; gl, ground lamella; MT, middle turbinate; s, septum.

Important endoscopic measurements from the anterior nasal spine

uncinate process	5.2 cms
anterior wall of sphenoid	6.1 cms
sphenoidal lumen	7.3 cms

Into the middle meatus open the frontal sinus and anterior/middle ethmoidal cells (anteriorly) and the maxillary sinus (posteriorly). The posterior ethmoidal sinuses open into the superior meatus, and the sphenoidal opening is into the spheno-ethmoidal recess situated postero-superiorly to the superior turbinate.

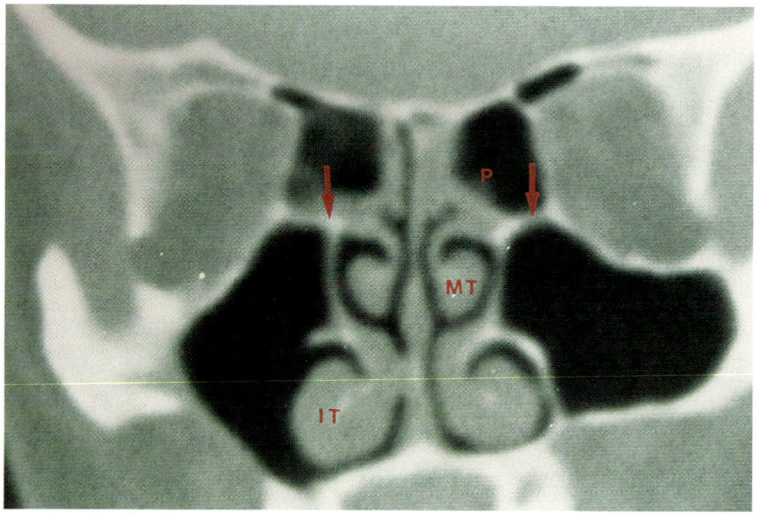

Fig. 3.7. Coronal CT scan view showing the ground lamellae on both sides (arrows). IT, inferior turbinate; MT, middle turbinate; P, posterior ethmoids.

Fontanelles

These are fibrous membranes stretched across the medial wall of the maxillary sinus. They are divided into anterior and posterior mainly by the uncinate process and the conchal process of the inferior concha. The accessory sinus openings are to be found in these membranes. In fact, if one or more openings on the lateral wall of the nose are seen on nasal endoscopy, they are always accessory ostia, and *not* the natural ostium of the maxillary sinus. Also, antro-choanal polyps exit the sinus through an accessory opening.

Agger Nasi Cells

These are the anterior-most ethmoidal cells, and form an intimate relation with the frontal recess. They lie just anterior to the superior attachment of the middle turbinate, and anterior to the frontal recess. These cells can dissect the lacrimal bone or the ascending process of the maxilla. Owing to their proximity to the frontal recess, it is wise to open these cells to access the frontal recess if there is any disease in the frontal sinus. Disease in this region may block the infundibulum and thereby the frontal recess. *A significant number of failed FESS procedures are due to inadequate clearance of this area.* Endoscopically, one often recognises these cells by a characteristic bulge on the lateral wall of the nose near the lateral attachment of the middle turbinate.

Chapter 4

Muco-Ciliary Concepts

Introduction

Unfortunately, the nose, with its convoluted internal architecture, is prone to obstructive disease. Although the muco-ciliary pathways work harmoniously in health, infection can rapidly throw this delicate mechanism into disarray. Most infections of the paranasal sinuses arise from a primary focus in the nose. Logically, therefore, recurrent disease implies an occult focus. The infection may also be secondary to trauma, of dental origin, or may occur via the blood stream. The key area of the lateral wall of the nose, the ostio-meatal complex, is most affected, thus blocking the openings of the maxillary and frontal sinuses. The most important site to be affected is the anterior ethmoidal complex, as the frontal and maxillary sinuses are dependent on them for their drainage. Messerklinger's work spanning over two decades established the genetically predetermined pathways of muco-ciliary clearance within the paranasal sinus system.

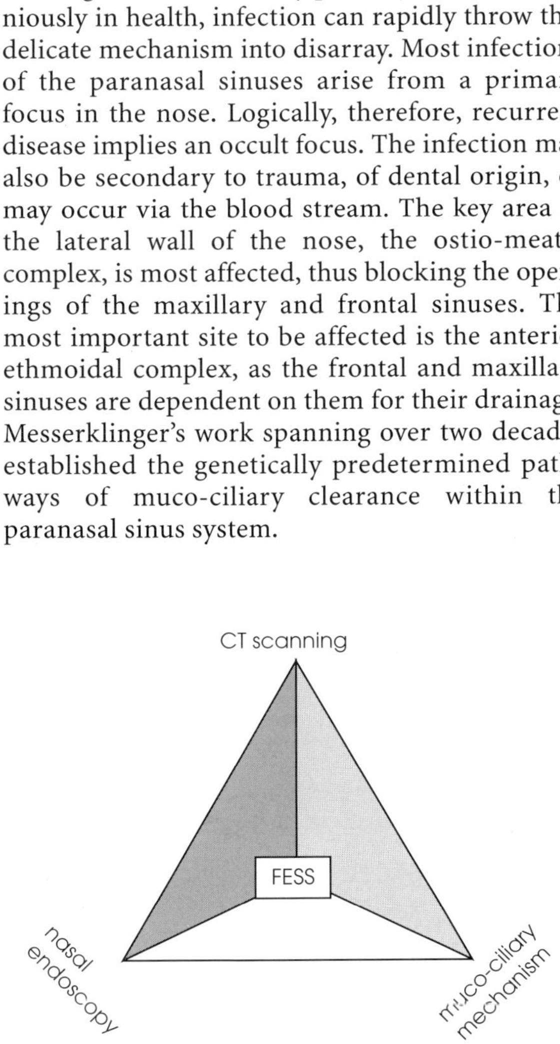

Fig. 4.1. The three factors fundamental to the concept of functional endoscopic sinus surgery (FESS).

Predisposing factors such as high septal deviations and/or spurs, medial/lateral rotation of the uncinate process, middle turbinate abnormalities such as knobbing, paradoxicity and conchosity go hand-in-hand with inflammatory and/or allergic processes to give rise to a vicious cycle of impaired ventilation and drainage of the sinuses. This knowledge, along with advances in imaging techniques and multi-angled endoscopic evaluation, led to the concept of FESS as it is known today.

Muco-Ciliary Clearance in the Nose and Paranasal Sinuses

General

Transport at a cellular level is integral to life itself. The movement of ions across cell membranes, the excretion of wastes from organs and the passage of nutrients all contribute to harmony and balance.

The nose, a point of entry for both air and foreign particles, relies on a very intricate system of muco-ciliary clearance (MCC) for its well-being. Two phylogenetically ancient systems blend to provide for this phenomenon: the production of mucus, which is present even in primitive algae, and the principle of ciliary locomotion, seen in certain protozoa.

Historical Aspects

The story of ciliary function investigation began in 1677, when Johannes Ham, a student of Leiden, first observed ciliary movements under a microscope. This was later confirmed by Van Leeuwenhoeck in seminal fluid. It took over a century, however, for Purkinje and Valentin to correlate this movement with the continuous cleansing action of ciliated epithelium. Martius in 1884 demonstrated ciliary motion stroboscopically and computed the beat frequency. The early half of the twentieth century saw a number of studies being carried out by pioneers in the field: Antweiler, Messerklinger, Negus and Proetz among others. Coloured particles and dyes (Hilding 1931, Tremble 1948, Evert 1965), and radioactive substances (Proctor and Wagner 1965) have all been introduced into the nasal cavity as markers for ciliary speed. Hilding successfully calculated the energy of ciliary epithelium in hen trachea. His contribution to animal and clinical studies has been monumental.

The introduction of the electron microscope provided new impetus to the investigation of ciliary phenomena. Jakus and Hall (1946) were responsible for demonstrating the longitudinal fibrillae within a single cilium. Further details were provided by Fawcett and Porter in 1954.

The relationship between nasal dysfunction and impaired mucociliary clearance has been described by Quinlan (1969), Puchelle (1981) and Sakakura (1985). Recurrent assaults on the lining of the nose lead to oedema and obstruction of the sinus openings, thus lowering intra-luminal oxygen tension, and encouraging the growth of anaerobic organisms. Exotoxin release then acts to further impede ciliary movement, this setting up a vicious cycle within the sinus. The aim of FESS is to re-establish normal drainage and ventilation channels by creating an easily accessible cavity. This encourages the muco-ciliary mechanism to regain harmonious function. Hence this form of surgery is known as "functional".

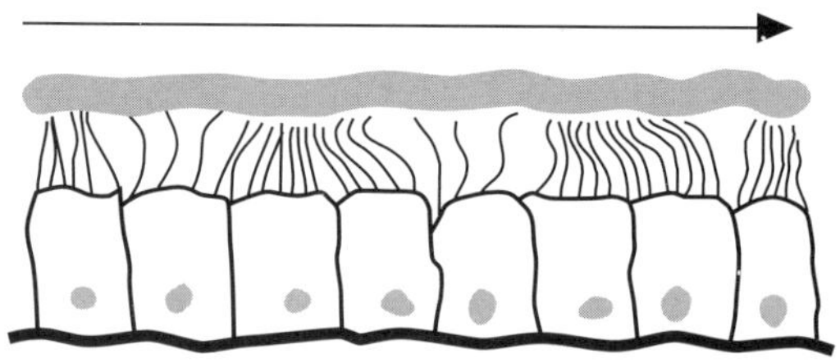

Fig. 4.2. Ciliated cells propelling the overlying mucous blanket.

Cilia

Phylogenetically, cilia are found in all animal species, including the nematodes. In the vertebrates, they are important in respiratory function, as well as the movement of spermatozoa and fallopian tube transport. Individual cilia move in a whip-like fashion, thus providing a lateral force to the overlying substance. When viewed in conjunction with their neighbours, this movement converts to one of continuous, wave-like motion.

The extent of ciliary epithelium in the airways is not constant, and is defined by environmental and genetic factors. A single ciliated epithelium cell carries between 50 and 300 cilia on its surface. The length of cilia in vertebrates is 3–8 μm, the diameter being 0.1–0.3 μm, with a narrowing towards the tip. The shaft contains nine peripheral tubules and two central single tubules. Energy for ciliary movement is ATP driven, and the average beat frequency is 5–20 Hz at 37 °C. Cilia are upright and rigid during the beat, and curved during the recovery phase. Mucus consists of 98% water, the rest being composed of albumin, globulin, glycoproteins and salts.

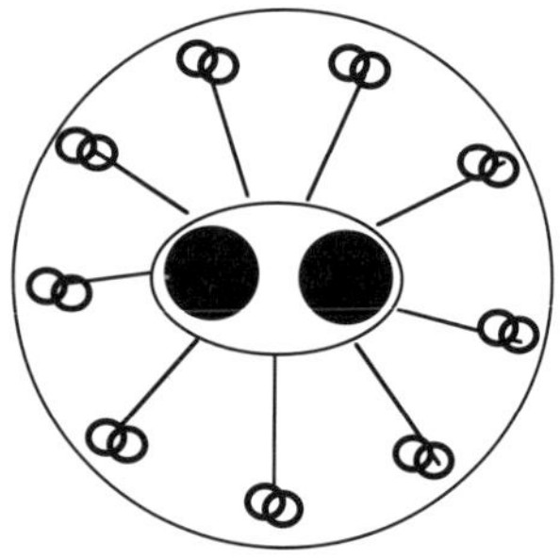

Fig. 4.3. Each cilium consists of the nine double microtubules surrounding two central microtubules.

All in all, MCC depends on an intact, metachronously beating ciliary epithelium, and an accurately balanced mucus layer of optimal viscoelasticity. Its physical properties thus seem to be more important than its biochemical properties.

MCC of the nose and paranasal sinuses is a highly directional activity. It is independent of body position, and is mainly directed backwards into the nasopharynx, except for a very small amount of forward movement at the very anterior end of the nasal septum. Within the maxillary antrum, MCC follows a star shaped pattern, originating from the floor and ascending into a spiral to end at the natural os. *Even in the presence of an inferior antrostomy opening, mucus still moves towards the natural ostium.* Messerklinger (1978) has shown that ethmoid and sphenoid clearance is directly through their respective ostia, whereas the frontal sinus shows circular traffic, only a small part of which exits via the fronto-nasal recess into the nose. Hence an increase in viscosity leads to almost immediate stasis in the frontal sinus.

Factors Influencing Mucociliary Clearance

Various parameters such as temperature and humidity affect MCC; the earliest studies in this regard are those of Engelmann (1877) and Dixon et al. (1905). Engelmann found increasing activity up to 45 °C, followed by a rapid decline. The optimal beat frequency lay between 30 and 40 °C. However, Proetz (1941) has stated that "the only natural enemy to cilia is excessive drying". Optimal relative humidity is around 90%. Neutral pH values are associated with good ciliary activity. Ciliary stasis rapidly sets in outside pH values of 7–10. Similarly, correct osmolarity and oxygen tension are essential for ciliary movements. Oxygenation is maintained both by diffusion from circulating blood, as well as from the surface itself. Interestingly, a mechanical load such as blood clot or mucus plug accelerates MCC.

A wide range of drugs and chemicals affect the mucociliary clearance. Smoking is known to have a deleterious effect, while sympathomimetic decongestant nose drops have been shown to retard ciliary function.

Pathophysiology of Sinusitis

Common pathogens found in the maxillary sinus are pneumococci, *Haemophilus influenzae* and other anaerobic organisms. Quite often there is underlying damage to the epithelium as a result of viral infection, this leads to exudate formation

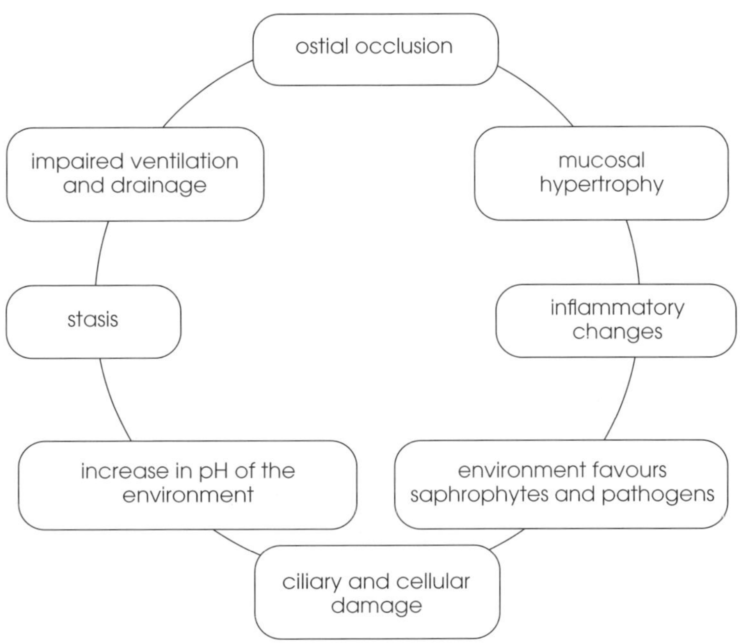

Fig. 4.4. The vicious circle of ostium closure and recurrent sinus infections.

and retention of secretions. The damaged mucosa and secretions form a favourable substrate for bacterial colonisation. As the oxygen tension in antral air is mainly dependent on the patency of the ostium and absorption by mucosa (Aust and Drettner 1974b), its closure leads to low partial pressure of oxygen PO_2 and high CO_2 levels and a rise in antral pH, thus affecting ciliary mechanisms (Aust and Drettner 1974b, Frederick and Braude 1974). Thus impaired ventilation due to ostial occlusion facilitates bacterial and cellular influence on the gas environment in such a way that heavy growth of anaerobes is initiated and maintained and a serous sinusitis is changed into a more purulent variety.

Clinical Assessment of Muco-Ciliary Function

Since Martius in 1884 used stroboscopy to study muco-ciliary function of the nose, a number of methods, both direct and indirect, have been used by other workers. Hilding in 1931 and Tremble in 1948 used dyes and coloured particles and timed their passage through the nasal cavities. Harper et al. (1962) resorted to radioactive technetium monitored by serial scanning. These methods, though sophisticated, often proved cumbersome for routine clinical use. Anderson (1974) introduced the use of saccharin placed on the nasal mucosa; by recording the time interval before the patient felt a sweet taste, an estimate of muco-ciliary clearance could be made. His technique has since been used by others in the field. Ginzel and Ellum (1980) and Hady (1983) modified this by using dyed saccharin in order to record dual observations.

In addition, researchers like King (1935), Flottes (1960), used an oil- or water-based contrast medium specifically to assess the muco-ciliary function of the maxillary sinus.

In our review of world literature we could not find a single instance of muco-ciliary studies being performed in patients of FESS using the saccharin test. Hence we undertook a prospective, controlled project involving 40 patients, and measured their "saccharin times" before and after surgery. Initially the injection technique was perfected on cadavers, and it was found that the solution could be easily instilled into the antrum via the membranous anterior fontanelle as an outpatient procedure.

Ten normal volunteers were initially chosen as a control study. Care was taken to eliminate those with nasal or sinus complaints. None of the control group suffered from any allergies.

Forty random patients with symptoms and signs of chronic sinusitis formed the test group. These patients had all suffered from chronic recurrent sinus disease (nasal obstruction, post-nasal drip, headache) for over a year. Diagnosis was confirmed by office endoscopy and coronal CT scans in all cases. All findings were recorded on a computer compatible proforma.

A standardised sterile solution of 25% saccharin was employed. All patients were tested in identical environmental conditions. They were instructed not to sniff forcibly, and to swallow at frequent intervals. The nature of the test solution was not revealed to the patient, they were merely asked to indicate any unusual taste they noticed.

Pre-operative assessment was carried out 24 hours prior to surgery, while post-operative recordings were done at 1, 3 and 6 months.

The procedure was carried out in the sitting position. Pre-operative recordings required limited local anaesthesia (4% xylocaine) in the region of the anterior fontanelle. A tuberculin syringe with a No.17 needle was employed. The needle was given a slight angulation in order to access the lateral nasal wall. 0.5 ml of saccharin were instilled into the antrum via the anterior fontanelle. In post-operative cases, the middle meatal antrostomy itself was used as an entry the antrum. No complications were noted during the study.

The control group exhibited a mean clearance time of 18 minutes (ranging from 12 to 25 minutes). In the patient group, pre-operative readings averaged at 46 minutes (36–62 minutes). Post-operative results at 1, 3 and 6 months are as shown in Fig 4.9. At 6 months, the reading was 23 minutes (range 18–42 minutes). This group consisted of 32 patients, eight having been lost to follow-up.

Our results showed a marked reduction in clearance times post-operatively, which was statistically significant which corresponded well with improvement in symptom profiles. We therefore propose this technique as a simple, safe and reli-

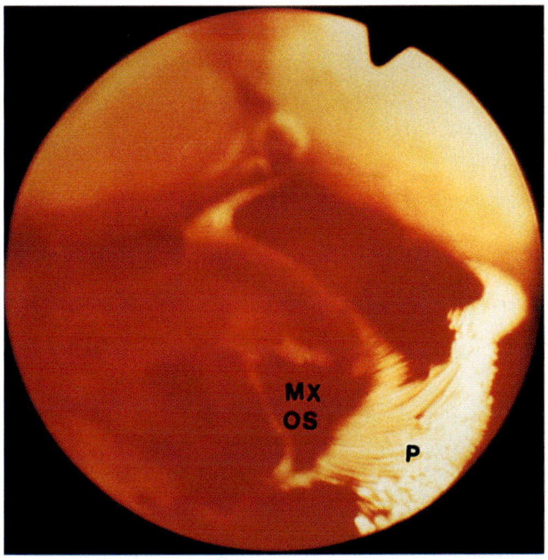

Fig. 4.5. Intra-operative picture of the left maxillary sinus showing pus (P) tracking towards the natural ostium (MX OS).

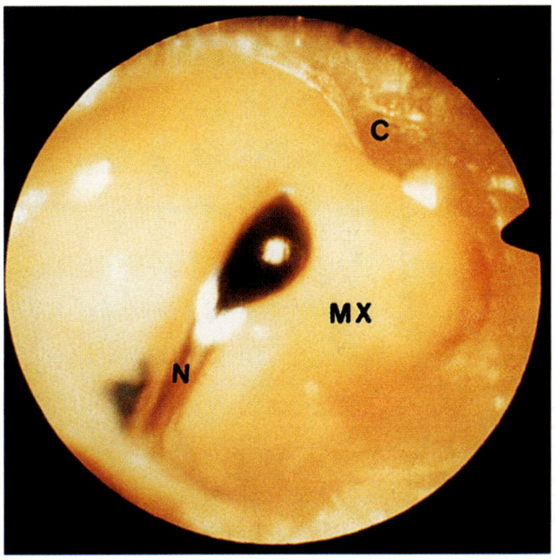

Fig. 4.7. Left maxillary antrum in the cadaver, showing instillation of a drop of dye through the anterior fontanelle. This was the first phase of our muco-ciliary clearance study (before actual patient trials). C, cannula; MX, maxillary sinus; N, needle.

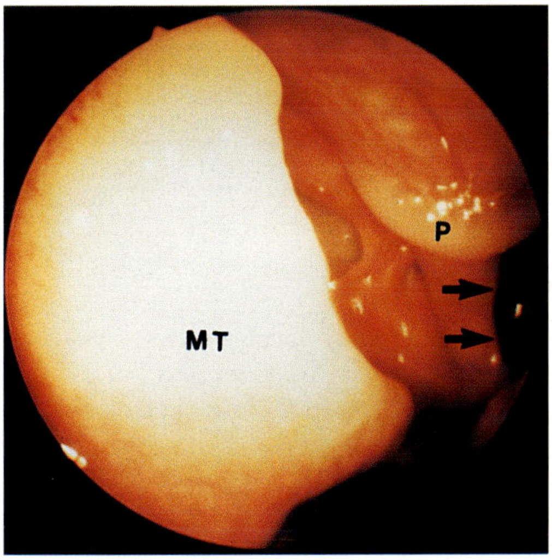

Fig. 4.6. Left-operated nasal cavity showing muco-pus (P) recirculating between a middle meatal antrostomy (arrows) and an accessory ostium. Care should be taken to remove the bridge of bone between the natural and the accessory ostium. MT, middle turbinate.

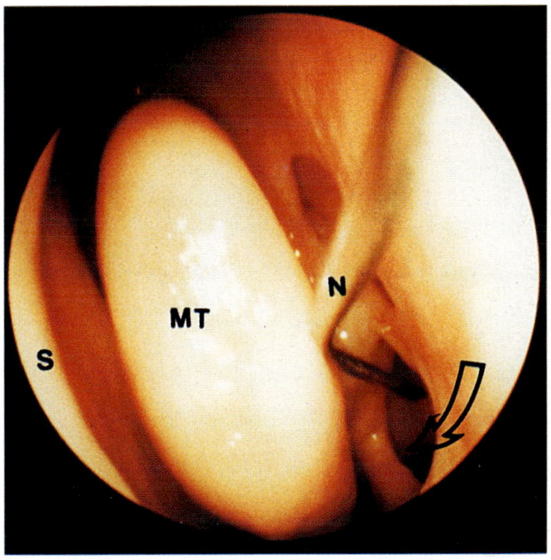

Fig. 4.8. Muco-ciliary study: patient trial. Post-operative view of the left middle meatal antrostomy (arrow) showing needle (N) being inserted into the maxillary sinus to instill standardised saccharin solution. S, septum; MT, middle turbinate.

able method of assessing results of surgery. Furthermore, the method could act as a pre-operative indicator of ciliary motility disorders, as prognosis in these cases is distinctly poor. The study shows a distinct delay in MCC time in chronic sinusitis cases as compared with controls. There is a definite improvement in these patients after FESS, almost reaching normal levels.

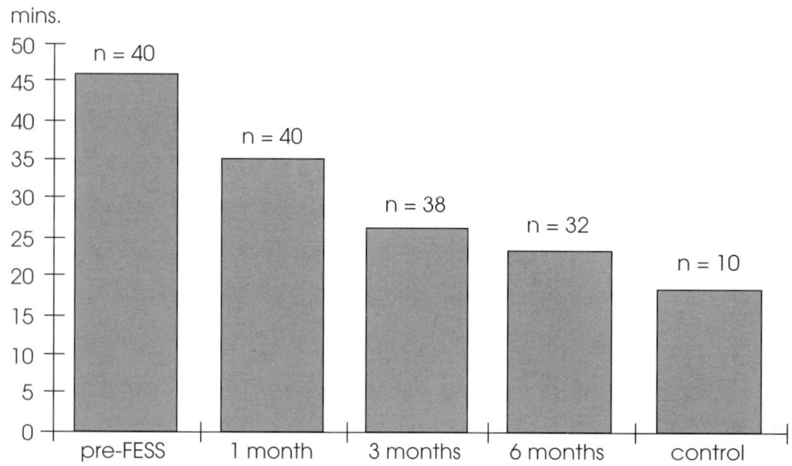

Fig. 4.9. Average clearance timings in control and pre- and post-FESS groups.

Saccharin proves to have several advantages: it is easily obtained, non-toxic and well recognised by the patient. The test itself is easily performed, reproducible, and requires no elaborate equipment.

In conclusion, as our understanding of the basic principles that control nasal and sinus function deepens, we come closer to achieving what is best for the patient. In addition to measuring MCC times, one could also focus on other parameters such as the ciliary beat frequency and electron microscope studies. Furthermore, appropriate tests are available for mucus secretion disorders such as cystic fibrosis and Young's syndrome which can also cause abnormal MCC.

Chapter 5

Office Nasal Endoscopy

Evolution of Nasal Endoscopy

It was Hirschmann in 1903 who used a modified cystoscope for diagnostic purposes in nose and sinus. Later on Reichert (1902), Valentin (1903) and Sargnon (1908) also used the endoscopic technique for the removal of foreign bodies and opening of cysts. The endoscope was inserted either through a dental socket or the anterior antral wall. Spielberg in 1922 described an alternative route to the sinuses through the inferior meatus. The scope of endoscopy in the nose and sinuses was enchanced with the introduction of better and brighter rod-based telescopes by Hopkins in the 1960s. With this development the diagnosis of nose and sinus pathologies became more precise and photographic archiving was made possible (Timm 1965, Messerklinger 1972, Draf 1973). The endoscopic evaluation of nose and sinus disease is well established across the globe. In addition to the diagnostic capabilities and surgery of the nose and sinuses the endoscope is used to perform endoscopic biopsies, to examine and irrigate the maxillary sinuses and carry out toilet post-operatively (Reynolds and Brandow 1975, Buiter 1976). Now the endoscope is used routinely for localisation of posterior epistaxis and subsequent cauterisation, excision of mucoceles, tumors of paranasal sinuses, orbital decompression and dacryocystorhinostomy(DCR).

We strongly recommend the endoscope's use even in those departments which do not routinely perform endoscopic sinus surgery, as it often reveals hidden pathologies. It is, in a sense, to sinus pathology what the microscope is to ear disease. If, for example, examination were to reveal ethmoid disease in a patient presenting with chronic sinusitis, it would be fruitless to carry out any form of surgery on the maxillary and/or frontal sinuses *per se*.

Instruments and Technique

Our routine technique uses a 4 mm, 30° Hopkins rod rigid endoscope. Although the 0° endoscope has been widely advocated for out-patient endoscopy, we would recommend its use only to the early stages of practice, when the surgeon is building up his depth perception and anatomical orientation. The very lie of the lateral wall requires an angled approach, especially when assessing the ostio-meatal complex.

The patient is examined in the sitting position. A side arm allows observation by trainee. The nose may be anaesthetised by application of cotton wicks soaked in 4% cocaine-adrenaline (care must be taken to squeeze out excess solution thoroughly to prevent absorption through the pharyngeal mucosa).

The nose is examined by three passes. The first pass traverses the floor of the nose, first examining the inferior meatus for previous antrostomy, etc. The endoscope is then guided further backwards towards the posterior choanae, examining the eustachian tube orifice, fossa of Rosenmullar and nasopharynx. Any pathology of and around the eustachian tube opening such as post-nasal discharge and mucosal oedema is noted. It is important to note that the anterior group of

First pass

inferior meatus	? naso-lacrimal duct opening previous antrostomy
floor of the nose post-nasal space eustachian tube orifice mucus channels	

Second pass

lateral wall of nose	agger nasi polyps accesory ostia uncinate process
middle meatus	hiatus semilunaris bulla ethmoides natural os ground lamella
middle turbinate deformity	

Third pass

superior turbinate/meatus
spheno-ethmoidal recess
sphenoidal ostium

sinuses drain in front of the opening, while secretions from the posterior group pass behind it. Rotating the endoscope allows examination of the contralateral tubal opening as well as the roof of the nasopharynx. The second and most important pass examines the ostio-meatal complex. It includes a detailed assessment of the agger nasi cells , middle turbinate and high deviation of septum. The endoscope then focuses on the uncinate process, and is gently guided into the middle meatus and rotated to bring the lateral wall of the nose into view. Here most frequently seen are the hiatus semilunaris, bulla ethmoidales, rarely a nasal view of the maxillary ostium, ground lamella and various pathologies. On the lateral wall accessory ostia (punched out appearance) may be found in the fontanelles. The third pass is made between the septum and the posterior part of the middle turbinate. The endoscope is directed superiorly to examine the superior turbinate and meatus, the spheno-ethmoidal recess and the sphenoidal ostium. Findings are recorded on a computer-compatible proforma. In paediatric noses where a 4 mm endoscope may be too large and hence traumatic, we employ a 2.7 mm endoscope.

Common Findings

Agger Nasi Cells. These most anterior ethmoid cells are usually pneumatised from the frontal recess and lacrimal bone. They may obstruct the frontal recess and if infected may spread infection into the frontal sinus. They are located immediately anterior to the middle turbinate and drain into the infundibulum.

Accessory Ostium. This is present in the fontanelle (anterior or posterior) in about 10%–40% of patients. Clinically they appear as "punched-out holes". They should not be mistaken for natural ostium. The bridge of tissue between accessory os and natural os should be removed to prevent recirculation of mucus into maxillary sinus.

Uncinate Process. This varies in length from 4 to 22 mm. It is a constant curved landmark on the lateral wall forming the antero-inferior boundary of the ethmoidal infundibulum. If it is enlarged, or rotated, it blocks the infundibulum.

Paradoxical Middle Turbinate. The normal curvature of the middle turbinate is convex medially; however, if the contour is reversed, it bulges laterally and hence obstructs the middle meatus. This predisposes to narrowing of the ostio-meatal complex (OMC) leading to infection.

Concha Bullosa. This is a pneumatisation of the middle turbinate with its own opening (Zuckerkandl, 1893). Quite often it acts as a reservoir of infection for recurrent ethmoiditis and blocks the middle meatus. It may also contribute to headaches.

Polyps and Discharge. Small polyps not seen on anterior rhinoscopy can be easily seen with an endoscope and the site and flow of muco-pus in the nasal cavity is diagnostic, as observed by Caldwell in 1893.

Bulla Ethmoidales. A large ethmoidal cell in the middle meatus may obstruct the OMC thus predisposing to recurrent infection and headaches.

Septal Spur. A high deviation of septum may also contribute to narrowing of the OMC.

Fig. 5.1. Fibre optic light source (250 watts) with 4 mm 0° and 30° endoscopes. This wattage allows for good photographic documentation.

Fig. 5.2. The endoscope is supported by the left thumb as it enters either nostril; the remaining four fingers of the left hand fix the head.

Fig. 5.3. First pass, showing the left nare (N) and the septum (S).

Fig. 5.4. Further inwards is the inferior turbinate (IT), the middle turbinate (MT) is seen in the distance. This is roughly the view one would obtain by anterior rhinoscopy. S, septum.

5.1

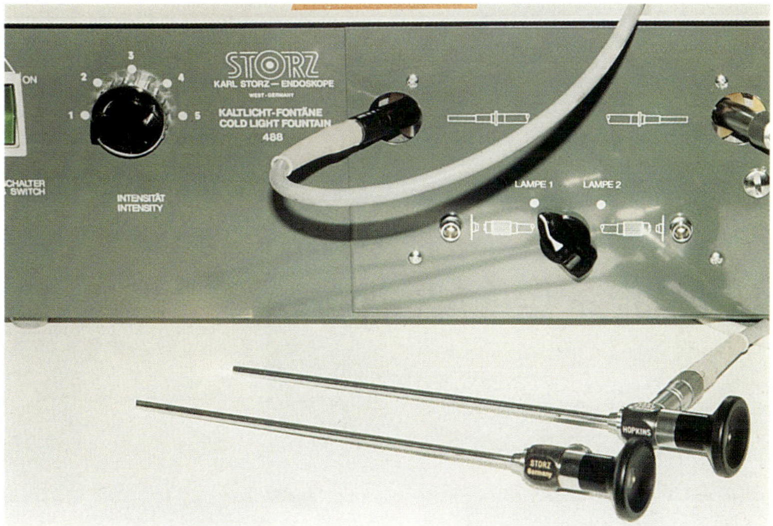

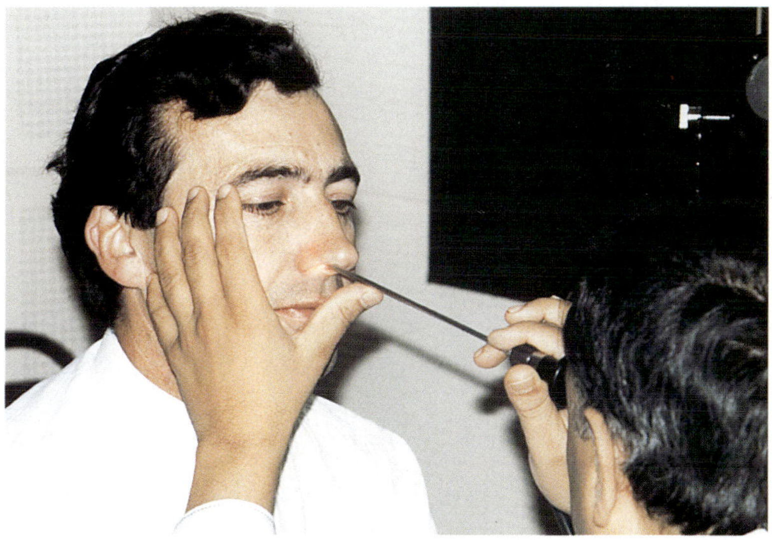

5.2

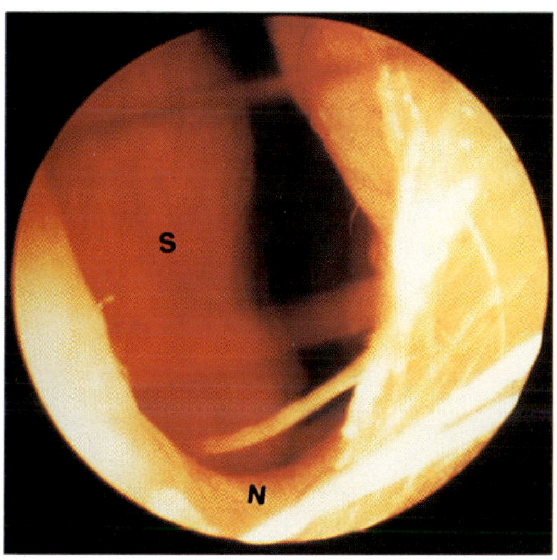

5.3

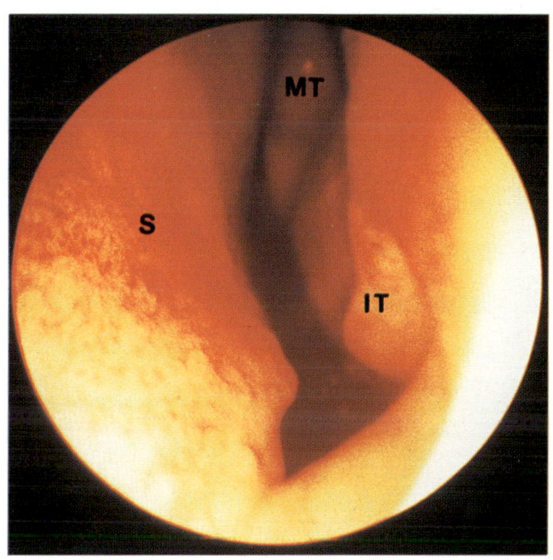

5.4

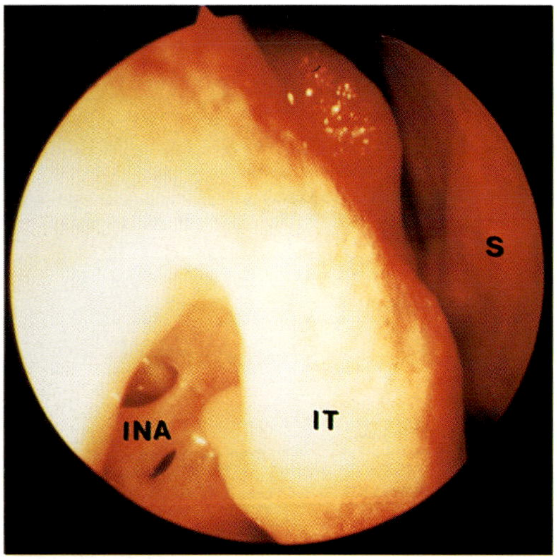

5.5

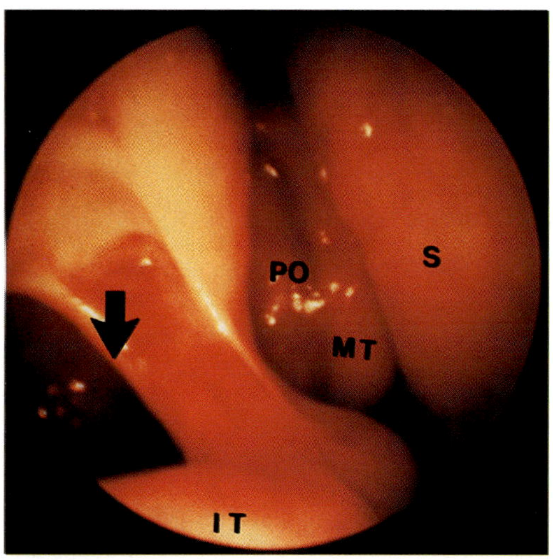

5.6

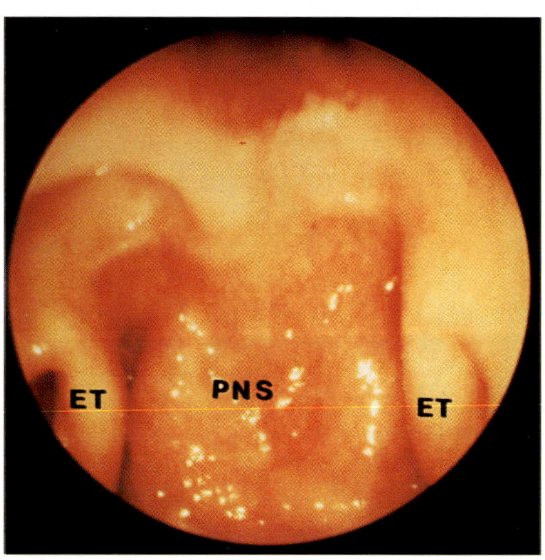

5.7

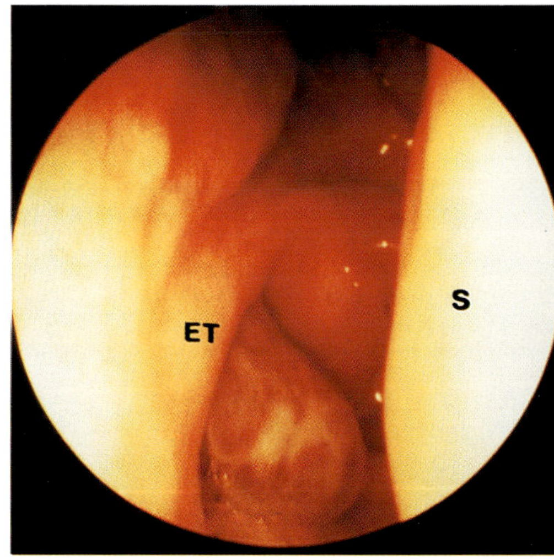

5.8

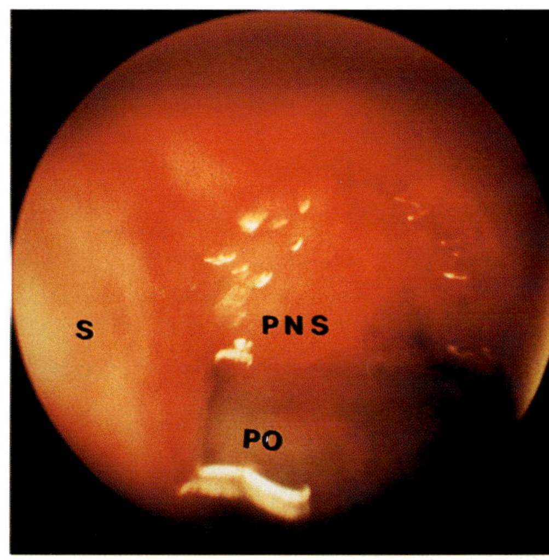

5.9

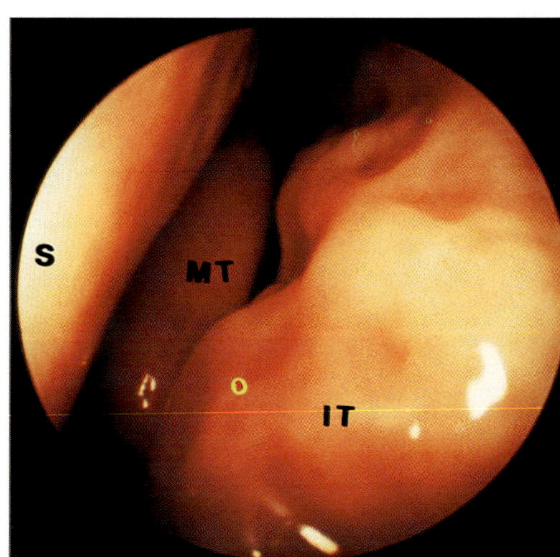

5.10

Fig. 5.5. The endoscope is guided into the inferior meatus to look for signs of previous surgery (INA, inferior nasal antrostomy). In this case, the antrostomy was quite stenosed, and the patient symptomatic.

Fig. 5.6. In this case, although the inferior-nasal antrostomy (arrow) was patent, there were polypoid (PO) changes not only inside the sinus but in the middle meatus. Once again, the patient was symptomatic. IT, inferior turbinate; MT, middle turbinate; S, septum.

Fig. 5.7. The first pass ends at the nasopharynx (PNS, post-nasal space). The eustachian tube openings (ET) are inspected. A 0° endoscope accesses the ipsilateral opening only, wheras a 30° endoscope may be rotated to visualise the contralateral side.

Fig. 5.8. Here the right first pass shows a small amount of viscous mucus emerging through the eustachian tube (ET) orifice. S, septum.

Fig. 5.9. An interesting finding. Examination of the left nasopharynx reveals an antro-choanal polyp (PO) protruding "around the bend" from the opposite side. S, septum, PNS, post-nasal space.

Fig. 5.10. The second pass begins with examination of the lateral wall of the nose. The endoscope is guided between the inferior turbinate (IT) and the septum (S) towards the middle turbinate (MT). Note the polypoidal inferior turbinate in this case.

Fig. 5.11. Very often (10%–40% of cases), an accessory ostium (AO) is seen in the anterior fontanelle. In fact, an ostium that is readily seen on the lateral wall (LW) is always an accessory one. MT, middle turbinate.

Fig. 5.12. The accessory ostium (AOS) in this case shows pus (P) tracking down from the maxillary sinus. Also note a large polyp (PO) on the floor of the nose. MT, middle turbinate; S, septum.

Fig. 5.13. The arrow demonstrates an early antro-choanal polyp protruding through the left accessory ostium. S, septum; MT, middle turbinate; AOS, accessory ostium.

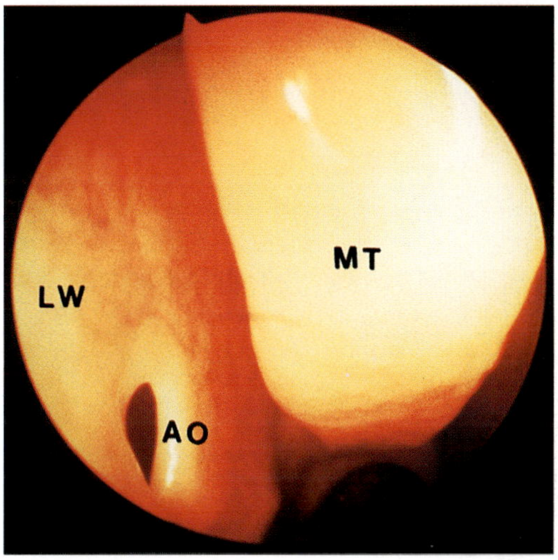

5.11

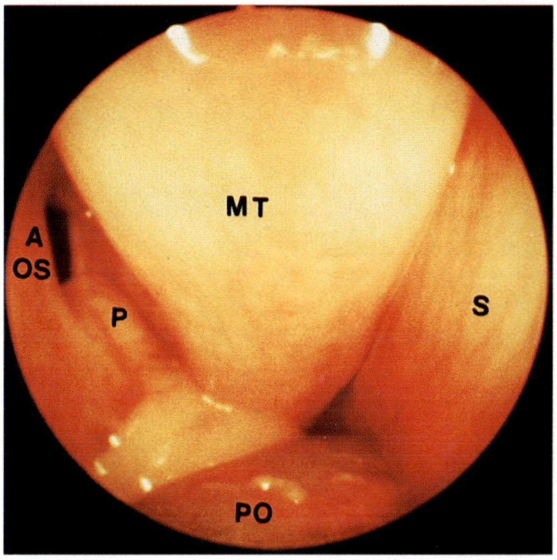

5.12

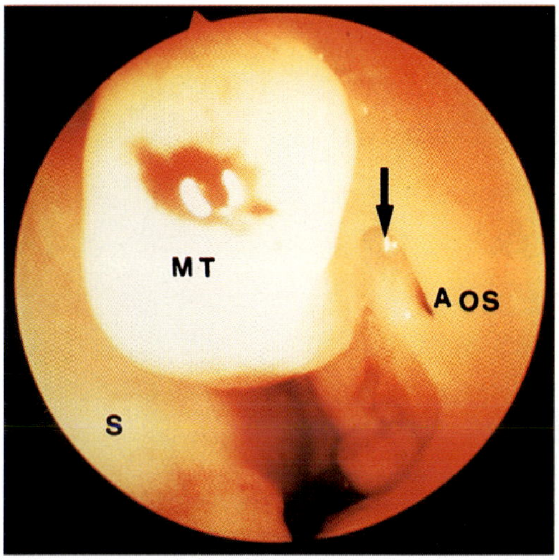

5.13

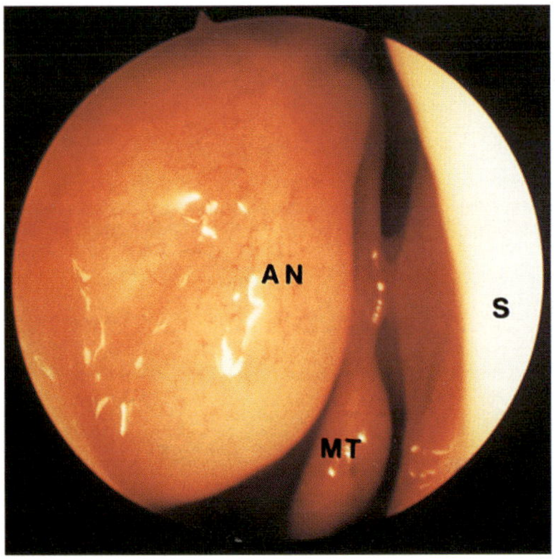

5.14

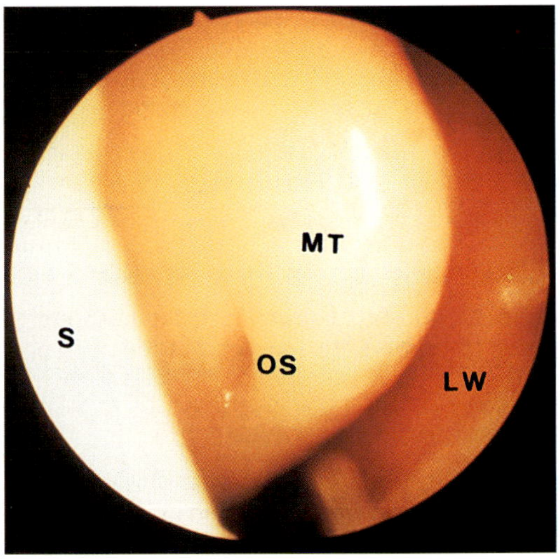

5.17

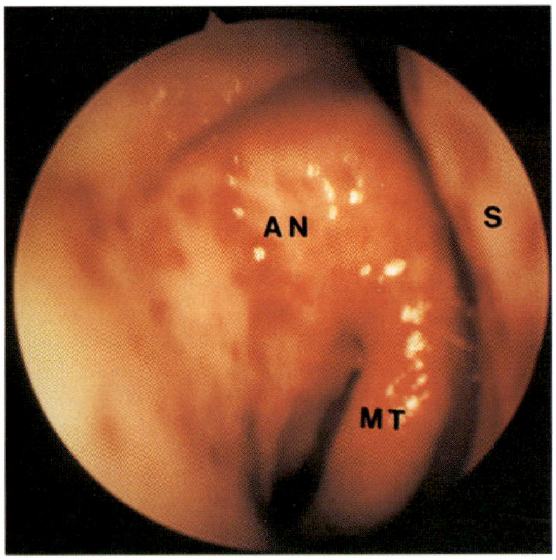

5.15

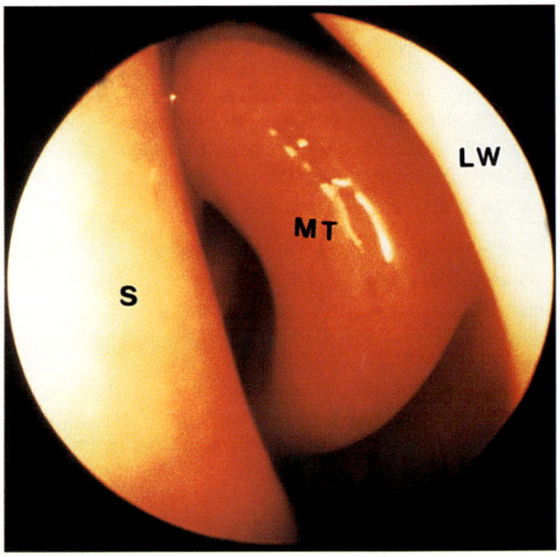

5.18

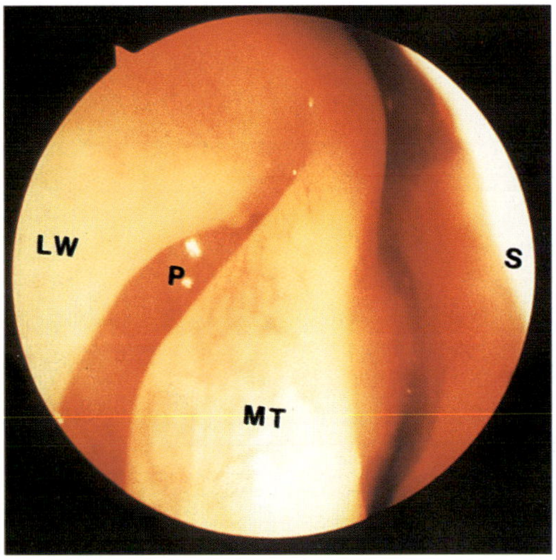

5.16

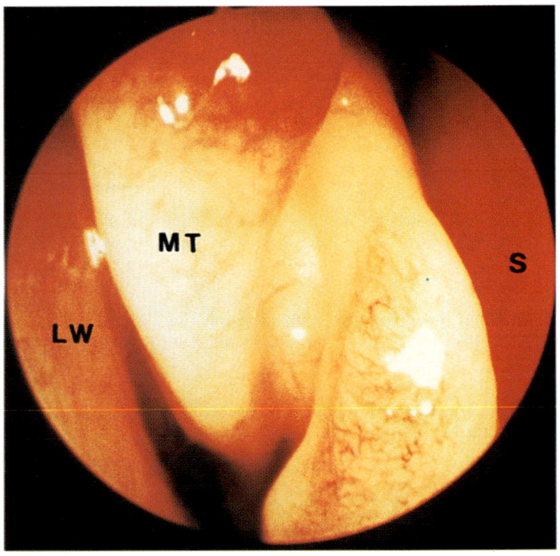

5.19

Fig. 5.14. Quite often, a bulge on the lateral wall at the attachment of the middle turbinate signifies agger nasi cells (AN). These are the most anterior ethmoid cells, often blocking the frontal recess. MT, middle turbinate, S, septum.

Fig. 5.15. Another view of the agger nasi cells (AN), chronic inflammation often causes a mottled appearance. MT, middle turbinate; S, septum.

Fig. 5.16. Endoscopic examination of the middle meatus often demonstrates minimal changes that could be causing the patient's symptoms. Here the meatus contains multiple, small polyps (P), not visible by anterior rhinoscopy. LW, lateral wall; MT, middle turbinate; S, septum.

Fig. 5.17. Deformities of the middle turbinate (MT) include pneumatisation (concha bullosa); sometimes an ostium (OS) is seen. If infected, discharge may be noticed too. S, septum; LW, lateral wall.

Fig. 5.18. The normal middle turbinate is convex medially, thus bowing away from the lateral wall (LW). Sometimes, however, this curvature is reversed, leading to a "paradoxicity" of the middle turbinate (MT). This impinges on the ostio-meatal complex. S, septum.

Fig. 5.19. At times the turbinate (MT) is bifid, and may contain air cells in one or both limbs. LW, lateral wall; S, septum.

Fig. 5.20. The middle turbinate may be bilobed (MT), giving rise to a false appearance of polyps at its anterior end, especially on anterior rhinoscopy. S, septum; LW, lateral wall.

Fig. 5.21. Endoscopy does not merely look at the lateral wall (LW); the convolutions of the septum must also be studied. Here the septum (S) shows a smooth deviation encroaching on a paradoxical middle turbinate (MT). This leads to crowding of the ostio-meatal complex.

Fig. 5.22. A sharp septal spur(s) juts out towards the middle meatus, thus causing stenosis of the ostio-meatal complex. Furthermore, apposition of mucosal surfaces causes muco-ciliary stasis often leading to oedema and polyps. MT, middle turbinate; LW, lateral wall.

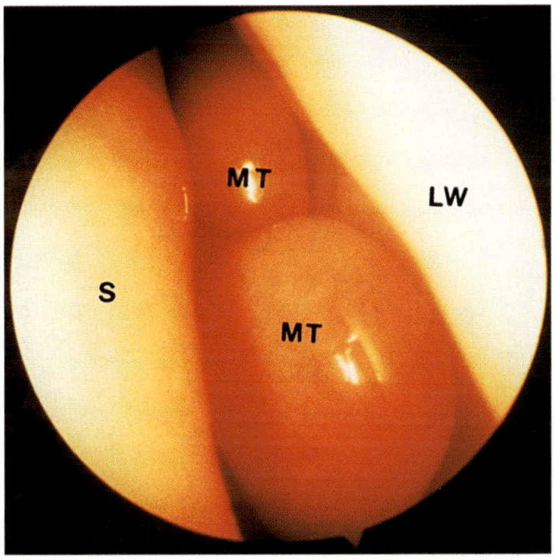

5.20

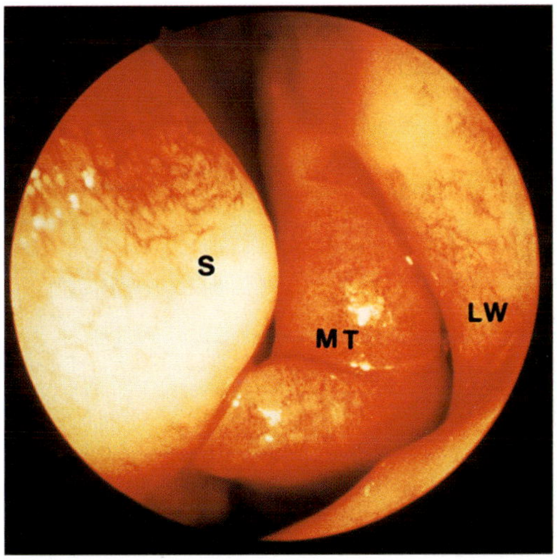

5.21

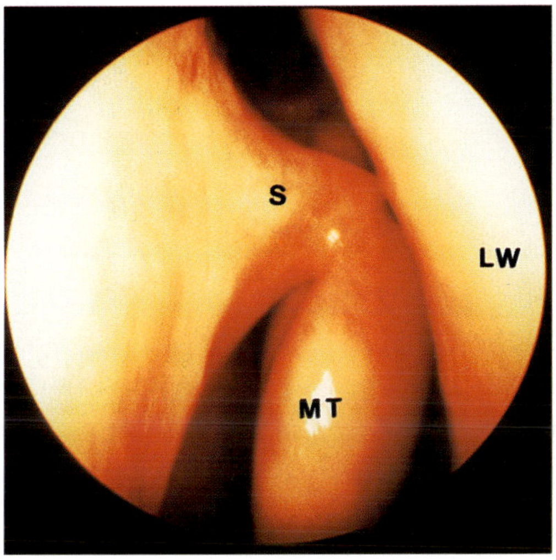

5.22

5.23

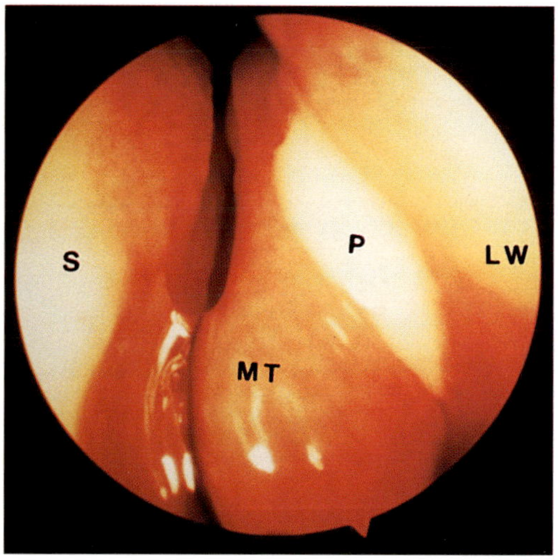

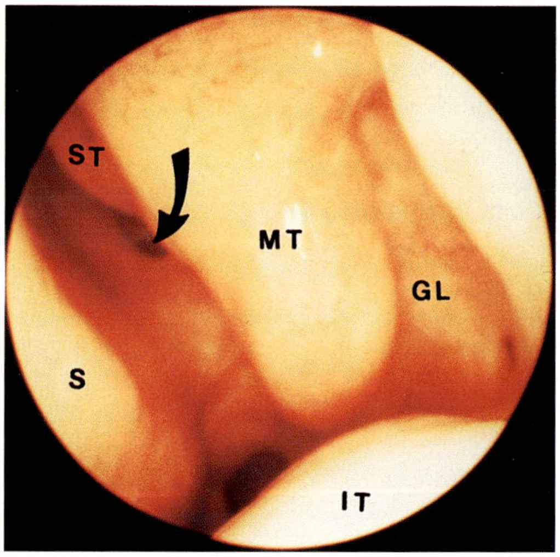

5.26

5.24

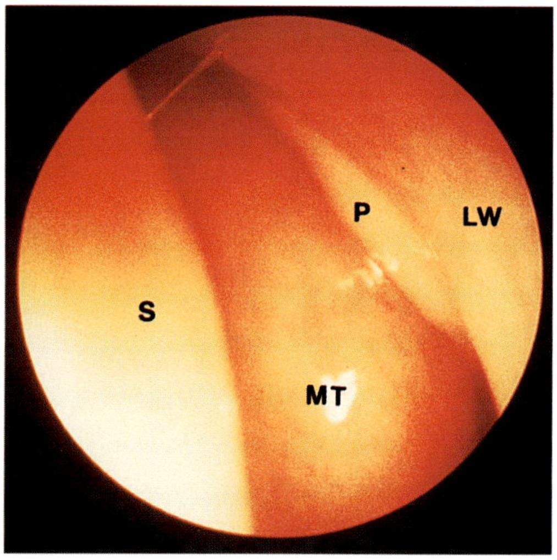

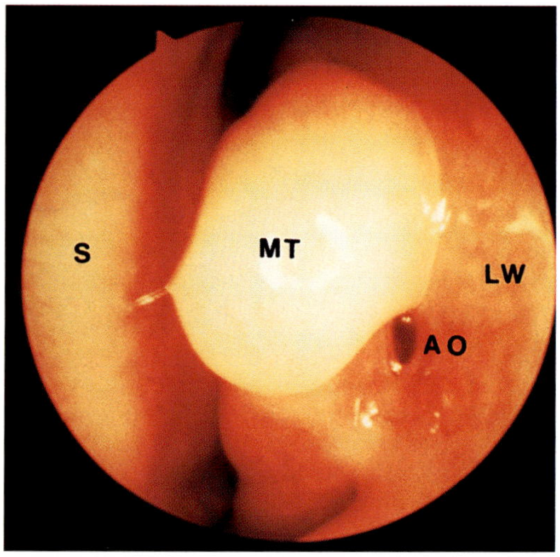

5.27

5.25

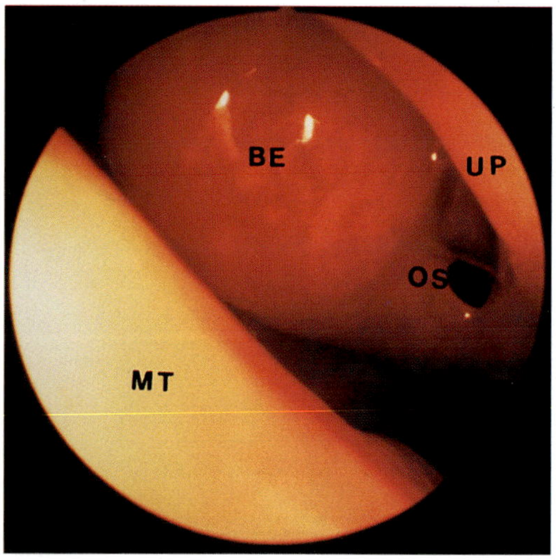

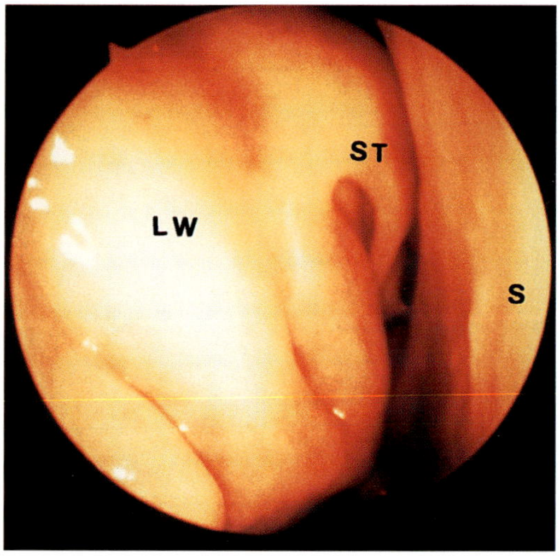

5.28

Fig. 5.23. The endoscope is now rotated into the middle meatus. Pus (P) may be seen tracking down the anterior meatal cleft, suggesting anterior sinus disease. S, septum; MT, middle turbinate; LW, lateral wall.

Fig. 5.24. Pus (P) emerging from the frontal recess is diagnostic of frontal sinusitis. S, septum; MT, middle turbinate; LW, lateral wall.

Fig. 5.25. As the endoscope enters the middle meatus, the accessory ostium (OS) comes into view. It is bounded anteriorly by the uncinate process (UP) and posteriorly by the bulla ethmoidales (BE). MT, middle turbinate.

Fig. 5.26. The posterior limit of the second pass is the ground lamella (GL) moving laterally across from the middle turbinate to the lamina papyracea. In addition, the superior turbinate (ST) and sphenoidal ostium (arrow) may be seen. S, septum; MT, middle turbinate; IT, inferior turbinate.

Fig. 5.27. Rarely, the middle turbinate (MT) may show a knobbing of its anterior end (rudimentary turbinate). Again, an accessory ostium (AO) is seen. S, septum; LW, lateral wall.

Fig. 5.28. Very rarely, the entire lateral wall (LW) may be grossly deformed, with no proper anatomical landmarks; this, of course, would make surgery hazardous. A CT is mandatory. ST, superior turbinate; S, septum.

Fig. 5.29. For a complete examination of the posterior group of sinuses, the endoscope is directed first backwards up to the posterior end of the inferior turbinate (IT), and then upwards. This brings the superior turbinate (ST) into view, along with the spheno-ethmoidal recess. PNS, post-nasal space; S, septum.

Fig. 5.30. Directing the 30° endoscope straight upwards will bring the sphenoidal recess into view (arrow). ET, eustachian tube; ST, superior turbinate; PNS, post-nasal space; S, septum.

Fig. 5.31. The third pass focuses on the superior turbinate (ST), its meatus (SM), and the spheno-ethmoidal recess. S, septum; LW, latera wall.

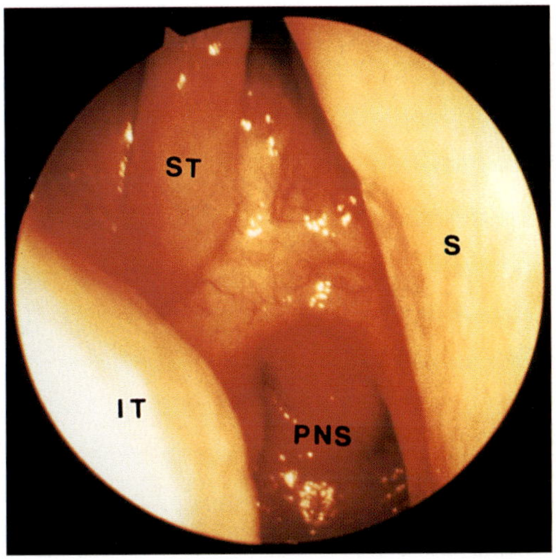

5.29

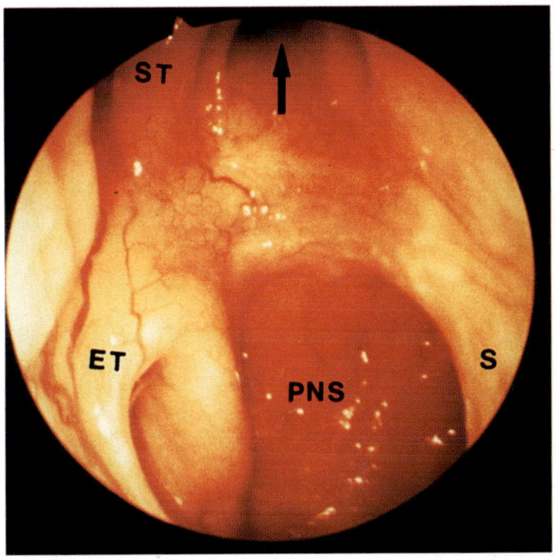

5.30

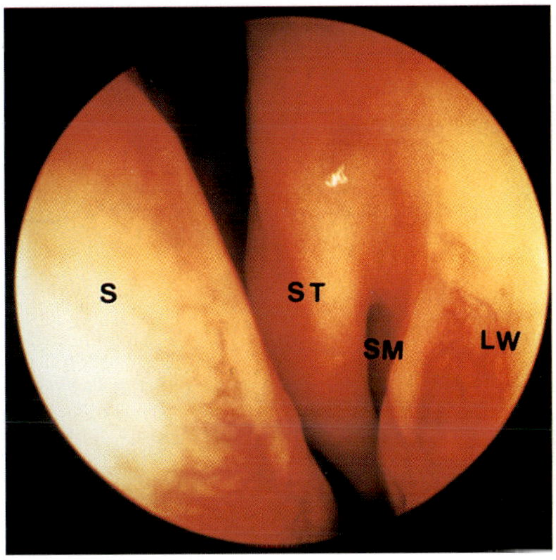

5.31

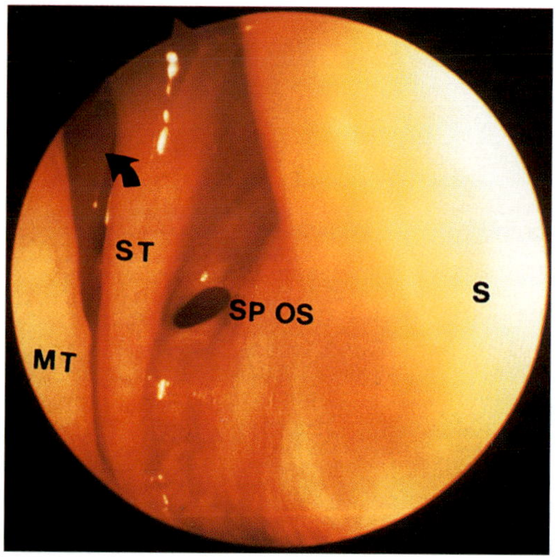

5.32

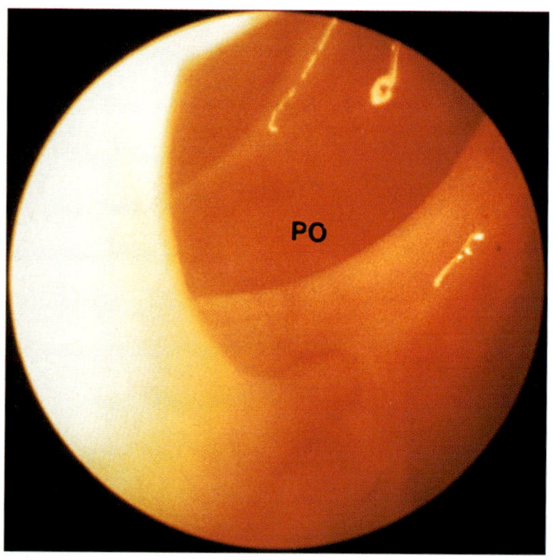

5.33

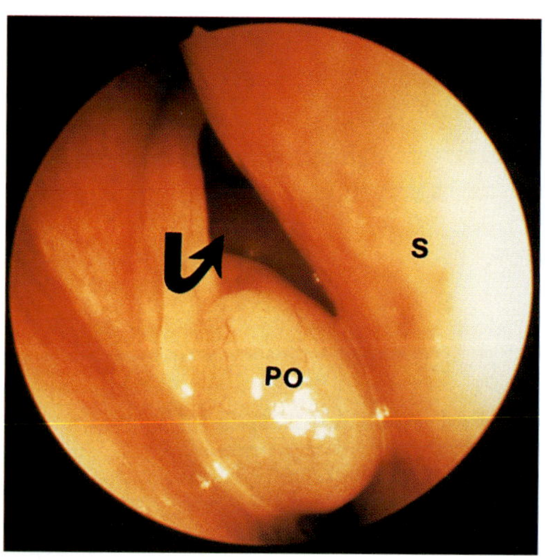

5.34

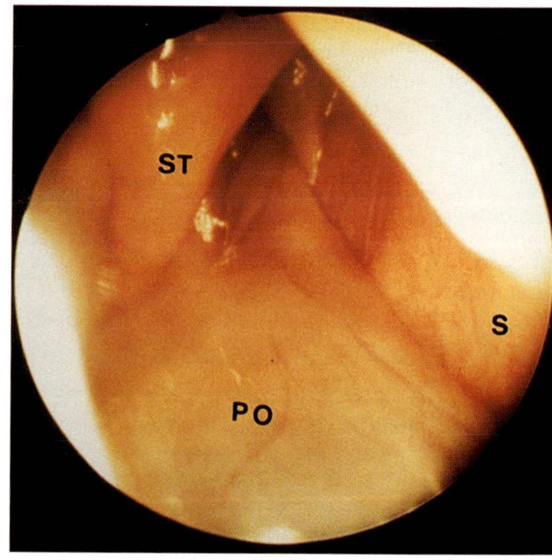

5.35

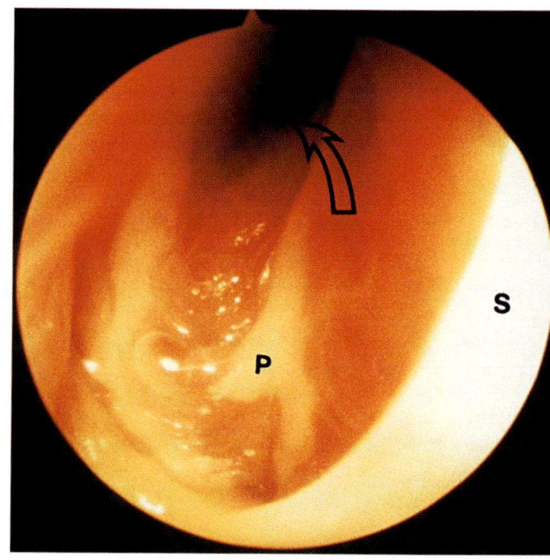

5.36

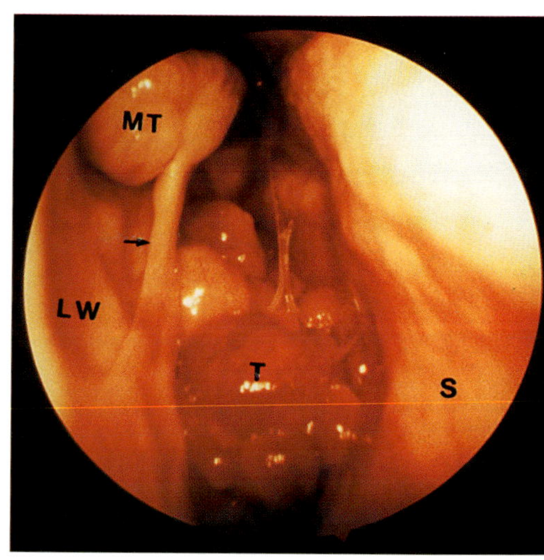

5.37

Fig. 5.32. Further examination shows the sphenoidal ostium (SP OS), the superior turbinate (ST) and its meatus (arrow). MT, middle turbinate; S, septum.

Fig. 5.33. Sometimes the sphenoidal ostium may be entered as in this view, showing an early polyp (PO). These patients often present as diagnostic enigmas until a throrough endoscopy reveals the true nature of the disease.

Fig. 5.34. Similarly, examination of the superior meatus (arrow) may also reveal a polyp (PO). This indicates a posterior group disease. S, septum.

Fig. 5.35. Pathology in the spheno-ethmoidal recess included polyps (PO), as seen emerging from the ostium. ST, superior turbinate; S, septum.

Fig. 5.36. Further to Fig. 5.35, pus (P) may be seen tracking down from the sphenoidal ostium (arrow). S, septum.

Fig. 5.37. Endoscopy may also be used to assess nasal tumours such as this inverted papilloma (T), which has recurred after excision several years ago. The arrow points to an adhesion due to previous surgery. LW, lateral wall; MT, middle turbinate; S, septum.

Fig. 5.38. This patient complained of nasal obstruction and crusting. Examination showed granulations (GR) on the floor of the nose, inferior surface of inferior turbinate (IT) and septum (S) a biopsy led to a diagnosis of sarcoidosis.

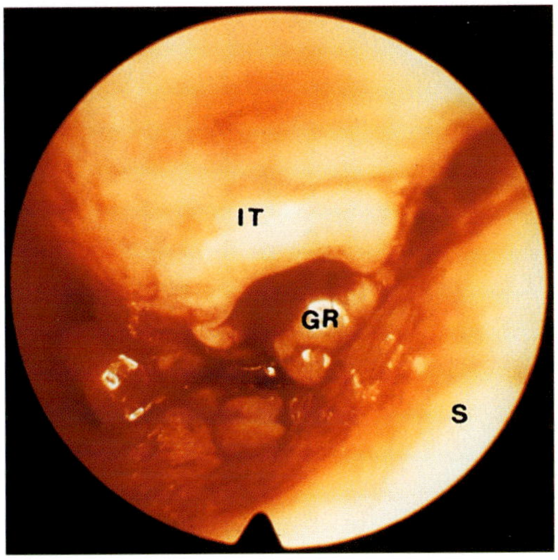

5.38

Chapter 6

Sinus Imaging

Introduction

One of the major developments in recent years has been sinus imaging with computed tomography (CT). It is now possible to study the intricate anatomy of paranasal sinuses and, more importantly, their relation to the surrounding vital structures such as skull base, lamina papyracea, internal carotid artery and optic nerve. It delineates key areas such as infundibulum and ostiomeatal complex. CT scanning is now mandatory in the pre-operative evaluation of sinus disease before undertaking functional endoscopic sinus surgery (FESS). The advent of CT scanning has significantly enhanced the scope of endoscopic sinus surgery by offering the surgeon more accurate information regarding sinus status than is possible with plain radiography.

Generally chronic sinus disease is evidenced by mucosal changes such as thickening, formation of polyps and presence of fluid levels in the sinuses.

CT Scan of the sinuses is mandatory in the following situations

- Children and adolescents
- All revision paranasal sinus surgery
- Gross nasal polyposis
- Extensive posterior ethmoid and/or sphenoid disease as evaluated on nasal endoscopy
- Distorted anatomy as seen on nasal endoscopy
- Mucoceles, tumors, etc.

Plain Radiography of the Paranasal Sinuses

Although CT scanning remains the investigation of choice for sinus disease, cost and availability often call for an almost reluctant reliance (on the part of the endoscopic surgeon) on plain films. Quite simply, bone throws a "shadow", while air is black, and any gradation between the two implies fluid or mucosal thickening. Thus it is a loss of translucency that characterises inflammation of the paranasal spaces. Assessment of the patient in the vertical position lets gravity enhance the information procured, as fluid levels (± menisci) are then reliably detected. Sinus views call for special fixation facilities of the skull to allow for standardisation of the patient/beam axis. Any deviation throws one set of sinuses away from the midline, causing a "white-out" and thus obscuring findings.

The routine views obtained are

- occipito-mental (Water's); this view demonstrates some important structures:
  - □ Sinuses (maxillary, postero-lateral wall, antero-lateral wall, frontal, ethmoid labyrinth
  - □ nasal septum
  - □ infraorbital foramen
  - □ zygoma
  - □ orbit

  If a fluid level is suspected, it may be confirmed by tilting the head by 20–40° to the side.
- occipito-frontal (Caldwell's), this film shows
  - □ sinuses (frontal, maxillary, [inferior segment] ethmoid labyrinth
- lateral view, this demonstrates
  - □ sinuses—sphenoid, frontal, maxillary, [posterior segment]
  - □ floor of the anterior cranial fossa

- orbit
- sella turcica

- oblique view, taken mainly for the posterior ethmoid cells and the optic foramina. The floor of the anterior cranial fossa is also seen clearly. The frontal sinuses are seen too.

The CT Scan

CT scanning of paranasal sinuses has proved the most dependable of pre-operative assessments of chronic sinus disease. By using wide and narrow window settings the bony details and soft tissue respectively can be shown precisely. Both coronal and axial scans are useful though coronal scans provide more useful information for the endoscopic surgeon.

Coronal scans have three distinct advantages over an axial scan:

1. The coronal scan illustrates progessively deeper structures encountered by the surgeon during the operation as he advances in an anterior to posterior direction, i.e. uncinate process, bulla ethmoidales, ground lamina, posterior ethmoids and sphenoid. In other words coronal scans synchronises with the steps of the operation.
2. It illucidates important relationships of the above structures to the lamina papyracea, and skull base.
3. The integrity of lamina papyracea is better appreciated in coronal scans.

It is our firm conviction that although CT is useful to elicit information about muco-periostial status, its true value lies in its ability to reflect accurately the surgical landscape of the core areas the surgeon is interested in. Both normal and variant anatomy (sinuses, ostia and intercommunicating air channels) guides one towards the cause of the patient's pathology. In our view, the scan serves as a "*road map*" for the surgeon as he negotiates the potentially hazardous clefts of the paranasal sinus unit.

We feel the primary advantage of the CT scan is the anatomical information it affords, and hence we read our films starting at the key area of the infundibular block, then moving outwards to include the para-infundibular block; finally noting the relationships of the anterior skull base and orbit to the sinuses (dehiscences of the internal carotid artery occur in a significant number of cases).

CT Scanning Technique

A coronal section CT scan is obtained in the following manner. The patient is placed prone on the scanner bed with the head hyperextended. The scanning plane is oriented so that the plane of sectioning is perpendicular to the infra-orbital meatal line. Scans are taken 3 mm apart (complete coronal section CT scan) and cover the region from the frontal sinuses to the sphenoid sinus. Bone windows (window width 4000), which are the same as those used for temporal bone imaging, are selected routinely for viewing. The selection of bone windows reduces the distortion resulting from metallic objects in comparison with soft tissue windows. It might be anticipated that a coronal section CT scan with soft tissue windows (window width 300) would be necessary in order to view soft tissue pathology; however, soft tissue/bony relationships are better demonstrated in a coronal sections with bone windows.

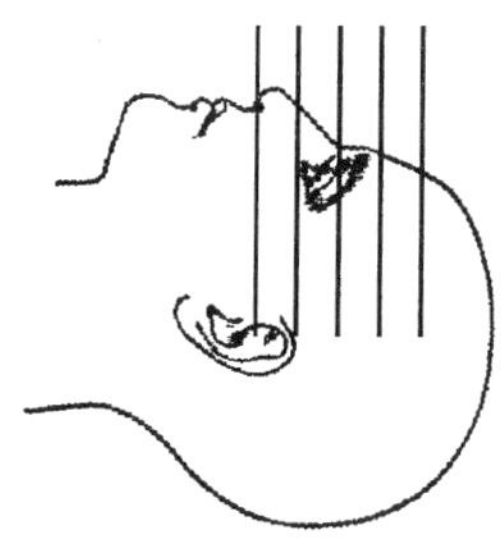

Fig. 6.1. Hyperextended head position for coronal CT scan of the sinuses.

CT Parameters

These films are used to obtain the following perspectives of the patients condition:

- the overall extent of pneumatisation of the paranasal sinus unit, and the extent of the disease, as reflected by haziness
- the *infundibular block*, consisting of the uncinate process, bulla ethmoidalis, the Haller cells and the middle meatus

- the *para-infundibular block*, consisting of the middle turbinate, agger nasi cells and the lamina papyracea
- the fovea ethmoidalis, or the floor of the anterior cranial fossa
- the optic nerve, and its relation to the posterior ethmoid cells
- the internal carotid artery in relation to the sphenoid sinus

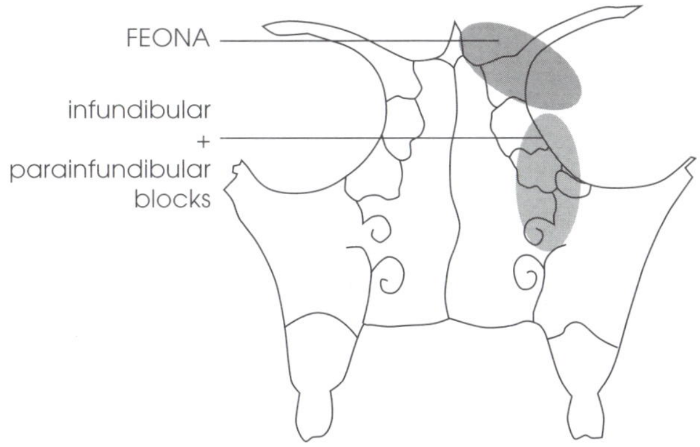

Fig. 6.2. FEONA, fovea ethmoidalis, optic nerve and internal carotid artery.

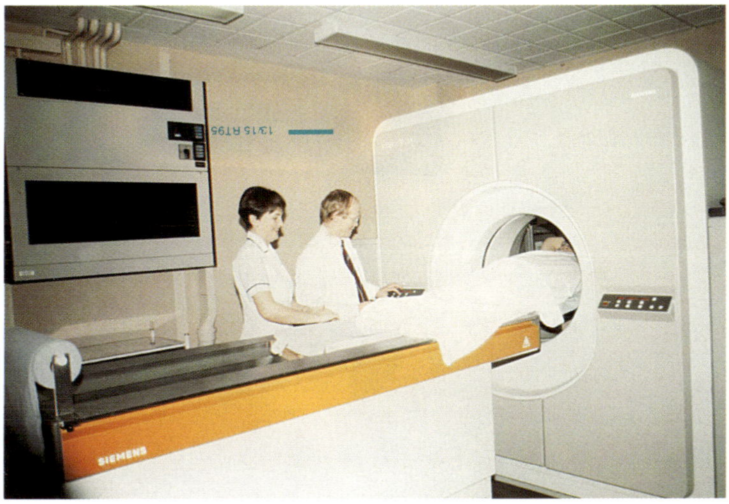

Fig. 6.3. View of a SIEMENS Somatom CR machine used for CT sinus imaging.

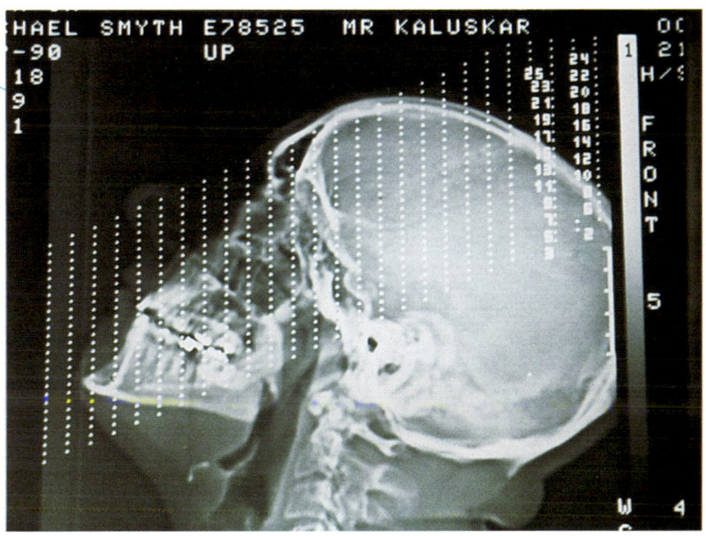

Fig. 6.4. Bone windows (3 mm) at 4-mm intervals are taken progressively from anterior to posterior for coronal CT scanning.

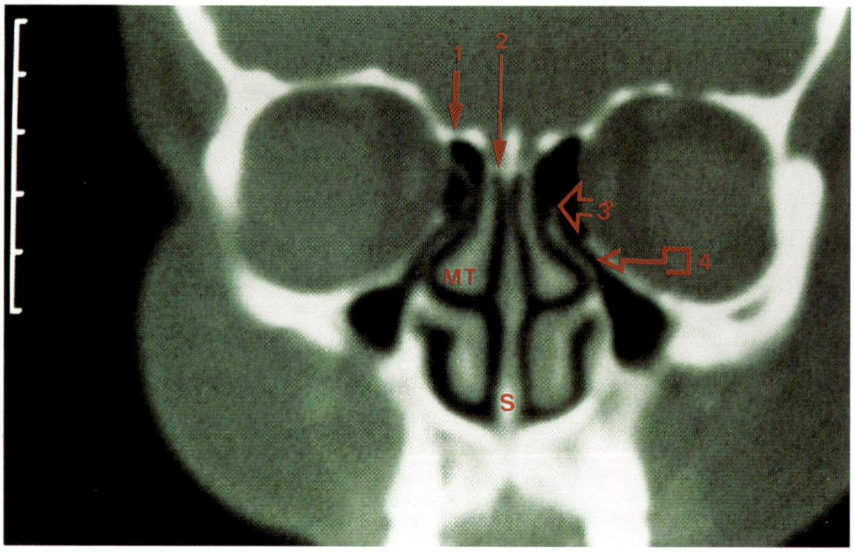

Fig. 6.5. Coronal sections reflect the surgeon's approach from anterior to posterior. More importantly, they reveal the key areas of the ostiomeatal complex as seen above (infundibular/para-infundibular blocks/FEONA). 1, the sloping roof (fovea) of the ethmoids; 2, level of cribriform plate; 3, upper limit of uncinate process; 4, infundibulum; S, septum; MT, middle turbinate.

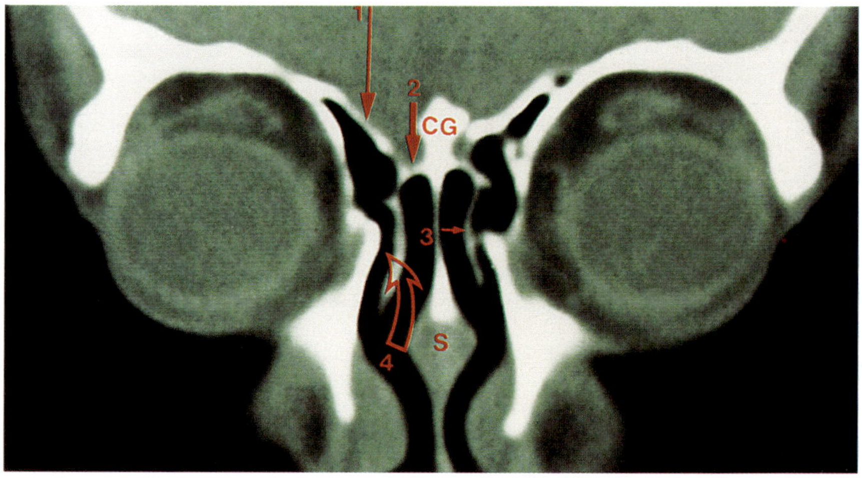

Fig. 6.6. In a more anterior section, the fronto-nasal recess (4) could be identified easily. Other structures include: 1, sloping roof of the ethmoids; 2, cribriform plate, which descends considerably into the nasal cavity; 3, superior attachment of the middle turbinate; CG, crista galli; S, septum.

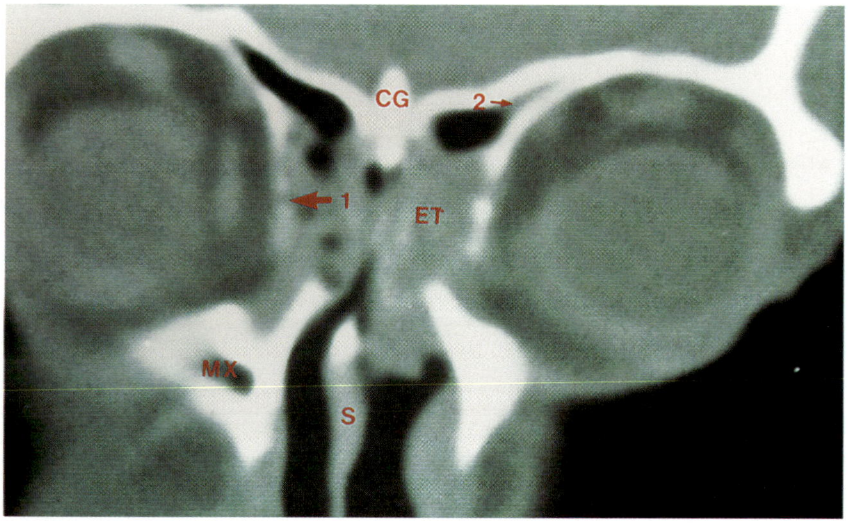

Fig. 6.7. Coronal cut through the anterior ethmoids (ET), showing extensive disease on both sides. 1, probable erosion of the lamina papyracea due to longstanding disease; 2, note disease in supra-orbital ethmoid cells; MX, maxillary sinus.

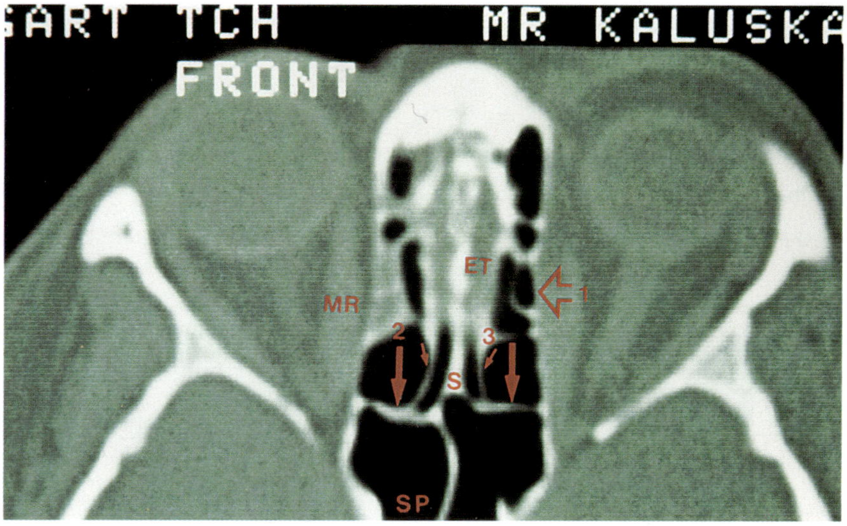

Fig. 6.8. An axial CT shows disease in the anterior and middle ethmoids. Although the relation of the orbit to the ethmoids is well seen, the important areas of the ostiomeatal complex are not seen adequately. 1, lamina papyracea separating the medial rectus (MR) from the ethmoids (ET); 2, lamellae separating the ethmoids from the sphenoids (SP) on either side; 3, upper attachments of the middle turbinate.

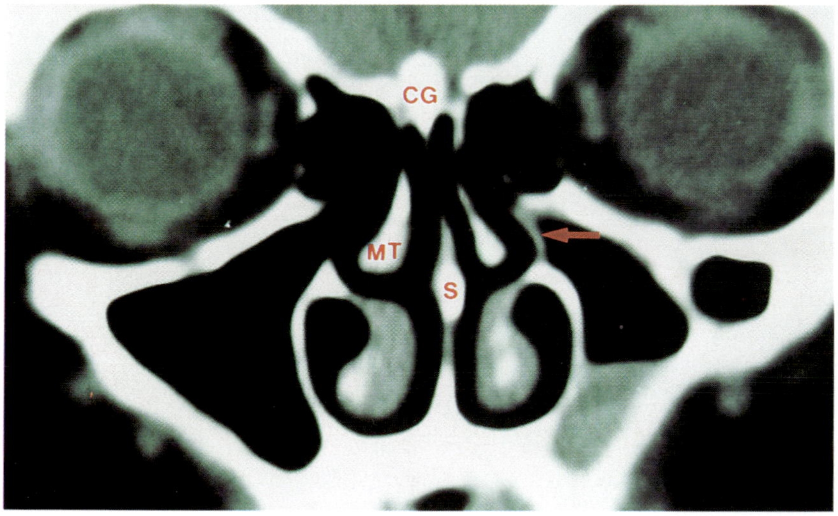

Fig. 6.9. Infundibular pathologies such as a laterally rotated uncinate process (arrow) cause stenosis of the ostio-meatal complex resulting in gross disease of the maxillary sinus. MT, middle turbinate; CG, crista galli; S, septum.

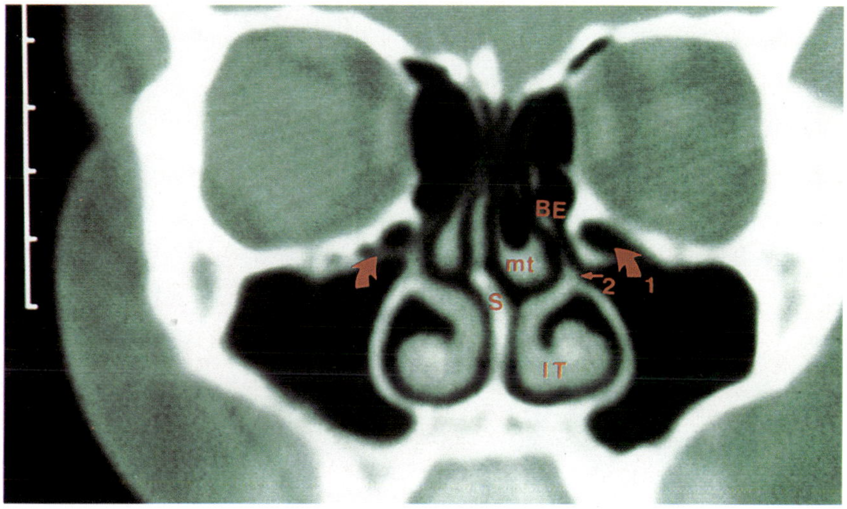

Fig. 6.10. Haller cells can be seen abutting the inferior orbital margin bilaterally (1). These are an uncommon finding, but quite significant in that they narrow the infundibulum. The uncinate process is seen on the left (2). BE, bulla ethmoidales; MT, middle turbinate; S, septum; IT, inferior turbinate.

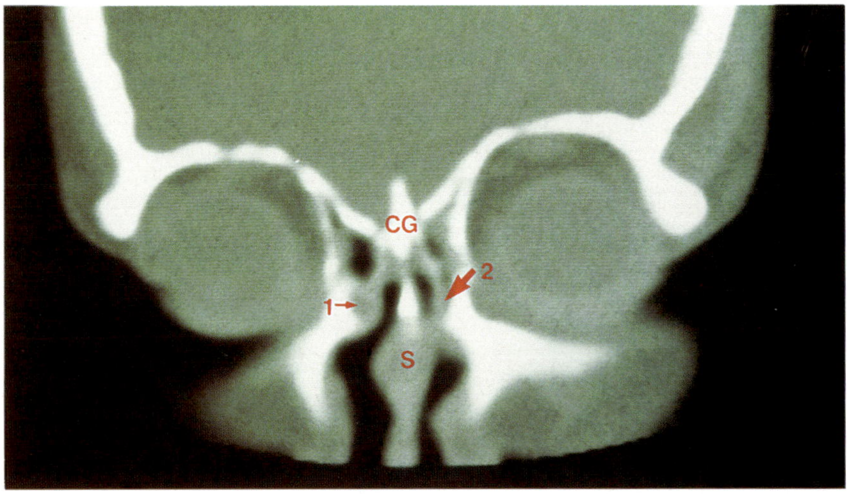

Fig. 6.11. Anterior coronal cut showing haziness in the right anterior ethmoids (1), and blockage of the fronto-nasal recess on the left (2). CG, crista galli; S, septum.

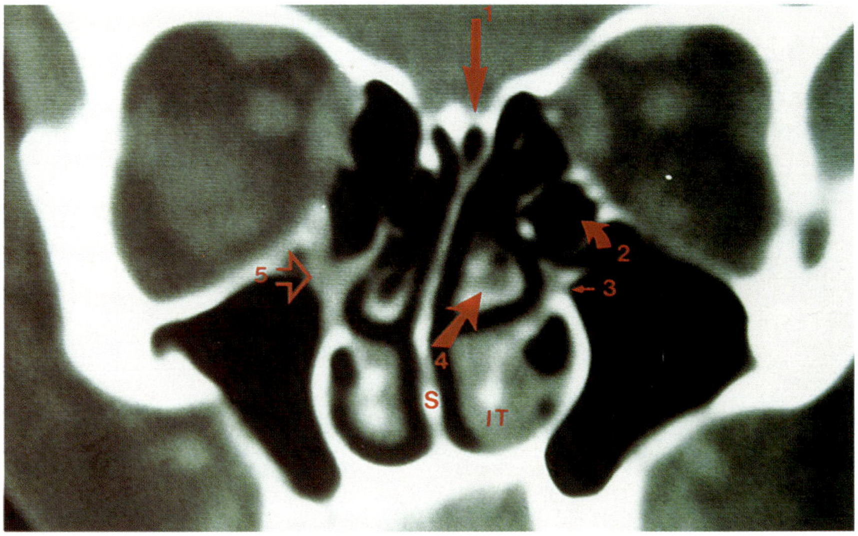

Fig. 6.12. CT scan showing several features: 1, cribriform plate; 2, large bulla ethmoidalis; 3, uncinate process; 4, haziness in a conchous left middle turbinate; 5, blocked infundibulum on the right. S, septum; IT, inferior turbinate.

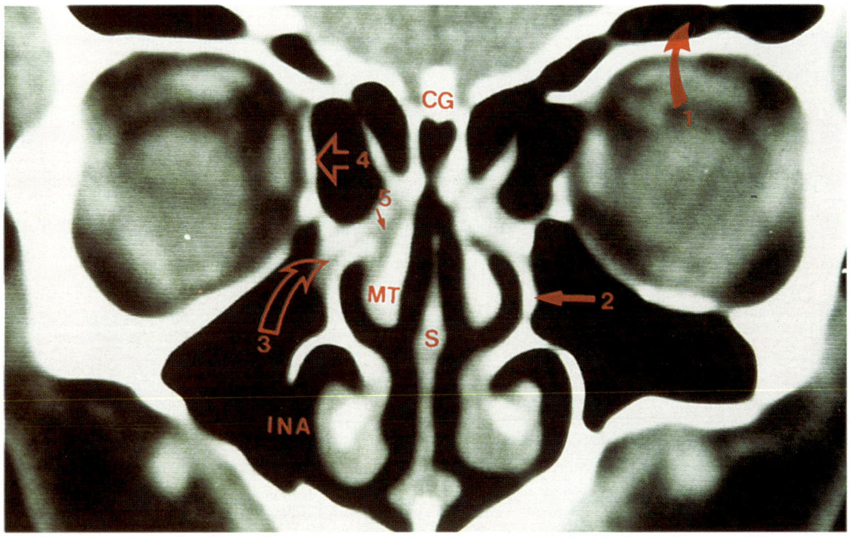

Fig. 6.13. Anterior cut showing several important features: 1, extensive supra-orbital pneumatisation; 2, laterally rotated uncinate process; 3, blocked, narrow infundibulum; 4, lamina papyracea; 5, anterior ethmoidal disease; INA, inferior nasal antrostomy. MT, middle turbinate, CG, crista galli; S, septum.

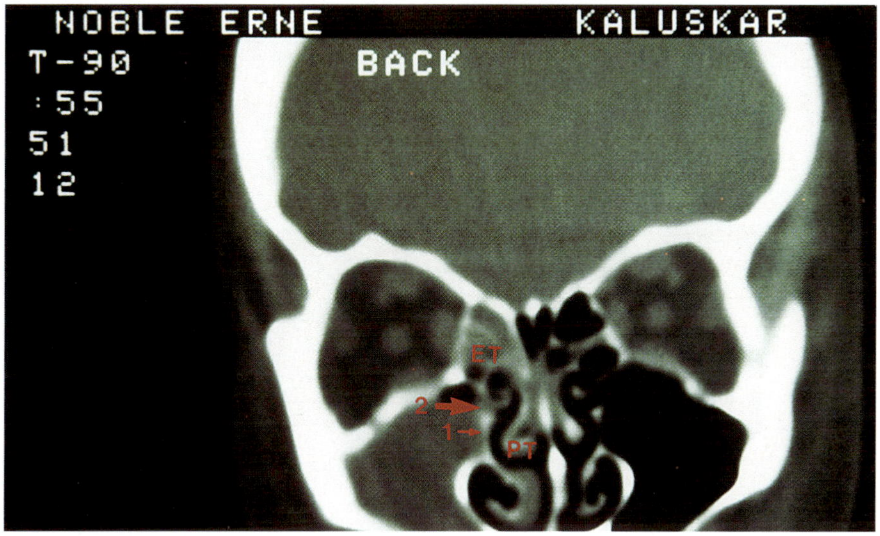

Fig. 6.14. Unusual, unilateral pan-sinusitis showing a paradoxical turbinate (PT) in association with ethmoidal (ET) and maxillary disease. 1, uncinate process; 2, blocked infundibulum.

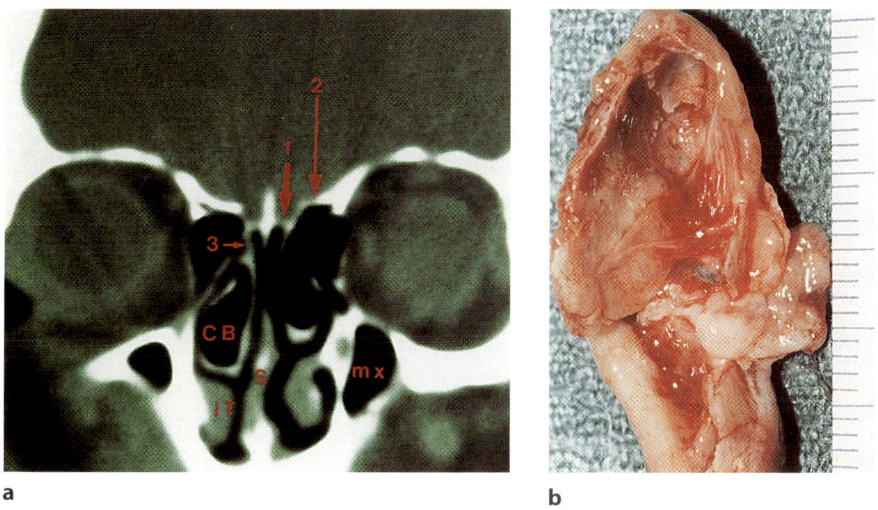

Fig. 6.15. **a** Right concha bullosa (CB), note its lateral wall encroaching on the OMC. 1, cribriform plate; 2, fovea ethmoidales; 3, superior attachment of the middle turbinate. IT, inferior turbinate; S, septum; MX, maxillary sinus. **b** Postoperative specimen of excised concha bullosa.

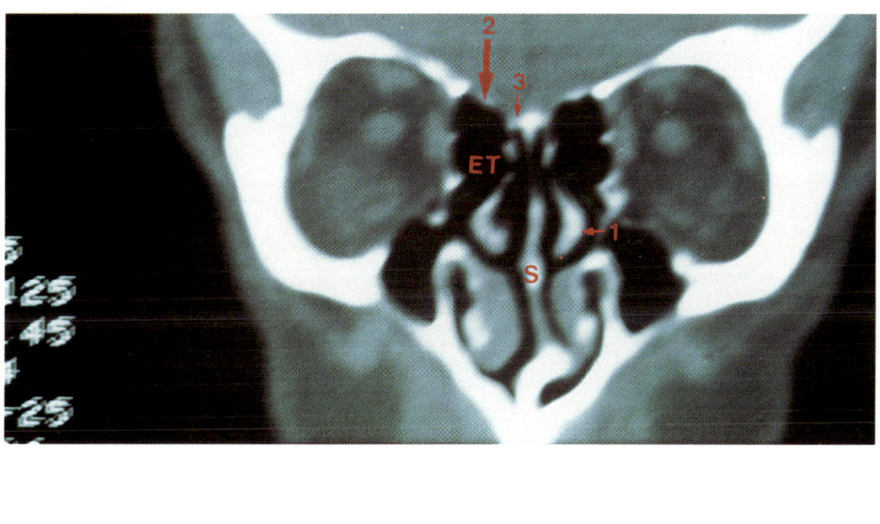

Fig. 6.16. Another common abnormality is the paradoxical turbinate, seen here bilaterally (1); 2, roof of the ethmoids; 3, cribriform plate. ET, ethmoids; S, septum.

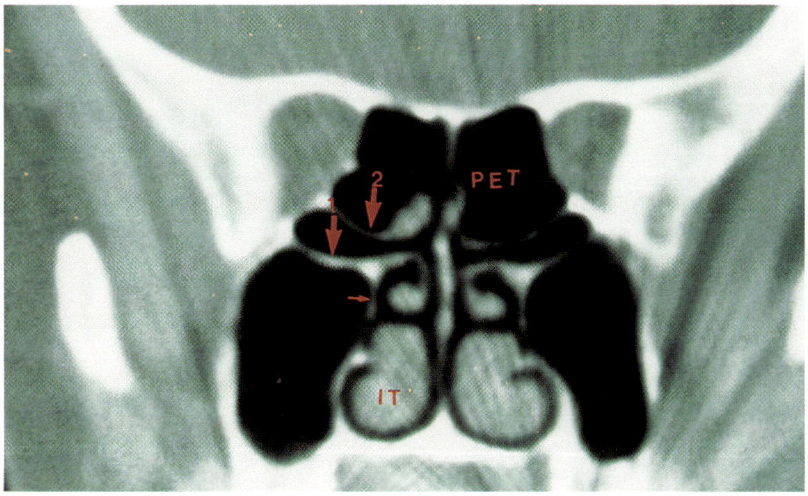

Fig. 6.17. Further posterior cuts show up a series of sequential ground lamellae, separating ethmoid cell planes. In this section, the main ground lamella (1) can be seen extending from the middie turbinate to the medial wall of the orbit. Just behind this is another lamella (2). Small arrow, uncinate process; PET, posterior ethmoids; IT, inferior turbinate.

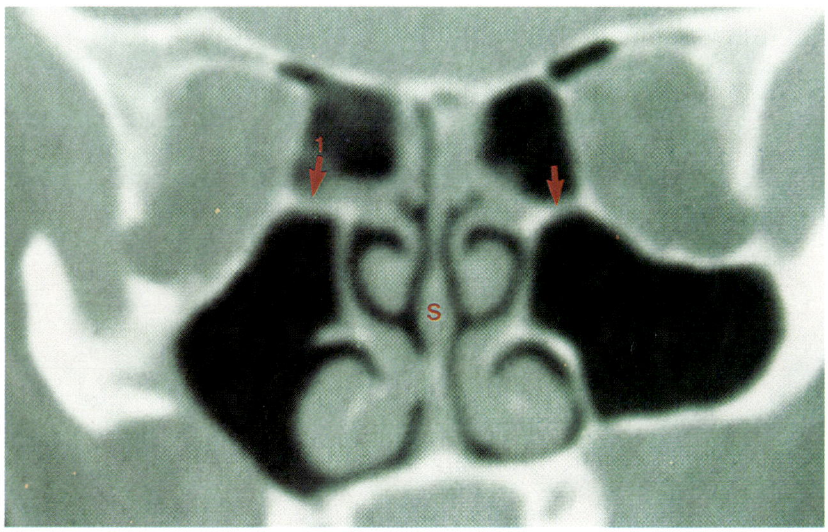

Fig. 6.18. The main ground lamella (1) is seen, note posterior cell disease. S, septum.

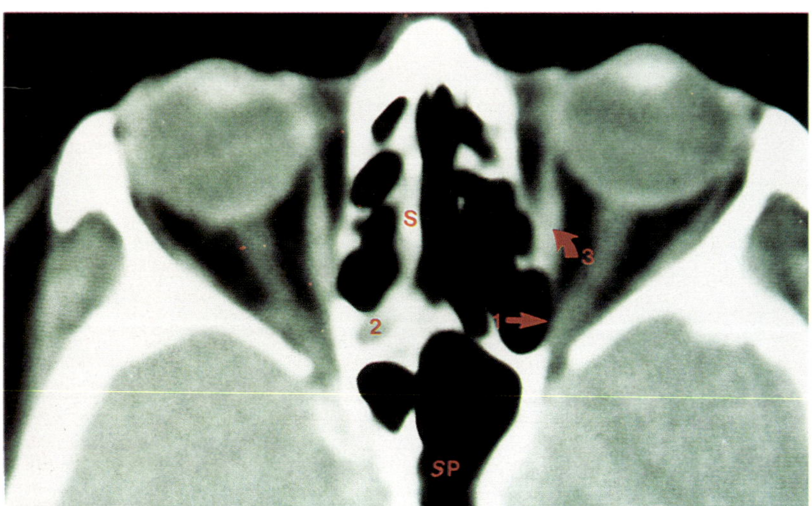

Fig. 6.19. Axial scan showing Onodi cell (lateral extension of the posterior ethmoidal cell) (1), thus exposing the optic nerve directly to the operative field. Note the posterior ethmoid disease on the opposite side (2). 3, medial rectus; SP, sphenoid sinus; S, septum.

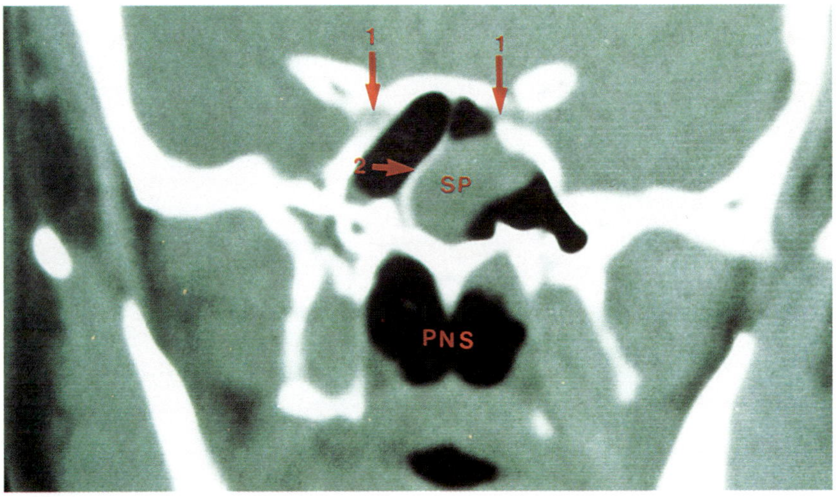

Fig. 6.20. Sphenoid cuts showing haziness in the left sphenoid sinus, along with dehiscences of both internal carotid arteries into the sphenoidal lumen (1). This is a not uncommon finding and, therefore, sphenoidal disease should be treated with extreme care. 2, inter-sphenoid septum. SP, sphenoid sinus; PNS, post-nasal space.

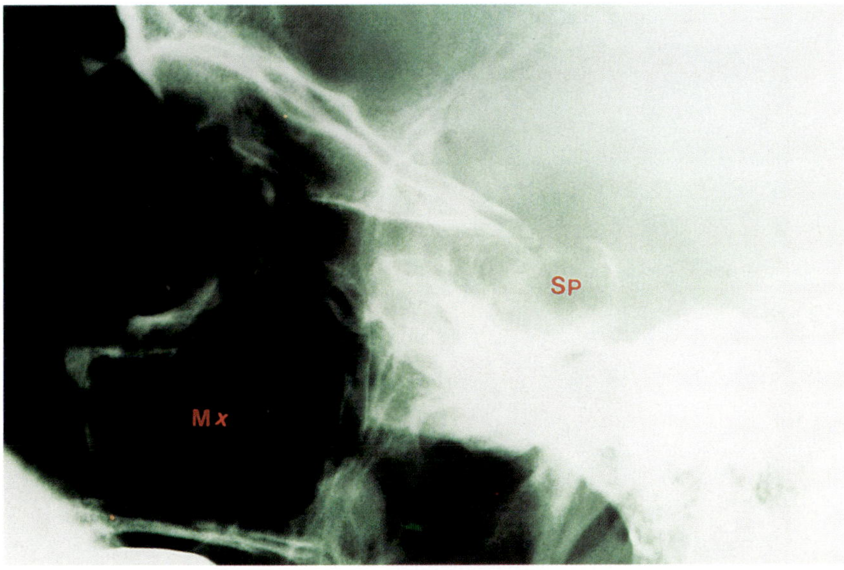

Fig. 6.21. This patient suffered from severe headaches, the only positive finding was a haziness of the sphenoidal sinus. MX, maxillary sinus; SP, sphenoid sinus.

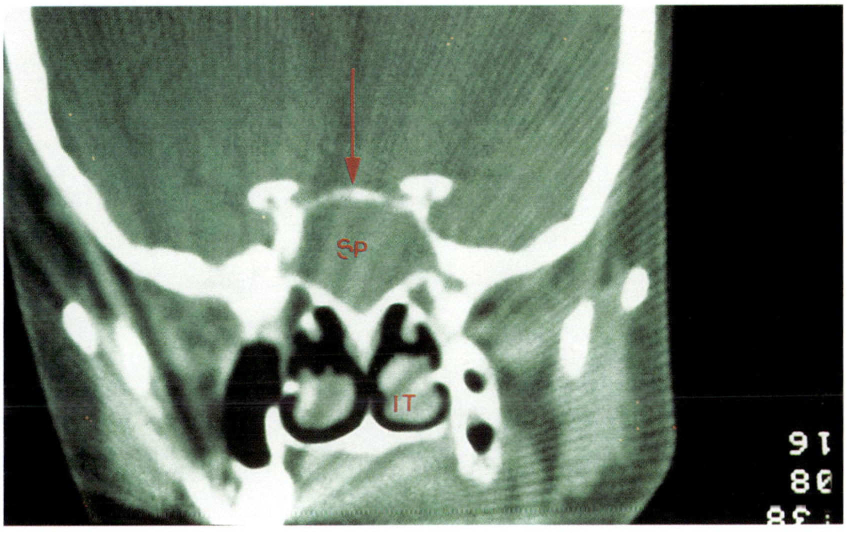

Fig. 6.22. The coronal CT scan revealed bowing of the posterior wall of the sphenoid sinus (arrow). Endoscopic removal confirmed a mucocoele. SP, sphenoid sinus; IT, inferior turbinate.

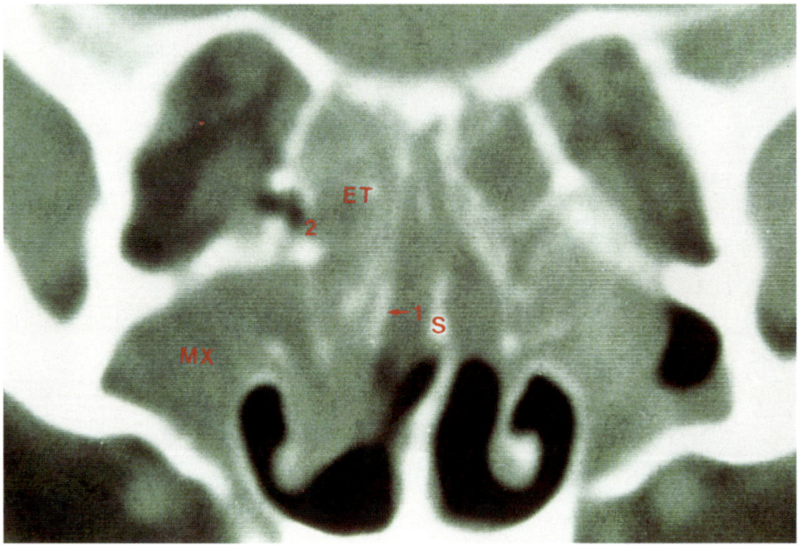

Fig. 6.23. This is a case of extensive, recurrent, bilateral pan-sinusitis. The patient underwent several polypectomies to no avail. There is complete distortion of surgical landmarks. Only remnants of the right middle turbinate could be seen (1), and there was an erosion of the lamina papyracea (2). CT results such as these should serve as warning lights, an external ethmoidectomy would be the preferred option for all except the truly experienced. MX, maxillary sinus; ET, ethmoids; S, septum.

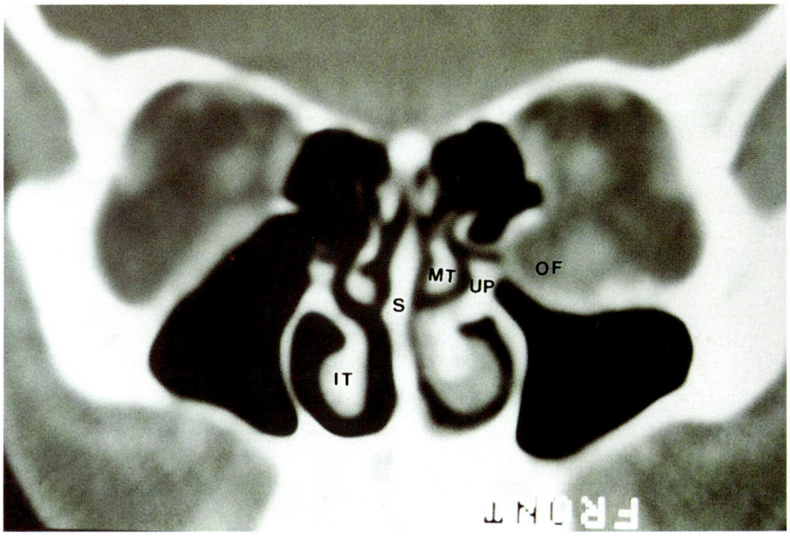

Fig. 6.24. Coronal CT scan showing inferior turbinate (IT), septum (S), middle turbinate (MT), uncinate process (UP) and orbital fat (OF) prolapsing into the infundibulum due to dehiscence of the lamina papyracea. The patient suffered from a blow-out orbital fracture.

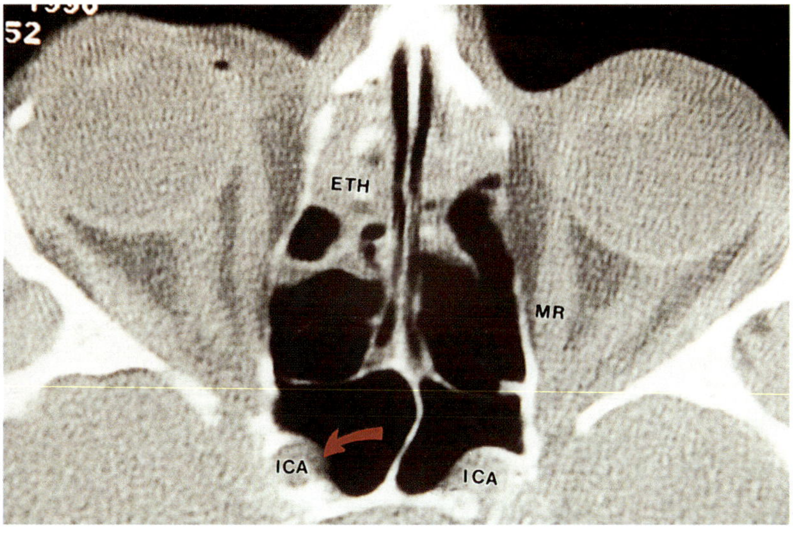

Fig. 6.25. Axial CT scan of the sinuses showing medial rectus muscle (MR), internal carotid artery (ICA) and dehiscence of the internal carotid artery in the lateral wall of the splenoid sinus (arrow). ETH, ethmoid sinuses showing mucosal disease.

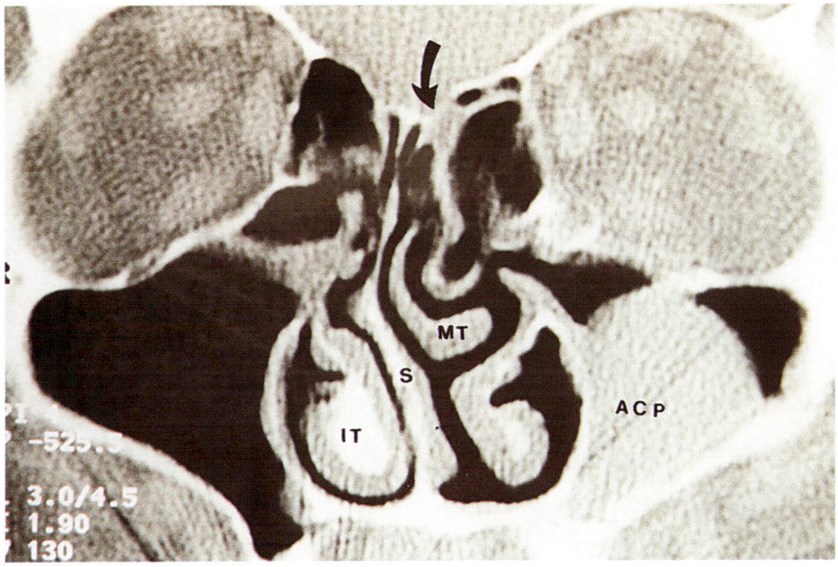

Fig. 6.26. Coronal CT scan of the sinuses. IT, inferior turbinate, S, septum; MT, middle turbinate; ACP, antrochoanal polyp. The arrow shows dehiscence of the lateral lamella of the cribiform plate.

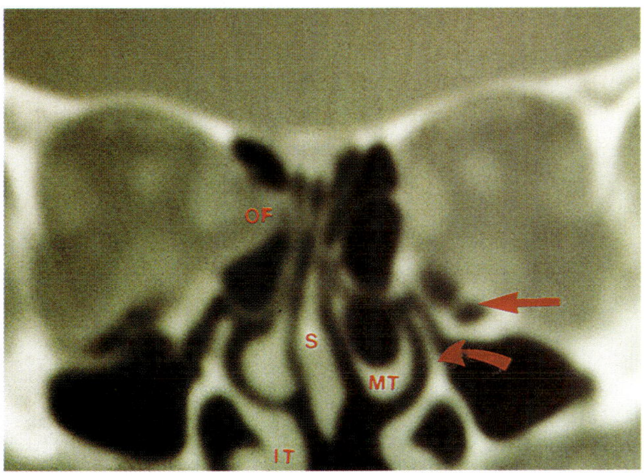

Fig. 6.27. Coronal CT scan. OF, prolapsed orbital fat in the ethmoid sinus due to dehiscence of the lamina papyracea; S, septum; MT, pneumatised middle turbinate (concha bullosa); IT, inferior turbinate; straight arrow, Haller cell; curved arrow, uncinate process.

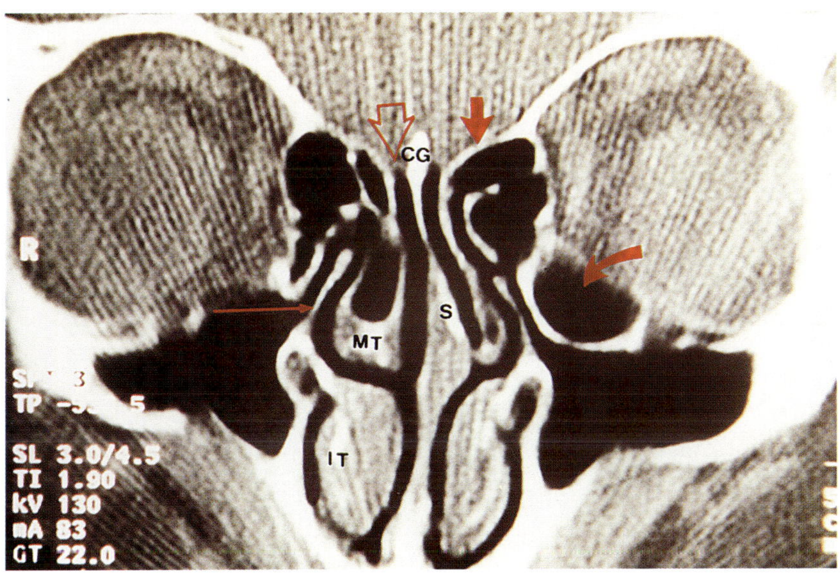

Fig. 6.28. Coronal CT scan of the sinuses. CG, crista galli; S, septum; MT, middle turbinate; IT, inferior turbinate; long arrow, uncinate process; hollow arrow, thin medial part of the fovea ethmoidales; short arrow, lateral part of the fovea ethmoidales; curved arrow, Haller cell, narrowing infundibulum.

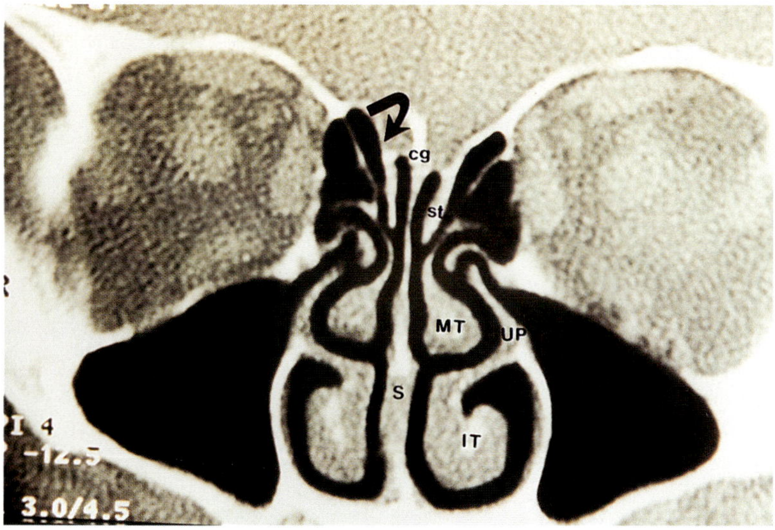

Fig. 6.29. Coronal CT scan. CG, crista galli; ST, superior turbinate; MT, middle turbinate; UP, uncinate process; IT, inferior turbinate; S, septum; arrow shows thin lateral lamella of the cribiform plate.

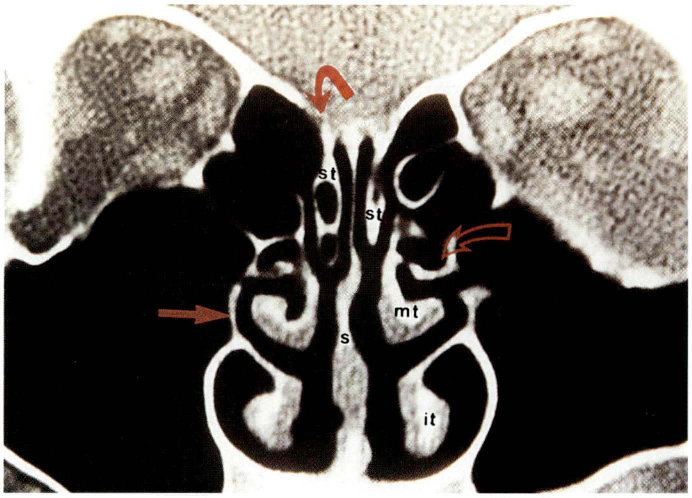

Fig. 6.30. Coronal CT scan of the sinuses. S, septum; IT, inferior turbinate; MT, middle turbinate; ST, superior turbinate (note pneumatisation); hollow arrow, bulla ethmoidales; curved arrow, dehiscent lateral lamella of the cribiform plate; straight arrow, uncinate process.

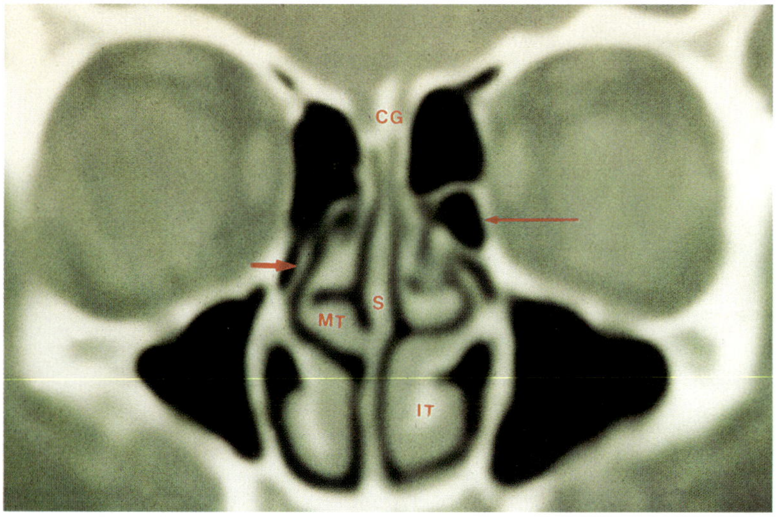

Fig. 6.31. Coronal CT scan. CG, crista galli; S, septum; MT, bifid middle turbinate, adherent to the septum; IT, inferior turbinate; short arrow, uncinate process; long arrow, bulla ethmoidales.

Why CT?

- diagnosis: confirmation or exclusion
- extent of pneumatisation
- anatomical relations with orbit/ anterior/cranial fossa
- structural abnormalities: uncinate process/concha bullosa/bulla ethmoidales/Haller cell/hypoplastic sinus, etc.
- "road map" for the surgeon

When CT?

- when the patient is not symptomatic
- treatment with antibiotics and decongestants for at least 3 weeks is recommended prior to CT

Frequent discussions regarding scans with the radiologist are recommended!

Plain films versus CT scans

On comparing the two investigations, we have found a good correlation (70%) between plain radiographs and CT scans for the maxillary sinus. The remaining 30% largely consists of patients in whom there is evidence of maxillary sinus disease on plain films, but their CT results are quite normal. This is probably due to the thickness of surrounding bone and soft tissue shadow giving an erroneous diagnosis of sinus infection on plain films. Similarly, the sphenoid sinus is clearly read on both plain films and CT scans (showing a correlation of 80% in our series); most of the remaining 20% are 'false' positive plain films.

Ethmoid sinus assessment proves more complicated. Fig. 6.32 shows the correlation between plain films and CT scans in a sample of 100 consecutive CT scan/plain film sets we analysed.

As can be seen, there is a significant risk of plain films not reflecting the true status of the sinuses in that they are under-read. Overall, 57 cases showed no correlation between the two modalities. Frontal sinus analysis produced a similarly mixed picture although the overall correlation was 67% (Fig. 6.33).

To sum up, the frontal and maxillary sinuses tend to be over-read on plain views, while the ethmoids tend to be under-read. As this is the key area of dysfunction in most cases of chronic rhinosinusitis requiring surgery, the value of plain radiographs as a pre-operative tool seems limited to gross anatomical discrepancies and acute infections with fluid levels.

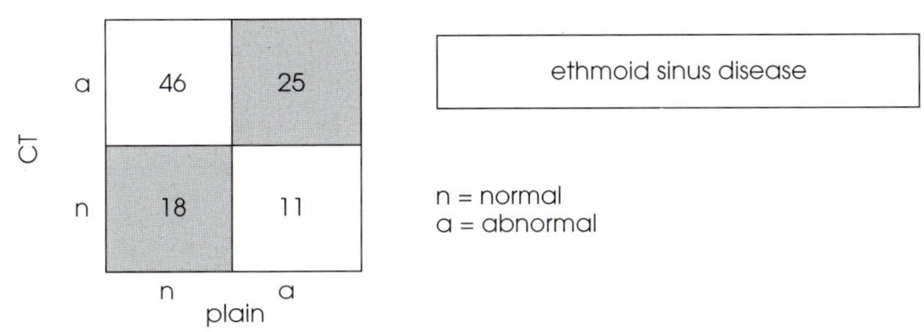

Fig. 6.32. Correlation between plain films and CT scans for ethmoid disease.

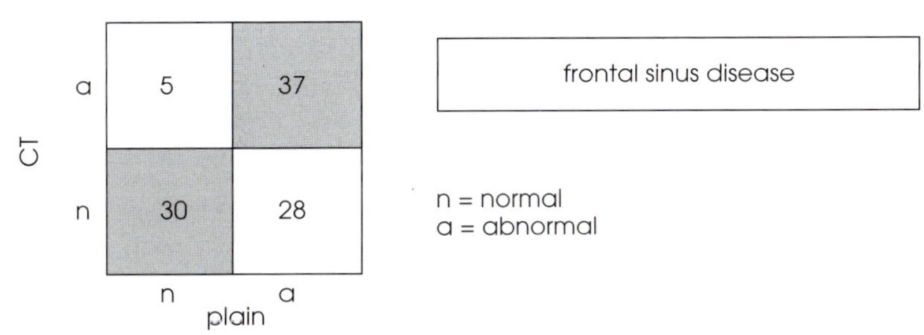

Fig. 6.33. Correlation between plain films and CT scans for frontal disease.

CT versus Operative Findings

In our experience, operative findings in the maxillary sinuses tend to correlate well with CT. For the ethmoids, however, we find that mucosal disease tends to be far more widespread than is detected on CT: It is essential, therefore, to be cautious regarding even seemingly insignificant changes in the infundibular and para-infundibular blocks.

Chapter 7

Medical Aspects of Sinusitis

Introduction

Sinusitis is an inflammation involving the mucous membrane of the paranasal sinuses. The cause is most often a respiratory virus, with secondary bacterial infection although a variety of other agents play a significant role. The clinical profile may range from a mild nasal stuffiness to severe, debilitating disease. Complications may be associated with significant morbidity and mortality.

Pathogenesis

Sinuses are lined with pseudostratified columnar ciliated epithelium, and drain through ostia into the nasal cavity. This drainage is dependent upon the functional patency of the ostium, which is more important than its anatomical size. Normal sinus clearance further depends on muco-ciliary integrity and action. Thus, any dysfunction of ostial patency, mucus production or ciliary action will increase the risk of sinusitis. Finally, hypoxia of the sinuses could further predispose to subsequent sinusitis.

The most crucial parameter in the pathogenesis of sinusitis appears to be the patency of the ostium. The normal size of the os varies for the different sinuses but may be as small as 1–2 mm (the usual diameter of an ethmoid cell ostium). The ostium may decrease further in size under certain normal circumstances such as being in the recumbent position. Other factors include

- inflammation and oedema associated with viral upper respiratory tract infections or allergic rhinitis
- local pathology such as:
  - nasal polyps
  - septal spurs/deviations
  - tumours
  - barotrauma
  - foreign bodies
  - dental abscesses
- focal pathology:
  - primary ciliary dyskinesias
  - Young's syndrome
  - cystic fibrosis
  - Kartagener's syndrome

In patients with these conditions, both the quality and the quantity of mucus production are deranged. For example, the mucus produced by cystic fibrosis patients is very viscid, and this probably accounts for the high incidence of sinusitis.

Muco-ciliary transport may be affected by ciliary dyskinesias – poor co-ordination of ciliary beat patterns may be found. Mucus is normally moved toward the ostia by the beating motion of the ciliated lining cells. Impairment of the nasal ciliary motility can be temporarily induced by viral infections, temperature changes, chemicals and certain drugs. Sinus infection causes a quantitative loss of cilia which is normally reversible, but if severe purulent rhino-sinusitis with scarring occurs this loss may be permanent, with ciliated epithelium replaced by squamous epithelium.

Immunoglobulins such as IgA, IgG and IgM are found in low concentrations in purulent secretions; thus, patients with hypogammaglobulinaemia are at risk for sinusitis.

Normal ventilation of the sinuses is important for mucosal aeration. There is a direct correlation between the ostial size and oxygen tension in the sinus. When the ostium is patent, the PO_2 is around 115 mmHg. The PO_2 falls with obstruction and may be reduced almost to zero in the setting of purulent sinusitis, with a concomitant increase in PCO_2. Anaerobic growth is encouraged by low oxygen tension. In patients with recurrent sinusitis, oxygen tension in the sinuses is reduced chronically *even when the sinus is infection-free*. Generalised causes of sinus compromise include:

- acquired immunodeficiency syndrome (AIDS), both T-cell and B-cell defects occur
- diabetes, which especially predisposes to sinus fungal infections

Bacteriology

Until recently, it was thought that the symptomatic paranasal sinus was sterile; with the advent of better aerobic and anaerobic laboratory techniques there is sufficient evidence to suggest that "normal flora" exist in the sinuses. Sinus aspiration often shows anaerobic isolates, predominantly *Bacteroides* sp. (over 50% *B. melaninogenicus*), anaerobic gram-positive cocci, and *Fusobacterium*. Predominant isolates among the aerobes are beta- and alpha-hemolytic strept ococci, *S. pneumoniae*, *Haemophilus influenzae* and *Staphylococcus aureus*.

Clinical Features of Chronic Sinusitis

The signs and symptoms of chronic sinusitis are often non-specific and difficult to differentiate from the common cold, influenza-type syndromes or allergic rhinitis. Purulent nasal discharge and facial pain or tenderness are the most common features of the disease. Additionally one might see disorders of smell, nasal congestion, cough or a history of recent upper respiratory tract infection.

The symptom profile in chronic sinusitis may be summarised as in Fig. 7.1.

Treatment

Although this book deals predominantly with functional endoscopic sinus surgery (FESS), we would like to emphasise that all patients must first be vigorously and adequately treated with medication. The principles of treatment are:

- eradicate infection (antibiotics)
- establish drainage and ventilation (systemic/local decongestants)
- treat underlying predisposing factors

Antibiotic therapy in sinusitis should, as always, be directed against the most common pathogens *Streptococcus pneumoniae*, *Haemophilus influenzae* and *Branhamella catarrhalis*. Empirical therapy is appropriate as sinus punctures are not routinely indicated except in complicated or refractory cases. Ampicillin, amoxicillin, cephalosporin group and trimethoprim/sulphamethoxazole have all been shown to be of use. The

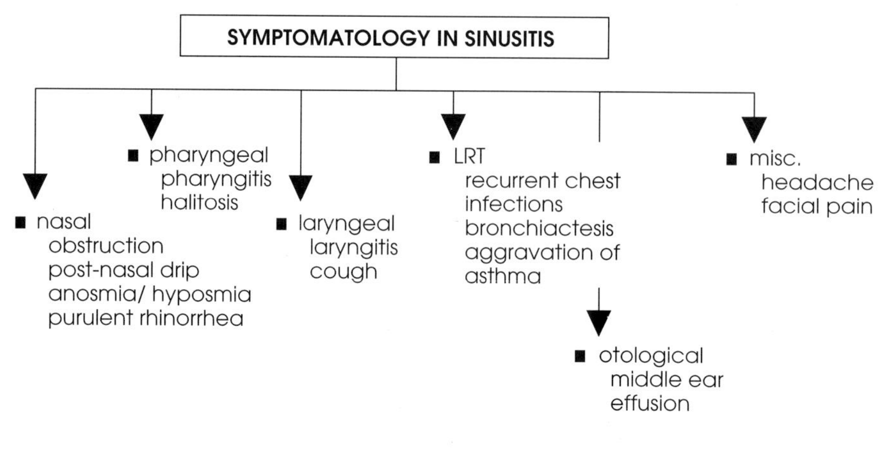

Fig. 7.1. Multiple symptoms of chronic sinusitis.

duration of the therapy should be at least 3 to 4 weeks. In addition to antibiotics, oral decongestants and steroid nasal sprays are commonly used. Inhalations help to loosen secretions. Antihistamines are of value in allergic patients. If a gram-negative result is returned by the laboratory, anaerobes and resistant gram-negative bacilli should be suspected, especially in chronic sinusitis. The cephalosporin group of drugs and metronidazole are of use in these cases. Mycotic sinusitis is a more morbid condition, often requiring surgical debridement plus systemic anti-fungals such as amphotericin B. FESS is considered in those patients who do not respond to medical management, and who have sinus disease as evinced by history of recurrent sinusitis, office endoscopy with or without CT scanning.

Acute Sinusitis: Summary

Causes

- upper respiratory tract infection
- entry of infected water during swimming/ diving
- dental extraction in the presence of root infection
- fractures involving the sinuses

Predisposing Factors

- local: obstruction of the nose, e.g. deviated septum, pathology of the ostio-meatal complex
- general: chilling, fatigue, atmospheric pollution

Clinical Features

Clinical features vary with the intensity of infection and efficacy of drainage.

- stabbing pain over the infected sinus, aggravated by bending or coughing
- muco-purulent discharge
- nasal obstruction (due to mucosal swelling)
- tenderness over the affected sinus
- oedema of the overlying soft tissue
- constitutional symptoms, e.g. pyrexia, malaise

Treatment

- systemic antibiotics (broad spectrum penicillin)
- analgesics
- restoring drainage of the affected sinus by systemic decongestants and/or irrigation. Endoscopic management of acute conditions should only be undertaken by an experienced practitioner.

Chapter 8

FESS Technique

Introduction

The two classic endoscopic approaches are those of Messerklinger (1985) and Wigand (1978). The Wigand technique calls for a back-to-front exenteration of disease from the paranasal sinuses, while the Messerklinger approach deals with the anterior group of sinuses unless there is definite evidence of posterior group disease in which case a complete fronto-ethmoido-sphenoidotomy and frontal sinusotomy could be performed. Both aim at re-establishing normal drainage channels, thereby reversing the diseased mucosa to normalcy; – the latter technique, however, is more widely practised the world over. We have based our technique on that of Messerklinger, as a vast majority of our patients present with early disease. The Wigand technique finds application in cases of massive polyposis or in revision surgery (where anatomical landmarks are distorted).

Both local and general anaesthetic are used for our patients. We prefer local anaesthesia, as there is significantly less bleeding. An awake patient is his own best guardian: any manoeuvring in the region of the orbit or the anterior cranial fossa elicits immediate pain in spite of adequate sedation and thorough topical anaesthesia! A general anaesthetic is therefore reserved for apprehensive patients, or those in whom a concomitant procedure such as septoplasty and/or rhinoplasty is required.

Local Preparation

Patients are operated on on a day-care basis, i.e. they are admitted on the morning of the operation, and leave that very evening.

The patient is positioned as shown in Fig. 8.1, and his or her clinical proforma and CT scan are made available. Sedation is achieved with Benzodiazepines derivative. The nose is prepared using serial cotton wool strip or soft universal strip-packs soaked in 4% cocaine + 1:1000 adrenaline. The cavity is filled from anterior to posterior and inferior to superior. On average, preparation takes 20–30 minutes in all. Once extensive decongestion is achieved, specific blocks are targetted towards the following:

- posterior end of middle turbinate *the spheno-palatine ganglion*
- junction of nasal bone and upper lateral cartilage *anterior ethmoidal nerves*
- middle meatus

The third block is an added precaution in case the spheno-palatine block is not satisfactory because of poor access to the region. Finally, the area of the uncinectomy on the lateral wall of the nose is

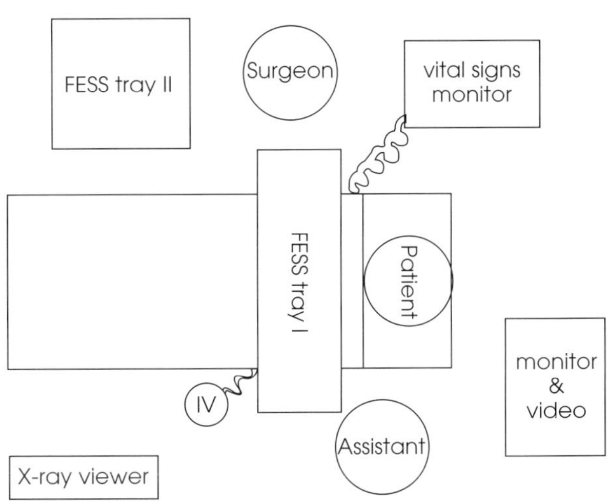

Fig. 8.1. Operating theatre set-up and layout for a patient operated on under local anaesthesia.

infiltrated with 0.5 ml of 1% lignocaine with 1:200,000 adrenaline so the anterior end of the middle turbinate is also infiltrated, especially when a concha bullosa needs to be tackled.

FESS Technique

Adequate access to the ostio-meatal complex is of vital importance to the surgeon, and hence any FESS procedure must begin only after correction of relevant septal deformities (caudal and/or high septal deviations), if any. Also, external architecture defects such as osteo-cartilaginous complex deformities should be taken care of by a functional rhinoplasty before endoscopic work commences.

The patient is placed in the 20° head-up position with the neck slightly flexed. This also brings the anterior cranial fossa into a more vertical plane. This means that when instruments are inserted posteriorly, they are less likely to be near the anterior cranial fossa. The procedure may be started using either the 0° or 30° endoscope; we prefer the former. As a first step, the middle turbinate is medialised gently, so that the middle meatus can be comfortably accessed.

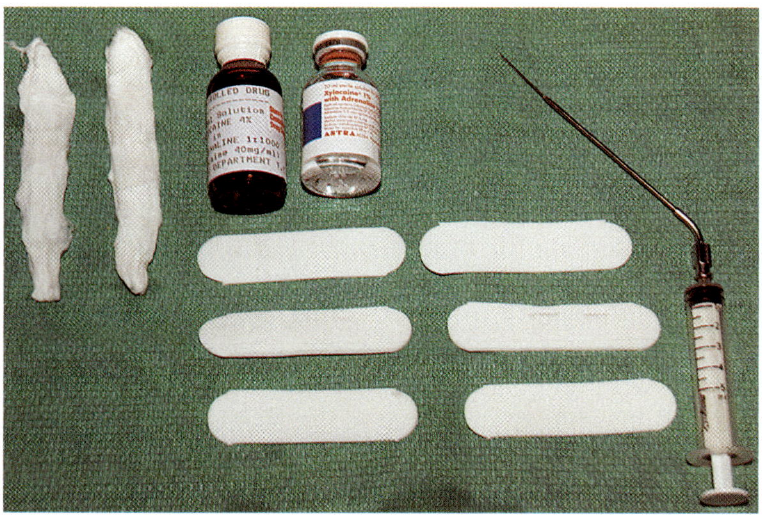

Fig. 8.2. A strip of cotton wool soaked in 4% cocaine/ adrenaline and carefully squeezed before use. In addition, 1% lignocaine with 1 : 200 000 adrenaline is infiltrated into the operative field.

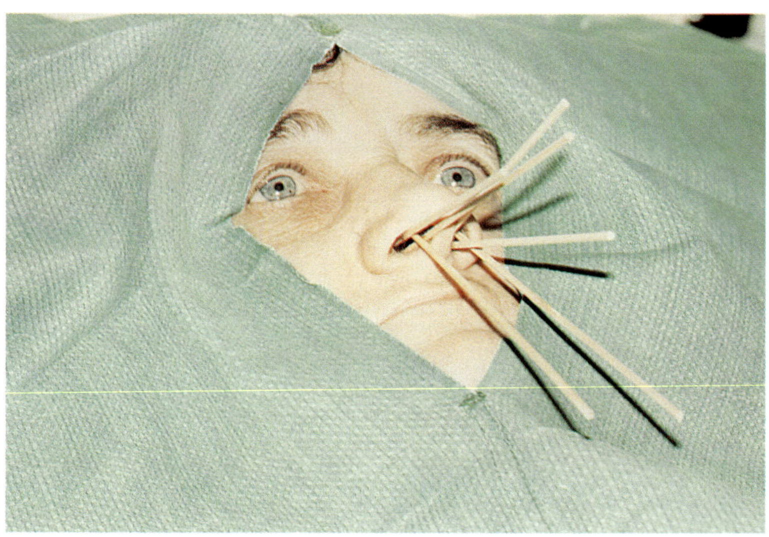

Fig. 8.3. After initial preparation of the nose, further blocks are applied with cotton wool applicators soaked and squeezed in 4% cocaine/adrenaline to the three sites (spheno-palatine nerves, anterior ethmoidal nerve, middle meatus) on each side.

Fig. 8.4. A local FESS in progress. The patient is given an intravenous sedative, and is monitored for vital functions.

Fig. 8.5. Often a concomitant procedure such as septorhinoplasty is performed in order of: 1, septoplasty; 2, hump removal; 3, tip rotation; 4, FESS; 5, osteotomies.

Fig. 8.6. Gentle medialisation of the middle turbinate is the first step of the operation. LW, lateral wall; UP, uncinate process; MT, middle turbinate; S, septum.

Fig. 8.7. Seen here is the site of incision (arrows), between the caudal border of the uncinate process (UP) and the cephalic border of the fronto-nasal process. LW, lateral wall; MT, middle turbinate; S, septum.

8.4

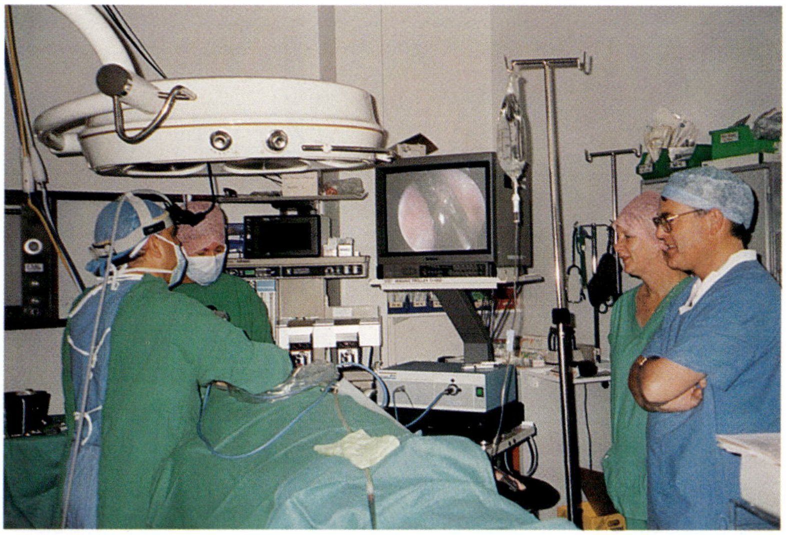

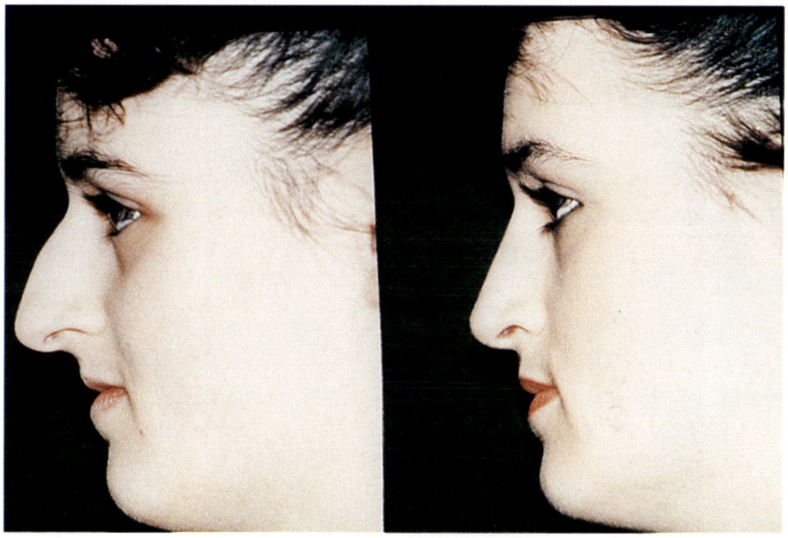

8.5

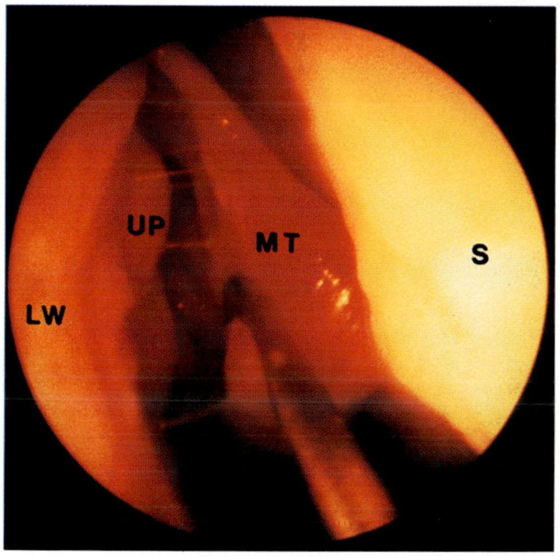

8.6

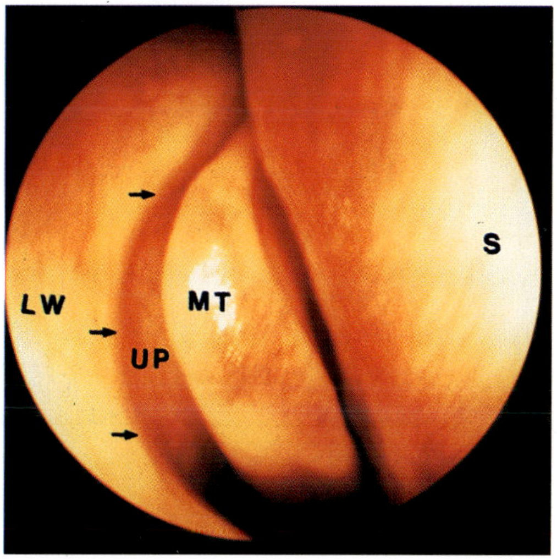

8.7

8.8

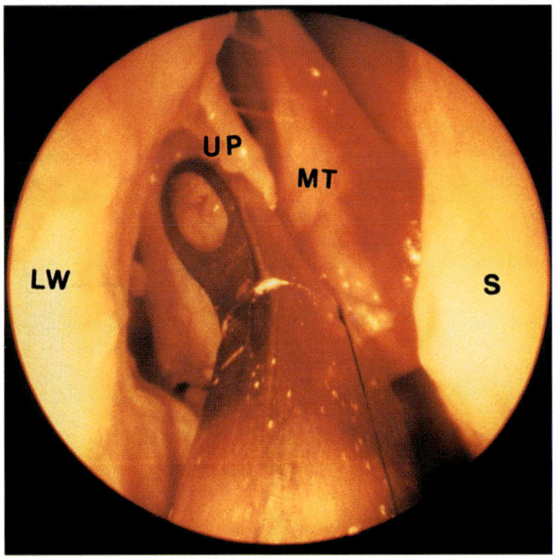

a

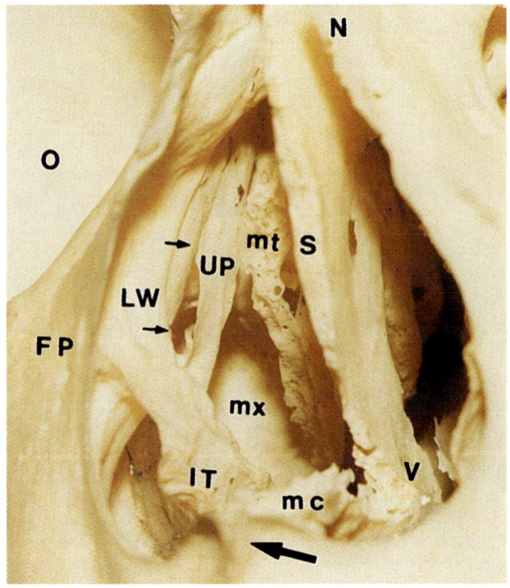

b

8.9

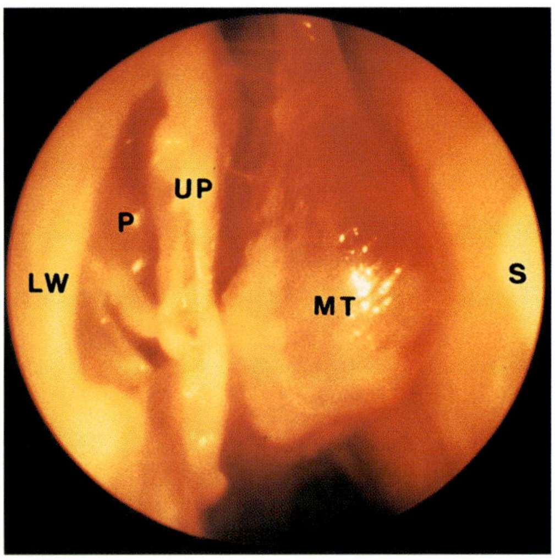

a

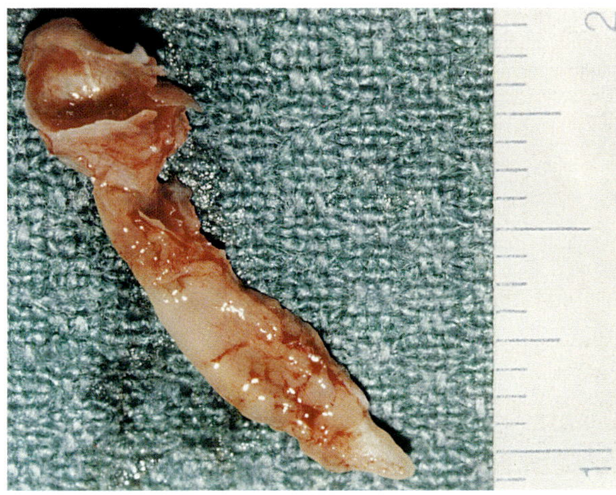

b

8.10

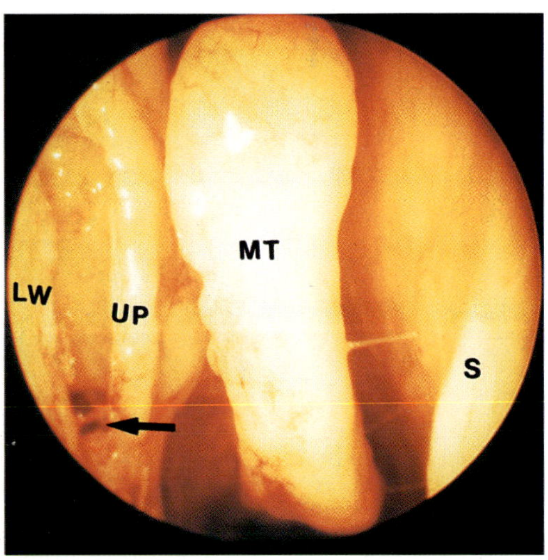

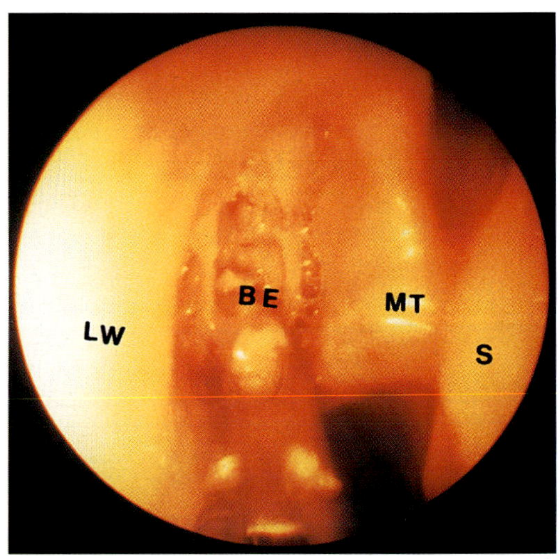

8.11

Fig. 8.8. a The incised uncinate process (UP) is grasped by means of an upward-cutting forceps at its superior and then its inferior ends, and mobilised gently medially, to detach it from the lateral nasal wall. b The intimate relations of the uncinate. LW, lateral wall; MT, middle turbinate; S, septum; O, orbit; FP, frontal process of maxilla; IT, inferior turbinate; MX, maxillary sinus; V, vomer; MC, maxillary crust; N, nasal bone.

Fig. 8.9. a Note the completely separated uncinate process (UP) from its superior to inferior ends, an infundibulotomy has been performed (polyps seen in the infundibulum). b Excised uncinate process with pneumatisation of its superior end. LW, lateral wall; P, polyp; MT, middle turbinate; S, septum.

Fig. 8.10. The initial incision must extend around the lower third of the uncinate process (UP), so as to expose the natural ostium (arrow). LW, lateral wall; MT, middle turbinate; S, septum.

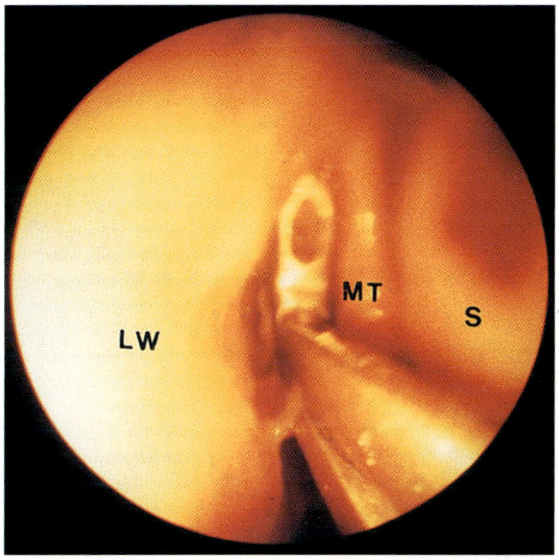

8.12

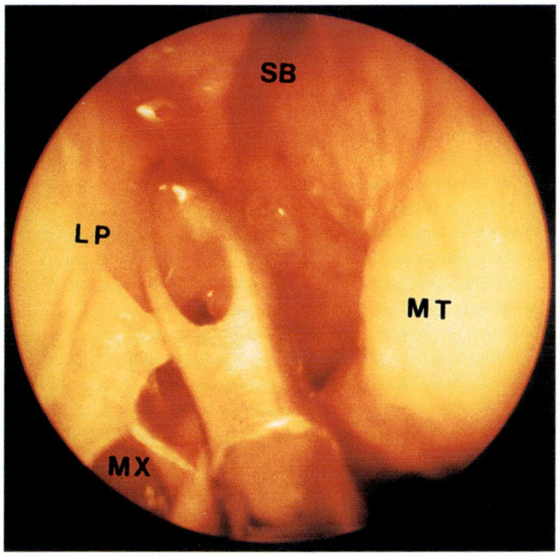

a

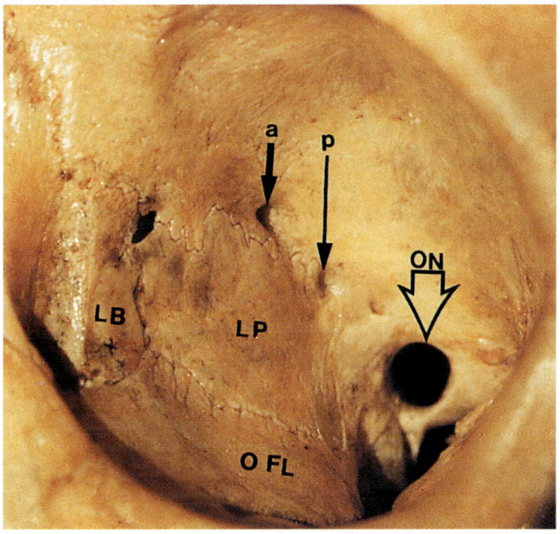

b

8.13

8.14

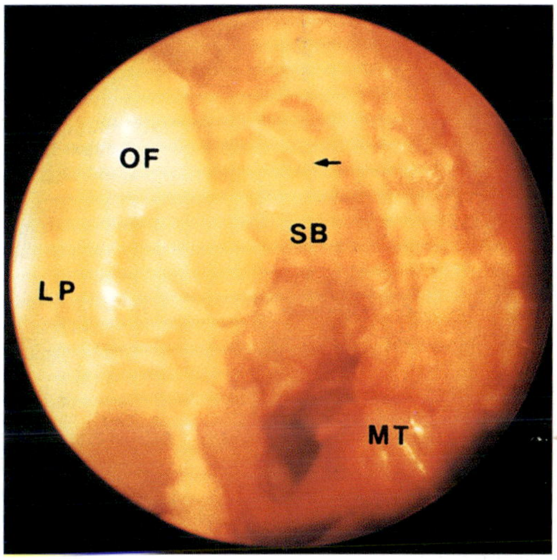

Fig. 8.11. Excision of the bulla, thus opening the middle ethmoidal cells. LW, lateral wall; BE, bulla ethmoidales; MT, middle turbinate; S, septum.

Fig. 8.12. Note the position of the forceps parallel to the lateral surface of the middle turbinate and lamina papyracea, with the middle turbinate constantly under vision. LW, lateral wall; MT, middle turbinate; S, septum.

Fig. 8.13. a Gentle removal of cells from the lamina papyracea. b The bony orbit showing ethmoidal arteries. a, anterior; p, posterior; ON, optic nerve; LP, lamina papyracea; SB, skull base; MT, middle turbinate; MX, maxillary sinus; LB, lacrimal bone; OFL, orbital floor.

Fig. 8.14. Identification of the skull base in the posterior aspects, note the olfactory nerves (arrow). OF, orbital fat (cadaver dissection). LP, lamina papyracea; SB, skull base; MT, middle turbinate.

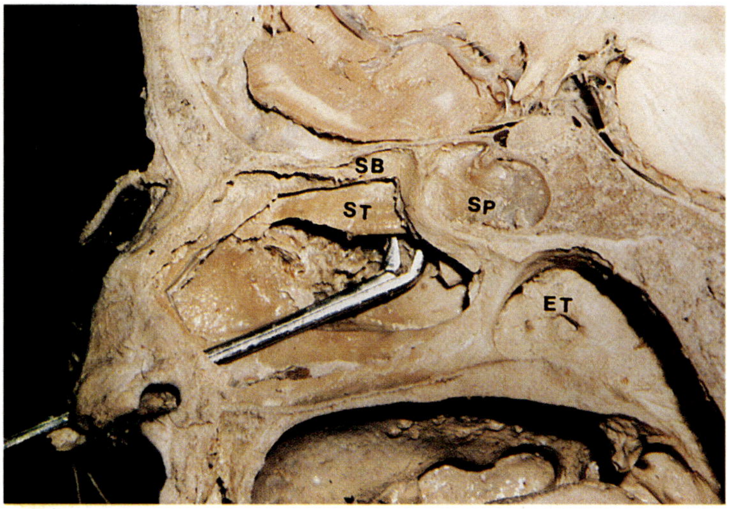

Fig. 8.15. Identification of the antero-inferior wall of the sphenoid sinus is important, so that its *infero-medial aspect* may be entered. SB, skull base; ST, superior turbinate; SP, sphenoidal sinus; ET, eustrachian tube.

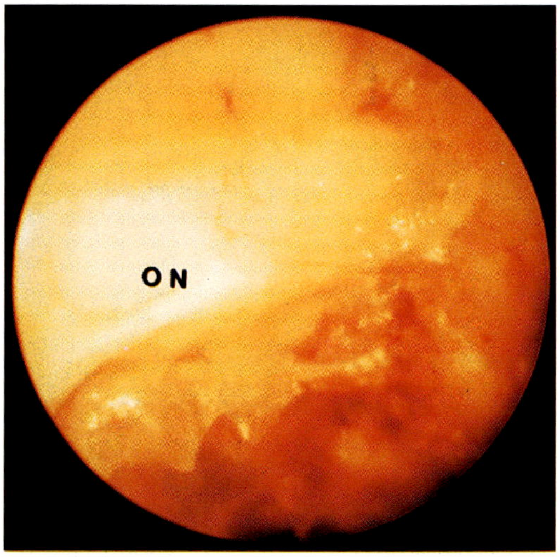

Fig. 8.16. The bulge of the optic nerve (ON) in the right sphenoid sinus.

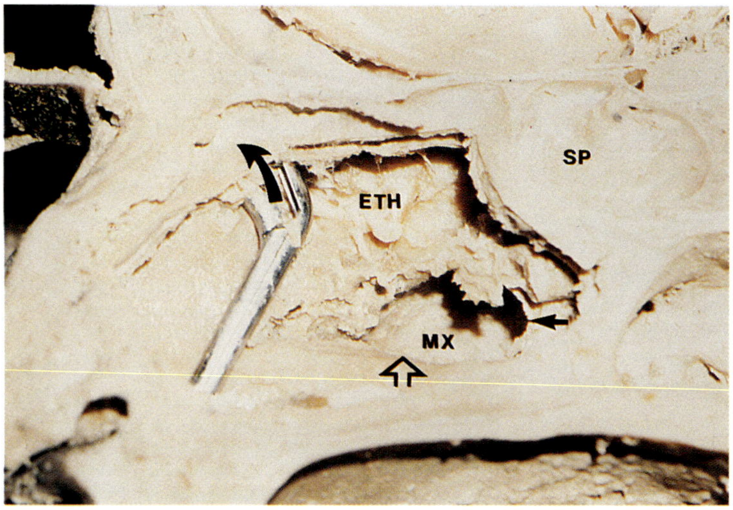

Fig. 8.17. After identifying the skull base posteriorly, dissection is carried out forwards to the frontal recess. ETH, excised ethmoid cells; MX, maxillary sinus; SP, sphenoid; arrow shows direction of frontal recess.

A sickle knife is used to carve out the uncinate process from its lateral attachment, from superior to inferior, and from anterior to posterior. The incision is parallel to the lateral border of the middle turbinate, and is taken in the groove usually seen on the lateral wall. The relation of the uncinate process to the orbit is of utmost importance: a laterally rotated process must be recognised on CT and tackled with due caution.

Uncinate Process

Inadequate removal of the upper part will lead to

1. difficulty in getting into frontal recess
2. may cause obstruction to the drainage of the frontal recess
3. adhesions may form which may obstruct frontal recess

Inadequate removal of the lower part will lead to

1. difficulty in getting into maxillary ostium
2. may cause obstruction of the infundibulum

The upward cutting (Weil–Blakely) forceps is used to grasp the uncinate process, and deliver it by gentle medial rotation by detaching superiorly and inferiorly. The base of the process is often left behind by beginners, leading to difficulties in identification of the natural os, and subsequent performance of an antrostomy. The uncinectomy allows a good view of the infundibulum, thus clearing the way for the next phase of the procedure.

The bulla (middle ethmoidal cell) is next removed using straight forceps. Careful dissection of the middle ethmoidal cells brings one to the roof of the ethmoid complex. The anterior ethmoid artery may be identified at this point, a useful landmark for the superior limit of the ethmoidectomy. The anterior ethmoid artery originates from the ophthalmic artery, a branch of the internal carotid system. It runs in a conduit along the skull base, and the junction of the middle and anterior ethmoidal cells. The medial limit of the dissection and the medial limit of the middle/anterior ethmoid complex is the upper attachment of the middle turbinate.

Guidelines for Surgery

- meticulous atraumatic technique
- minimise bleeding – mucosal & arterial
- use 0° endoscope
- 30° and 70° 'scopes may cause distortion, trauma and bleeding
- identify landmarks: uncinate process/lamina papyracea/ground lamella/skull base
- approach should be directed medially and downwards as the endoscope is advanced

Operative Landmarks

- middle turbinate
- uncinate process
- lamina papyracea
- ground lamella
- skull base
- upper border of posterior choana
- anterior wall of sphenoid

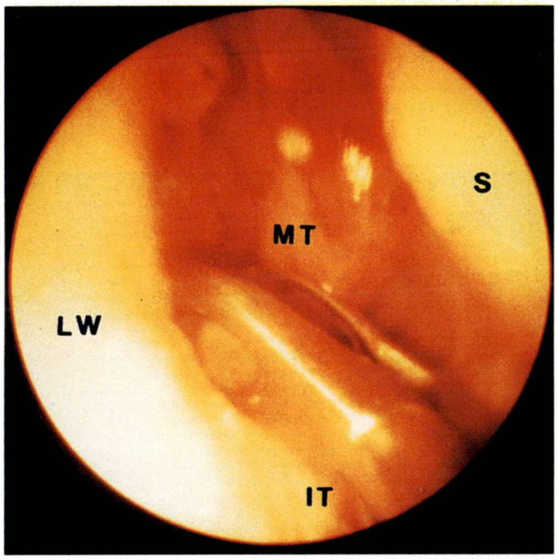

Fig. 8.18. The first step for a middle meatal antrostomy is identification of the natural ostium at the lower border of the middle turbinate (MT) (if not already done). LW, lateral wall; IT, inferior turbinate; S, septum.

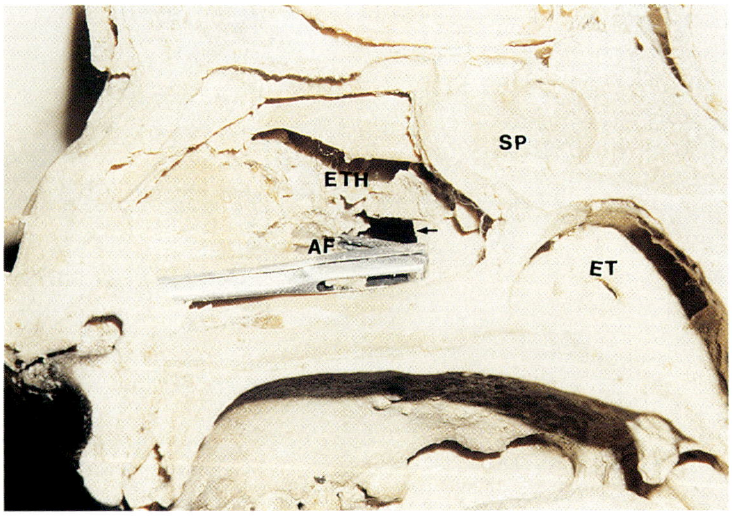

8.19

Fig. 8.19. The middle meatus is enlarged at the expense of the anterior fontanelle (AF) (antero-inferiorly). ETH, excised ethmoid cells; SP, sphenoid; ET, eutachian tube.

8.20

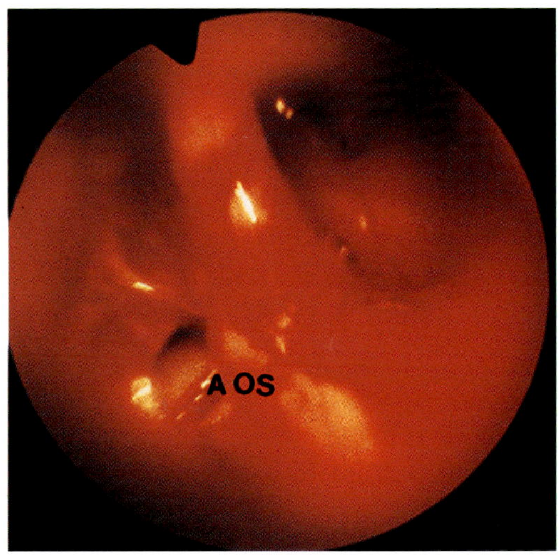

Fig. 8.20. To prevent recirculation of mucus, the natural ostium must be connected to an accessory one if present. Shown is an enlarged middle meatal antrostomy along with an accessory ostium AOS).

Fig. 8.21. **a** Wedge resection of MT, note approximately 0.5–1 cm of the anterior end of the middle turbinate is excised. **b** The excised wedge of the anterior end of the middle turbinate (MT), pneumatisation of the anterior end is seen. S, septum; LW, lateral wall.

8.21

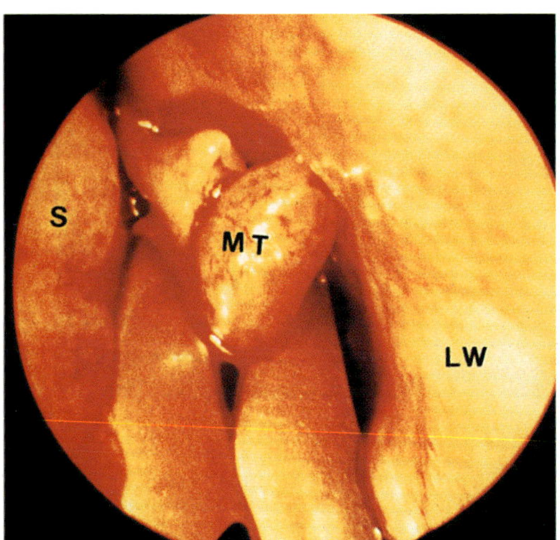

a

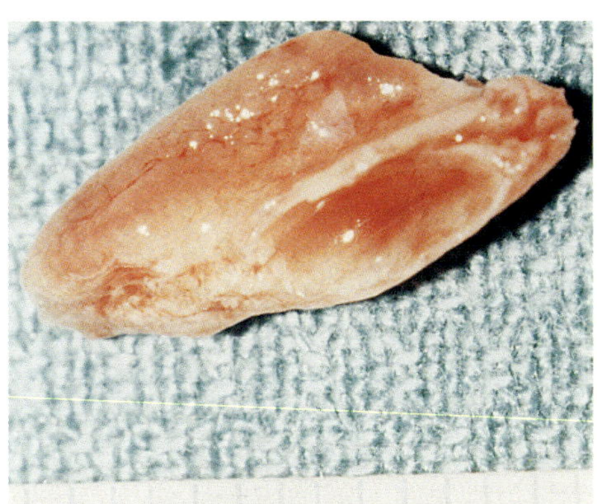

b

There is good reason to stay away from this superior junction of the middle turbinate with the skull base. The cribriform plate is thin here, and easily traumatised because of multiple perforations by the olfactory nerve. The ethmoid roof is hard throughout its length, but at its medial-most aspect it turns down sharply to join the cribriform plate, where it is quite thin.

The lamina papyracea is now identified, and kept in perspective throughout the operation. The ethmoid air cell complex is debulked gently, avoiding any tugging motions. All exenterated material is immersed in water to check for adipose tissue, a warning sign of breach of the lamina papyracea. Multiple small bites are better than a single large one. All forceps movements should be parallel to the lamina.

The skull base is identified posteriorly where it is at its thickest. The ground lamella is a thin plate of bone which connects the middle turbinate to the lamina paparycea. The perforation is always done medially to avoid traumatising the medial orbital wall. Once the larger posterior ethmoidal cells are identified, the endoscope should follow a gentle curve downwards and medially, *not upwards or laterally*. This prevents inadvertent entry into the cranium, or damage to the optic nerve.

If the posterior ethmoidal cells are seen to be clear, the procedure stops here. If, however, there is disease in this area, it is important to exenterate all cells with gentle biting movements from backwards to forwards. The antero-medial section of the roof is extremely thin; there is a distinct risk of breaching this boundary (danger area). Next the fourth important landmark is identified, the antero-inferior wall of the sphenoid sinus. This is opened and enlarged (sphenoidotomy). The sphenoid sinus is best opened at its inferior and medial aspect. The manoeuvre is carried out keeping in view the posterior choana which is 1 to 1.5 cm from the anterior sphenoidal wall. The sinus is normally at a distance of 7 cm from the anterior nasal spine. At 9 cm the tip of the forceps lies within the lumen. As a dehiscence of the internal carotid artery occurs in a quarter of patients, no instrumentation should be done within the sinus.

The middle meatal antrostomy follows, and we find that an incision along the caudal border of the uncinate process right up to the inferior end allows for better identification of the natural ostium. Landmarks locating the natural ostium are of crucial importance. The ostium usually lies at the same level as the inferior margin of the middle turbinate anterior to the bulla ethmoidales. Theoretically, the ostium is sited at the mid-point of the middle turbinate; however, this degree of depth perception is often difficult during an endoscopic procedure.

Finally, the close relationship of the ostium to the inferior orbital wall must always be kept in mind, and any evidence of lamina papyracea dehiscence must be looked for. The ostium is almost always in the same sagittal plane as the lamina papyracea, which by now should have been identified as a part of the ethmoidectomy. A helpful tip for identification of the os is the presence of air bubbles at the site. In difficult cases, as in gross sinus pathology, distorted anatomy or recurrent disease, we advise a trocar and cannula be passed through the inferior meatus into the sinus cavity, so that it may be irrigated under endoscopic vision. This helps to locate the position of the ostium. It should be remembered that fluid may exit the sinus through an accessory ostium as well, thus misleading the surgeon!

After identifying the ostium, it is denuded of its surrounding bone using upward-cutting forceps, and the antrostomy is now enlarged at

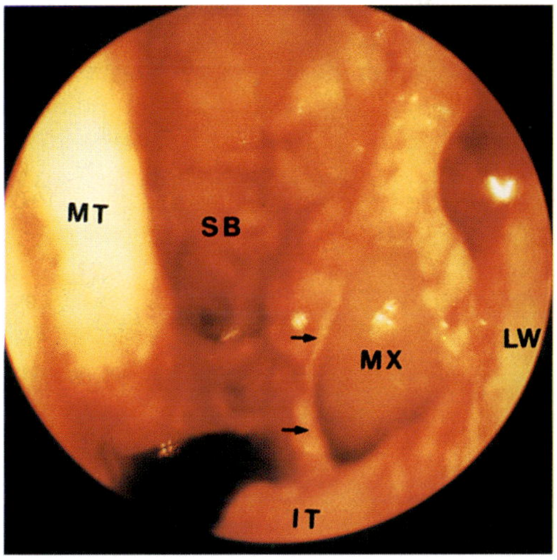

Fig. 8.22. Operated cavity as seen at the end of the procedure (left side). The arrows show mucosa everted over the antrostomy margin. MT, middle turbinate; SB, skull base; MX, maxillary sinus; LW, lateral wall; IT, inferior turbinate.

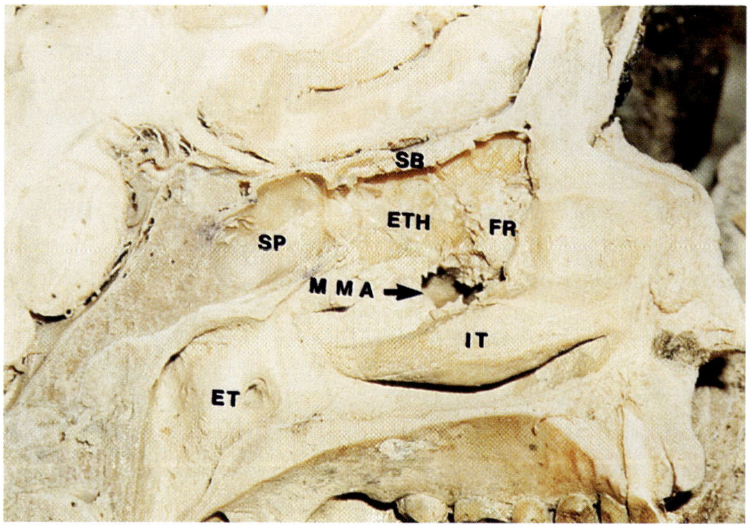

Fig. 8.23. Similar view to Fig. 8.22 in a cadaver. The relationship of the middle meatal antrostomy (MMA) to the inferior turbinate is seen. SP, sphenoid sinus; ET, eustachian tube; ETH, ethmoid canity; FR, frontal recess; IT, inferior turbinate; SB, skull base.

the expense of the antero-inferior fontanelle. A factor of paramount importance at this stage is that anterior sacrifice of bone must be stopped when the thickness of the fronto-nasal process of the maxilla is encountered or else unnecessary trauma may result in excessive bleeding. Also, the nasolacrimal duct may be damaged in this region, leading to epiphora in the post-operative period.

Differences between Accessory Os. & Natural Os

Accessory ostium	**Natural ostium**
located in anterior and/or posterior fontanelle, sometimes multiple	located in the lower part of the infundibulum related to uncinate process anteriorly and bulla ethmoidales posteriorly
easy to identify	difficult to identify: can be found at the level of lower border of middle turbinate or upper border of inferior turbinate
circular or punched out appearance	usually oval-shaped
lies in saggital plane	lies more obliquely
present in 10% to 40% of patients	always present

Another tip concerns the opening itself; in order to prevent stenosis, it is prudent to "tuck in" the mucosa around the ostium as a final step.

Often the natural ostium is not easy to identify intra-nasally due to:

- thick periosteal bone
- scarring from prior surgery
- blood obscuring the operative field
- hypoplastic sinus
- gross pathology (polyposis)

In order to circumvent these problems, we employ a modified Messerklinger technique, combining intra-antral and intra-nasal views (CAMMA: Combined Approach Middle Meatal Antrostomy). Having exposed the ostio-meatal complex, a 0° endoscope is inserted through a sublabial cannula into the sinus, and the medial wall of the sinus is brought into view. Working through the nose with the other hand, the curved upward cutting forceps are used to enlarge the natural ostium. Since this is the only instrument in the nose at this time, the range of movements is quite free; in addition, the forceps teeth can be directly visualised through the endoscope as they gradually fashion the middle meatal antrostomy at the expense of the antero-inferior wall of the ostium. If an accessory ostium has been identified, it must be communicated with the natural ostium as a part of the middle meatal antrostomy.

The patient leaves the hospital after 24 hours if a concomitant operation such as septoplasty or

> **Middle meatal antrostomy (infundibuloplasty): salient points**
>
> **FACTS**
> - "MMA" is not synonymous with FESS!
> - a "hole" in the fontanelle is NOT "MMA"
> - a functional MMA is more important than the physical size of the MMA
>
> **Technique of MMA**
> 1. remove lower end of the uncinate process
> 2. preserve bulla ethmoidales until natural ostium is identified
> 3. remove stenotic areas of the ethmoid around natural ostium
> 4. remove any pathology around natural ostium with preservation of mucosa
> 5. connect natural ostium with the accessory ostium

septo-rhinoplasty has been performed. Patients operated on under a local anaesthetic are fit to go home the same evening, while FESS patients who have had a general anaesthetic go home the following day.

All in all, the CAMMA technique proves to be a safe method of visualising the natural ostium from within. It leaves the nasal cavity free for instrumentation, an important point for beginners who may find difficulty working in a limited space. The routine technique calls for three instruments (endoscope, cannula and forceps) being passed into the nose simultaneously, each about 4 mm in diameter. The CAMMA procedure cuts this down to a minimum of one instrument. Furthermore, it is known that leaving a bridge of tissue between the natural and accessory ostia leads to a local recirculation of mucus and constant reinfection of the maxillary sinus. The technique permits direct visualisation of the ostia, thus facilitating their communication.

The final step is clearance of the frontal recess and agger nasi cells if there is disease in the frontal sinus. This is performed at a late stage in order to prevent soiling of the operative field by the almost certain oozing that takes place. For this procedure, a 30° or 70° 4 mm endoscope is recommended. Upward cutting forceps are used to enter the frontal recess at the attachment of the middle turbinate to the lateral wall of the nose. It is not always easy to find the frontal recess, as it may be surrounded by a cluster of anterior ethmoidal cells (agger nasi). However, in this situation, usually the *most medial cell* is the one which will lead into the frontal recess and ultimately into the frontal sinus.

We find exenteration of this rather inaccessible region is best performed by use of ophthalmic upward-cutting punch forceps. It is important to clear the anterior ethmoids, as one of the most common causes of recurrence and persistence of symptoms is the presence of disease in this region. Unfortunately, until adequate dexterity is achieved, one tends to overlook this part of the procedure as it is rather cumbersome working

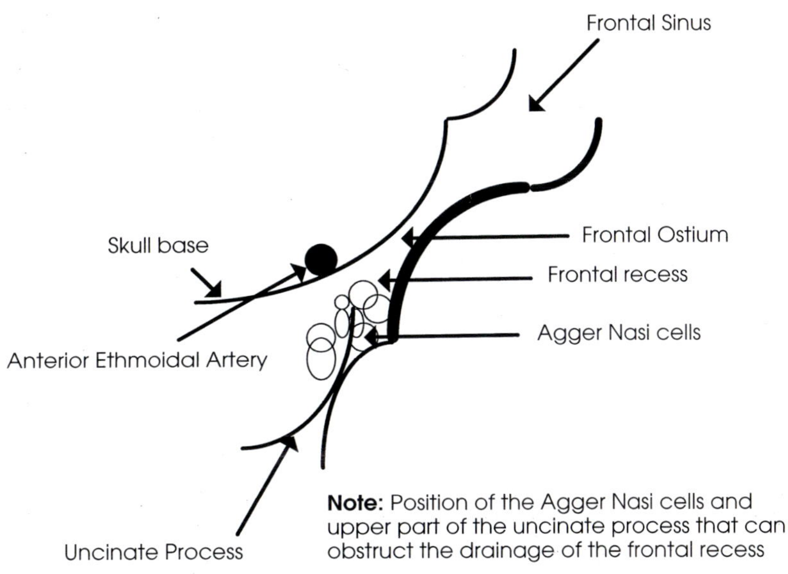

Fig. 8.24. Surgical anatomy of the frontal recess.

through 30° and 70° endoscopes; however, it must be remembered that any compromise at this stage might well bring the patient back with residual disease.

Surgery of the frontal recess

Difficulties

- Difficult surgical anatomy to understand on the operation table
- Dangerous area as it is close to skull base and orbit
- Surgical area is difficult to view
- 30° and 70° degree endoscopes need to be used. These endoscopes cause distortion and foreshortening of the surgical field and are difficult to handle
- Special instruments are needed
- It is extremely important to preserve the mucosa otherwise stenosis readily occurs
- Osteoneogenesis is common in this area
- Surgeon must learn the technique on cadavers

Key points in the surgical technique

1. Identify skull base and anterior ethmoidal artery
2. Use 30° and 70° degree endoscopes
3. Keep attachment of the middle turbinate and cribiform plate in sight all the time
4. Work in slightly lateral direction
5. Always work anteriorly
6. Never work anterior to posterior direction
7. Gently palpate with ostium seeker or suction cannula for "false" ceiling (egg shell consistency of the aggar nasi cells)
8. Always preserve mucosa
9. If stent is to be used never insert the stent tightly which may cause mucosal necrosis and stenosis

Anatomical relations of the aggar nasi cells

- Laterally: nasal and lacrimal bones
- Anteriorly: frontal process of maxilla
- Superiorly: frontal recess and frontal sinus
- Inferiorly and medially: uncinate process
- Posteriorly: ethmoidal infundibulum

Note: location and extent of pneumatisation vary widely

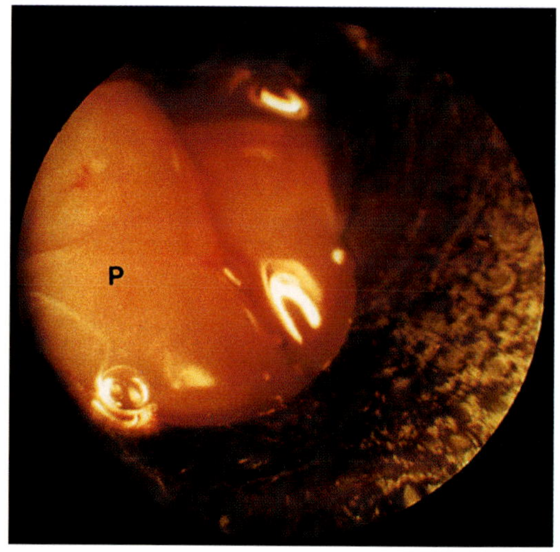

Fig. 8.25. Quite often, the maxillary sinus is full of thickened mucosa and polyps, and identification of the ostium is quite difficult. Not much is seen on antroscopy except polyp (P).

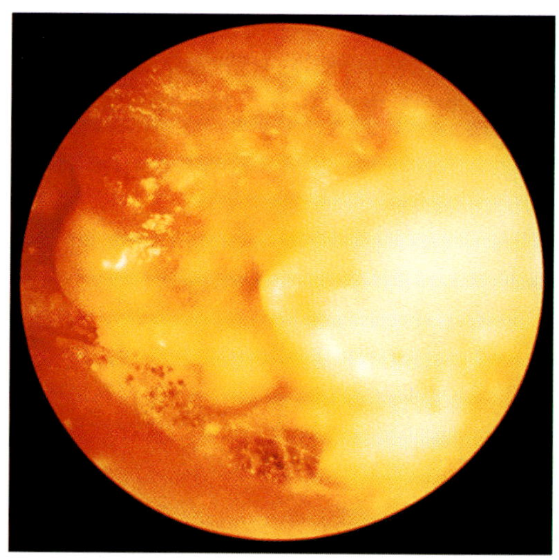

Fig. 8.26. Scarring, muco-pus or granulation tissue can obscure the view of maxillary natural ostium, as seen on antroscopy.

Fig. 8.27a–h. CAMMA technique. **a** Identify the canine fossa always lateral to the canine tooth. **b** A small stab is made through the mucous membrane. **c** The trocar and cannula are inserted through the canine fossa. **d** The endoscope is inserted through the sub-labial cannula into the sinus, while instrumentation is performed through the nose under direct vision. (From *Laryngoscope*, vol. 102, June 1992, with permission). **e** Backward cutting forceps seen enlarging the natural ostium under direct vision. P, polyps; PW, posterior wall of maxillary sinus. **f** Cysts may also be excised under direct vision. PW, posterior wall of maxillary sinus; c, cyst. **g** The sinus is irrigated, and outflow assesed through the nose endoscopically to check the size of newly fashioned antrostomy. **h** The stab incision may be closed with a single stitch.

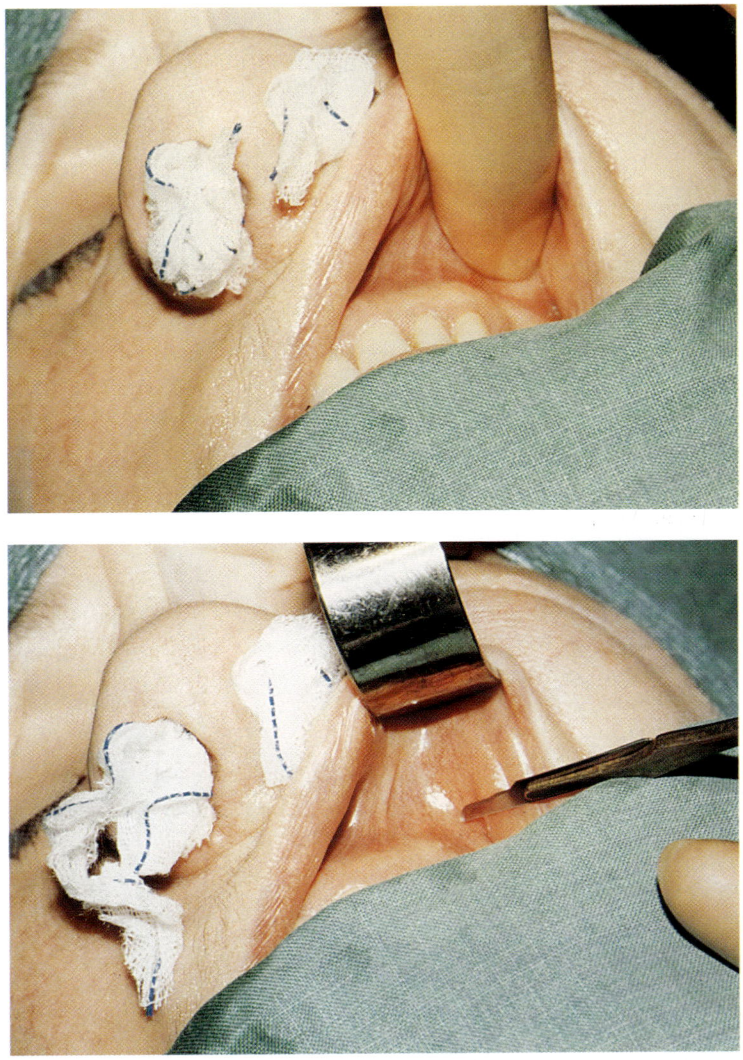

a

b

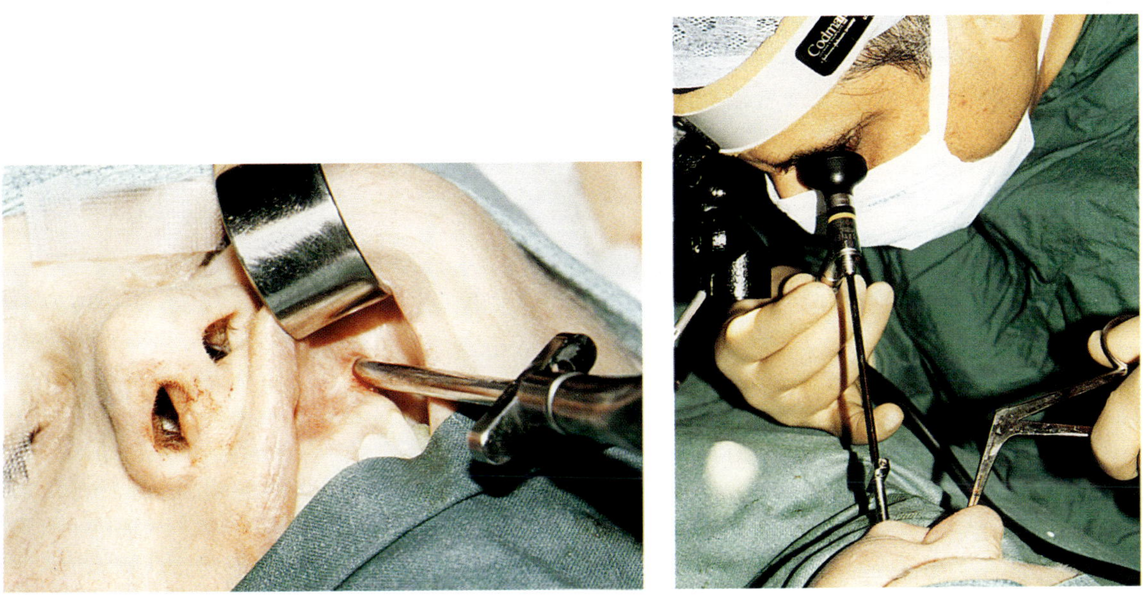

c

d

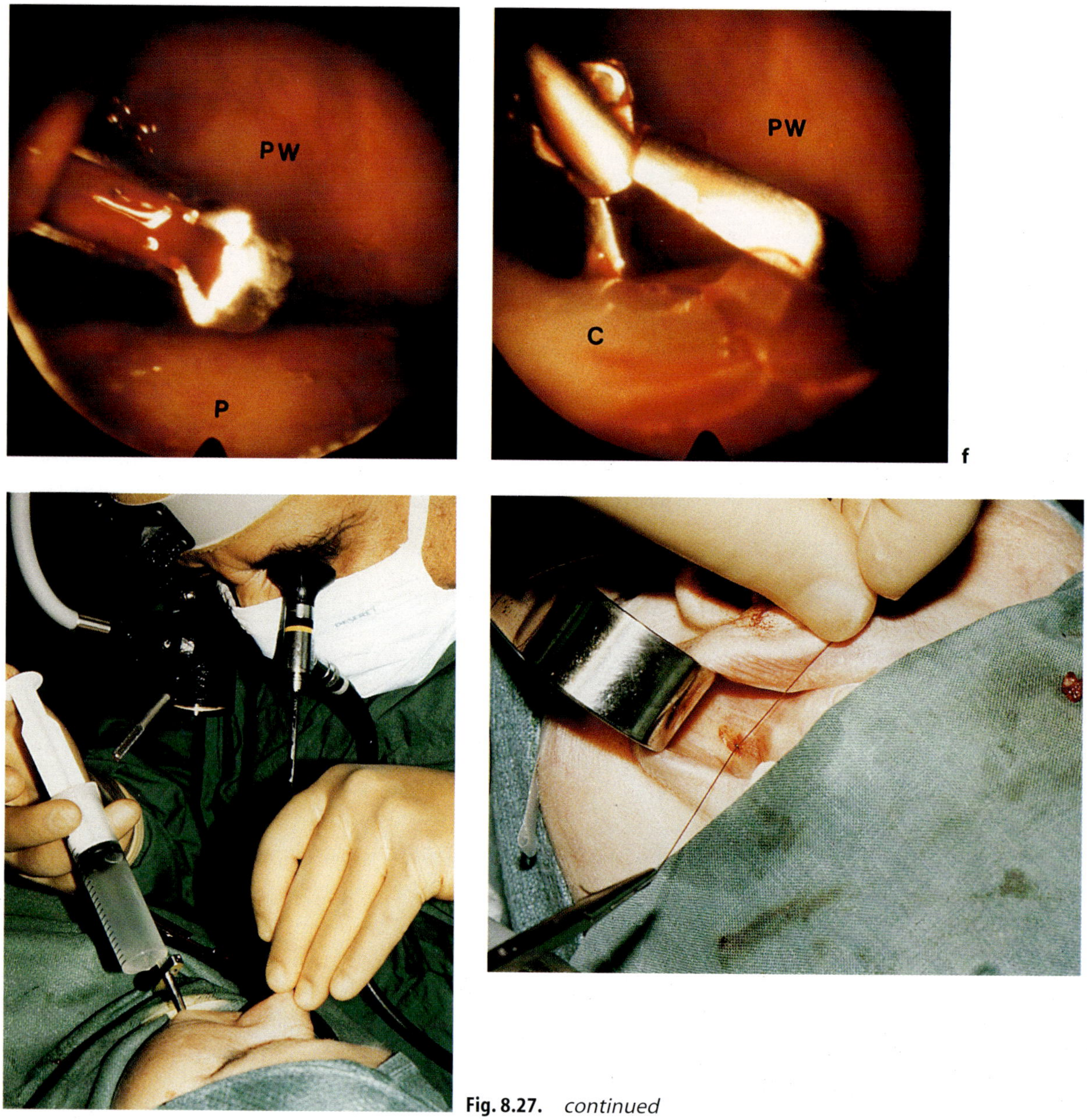

Fig. 8.27. *continued*

It is widely recognised that the post-operative period in FESS is as important as the operation itself. The commonest complication is the formation of adhesions between the middle turbinate and the lateral wall of the nose, which impede both mucous drainage and ventilation, as well as access to the operated field.

With this in mind, we have further modified the Messerklinger technique to include a wedge resection of the antero-inferior end of the middle turbinate, thus widely opening the ostio-meatal complex. This simple manoeuvre (shown in Fig. 8.28) affords easy passage to the surgeon during post-operative endoscopic cavity care, and minimises adhesions by preventing apposition of raw surfaces.

After the FESS has been performed, angled endoscopic scissors are used to make a cut in the anterior end of the middle turbinate (in the postero-inferior direction). A second cut joins the

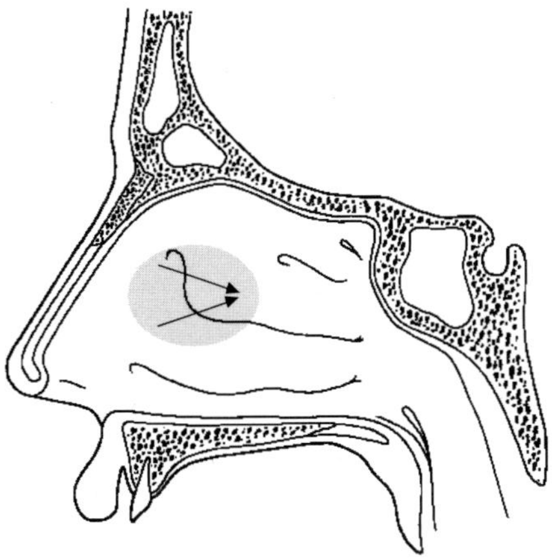

Fig. 8.28. Techniques of wedge resection of antero-inferior end of the middle turbinate. Upper arrow shows a cut in the posterior-inferior direction; lower arrow shows a cut in the infero-superior direction.

first from inferior to superior, thus freeing a small wedge approximately one sixth the length of the turbinate. Bleeding is usually minimal, and could be controlled by packing the nose for a few minutes. Under local anaesthetic, the procedure must be preceded by injection of lignocaine/adrenaline into the anterior end of the middle turbinate.

Our experience shows that significant number of middle turbinates do contain conchal cells, often showing mucosal thickening and/or pus, in spite of a radiologically normal turbinate on CT. It has been suggested that middle turbinate pathology may account for non-specific facial pains; this is borne out by our findings. Therefore, wedge resection also opens out an overt or latent concha bullosa.

Surgery of the Middle Turbinate

An enlarged middle turbinate due to either hypertrophied mucous membrane or to pneumatisation (concha bullosa) causes obstruction of the normal flow of air and/or mucus. Endoscopic visualisation under topical anaesthesia may reveal mucosal oedema in the middle meatus, but often there may be no overt evidence of disease. Plain films of the paranasal sinuses again do not aid in the diagnosis. On the other hand, CT is quite effective in elucidating the extent of disease at an early stage. Additional findings such as mucosal swelling and blockage of the structures comprising the lateral nasal wall or ostio-meatal complex are often revealed (stenosis of the infundibular block).

Surgical management of a conchous middle turbinate must be based on its size and shape, and its relationship to surrounding structures (including areas of mucosal apposition). The nasal septum, the uncinate process and the anterior and middle ethmoidal sinuses all may be impinged upon by the turbinate laterally. If the septum itself is thickened superiorly, or is deviated towards the enlarged middle turbinate, it may be necessary to shave a wafer-thin layer off its cephalic cartilaginous portion and perpendicular plate by a septoplasty. Frequently, this itself improves nasal function. We excise the turbinate by wedge resection at its anterior end, as described earlier. In fact, a pilot study carried out by us revealed pus/polyps in middle turbinates that appeared otherwise normal, both endoscopically and radiologically. We now perform the procedure almost routinely. An attractive side-effect is the decrease in formation of adhesions between the lateral turbinate surface and the lateral nasal wall, although stringent post-operative care is still mandatory.

The Caldwell–Luc Procedure Revisited

The classical trans-buccal sino-rhinostomy operation was first described for the relief of chronic sinusitis by George Caldwell of New York in 1893. A few years later (1897), Luc in Paris proposed a similar technique. History bestowed the procedure with the combined names of Caldwell and Luc.

With the introduction of FESS, the traditional Caldwell–Luc procedure (CLP) has come under increasing scrutiny the world over. Long-term results evaluated at several centres indicate poor patient satisfaction, and persistence of disease as demonstrated by both endoscopy and CT. In fact, a large proportion of these patients have had to undergo FESS clearance. The reason for this is clear. The Caldwell–Luc technique involves accessing the maxillary sinus via the sublabial route, currettage of the sinus lining, and creation of an inferior meatal intra-nasal antrostomy. Thus the technique focuses only on exenteration of maxillary disease; furthermore, the inferior meatal antrostomy most often proves an inadequate answer to obstructive disease at the natural ostium. Today we know that the muco-ciliary pathways of the sinus are almost certainly genetically predetermined, and follow a route towards the natural middle meatal opening. Furthermore, chronic sinus disease almost always resides in the anterior ethmoids, which are not tackled at all during a CLP. Ironically, it was Caldwell who said, "the semilunar fold acts as an imperfect valve to the maxillary sinus, and incidentally guides fluids from the higher cells into the antrum, as I have repeatedly demonstrated on the cadaver. For this reason, the diagnosis of empyema of the antrum is not sufficient until the frontal and anterior ethmoid cells have been excluded. The antrum may be the *receptacle* (not the origin) of pus, or become involved secondarily". This was as far back as 1893, the very year George Caldwell described the technique that bears his name. Restricted as he was with poor instrumentation, he was magnanimous enough to acknowledge his own limitations (and thereby the technique's), a fact often forgotten by modern-day proponents of the CLP.

In view of this ongoing debate, we reviewed our own series of 36 CLPs done over the last decade. In our cumulative experience, the incidence of persistent post-operative symptoms was 70%. A breakdown of these symptoms is as shown in Figs 8.29 and 8.30.

A fifth of these (20%) had patent antro-nasal windows at presentation. In the other cases (80%), the window was either completely occluded, or severely stenosed. Nasal endoscopy and CT scanning was performed, revealing an anterior ethmoid focus of disease *in all cases*. Our endoscopic findings ranged from common middle meatal obstructive factors such as turbinate para-

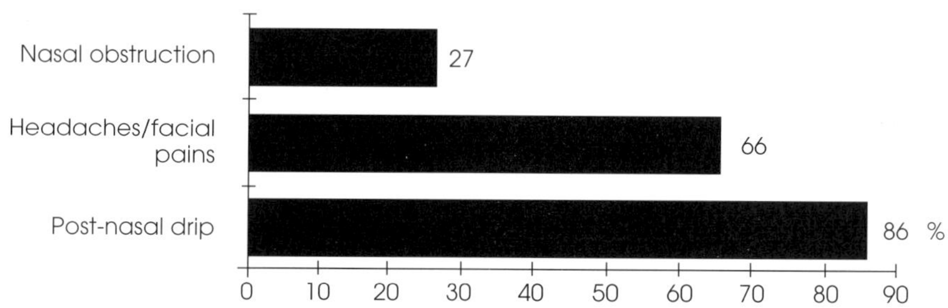

Fig. 8.29. Symptom profile of Caldwell–Luc patients ($n = 36$).

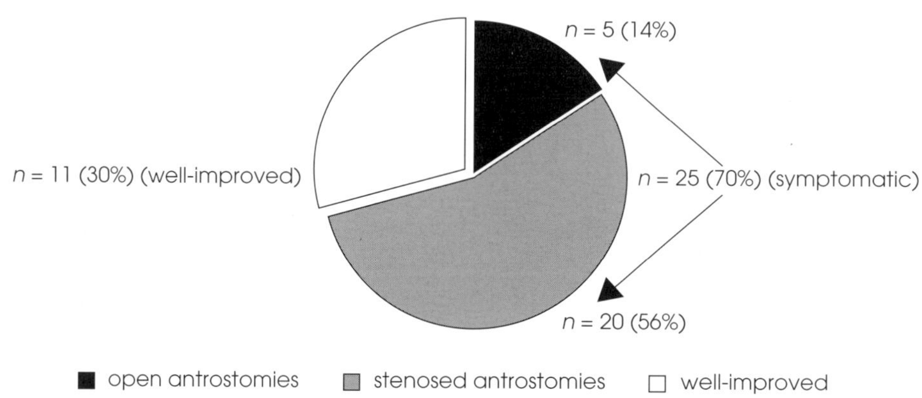

Fig. 8.30. Clinical findings of INA and outcome of our series of CLP patients.

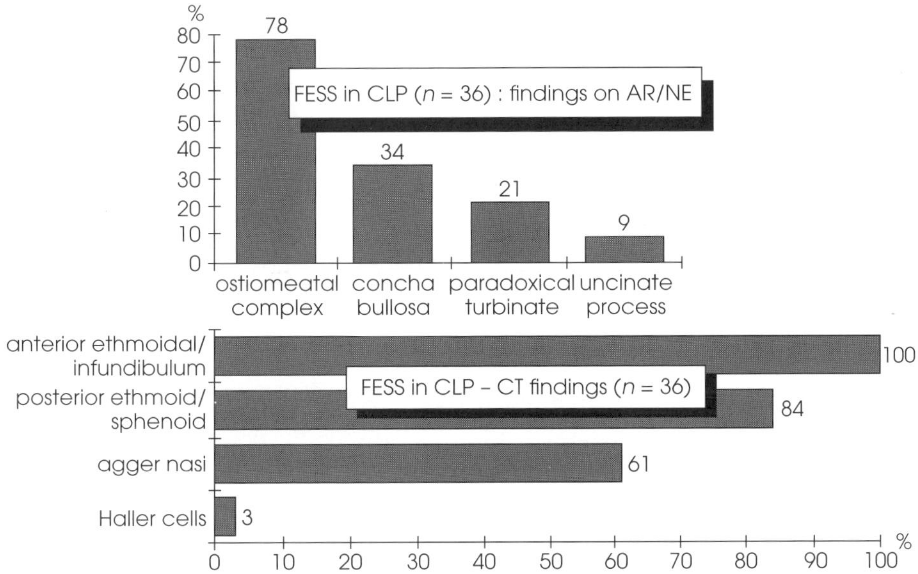

Fig. 8.31. Anterior rhinoscopy (AR), nasal endoscopic (NE) and CT findings in Caldwell–Luc patients (CLP) undergoing functional endoscopic sinus surgery (FESS).

doxicity, turbinate hyper-pneumatisation and medial/lateral rotation of the uncinate process to rarer turbinate malformations. CT confirmed endoscopic findings, and also showed anterior ethmoidal cell clouding and/or a narrow infundibulum, amongst other pathologies. These findings are displayed in Fig. 8.31.

As most studies so far have attempted to evaluate CLP results subjectively, we felt the need to develop an objective, easily performed, reproducible technique for CLP evaluation. Sterile saccharin was used to evaluate muco-ciliary clearance time. In those cases where an the inferior meatal antrostomy was patent (open-window group), the solution was instilled through the antrostomy itself. In cases of stenosis or closure (closed-window group), the anterior fontanelle was used to access the sinus. Details of the actual technique may be found on page 18.

Our study confirmed that muco-ciliary timings were elevated in all of the 12 CLP patients examined, irrespective of the status of their antro-nasal windows. Pre-FESS times in this group averaged 50 minutes, while 6 month reviews revealed a

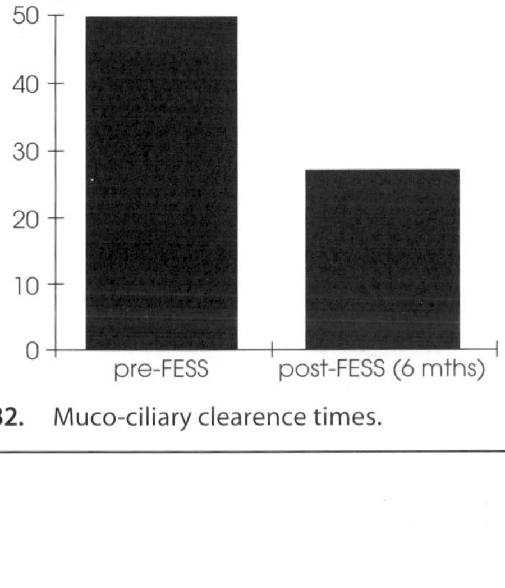

Fig. 8.32. Muco-ciliary clearence times.

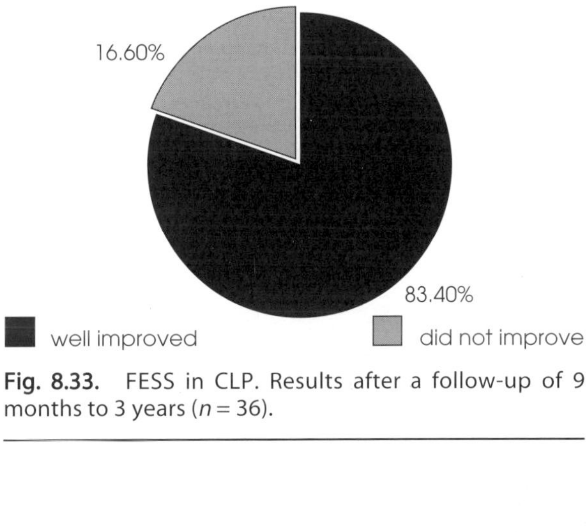

Fig. 8.33. FESS in CLP. Results after a follow-up of 9 months to 3 years ($n = 36$).

markedly reduced time of 27 minutes in nine patients (see above). Three patients had persistently delayed saccharin times – two of them were at 34 and 36 minutes – these might have had underlying mucosal abnormalities. The third showed no decrease in clearance at 6 months, and also had no relief from her symptoms. She was diagnosed to be aspirin-sensitive. All the other patients, however, improved symptomatically. This leads us to believe that the CLP does not control sinus disease in most patients. Our overall results are shown in Fig. 8.33.

No doubt one is averse to giving up such traditional methods of dealing with sinus disease – old habits die hard ! However, as one looks back at one's results over time, it becomes increasingly clear that FESS offers a better alternative to these patients, by tackling the disease more physiologically. The technique takes time to learn, and requires more expensive instruments; however, a systematic approach based on surgical anatomy, cadaver dissections and handling the endoscope would bear fruit in time.

Advantages of FESS

- targetted at exenterating the cause and not the sequelae of disease

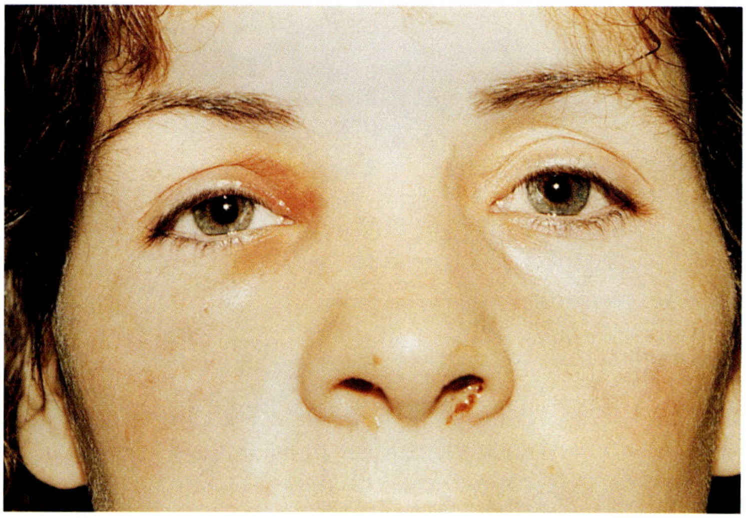

Fig. 8.34. If there is a dehiscence of the lamina papyracea, or a breach during the procedure, mild ecchymosis may result post-operatively, this settles in 3–4 days.

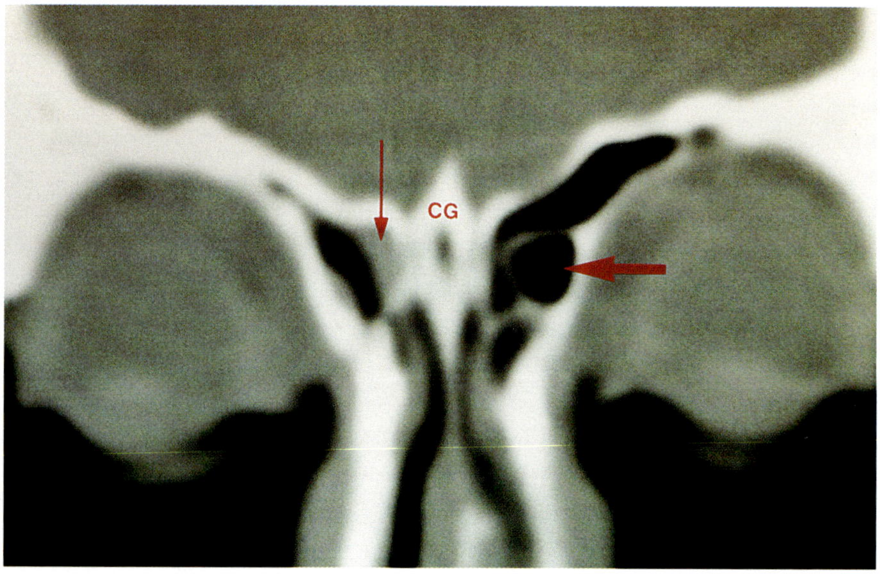

Fig. 8.35. Coronal CT scan. CG, crista galli; thin arrow shows mucosal disease in the frontal recess; thick arrow shows frontal cell obstructing frontal recess.

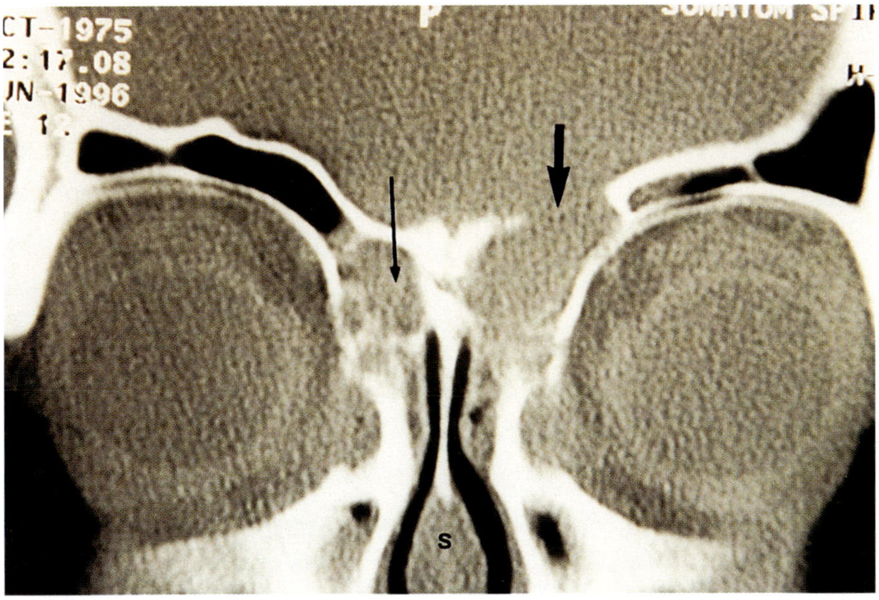

Fig. 8.36. Coronal CT scan showing bilateral frontal recess disease. Thin arrow shows mucosal disease in the frontal recess; thick arrow shows dehiscence of the superior wall.

- re-establishes normal ciliary function by drainage and ventilation of the sinuses
- can be performed under local anaesthesia on a day surgery basis and hence is cost-effective
- excellent visualisation of the operative area with multi-angled endoscopes

Disadvantages of FESS

- meticulous atraumatic technique is required for optimum results
- angulated endoscopes may cause disorientation, increasing the chances of complications
- use of endoscopes may give a false sense of security, the endoscope is not a substitute for surgical expertise but a useful tool
- because of the “microscopic view”, the surgery may extend beyond required limits

Chapter 9

The Ethmoid Cavity: Post-Operative Care

The final ethmoid cavity is bounded by the lateral wall of the nose, the lamina papyracea, skull base and the middle turbinate. Posteriorly the cavity is related to the sphenoid, while antero-inferiorly it communicates with the nasal cavity. Into this space open the enlarged middle meatal antrostomy and the fronto-nasal recess. Thus the cavity acts as a portal for drainage from the maxillary and frontal sinuses.

It is of paramount importance to maintain the patency of the ethmoid cavity. We routinely apply ointment in the ethmoid cavity at the end of the operation (Soframycin/hydrocortisone). However, synaechia could remain a major problem in cavity care, leading to stasis, reinfection and poor surgical outcome. A correlation could be made between the mastoid and ethmoid cavity. Just as even a well-fashioned mastoid bowl requires assiduous toilet, an endoscopic ethmoidectomy must be regularly cleaned out until epithelialisation occurs.

The success of the procedure depends on meticulous care of the resultant ethmoid cavity. Nothing can replace regular follow-up till the cavity has healed satisfactorily. We see all post-operative patients in a special "Endoscopy Clinic". This is so that one might devote adequate time to their management, and a well-defined protocol could be established. Following discharge, the patient is seen twice-weekly for at least 2 weeks and weekly thereafter until the cavity begins to epithelialise. Every patient undergoing FESS with an instant septoplasty and/or rhinoplasty are advised to use warm alkaline nasal douch two or three times a day. Alkaline solution being mucolytic it also clears blood clots and crust from the the ethmoid cavity and enhances the epithelialisation. Our experience shows that this is very well tolerated by the patients as they tend to breathe more freely after nasal douch.

In endoscopic clinic the patient has his or her nose anaesthetised with 4% cocaine/adrenaline, and endoscopic suction clearance is performed. Blood clots, crusts, sequestrum and thickened mucosa are removed, as are small, residual polypi. The fibrous connections between the lateral wall and the turbinate are broken down, so as to prevent adhesions. Finally, the cavity is painted with an antibiotic/ steroid ointment. Post-operative antibiotics are not used routinely. They may be indicated if the procedure reveals extensive infection. In these cases, a broad-spectrum antibiotic is administered.

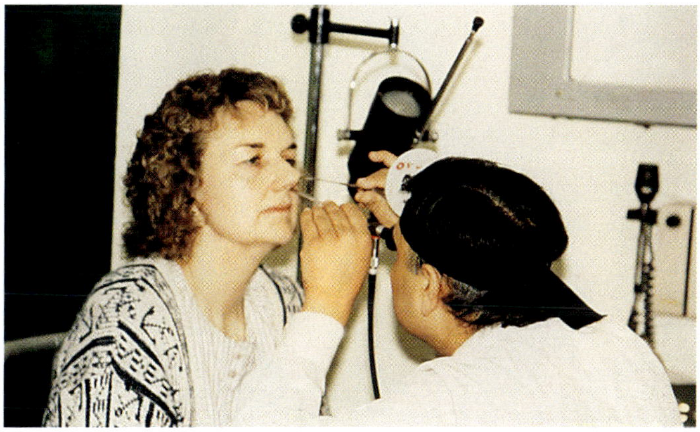

Fig. 9.1. The patient is seen in the endoscopy clinic after anaesthetic has been applied to the ethmoid cavity prior to endoscopic clearance.

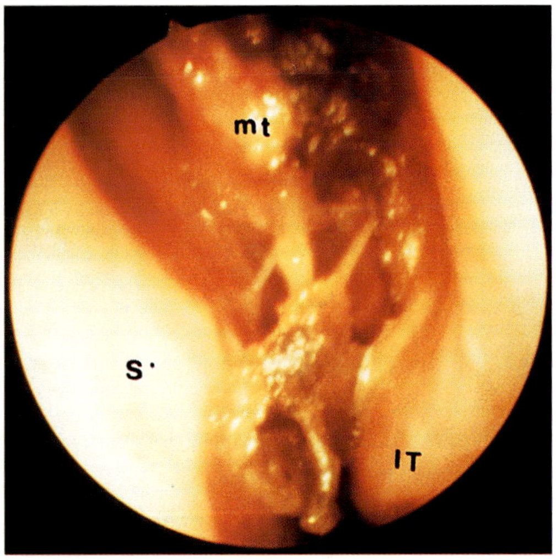

9.2

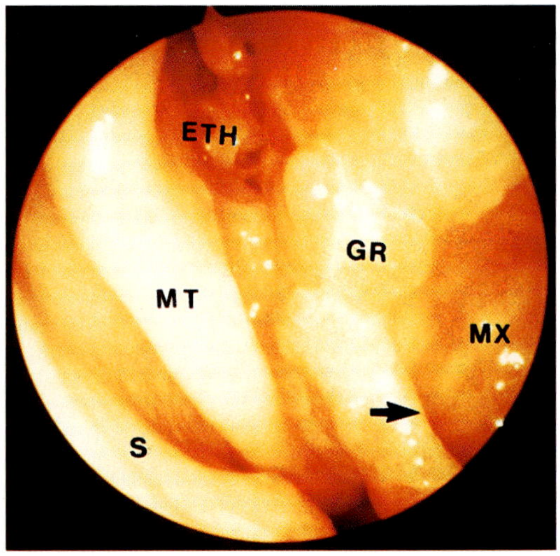

9.5

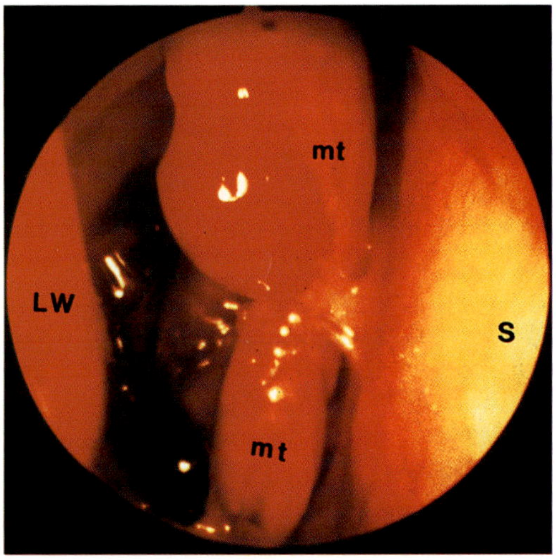

9.3

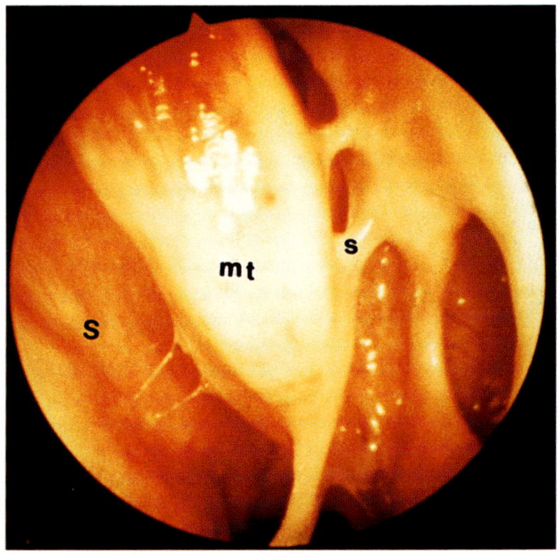

9.6

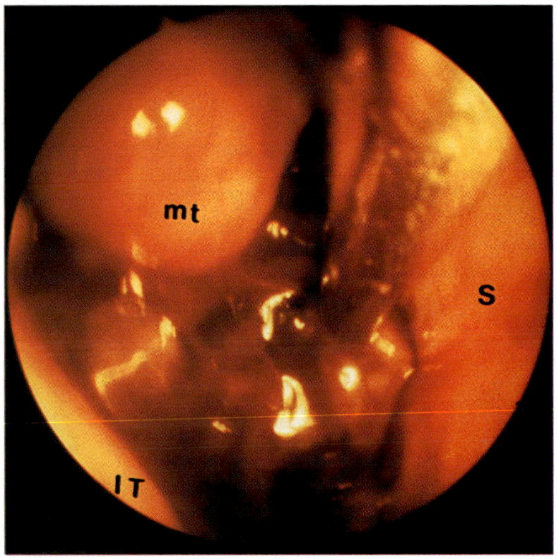

9.4

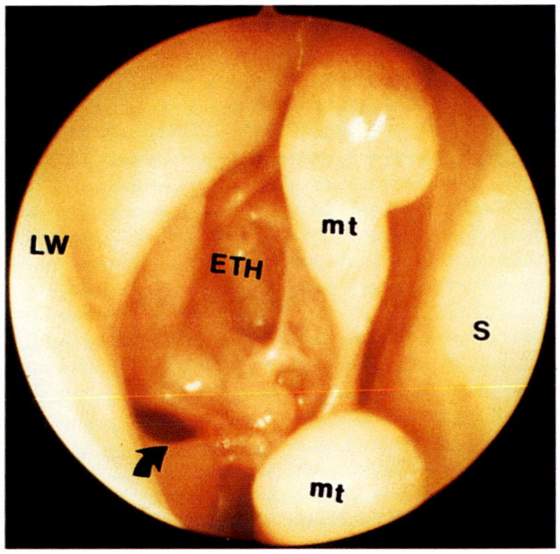

9.7

Fig. 9.2. Post-operative ethmoid cavity after 3 days, containing blood clots and crusts. Proper cavity care is essential. S, septum; MT, middle turbinate; IT, inferior turbinate.

Fig. 9.3. Wedge resection of the middle turbinate (MT) allows for proper access to the cavity (first post-operative week). LW, lateral wall; S, septum.

Fig. 9.4. Suction clearance of the ethmoid cavity, mucus and clots are removed. IT, inferior turbinate; MT, middle turbinate; S, septum.

Fig. 9.5. Four weeks post-operatively, note the granulations (GR) on the edge of the antrostomy (arrow), and in the maxillary sinus. This usually settles over a period of time. S, septum; MT, middle turbinate; MX, maxillary sinus; ETH, ethmoid cavity.

Fig. 9.6. Post-operative cavity after 6 months. Note small synaechiae (s), but large MMA with healthy sinus mucosa. S, septum; MT, middle turbinate.

Fig. 9.7. One year post-operatively, the ethmoid cavity (ETH) and MMA (arrow) are healthy. The unlocked ostio-meatal complex and ethmoid cavity are seen. LW, lateral wall; MT, middle turbinate; S, septum.

Fig. 9.8. Three years post-operatively, a small adhesion (A) has formed over the MMA (arrow); however, the cavity is clear. A small retention cyst (C) is noted. ETH, ethmoid cavity; S, septum.

Fig. 9.9. A classical MMA (small arrow) and widened frontal recess (curved arrow) at 4 years is the result of meticulous follow-up. MT, middle turbinate; MX, maxillary sinus.

Fig. 9.10. If care is not exercised, adhesions would form as shown. LW, lateral wall; MT, middle turbinate; S, septum.

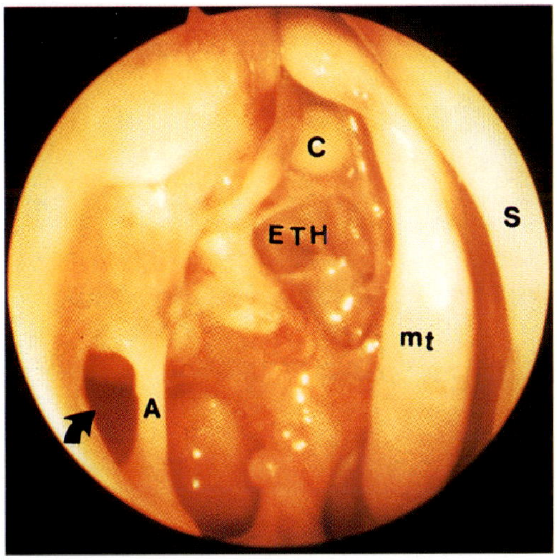

9.8

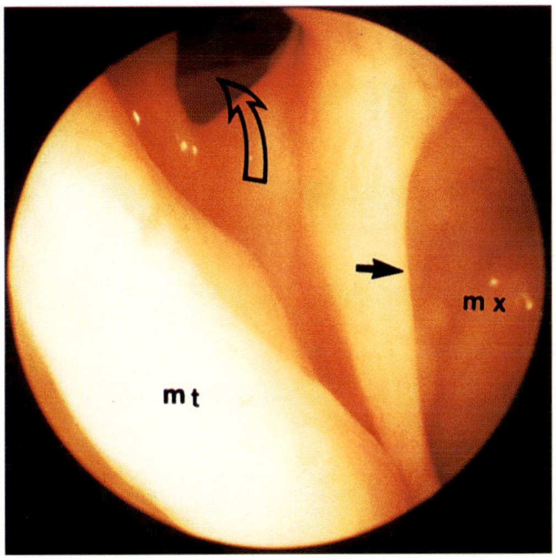

9.9

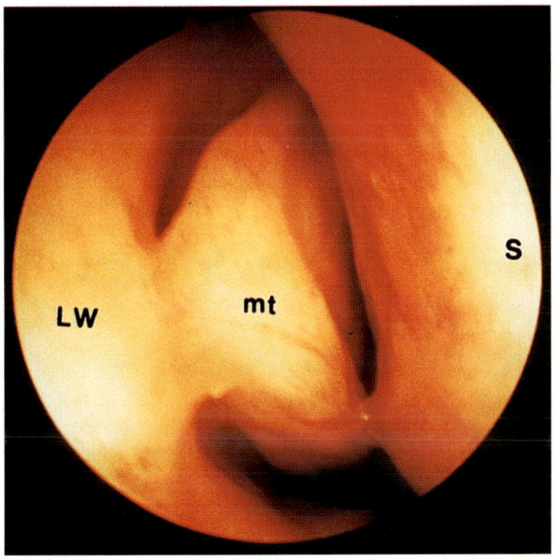

9.10

Chapter 10

Headaches and Facial Pains

Sinus disease almost always starts in the middle meatus. Both secretion and ventilation have to take place through the very narrow infundibular isthmus. Any narrowing or blocking of these spaces may lead to retention of secretions and/or poor ventilation, thus predisposing to chronic infection. At times, localised areas of contact of apposing mucosal surfaces in the key areas of the anterior ethmoids may lead to such a blockage and alter nasal function. Ciliary beating stops at the site of mucous membrane contact, thus causing stasis and providing ideal conditions for viral and bacterial infections. These localised mucosal factors may be summarised as

- constant intense mucosal contact
- poor ventilation or non-ventilation of the sinus with resulting hypoxia
- pressure from proliferating polyps

The sinuses may cause pain by the following mechanisms:

- Mechanical, due to blockage of sinus ostia. Here, the pain is poorly localised to the areas of Vth nerve distribution. It may vary in degree. Wolf (1955) has shown that the mucous membrane of the nose and sinuses has varying degrees of sensitivity, so that stimulation of the turbinates causes the least amount of pain, while instrumentation of the frontal recess causes moderate pain. The sinus ostia produce severe pain on stimulation. The two main stimuli that cause pain are temperature and pressure, so that even a slight touch with cotton wool in the region of the ostium could cause severe pain.
- Functional, due to affection of cilia, and mucosal oedema. Precipitating factors include exotoxin producing organisms and nasal irritants (e.g. noxious fumes) and allergens (e.g. pollens).

The occurrence of headaches and facial pains as a result of inflammation of the mucous membrane of the nose and sinuses in acute sinusitis is well known. In this situation, there is a blockage of the natural ostium, retention of secretions and mucosal irritation leading to hypoxia and a sensation of pain in the distribution of the Vth nerve. Furthermore, Stammberger (1988) also showed that negative pressure due to malventilation leads to bleeding into the infected sinus, and initiates pain. With the development of multi-angled endoscopes, it is now possible to explore the clefts of the nasal cavity, and to assess various structures in the region of the ostio-meatal complex. For example, areas of mucosal contact between two apposing surfaces lead to ciliary stasis and changes in the microvascular supply, causing reflex engorgement of the tissues and thus chronic inflammation and local increase in the concentration of vasoactive amines.These vasoactive amines are potent biotransmitters such as noradrenaline and acetylcholine; these act by inducing pain, or lowering pain threshold. They also reach blood vessels, and affect nasal secretions. Uddman et al. (1983) discovered a further transmitter, substance P. This is another potent neurotransmitter for sensory neurons, which acts locally in nasal mucosa as a strong vasodilator, and induces plasma extravasation, hypersecretion and smooth muscle contraction (this may account for the lower respiratory tract symptoms). Substance P further triggers oedema by simultaneous release of histamine from the mast cells.

Two further kinins exist in the submucosa of the nasal mucous membrane around blood vessels and glands. These are tachykinins and neurokinins. These also co-exist with substance P in C fibres, which carry the sensation of pain.

The work of Aust and Drettner (1974a) implicates three sinus factors in the production of pain: the size of the ostium, the sinus volume and the degree of nasal breathing. Functional sinus ostial size is directly proportional to oxygen exchange, while sinus volume is inversely proportional. This means that a small change in the diameter of the ostium, such as that due to mucosal thickening or blockage by polyps, has a dramatic effect on the exchange of oxygen in the maxillary sinus. Furthermore, oxygen exchange is twice as fast during nasal respiration as it is with the nostrils closed.

In all cases of possible sinus pain, a differential diagnosis must be kept in mind:

- Temporo-mandibular dysfunction
- tension headaches
- migraine
- brain tumours
- post-head injury (including whiplash)
- cervical arthritis
- neuralgias

CT scanning enables us to discover disease otherwise hidden from the eye. Some of the typical and more frequent abnormalities predisposing to headaches and/or recurrent sinusitis are listed in Table 10.1.

Table 10.1. Abnormalities predisposing to headaches and/or recurrent sinusitis

Septal abnormalities	Ethmoidal labryrinth	Uncinate process
High deviated nasal septum	Agger nasi cells	Medial rotation
Spur onto ostio-meatal complex	Middle turbinate concha bullosa paradoxical turbinate	Lateral rotation Contact with middle turbinate
Apposition of septum to superior turbinate	Large ethmoid bulla filling middle meatus	Pneumatisation
	Mucosal apposition	

A detailed analysis of clinical findings on anterior rhinoscopy and nasal endoscopy is shown in Fig. 10.1.

Not all these conditions are disease states *per se*, but they are factors that may reduce the already narrow spaces of the anterior ethmoids, and thus give rise to areas of mucosal contact, secretion retention and poor ventilation, and/or infection of larger sinuses. This promotes polypoid regeneration of apposing mucosal surfaces.

> **Key Points**
>
> - sinus pathophysiology as a cause of headaches and facial pains is well established
> - "headache" patients must undergo a thorough nasal endoscopic examination
> - CT findings such as concha bullosa, large bulla ethmoidales and a crowded ostio-meatal complex in conjunction with history and endoscopy are indications for surgery
> - Endoscopic sinus surgery is the treatment of choice in such patients
> - *The nose and sinus should be considered as one functional unit*

All these conditions, even when very small and circumscribed, may have one dominating clinical symptom of headache.

Coronal CT scanning was performed in all cases; it confirmed the clinical findings seen above. In addition, Fig. 10.2 shows disease in the agger nasi and Haller cells; 10% of our CT's were reported as normal, no mucoperiosteal thickening was noted.

The character of conchal pain is described as intermittent, and is said to be due to cyclic congestion of the lining mucous membrane. As a result of inflammation, a swelling of the middle turbinate presses on to surrounding structures such as the septum, producing pain over the medial canthus, the supraorbital region and occasionally the side of the face. The pain is mediated via branches of cranial nerve V2, and via peripheral interconnections with V1. Such headaches, unlike migraines, are not associated with auras, and may often occur in a dry nose. Moreover, they are of considerably shorter duration than migrainous headaches, and respond well to vasoconstrictor agents. The diagnosis is made from the character of the pain, and is confirmed by immediate relief with the application of a vasoconstrictor to the points of contact between septum and the lateral nasal wall. These contact points should be further evaluated by a coronal CT scan.

In keeping with these concepts, we analysed our data of 60 patients who presented with headaches and/or facial pains as a *predominant* feature. They ranged in age from 16 to 54 years (average age 34 years). The male to female ratio was 32 : 28, and follow-up extended from 8 months to 6 years.

The results are shown in Fig. 10.3.

As can be seen from Fig. 10.3, the majority of patients were well improved (very pleased with results) or improved (with intermittent symptoms, 52 in all). We do not know why eight patients did not do well.

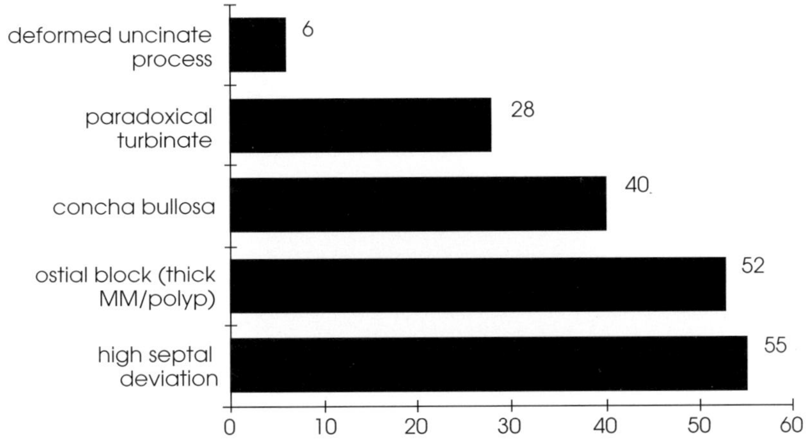

Fig. 10.1. Clinical findings on anterior rhinoscopy and nasal endoscopy in 60 patients. Figures given are percentages.

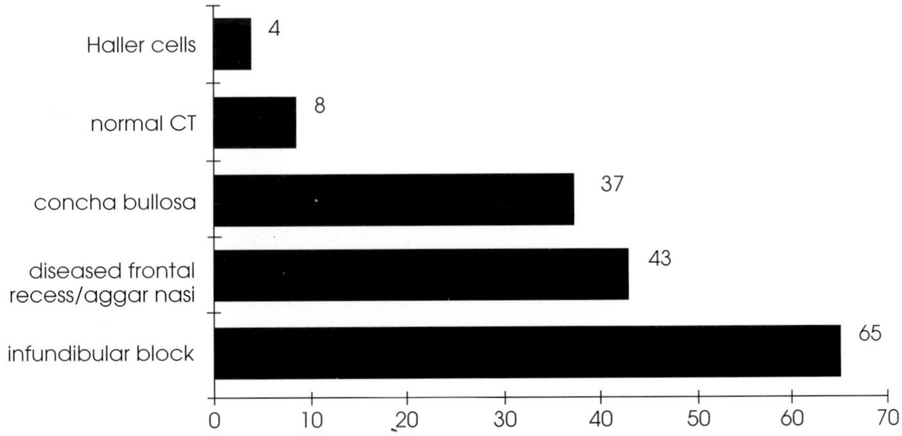

Fig. 10.2. Findings on CT scan in 60 patients. Figures given are percentages.

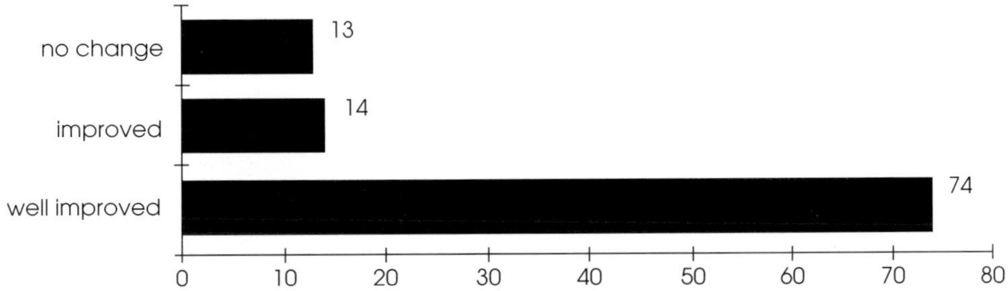

Fig. 10.3. Results of FESS in treating headaches and facial pains in 60 patients. Figures given are percentages.

Chapter 11

Complications in FESS

Complications in FESS take on special significance, as a minimal indiscretion or breach of surgical landmarks may result in severe morbidity including intra-orbital and intra-cranial complications. Nowhere is the axiom "prevention is better than cure" more valid than in this intricately contoured region. Careful training and planning are vital to a successful outcome. We have classified the complications as described in Fig. 11.1.

The most troublesome complication during surgery is *bleeding*. Usually it is not severe enough to cause the procedure to be abandoned, but therein lies the danger, the surgeon proceeds with compromised vision, making his task more difficult, and risking complications. Bleeding may occur from one of two sources: a generalised mucosal ooze due to clumsy instrumentation, or a brisk, arterial bleed due to injury to the anterior or posterior ethmoidals, or branches of the spheno-palatine artery, packing usually controls the bleeding. The posterior septal artery, which supplies the posterior middle turbinate and septum, can bleed vigorously, and is controlled with cautery and packing. If the patient has had a recent insult to his respiratory lining by an upper respiratory tract infection, it is wise to institute a course of systemic antibiotics and decongestants before taking him up for surgery. Bleeding diatheses must be excluded, and it is important to note that patients with aspirin intolerance often exhibit an increased clotting time.

Pre-operative nasal packing must be gentle, it is often at this stage that an impatient surgeon sets off a potential problem. We "walk our pack in", i.e. starting anteriorly within the nasal vestibule, several packs are placed sequentially, towards and finally into the middle meatus. In addition, a pack is compactly positioned against the roof of the nasal cavity, thus desensitising branches of the ethmoidal nerves. We recommend cotton wool strips or soft universal strips instead of ribbon gauze. An average of five strips are used on each side, each staying in place for several minutes.

Vessel trauma poses a more serious threat to the procedure, if only because ligation or even coagulation is impossible. Packing and waiting is the rather frustating answer. The danger point for anterior ethmoidal damage during FESS is when the sinus lateralis is entered above the ethmoid

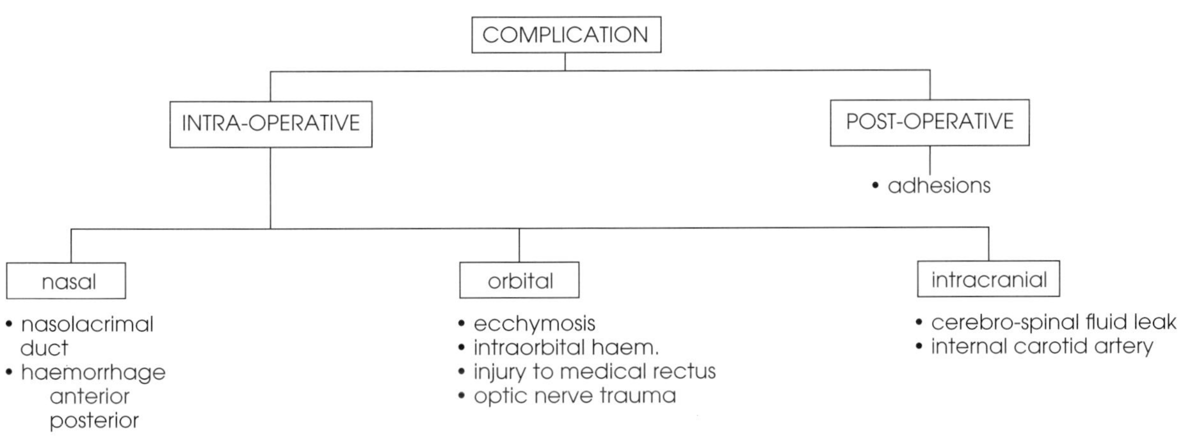

Fig. 11.1. Classification of complications in FESS.

bulla. It is here that the artery is seen to arch across from the lamina papyracea into the ethmoidal cells. The posterior ethmoidal is sometimes encountered after entering the ground lamella, especially if the endoscope is not forwarded in a parabolic fashion. Of the sphenopalatine artery, it may be said "the last, but most certainly not the least". As it exits the sphenopalatine foramen, it divides to supply the lateral wall of the nose. The anterior branch may be grasped and injured while enlarging the posterior limits of the middle meatal antrostomy. In addition, a mild *peri-orbital ecchymosis* may occur with compromise of the lamina papyracea, either due to the surgeon, or due to inherent dehiscence.

The opening of the *naso-lacrimal duct* is securely tucked away under the inferior turbinate, and rarely enters the surgical field. However, while resecting the anterior fontanelle with the backward-cutting forceps, there is a distinct risk to the duct, which lies only a few millimetres away. This is especially important in children. It is prudent to stop as soon as thick bone is encountered. Post-operatively if the patient develops epiphorae, an ophthalmologist's opinion should be requested.

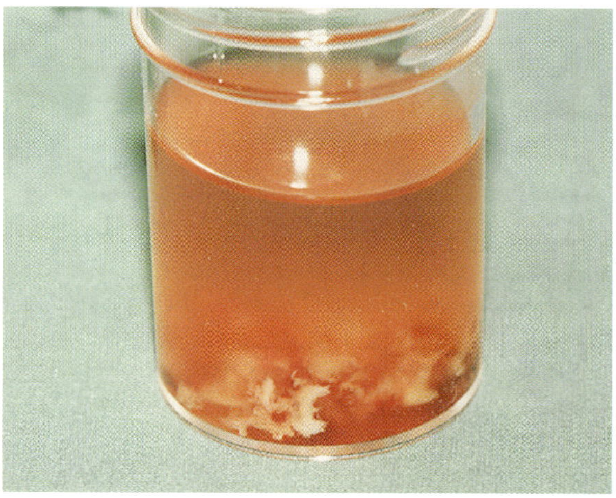

11.2

Fig. 11.2. Place all material removed from the ethmoid cavity into water, it must sink, only fat and brain tissue does not.

Fig. 11.3. In case of any doubt as regards breach of the lamina, gentle pressure should be applied to the eyeball to see the transmitted movements through the endoscope, and caution exercised.

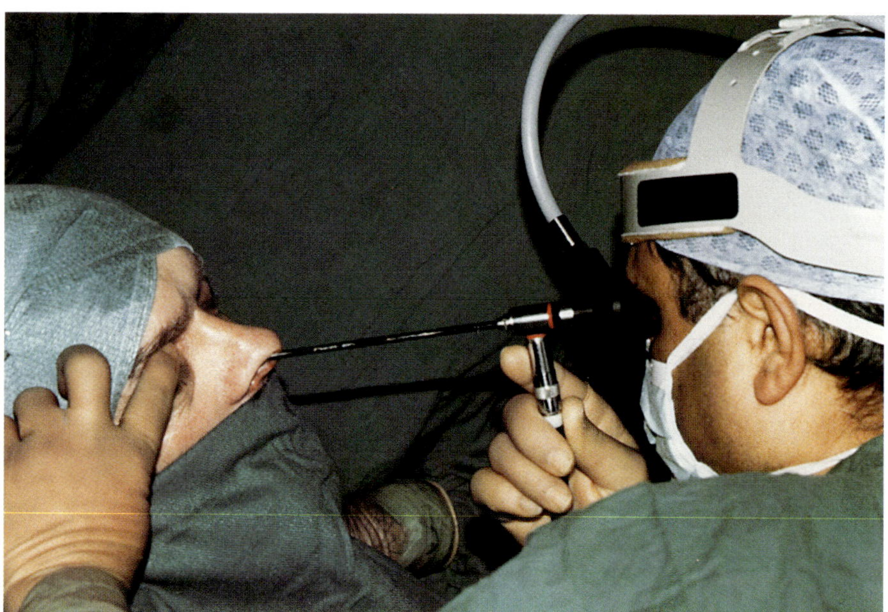

11.3

11.4

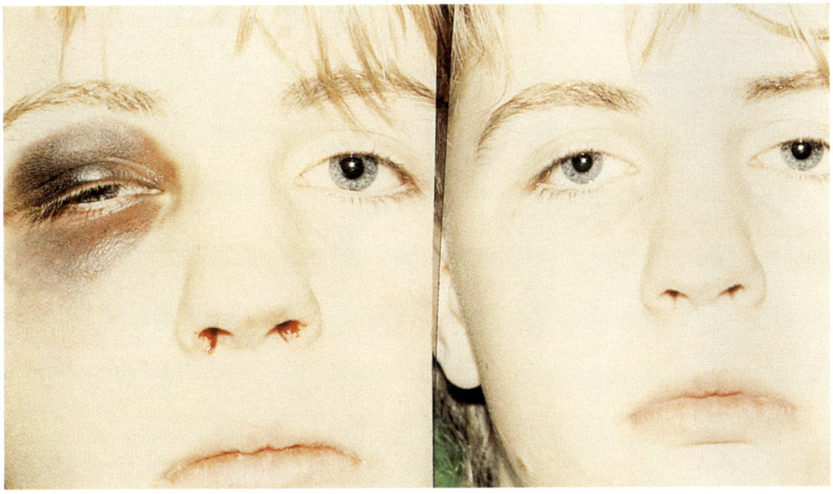

Fig. 11.4. With exposure of the orbital periosteum, a mild ecchymosis may occur in the immediate post-operative period – close observation must be maintained. The ecchymosis settles in 3 to 4 days.

Fig. 11.5. Patients should be instructed not to blow their noses for several days, as this may lead to a surgical emphysema (around the orbit on the right, arrow). Again, this settles rapidly.

Fig. 11.6. Acute sinusitis causing intra-orbital complications such as cellulitis and/or abscess should be treated endoscopically only by the experienced surgeon; otherwise conventional methods are employed.

Fig. 11.7. The same patient post-operatively.

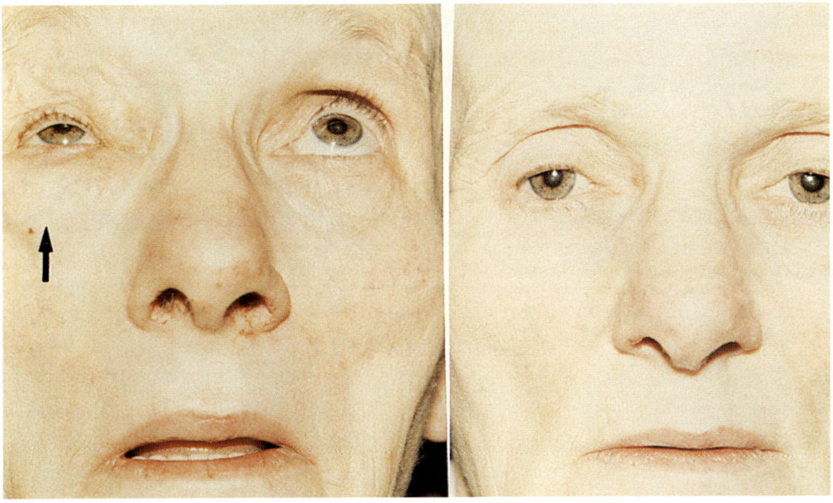

11.5

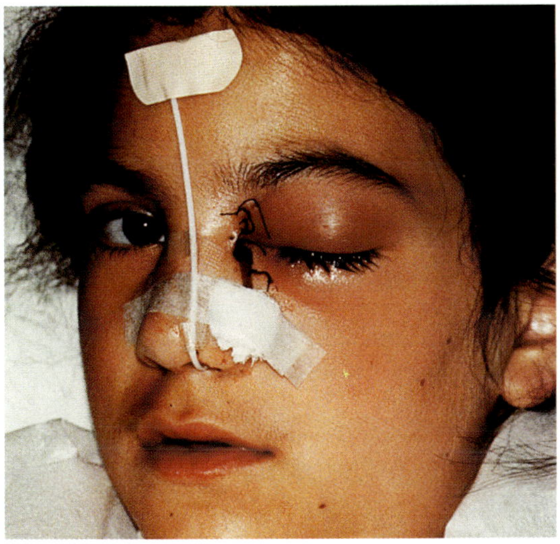

11.6

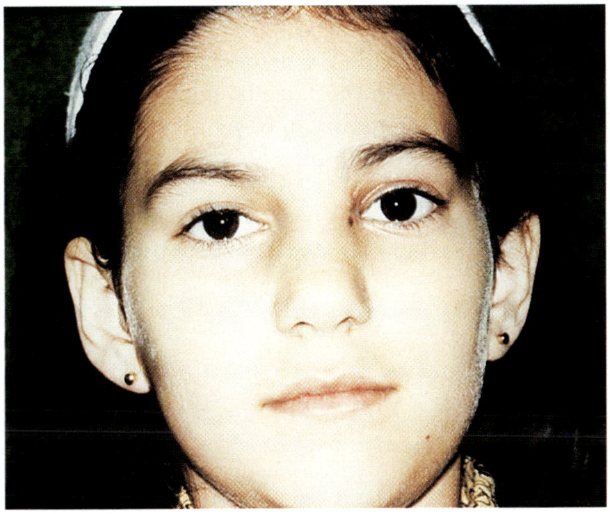

11.7

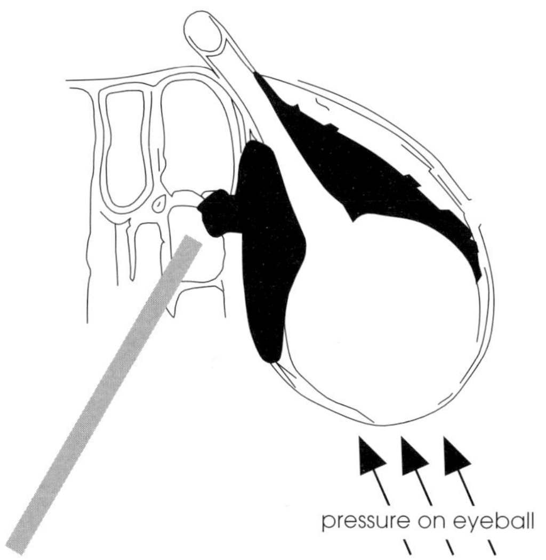

Fig. 11.8. Pressure on the eyeball shows transmitted movements on the exposed orbital fat examined through the nose with an endoscope.

Yet another intra-operative danger zone is the *orbit*, which might easily be damaged when dealing with the ethmoidal labyrinth. Simple breach of the lamina papyracea is not in itself a problem; the out-pouching of yellow, globular fat is easily recognised and confirmed by gently pressing the eyeball. The sign involves gentle pressure on the eyeball, simultaneously observing the medial orbital wall endoscopically, when pressure transmited through the dehiscence causes protrusion of orbital fat into nasal cavity. (Fig 11.8)

Another simple check is to immerse all endoscopically removed tissue into water, when fat will be seen to float, thus alerting the surgeon to possible danger. As ethmoid disease is exenterated from the region of the lamina papyracea, it is important to keep in mind the fact that the lamina lies in the same plane (more or less) as the lateral wall of the nose.

Unfortunately, in spite of careful technique, complications in this area have been known to occur, including an anecdotal case of bilateral total blindness.

Blindness may be temporary or permanent, the latter occurs as a result of direct optic nerve injury or prolonged intra-ocular pressure due to haematoma, causing neural damage. Direct injury to the optic nerve is irreversible, whereas increased intra-ocular pressure could be reversed. Major orbital injury occurs as a result of not heeding a breach in the lamina papyracea, and in most cases the peri-orbita too.

Prevention of this catastrophic complication begins pre-operatively, when evaluating the patient for the first time, and carries over to operative and post-operative phases. Patients who have extensive disease and/or have had previous interventions warrant caution and careful evaluation. Intra-operatively the patient's eyes are always left uncovered, and are closely observed. Eye movements during surgery, ecchymosis, proptosis and pupillary change are all signs which indicate the orbit has been entered. If pressure on the eyeball with transmitted movements is encountered with protrusion of orbital fat, the patient should be observed closely for signs of increased intra-ocular pressure. Patients who develop signs consistent with entry into the orbital cavity, e.g. ecchymosis, lid oedema or proptosis, should be monitored very closely. Intra-operative insult to the orbit should be treated immediately with mannitol 1–2 gm intravenously infused over 30–60 minutes, along with high doses of intravenous steroids; an ophtalmology consultation is obtained, and orbital massage is begun immediately. If there is impairment of vision with evidence of intra-orbital haematoma (evinced by severe pain, ecchymosis and proptosis), a lateral canthotomy or a Lynch incision with external ethmoidectomy is performed. The peri-orbita should be incised to assist drainage and decompression. Neural injury can occur if the vascular supply to the optic nerve is compromised for 60–90 minutes.

Delayed proptosis and increased orbital pressure can also occur. Post-operatively, it is important that the nursing staff be aware of possible complications and should be specifically instructed to contact the surgeon immediately in case of danger signals such as proptosis and pain in the eye.

All in all, management of lamina papyracea injury is either by abandoning the procedure (advised for beginners), or by proceeding extremely carefully (by advanced surgeons). Occasionally, the lamina itself may be dehiscent, and may pose the same problems.

Intra-orbital haemorrhage is of grave significance. On waking up, the patient complains of severe pain in the affected eye. A gradual proptosis occurs, and vision may be compromised, and the pupil reacts sluggishly. Immediate steps must be taken to reduce the intra-orbital pressure, and to decompress the optic nerve via a lateral canthotomy. If this does not succeed, an external decompression is resorted to. In addition to these measures, the patient must be given mannitol, high doses of steroids and diuretics.

Diplopia may occur due to direct or indirect injury to the medial rectus muscle or its neurovascular supply. This may be due to direct injury, cautery or laser. Diplopia may be related to oedema and is temporary; if a persistent diplopia develops, the treatment is strabismus surgery, which may not resolve the condition completely.

Subcutaneous orbital emphysema may result from the entry of air into the lamina papyracea, usually due to a Valsalva manoeuvre (vomiting, nose blowing). This is temporary, and resolves within a week.

Skull base trauma leads to a *cerebrospinal fluid (CSF) leak* just above the sphenoid ostia or sphenoidotomy openings (more commonly lateral lamella of the cribriform plate). The cribriform plate and fovea ethmoidales are other sites of injury. CSF leaks are usually detected endoscopically, and can be plugged directly with post-auricular temporalis fascia and muscle. Post-operative CSF leaks will often close spontaneously, but such patients require antibiotics and close monitoring for continuing leaks. Persistent leaks require neurosurgical assessment and a craniotomy may be needed. Sphenoid sinus breach can occur from forceps biting on the superior wall. Leakage may be controlled with packing, fibrin glue combining thrombin, calcium chloride and cryoprecipitate with a muscle plug has been used to close the leak under endoscopic guidance. An important guide in FESS is the middle turbinate, working along its lateral surface avoids injury to the cribriform plate. The base of the skull is often yellowish in colour, delineating it from the ethmoid complex. Also, in the majority of cases, the ethmoid and sphenoid anterior wall are of egg-shell consistency. When great difficulty is encountered during removal of disease or extremely hardened bone is apparent, it is prudent to withdraw and re-evaluate surgical anatomy. Patients with chronic disease will often develop thickened bone; also, those with repeated interventions may have distorted anatomy and great care has to be taken in these cases.

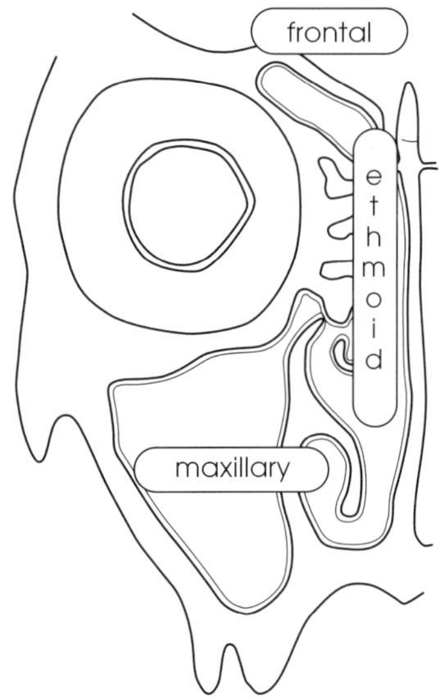

Fig. 11.9. Note the proximity of the paranasal sinuses to the orbit.

Post-operative complications include *synaechia*, without doubt the most common problem occuring with endoscopic sinus surgery. Because the middle turbinate is often preserved, its raw lateral surface apposes the lateral wall of the nose, and scarring can occur. Therefore, meticulous post-operative care is necessary. The surgeon should lyse all adhesions, which usually occur anteriorly. It should be noted that the middle turbinate has a natural tendency to move laterally, contributing to synaechia formation. If persistent adhesions block the middle meatus, a revision procedure is necessary. Finally, antrostomy closure may occur if the opening is not wide enough; lining the opening with mucosa helps in maintaining patency.

High Risk Areas in Endoscopic Sinus Surgery: Prevention of Complications

T. Ohinishi

Endoscopic sinus surgery is an excellent modality for treatment of chronic or recurrent sinusitis. It enables minimally invasive surgery for precise resection of disease within the paranasal sinuses for a better functional recovery. The only major drawback of this technique is the occurrence of potentially very serious surgical complications in a very small number of cases. Prevention of complications in endoscopic sinus surgery is an essential part of this technique.

The prevention of complications begins before the operation with study of the CT scan preferably with the radiologist.

CT Scan: Coronal and Axial

Routine examination of the CT scan of the sinuses is important as the patient may have anatomical variations in the

- paranasal sinuses
- base of the skull
- in the course of the arteries and nerves

The following anatomical sites must be carefully examined by the surgeon prior to the operation:

1. The roof of the ethmoid sinuses. The most frequent site of surgical injury is the lateral lamina of the cribriform plate as it is extremely thin and injury can lead to a CSF leak. This slanting lateral wall usually forms a gradual slope towards the roof of the ethmoid sinus.
2. The cribriform plate. In some cases the cribriform plate extends laterally into the ethmoid sinus creating a right-angled corner within the sinus. The thin bony wall in this area is often seen with a small bony defect and dehiscence. Simple stripping of the mucoperiosteum may compromise the defence system against infections. Moreover the dura mater in this area is intimately adherent to the thin bony wall and simple elevation of bone in this area may tear the dura mater creating a CSF leak.
3. The lamina papyracea. This is sometimes very thin, fragile, or dehiscent as a result of previous operation and/or longstanding multiple nasal polyps.
4. Dehiscence and unusual thinning as the patient may have had previous surgical interventions and resulting bony defects. A meningoencephalocele may present as a large nasal polyp through a defect in the fovea ethmoidalis or cribriform plate. The level of the anterior and posterior ethmoidal artery and its protrusion in the ethmoid sinus can be seen in the coronal view. Any difference in the height of the right and left anterior ethmoidal artery should be noted and kept in mind during operation.

In the axial CT scan the surgeon should pay particular attention to the optic nerve in relation to the posterior ethmoids where it is more likely to be damaged than the sphenoid sinus.

Haemostasis

Proper haemostasis is very important to obtain a clear operative field, thus allowing proper identification of various surgical landmarks and thereby preventing complications. Effective haemostasis should be achieved by adequate preparation of the nose and meticulous infiltration around the anterior and posterior ethmoidal arteries.

Through the microscopic observation Ohinishi (1981) has identified five main areas of the roof of the ethmoid sinus where surgical complications are likely to occur during endoscopic sinus surgery:

1. Medial wall of the ethmoid sinus or lateral lamina of the cribriform plate
2. Along the course of the anterior ethmoidal artery or nerve
3. Around the anterior origin of the middle turbinate or conchal lamina
4. Antero-lateral aspect of the roof of the ethmoid sinus
5. Area around the posterior ethmoidal artery or nerve.

Figures 11.10 to 11.14 show operating microscopic findings of the roof of the ethmoid sinus with natural dehiscence, indicating that defect or dehiscence in the roof of the ethmoid sinus is not a rare occurrence.

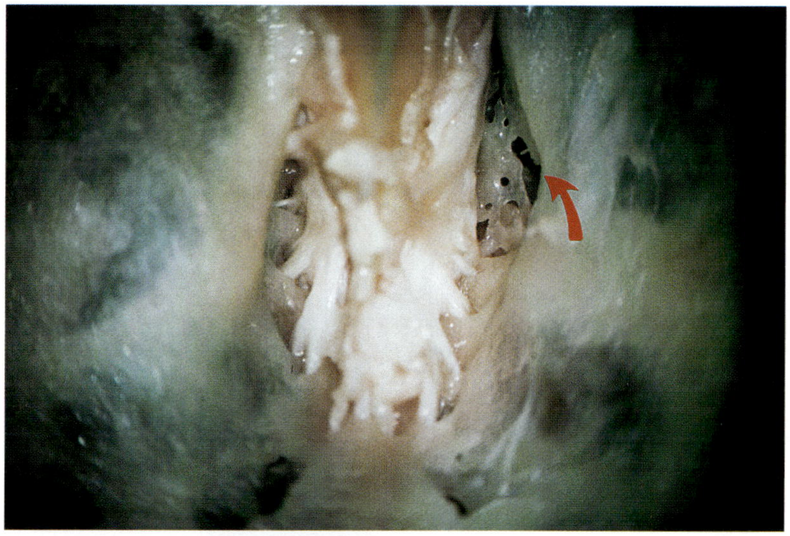

Fig. 11.10. Lamina cribrosa and the lateral wall: view from the intracranium, note multiple dehiscence in the lateral lamina of the right cribriform plate and multiple perforations for the olfactory fila. Top of the picture is anterior. Arrow shows multiple dehiscence in the lateral lamina of the right cribriform plate.

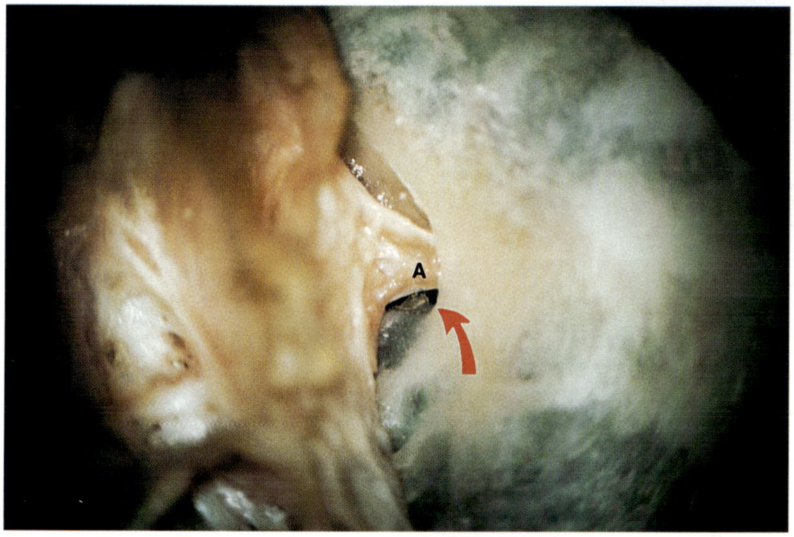

Fig. 11.11. Insertion of the right anterior ethmoidal artery (A). View from the intracranium. Note a defect of bone around the passage of the right anterior ethmoidal artery. Top of the picture is anterior. Arrow shows dehiscence around right anterior ethmoidal artery.

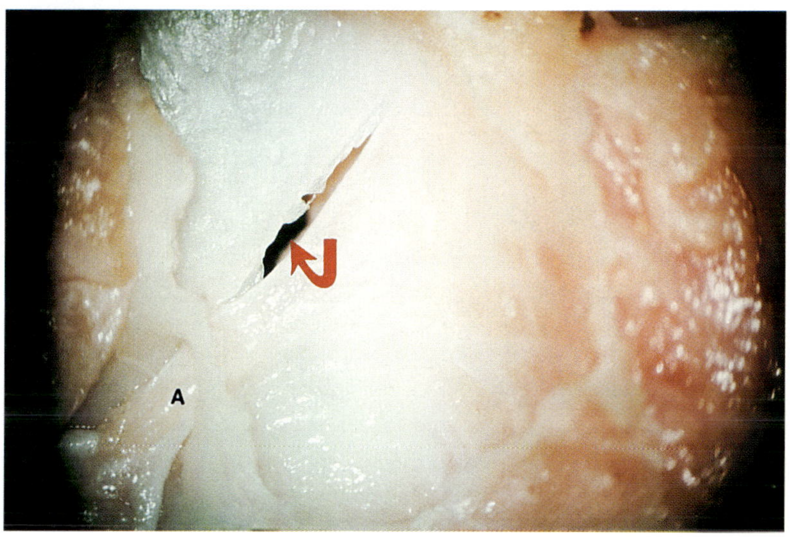

Fig. 11.12. Roof of the right ethmoid sinus, view from the sinus side. Note the course of the anterior ethmoidal artery (A) from the right orbit, through the lamina papyracea, to the ethmoid sinus and the defect of bone along the entire course of the anterior ethmoidal artery within the ethmoid sinus. Top of the picture is anterior. Arrow shows dehiscence along the entire course of the anterior ethmoidal artery.

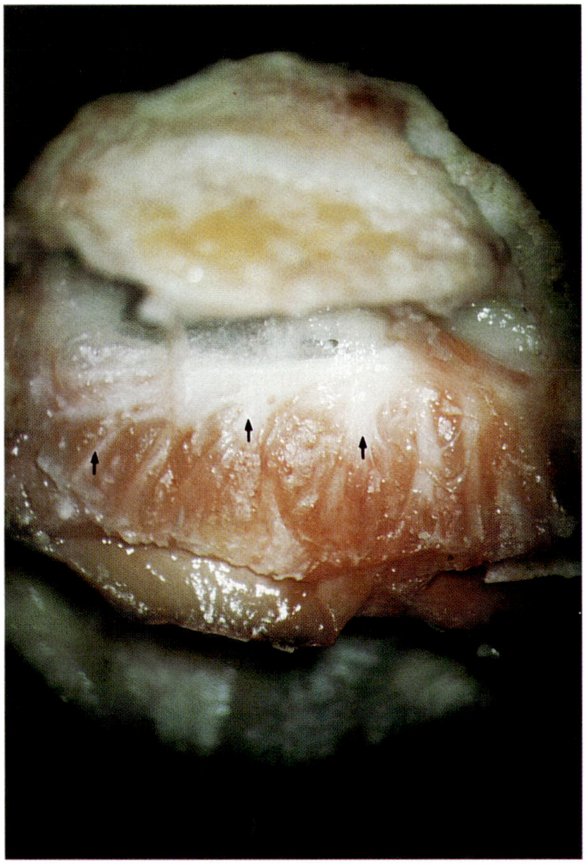

Fig. 11.13. Upper lateral wall of the right nasal cavity, note abundant distribution of the olfactory fila (arrow) on the upper nasal turbinate and the lateral wall as seen after resection of roof of the ethmoid sinus. Right of the picture is anterior.

High-Risk Areas

Relatively high rates of serious complications have been reported by Stankeiwicz (1987, 1989), Freedman and Kern (1979) and Maniglia (1989) due to violation of the orbit and skull base. Awareness of each high risk area during sinus surgery is of paramount importance in the prevention of complications.

I have identified the following five high-risk areas that would be encountered during endoscopic sinus surgery:

Lamina papyracea. The lateral wall of the ethmoid sinus consists of lamina papyracea, the thin medial wall of the orbit, which is a convex bulge towards the cavity of the ethmoid sinus. Figure 11.15 shows an endoscopic view of the lamina when fully exposed. This is the site of relatively frequent injuries during ethmoid sinus operation. Surgical injury of the lamina papyracea may cause a simple prolapse of fatty tissue, diplopia when the medial rectus is injured, intra-orbital haemorrhage or infectious and in rare cases blindness. Complications are more severe when the lamina papyracea is traumatised near the apex as the optic nerve may be damaged. Intraorbital hematoma can cause blindness by extension to the optic nerve or compression of the central retinal artery.

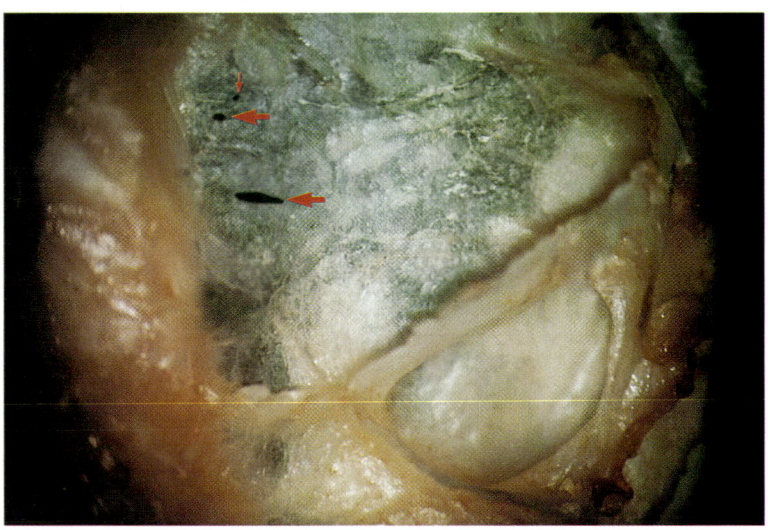

Fig. 11.14. Roof the right ethmoid sinus after the dura mata and the mucoperiosteum have been removed. View from sinus cavity, note the thin bony roof of the right posterior ethmoid sinus and three small dehiscences (arrow).

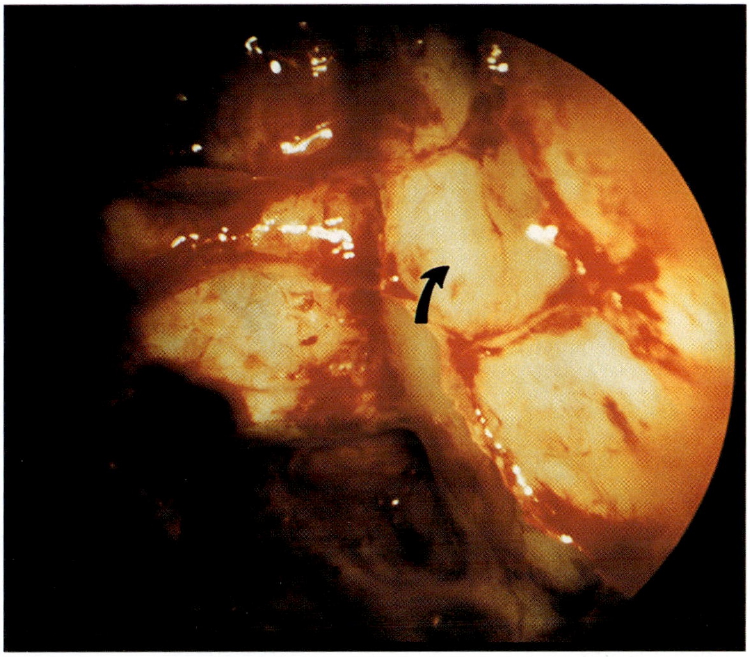

Fig. 11.15. Lamina papyracea (arrow), left ethmoid sinus.

The roof of the ethmoid sinus around the anterior ethmoidal artery. The anterior ethmoidal artery usually travels beneath the roof of the ethmoid sinus along with the nerve. There are elevations of the bony wall at the medial and lateral ends of the bony canal containing the artery. When injured it may lead to orbital haematoma in the lateral wall and CSF leakage or intracranial infections in the medial wall. Figure 11.16 shows a case where the anterior ethmoidal artery and nerve, without bony canal, course below the roof of the ethmoid sinus. This area has the unique anatomical features characterised by thin bony walls, a narrow cavity that tapers towards the ostium of the frontal sinus. Even in minor injury to this thin wall may tear the dura mater and can cause CSF leak.

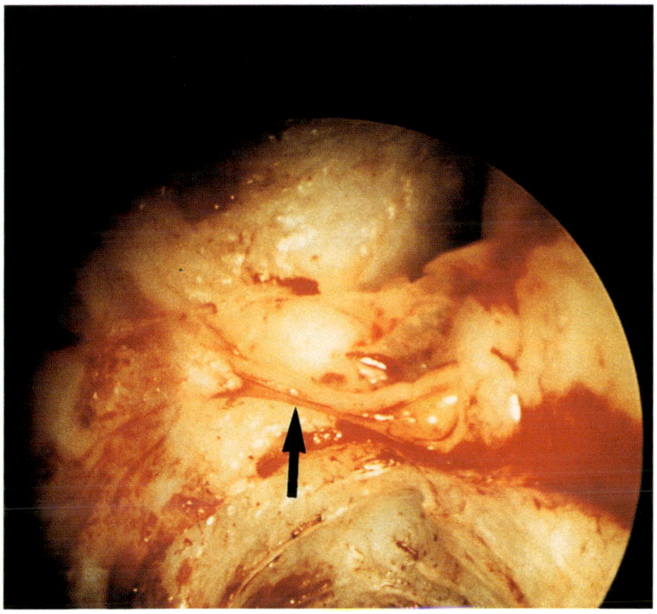

Fig. 11.16. Anterior ethmoidal artery and nerve (arrow), left ethmoid sinus.

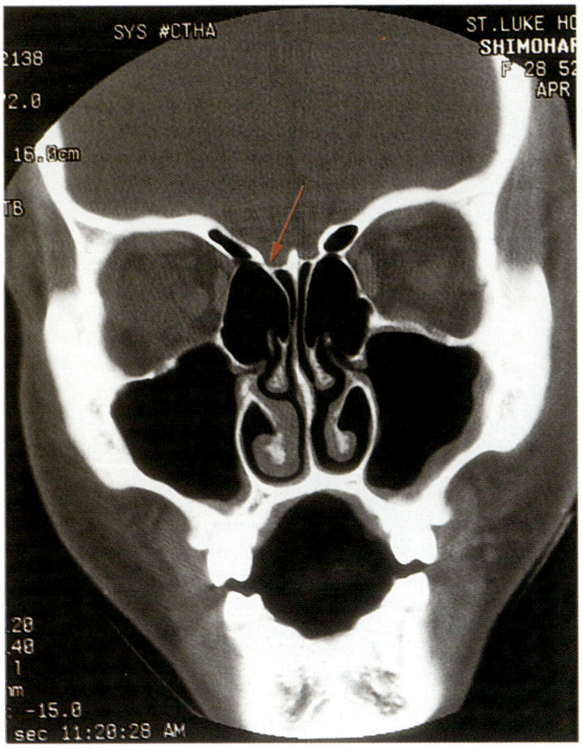

Fig. 11.17. CT findings of the lateral lamina of the cribriform plate (arrow) and the anterior ethmoidal arteries.

Lateral lamina of the cribriform plate. The medial wall of the anterior ethmoid sinus usually extends upwards above the level of the lamina cribrosa, which is the lateral wall of the olfactory bulb. This wall often bulges into the sinus cavity as is shown in the CT scan of the paranasal sinuses in Fig. 11.17. This bulging is a likely sight of surgical trauma during removal of disease in the frontal recess area. After penetrating the lamina cribrosa, the olfactory filaments distribute abundantly in the bony plates of the superior and middle tribunate which are the medial wall of the anterior ethmoid sinus. Because of these penetrations the medial wall of the anterior ethmoid sinus occasionally has bony dehiscence which may serve as a route to intra-cranial infections when damaged. Figure 11.18 shows an endoscopic view of the lateral lamina of the cribriform plate.

Ethmoid roof near posterior ethmoidal artery. The posterior ethmoidal artery travels in the roof of the posterior ethmoid sinus. The artery usually courses above the bony roof which is thicker than the roof of the anterior ethmoid sinus, and is hidden behind the bony plate. In some cases the artery, with its surrounding thin bony canal, travels beneath the bony roof. Damage to this artery may cause significant bleeding. Figure

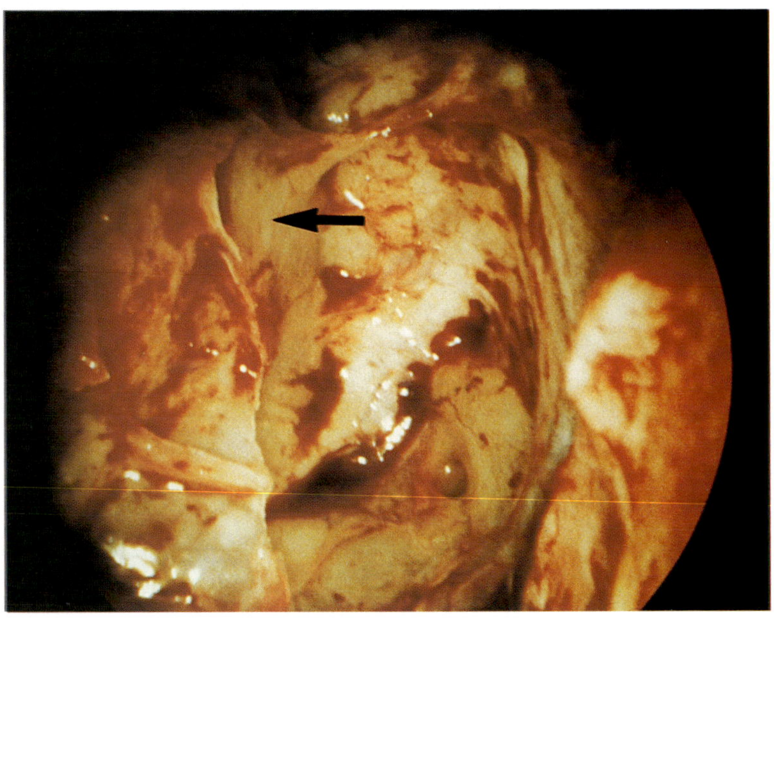

Fig. 11.18. Lateral lamina of the cribriform plate (arrow), left ethmoid sinus.

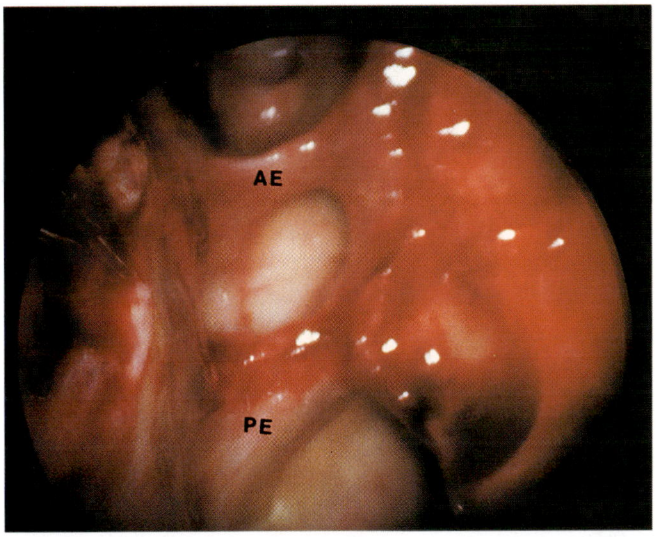

Fig. 11.19. Posterior ethmoidal artery (PE) with the anterior ethmoidal artery (AE), right ethmoid sinus.

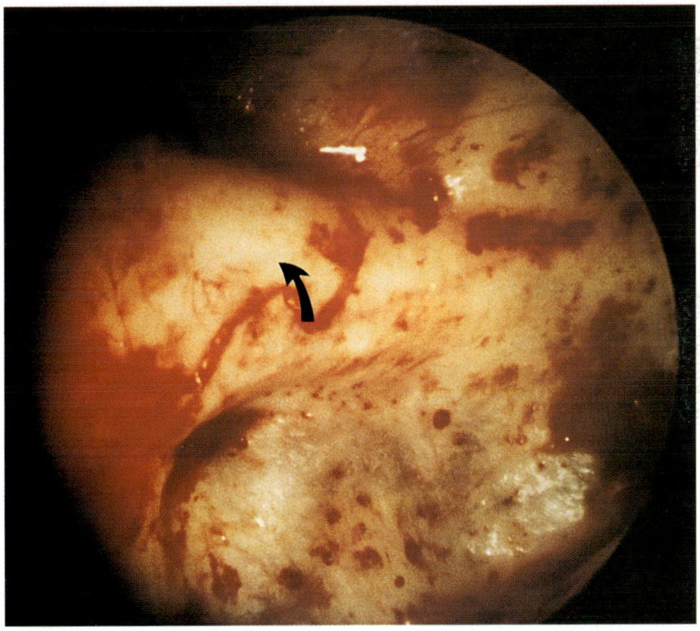

Fig. 11.20. Optic canal (arrow) at the bordering area between the sphenoid and ethmoid sinuses, right sphenoid sinus.

11.19 shows a case where a posterior ethmoidal artery travels underneath the sinus roof, posing a risk of injury.

The bordering area between the sphenoid and posterior ethmoid sinuses. The bony bulge of the optic nerve can be seen in this area during surgery. Figure 11.20 shows such a bulging of the optic canal in the sphenoid sinus. However, injury to the nerve may occur at the area bordering the sphenoid and ethmoid sinuses where the optic canal is obscure. An attempt to remove the thick buttress between the two sinuses poses a risk of injury to the nerve that may lie superficially to the sinus wall. The surgeon should also be aware of Onodi cells (excessive pneumatisation of the posterior ethmoidal cell) where the optic canal protrudes into the posterior ethmoid sinus. The internal carotid artery courses along the lateral surface of the sphenoid bone within the cavernous sinus causing a slight bony bulge into the sinus cavity. Injury to the artery may produce profuse and life-threatening bleeding into the sinus and the intracranial space. Craniotomy and clamping the artery may become necessary to

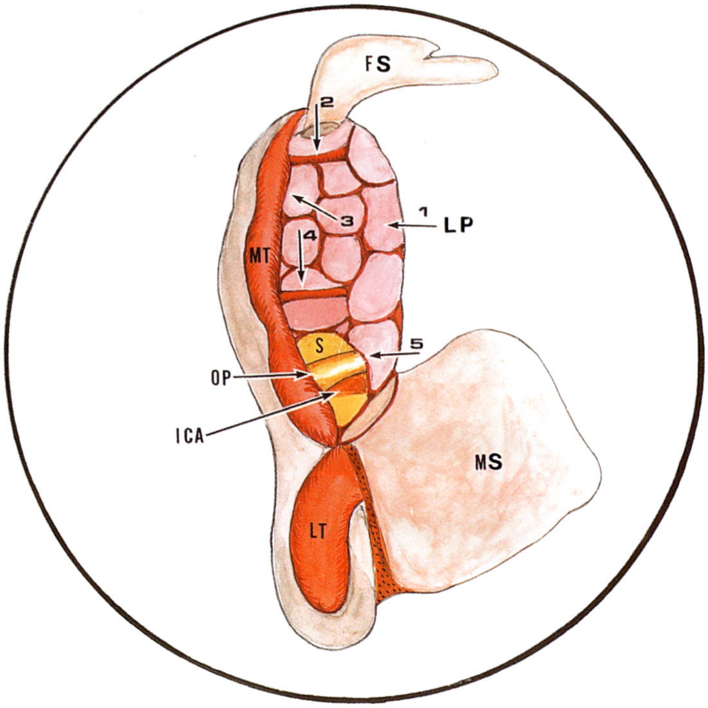

Fig. 11.21. Schematic drawing showing the high-risk areas in the left ethmoid sinuses. FS, frontal sinus; MT, middle turbinate; OP, optic nerve; ICA, internal carotid artery; S, sphenoid; IT, inferior turbinate; MS, maxillary sinus; LP(1), lamina papyracea; 2, anterior ethmoidal artery; 3–4, anterior ethmoidal cells; S, posterior ethmoidal cells.

prevent a catastrophe. Figure 11.21 shows a schematic drawing of the five-risk areas in the left ethmoid sinus.

Conclusion

The use of endoscope in sinus surgery has enabled us to visualise minute details of the structures of the paranasal sinuses and at the same time allows us to perform meticulous surgery (Ohinishi *et al.* (1990). Our previous study indicated that serious surgical complications in endoscopic sinus surgery are likely to occur when the surgeon is in haste to complete the operation owing to other commitments. As Ohnishi *et al.* (1993) indicated, a composed and calm attitude in the surgeon is an important requirement for sure and safe endoscopic sinus surgery. Perhaps most important is that the surgeon should be able to see in his or her mind the vital structures on the other side of the thin bony walls being worked upon.

Bibliography

Freedman HM, Kern EB (1979) Complications of intranasal ethmoidectomy: a review of 1000 consecutive operations. Laryngoscope 89: 421–434

Maniglia AJ (1989) Fatal and major complications secondary to nasal and sinus surgery. Laryngoscope 99: 276–283

Ohnishi T (1981) Bony defects and dehiscences of the roof of the ethmoid cells. Rhinology 19: 195–202

Ohnishi T, Esaki S, Iwasaki M; Tachibana T (1990a) Endoscopic microsurgery of the ethmoid sinus. Am J Rhinol 4: 119–127

Ohnishi T, Esaki S, Iwasaki M; Tachibaba T (1990b) Endoscopic microsurgery of the ethmoid sinus. Am J Rhinol 4: 119–127

Ohnishi T, Tachibana T, Kaneko Y, Esaki S (1993) High-risk areas in endoscopic sinus surgery and prevention of complications. Laryngoscope 103: 1181–1185

Stankiewiez JA (1987) Complications of endoscopic intranasal ethmoidectomy. Laryngoscope 97: 1270–1273

Stankiewiez JA (1989) Complications in endoscopic intranasal ethmoidectomy: an update. Laryngoscope 99: 686–690

Takahashi R (1944) Clinicoanatomical studies of the canalis orbitocranialis and canalis orbit-ethmoidalis in relation to the ethmoid cells. ORL Soc 50(3): 224–240. [English version: A collection of ear nose and throat studies. Department of Otolaryngology, The Jikei University School of Medicine, Tokyo 1971, pp 174–191

Chapter 12

Prevention of Pitfalls in FESS

The prevention of pitfalls in FESS begins not at the time of operation, but as soon as the patient presents himself to the endoscopic surgeon. Such measures can be conveniently divided into three distinct phases: pre-operative, intra-operative and post-operative.

Pre-operative Measures

There are three pre-operative measures which should be taken: History taking, nasal endoscopy and a CT scan.

As in most doctor–patient interactions, history taking is the most important aspect of patient diagnosis. A positive history of allergies, aspirin sensitivity, bleeding tendencies and recent infections must be inquired into prior to operation. Patients with specific or generalised atopy do less well following FESS, and must be warned accordingly. Also, indications for surgery in this group must be based on anatomical and/or radiological assessment of the ostio-meatal complex, and not purely on symptoms. Medical therapy must be rigorously pursued before a decision to operate is taken. Aspirin sensitivity was first described by Widal in 1922, but has been in the limelight only in recent years. It presents as part of a triad of symptoms, along with nasal polyposis and asthma. Prolonged bleeding times may also be noted. The underlying aetiology is thought to be a primary connective tissue disorder in the sensitive individual. Here again results are poorer, and no false hopes should be raised pre-operatively. Operating in a recently infected patient invites trouble, as generalised congestion and mucosal oedema cause bleeding into the narrow surgical confines, thus making the procedure difficult and hazardous.

A few salient points must be mentioned for out-patient nasal endoscopy. The 4 mm 0° and 30° endoscopes are advocated routinely. The first and second passes could be performed without anaesthetising the nose, but it is recommended that examination of the spheno-ethmoidal recess and sphenoid ostium be performed with local anaesthesia such as 4% cocaine in 1 : 1000 adrenaline.

We have had cases where the patient mentions some form of "nose operation" – endoscopy reveals an inferior meatal antrostomy, at times stenosed, or associated with residual antral disease, hence a routine examination of the inferior meatus is important. Similarly, close examination may reveal an apparent polyp to be a bifid anterior end of the middle turbinate; a medially rotated uncinate process may mimic a polyp in the middle meatus.

CT scanning should not be performed in the presence of recent rhinitis and/or sinusitis, as a false picture of mucosal oedema may be presented. Also, constant dialogue between the surgeon and radiologist is necessary, especially in regard to uncinate process position, the relation of the ethmoid roof to the nasal cavity and dehiscences of the lamina papyracea, optic nerve or internal carotid artery. So also the extent of disease should be assessed.

Intra-operative Measures

Local versus general anaesthetic and ease of access are intra-operative factors which should be considered.

Local anaesthetic is ideal, as it allows the patient to return home after a relatively safe, bloodless procedure. The patient inevitably alerts the surgeon if he approaches the lamina papyracea and/or skull base; however, the surgeon must anaesthetise the nose thoroughly, or the patient becomes unco-operative and the procedure tedious. Associated sepal deviations and/or external nasal deformities require a general anaesthetic.

The initial incision should be deep, and should follow the groove between the lateral wall of the nose and the caudal border of the uncinate process. A superficial incision will only lead to mucosal bleeding, and a proper infundibulotomy will not be achieved. The incision must extend up to the lower part of the uncinate process so as to facilitate a middle meatal antrostomy at a later stage in the operation. Similarly the upper part of the uncinate process must be removed properly, or else one would encounter difficulty in accessing frontal recess.

After the bulla has been removed, the lamina papyracea is identified and the forceps always kept parallel to its plane. Next comes the ground lamella, which should be identified and breached medially to gain access to the posterior cells. At this stage, the skull base must be identified by its thickness, and then skeletonised anteriorly by removing the remaining ethmoidal cell stumps. The ethmoid cavity has now been created; fine touches to its contours provides a well exteriorised cavity.

A middle meatal antrostomy often brings the surgeon up against the most diseased part of the nasal cavity, with polyps and discharge (in close proximity to the orbit) frequently distorting the view. Identification of the ostial site is easier when the lower part of the uncinate process has been completely excised. The level of the antrostomy is usually at the lower margin of the middle turbinate, just above the upper margin of the inferior turbinate. The upwards-cutting forceps must be directed downwards and laterally. For beginners, the combined approach (CAMMA) may be used. Caution must be exercised when advancing posteriorly, as branches of the spheno-palatine artery may be encountered. Post-operative stenosis is prevented by turning a mucosal flap over the antrostomy edge.

The fronto-nasal recess is left until last. A 70° endoscope is used to clear this area. As the ethmoidectomy has already been done at this stage, the almost unavoidable trickle of blood does not hamper the surgeon in his procedure. If oozing is excessive at any stage of the technique, the cavity may be packed with 4% cocaine-adrenaline strips, and the opposite side may be operated on.

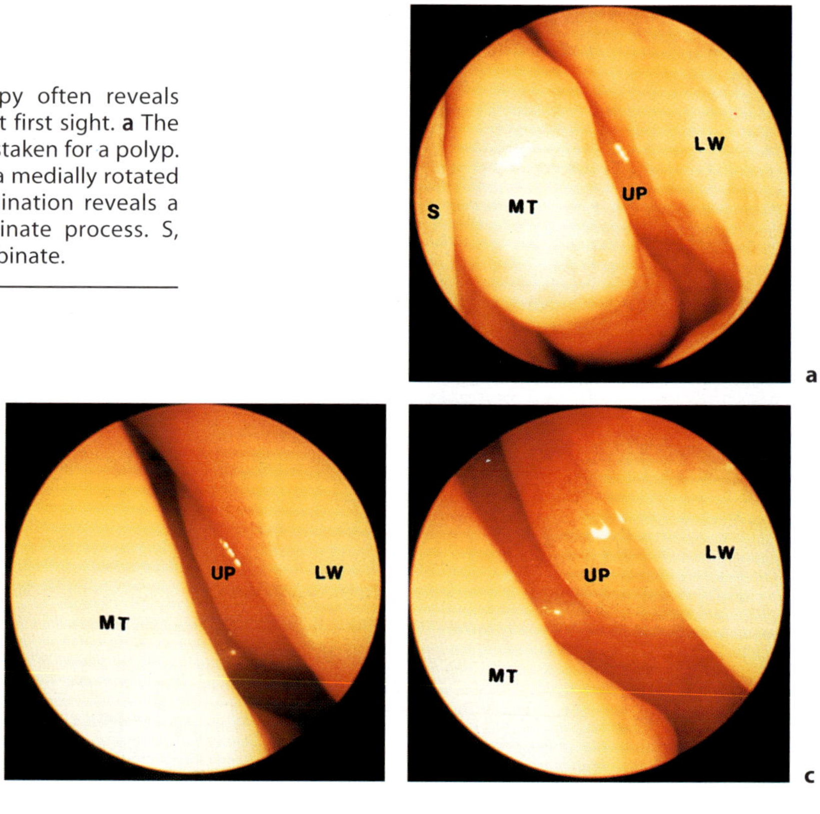

Fig. 12.1a–c. Proper nasal endoscopy often reveals abnormalities which appear different at first sight. **a** The uncinate process (UP) could well be mistaken for a polyp. **b** Closer examination shows this to be a medially rotated uncinate process. **c** Even closer examination reveals a congested and medially rotated uncinate process. S, septum; LW, lateral wall; MT, middle turbinate.

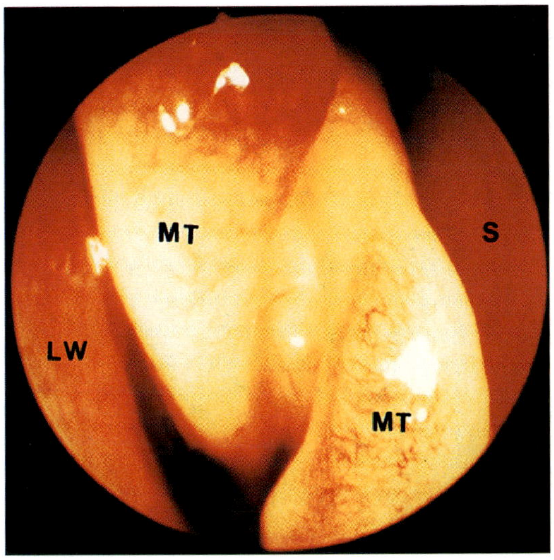

12.2

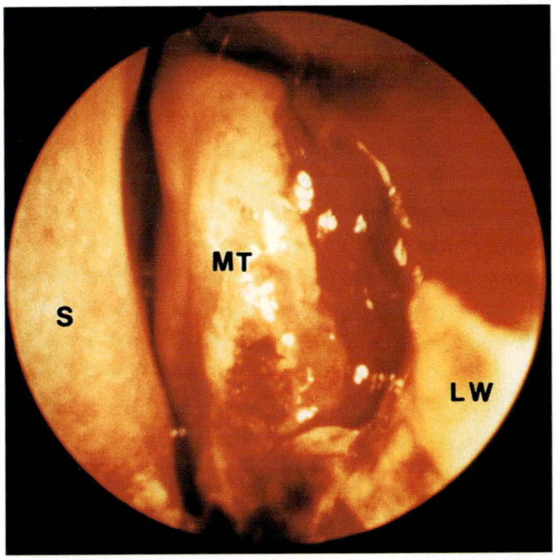

12.3

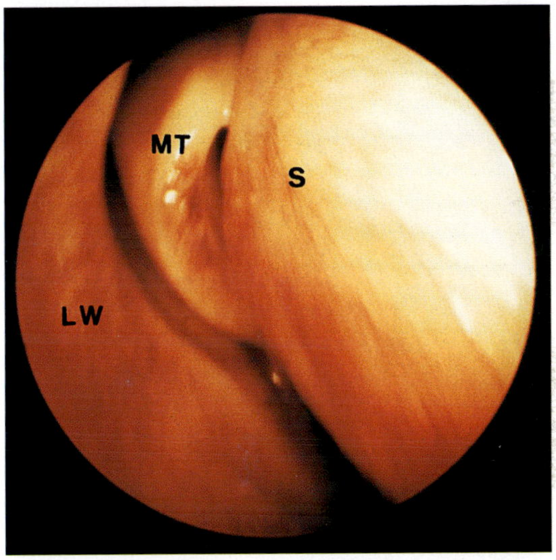

12.4

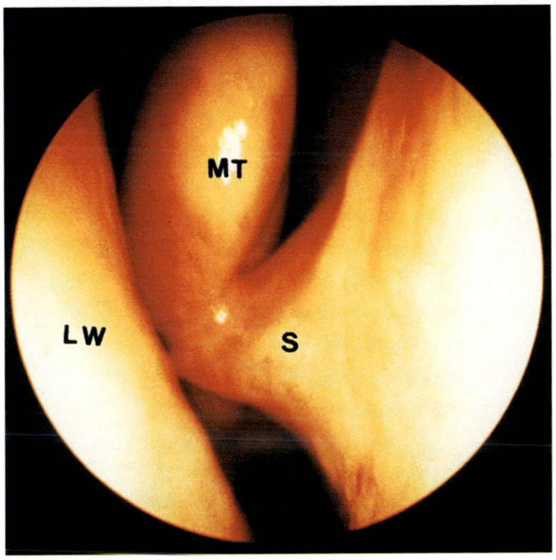

12.5

Fig. 12.2. Similarly to Fig. 12.1, this middle turbinate shows a bifid anterior end in its longitudinal axis. Clinically, this might mimic a polyp. LW, lateral wall; MT, middle turbinate; S, septum.

Fig. 12.3. A traumatic start causing mucosal bleeding can obscure the field, increasing the risk of complications. This should be avoided at all costs, a gentle technique is important. S, septum; MT, middle turbinate; LW, lateral wall.

Fig. 12.4. A high deviation of the nasal septum associated with lateral rotation of the uncinate and a paradoxical turbinate is a bad combination for the novice as the operative area is very restricted. In this case, a septoplasty should be performed first. LW, lateral wall; MT, middle turbinate; S, septum.

Fig. 12.5. A septal spur would also restrict access to the middle meatus, and trauma may lead to post-operative adhesions. LW, lateral wall; MT, middle turbinate; S, septum.

12.6

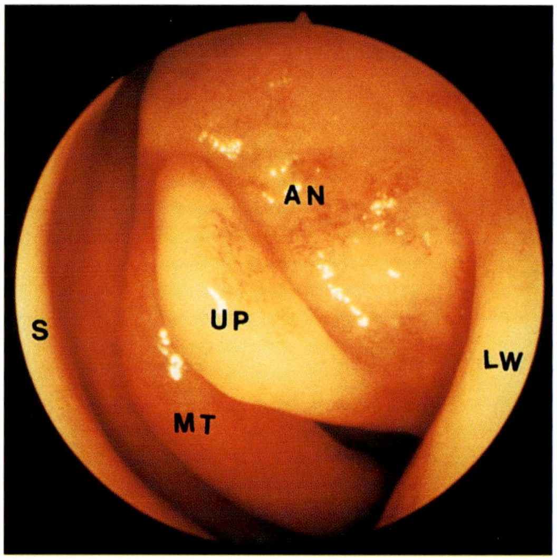

12.9

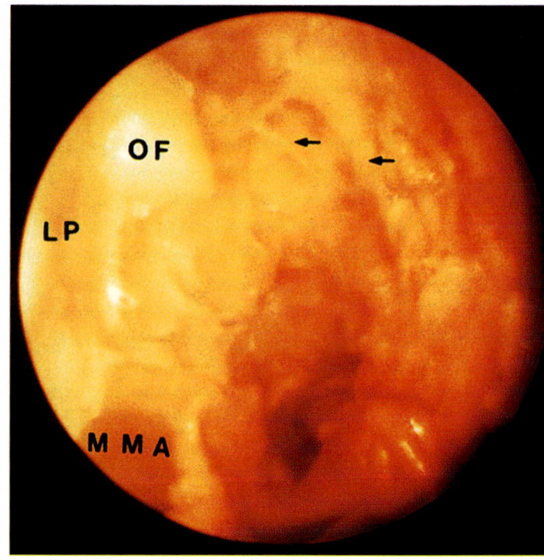

12.7

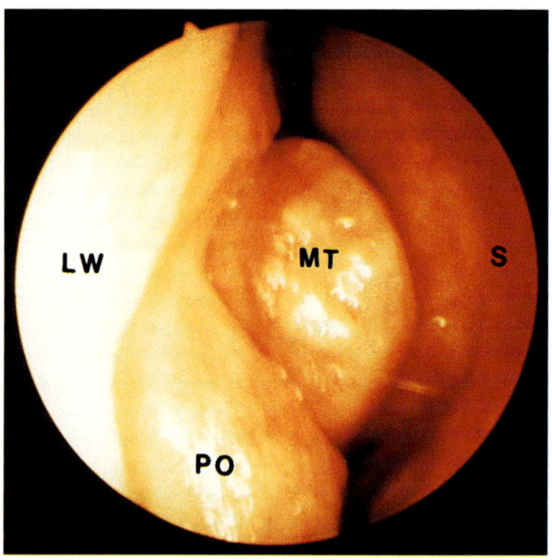

12.8

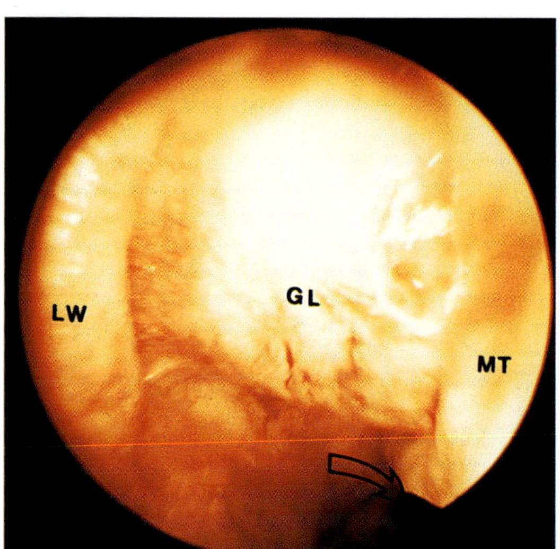

Fig. 12.6. In this case, a medially rotated uncinate process allows for easy identification of surgical landmarks, a good case for the initiate. S, septum; MT, middle turbinate; UP, uncinate process; AN, agger nasi cells; LW, lateral wall.

Fig. 12.7. In a situation where polyps (PO) protrude from the middle meatus, they should be removed gently with minimal trauma to create access to the uncinate process. LW, lateral wall; MT, middle turbinate; S, septum.

Fig. 12.8. In this case, the ground lamella (GL) is covered with thick, polypoid mucosa, and there is a danger of losing one's orientation, and of entering the posterior ethmoids unknowingly. Arrow, post-nasal space; LW, lateral wall; MT, middle turbinate.

Fig. 12.9. The skull base is identified by its nature (thick) and by small olfactory filaments running across (arrows) and the anterior ethmoidal artery in its bony canal. Orbital fat (OF) is seen as yellow in comparison with its surrounding mucosa. If pulled, it does not detach easily. LP, lamina papyracea; MMA, middle meatal antrostomy.

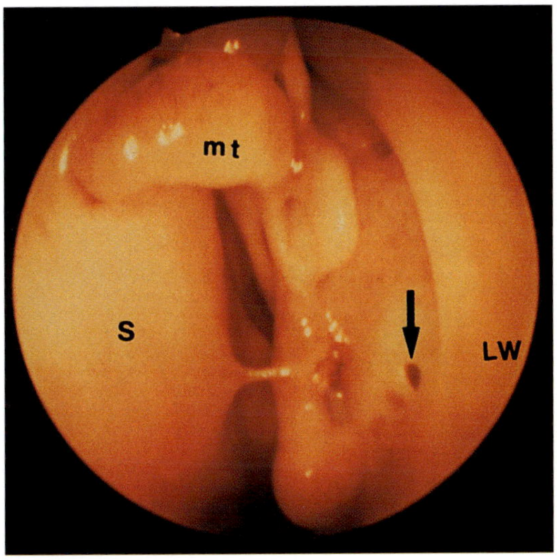

Fig. 12.10. In this revision case the normal anatomy is totally distorted due to scar tissue and incomplete excision. Such cases must not be taken lightly, and a thorough pre-operative evaluation with CT scan is required. Arrow shows the site of a probable attempt at MMA. mt, middle turbinate remnant; LW, lateral wall; S, septum.

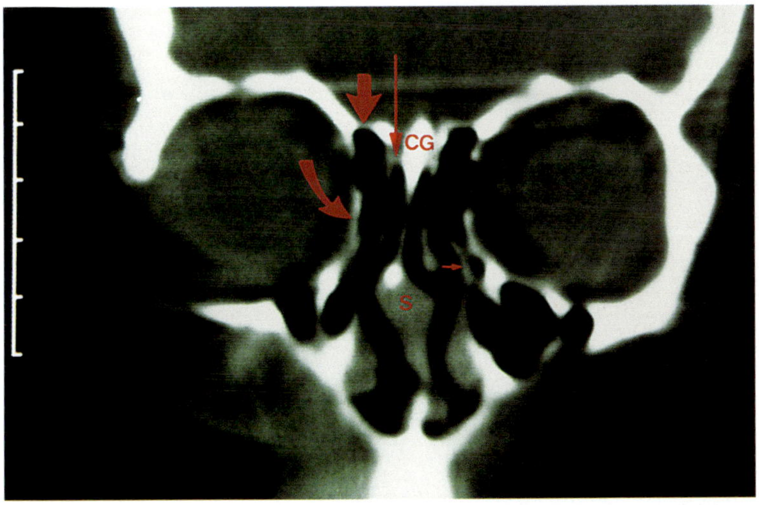

Fig. 12.11. The surgeon must know radiological anatomy of the paranasal sinuses, especially their relation to the skull base (long thin arrow). Thick arrow, fovea ethmoidales; small arrow, uncinate process; curved arrow, lamina papyracea; CG, crista gelli; S, septum.

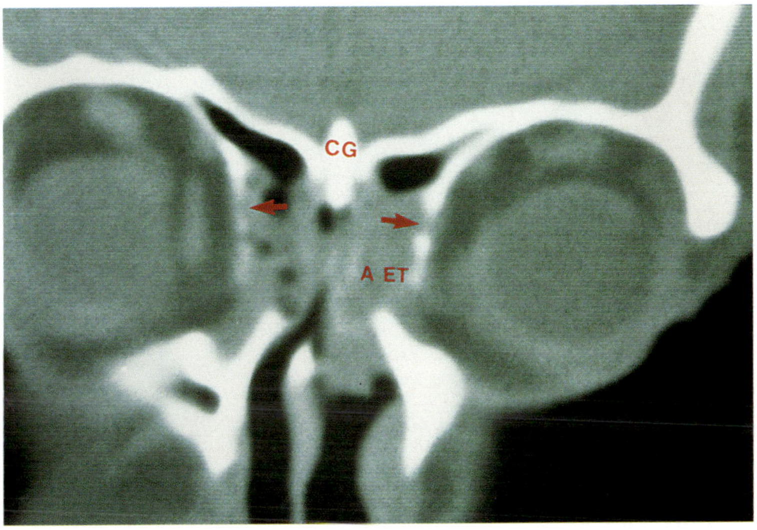

Fig. 12.12. Note the dehiscence of the lamina papyracea (arrows) due to chronic nasal polyposis and multiple operations. A ET, anterior ethmoids; CG, crista gelli.

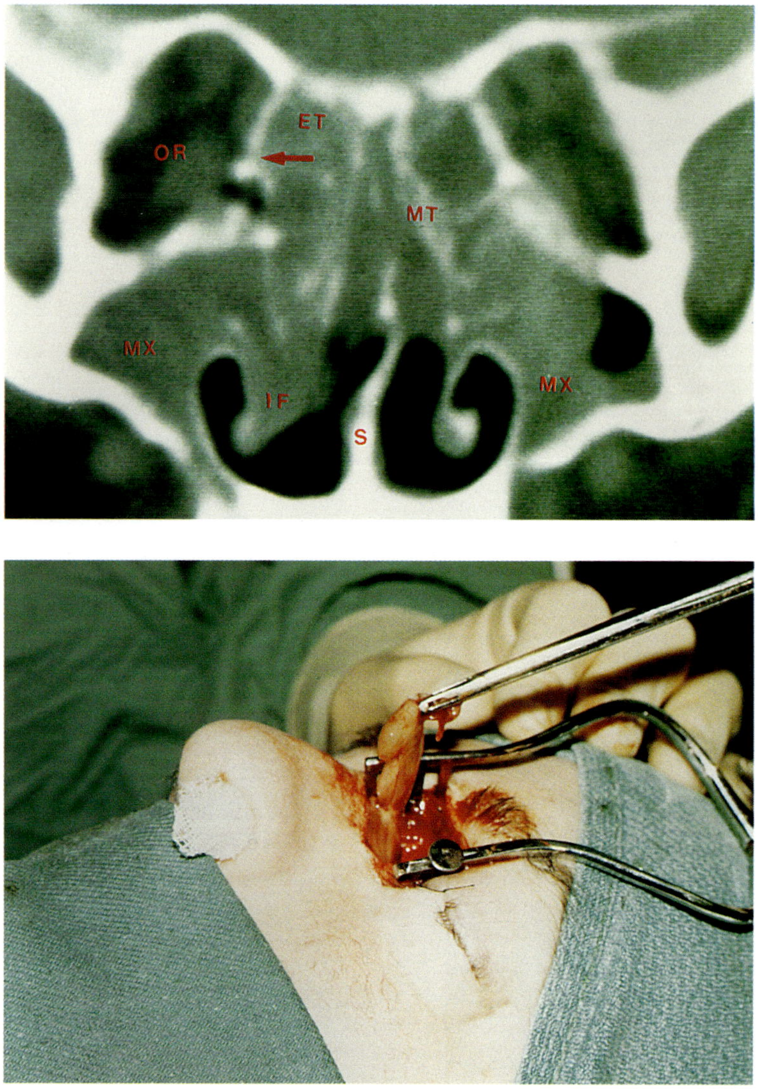

a

b

Fig. 12.13. **a** In cases of bilateral, extensive polyposis and multiple operations with distorted anatomy as seen on this coronal CT scan, FESS must be undertaken with extreme care. The arrow shows dehiscence of the lamina papyracea on the right. **b** In this case, an external ethmoidectomy may be the operation of choice. OR, orbit; MX, maxillary sinus; IF, inferior turbinate; MT, middle turbinate; ET, ethmoid; S, septum.

Post-operative Measures

The most important post-operative pitfall is inadequate after-care. In order for the newly created ethmoid cavity to heal satisfactorily, regular and frequent endoscopic toilet is absolutely mandatory. If this is not done, the surgeon may have the unfortunate experience of seeing a potentially good procedure turn into a failure. One of the earliest indicators of a mismanaged cavity is the development of adhesions. This usually occurs in the first few post-operative days. This is prevented by causing minimal tissue trauma during the operation, either with the endoscope or with instruments. The middle turbinate is gently medialised at the start of the procedure, thus preventing apposition of raw mucosal surfaces should they occur as a result of trauma. Wedge resection of the middle turbinate further modifies the ethmoid cavity, creating an anterior portal and allowing for easy access post-operatively, through which adhesions may be broken. The middle meatal antrostomy may also be compromised by the overgrowth of fibrous tissue. Furthermore, a mucosal flap should be turned inwards to line the antrostomy edges. In cases with an underlying allergy, treatment must be maintained in the post-operative period.

Chapter 13

FESS Data Analysis

Background

An analysis of our results was made possible by using a computer-compatible proforma as shown in Appendix B. Out-patient assessment, operative findings and steps and post-operative follow-up were all recorded. Symptoms were divided into the classical triad of nasal obstruction (mild, moderate, severe; right, left, bilateral), post-nasal drip and headaches and facial pains. Sequelae such as pharyngitis, laryngitis, otitis media and hyposmia were also noted. A history of asthma and aspirin intolerance were carefully recorded. Out-patient evaluation included anterior rhinoscopy, nasal endoscopy, plain radiographs and CT scans. Operative notes indicated internal architecture abnormalities and a sequential record of the procedure, while post-operative follow-up at 3 months, and then at 1, 2 or three years assessed symptoms and endoscopic findings.

This chapter presents our results in graphical fashion. It will afford the reader a clear picture of the statistics generated by our patient pool, and emphasise the need for careful and thorough documentation. We find analyses such as these act as a "rear-view mirror", telling us where we might have gone wrong, and guiding us to improvement in the future.

Over a period of 5 years, a total of 532 cases of FESS were carried out. The age range was between 7 and 87 years, with an average of 33.3 years. Just over half (51%) of patients were male. Bilateral procedures were performed in 511 patients (96%, 1022 procedures). The following results have been compiled from all patients with a minimum follow-up period of six months (n=478) and a maximum follow-up of 4 years.

An allergic history was elicited in 19% of cases (most of these patients were sensitive to common allergens such as house dust and pollen). Four cases of aspirin sensitivity were treated. Secondary symptoms such as sore throats, laryngitis, middle ear effusions and smell disorders accounted for 38% of cases. A number of patients admitted to taste and smell disorders of varying degrees.

The cardinal symptoms recorded were nasal obstruction, post-nasal drip and headaches and/or facial pains. These were further graded as mild, moderate and marked. Over 60% of patients had marked symptoms.

Anterior rhinoscopy showed a deviated nasal septum in 63% of patients. These deviations, both caudal and cephalic, were often the cause of inadequate access to the ostio-meatal complex. In 76% of these, a septoplasty/rhinoplasty was carried out. Figure 13.1 summarises our anterior rhinoscopy/nasal endoscopy or operative findings.

The following trends emerged during our study, and are worth noting

Unilaterality of Findings

Only a small number, 19 (4%) of patients had unilateral symptoms. Nasal endoscopy in these cases confirmed the presence of unilateral findings such as a conchous middle turbinate, rotation of the uncinate process, etc. These findings were further confirmed on a coronal CT scan.

Concha Bullosa

We found concha bullosae in 38% of cases, 20% of these (36 patients) showed both middle turbinates to be conchous. This figure is slightly higher than that seen in other studies, possibly due to the fact that we perform a routine wedge resection of the middle turbinate in almost all cases, hence occult conchosities are brought to light. Quite often the conchosity was in the posterior segment of the turbinate. *Almost all these patients had headaches and/or facial pains as a presenting symptom.*

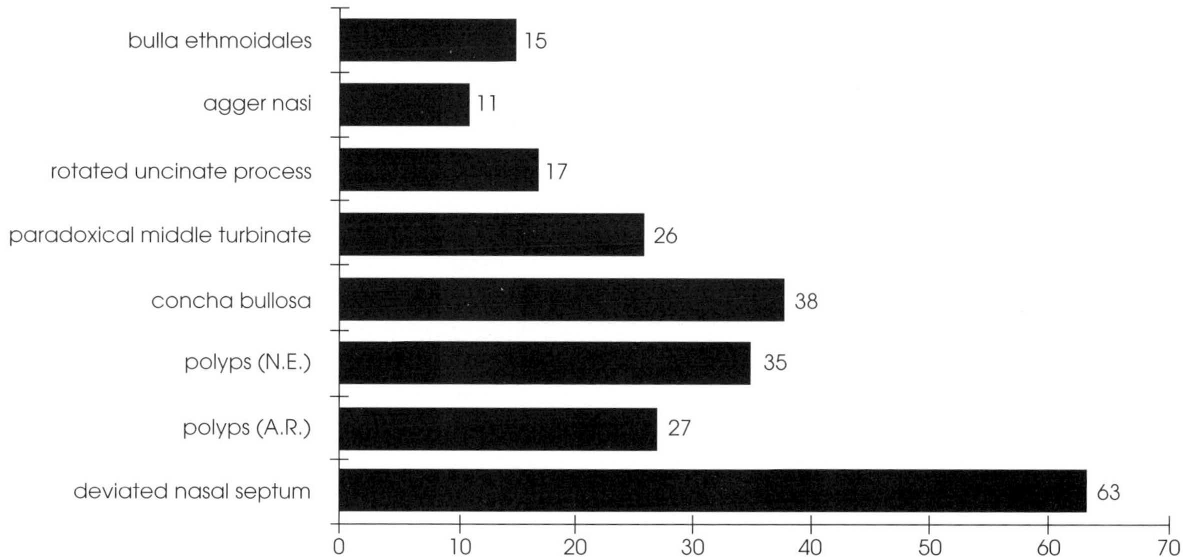

Fig. 13.1. Anterior rhinoscopy/nasal endoscopy findings in 478 patients. Figures given are percentages; the figure for polyps (N.E) indicates findings not noted on anterior rhinoscopy.

Excising the lateral wall of the conchal sinus or wedge resection relieved the majority of these patients.

CT Findings

A small number of our patients (c. 10%) had clear CT scans in spite of being symptomatic. Once again, most of these patients suffered headaches and facial pains. Nasal endoscopy in these patients showed mild rotation of the uncinate process, muco-pus in the middle meatus, or a septal spur impinging on the middle turbinate. These findings led us to undertaking FESS in these cases.

As discussed in Chapter 6, we found plain radiograph correlation of the maxillary sinuses to have a high index of accuracy, while ethmoid assessment was poor. The key areas of the ethmoid infundibulum and the frontal recess lent themselves to assessment only by well-taken coronal CT scans. As far as comparison between CT and operative findings was concerned, we found a majority of patients with more extensive disease than that seen on film, a factor that must be kept in mind when reading the CT scan.

Post-operative Results

All FESS patients were followed up in a special Endoscopy Clinic. They were called in (initially on a twice-weekly basis, then weekly), until the cavity showed good epithelialisation, and crusting was minimal. At the clinic, they underwent rigorous nasal toilet sometimes under local anaesthesia. Their symptoms were noted on the endoscopic proforma at 3 and 6 months and then annually. The three cardinal symptoms showed the following trends over time (Fig. 13.2).

Long-term follow-up of the well-improved group (Fig. 13.3) showed nasal obstruction settled in 96.2%, headaches and facial pains in 94.4% and post-nasal drip in 92.1% (Fig. 13.4).

As can be seen in Fig. 13.4, nasal obstruction seems to be the most improved. It must be noted that obstruction may not always be due to polyps and/or septal deviations, a congestion of the nasal mucosa secondary to chronic sinusitis most often contributes too. Almost all patients who had headaches and/or facial pain had obstruction of the ostio-meatal complex. Post-nasal catarrh seems to be the most refractory symptom of the three; this may be due to the fact that hyper-

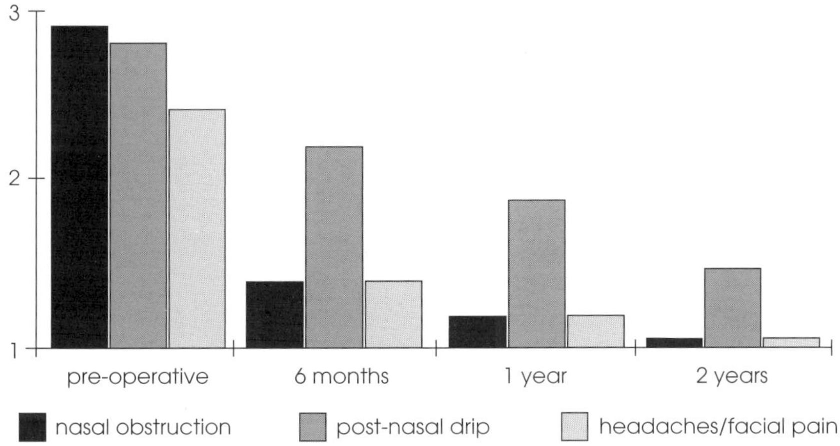

Fig. 13.2. Symptom profile in 478 patients.

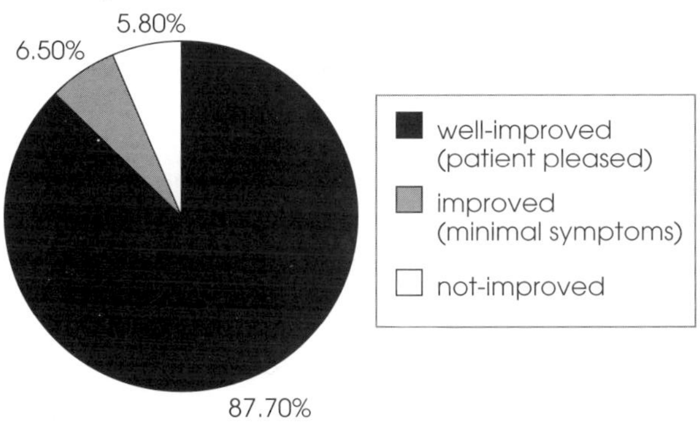

Fig. 13.3. Results of long-term follow-up of the well-improved group.

trophic sinus mucosa takes a longer time to revert back to normal; in addition, a parasympathetic overactivity might be implicated in those cases where a significant improvement was not seen.

The second group (improved – minimal symptoms, Fig. 13.3) showed a high degree of allergic symptoms, such as watering and itching of eyes, excessive, clear rhinorrhea and sneezing. These patients could have either perennial or seasonal rhinitis. The offending agent for the former group, when investigated, was most often house dust and house dust mite. The latter group showed a high incidence of sensitivity to grass pollens. Some cases of asthma or lower respira-

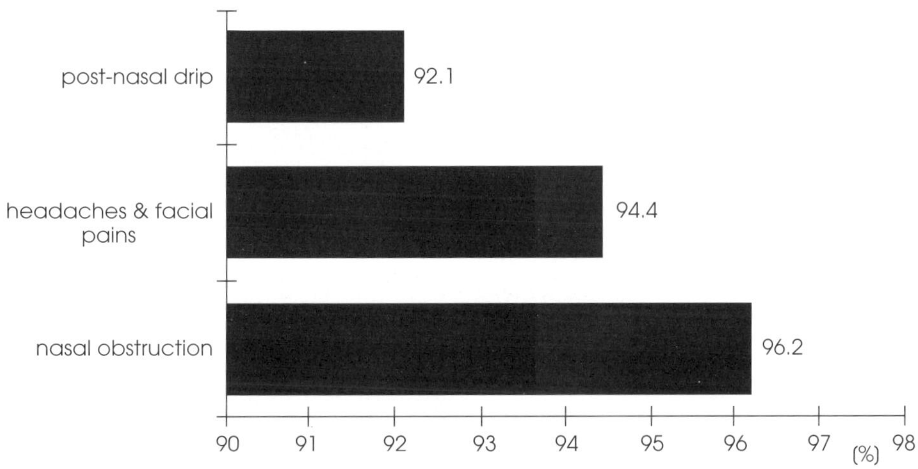

Fig. 13.4. Symptom profile in the well-improved group at 3 years ($n = 416$).

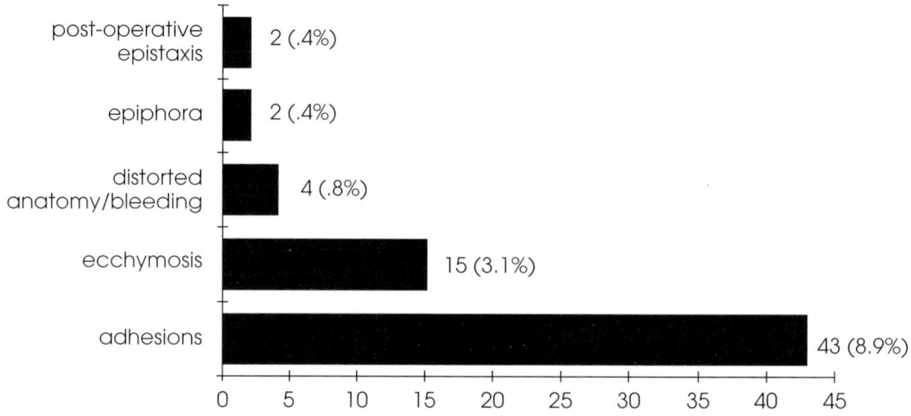

Fig. 13.5. Complications in our series ($n = 478$ patients).

tory tract infections did not therefore do as well as might have been expected.

Finally, there were a small number of patients whose symptoms remained unchanged following surgery (n=28, Fig. 13.3). Eleven of these had residual disease, especially in the posterior ethmoids/sphenoid. The bulk of these were in the first year; four went on to have external ethmoidectomies, while the remainder had endoscopic revision. All our cases of aspirin sensitivity (n=6) did badly. The classical course in these patients is an early improvement, lasting a few months at the most, followed by recurrence of disease and symptoms. One patient suffered from Kartagener's syndrome. These procedures were essentially carried out to facilitate drainage. Six cases had persistent adhesions blocking the middle meatal antrostomies and causing obstruction of the ethmoid cavities. In the remaining four patients, no obvious cause could be found; endoscopically, they had clean cavities – autoimmune mechanisms or hyper-reactivity of the nasal mucosa may play a role here. (Conversely, we have had patients with less than perfect cavities, who had near-total improvement in their symptoms.)

Complications in our Series

The most common complication both in our experience as well as in reported studies is *not* intra-operative, rather it is the development of post-operative adhesions (Fig. 13.5). In our series, 43 of 478 cases had some degree of adhesions. These ranged from some completely asymptomatic adhesions (37 cases) to those that obstructed middle meatal antrostomies and/or the ethmoid cavity, resulting in poor ventilation and drainage, requiring further surgery (6 cases). The adhesions could be unilateral or bilateral. The site of adhesions was either in the "first pass area" (i.e. between the inferior turbinate and the septum), or in the "second/third pass areas" (between the middle turbinate and the lateral wall of the nose). This must be considered against the background that 61% of our patients underwent a concomitant septoplasty or septorhinoplasty. This, naturally, would also contribute to the formation of adhesions. Our experience shows that some patients develop fibrous adhesions in spite of meticulous post-operative care. This is perhaps due to variable mucous membrane healing in different patients. The golden rule we observe is that if the adhesions are not causing any symptoms, they should be left well alone.

The second most common complication was post-operative ecchymosis (15 cases). This occurred in the earlier cases, and was predominantly right-sided. This, we feel, is due to the difficulty encountered by a right-handed surgeon in accessing the right natural os. The discolouration was only temporary, and settled in a few days. It is important to note that some of these cases had a dehiscent lamina papyracea.

Two of our cases had epiphorae, one settled after 3 weeks, and was probably due to post-operative oedema of the naso-lacrimal duct, while one did not settle after 18 months. Investigations showed scarring in the lower naso-lacrimal duct and she may need to undergo a dacryocystorhinostomy.

During the actual operation, we had significant bleeding in two cases, necessitating abandonment of the procedure, no aetiology was elicited. Two had partial procedures due to technical difficulties, one had a hypoplastic sinus. Finally, two patients bled profusely towards the end of surgery, and required post-nasal packing for 48 hours. Recovery thereafter was uneventful.

We have tried to reduce complications of adhesions by inserting a Merocel pack into the middle meatus, using splint in the ethmoid cavity, and even suturing the middle turbinate to the septum. However, our experience now shows that these measures have somewhat limited value in preventing adhesions. Only the diligent post-operative care of the ethmoid cavity has been successful in reducing adhesions significantly in recent years.

We had no serious intraorbital complications such as injury to medial rectus or optic nerve, nor any intracranial complications.

Chapter 14

Tips for Beginners

The age old chestnut, "practice makes perfect" is not, unfortunately, good advice for the initiate in FESS. Especially in today's age of medico-legal monitoring and heightened patient awareness, mistakes in this intricate region may prove very costly indeed. It must be remembered that though the surgeon tends to think in terms of statistics and series, for the patient an operation is a 100% event; it either succeeds or fails, there are usually no in-betweens. How, then, should one prepare oneself for this potentially hazardous but richly rewarding journey?

Training for FESS begins not in the operating theatre, but in the anatomy department, where an in-depth understanding of the surgical anatomy of both "hard" and "soft" parts of the nose and sinuses is gained. The next step is cadaveric dissection, and then on to the out-patient clinic, where a constant flow of patients is usually assured, and handling of endoscopes can be learnt. It is at this early stage that the initiate has to face the extremely practical problem of choice of instrument, which endoscope should he buy? As these instruments are expensive, we strongly recommend the 30° endoscope to start with, as it serves the function of both diagnosis and operation remarkably well. If, however, this is not the case, we would then recommend beginning with a 0° 4 mm endoscope, as it allows one to develop an undistorted depth perception and orientation of the internal nasal architecture. We would suggest the 30° instrument as the next in line, and finally the 70° endoscope for advanced users.

The endoscope must be considered an integral part of the clinical examination of any patient complaining of nose and/or sinus symptoms; it is not a second-level tool to be used only in difficult cases, or in those where anterior rhinoscopy shows no clear-cut pathology. One is often astonished at the hidden pathologies that are often brought to light during endoscopic examination. The examination must be disciplined and thorough: all three passes must be undertaken irrespective of the presence or absence of pathology. Finally, the importance of careful local anaesthesia cannot be over-emphasised. Findings must be carefully documented, preferably using a clinical profile sheet, such as the one we use (see Appendix B).

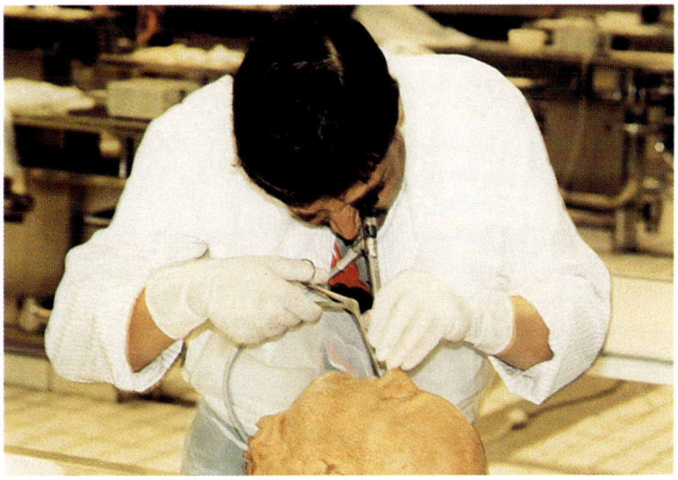

Fig. 14.1. The only way to master the technique is to dissect as many cadavers as possible.

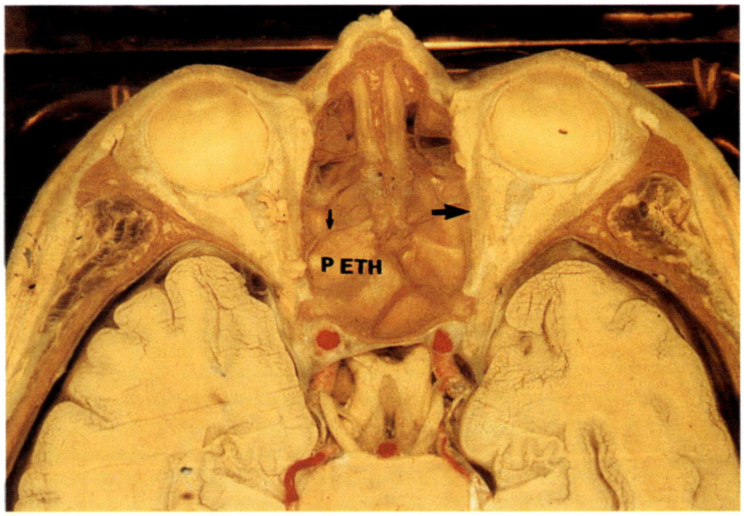

Fig. 14.2. As one advances posteriorly, one comes closer to the medial rectus, optic nerve and internal carotid artery in the posterior ethmoids and sphenoid. Also, the ethmoid cavity gets narrower; appropriate instruments should be used, and extreme care exercised. P ETH, posterior ethmoids; thick arrow, medial rectus; thin arrow, ground lamella.

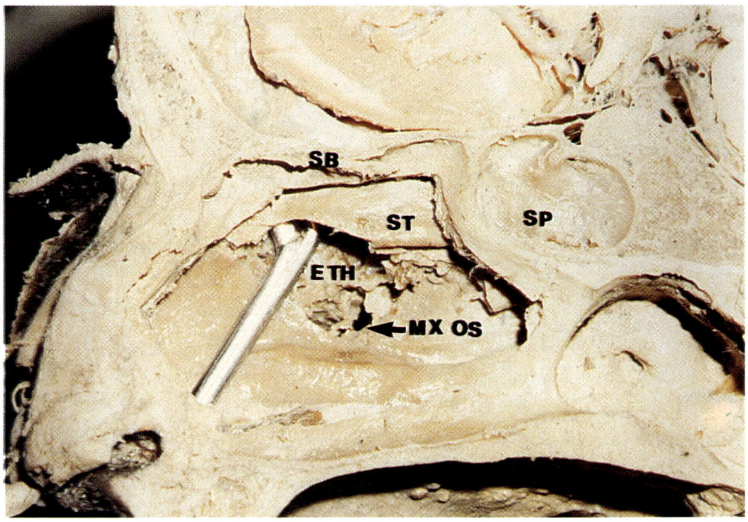

Fig. 14.3. As the instrument and the endoscope are advanced posteriorly, the dissection should follow inferiorly and medially, or else one might enter the cranium. SB, skull base; ETH, ethmoids; MX OS, maxillary ostium; ST, superior turbinate; SP, sphenoid sinus.

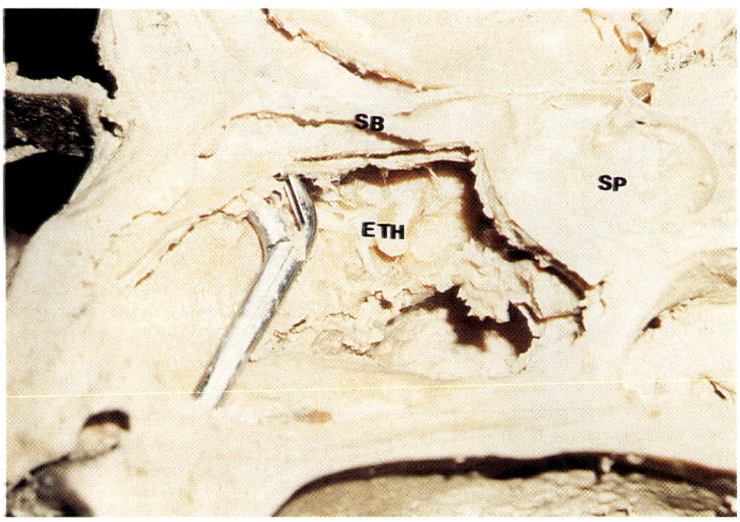

Fig. 14.4. Great care should be taken while clearing the frontal recess, any movements medially at this level will push the instrument through the thin olfactory fossa into the cranium. Hence, movements should be directed laterally. SB, skull base; ETH, ethmoids; SP, sphenoid sinus.

As the beginner acquires real-time experience of nasal endoscopy, so must he also acquaint himself with the theoretical aspects of disease in this region. This knowledge is not to be gleaned merely from textbooks, any number of hours spent in dissection and the study of the skull bones would be on too few!

The next step is perhaps the most crucial one. By now the beginner should have a clear, three-dimensional concept of the surgical anatomy of the nose and sinuses, and be comfortable with the handling of straight and angled endoscopes, he is ready for cadaveric dissection, which we consider to be absolutely mandatory, and there are no short-cuts. A useful tip at the end of the cadaveric dissection is to deliberately breach the boundaries of the cavity in order to see the relationships of the orbit and the intracranial structures. Furthermore, one of the most important skills is the ability to interpret CT scans of the nose and paransal sinuses. Constant dialogue with the radiologist, and correlation of clinical findings with radiology reports helps to a large extent. It is often the endoscopist's job to indicate his requirements to the radiologist, who otherwise tends to give only a gross reading of his films.

Initial surgery must be confined to patients with anterior group disease, i.e. the anterior/middle ethmoids, maxillary and frontal sinuses. Revision cases or extensive disease must be be avoided, as excessive scarring and distorted anatomy could pose hazards. High septal deviation patients should have their septum corrected before the actual FESS, we often perform septoplasties or rhinoplasties in order to achieve clear access to the lateral wall of the nose as a one-stage procedure.

It is advisable to begin with cases under local anaesthesia, as there is significantly less bleeding, and hence the various surgical landmarks could be easily identified. Also, the patient is aware of the procedure, and therefore warns the surgeon of pain (suggesting proximity to the orbital periosteum and skull base). The single most irritating factor for a beginner is bleeding, often caused inadvertently at the time of packing itself, or later due to coarse movements within the nose. Surgical technique overall, must be meticulous and atraumatic as in ear surgery, and frequent use of suctioning or rough movements with the endoscope or instruments should be avoided. Not only does blood obscure the field, it also stains the end of the endoscope, leading to frequent withdrawals and thereby excessive manipulation. If there is mucosal bleeding, it is prudent to pack the nose with cocaine-adrenaline strips, and wait for a few minutes. Fogging is prevented by dipping the endoscope in warm water or anti-fog solution, or by the rather ingenious method of gently touching it to the inferior turbinate. A constant watch must be kept on the eyes, as any transmitted movements could be picked up; towards this end, it is vital that the eyes remain uncovered all the time. The patient's CT scans must always be on hand, so that they can be used as a "road-map" during the operation.

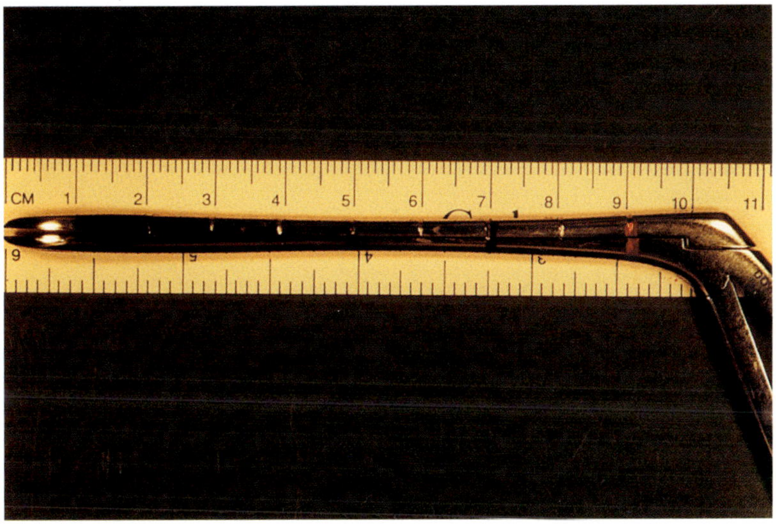

Fig. 14.5. Always be aware of the depth of the instrument within the nasal cavity from the anterior nasal spine. At 7 cm is the anterior wall of the sphenoid, and at 9 cm one is within the sphenoid sinus. Marking the forceps shaft is a useful trick.

As the learning curve is traversed, posterior sinus involvement is tackled more frequently, and cases may be operated on under a general anaesthetic. Only when the surgeon is at an advanced stage of expertise should recurrent or revision cases or extensive disease be taken on, gradually undertaking new horizons in endoscopic sinus surgery.

Ten commandments in endoscopic sinus surgery

- acquire an intimate knowledge of the surgical anatomy of the nose and sinuses, especially of the ostio-meatal complex
- evaluate patients thoroughly with nasal endoscopy and CT scan pre-operatively
- develop good endoscopic diagnostic skills using the 0° endoscope and acquire familiarity with the instruments
- perform several cadavaric dissections in order to master the technique
- remember, local anaesthesia permits a better field and is also safer. The patient simply would not allow the surgeon to enter the orbit or skull base
- avoid unnecessary trauma to the nasal mucosa to minimise bleeding
- use the CAMMA technique in the initial stages to gain better visualisation of the operative area
- do not select cases with extensive disease or a history of multiple interventions in the past
- when in doubt due to distorted anatomy, choose an external approach
- meticulous post-operative care is essential for preventing adhesions

Broadening the Horizons

As the endoscope comes into wider use, its applications extend to include –

- posterior epistaxis
- mucoceles
- acute sinusitis not responding to medical treatment
- orbital cellulitis and abscess
- dacryocystorhinostomy
- inverted papilloma
- CSF leak
- frontal sinus osteomas
- orbital decompression for Graves disease
- optic nerve decompression
- pituitary tumors
- foreign bodies: nose and sinuses
- olfactory disorders: hyposmia, dysosmia
- endoscopic management of choanal atresia
- endoscopic vidian neurectomy
- endoscopic management of intranasal and eustachian tube adenoids

Endoscopic Approach to Posterior Epistaxis

The role of nasal endoscopy in the evaluation of chronic sinus disease in now universally accepted in patients complaining of nasal and/or sinus symptoms (Kaluskar 1992). As one develops the expertise in handling the multiangled endoscopes and becomes familiar with the normal and abnormal anatomy of the nasal cavity, the use of nasal endoscopy can be further extended to embrace a common clinical emergency, epistaxis, with most gratifying results. Epistaxis affects about 10% of the population showing significant episodes (Weiss 1972; Shaheen 1975). Rarely it can be fatal. (Hara 1962)

Successful management of epistaxis is directly related to the location of the bleeding point. The more common anterior epistaxis is the least dangerous as it is easily located and cauterised. On the other hand posterior epistaxis may be quite severe, and more difficult to identify and control. It usually occurs in the elderly, who are intolerant to haemodynamic change and are often associated with hypertension, arteriosclerosis and various coagulopathies. The conventional treatment calls for a nasal packing, both anterior and posterior, antibiotics and a prolonged hospitalisation with increased morbidity. This results in significant pain in the nose and swallowing difficulties due to postnasal packing.

In recent years, one of the most valuable applications of the nasal endoscope has been its use in dealing with posterior epistaxis. This has shown excellent efficacy approaching 90% (Wurman 1992), a minimal complication rate (McGarry et al. 1991) and higher degree of patient satisfaction (O'Leary and Stickney 1992).

Posterior epistaxis occurs in an inaccessible area of the nasal cavity. The main pathological change in the arteries in the nasal cavity is increased fibrous tissue in the muscularis layer of the arterial wall (Shaheen 1975) and it may be that the rigid skeleton of the arterial wall prevents its contraction. Our experience in nasal endoscopy for posterior epistaxis has revealed the source of bleeding point as follows:

- the posterior part of the lateral wall near the sphenopalatine foramen
- from the posterior end of the inferior turbinate
- the middle turbinate and its medial surface
- from the floor beneath the inferior turbinate
- the middle and posterior part of the septum

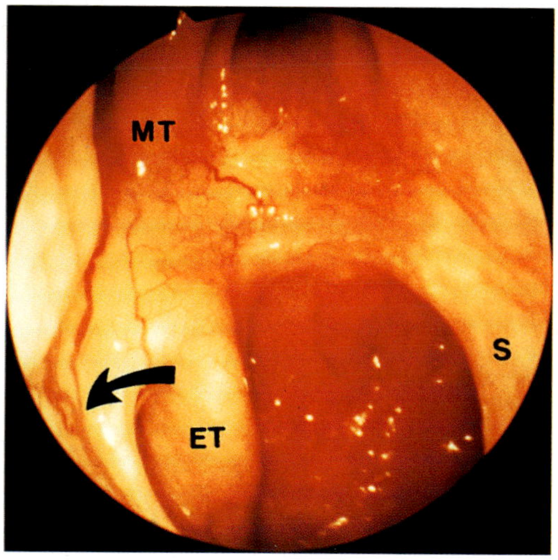

Fig. 15.1. Endoscopic view of Woodruff's plexus (arrow). ET, eustachian tube; S, septum; MT, posterior end of the middle turbinate.

- from the post-nasal space especially from Woodruff's plexus (Fig. 15.1) (Woodruff 1949).

In our experience bleeding from the Woodruff's plexus is usually a small trickle compared with the severe epistaxis from one of the branches of the sphenopalatine artery or anterior ethmoidal artery. The source of bleeding in posterior epistaxis has almost always been a single spurting blood vessel rather than engorged multiple blood vessels seen in anterior epistaxis.

Treatment Modalities of Posterior Epistaxis

Traditionally methods of treatment of posterior epistaxis are:

- anterior and posterior nasal packing with bismuth iodine paraffin paste (BIPP)
- various balloon catheters
- arterial ligation
  - internal maxillary artery
  - anterior and posterior ethmoidal arteries
  - external carotid artery
- arterial embolisation

In practice most surgeons resort to anterior and posterior nasal packing usually performed under general anaesthesia. In intractable cases anterior ethmoidal or external carotid ligation is performed depending upon the perceived source of bleeding. The drawbacks of the antero-posterior packing are as follows:

- prolonged admission, greater than 5 days on average
- High complication rate of 68% in these patients as shown by Wang et al. 1981 and Fairbanks 1986. There is certainly an increased morbidity in the post-operative period with significant pain and discomfort. Jensen et al. (1991) have shown the complications following anterior and posterior nasal packing such as sinusitis, otitis media, aspiration, toxic shock, hypoxia/hypercapnia, myocardial infarction, cardiovascular accidents, alar and columellar necrosis and scarring, development of septal perforations and adhesions.

The alternative method of posterior endoscopic cauterisation in the case of epistaxis affords the following advantages:

- its efficacy is in the region of 90%
- it does not require nasal packing
- it can be carried out under local anaesthesia in an out-patient clinic and has an excellent patient tolerance
- most patients are treated as out-patients
- even if admission becomes necessary, the mean in-patient stay is significantly reduced
- long-term side-effects of nasal packing are abolished
- it affords an early screening method for more extensive surgery as the potential severity of bleeding is obvious.

The side-effects of posterior endoscopic cautery are few. Wurman (1988) reported transient palatal anaesthesia and considered the theoretical risk of damaging the eustachian tube orifice with scar tissue in overenthusiastic cautery. However, these are avoidable complications.

We have now successfully treated 18 patients with posterior epistaxis including one case of haemangioma on the posterior part of the inferior turbinate (Fig. 15.2). Only five patients needed hospital admission, three for overnight observation due to social circumstances and two needed 3 days admission as they were post-operative (FESS) posterior epistaxis. All others, (13 patients) were treated as out-patients following endoscopic electric cauterisation under local anaesthesia.

Posterior endoscopic cauterisation of the source of bleeding is, initially, a cumbersome procedure as the endoscope and suction or cautery equipment has to be introduced into the nasal cavity. However, this can be overcome as one gains experience in handling the endoscopes. Nasal anatomy may make the passage of endoscopes a little difficult and an occasional septal correction may be required. In cases of septal deflections, we have used 2.7-mm endoscopes for successful posterior epistaxis cauterisation without the need for septal surgery.

15.2

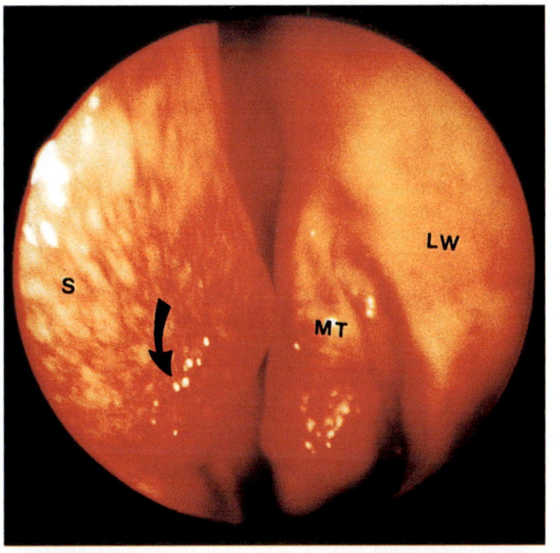

15.3

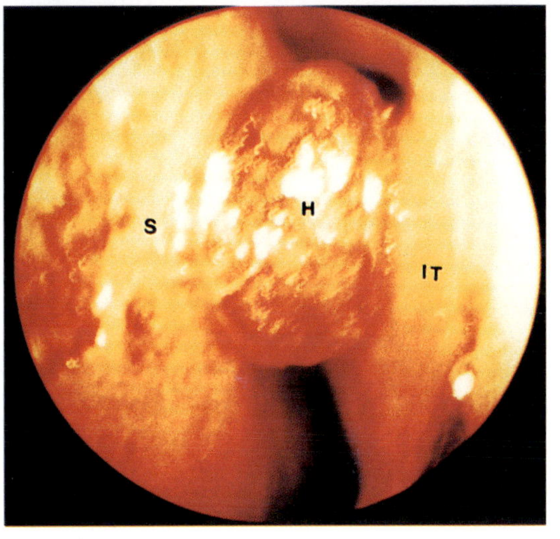

Fig. 15.2. An elderly patient was admitted with profuse posterior epistaxis. Thorough cocainisation of the nose and careful nasal endoscopy revealed a bleeding point far back on the nasal septum. This was cauterised with suction/cautery under local anaesthetic, thus avoiding a post-nasal pack. S, septum; MT, middle turbinate; LW, lateral wall; arrow, engorged blood vessels.

Fig. 15.3. A haemangioma on the posterior end of the left inferior turbinate. Multiple attempts at anterior nasal packing proved unsuccessful; however, nasal endoscopy identified the cause. S, septum; H, haemangioma; IT, inferior turbinate.

The equipment (see Fig. 15.4) used for posterior nasal endoscopy is quite common; for example:

- 4 mm and sometimes 2.7 mm 0° endoscope
- suction equipment, preferably with diathermy
- standard electric cautery equipment
- silver nitrate sticks in case of continuous oozing, e.g. Woodruff's plexus

Technique of Endoscopic Cauterisation

The patient is placed in a semi-sitting position and is given a bowl to clear blood clots into from the back of the throat. Initially 4% cocaine with 1 : 1000 adrenaline is applied on the soft Unicergel or cotton wool pledges to the nasal cavity which is actively bleeding. Usually, after a few repeated insertion of cocaine adrenaline the nose is well anaesthetised but, if necessary, anaesthesia could be supplemented by a sphenopalatine block with 1% lignocaine and 1 : 200 000 adrenaline injected at the posterior end of the middle turbinate. Sometimes a local injection of lignocaine can be made on the septum at the site of the bleeding point. A meticulous examination of the nasal cavity including various meati should be made and any blood clots removed. A bleeding point is usually seen in the form of markedly engorged blood vessels on turbinates, a large single vessel on the lateral wall, or a nipple-like projection

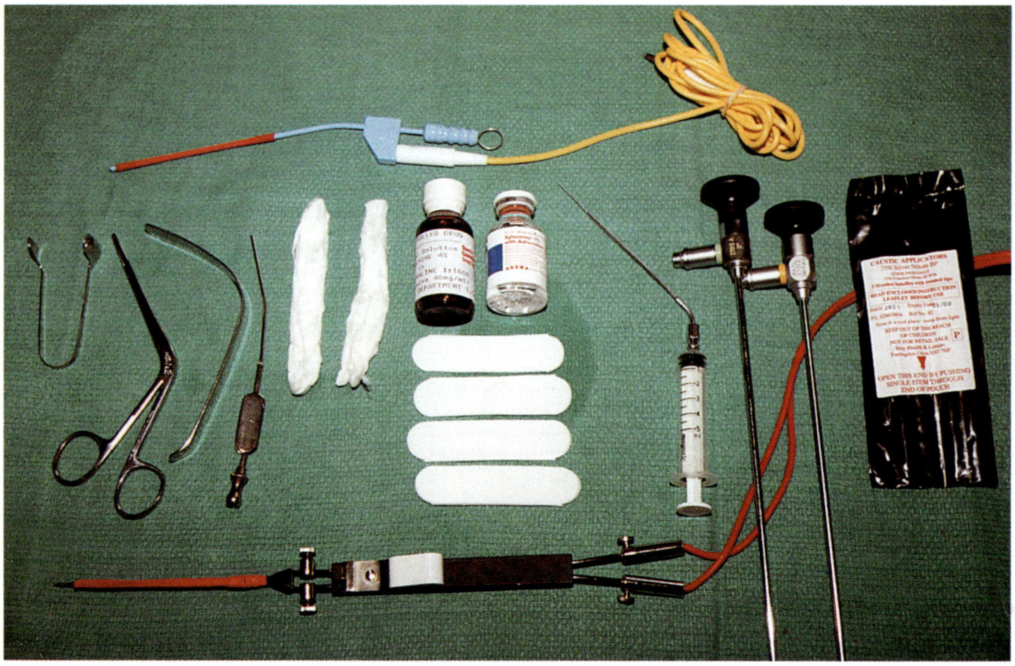

Fig. 15.4. Equipment for endoscopic cauterisation of posterior epistaxis. Note the suction cautery with insulation at the top of the photograph and the electric cautery at the bottom; 4% cocaine with 1: 1000 adrenaline and 1% lignocaine with 1: 200 000 adrenaline and endoscopic needle in the middle; far right are 4-mm 0° and 30° endoscopes and silver nitrate sticks.

usually on the septum is easily seen. The bleeding point is then electro-cauterised under endoscopic control.

We strongly recommend that nasal endoscopes should be routinely used in preference to the blind insertion of nasal packing or balloons. The morbidity and mortality associated with this most common ENT emergency can be greatly reduced. A thorough nasal endoscopy after topical anaesthesia, location of bleeding point and cauterisation should be the first-line treatment in all cases of posterior epistaxis. Nasal endoscopy in cases of posterior epistaxis is a significant major advancement for precise location of a bleeding point.

Mucoceles of the Paranasal Sinuses

A mucocele arises from the retention and accumulation of thick, mucoid secretions within a paranasal sinus, giving rise to expansion, thinning or lysis of one or more walls of the sinus. When mucoceles are infected, they form mucopyoceles. Mucoceles are lined by the epithelium. Sometimes they are misdiagnosed due to alarming neurological symptoms and signs. They need proper radiological investigation and once diagnosed the treatment is usually of marsupialisation and an adequate drainage. Approximately 65% of mucoceles occur in frontal sinuses, 30% in anterior ethmoids, (Howarth 1921; Rogers 1976). Sphenoidal and posterior ethmoidal mucoceles are rare. Maxillary involvement occurs in about 5% to 10% of cases. (Zizmor, 1974).

In his biography of Francis I, Francis Hackett (1934) recounts an illness from which the French King suffered while a prisoner in Spain in 1524. The illness described accurately follows the clinical course that an infected mucocele can take when untreated. The first documented case was reported by Dezeimerisn in 1725. However, Langenbeck in 1818 wrote the detailed description of a frontal sinus mucocele calling it a Hydatid. It was a French pathologist, Berthon in

1880, who reported the first treatment of a frontal mucocele. Onodi in 1901 described the histological characteristics of these lesions. In 1909 Rollet suggested the term mucocele. Herrnheiser presented the first detailed description of the radiological appearances of these lesions in 1926.

Aetiology

The aetiology of the mucocele is primarily that of obstruction and inflammation of the sinus due mainly to the following factors in a majority of patients:

- chronic sinusitis
- nasal polyps
- trauma either accidental or operative
- neoplasms

In a small number of patients the nature of obstruction is not clear. In these cases it may relate to anatomic variations and absence of alternative pathways. Histological examination of the mucocele confirms the chronic inflammatory process where osteolysis, new bone formation and sclerosis has been identified.

Clinical Features

Mucoceles are slow-growing lesions and hence do not cause symptoms and signs until there is involvement of the surrounding structures. Symptoms vary in general and clinical features depends upon the size, location and presence of secondary infection. The following are the common symptoms in case of fronto-ethmoid mucoceles. Slowly growing, painless, fluctuant swelling in the forehead or fronto-nasal region of the orbit just above the inner canthus. This may ultimately cause upper eye lid ptosis and proptosis with inferolateral displacement of the eye (Figs. 15.5, 15.6). Headaches, pain in the region of the swelling, diplopia, blurring of the vision and epiphora may occur. Progressive compression of the orbital contents may lead to visual impairment and even optic atrophy. Endoscopic examination may reveal bulging in the region of the middle meatus. CT scan both coronal and axial (Figs. 15.7–15.9) usually clinches the diagnosis by showing radio opacity, bony destruction, marginal bone sclerosis and loss of scalloping in case of frontal sinus mucocele. Thorough CT assessment is essential to decide a surgical approach. MRI would be useful to differentiate mucocele from neoplasms and meningoceles.

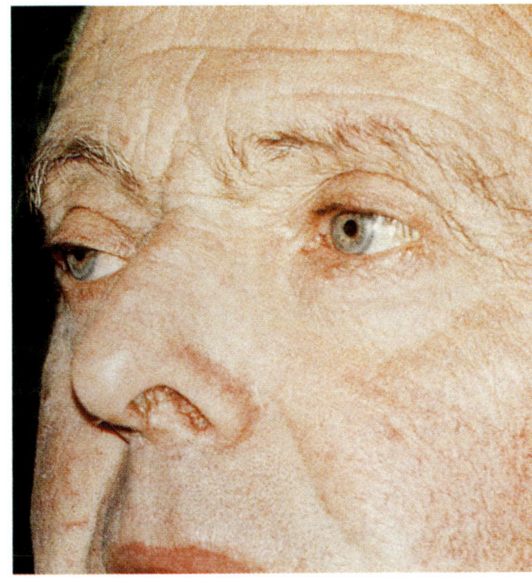

Fig. 15.5. Right large fronto-ethmoid mucocele in a 60-year-old man. Note the right proptosis. The patient had had several sinus operations in the past.

Sphenoid sinus mucoceles are uncommon and usually present with headache and opthalmological symptoms. The headaches are localised to the vertex, forehead or around the eye. Inflammation or oedema of the optic canal is usually responsible for the visual disturbances rather than direct compression.

Treatment

Prior to the advent of endoscopic surgery, patients were treated by external approach either by the Lynch–Howarth or osteoplastic procedures requiring several days in the hospital. Both these approaches have been associated with high degree of recurrence. A non-invasive endoscopic approach is much preferable in suitable cases where the mucoceles can be widely marsupialised under direct vision, avoiding external incisions and post-operative morbidity. In more laterally placed mucoceles a combined approach may be necessary.

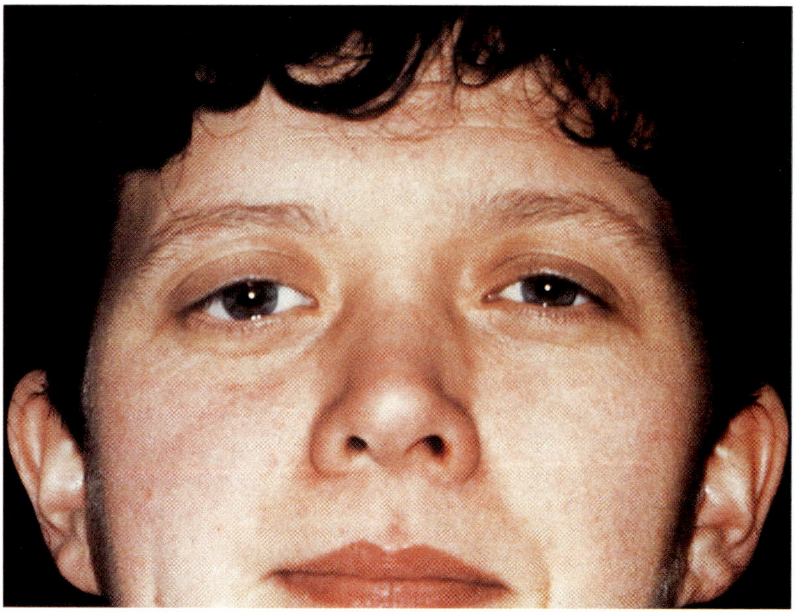

Fig. 15.6. A 32-year-old woman complaining of right epiphora and swelling of the right eye. A diagnosis of fronto-ethmoid mucocele was made which was marsupialised through the endoscope.

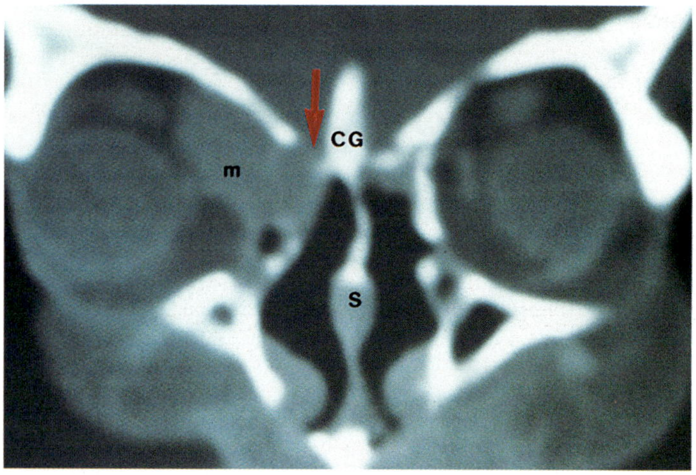

Fig. 15.7. Coronal CT scan of the sinuses. The arrow shows destruction of the bone due to mucocele. Also note the complete absence of both middle turbinates. M, fronto-ethmoid mucocele; CG, crista galli; S, septum.

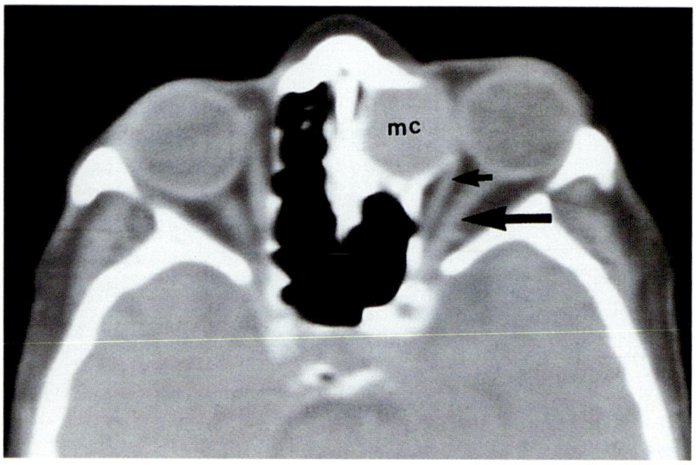

Fig. 15.8. Axial CT scan showing mucocele of the ethmoid sinus causing minimal unilateral proptosis (mc); medial rectus (short arrow); optic nerve (long arrow).

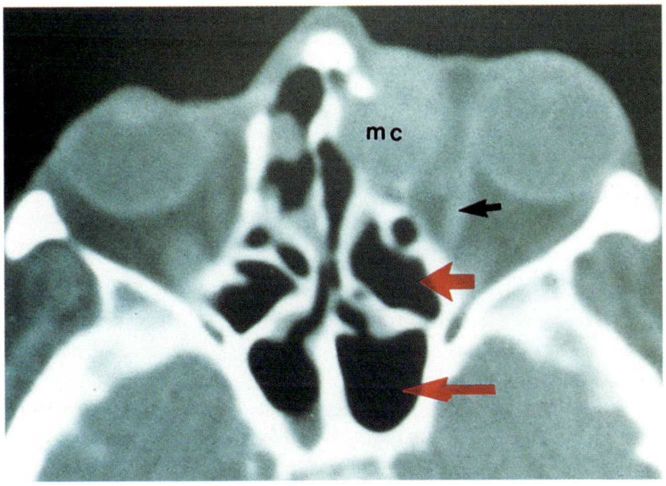

Fig. 15.9. Axial CT scan showing mucocele of the fronto-ethmoid sinus causing unilateral proptosis (mc); medial rectus muscle (small arrow); posterior ethmoid sinus (short arrow); sphenoid sinus (long arrow).

The operation is performed either under local or general anaesthesia as a day-case procedure. After topical decongestion of the nasal cavity a 4-mm, 0° endoscope is used and after identifying surgical landmarks the inferior expansion of the mucocele is uncapped with a sickle knife. The opening of the mucocele is then widely enlarged with an upward-cutting forceps and the turbid inspissated mucus is removed. If a mucocele is not seen in the nasal cavity then uncinectomy, anterior ethmoidectomy with identification of the skull base may need to be performed to gain access to the wall of the mucocele. A close scrutiny of the CT scan of the mucocele will help the surgeon to have proper exposure of the mucocele. Post-operative cavity care is the same as after any ethmoidectomy. The proptosis usually settles over a period of few weeks following surgery.

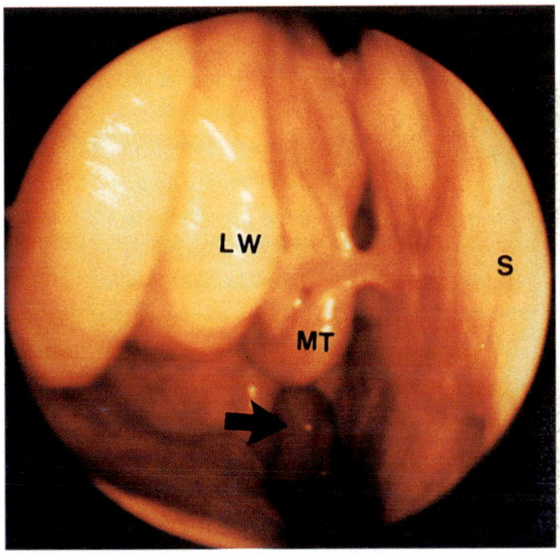

Fig. 15.10. Endoscopic view of the right nasal cavity in a patient who had had multiple sinus surgeries in the past. LW, prolapsed lateral wall of the nose with protrusion of the mucocele; MT, remnants of the middle turbinate; S, septum; post-nasal space (arrow).

Endoscopic Approach to Acute Sinusitis and Orbital Complications

Acute sinusitis most commonly responds to oral antibiotics and decongestants. However, there are cases in which the disease is disturbingly persistent and there is a further danger of infection spreading into the orbit or to the cranium.

The orbital complications such as orbital cellulitis and abscess most commonly arise from ethmoiditis. Patients who do not rapidly respond to intravenous antibiotics and decongestants need urgent surgical drainage. The problems during operation in acute cases are a congested and oedematous mucous membrane with vasodilatation and hence an increased tendency to bleed during operation. The surgical landmarks are distorted and there is an increased risk of complications, e.g. CSF leak during operation due to disease, inflammation and increased bleeding. Needless to emphasise that such an operation should only be undertaken by an experienced endoscopic surgeon (Fig. 11.7), otherwise an external approach should be performed for surgical drainage.

Chandler in 1970 proposed a classification of orbital complications of sinusitis as follows:

- inflammatory oedema
- orbital cellulitis
- subperiosteal abscess
- orbital abscess
- carvernous sinus thrombosis

The first three conditions are amenable to endoscopic ethmoidectomy and a surgical drainage; the last two, being more advanced and serious, should be approached by a combined technique of both endoscopic ethmoidectomy and an external ethmoid exploration and orbital decompression.

Lasers in FESS

Laser is an acronym for Light Amplification by Simulated Emission of Radiation. Although lasers have been used in medicine since the 1960s they have only recently been employed as a adjunct to endoscopic sinus surgery. Their action is threefold: cutting, coagulation and vaporisation; all three actions are of use in FESS. The KTP/532 laser has the added advantage of fibre optic delivery systems, where the quartz fibre that transmits the laser energy can be passed through an endoscope, or a suitable hand-held instrument. This permits use of the laser right along the nasal cavity. The specific advantages of using KTP/532 in FESS are as follows:

- minimal bleeding, thereby allowing clear visualisation of surgical landmarks, making the surgery safe
- minimal trauma
- can be used under local anaesthetic
- green visible beam, therefore no need for an aiming beam as with other lasers
- as the KTP/532 wavelength is highly absorbed by the haemoglobin, it is a better coagulant.

Thus we find the KTP to be valuable for:

- nasal polyposis, where polyps could be evaporated with minimal bleeding
- adhesions, which can be divided with minimal risk of causing scar tissue
- using the property of "scatter" (dispersion of laser energy due to distance between the end of the probe and the target tissue), turbinate diathermy may be performed with minimum trauma to the surface lining, and maximum damage to the sub-mucosa
- evaporation of the posterior end of the inferior turbinate ("mulberry ends")
- excision of a concha bullosa
- wedge resection of middle turbinate

The disadvantages are:

- cost of equipment
- need for safety measures for all concerned as the laser gives out an extremely intense light, it can damage the retina. Hence, special glasses to filter the beam should be worn by all (including the patient).

We have so far performed 112 cases of endoscopic sinus surgery with KTP Laser, which included nasal polyposis, excision of concha bullosa, turbinoplasties, FESS revision cases, middle meatal antrostomies and excision of posterior ends of middle turbinates with great success.

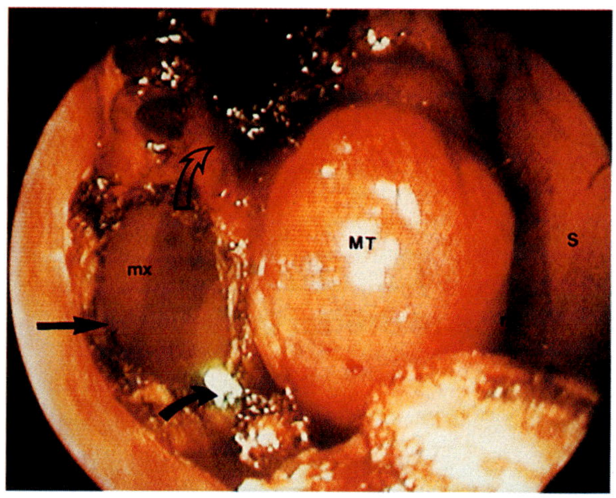

Fig. 15.11. Endoscopic view of a right revision FESS with a KTP/S32 laser. Straight arrow, middle meatal antrostomy; curved arrow, 600 micron quartz fibre transmitting laser creating a MMA; hollow curved arrow, ethmoid cavity, MT, middle turbinate; S, septum; MX, maxillary sinus.

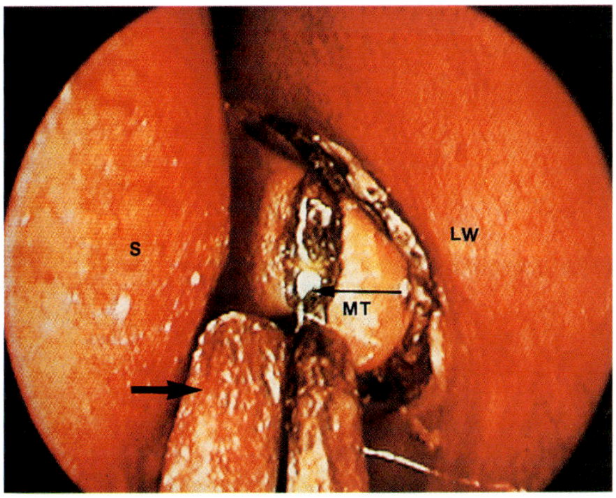

Fig. 15.12. Endoscopic view of the left nasal cavity during a KTP/S32 laser revision. FESS: excision of the anterior part of the middle turbinate. S, septum; MT, adherent middle turbinate to the lateral wall (LW) of the nose; thick arrow, suction part of the laser microstat long arrow, 600 micron quartz laser fibre. Note the complete absence of bleeding.

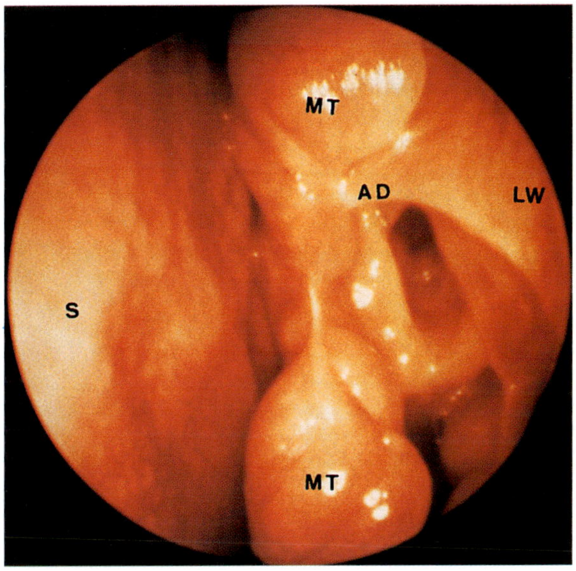

Fig. 15.14. Revision FESS: left nasal cavity. S, septum; MT, excised middle turbinate from previous surgery; LW, lateral wall; AD, adhesions between middle turbinate and lateral wall filling the ethmoid cavity.

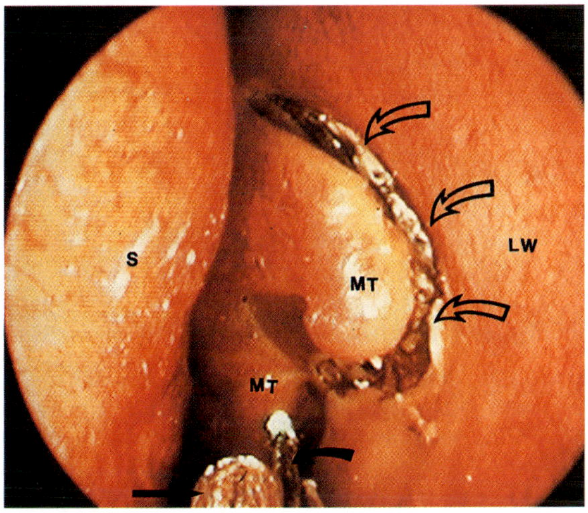

Fig. 15.13. Endoscopic view of the left nasal cavity showing KTP/S32 laser revision FESS. Laser uncinectomy has been performed (three hollow arrows). S, septum; MT, partially excised and adherent middle turbinate; straight arrow, suction part of the laser microstat; curved arrow, 600 micron quartz fibre transmitting laser.

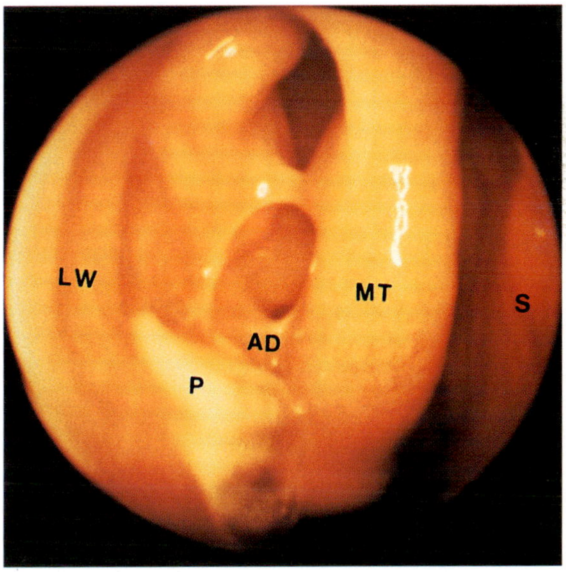

Fig. 15.15. Revision FESS: right nasal cavity. AD, adhesions between the middle turbinate (MT) and the lateral wall (LW) filling the ethmoid cavity; P, pus exuding from the maxillary sinus; S, septum.

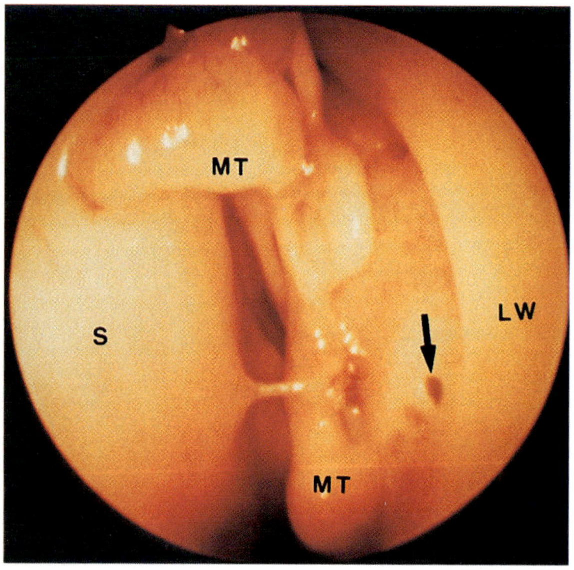

Fig. 15.16. Revision FESS: left nasal cavity. S, septum; MT, remnants of the middle turbinate following two previous sinus operations; LW, lateral wall; arrow shows stenosed middle meatal nasal antrostomy.

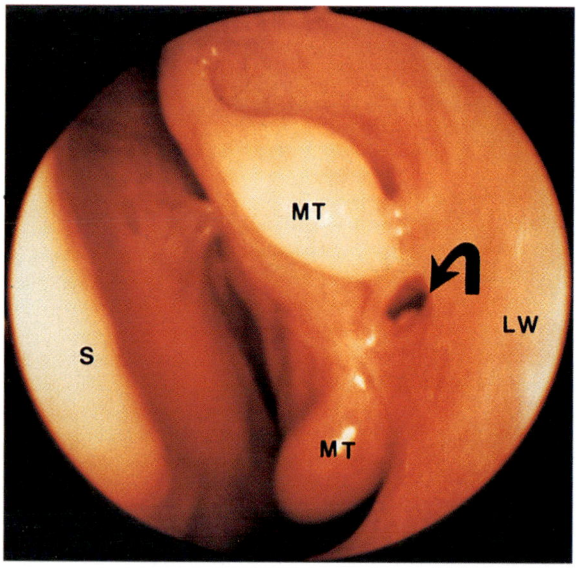

Fig. 15.17. Revision FESS: Left nasal cavity. S, septum; MT, middle turbinate adherent to the lateral wall (LW) after previous partial excision; arrow shows stenosed middle meatal antrostomy.

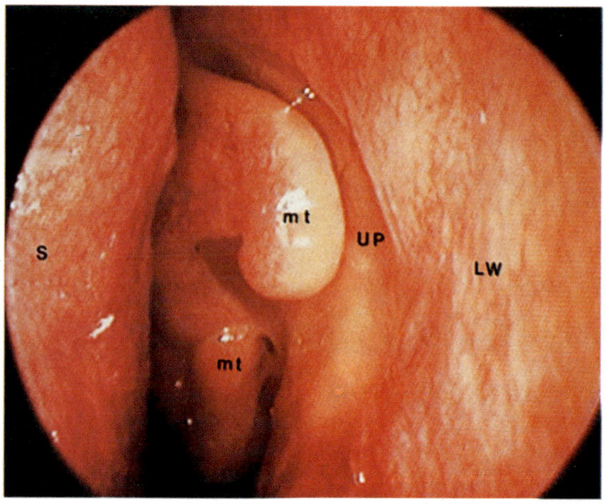

Fig. 15.18. Revision FESS: left side of nasal cavity. S, septum; MT, middle turbinate partially excised and adherent to the lateral wall (LW); UP, uncinate process.

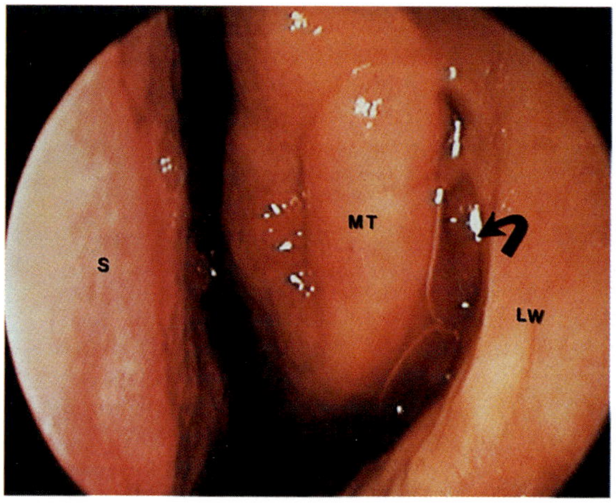

Fig. 15.19. Revision surgery: left side of nasal cavity. S, septum; MT, middle turbinate adherent to the lateral wall (LW); arrow shows polypoidal disease in the middle meatus.

Endoscopic Orbital Decompression

Orbital decompression for dysthyroid orbitopathy has been performed in the past either through the antrum or externally. With the advent of intranasal endoscopes it is now possible to decompress the medial and inferior wall of the orbit, thereby relieving pressure on the optic nerve. This clearly avoids the morbidity of the external ethmoidectomy or Caldwell-Luc antrotomy.

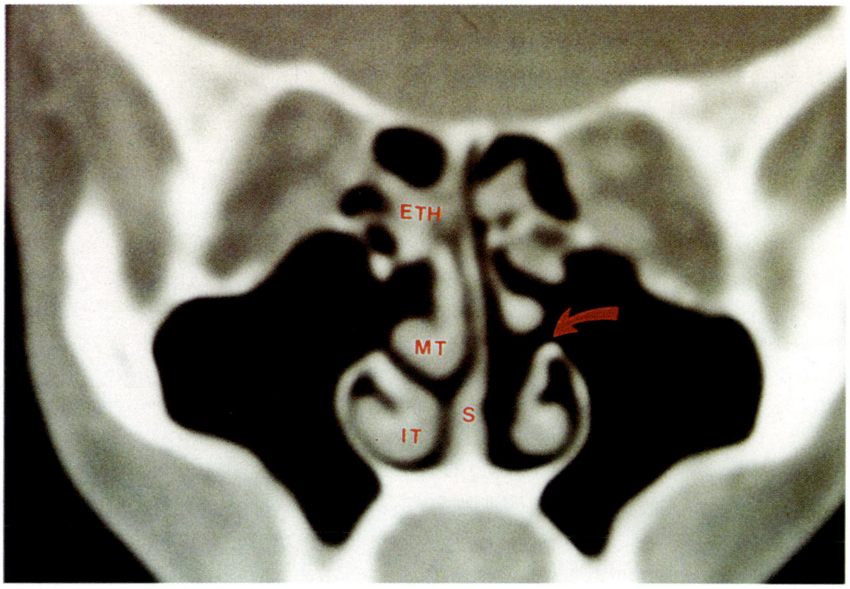

Fig. 15.20. Revision FESS. Coronal CT scan showing mucosal disease in the posterior ethmoid sinuses (ETH); middle turbinate (MT); septum (S); inferior turbinate (IT); middle meatal antrostomy (arrow).

Endoscopic orbital decompression is performed under general anaesthesia. A complete sphenoethmoidectomy is performed to skeletonize the medial wall of the orbit and the skull base. Anteriorly the frontal recess and posteriorly the orbital apex are identified. A large middle meatal antrostomy is performed from the posterior margin of the nasolacrimal duct anteriorly to the posterior limit of the maxillary sinus. Inferiorly the antrostomy extends to the attachment of the inferior turbinate. The exposed medial wall of the orbit is then removed with a blunt hook or similar instrument. The lateral limit for the removal of the bone of the orbital floor is the position of the infraorbital nerve seen through the large antrostomy. Care is taken to preserve the small bony area of the medial wall of the orbit near the frontal recess or else the soft tissues will obstruct the frontal recess.

When removal of the bony orbital wall is complete, the periorbita is incised with a sickle knife in linear strokes from posterior to the anterior. Further fibrous periorbital bands are excised to allow the orbital contents to prolapse into the maxillary sinus and into the nasal cavity. The extent of decompression is assessed by palpating the eyeball and looking for herniated orbital contents in the nose through the endoscope. The post-operative care is same as for any other sphenoethmoidectomy.

Endoscopic Dacryocystorhinostomy (DCR)

As the endoscope broadens our horizons, it leads us to a structure that is traditionally within the purview of the ophthalmologist, the orbit. When one considers that two out of the three walls of the eye socket abut the nasal and sinus cavities, procedures involving this structure can logically also be handled by the otolaryngologist. Endoscopic control allows for tackling disease at the lower end of the naso-lacrimal apparatus, which is the most common site of affection. It has made the technique an even more aesthetic and successful procedure in the hands of the ear, nose and throat surgeon. Furthermore, an anterior ethmoid focus of disease may be tackled simultaneously. Indeed, a significant number of dacryocystitis cases are caused by infection in the anterior ethmoidal cells.

The advantages include:

- internal approach; therefore no external incision
- can be performed under local anaesthesia (day-case surgery)
- good success rate comparable with the external approach

The operation is performed for a congenital or acquired stenosis, producing chronic dacryocystitis. Investigations include a coronal CT scan to rule out anterior ethmoid disease and dacryocystogram. Cannulation is then performed to locate the site of obstruction; it may be further supplemented by a retinal light probe.

The operation is performed by an initial uncinectomy, followed by exenteration of the anterior ethmoidal cells. The lacrimal sac is then identified just in front of the anterior end of the middle turbinate, by removing the overlying bone. The sac may be illuminated by passing a light fibre along the inferior canaliculus. Once this is done, the medial wall of the sac is excised and left open, or a silastic tube may be inserted via the canaliculus into the sac and brought out through the nose to be secured for a period of 2–3 months. Published reports suggest excellent results with minimum morbidity and complications.

Vidian Neurectomy

This operation is based on the fact that the autonomic nerve supply to the nasal cavity is through sympathetic and parasympathetic fibres, carried by the vidian nerve. The vidian nerve is formed from the union of pre-ganglionic parasympathetic fibres from the greater superficial petrosal nerve with post-ganglionic fibres containing the deep petrosal nerve. The nerve runs across the pterygo-palatine fossa to connect with the spheno-palatine ganglion, from where post-ganglionic parasympathetic fibres emerge to supply the nasal mucosa. They exert their action via the release of acetylcholine.

Vasomotor instability is manifested mainly by rhinorrhea, excessive sneezing, nasal obstruction and headaches. Rhinorrhea is the most characteristic symptom, and is due to excessive overactivity of the seromucinous glands. This, together with increased capillary permeability and vascular dilatation results in copious, clear and watery secretions of low mucin content. Nasal obstruction occurs because of enormous dilatation and oedema of the inferior turbinate. Headaches follow because of insufficient ventilation. Racial, constitutional and emotional factors play an important part in the aetiology of this condition.

Surgical Anatomy

The pterygoid or vidian canal is bounded above by the undersurface of the body of the sphenoid, and below by the medial pterygoid plate. The anterior end opens into the pterygo-maxillary fissure, lateral to the posterior border of the spheno-palatine foramen. Its posterior end opens into the foramen lacerum. The nerve traverses the pterygoid canal and joins the spheno-palatine ganglion. It is important to note that the ganglion lies between the mucosa and the periosteum, while the vidian nerve lies between the periosteum and bone (Minnis, 1971). The average distance between the posterior border of the septum and the anterior opening of the vidian canal is 18.6 mm, while the distance from the mouth of the vidian canal to the foramen rotundum is 6.5 mm. The distance between the pterygoid canal from the posterior border of the spheno-palatine foramen is 2 mm. The average dimensions of the pterygoid canal are 1.5–2.5 mm diameter and 19–20 mm length.

Vidian neurectomy has been a well-established procedure in the treatment of vasomotor rhinitis. Several routes to the vidian nerve have been described, including the trans-antral, trans-palatal and trans-nasal. The endoscope may be used to access the spheno-palatine foramen and thereby the pterygoid canal via the trans-nasal route.

Technique

Under general anaesthetic, a 4 mm angled (30°) endoscope is used to visualise the posterior end of the middle turbinate, which may need lateralisation to allow better access. The mouth of the spheno-palatine foramen is then located by one of three methods (Khasgiwala 1972):

- just posterior to or slightly above the posterior end of the middle turbinate
- at the junction of the superior and lateral walls of the nasal cavity, 12–14 mm antero-superior to the superior border of the posterior choana
- the superior turbinate acts as a “pointer” to the spheno-palatine foramen

The site of the foramen is confirmed with an insulated, 15 cm blunt probe, and an incision is made over the posterior (relatively avascular) lip with a sickle knife. The muco-periosteum is dissected forwards to expose the foramen under endoscopic control. The spheno-palatine artery is often encountered here; however, it rarely bleeds as it has its own covering and may be kept away from the operative field. The probe is further advanced into the foramen in a posterior direction, thus separating the periosteum off the lateral wall of the sphenoid process of the palatine bone. The opening of the pterygoid canal is reached 5–6 mm posterior and lateral to the spheno-palatine foramen. Finally, a slim cautery/probe is inserted into the dissected foramen and used to cauterise the vidian nerve. The mucosa is then replaced; no packing is normally necessary. Both sides are done at one sitting.

The procedure is a distinct step forward from the gross or microscopic trans-nasal approaches, as it allows for angled visualisation of the lateral nasal wall, and better visualisation of the pterygo-palatine foramen. Some published results show a high degree of efficacy of the technique.

Appendix

Appendix A. Instruments

Instruments Used in Office Nasal Endoscopy

1. Hopkins straight telescope, 0°, 4 mm, fibre optic light transmission incorporated
2. Hopkins forward oblique telescope, 30°, 4 mm
3. Cold Light Fountain, 220 VAC, 40–60 Hz.
   150 watt halogen lamp
   colour temperature approximately 3400 K
   with two light outlets
   light intensity adjustable
4. Fibre optic light cable, size 3.5 mm, length 230 cm
5. Wittmoser Articulated Optical Arm with 4 joints, length 820 mm. Used for optically connecting the endoscopic telescope through a still or video camera, and for simultaneous observation by the operator and observer
6. Special zoom lens with adapter to Hopkins telescope for still photography

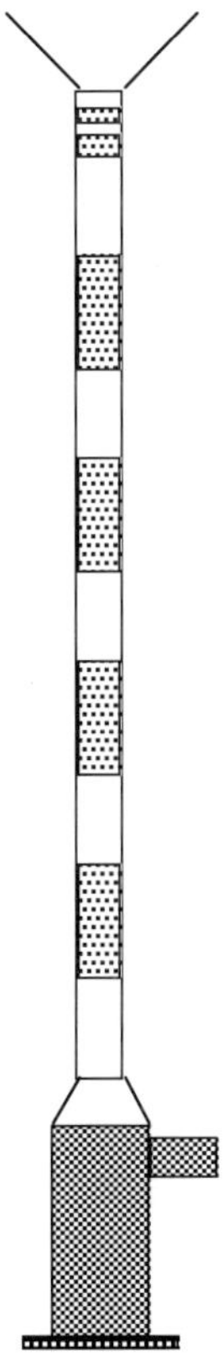

Fig. A.1. The Hopkins rod lens system affords a smaller diameter, a larger viewing angle and a brighter image.

Instruments Used in FESS

1. Hopkins straight telescope, 0°, 4 mm, fibre optic light transmission incorporated
2. Hopkins forward oblique telescope, 30°, 4 mm
3. Hopkins wide-angle lateral telescope, 70°, 4 mm
4. Hopkins wide-angle lateral telescope, 2.7 mm, 0°
5. Hopkins wide-angle lateral telescope, 2.7 mm, 30°
6. Anti-fog solution
7. Endoscope sleeve (for good grip and precision of movements)
8. Killian's fibre optic nasal speculum, 7.5 cm
9. Freer elevator, double ended, 20 cm
10. Angular long needle (for local anaesthetic)
11. Sickle knife, 19 cm, pointed
12. Blakesly Weil ethmoid forceps, 19 cm, size No. 1, size No. 4
13. Blakesly Weil ethmoid forceps, upturned, 19 cm, size No.2
14. Blakesly Weil ethmoid forceps, upturned, 90°, size No.2 (for frontal recess and middle meatal antrostomy)

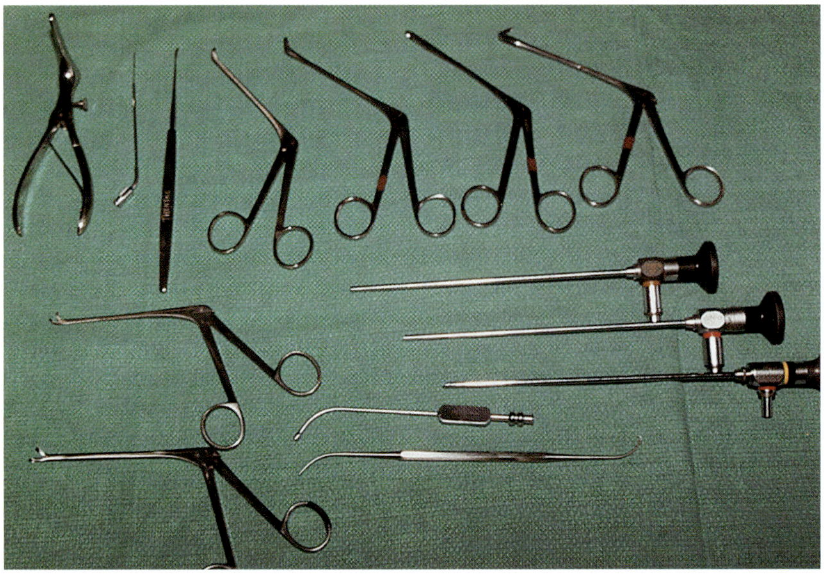

Fig. A.2. Instrument set commonly used for FESS. Top, from left to right: Killian's fibre optic speculum, endoscopic needle, sickle knife, small and large 45° upward-cutting forceps, straight forceps, backward-cutting forceps. Bottom, from left to right: "cut through" straight and upward-cutting forceps, suction cannula, maxillary and frontal ostium seeker, three 4-mm endoscopes (0°, 30° and 70°).

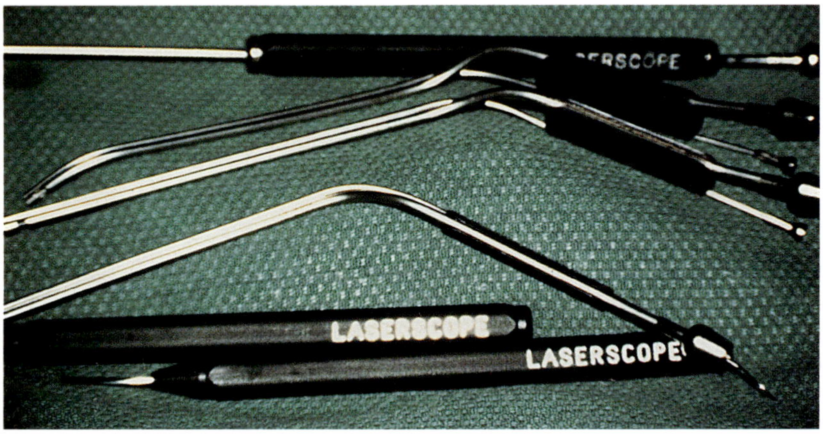

Fig. A.3. A set of KTP 532 laser delivery instruments for endoscopic sinus surgery.

15. Stammberger backward-cutting forceps (for middle meatal antrostomy)
16. Endoscopic scissors
17. Angled suction
18. Trocar and cannula (for maxillary sinoscopy)
19. Giraffe forceps

Care of Instruments

Endoscopy instruments are expensive, and should be handled with care! All steel instruments can be autoclaved. The endoscopes and fibre optic cables are chemically sterilised with CIDEX, although autoclavable endoscopes have recently became available.

Appendix B. Documentation

Proforma

In order to keep accurate records of all our patients, we developed computer-compatible proformas for pre-, intra- and post-operative notes. The proformas also display key areas visually, in order to give the surgeon a rough pictorial representation of the disease. The graphic on Chart 2, for example, depicts the site and extent of disease which could then be accurately monitored post-operatively. Each piece of information is given a number, so that it may be processed on a database.

Functional Endoscopic Sinus Surgery Chart 1

Age _______ years
Sex 0 (male)
1 (female)

Consultant - S.K. Kaluskar, M.S.,F.R.C.S.,D.L.O.(Eng.)
Tyrone County Hospital, Omagh, Northern Ireland

name ____________
address ____________
____
____
hosp. no. ____________
d.o.birth ________ *sticker if available*

Catalogue number | 9 / |

Date of operation ____|____|____

Surgeon / s – – – – – – – – –
Anaesthetist – – – – – – – – –

Pre-Operative Assessment

Grades:
1 - mild, 2 - moderate, 3 - severe

Symptoms
nasal obstruction
2,3,4 R / L / bilateral
5,6,7 Grade: 1 2 3
post-nasal drip
8,9,10 Grade: 1 2 3
headaches & facial pains
11,12,13 Grade: 1 2 3

h/o allergies Y/N 14,15
allergens ____________________

asthma 16
ASA intolerance 17

allergic profile
18,19,20 Grade : 1 2 3
21 sneezing
22 rhinorrhea
23 nasal irritation

smell
anosmia 24
hyposmia 25
parosmia 26

misc.
sore throats 27
laryngitis 28
LRTI 29
otitis media 30

Anterior Rhinoscopy
septal deviation
31,32,33 - R / L / bilateral
34,35 - caudal / high
polyps
36,37,38 - R / L / bilateral
39,40 - single / multiple

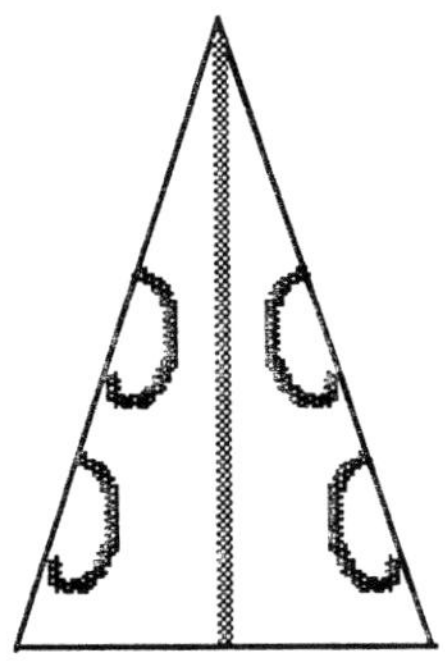

N. Endoscopy

high septal deviation - R / L / bilateral 41,42,43

Sphe.- eth. recess 57a - uncinate process (R / L) 44,45
- concha bullosa (R / L) / paradoxical turb. (R / L) 46,47,48,49

Acc. Os 57b - bulla ethmoidales (R / L) / agger nasi (R / L) 50,51,52,53
- m. meatus : polyps (R / L) / mucopus (a / p) 54,55,56,57

Past History - Caldwell Luc / IN Antrostomy / Polypectomy / 58,59,60,
Sinusitis / Chest Complaints 61,62

Functional Endoscopic Sinus Surgery Chart (2)

Plain Xrays	grade R/L	cyst / fl. level R/L
maxillary		
frontal		
ethmoid		
sphenoid		

63 to 78

clear 0, slight haziness 1, marked haziness 2

CT scan	grade R/L	other R/L
maxillary		
frontal		
ethmoid		
sphenoid		

79 to 94

clear 0, haziness 1

extent of disease

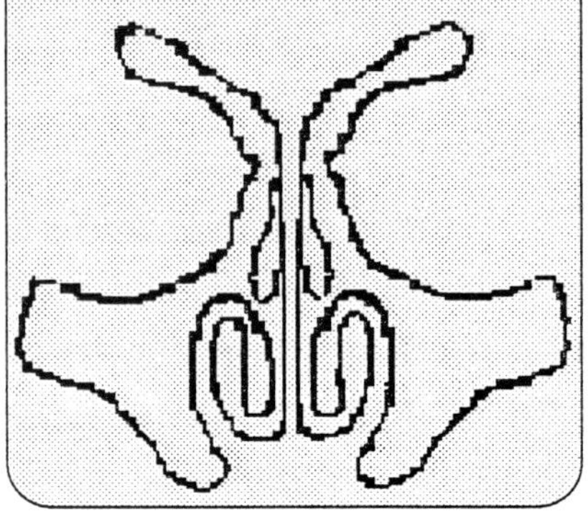

key areas on CT scan

Rt. Lt.

u.p.
bulla ethmoidales
ethmoid infundibulum
m.t.
frontal recess
ant. eth.
post.eth. 95 to 106
sphenoid
fovea eth.
l.p.
o.n.(O.cell)
ica

Additional Remarks :

Operative Notes

- **LA / GA** 107, 108
- unilateral / bilateral 109, 110

approach
- nasal / combined 111, 112

ethmoidectomy 113, 114
- anterior / frn.sinusotomy / mma
- posterior / sphenoidotomy / fro.sin./mma 115, 116

Mx. ostium 117, 118
- blocked - thick muco. / polyps / granu.

-sinus mucosa 119, 120
nor. / thick. / polypoidal / cysts 121, 122

- associated procedures
- septo / rhinoplasty - 123, 124
- ear 125

- any complications 126

POST OP. FOLLOW UP

no change 1, mild improvement 2, good improvement 3, no sym.4
nasal endoscopy : well healed 1, disease present 2

follow - up	3 mon.	6 mon.	1 yr.	18 mon.	2 yr.	3 yr.	4 yr.	5 yr.
nasal obst.								
P.N.D.								
headaches - f.pain								

SKK / ASK

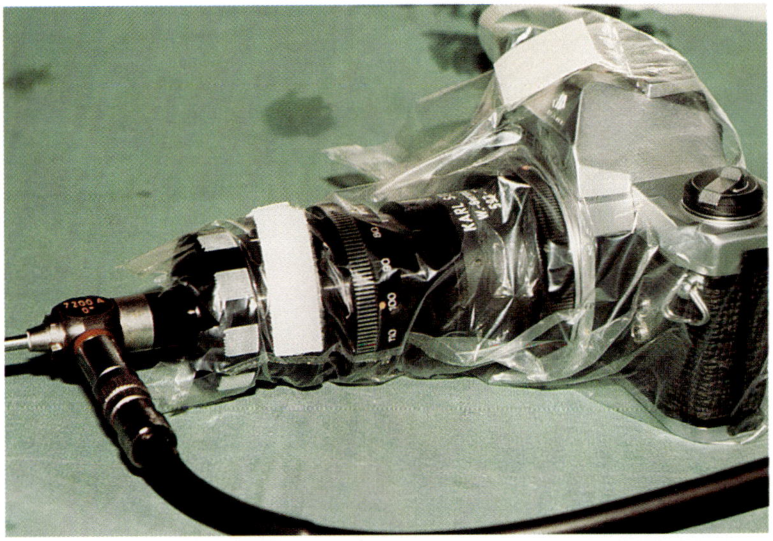

Fig. A.4. Single Lens Reflex camera with a special zoom lens for endoscopic still photography.

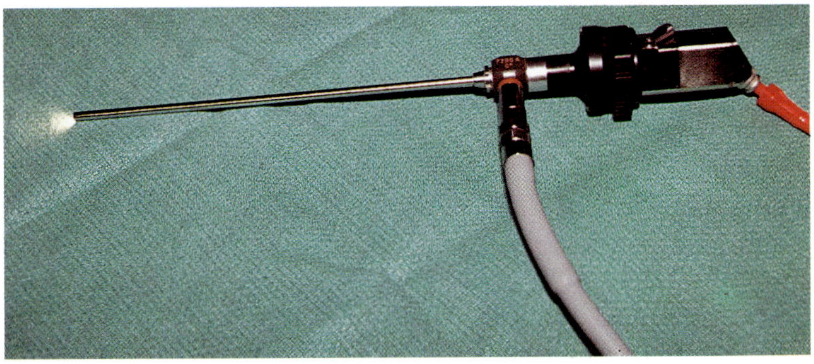

Fig. A.5. A 4-mm 0° endoscopic CCD video camera with C-mount.

Still Photography

1. Single lens reflex camera
2. Zoom lens, *f* 70–140, with adapter to Hopkins telescope
3. Cold light fountain, 150 or 250 watt
4. Ektachrome slide daylight film (high-quality prints may be made)
5. Customised sterile camera drape (for intraoperative photography)

Video Photography

1. CCD video camera
2. High resolution colour monitor
3. Super VHS video recorder
4. Pro-grade S-VHS video tapes
5. Professional editing suite with high-resolution monitors using a Commodore Amiga with title software

Bibliography

Anderson I, Camner P, Jenson PL, Philipson K, Proctor DF (1974) A comparison of nasal and tracheobronchial clearance. Arch Environ Health 29: 290–293

Aust R, Drettner B (1974a) The functional size of the human maxillary ostium in vivo. Acta Otolaryngol 78: 432–435

Aust R, Drettner B (1974b) Oxygen tension in human maxillary sinus under normal and pathological conditions. Acta Otolaryngol 78: 264–269

Berthon E (1880) Essai sur less abcés et hydropiris des sinus frontaux. Thesis, Paris

Bryant FL (1960) Management of chronic maxillary sinusitis in children. J La State Med Soc 112: 390–393

Buiter CT (1976) Endoscopy of the upper airway. Elsevier, Amsterdam

Buiter CT, Straatman NJA (1981) Endoscopic antrostomy in the nasal fontanelle. Rhinology 19: 17–24

Chandler JR, Langenbrunaer DJ, Stevens ER (1970) The pathogenesis of orbital complications and acute sinusitis. Laryngoscope 80: 1414–1428

Dedola GL (1987) Computed tomography of the ethmoid labyrinth and adjacent structures. Ann Otol Rhinol Laryngol 96: 239–250

Deitmer TH (1989) Physiology and pathology of mucociliary system. Adv 43 Otorhinolaryngol

Dezeimeris, cited by Berthon (1880)

Draf W (1978) Endoscopy of the paranasal sinuses. Actas, XI World Congress for ORL, Buenos Aires

Draf W (1982) Die chirurgische Behandlung entzündlicher Erkrankungen der Nasennebenhöhlen. Arch Otorhinolaryngol 235: 133–305, 367–377

Draf W (1983) Endoscopy of the paranasal sinuses. Springer, Berlin Heidelberg New York

Evert G (1965) On the mucous flow rate in the human nose. Acta Otolaryngol (Stockh) [Suppl] 200: 1

Fairbanks DFN (1986) Complications of nasal packing. Otolaryngol Head Neck Surg 94: 412–415

Fawcett D, Porter K (1954) Study of the fine structure of ciliated epithelia. J Morphol 94: 221

Flottes L, Clerc P, Riu R, Devilla F (1960) La physiologie des sinus. Libraire Arnette, Paris

Ginzel A, Ellum P (1980) Nasal mucociliary clearance in patients with septal deviation. Rhinology 18: 177–181

Grünwald L (1925) Deskriptive und topographische Anatomie der Nase und ihrer Nebenhöhlen In: Denker A, Kahler O (eds) Handbuch der Hals-Nasen-Ohrenheilkunde. Springer, Berlin Heidelberg New York, pp 1–95

Hackett F (1934) Francis I. Doubleday, New York, 613–616

Hady MRA, Shehata O, Hassan R (1983) Nasal mucociliary function in different diseases of the nose. J Laryngol Otol 97: 497–502

Hansen R (1968) Headaches of nasal origin. Laryngoscope 78: 1164–1171

Hara MJ (1962) Severe epistaxis. Arch Otolaryngol Head Neck Surg 75: 258–269

Harper OV, Andros G, Lathrop KA (1962) Observation on the use of 6 hours 99^{m} technetium as a tracer in biology and medicine. Argonne Cancer Res Hosp 18: 76–80

Herrnheisser G (1926) Der Rontegenbefund bei der mucocele oder pycele der Stirnhohle und der Siebbeinzellen. Z Hals Nasen 14: 319–326

Hilding AC (1931) Ciliary activity and course of secretion currents of the nose. Proc Mayo Clin 6: 285

Hilding AC (1932) Physiology of drainage of nasal mucous: experimental work on accessory sinuses. Am J Physiol 100: 664

Hilding AC (1941) Experimental sinus surgery: effects of operation windows on normal sinuses. Ann Otol Rhinol Laryngol 50: 379–392

Hilding AC (1944) The role of ciliary action in production of pulmonary atelectasis, vacuum in the paranasal sinuses, and in otitis media. Ann Otol 52: 816–833

Hilding AC (1950) Physiologic basis of nasal operations. Calif Med 72: 103–107

Howarth WC (1921) Mucocele and pyocele of the accessory nasal sinuses. Lancet II: 744–746

Illum P, Jeppesen F (1972) Sinoscopy: endoscopy of the maxillary sinus. Technique, common and rare findings. Acta Otolaryngol [Stockh.] 73: 506–512

Jakus MA, Hall CE (1946) Electron microscope observations of trichocysts and cilia of paramecium. Biol Bull 91: 131–146

Jensen PF, Kristensen S, Juul A, Johannessen NW (1991) Episodic nocturnal hypoxia and nasal packing. Clin Otolaryngol 16: 433–435

Kaluskar SK, Patil NP (1992a) The role of out-patient nasal endoscopy in the evaluation of chronic sinus disease. (editorial). Clin Otolaryngol 17: 193–194

Kaluskar SK, Patil NP (1992b) Combined approach middle meatal antrostomy (CAMMA) Laryngoscope 102: 709–711

Kaluskar SK, Patil NP (1992c) Clinical atlas of ear, nose and throat diseases. Bhalani Publishing House, India

Kaluskar SK, Patil NP (1992d) Office nasal endoscopy in the evaluation of chronic sinus disease. Karl Storz.

Kaluskar SK, Patil NP, Sharkey AN (1993) The role of CT in functional endoscopic sinus surgery. Rhinology 31: 49–52

Kaluskar SK, Leydon P (1994) Endoscopic approach to posterior epistaxis. Proc Irish Otolaryngol Soc 1994: 39–42

Kaluskar SK (1995) Endoscopic approach to orbital complications. Presented at Functional Endoscopic Sinus Surgery Meeting, April 1995, Omagh, N. Ireland

Kaluskar SK (1995) KTP/532 Laser in Functional Endoscopic Sinus Surgery. Presented at Advanced Symposium on Sinus Surgery, August 1995, Cairns, Australia

Kaluskar SK (1996) KTP/532 Laser in Revision Endoscopic Sinus Surgery. Presented at World Association of Laser Therapy, June 1996, Jerusalem, Israel

Kaluskar SK (1996) Endoscopic approach to posterior epistaxis. J Min Invasive Ther All Technol 5: 75–77

Karenfelt C, et al. (1978) The role of local gas composition in pathogenesis of maxillary sinus empyema Acta Otolaryngol 85: 116–121

Kennedy DW, Goodstein ML, Miller NR, Zinreich SJ (1990) Endoscopic transnasal orbital decompression. Arch Otolaryngol Head Neck Surg 116: 275–282

Khasgiwala CK (1972) Vidian neurocoagulation via transseptal approach. MS thesis, University of Gujarat, India

King E (1935) A clinical study of the functioning of the maxillary sinus mucosa. Ann Otol 44: 480–482

Langenbeck CJM (1819) Neue Bibliohek fur die Chirugie and opthalmologie. Bey DanBrudern Hahn, Hannover

Langraf-Favre FF (1974) Pain of nasal origin. Rhinology 12: 73–78

Lynch RC (1921) The technique of a radical frontal sinus operation which has given me the best results. 31: 1–5

Mattox DE, Kennedy DW (1990) Endoscopic management of cerebrospinal fluid leaks and cephaloceles. Laryngoscope 100: 857–862

McGarry GW (1991) Nasal endoscope in posterior epistaxis, a preliminary evaluation. J Laryngol Otol 105: 428–431

Messerklinger W (1978) Endoscopy of the nose. Urban & Schwarzenberg, Munich

Mosher HP (1912) The applied anatomy and the intranasal surgery of the ethmoidal labyrinth. Trans Am Laryngol Assoc 34: 25–39

Mosher HP (1929) The surgical anatomy of the ethmoidal labyrinth. Am Acad Ophthalmol Otolaryngol 376–410

Myerson MC (1932) The natural orifice of the maxillary sinus. Anatomic Studies. Arch Otolaryngol 15: 80–91

O'Leary, Stickney K, Makielski K, Weymuller EA Jr (1992) Rigid endoscopy for the control of epistaxis. Arch Otolaryngol Head Neck Surg 118: 966–967

Onodi A (1901) Das Verhaltniss der kieferhole zurkeilbein hohle und zu der vorderen Siebbeinzellen. Arch Laryngol Rhinol 11: 391–395

Proctor DF, Wagner HN (1965) Clearance of particles from the human nose. Arch Environ Health 11: 366–371

Proetz AW (1941) Essays on the applied physiology of the nose. Annals Publ. St Louis

Puchelle E, Aug F, Zahm JN, Bertrand A (1981) Comparison of three methods for measuring nasal mucociliary clearance in man. Acta Otolaryngol 91: 247–303

Quinlan M, Salmon S, Swift D, Wagner H, Proctor DF (1969) Measurement of mucociliary function in man. Am Rev Respir Dis 99: 13

Rogers JH (1976) Mucocele of the paranasal sinuses. Otolaryngol Clin North Am 9: 233

Rollet M (1896) Mucocele de l'angle superointern des orbites. Lyon Med 81: 573–575

Rollet M (1909) In: Lagrange F, Valude E (eds) Encyclopedie francaise d'opthalmology. Doin, Paris, 588

Ryan RE et al. (1979) Headache of nasal origin. Headache 19: 173–179

Sakakura Y, Majuma Y, Saida S, Ukai K, Mioshi Y (1985) Reversibility of reduced mucociliary clearance in chronic sinusitis. Clin Otolaryngol 10: 79–83

Shaheen OH (1975) Arterial epistaxis. J Laryngol Otol 89: 17

Sluder G (1927) Nasal neurology: headache and eye disorders. Kimpton, London 1927

Stammberger H (1985) Endoscopic surgery for mycotic and chronic recurring sinusitis. Ann Otol Rhinol Laryngol 94 (suppl 119): 1–11

Stammberger H (1986a) Nasal and paranasal sinus endoscopy: a diagnostic and surgical approach to recurrent sinusitis. Endoscopy 6: 213–218

Stammberger H (1986b) Endoscopic endonasal surgery: concepts in treatment of recurring rhinosinusitis. 1. Anatomic and pathophysiologic considerations. 11. Technique. Otolaryngol Head Neck Surg 94: 143–156

Stammberger H, Zinreich SJ, Kopp W, Kennedy DW, Johns ME, Rosenbaum AE (1987) Zur operativen Behandlung der chronisch-rezidivierenden Sinusitis – Caldwell-Luc versus funktionelle endoskopische Technik. Head Neck Otolaryngol 35: 93–105

Terrier G, Baumann RP, Pidoux JM (1976) Endoscopic and histopathological observations of chronic maxillary sinusitis. Rhinology 14: 129–132

Tremble GE (1948) Clinical observations in movement of nasal cilia. Laryngoscope 206–224

Waitz G, Wigand ME (1992) Results of endoscopic sinus surgery for the treatment of inverted papillomas. Laryngoscope 102: 917–922

Wang et al. (1981) Posterior epistaxis. Comparison of treatment. Otolaryngol Head Neck Surg 89: 1001–1006

Weiss N (1990) Otolaryngology: an illustrated history. Butterworth, London

Weiss NS (1972) Relation of high blood pressure to headache epistaxis, and selected other symptoms. N Engl J Med 287: 631

Wigand ME (1981) Transnasal ethmoidectomy under endoscopical control. Rhinology 19: 7–15

Wigand ME, Hosemann W (1985) Endoscopic ethmoidectomy for chronic sinubronchitis. In: Myers E (ed) New dimensions in otorhinolaryngology: head and neck surgery, vol. 1. Elsevier, Amsterdam, pp 549–552

Wigand ME, Steiner W, Jaumann MP (1978) Endonasal sinus surgery with endoscopical control: from radical operation to rehabilitation of the mucosa. Endoscopy 10: 255–260

Woodruff GH (1949) Cardiovascular epistaxis and the nasopharyngeal plexus. Laryngoscope 59: 1238–1247

Wurman LH, Garry Sack J, Flannery JV, Paulson TO (1988) Selective endoscopic electrocautery for posterior epistaxis. Laryngoscope 98: 1348–1349

Wurman LH, Sack JG, Flannery JV Jr, Lipsman RA (1992) The management of epistaxis. Am J Otolaryngol 193–209

Zinreich SJ, Kennedy DW, Rosenbaum AE, Gayler BW, Kumar AJ, Stammberger H (1987) CT of nasal cavity and paranasal sinuses: imaging requirements for functional endoscopic sinus surgery. J Radiol 163: 769–775

Zizmor J (1974) Mucocele of the paranasal sinuses, Can J Otolaryngol [Suppl] 1

Index

P

S

T

U

V

W